T0296698

The Phylogenetic Handbook

A Practical Approach to Phylogenetic Analysis and Hypothesis Testing

Second Edition

Edited by

Philippe Lemey
Katholieke Universiteit Leuven, Belgium

Marco Salemi
University of Florida, Gainesville, USA

Anne-Mieke Vandamme
Katholieke Universiteit Leuven, Belgium

CAMBRIDGE
UNIVERSITY PRESS

CAMBRIDGE
UNIVERSITY PRESS

University Printing House, Cambridge CB2 8BS, United Kingdom

One Liberty Plaza, 20th Floor, New York, NY 10006, USA

477 Williamstown Road, Port Melbourne, VIC 3207, Australia

314-321, 3rd Floor, Plot 3, Splendor Forum, Jasola District Centre, New Delhi - 110025, India

79 Anson Road, #06-04/06, Singapore 079906

Cambridge University Press is part of the University of Cambridge.

It furthers the University's mission by disseminating knowledge in the pursuit of education, learning and research at the highest international levels of excellence.

www.cambridge.org
Information on this title: www.cambridge.org/9780521730716

© Cambridge University Press 2009

First published 2009
8th printing 2017

A catalogue record for this publication is available from the British Library

ISBN 978-0-521-87710-7 Hardback
ISBN 978-0-521-73071-6 Paperback

The Phylogenetic Handbook

A Practical Approach to Phylogenetic Analysis and Hypothesis Testing

Second Edition

Edited by

Philippe Lemey
Katholieke Universiteit Leuven, Belgium

Marco Salemi
University of Florida, Gainesville, USA

Anne-Mieke Vandamme
Katholieke Universiteit Leuven, Belgium

CAMBRIDGE
UNIVERSITY PRESS

CAMBRIDGE
UNIVERSITY PRESS

University Printing House, Cambridge CB2 8BS, United Kingdom

One Liberty Plaza, 20th Floor, New York, NY 10006, USA

477 Williamstown Road, Port Melbourne, VIC 3207, Australia

314-321, 3rd Floor, Plot 3, Splendor Forum, Jasola District Centre, New Delhi - 110025, India

79 Anson Road, #06-04/06, Singapore 079906

Cambridge University Press is part of the University of Cambridge.

It furthers the University's mission by disseminating knowledge in the pursuit of education, learning and research at the highest international levels of excellence.

www.cambridge.org
Information on this title: www.cambridge.org/9780521730716

© Cambridge University Press 2009

First published 2009
8th printing 2017

A catalogue record for this publication is available from the British Library

ISBN 978-0-521-87710-7 Hardback
ISBN 978-0-521-73071-6 Paperback

Cambridge University Press has no responsibility for the persistence or accuracy of URLs for external or third-party internet websites referred to in this publication, and does not guarantee that any content on such websites is, or will remain, accurate or appropriate.

Contents

3 Multiple sequence alignment   68

Contributors

Guy Bottu
Belgian EMBnet Node
Brussels, Belgium

Alexei Drummond
Department of Computer Science
University of Auckland
Private Bag 92019
Auckland, New Zealand

Simon Frost
Antiviral Research Center
University of California
150 W Washington St, Ste 100
San Diego, CA 92103, USA

Des Higgins
Conway Institute
University College Dublin
Ireland

Katharina T. Huber
School of Computing Sciences
University of East Anglia
Norwich, UK

John P. Huelsenbeck
Department of Integrative Biology
University of California at Berkeley
3060 Valley Life Sciences Bldg
Berkeley, CA 94720-3140, USA

Sergei Kosakovsky Pond
Antiviral Research Center
University of California
150 W Washington St, Ste 100
San Diego, CA 92103, USA

Mary Kuhner
Department of Genome Sciences
University of Washington
Seattle (WA), USA

Philippe Lemey
Rega Institute for Medical Research
Katholieke Universiteit Leuven
Leuven, Belgium

Darren Martin
Institute of Infectious Disease and Molecular
    Medicine
Faculty of Health Sciences
University of Cape Town
Observatory 7925
South Africa

Vincent Moulton
School of Computing Sciences
University of East Anglia
Norwich, UK

Fred R. Opperdoes
C. de Duve Institute of Cellular Pathology
Universite Catholique de Louvain
Brussels, Belgium

Art Poon
Antiviral Research Center
University of California
150 W Washington St, Ste 100
San Diego, CA 92103, USA

David Posada
Department of Biochemistry
Genetics and Immunology
University of Vigo
Spain

Oliver Pybus
Department of Zoology
University of Oxford
South Parks Road
Oxford OX1 3PS, UK

Andrew Rambaut
Institute of Evolutionary Biology
University of Edinburgh
Ashworth Laboratories
Kings Building
West Mains Road
Edinburgh EH3 9JT, UK

Allen Rodrigo
School of Biological Sciences
University of Auckland
New Zealand

Fredrik Ronquist
Department of Entomology
Swedish Museum of Natural History
Box 50007, SE-104 05 Stockholm
Sweden

Marco Salemi
Department of Pathology, Immunology, and
    Laboratory Medicine
University of Florida
Gainesville, Florida
USA

Mika Salminen
HIV Laboratory
National Public Health Institute
Department of Infectious Disease
    Epidemiology
Helsinki, Finland

Heiko Schmidt
Center for Integrative Bioinformatics Vienna
    (CIBIV)
Max F. Perutz Laboratories (MFPL)
Dr. Bohr Gasse 9
A-1030 Wien, Austria

Beth Shapiro
Department of Biology
The Pennsylvania State University
326 Mueller Lab
University Park, PA 16802
USA

Korbinian Strimmer
Institute for Medical Informatics Statistics
    and Epidemiology (IMISE)
University of Leipzig
Germany

Jack Sullivan
Department of Biological Science
University of Idaho
Idaho, USA

David L. Swofford
Center for Evolutionary Genomics
Institute for Genome Sciences & Policy
Box 90338
Duke University
Durham, NC 27708
USA

Anne-Mieke Vandamme
Rega Institute for Medical Research
Katholieke Universiteit Leuven
Leuven, Belgium

Yves Van de Peer
VIB / Ghent University
Bioinformatics & Evolutionary Genomics
Technologiepark 927
B-9052 Gent, Belgium

Paul van der Mark
School of Computational Science
Florida State University
Tallahassee, FL 32306-4120, USA

Marc Van Ranst
Rega Institute for Medical Research
Katholieke Universiteit Leuven
Leuven, Belgium

Arndt von Haeseler
Center for Integrative Bioinformatics
   Vienna (CIBIV)
Max F. Perutz Laboratories (MFPL)
Dr. Bohr Gasse 9
A-1030 Wien, Austria

Xuhua Xia
Biology Department
University of Ottawa
Ottawa, Ontario
Canada

Foreword

"It looked insanely complicated, and this was one of the reasons why the snug plastic cover it fitted into had the words DON'T PANIC printed on it in large friendly letters."

Douglas Adams
The Hitch Hiker's Guide to the Galaxy

As of February 2008 there were 85 759 586 764 bases in 82 853 685 sequences stored in GenBank (*Nucleic Acids Research*, Database issue, January 2008). Under any criteria, this is a staggering amount of data. Although these sequences come from a myriad of organisms, from viruses to humans, and include genes with a diverse arrange of functions, it can all, at least in principle, be studied from an evolutionary perspective. But how? If ever there was an invitation panic, it is this. Enter *The Phylogenetic Handbook*, an invaluable guide to the phylogenetic universe.

The first edition of *The Phylogenetic Handbook* was published in 2003 and represented something of a landmark in evolutionary biology, as it was the first accessible, hands-on instruction manual for molecular phylogenetics, yet with a healthy dose of theory. Up until this point, the evolutionary analysis of gene sequence was often considered something of a black art. *The Phylogenetic Handbook* made it accessible to anyone with a desktop computer.

The new edition *The Phylogenetic Handbook* moves the field along nicely and has a number of important intellectual and structural changes from the earlier edition. Such a revision is necessary to track the major changes in this rapidly evolving field, in terms of both the new theory and new methodologies available for the computational analysis of gene sequence evolution. The result is a fine balance between theory and practice. As with the First Edition, the chapters take us from the basic, but fundamental, tasks of database searching and sequence alignment, to the complexity of the coalescent. Similarly, all the chapters are written by acknowledged experts in the field, who work at the coal-face of developing new methods and using them to address fundamental biological questions. Most of the authors are also remarkably young, highlighting the dynamic nature of this discipline.

The biggest alteration from the First Edition is the restructuring into a series of sections, complete with both theory and practice chapters, with each designed to take the uninitiated through all the steps of evolutionary bioinformatics. There are also more chapters on a greater range of topics, so the new edition is satisfyingly comprehensive. Indeed, it almost stands alone as a textbook in modern population genetics. It is also pleasing to see a much stronger focus on hypothesis testing, which is a key aspect of modern phylogenetic analysis. Another welcome change is the inclusion of chapters describing Bayesian methods for both phylogenetic inference and revealing population dynamics, which fills a major gap in the literature, and highlights the current popularity of this form of statistical inference.

The Phylogenetic Handbook will calm the nerves of anyone charged with undertaking an evolutionary analysis of gene sequence data. My only suggestion for an improvement to the third edition are the words DON'T PANIC on the cover.

Edward C. Holmes
June 12, 2008

Preface

The idea for *The Phylogenetic Handbook* was conceived during an early edition of the Workshop on Virus Evolution and Molecular Epidemiology. The rationale was simple: to collect the information being taught in the workshop and turn it into a comprehensive, yet simply written textbook with a strong practical component. Marco and Annemie took up this challenge, and, with the help of many experts in the field, successfully produced the First Edition in 2003. The resulting text was an excellent primer for anyone taking their first computational steps into evolutionary biology, and, on a personal note, inspired me to try out many of the techniques introduced by the book in my own research. It was therefore a great pleasure to join in the collaboration for the Second Edition of *The Phylogenetic Handbook*.

Computational molecular biology is a fast-evolving field in which new techniques are constantly emerging. A book with a strong focus on the software side of phylogenetics will therefore rapidly grow a need for updating. In this Second Edition, we hope to have satisfied this need to a large extent. We also took the opportunity to provide a structure that groups different types of sequence analyses according to the evolutionary hypothesis they focus on. Evolutionary biology has matured into a fully quantitative discipline, with phylogenies themselves having evolved from classification tools to central models in quantifying underlying evolutionary and population genetic processes. Inspired by this, the Second Edition provides a broader coverage of techniques for testing models and trees, detecting recombination, the analysis of selective pressure and genealogy-based population genetics. Changing the subtitle to *A Practical Approach to Phylogenetic Inference and Hypothesis Testing* emphasizes this shift in focus. Thanks to novel contributions, we also hope to have addressed the need for a Bayesian treatment of phylogenetic inference, which started to gain a great deal of popularity at the time the content for the First Edition was already fixed.

Following the philosophy of the First Edition, the book includes many step-by-step software tutorials using example data sets. We have not used the same data sets throughout the complete Second Edition; not only is it difficult to find data sets that consistently meet the assumptions or reveal interesting aspects of all the methods described, but we also feel that being confronted with different data with their own characteristics adds educational value. These data sets can be retrieved from www.thephylogenetichandbook.org, where other useful links listed in the book can also be found. Furthermore, a glossary has been compiled with important terms that are indicated in italics and boldface throughout the book.

We are very grateful to the researchers who took the time to contribute to this edition, either by updating a chapter or writing a novel contribution. I hope that my persistent pestering has not affected any of these friendships. We would like to thank Eddie Holmes in particular for writing the Foreword to the book. It has been a pleasure to work with Katrina Halliday and Alison Evans of Cambridge University Press. We also wish to thank those who supported our research and the work on this book: the Flemish "Fonds voor Wetenschappelijk Onderzoek", EMBO and Marie Curie funding. Finally, we would like to express our thanks to colleagues, family and friends onto whom we undoubtedly projected some of the pressure in completing this book.

Philippe Lemey

Section I

Introduction

Basic concepts of molecular evolution

Anne-Mieke Vandamme

1.1 Genetic information

It was the striking phenotypic variation of finches in the Galapagos Islands that inspired Darwin to draft his theory of evolution. His idea of a branching process of evolution was also consistent with the knowledge of fossil researchers who revealed phenotypic variation over long periods of time. Today, evolution can be observed in real time by scientists, with the fastest evolution occurring in viruses within months, resulting, for example, in rapid development of human immunodeficiency virus (HIV) drug resistance. The phenotype of living organisms is always a result of the genetic information that they carry and pass on to the next generation and its interaction with the environment. Thus, if we want to study the driving force of evolution, we have to investigate the changes in the genetic information.

The genome, carrier of this genetic information, is in most organisms deoxy-ribonucleic acid (**DNA**), whereas some viruses have a ribonucleic acid (**RNA**) genome. Part of the genetic information in DNA is transcribed into RNA: either mRNA, which acts as a template for **protein** synthesis; rRNA, which together with ribosomal proteins constitutes the protein translation machinery; tRNA, which offers the encoded amino acid; or small RNAs, some of which are involved in regulating expression of genes. The genomic DNA also contains elements, such as *promotors* and *enhancers*, which orchestrate the proper transcription into RNA. A large part of the genomic DNA of eukaryotes consists of genetic elements such as introns or alu-repeats, the function of which is still not entirely clear. Proteins, RNA, and to some extent DNA, constitute the phenotype of an organism that interacts with the environment.

DNA is a double helix with two antiparallel *polynucleotide* strands, whereas RNA is a single-stranded polynucleotide. The backbone in each DNA strand

The Phylogenetic Handbook: a Practical Approach to Phylogenetic Analysis and Hypothesis Testing, Philippe Lemey, Marco Salemi, and Anne-Mieke Vandamme (eds.). Published by Cambridge University Press. © Cambridge University Press 2009.

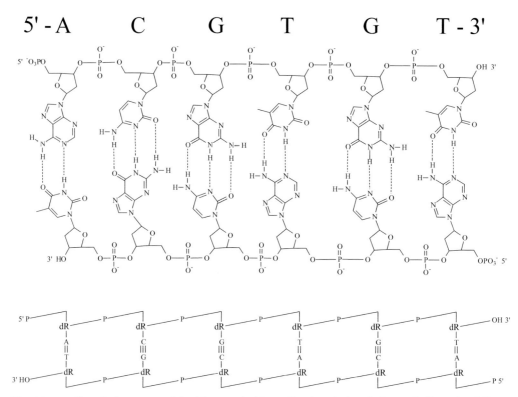

Fig. 1.1   Chemical structure of double-stranded DNA. The chemical moieties are indicated as follows: dR, deoxyribose; P, phosphate; G, guanine; T, thymine; A, adenine; and C, cytosine. The strand orientation is represented in a standard way: in the upper strand 5′–3′, indicating that the chain starts at the 5′ carbon of the first dR, and ends at the 3′ carbon of the last dR. The one letter code of the corresponding genetic information is given on top, and only takes into account the 5′–3′ upper strand. (Courtesy of Professor C. Pannecouque.)

consists of deoxyriboses with a phosphodiester linking each 5′ carbon with the 3′ carbon of the next sugar. In RNA the sugar moiety is ribose. On each sugar, one of the four following bases is linked to the 1′ carbon in DNA: the *purines*, **adenine** (**A**), or **guanine** (**G**), or the *pyrimidines*, **thymine** (**T**), or **cytosine** (**C**); in RNA, thymine is replaced by **uracil** (**U**). Hydrogen bonds and base stacking result in the two DNA strands binding together, with strong (triple) bonds between G and C, and weak (double) bonds between T/U and A (Fig. 1.1). These hydrogen-bonded pairs are called *complementary*. During DNA duplication or RNA transcription, DNA or RNA polymerases synthesize a complementary 5′–3′ strand starting with the lower 3′–5′ DNA strand as template, in order to preserve the genetic information. This genetic information is represented by a one letter code, indicating the 5′–3′ sequential order of the bases in the DNA or RNA (Fig. 1.1). A nucleotide sequence is thus represented by a contiguous stretch of the four letters A, G, C, and T/U.

Table 1.1 Three- and one-letter abbreviations of the 20 naturally encoded amino acids

Amino acid	Three-letter abbreviation	One-letter abbreviation
Alanine	Ala	A
Arginine	Arg	R
Asparagine	Asn	N
Aspartic acid	Asp	D
Cysteine	Cys	C
Glutamic acid	Glu	E
Glutamine	Gln	Q
Glycine	Gly	G
Histidine	His	H
Isoleucine	Ile	I
Leucine	Leu	L
Lysine	Lys	K
Methionine	Met	M
Phenylalanine	Phe	F
Proline	Pro	P
Serine	Ser	S
Threonine	Thr	T
Tryptophan	Trp	W
Tyrosine	Tyr	Y
Valine	Val	V

In RNA strands encoding a protein, each triplet of bases is recognized by the ribosomes as a code for a specific amino acid. This translation results in polymerization of the encoded amino acids into a protein. Amino acids can be represented by a three- or one-letter abbreviation (Table 1.1). An amino acid sequence is generally represented by a contiguous stretch of one-letter amino acid abbreviations (with 20 possible letters).

The **genetic code** is universal for all organisms, with only a few exceptions such as the mitochondrial code, and it is usually represented as an RNA code because the RNA is the direct template for protein synthesis (Table 1.2). The corresponding DNA code can be easily reconstructed by replacing the U by a T. Each position of the triplet code can be one of four bases; hence, 4^3 or 64 possible triplets encode 20 amino acids (61 *sense* codes) and 3 stop codons (3 *non-sense* codes). The genetic code is said to be degenerated, or redundant, since all amino acids except methionine have more than one possible triplet code. The first codon for methionine *downstream* (or 3′) of the ribosome entry site also acts as the start codon for the translation of a protein. As a result of the triplet code, each

Table 1.2 The universal genetic code

	U		C		A		G		
	Codon	Amino acid	Codon	Amino acid	Codon	Amino acid	Codon	Amino acid	
U	UUU	Phe	UCU	Ser	UAU	Tyr	UGU	Cys	U
	UUC	Phe	UCC	Ser	UAC	Tyr	UGC	Cys	C
	UUA	Leu	UCA	Ser	UAA	STOP	UGA	STOP	A
	UUG	Leu	UCG	Ser	UAG	STOP	UGG	Trp	G
C	CUU	Leu	CCU	Pro	CAU	His	CGU	Arg	U
	CUC	Leu	CCC	Pro	CAC	His	CGC	Arg	C
	CUA	Leu	CCA	Pro	CAA	Gln	CGA	Arg	A
	CUG	Leu	CCG	Pro	CAG	Gln	CGG	Arg	G
A	AUU	Ile	ACU	Thr	AAU	Asn	AGU	Ser	U
	AUC	Ile	ACC	Thr	AAC	Asn	AGC	Ser	C
	AUA	Ile	ACA	Thr	AAA	Lys	AGA	Arg	A
	AUG	Met	ACG	Thr	AAG	Lys	AGG	Arg	G
G	GUU	Val	GCU	Ala	GAU	Asp	GGU	Gly	U
	GUC	Val	GCC	Ala	GAC	Asp	GGC	Gly	C
	GUA	Val	GCA	Ala	GAA	Glu	GGA	Gly	A
	GUG	Val	GCG	Ala	GAG	Glu	GGG	Gly	G

The first nucleotide letter is indicated on the left, the second on the top, and the third on the right side. The amino acids are given by their three-letter code (see Table 1.1). Three stop codons are indicated.

contiguous nucleotide stretch has three reading frames in the 5′–3′ direction. The complementary strand encodes another three reading frames. A reading frame that is able to encode a protein starts with a codon for methionine, and ends with a stop codon. These reading frames are called **open reading frames** or **ORFs**.

During duplication of the genetic information, the DNA or RNA polymerase can occasionally incorporate a non-complementary nucleotide. In addition, bases in a DNA strand can be chemically modified due to environmental factors such as UV light or chemical substances. These modified bases can potentially interfere with the synthesis of the complementary strand and thereby also result in a nucleotide incorporation that is not complementary to the original nucleotide. When these changes escape the cellular repair mechanisms, the genetic information is altered, resulting in what is called a *point mutation*. The genetic code has evolved in such a way that a point mutation at the third codon position rarely results in an amino acid change (only in 30% of possible changes). A change at the second codon position always, and at the first codon position mostly (96%), results in an amino acid change. Mutations that do not result in amino acid changes are called silent or *synonymous mutations*. When a mutation results in the incorporation of a different amino acid, it is called non-silent or *non-synonymous*. A site within a coding triplet is said to be *fourfold degenerate* when all possible changes at that site are synonymous (for example "CUN"); *twofold degenerate* when only two different amino acids are encoded by the four possible nucleotides at that position (for example, "UUN"); and *non-degenerate* when all possible changes alter the encoded amino acid (for example, "NUU").

Incorporation errors replacing a purine (A, G) with a purine and a pyrimidine (C, T) with a pyrimidine occur more easily because of chemical and steric reasons. The resulting mutations are called *transitions*. *Transversions*, purine to pyrimidine changes and the reverse, are less likely. When resulting in an amino acid change, transversions usually have a larger impact on the protein than transitions, because of the more drastic changes in biochemical properties of the encoded amino acid. There are four possible transition errors (A ↔ G, C ↔ T), and eight possible transversion errors (A ↔ C, A ↔ T, G ↔ C, G ↔ T); therefore, if a mutation occurred randomly, a transversion would be two times more likely than a transition. However, the genetic code has evolved in such a way that, in many genes, the less disruptive transitions are more likely to occur than transversions.

Single nucleotide changes in a particular codon often change the amino acid to one with similar properties (e.g. hydrophobic), such that the tertiary structure of the encoded protein is not altered dramatically. Living organisms can therefore tolerate a limited number of nucleotide point mutations in their coding regions. Point mutations in non-coding regions are subject to other constraints, such as conservation of binding places for proteins, conservation of base pairing in RNA tertiary structures or avoidance of too many homopolymer stretches in which polymerases tend to stutter.

Errors in duplication of genetic information can also result in the **deletion** or **insertion** of one or more nucleotides, collectively referred to as *indels*. When multiples of three nucleotides are inserted or deleted in coding regions, the reading frame remains intact and one or more amino acids are inserted or deleted. When one or two nucleotides are inserted or deleted, the reading frame is disturbed and the resulting gene generally codes for an entirely different protein, with different amino acids and a different length from the original protein. The consequence of this change depends on the position in the gene where the change took place. Insertions or deletions are therefore rare in coding regions, but rather frequent in non-coding regions. When occurring in coding regions, indels can occasionally change the reading frame of a gene and make another ORF of the same gene accessible. Such mutations can lead to acquisition of new gene functions. Having small genomes, viruses make extensive use of this possibility. They often encode several proteins from a single gene by using overlapping ORFs. Another type of mutation that can change reading frames or make accessible new reading frames is mutations in splicing patterns. Eukaryotic proteins are encoded by coding gene fragments called **exons**, which are separated from each other by **introns**. Joining the introns is called **splicing** and occurs in the nucleus at the pre-mRNA level through dedicated spliceosomes. Mutations in splicing patterns usually destroy the gene function, but can occasionally result in the acquisition of a new gene function. Viruses have used these mechanisms extensively. By alternative splicing, sometimes in combination with the use of different reading frames, viruses are able to encode multiple proteins by a single gene. For example, HIV is able to encode two additional regulatory proteins using part of the coding region of the *env* gene by alternative splicing and overlapping reading frames.

When parts of two different DNA strands are combined into a single strand, the genetic exchange is called **recombination**. Recombination has a major effect on the genetic make-up of organisms (see Chapter 15). The most common form of recombination happens in eukaryotes during *meiosis*, when recombination occurs between *homologous chromosomes*, shuffling the *alleles* for the next generation. Consequently, recombination contributes significantly to evolution of diploid organisms. More details on the process and consequences of recombination are provided in Chapter 15.

Another form of genetic exchange is lateral gene transfer, which is a relatively frequent event in bacteria. A dramatic example of this is the origin of eukaryotes arising from bacteria acquiring other bacterial genomes that evolved into organelles such as mitochondria or chloroplasts. The bacterial predecessor of mitochondria subsequently exchanged many genes with the "cellular" genome. Substantial parts of mammal genomes are "littered" with endogenous retroviral sequences, with the "fusion" capacity of some retroviral envelope genes at the origin of the placenta. Every retroviral infection results in lateral gene transfer, usually only in somatic cells.

Genetic variation can also be caused by **gene duplication**. Gene duplication results in genome enlargement and can involve a single gene, or large genome sections. They can be partial, involving only gene fragments, or complete, whereby entire genes, chromosomes (*aneuploidy*) or entire genomes (*polyploidy*) are duplicated. Genes experiencing partial duplication, such as domain duplication, can potentially have a greatly altered function. An entirely duplicated gene can evolve independently. After a long history of independent evolution, duplicated genes can eventually acquire a new function. Duplication events have played a major role in the evolution of species. For example, complex body plans were possible due to separate evolution of duplications of the homeobox genes (Carroll, 1995), and especially in plants, new species are frequently the result of polyploidy.

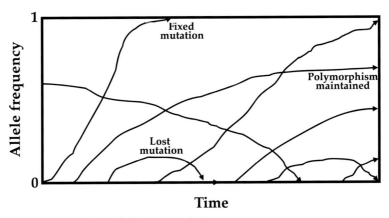

Fig. 1.2   Loss or fixation of an allele in a population.

1.2 Population dynamics

Mutations in a gene that are passed on to the offspring and that coexist with the original gene result in **polymorphisms**. At a polymorphic site, two or more variants of a gene circulate in the population simultaneously. Population geneticists typically study the dynamics of the frequency of these polymorphic sites over time. The location in the genome where two or more variants coexist is called the *locus*. The different variants for a particular locus are called *alleles*. Virus genomes, in particular, are very flexible to genetic changes; RNA viruses can contain many polymorphic sites in a single population. HIV, for example, does not exist in a single host as a single genomic sequence, but consists of a continuously changing swarm of variants sometimes referred to as a *quasispecies* (Eigen & Biebricher, 1988; Domingo *et al.*, 2006). Although this has become a standard term for virologists, the quasispecies theory has specific mathematical formulations and to what extent virus populations comply with these is the subject of great debate. The high genetic diversity mainly results from the rapid and error prone replication of RNA viruses. *Diploid* organisms always carry two alleles. When both alleles are identical, the organism is *homozygous* at that locus; when the organism carries two different alleles, it is *heterozygous* at that locus. Heterozygous positions are polymorphic.

Evolution is always a result of changes in *allele frequencies*, also called *gene frequencies* whereby some alleles are lost over time, while other alleles sometimes increase their frequency to 100%, they become *fixed* in the population (Fig. 1.2). The rate at which this occurs is called the **fixation rate**. The long-term evolution of a species results from the successive fixation of particular alleles, which reflects fixation of mutations. Terms like **fixation rate, mutation rate, substitution rate** and **evolutionary rate** have been used interchangeably by several authors, but they can refer to markedly different processes. This is particularly so for mutation rate, which should preferably be reserved for the rate at which mutations arise at the DNA level, usually expressed as the number of nucleotide (or amino acid) changes per site per replication cycle. Fixation rate, substitution rate, and the rate of molecular evolution are all equivalent when applied to sequences representing different species or populations, in which case they represent the number of new mutations per unit time that become fixed in a species or population. However, when applied to sequences representing different individuals within a population, the interpretation of these terms is subtly altered, because not all observed mutational differences among individuals (polymorphisms) will eventually become fixed in the population. In these cases, fixation rates are not appropriate, but substitution rate or the rate of molecular evolution can still be used to represent the rate at which individuals accrue genetic differences to each other over time (under the selective regime acting on this population). To summarize this from a phylogenetic perspective, the differences in nucleotide or amino acid sequences between taxa are generally called substitutions (although recently generated mutations can be present on terminal branches of trees). If these taxa represent different species or populations, the substitutions will be equivalent to fixation events. If the taxa represent different individuals within a population, branch lengths measure the genetic differences that accrue within individuals, which are not, but ultimately may lead to, fixation events.

The rate at which populations genetically diverge over time is dependent on the underlying mutation rate, the **generation time**, the time separating two generations, and on evolutionary forces, such as the fitness of the organism carrying the allele or variant, positive and negative selective pressure, population size, genetic drift, reproductive potential, and competition of alleles. If a particular allele is more fit than others in a particular environment, it will be subject to **positive selective pressure**; if it is less fit, it will be subject to **negative selective pressure**. An allele can confer a lower fitness to the homozygous organism, while heterozygosity of both alleles at this locus can be an advantage. In this case, polymorphism is advantageous and will be maintained; this is called **balancing selection** (heterozygote is more fit than either homozygote). For example, humans who carry the hemoglobin S allele on both chromosomes suffer from sickle-cell anaemia. However, this allele is maintained in the human population because heterozygotes are, to some extent, protected against malaria (Allison, 1956). Fitness of a variant is always the result of a particular phenotype of the organism; therefore, in coding regions, selective pressure always acts on mutations that alter function or stability of a gene or the amino acid sequence encoded by the gene. Synonymous mutations could at first sight be expected to be neutral since they do not result in amino acid changes. However, this is not always true. For example, synonymous changes can alter

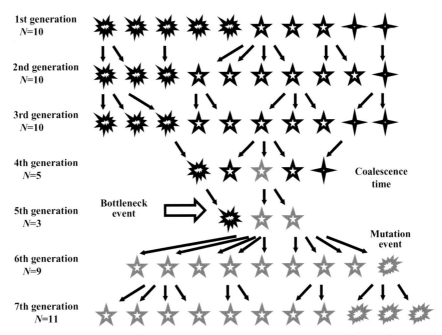

Fig. 1.3   Population dynamics of alleles. Each different symbol represents a different allele. A mutation event in the sixth generation gives rise to a new allele. The figure illustrates fixation
and loss of alleles during a bottleneck event, and the concept of coalescence time (tracking
back the time to the most recent common ancestor of the grey individuals). N: population
size.

RNA secondary structure and influence RNA stability; also, they result in the usage
of a different tRNA that may be less abundant. Still, most synonymous mutations
can be considered selectively neutral.

The rate at which a mutation becomes fixed through *deterministic* or *stochastic*
forces depends on the **effective population size** (N_e) of the organism. This can
be defined as the size of an idealized population that is randomly mating and
that has the same gene frequency changes as the population being studied (the
"census" population). The effective population size is smaller than the overall
population size (N), when a substantial proportion of a population is producing
no offspring, when there is inbreeding, in cases of population subdivision, and when
selection operates on linked viral mutations. The effective population size is a major
determinant of the dynamics of the allele frequencies over time. When the (effective)
population size varies over multiple generations, the rates of evolution are notably
influenced by generations with the smallest effective population sizes. This may
be particularly true if population sizes are greatly reduced due to catastrophes,
or during migrations, etc. (Fig. 1.3). Such events can significantly affect genetic
diversity and are called **genetic bottlenecks**. Two individual lineages merging into

a single ancestor as we go back in time is referred to as a *coalescent event*. In general, the most recent common ancestor of the extant generation never traces back to the first generation of a population. In Fig. 1.3, all individuals of the seventh generation have one common ancestor in the fourth generation (tracing back the gray individuals), which is called the coalescence time of the extant individuals.

An entirely *deterministic* evolutionary pattern would require that changes in allele or gene frequencies depend solely on the reproductive fitness of the variants in a particular environment and on the environmental conditions. In such a situation the gene frequencies can be predicted if the fitness and environmental conditions are known. In deterministic evolution, changes other than environmental conditions, such as chance events, do not influence allele/gene frequencies. This can only hold true if the effective population is infinitely large. **Natural selection**, the effect of positive and negative selective pressure, accounts entirely for the changes in frequencies. When random fluctuations determine in part the allele frequencies, allele/gene frequencies cannot be predicted exactly. In such a *stochastic* model, one can only determine the probability of frequencies in the next generation. These probabilities still depend on the reproductive fitness of the variants in a particular environment and on the environmental conditions. However, chance events also play a role in populations of limited size. Consequently, only statistical statements about allele/gene frequencies can be made. Random ***genetic drift***, therefore, contributes significantly to changes in frequencies under a stochastic model. The smaller the effective population size, the larger the effect of chance events and the more the mutation rate is determined by genetic drift rather than by selective pressure.

Evolution is never entirely deterministic or entirely stochastic. Depending on the interplay of effective population size and the distribution of selective coefficients, the evolution of allele/gene frequencies is more affected by either natural selection or genetic drift. Genetic mutations are always random, but they can sometimes result in an adaptive advantage. In this case, positive selective pressure will increase the frequency of the *advantageous mutation*, eventually leading to fixation after fewer generations than expected for a neutral change, provided the effective population size is large enough. A mutation under negative selective pressure can become fixed due to random genetic drift when it is not entirely deleterious, but this generally requires more generations than expected for a neutral change. Non-synonymous mutations result in a phenotypic change of an organism, and are subject to selective pressure if they change the interaction of that organism with its environment. As explained above, synonymous mutations are usually neutral and therefore become fixed due to genetic drift. The effect of positive and negative selective pressure can be investigated by comparing the synonymous and non-synonymous substitution rate (see also Chapters 13 and 14).

Darwin realized that the factors that shaped evolution were an environment with limited resources, inheritable variations among organisms that influenced fitness, competition between organisms, and natural selection. In his view, the survival of the fittest was the result of these factors and the major force behind the origin of species (Darwin, 1859). Only in the twentieth century, after the rediscovery of Mendelian laws, was it realized that the sources of variation on which selection could act were random mutations. From this perspective, random mutations result in genetic variation, on which natural selection acts as the dominant force in evolution. Advantageous mutations that have a strong fitness impact become fixed due to positive selective pressure, mutations that result in a disadvantage are eliminated, and neutral mutations result in polymorphisms that can be maintained for a long period in a population. Changes in the environment can change the fate of neutral mutations into advantageous or disadvantageous mutations resulting in subsequent fixation or elimination. The phenotypic effects of *adaptive evolution* can be more easily appreciated when long time periods are considered (such as adaptive evolution in finches on the Galapagos islands), or when changes in the environment drastically change the fitness of the existing population (such as when new adaptive niches become available upon extinction of species). In adaptive evolution, a nucleotide or amino acid substitution is mainly the result of a process of positive selection. The surviving organisms increase their fitness, and become more and more adapted to the environment.

According to the **neutral theory of molecular evolution**, stochastic fixation of mutations is the most important driving force behind substitutions. Kimura advocated that the majority of genetic substitutions were the result of random fixation of neutral or nearly neutral mutations (Kimura, 1968). Positive selection does operate, but the effective population size is, in general, so small in comparison to the magnitude of the selective forces that the contribution of positive selection to evolution is too weak to shape the genome significantly. According to the neutral theory, only a small minority of mutations become fixed because of positive selection. Organisms are, in general, so well adapted to the environment that many non-synonymous mutations are deleterious and, therefore, quickly removed from the population by negative selection. Stochastic events predominate and substitutions are mainly the results of random genetic drift.

The fitness of an individual can be represented in an adaptive landscape where the x–y coordinates are considered as genetic variation, and the height represents the fitness of the genetic variant (Fig. 1.4a). The landscape has fitness valleys and fitness peaks, where each individual (with its individual genetic make-up) occupies a dot in the landscape. As long as a species, represented by a population of organisms (swarm of dots in the landscape), is not wandering close to the adaptive peak, adaptive evolution will drive the population upwards, selecting those individuals

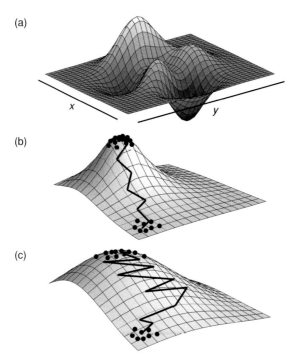

(a)

(b)

(c)

Fig. 1.4   Illustration of a fitness landscape with *x–y* coordinates representing genotype space and peaks representing optimal genotypes (a). (b) and (c) represent two different paths for populations climbing an adaptive peak with different slope.

with mutations advantageous for the particular environment (Fig. 1.4b). How erratic that path is depends on the effective population size and thus on the amount of genetic drift, and on the steepness of the slope towards the adaptive peak (Fig. 1.4b vs. 1.4c). Once this peak has been reached, neutral evolution will predominate and mutations are fixed, mainly because of random genetic drift. In this case, the smaller the effective population size, the faster mutations become fixed. Also, the sharper the fitness peak in the landscape, the lower the genetic variation in the population (Fig. 1.4b vs. 1.4c). Changes in the environment can reshape the adaptive landscape and again stimulate adaptive evolution. To what extent adaptive or neutral evolution acts upon the genetic make-up of an organism can be investigated with specific tools that will be explained in detail in Chapters 13 and 14.

1.3 Evolution and speciation

Two different populations of a single species can evolve into two different daughter species if they lose the capacity to interbreed due to accumulation of substitutions

Table 1.3 Evolutionary forces affecting speciation

Force	Variation within populations	Variation between populations
Mutation	+	+
Inbreeding or genetic drift (N_e is small)	−	+
Migration (↔ isolation)	+	−
Selection		
Positive/negative	+/−	+/−
Balancing	+	−
Incompatible	−	+

N_e: effective population size.

that prevent mating or fertile offspring. Evolutionary forces favoring speciation are those that decrease variation within a population, and increase the variation between populations (Table 1.3). The more mutations that arise in each population, the larger the variation within, but also between, populations. In a diploid population with $2N$ copies per gene, the frequency of a newly generated mutation is $1/(2N)$. The frequency of a new mutation is equal to its probability of fixation by genetic drift only. If the population size is smaller, the contribution of random genetic drift is larger. It takes, on average, $4N$ generations before a new polymorphism becomes fixed solely due to genetic drift. Therefore, the smaller the population size, the higher the chances that genetic drift will diversify two populations from each other. However, **migration** prevents speciation since it exchanges the new mutations between the two populations; conversely, isolation favors speciation. The *migration rate* is expressed in fraction of migrants, usually immigrants, in a population per generation. Even limited migration can prevent speciation. Migration prevents loss of heterozygosity if the number of migrants per generation is larger than 1. Within a population, individuals are under similar selective pressure, which reduces genetic variation. Different populations can experience different positive selective pressure, driving the populations apart and favoring speciation. Negative selective pressure favors the conservation of the parent phenotype, keeping both populations more similar. Balancing selection also maintains gene distributions and thus prevents speciation, whereas incompatible selection, where the heterozygote is less fit, will favor speciation.

Speciation is classified according to its driving evolutionary forces. If *reproductive isolation* between populations is caused mainly by geographical isolation (a new geographic barrier, or a founder effect of a population moving beyond a geographical barrier, etc.), it is classified as **allopatric speciation**. Allopatric speciation follows primarily the neutral model of evolution. Adaptation to the new geographic environment through adaptive evolution was not the cause of the speciation. **Sympatric speciation** happens when **disruptive selection** results in adaptation to different ecological niches in two populations within reproductive distance. The magnitude of the positive selective pressure must be very large to result in a fast morphological change preventing interbreeding. The impact of genetic drift can be large and usually contributes to differentiation between the two daughter species. **Parapatric speciation** lies somewhere in-between, when reproductive isolation results from a gradient of allele/gene frequencies over a rather homogenous geographic area, with often hybrid zones between the two species. In all cases, mutations preventing interbreeding are an important factor.

Having discussed evolutionary processes and speciation, it is useful to point out that the term phylogeny was originally used for evolutionary relationships between species, whereas **genealogy** was the preferred term to describe the shared ancestry within a population. However, the literature does not strictly adhere to these definitions.

1.4 Data used for molecular phylogenetics

To investigate the evolution and relationships among genes and organisms, different kinds of data can be used. The classical way of estimating the relationship between species is to compare their morphological characters (Linnaeus, 1758). Taxonomy is still based largely on morphology. The increasingly available molecular information, such as nucleotide or amino acid sequences, and restriction fragment length polymorphisms (RFLPs), can also be used to infer phylogenetic relationships. Whether the morphological or molecular approach is preferable for any particular evolutionary question has been hotly debated during the last decennia (e.g. Patterson *et al.*, 1993). However, the use of molecular data for inferring phylogenetic trees has now gained considerable interest among biologists of different disciplines, and it is often used in addition to morphological data to study relationships in further detail. For extinct species, it is difficult or impossible to obtain molecular data, and using morphological characteristics of mummies or fossils is usually the only way to estimate their relationships. However, organisms such as viruses do not leave fossil records. The only way to study their past is through the phylogenetic relationships of existing viruses. In this book, we introduce the concepts, mathematics, and techniques to infer phylogenetic trees from molecular data and to test hypotheses about the processes that have shaped nucleotide and amino acid sequences.

According to evolutionary theory, all organisms evolved from one common ancestor, going back to the origin of life. Different mechanisms of acquiring variation have led to the biodiversity of today. These mechanisms include mutations, duplication of genes, reorganization of genomes, and genetic exchanges such as recombination, reassortment, and lateral gene transfer. Of all these sources, mutations (i.e. point mutations, insertions, and deletions) are most often used to infer relationships between genes. Basically, phylogenetic methods consider the similarity among the genes, assuming that they are **homologous** (i.e. they share a common ancestor). Although it is assumed that all organisms share a common ancestor, over time the similarity in two genes can be eroded so that the sequence data themselves do not carry enough information about the relationship between the two genes and they have accumulated too much variation. Therefore, the term homology is used only when the common ancestor is recent enough for the sequence information to have retained enough similarity for it to be used in phylogenetic analysis. Thus, genes are either homologous or they are not. Consequently, such an expression as 95% homology does not exist, rather one should use the concept of 95% *similarity*.

When two sequences are compared, one can always calculate the percentage similarity by counting the number of identical nucleotides or amino acids, relative to the length of the sequence. This can be done even if the sequences are not homologous. DNA is composed of four different types of residues: A, G, C, and T. If gaps are *not* allowed, on average, 25% of the residues in two randomly chosen aligned sequences would be identical. If gaps *are* allowed, as many as 50% of the residues in two randomly chosen aligned sequences can be identical, resulting in 50% similarity by chance. For proteins, with 21 different types of codons (i.e. 20 amino acids and 1 terminator), it can be expected that two random protein sequences – after allowing gaps – can have up to 20% identical residues. Any input can be used to construct phylogenetic trees, but when using sequences with less than 60% similarity for nucleotide sequences, or less than 25% similarity for amino acid sequences, such trees are of little value. In general, the higher the similarity, the more likely the sequences are to be homologous.

Taxonomic comparisons show that the genes of closely related species usually only differ by a limited number of point mutations. These are usually found in the third (often redundant) codon positions of ORFs, resulting in a faster evolutionary rate at the third codon position compared with that at the first and second codon position. The redundancy of the genetic code ensures that nucleotide sequences usually evolve more quickly than the proteins they encode. The sequences may also have a few inserted or deleted nucleotides (i.e. indels). Genes of more distantly related species differ by a greater number of substitutions. Some genes are more conserved than others, especially those parts encoding, for example, catalytic sites or the core of proteins. Other genes may have insufficient similarity for phylogenetic purposes. Distantly related species often have discernible sequence relatedness only in the genes that encode enzymes or structural proteins. These similarities can be very distant and involve only short segments (i.e. motifs), interspersed with large regions with no similarity and of variable length, which indicates that many mutations and indels have occurred since they evolved from their common ancestor. Some of the proteins from distantly related species may have no significant sequence similarity but clearly similar secondary and tertiary structures. Protein sequence similarity is lost more quickly than secondary and tertiary structure. Thus, differences between closely related species are assessed most sensitively by analysis of their nucleotide sequences. More distant relationships, between families and genera, are best analyzed by comparing amino acid sequences, and may be revealed only by parts of some genes and their encoded proteins (see also Chapter 9).

Phylogenetic analysis establishes the relationships between genes or gene fragments, by inferring the common history of the genes or gene fragments. To achieve this, it is essential that homologous sites be compared with each other (*positional homology*). For this reason, the homologous sequences under investigation are aligned such that homologous sites form columns in the **alignment**. Obtaining the correct alignment is easy for closely related species and can even be done manually. The more distantly related the sequences, the less straightforward it is to find the best alignment. Therefore, alignments are usually constructed with specific software packages that implement particular algorithms. This topic is discussed extensively in Chapter 3.

Many popular algorithms start by comparing the sequence similarity of all sequence pairs, aligning first the two sequences with the highest similarity. The other sequences, in order of similarity, are added progressively. The alignment continues in an iterative fashion, adding gaps where required to achieve positional homology, but gaps are always introduced at the same position for all members of each growing cluster. Alignments obtained in this way are optimal only with respect to the clusters of sequences, there is no global optimization for the individual sequences in the overall alignment. When during the clustering procedure, gaps have been added to the cluster, affecting each individual sequence of the cluster at the common position, the resulting alignment can often be improved by manual editing. Obtaining a good alignment is one of the most crucial steps towards a good phylogenetic tree. When the sequence similarity is so low that an alignment becomes too ambiguous, it is better to delete that particular gene fragment from the alignment. Columns with gaps at the beginning and end of a sequence, representing missing sequence data for the shorter sequences, have to be removed to consider equal amounts of data for all sequences, unless the software used can deal with such missing data. Often, columns in the sequence alignment with deletions and insertions for the majority of the sequences are also removed from the analysis (see Chapter 3).

For a reliable estimate of the phylogenetic relationship between genes, the entire gene under investigation must have the same history. Recombination events within the fragment under investigation will affect phylogenetic inference (see Chapter 15). Recombination outside the fragment of interest does not disturb the tree; and a different clustering of two consecutive fragments can even be used to investigate recombination.

Genes originating from a duplication event recent enough to reveal their common ancestry at the nucleotide or amino acid level are called **paralogous**. Comparing such genes by phylogenetic analysis will provide information about the duplication event. Homologous genes in different species that have started to evolve independently because of the speciation are called **orthologous**. Comparing such genes by phylogenetic analysis will provide information about the speciation event. Therefore, to prevent flawed conclusions on speciation events, it is important to know *a priori* whether the genes are orthologous or paralogous.

Evolution under similar selective pressures can result in **parallel** or **convergent evolution**. When two enzymes evolved to have similar function, the similar functional requirements can result in a similar active site consisting of the same or similar amino acids. This effect can result in the two sequences having higher similarity than expected by chance, which can be mistaken for homology. Other events can result in a higher similarity of two sequences than the similarity expected from their evolutionary history. Sequence reversals occur when a substitution reverts back to the original nucleotide, multiple hits when a substitution has occurred several times at the same nucleotide, and parallel substitutions when the same substitution happened in two different lineages. Such events can lead to **homoplasy** in the alignment, and they can confound the linear relationship between the time of evolution and sequence divergence.

Currently, sequence information is stored in databases such as the National Center for Biotechnology Information (NCBI), the National Library of Medicine (NLM), the European Molecular Biology Laboratory (EMBL) or the DNA Database of Japan (DDJ). A search for homologous sequences based on similarity scores can be done in various ways in different databases. Some organizations provide a search service via the worldwide web (e.g. NCBI Blast). However, no search method is perfect and related sequences may be missed. Information on search engines is provided in Chapter 2.

1.5 What is a phylogenetic tree?

Evolutionary relationships among genes and organisms can be illustrated elegantly using a phylogeny, which is comparable to a pedigree and shows which genes or organisms are most closely related. The various diagrams used for depicting

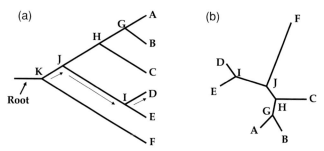

Fig. 1.5   Structure of (a) a rooted and (b) an unrooted phylogenetic tree. Both trees have the same topology. A rooted tree is usually drawn with the root to the left. A, B, C, D, E, and F are external nodes or operational taxonomic units. G, H, I, J, and K are internal nodes or hypothetical taxonomic units, with K as root node. The unrooted tree does not have a root node. The lines between the nodes are branches. The arrow indicates the direction of evolution in the rooted tree (e.g. from root K to external node D). The direction of evolution is not known in an unrooted tree.

these relationships are called phylogenetic trees because they resemble the structure of a tree (Fig. 1.5), and the terms referring to the various parts of these diagrams (i.e. *root*, *branch*, *node*, and *leaf*) are also reminiscent of trees. *External (terminal) nodes* or *leaves* represent the *extant (existing) taxa* and are often called *operational taxonomic units* (*OTUs*), a generic term that can represent many types of comparable **taxa** (e.g. a family of organisms, individuals, or virus strains from a single species or from different species). Similarly, *internal nodes* may be called *hypothetical taxonomic units* (*HTUs*) to emphasize that they are the hypothetical progenitors of OTUs. A group of taxa that share the same branch have a ***monophyletic*** origin and is called a **cluster**. In Fig. 1.5, the taxa A, B, and C form a cluster, have a common ancestor H, and, therefore, are of monophyletic origin. C, D, and E do not form a cluster without including additional strains and are not of monophyletic origin, they are called ***paraphyletic***. The branching pattern – that is, the order of the nodes – is called the **topology** of the tree.

An ***unrooted*** tree only positions the individual taxa relative to each other without indicating the direction of the evolutionary process. In an unrooted tree, there is no indication of which node represents the ancestor of all OTUs. To indicate the direction of evolution in a tree, it must have a root that indicates the common ancestor of all the OTUs (Fig. 1.5). The tree can be *rooted* if one or more of the OTUs form an ***outgroup*** because they are known as, or believed to be, the most distantly related of the OTUs (i.e. *outgroup rooting*). The remainder then forms the *ingroup*. The root node is the node that joins the ingroup and outgroup taxa and thus represents their common ancestor. It is still possible to assign a root even in the case where there is no suitable outgroup available, for example, because all available outgroup taxa are related too distantly and the alignment with the

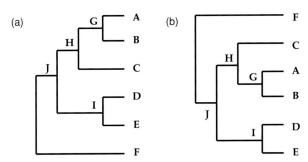

Fig. 1.6   Structure of a rooted phylogenetic tree. This is the same tree as in Fig. 1.5 but in a different style. Both trees (a) and (b) have identical topologies, with some of the internal nodes rotated.

outgroup is too ambiguous. Assuming that the rate of evolution in the different lineages is similar, the root will then lie at the midpoint of the path joining the two most dissimilar OTUs (***midpoint rooting***).

When trying to root a tree, it is not advisable to choose an outgroup that is related too distantly to the ingroup taxa. This may result in topological errors because sites may have become saturated with multiple mutations, implying that information at these sites may have been erased. On the other hand, an outgroup that is related too closely to the taxa under investigation is also not appropriate; it may not represent a "true" outgroup. The use of more than one outgroup generally improves the estimate of the tree topology. As mentioned previously, midpoint rooting could be a good alternative when no outgroups are available, but only in the case where all branches of the tree have roughly similar evolutionary rates.

Various styles are used to depict phylogenetic trees. Figure 1.6 demonstrates the same tree as in Fig. 1.5, but in a different style. Branches at internal nodes can be rotated without altering the topology of a tree. Both trees in Fig. 1.6 have identical topologies. Compared with tree (a), tree (b) was rotated at nodes J and H.

Frequently, biologists are interested in the time of the common origin of a gene or the divergence of a group of taxa. Phylogenetic analysis provides useful tools to calculate the *time to the most recent common ancestor* (*TMRCA*) for all the extant alleles/genes. Divergence time calculations are often used when investigating the origin of a species. In such calculations, one can consider a between-species approach, or a within-species approach.

Since phylogenetic trees represent the evolutionary history of a particular gene or DNA sequence, they can be called *gene trees*. Whether these gene trees reflect the relationships among species and can be considered a *species tree* depends on whether the aligned genes are orthologous or paralogous genes as explained above. To establish evolutionary relationships between species, orthologous genes have

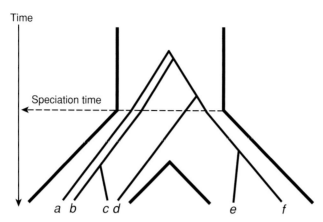

Fig. 1.7   A schematic representation of a lineage sorting at speciation. Six extant alleles (*a–f*) and their evolutionary relationships are represented for the two daughter species. In several cases, the origin of some polymorphisms predates the speciation event and due to the particular *lineage sorting* scenario, the gene tree will not match the history of speciation.

to be compared. When dealing with a gene that has polymorphic sites in the parent and daughter species, the nodes never really reflect the speciation event, but merely separation between different alleles. In the case where different alleles are investigated in daughter species and some sequence variation already existed in the parent species, separation of different alleles may both precede and follow upon speciation.

When analyzing alleles within a species, the divergence time will depend on the extinction dynamics of alleles after speciation. Because of the loss of alleles after speciation, the divergence time for the different alleles within a species can be more recent than the actual speciation time (e.g. alleles *e* and *f* in Fig. 1.7). For example, the divergence time of human mitochondrial DNA, which is inherited through the female line, is calculated to be around 200 000 years ago (Vigilant *et al.*, 1991; Ingman *et al.*, 2000). The divergence time for the Y chromosome is around 70 000 years ago (Dorit *et al.*, 1995; Thomson *et al.*, 2000), yet human speciation was not at a different time for women than for men. The estimated dates are the divergence times for the two different genes analyzed, whose polymorphic origins do not necessarily have to be simultaneously. When the origin of the polymorphisms predates speciation, the divergence time of the existing alleles of a species can even precede speciation (e.g. alleles *a*, *b*, *c*, and *d* in Fig. 1.7). In addition, different genes can have different divergence times when recombination has occurred among loci. In this case, each locus is a different "run" of the coalescent process (see Chapter 15). Divergence times may also vary among genes because they can have different allelic copy numbers (i.e. autosomal vs. sex-linked vs. mitochondrial loci). For closely related species, ancestral diversity gives rise to **lineage sorting**, so the gene tree phylogeny may not exactly match the true history of speciation (see Fig. 1.7). Whether the divergence time of existing alleles precedes or follows speciation will also be dependent on the effective population size.

To estimate divergence times, we generally assume that sequence divergence accumulates at a roughly constant rate over time; this assumption is referred to as the ***molecular clock*** hypothesis. When the molecular clock holds, all lineages in the tree have accumulated substitutions at a similar rate (see also Chapter 11). However, the evolutionary rate is dependent on many factors, including the underlying mutation rate, metabolic rates in a species, generation times, population sizes, and selective pressure. Therefore, real molecular data frequently violates a strict molecular clock assumption. Chapter 11 discusses how to perform statistical tests that evaluate how the evolutionary rates along the branches in a tree deviate from a uniform rate. It is interesting to note that ***relaxed molecular clock*** models have recently been developed, which can accommodate rate variation when estimating divergence dates.

The ancestor–descendant relationship between species (or groups of organisms) can be represented using a ***cladogram***, which is not necessarily based on phylogenetic analysis. Cladograms can be drawn based on morphological characters of fossils, and the branches can be calculated from independent methods such as radiocarbon dating. A ***phylogram*** depicts the phylogenetic relationships between a group of taxa with branch lengths representing evolutionary distances, which were inferred using a phylogenetic approach.

1.6 Methods for inferring phylogenetic trees

Reconstructing the phylogeny from nucleotide or amino acid alignments is not as straightforward as one might hope, and it is rarely possible to verify that one has arrived at the "true" conclusion. Although there are many methods available, none of them guarantees that the inferred phylogenetic tree is, in fact, the "true" phylogenetic tree.

The methods for constructing phylogenetic trees from molecular data can be grouped according to the kind of data they use, **discrete character** states or a **distance matrix** of pairwise dissimilarities, and according to the algorithmic approach of the method, either using a *clustering algorithm* usually resulting in only one tree estimate, or using an *optimality criterion* to evaluate different tree topologies.

Character-state methods can use any set of discrete characters, such as morphological characters, physiological properties, restriction maps, or sequence data. When comparing sequences, each sequence position in the aligned sequences is a "character," and the nucleotides or amino acids at that position are the "states." Usually, all character positions are analyzed independently; so each alignment column is assumed to be an independent realization of the evolutionary process. Character-state methods retain the original character status of the taxa and therefore, can be used to reconstruct the character state of ancestral nodes.

In contrast, *distance-matrix* methods start by calculating some measure of the dissimilarity of each pair of OTUs to produce a pairwise distance matrix, and then infer the phylogenetic relationships of the OTUs from that matrix. These methods seem particularly well suited for rapidly analyzing relatively large data sets. Although it is possible to calculate distances directly from pairwise aligned sequences, more consistent results are obtained when all sequences are aligned. Distance-matrix methods usually employ an **evolutionary model**, which corrects for multiple hits. When two sequences are very divergent, it is likely that, at a certain position, two or more consecutive mutations have occurred. Because of multiple events, two sequences are related more distantly than can be deduced from the actual percentage differences. Mathematical models allow for correcting the percentage difference between sequences. This results in **genetic** or **evolutionary distances**, which are always bigger than distances calculated by direct comparison of sequences (also called **p-distance**, see Chapter 4). Distance methods discard the original character state of the taxa. As a result, the information required to reconstruct character states of ancestral nodes is lost. The major advantage of distance methods is that they are generally computationally inexpensive, which is important when many taxa have to be analyzed.

Tree-evaluation methods employ an *optimality* or *goodness-of-fit criterion* and examine different tree topologies for a given number of taxa in the search for the tree that optimizes this criterion. **Maximum likelihood** methods (discussed later in this chapter and in Chapter 6) have the advantage of using a statistical criterion because they consider the probability that a tree gave rise to the observed data (i.e. the aligned sequences) given a specific evolutionary model. This allows the investigator to compare the relative support for different phylogenetic trees in a statistical framework (see Chapter 12). Unfortunately, an exhaustive search exploring all possible trees is usually not possible since the number of possible trees, and thus the computing time, grows explosively as the number of taxa increases; the number of bifurcated rooted trees for n OTUs is given by $(2n-3)!/(2^{n-2}(n-2))!$ (Table 1.4). This implies that, for a data set of more than 10 OTUs, only a subset of possible trees can be examined. In this case, various *heuristic* strategies have been proposed to search the "tree space," but there is no algorithm that guarantees to find the best tree under the specified criterion.

The **clustering** methods avoid the problem of having to evaluate different trees by gradually clustering taxa into a single tree. These tree-construction methods employ

Table 1.4 Number of possible rooted and unrooted trees for up to 10 OTUs

Number of OTUs	Number of rooted trees	Number of unrooted trees
2	1	1
3	3	1
4	15	3
5	105	15
6	954	105
7	10 395	954
8	135 135	10 395
9	2 027 025	135 135
10	34 459 425	2 027 025

Table 1.5 Classification of phylogenetic analysis methods and their strategies

	Optimality search criterion	Clustering
Character state	Maximum parsimony (MP)	
	Maximum likelihood (ML)	
	Bayesian inference	
Distance matrix	Fitch–Margoliash	UPGMA
		Neighbor-joining (NJ)

various algorithms to construct a tree topology, which differ in their methods of determining the relationships between OTUs, in combining OTUs into clusters, and in computing branch lengths. They are usually fast and can compute phylogenies for large numbers of OTUs in a reasonable time. Because they produce only one tree, we have no idea how well the data support alternative hypotheses. However, various statistical procedures have been proposed to evaluate the robustness of clusters in the tree. The majority of distance-matrix methods use clustering algorithms to compute the "best" tree, whereas most character-state methods employ an optimality criterion.

Table 1.5 classifies phylogenetic tree-construction and tree-analysis methods according to the data and algorithmic strategy used: character state or distance-matrix, tree-evaluation, or clustering methods. All methods use particular assumptions, which are not necessarily valid for the evolutionary process underlying the sequence data. Therefore, it is important to be aware of the assumptions that were made when evaluating the tree generated by each method. The different methods and their assumptions are explained in detail in the following chapters.

UPGMA is an abbreviation of ***unweighted pair group method with arithmetic means***. This is probably the oldest and simplest method used for reconstructing phylogenetic trees from distance data. Clustering is done by searching for the smallest value in the pairwise distance matrix. The newly formed cluster replaces the OTUs it represents in the distance matrix, and distances between the newly formed cluster and each of the remaining OTUs are calculated. This process is repeated until all OTUs are clustered. In UPGMA the distance of the newly formed cluster is the average of the distances of the original OTUs. This process of averaging assumes that the evolutionary rate from the node of the two clustered OTUs to each of the two OTUs is identical. The whole process of clustering thus assumes that the evolutionary rate is the same in all branches, which is frequently violated. Therefore, UPGMA tends to give the wrong tree when evolutionary rates differ along the different branches (see also Chapter 5).

The ***Neighbor-joining*** (***NJ***) method constructs a tree by sequentially finding pairs of neighbors, which are the pairs of OTUs connected by a single interior node. The clustering method used by this algorithm is quite different from the one described above, because it does not attempt to cluster the most closely related OTUs, but rather minimizes the length of all internal branches and thus the length of the entire tree. The NJ algorithm starts by assuming a star-like tree that has no internal branches. In the first step, it introduces the first internal branch and calculates the length of the resulting tree. The algorithm sequentially connects every possible OTU pair and finally joins the OTU pair that yields the shortest tree. The length of a branch joining a pair of neighbors, X and Y to their adjacent node is based on the average distance between all OTUs and X for the branch to X, and all OTUs and Y for the branch to Y, subtracting the average distances of all remaining OTU pairs. This process is then repeated, always joining two OTUs (neighbors) by introducing the shortest possible internal branch (see also Chapter 5).

The *Fitch–Margoliash* method is a distance-matrix method that evaluates all possible trees to find the tree that minimizes the differences between the pairwise genetic distances and the distance represented by the sum of branch lengths for each pair of taxa in the tree.

Maximum parsimony (***MP***) aims to find the tree topology for a set of aligned sequences that can be explained with the smallest number of character changes (i.e. substitutions). For a particular topology, the MP algorithm infers for each sequence position the minimum number of character changes required along its branches to explain the observed states at the terminal nodes. The sum of this score for all positions is called the *parsimony length* of a tree and this is computed for different tree topologies. When a reasonable number of topologies have been evaluated, the tree that requires the minimum number of changes is selected as the maximum parsimony tree (see Chapter 8). Maximum parsimony is very sensitive to the "*long-branch attraction*" problem since there is no way to correct for multiple hits. MP always assumes that a common character is inherited directly from a common ancestor and thus always underestimates the real divergence between distantly related taxa.

Maximum likelihood (ML) is similar to the MP method in that it examines different tree topologies and evaluates the relative support by summing over all sequence positions. ML algorithms search for the tree that maximizes the probability of observing the character states, given a tree topology and a model of evolution. For a particular tree, the *likelihood* calculation involves summing over all possible nucleotide (or amino acid) states in the ancestral (internal) nodes. Numerical optimization techniques are used to find the combination of branch lengths and evolutionary parameters that maximizes the likelihood. Depending on the search algorithm, the likelihood of a number of tree topologies is searched with this criterion, and the tree yielding the highest likelihood is chosen as the best tree. Unfortunately, obtaining the likelihood of a tree can be computationally very demanding (see Chapter 6).

Bayesian methods are character-state methods that use an optimality criterion, but they are conceptually very different from MP and ML in that they do not attempt to search only for the single best tree. Bayesian methods also employ the concept of likelihood, but by targeting a probability distribution of trees, they search for a set of plausible trees or hypotheses for the data. This *posterior distribution* of trees inherently holds a confidence estimate of any evolutionary relationship. Bayesian methods require the researcher to specify a prior belief, which is formalized as a *prior distribution* on the model parameters, i.e. substitution model parameters, branch lengths, and tree topology. The relative evidence present in the data is then used to evaluate how one should update his/her prior belief. If no appropriate biological information is available, the prior belief is preferably vague or uninformative. In Bayesian phylogenetic inference, a uniform prior on topology is the most objective (e.g. every tree topology is assumed to be equally likely before looking at the data). Posterior probabilities are obtained by exploring tree space using a sampling technique, called *Markov chain Monte Carlo* (*MCMC*). This sampling method starts by simulating a random set of parameters and proposes a new "state," which is a new set of parameters, by changing the parameters to some extent using random operators. In each step, the likelihood ratio and prior ratio is calculated for the new state relative to the current state. When the combined product is better, the parameters are accepted and a next step is proposed; if the outcome is worse, the probability that the state is rejected is inversely proportional to how much worse the new state is. After an initial convergence to a set of probable model/tree solutions ("*burn-in*", which needs to be discarded), it is hoped that this stochastic algorithm samples from the "posterior" probability distribution.

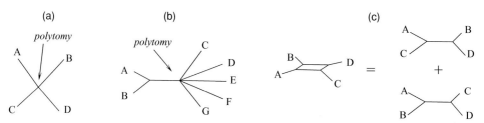

Fig. 1.8   Non-bifurcating trees and networks; arrows indicate polytomy. (a) Star-like (or multifurcating) tree. (b) Tree with an internal polytomy. (c) Networks representation: the network on
the left is one way of displaying simultaneously the two conflicting tree topologies on the
right.

The frequency by which a particular tree topology is sampled is then proportional
to its posterior probability. The results are usually presented as summary features
of the samples, for example, the mean or median for continuous parameters,
but for trees, a consensus tree or *maximum a posteriori* tree can be presented.
Bayesian methods are computer intensive, but from a single MCMC, run support
values for the clusters in a tree can be derived (Chapter 7). When the tree is
only a **nuisance parameter** in obtaining estimates for other parameters, Bayesian
inference provides a natural way of taking phylogenetic uncertainty into account
(e.g. Chapter 18).

There have been several reports comparing different algorithms on different data
sets, either simulated or empirical. As extensively explained in the rest of this book,
some methods clearly perform better than others on empirical or simulated data
sets. When judging the reliability of different methods, one always has to bear in
mind that different tree-inference algorithms are based on different assumptions,
and searching for a tree implies already an assumption about the evolutionary
process (see below). The use of statistical methods assists in assessing the reliability
of certain clusters (or tree topologies, see Chapter 12). However, they are also
dependent on the phylogeny method used and suffer from the same systematic bias
(i.e. when the assumptions of the method are not met by the data). They should
be regarded as ways to explore the "neighborhood" of a point estimate obtained
by any estimator. A widely used technique to evaluate tree topologies is **bootstrap
resampling**, which will be explained in detail in the next few chapters.

1.7 Is evolution always tree-like?

The algorithms discussed in the previous section usually generate *strictly bifurcating trees* (i.e. trees where any internal node is always connected to only three
other branches; see Fig. 1.5). This is the standard way of representing evolutionary

relationships among organisms, but it presumes that, during the course of evolution, any ancestral sequence (internal nodes of the tree) can give rise to only two separate lineages (leaves). However, there are phenomena in nature, such as the explosive evolutionary radiation of HIV or HCV, that might be best represented by a multifurcating tree, such as the one shown in Fig. 1.8a, or by a tree that allows some degree of multifurcation (Fig. 1.8b). Multifurcations on a phylogenetic tree are also known as **polytomies**, and can be classified as *hard polytomies* and *soft polytomies*. Hard polytomies represent explosive radiation in which a single common ancestor almost instantly gave rise to multiple distinct lineages at the same time. Hard polytomies are difficult to prove and it is even questionable as to whether they actually do occur (for a detailed discussion, see Li, 1997, and Page & Homes, 1998). Soft polytomies, on the other hand, represent unresolved tree topologies. They reflect the uncertainty about the precise branching pattern that best describes the data. Finally, there are situations – for example, in the case of recombination – in which the data seem to support two or more different tree topologies to some extent. In such cases, the sequences under investigation may be better represented by a network, such as the one depicted in Fig. 1.8c. Networks are covered in detail in Chapter 21.

FURTHER READING

Griffiths A. J. F., Miller, J. H., Suzuki, D. T., Lewontin, R. C., & Gelbart, W. M. (2005). *An Introduction to Genetic Analysis*. New York, USA: W. H. Freeman.

Hillis, D. M., Moritz, C., & Mable, B. K. (1996). *Molecular Systematics*. Sunderland, MA, USA: Sinauer Associates, Inc.

Li, W.-H. (1997). *Molecular Evolution*. Sunderland, MA, USA: Sinauer Associates, Inc.

Li, W.-H. & Graur, D. (1991). *Fundamentals of Molecular Evolution*. Sunderland, MA, USA: Sinauer Associates, Inc.

Nei, M. (1987). *Molecular Evolutionary Genetics*. New York, USA: Columbia University Press.

Page, R. D. M. & Homes, E. C. (1998). *Molecular Evolution. A Phylogenetic Approach*. Oxford, UK: Blackwell Science Ltd.

Section II

Data preparation

Sequence databases and database searching

THEORY

Guy Bottu

2.1 Introduction

Phylogenetic analyses are often based on sequence data accumulated by many investigators. Faced with a rapid increase in the number of available sequences, it is not possible to rely on the printed literature; thus, scientists had to turn to digitalized databases. Databases are essential in current bioinformatic research: they serve as information storage and retrieval locations; modern databases come loaded with powerful query tools and are cross-referenced to other databases. In addition to sequences and search tools, databases also contain a considerable amount of accompanying information, the so-called **annotation**, e.g. from which organism and cell type a sequence was obtained, how it was sequenced, what properties are already known, etc. In this chapter, we will provide an overview of the most important publicly available sequence databases and explain how to search them. A list of the database URLs discussed in this section is provided in Box 2.1.

To search sequence databases, there are basically three different strategies.

– To easily retrieve a known sequence, you can rely on unique *sequence identifiers*.
– To collect a comprehensive set of sequences that share a taxonomic origin or a known property, the annotation can be searched *by keyword*.
– To find the most complete set of *homologous* sequences a search *by similarity* of a selected query sequence against a sequence database can be performed using tools like BLAST or FASTA.

The Phylogenetic Handbook: a Practical Approach to Phylogenetic Analysis and Hypothesis Testing,
Philippe Lemey, Marco Salemi, and Anne-Mieke Vandamme (eds.). Published by Cambridge
University Press. © Cambridge University Press 2009.

Box 2.1 URLs of the major sequence databases and database search tools

ACNUC: *http://pbil.univ-lyon1.fr/databases/acnuc/acnuc.html*
BioXL/H: *http://www.biocceleration.com/BioXLH-technical.html*
BLAST: *http://www.ncbi.nlm.nih.gov/blast/*
DDBJ and DAD: *http://www.ddbj.nig.ac.jp/*
EMBL: *http://www.ebi.ac.uk/embl/*
EMBL Sequence Version Archive: *http://www.ebi.ac.uk/cgi-bin/sva/sva.pl*
Ensembl: *http://www.ensembl.org/*
Entrez: *http://www.ncbi.nlm.nih.gov/Entrez/*
fastA: *http://fasta.bioch.virginia.edu/fasta/*
GenBank: *http://www.ncbi.nlm.nih.gov/Genbank/*
Gene Ontology: *http://www.geneontology.org/*
HAMAP: *http://www.expasy.org/sprot/hamap/*
HCV database: *http://hcv.lanl.gov/*
HIV database: *http://hiv-web.lanl.gov/*
HOGENOM: *http://pbil.univ-lyon1.fr/databases/hogenom.html*
HOVERGEN: *http://pbil.univ-lyon1.fr/databases/hovergen.html*
IMGT/HLA: *http://www.ebi.ac.uk/imgt/hla/*
IMGT/LIGM: *http://imgt.cines.fr/*
MPsrch, Scan-PS, WU-BLAST and fastA at EBI: *http://www.ebi.ac.uk/Tools/
    similarity.html*
MRS: *http://mrs.cmbi.ru.nl/mrs-3/*
NCBI Map Viewer: *http://www.ncbi.nlm.nih.gov/mapview/*
ORALGEN: *http://www.oralgen.lanl.gov/*
PDB: *http://www.rcsb.org/*
PRF/SEQDB: *http://www.prf.or.jp/*
RefSeq: *http://www.ncbi.nlm.nih.gov/RefSeq/*
Sequin: *http://www.ncbi.nlm.nih.gov/Sequin/*
SRS: *http://www.biowisdom.com/navigation/srs/srs*
SRS server of EBI: *http://srs.ebi.ac.uk/*
SRS list of public servers: *http://downloads.biowisdomsrs.com/publicsrs.html*
Taxonomy: *http://www.ncbi.nlm.nih.gov/Taxonomy/*
TIGR: *http://www.tigr.org/*
UCSC Genome Browser: *http://genome.ucsc.edu/*
UniGene: *http://www.ncbi.nlm.nih.gov/UniGene/*
UniProt at EMBL: *http://www.ebi.ac.uk/uniprot/*
UniProt at SIB: *http://www.expasy.uniprot.org/*
VAST: *http://www.ncbi.nlm.nih.gov/Structure/VAST/vastsearch.html*
WU-BLAST: *http://blast.wustl.edu/*

2.2 Sequence databases

2.2.1 General nucleic acid sequence databases

There are parallel efforts in Europe, USA, and Japan to maintain public databases with all published nucleic acid sequences:

- **EMBL** (European Molecular Biology Laboratory) database, maintained at the EMBL-EBI (European Bioinformatics Institute, Hinxton, England, UK).
- **GenBank**, maintained at the NCBI (National Center for Biotechnology Information, Bethesda, Maryland, USA).
- **DDBJ** (DNA Data Bank of Japan), maintained at the NIG/CIB (National Institute of Genetics, Center for Information Biology, Mishima, Japan).

In the early 1980s the database curators scanned the printed literature for novel sequences, but today the sequences are exclusively and directly submitted by the authors through World Wide Web submission tools (or by email after preparation of the data using the `Sequin` software). There is an agreement between the curators of the three major databases to cross-submit the sequences to each other. The databases contain RNA as well as DNA sequences but, by convention, a sequence is always written as DNA, that is with Ts and not with Us. Note that often, but not always, sequence analysis software treats U and T without making a distinction. Modified bases are replaced by their "parent" base ACGT but the modification is mentioned in the annotation.

The databases have the notion of *release.* At regular intervals (2 months for GenBank, 3 months for the two others) the database is "frozen" in its current state to make a "release." The sequences that are submitted since the last release are available as "daily updates." On June 15, 2007, GenBank release 160.0 contained more than 77 248 690 945 nucleotides in 73 078 143 sequences. Over the years, the sequence databases have grown at an enormous rate (Fig. 2.1).

Each sequence has a series of unique identifiers (see Fig. 2.2), which allow easy retrieval:

Entry name, locus name or identifier (ID)

The ID was originally designed to be mnemonic (e.g. ECARGS for the ArgS gene of *E. coli*) but, facing the rapid accumulation of anonymous fragments, many sequences now have an ID that is simply identical to their AC. The ID can change from release to release (in the face of new information of what the sequence is) and can be different in the three databases. Note that the EMBL has decided to drop the use of IDs from June 2006 on, but GenBank and DDBJ still use them.

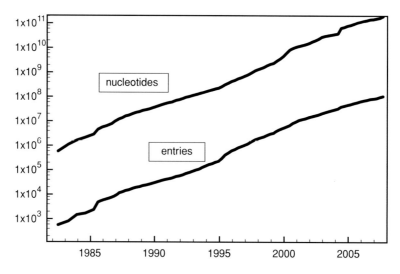

Fig. 2.1   Growth (on a logarithmic scale) of the number of nucleotides and entries in the public EMBL database (non-redundant) from June 1982 to September 2007.

```
LOCUS       HUMBMYH7                28452 bp     DNA      linear    PRI 30-OCT-2002
DEFINITION  Homo sapiens beta-myosin heavy chain (MYH7) gene, complete cds.
ACCESSION   M57965 M30603 M30604 M30605 M57747 M57748 M57749
VERSION     M57965.2  GI:24429600
```

Fig. 2.2   First lines of a GenBank entry. Note the different identifiers: the entry name HUMBMYH7, the primary accession number M57965 (followed by 6 secondary accession numbers), the version number M57965.2 and the GI number.

Accession number (AC)

Contrary to the ID, the AC remains constant between releases and there exists an agreement between the managers of the three databases to give the same AC to the same sequence. Besides its *primary accession number*, a sequence can have *secondary accession numbers*; the reason is that, when several sequences are merged into one, the new sequence gets a new AC but inherits all the old ACs. Because of its stable and universal nature the AC is thus the most useful for retrieving a sequence.

Version number

The version number is made from the AC and a number that is incremented each time the sequence is modified. The version number is thus useful if it is necessary to find back exactly the sequence that was used for a particular study. To allow the retrieval of "old sequences" the EBI has set up a "*Sequence Version Archive*," which you can search by SV or by AC (to retrieve the last entry that had that primary accession number). Note, however, that the version number was introduced in the beginning of 1999 and does not cover the period before.

GenInfo number (GenBank only)

Each sequence treated by the NCBI is given a GI number, which is unique over all NCBI databases.

In addition to these identifiers, each coding sequence (CDS) identified in a database entry gets a *protein identifier* (protein_id), which is shared by the three databases and followed by a version number; in GenBank each CDS has also its own GI number.

The databases can be downloaded by anonymous ftp and can also be searched on-line. GenBank can be searched as part of the Entrez Nucleotide table. EMBL and DDBJ are available in the SRS servers of EBI and NIG respectively (see below).

In some interfaces, the ***Expressed Sequence Tags*** (ESTs) and ***Genome Survey Sequences*** (GSSs) sections of the databases are presented in separate sections because they make up huge collections of sequences with a high error rate ($\sim$1%) and often have a poor annotation. ESTs are basically mRNA fragments, obtained by extracting RNA followed by reverse transcription and single-pass automated DNA sequencing. GSSs are single-pass fragments of genomic sequence.

The three databases have two "annexes":

Whole Genome Shotgun (WGS) sequences

Sequences from genome sequencing projects that are being conducted by the whole genome shotgun method. They do not have a stable accession number. At regular time intervals, all the WGS sequences for one organism are removed from the database and replaced by a new set.

Third Party Annotations (TPA)

Sequences with improved documentation and/or sequences constructed by assembling overlapping entries, submitted by a person who is not the original author of the sequence(s).

2.2.2 General protein sequence databases

Proteins can be sequenced using the classic Edman degradation method, using modern methods based on peptide mass spectrometry, or their sequence can be deduced from the 3-D structure, as determined by X-ray crystallography or NMR. The bulk of the data in the protein databases is, however, obtained by sequencing and translating coding DNA/RNA. In this respect, it is important to be aware that the databases can, and do, contain errors (translations of open reading frames that do not correspond to real proteins, frame-shifting sequencing errors, erroneously assigned intron–exon boundaries, etc.).

Contrary to the situation for nucleic acids, which are stored in three nearly identical databases, proteins are stored in databases that differ considerably in content and quality. The main source of information is the **UniProt** knowledgebase, which is meant to contain all protein sequences that have ever been published, with exclusion, however, of minor variants of the same sequence, short fragments, and "dubious" sequences. It contains a collection of translations of open reading frames extracted from the annotation of the EMBL sequences and also submissions from authors who have performed "real protein sequencing". The sequences in UniProt are accompanied by an excellent annotation, including cross-references to many other databases and a tentative assignment of motifs, based on automated database searches, which contributes significantly to the value of this database. UniProt is composed of two parts, **UniProt/SwissProt** (maintained by the team of Prof. Amos Bairoch at the University of Geneva, Switzerland, in collaboration with the EMBL-EBI) and **UniProt/TrEMBL** (maintained at the EMBL-EBI). TrEMBL is a computer-generated database with little human interference; SwissProt is curated by human annotators, who manually add information extracted from the literature. Currently, EMBL translations always enter TrEMBL first and get an accession number. After their annotation has been verified and enriched by the curators, they are removed from TrEMBL and enter SwissProt while keeping their AC; author submissions directly go to SwissProt. SwissProt sequences have an ID of type XXXXX_YYYYY, where XXXXX stands for the nature of the protein and YYYYY is a unique organism identifier (e.g. TAP_DROME for the target of Poxn protein of *D. melanogaster*), TrEMBL sequences have a provisional ID composed of the AC and the organism identifier.

For the purpose of BLAST similarity searches, EBI also provides the following collections:

UniProt splice variants, useful in case the query sequence matches only a "variant" and not the "representative" sequence present in SwissProt.

UniRef100, a subset from UniProt, chosen so that no sequence is identical to or is a subfragment of another sequence; useful for faster searching.

UniRef90 and **UniRef50**, still smaller, are made from UniRef100 by excluding sequences shorter than 11 amino acids and by selecting sequences so that no sequence has more than 90% and 50% identity respectively with another (see also Chapter 9).

In addition to UniProt the following databases are also available:

GenPept: collection of open reading frames extracted from GenBank, maintained at the NCI-ABCC (National Cancer Institute, Advanced Biomedical Computing Center, Frederick, Maryland, USA).
DAD: collection of open reading frames extracted from DDBJ, maintained at the NIG/CIB.
PRF/SEQDB (Protein Resource Foundation), maintained at the PRF (Osaka, Japan), contains both translations and proteins submitted by authors.

2.2.3 Specialized sequence databases, reference databases, and genome databases

In addition to the general databases, there exist a large number (more than 100) of specialized databases. They can be constructed based on the general databases; they can accept author submissions and/or are generated by automated genome annotation pipelines. They offer one or more of the following advantages:

– The data form a well-defined searchable set of sequences. Searching a specialized database instead of a general one might yield the same result, but in less time and less "polluted" by background noise.
– The database is often "cleaned," containing less errors and less redundant entries.
– Sometimes the annotation has been standardized, so that you can easily find all the sequences you want without having to repeat the search using alternative keywords.
– The annotation is usually vaster and of better quality than the one found in the general databases. Moreover, some databases have the sequences already clustered into families and/or contain tables with sequence-related data, like genomic maps.

In Box 2.2 we provide a (necessarily incomplete and subjectively selected) list with the most interesting specialized databases.

2.3 Composite databases, database mirroring and search tools

2.3.1 Entrez

None of the general databases available is really complete. To make up for this deficiency, efforts have been undertaken to make composite databases. The most popular one is the Entrez database maintained by NCBI. Originally, it contained only sequences and abstracts from the medical literature, but in the course of the years new tables have been added (there are currently more than 30), making it an integrated database of data related to molecular biology. The individual tables are interlinked (Fig. 2.3), so that you can search, for example, for a protein sequence and then, by simply making a link, obtain the nucleic acid sequence of the gene that codes for it, its 3-D structure, the abstract of the reference that describes its sequencing, etc. Box 2.3 provides more detail on a selection of ENTREZ databases.

Box 2.2 A selection of secondary databases

RefSeq: collection of "reference" sequences, maintained at the NCBI. One can find in RefSeq an entry for each completely sequenced genomic element (chromosome, plasmid or other), as well as genomic sequences from organisms for which a genome sequencing project is underway, and also genomic segments corresponding to complete genes (as referenced in the Gene table of Entrez, see Box 2.3), together with their transcript (mRNA or non-coding RNA) and protein product. Some RefSeq entries are generated fully automatically, others are generated with some amount of human intervention (curated). The nature of a RefSeq Entry can be distinguished by its accession number, e.g. NC_* for (curated) complete genomic element or NT_* for (automated) intermediate assembly from BAC.

Some databases with automatically assembled and annotated eukaryotic genomes allow the user to navigate through genomic maps and extract well-defined portions of chromosome:

Ensembl: maintained at the EMBL-EBI, in collaboration with the Wellcome Trust Sanger Institute (Hinxton, England, UK)
NCBI Map Viewer: maintained at the NCBI
UCSC Genome Browser: maintained at the University of California, Santa Cruz (USA)

The Institute for Genomic Research, Rockville, Maryland, USA (**TIGR**): various genomes, mainly of microbes. The TIGR website also contains a collection of hyperlinks to other sites with information about genomes.

Databases with clustered sequences:

UniGene: collection of sequences grouped by gene, extracted from GenBank (for the moment only for a restricted number of organisms), maintained at the NCBI and accessible as a table of Entrez, see Box 2.3
Homologous Vertebrate Genes (**HOVERGEN**) database: vertebrate sequences (coding sequences extracted from EMBL and proteins extracted from UniProt), classified by protein family, maintained at the Université Claude Bernard (Lyon, France)
Homologous sequences from complete Genomes (**HOGENOM**) database: sequences from completely sequenced genomes (coding sequences extracted from EMBL and proteins extracted from UniProt), classified by protein family, maintained at the Université Claude Bernard (Lyon, France)
High-quality Automated and Manual Annotation of microbial Proteomes (**HAMAP**) database: protein sequences from completely sequenced prokaryotic genomes (eubacteria, archaebacteria, and also plastids), extracted from UniProt, classified by protein family and accompanied by expert annotation, maintained at the University of Geneva (Switzerland)

Databases with sequences related to sexually transmitted diseases, maintained at the LANL (Los Alamos National Laboratory, New Mexico, USA):

HIV database: HIV and SIV sequences
HCV database: Hepatitis C and related flavivirus sequences
ORALGEN: sequences from Herpes simplex virus and other oral pathogens

Immunogenetics **IMGT**: collection of sequence databases of immunological interest:

Laboratoire d'ImmunoGénétique Moléculaire (**IMGT/LIGM**): subset of EMBL with the immunoglobulin and T-cell receptor genes, maintained by the group of Professor Marie Paul Lefranc at the Laboratoire d'ImmunoGénétique, University of Montpellier (France). The documentation is enriched in three levels (the keywords are standardized and then the feature list is completed, first automatically and later by experts).

Human histocompatibility locus A (**IMGT/HLA**): contains genes for the human HLA class I and II antigens; it is maintained at the Anthony Nolan Research Institute (London, UK) and the sequences are named and documented according to the WHO HLA Nomenclature. It accepts direct submissions by authors.

Protein Data Bank (**PDB**): 3-D structures, as determined by X-ray crystallography or NMR, of proteins, nucleic acids, and their complexes, maintained by the Research Collaboratory for Structural Bioinformatics. Some sites offer a search against the protein or nucleic acid sequences corresponding to the PDB entries.

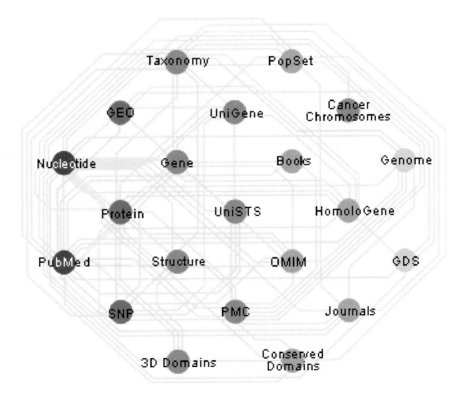

Fig. 2.3   Relationships among the different ENTREZ databases.

Box 2.3 A selection of ENTREZ databases

Nucleotide: A complete and non-redundant ("nr") nucleic acid sequence database. It contains, in that order, RefSeq, GenBank, EMBL, DDBJ and sequences from PDB. Significant effort has been made to remove duplicates. The NCBI also offers a BLAST search of your sequence against this "nr" database. Furthermore, the NCBI performs at regular intervals a BLAST search of each sequence against the complete database, so that Entrez can provide a "Related Sequences" link that allows retrieving highly similar sequences.

Protein: Idem for protein sequences, composed of RefSeq, translations from open reading frames found in GenBank, EMBL and DDBJ, sequences from the PDB, UniProt/SwissProt, PIR and PRF.

UniGene: Collection of GenBank sequences grouped by gene (for the moment only for a restricted number of organisms), as already mentioned above Box 2.2 about specialized databases.

Structure: Molecular Modeling Database (MMDB): 3-D structures of proteins, nucleic acids, and their complexes, derived from the PDB database, but in improved format. The NCBI has a tool for comparing 3-D structures, Vector Alignment Search Tool (VAST); they offer a search of a query protein structure against MMDB, as well as inside MMDB information about very similar structures. NCBI also provides a specialized tool for viewing the structures, Cn3D, which can be installed as a helper application ("plug-in") in a Web browser.

Gene: Information about genetic loci (including the alternative names of the gene, the phenotype, the position on the genetic map and cross-references to other databases), corresponding to the organisms and sequences in RefSeq.

PopSet: Database of multiple sequence alignments of DNA sequences of different populations or species, submitted as such to GenBank and used for phylogenetic and population studies.

PubMed: MEDLINE is a database with abstracts from the Medical Literature (starting from 1966), maintained at the National Library of Medicine (Bethesda, Maryland, USA). PubMed comprises MEDLINE plus not yet annotated references and references to non-life-science articles published in MEDLINE journals. The WWW interface from PubMed provides hyperlinks to on-line versions of some journals.

Online Mendelian Inheritance in Man (OMIM): A collection of really neat literature reviews. There is one entry for each identified human gene and for each trait shown to be hereditary in man. OMIM is maintained at the Johns Hopkins University School of Medicine (Baltimore, Maryland, USA) and began as the personal work of Dr. Victor McKusick, but in the meantime a whole team of experts has taken over the work.

Taxonomy: Contains an entry for each species, subspecies or higher taxon for which there is at least one sequence in GenBank/EMBL/DDBJ. The entry has a standard

name, which is used to make the "Organism" fields of the databases, as well as a
series of alternative names. At present, the Taxonomy nomenclature is the standard
for the three main nucleic acid databases and for UniProt.
   Medical Subject Headings (**Mesh**): A collection of standardized keywords, used by the
NLM to index MEDLINE.

2.3.2 Sequence Retrieval System (SRS)

An alternative to searching databases at their maintenance site or turning to "the
big site that has everything" is to have a local site with local copies of a series of
often used databases. It is not trivial to install "mirrors" of a variety of different
databases in one place under a common interface, since these databases not only
have a different design, but they are also maintained using different incompatible
and often expensive commercial software tools. A solution is to export/import
the databases as eXtensible Markup Language (XML), text interwoven with "tags"
that ease parsing, or even as a simple "flat file" in American Standard Code for
Information Interchange (ASCII) and use software that can generate searchable
indexes with keywords. At the time of this writing, the most popular and most
elaborate tool to do this is SRS. It was originally developed by Thure Etzold and
coworkers at the EMBL/EBI. Its copyright now belongs to the private enterprise,
BioWisdom Ltd. (Cambridge, UK), but academic users can still obtain a free license.
SRS can handle XML as well as simple ASCII; the commercial version, which has
extended capabilities, can also integrate databases under native RDBS, and has a
tool to automate the update of the databases (Prisma). The different databases can
be interlinked in a way similar to the tables of Entrez. To help the user perform
their query, version 8 of SRS has databases and links between databases organized
according to a *concept-oriented database* **ontology** (so that you can, for example,
search for "Macromolecular Structure" without having to know explicitly that you
need to select the PDB). There is also a *synonym finder* that proposes alternatives for
keywords. SRS can submit sequences to analysis tools (BLAST, CLUSTAL, EMBOSS, etc.)
and can present the results on-the-fly as a searchable database and/or in a graphical
survey. There are now more than 40 publicly accessible SRS servers worldwide; Bio-
Wisdom maintains a list of these sites and their collection of available databases (see
Box 2.1).

   Note that, at the moment of this writing, the future of SRS is somewhat uncertain
because of the recent sale of SRS by Lion Ltd. to BioWisdom Ltd. Alternatives do
exist, however; among the many available we can cite the great classic ACNUC of

the Université Claude Bernard (Lyon, France) and the very promising MRS of the
University of Nijmegen (the Netherlands).

2.3.3 Some general considerations about database searching by keyword

While searching databases, it is important to understand that the search results
depend in the first place on what has been written in the database by the submitting
authors and by the curators. Notwithstanding ongoing efforts, databases are only
partly standardized. So, you can find "HIV-1" as well as "HIV1." If you are looking
for proteases, sometimes you find the word "protease" in the text, but sometimes
only the name of the particular protease, like "trypsin" or "papain." Even worse,
there can be typographical errors like "psuedogene" instead of "pseudogene."
Furthermore, search results also depend on what has been put in the alphabetic
indexes used to search the database. So, it serves no purpose to search for "reverse
transcriptase" if "reverse" and "transcriptase" have been placed in the index as
separate keywords. It is a good idea to repeat your query several times with different
search terms. After you have performed some searches, you will get a better feeling
of what to expect in the databases. In this respect, the utility of some specialized
databases with standardized keywords has already been mentioned (see above).
The Taxonomy database in particular, which is part of Entrez, is useful to search by
keyword. In Entrez, or in an SRS server that holds a copy of the Taxonomy database,
it is possible to search for an organism name, eventually to browse up and down
in the hierarchy of the taxonomy, and then make the link to a sequence database.
While the "Organism" field has already been standardized a long time ago, a similar
effort is under way to standardize the "Keywords" fields of some databases, using
the *Gene Ontology* (GO).

A few more notes about composing queries:

- With Entrez and SRS (and in nearly all database search systems), a search is case-insensitive (there is no difference in using capitals or small letters).
- Search systems usually allow the use of *wild cards*: ? and * stand, respectively, for one character and for any string of characters, including nothing. SRS and Entrez automatically append a wildcard * at the end of each search word longer than one letter; you can turn this feature off by putting the search term between quotes. Entrez, however, only allows wild cards at the end of a search term.
- You can compose a query with logical combinations of search terms. Entrez will recognize AND, OR, and NOT (means *but not*); SRS uses the symbols & | ! instead. You can nest logical expression with parentheses "( )."
- We already mentioned the problem of search terms containing spaces. SRS has a special feature: when you type, for example, **protein c** it will in fact search **protein* & c | protein c*** unless you type explicitly with quotes **"protein c"**.

2.4 Database searching by sequence similarity

2.4.1 Optimal alignment

Relationships between homologous sequences are commonly inferred from an **alignment**. An alignment has bases/amino acids that occupy corresponding positions written in the same column. Deletions/insertions are indicated by means of a gap symbol, usually a ' - ' (or sometimes a ' . '). From any alignment a *similarity score* can be computed, which tells something about the quality of the alignment. This score is computed by summing up a comparison score for each aligned pair of bases/amino acids and by subtracting a penalty for each introduced gap. For nucleic acids one generally uses a simple match/mismatch score, although more complex scoring schemes like **transition/transversion** scoring have been proposed. For proteins **log odds matrices** are usually employed, like the classic **PAM** (Dayhoff *et al.*, 1978) and **BLOSUM** substitution matrices (Henikoff & Henikoff, 1992; see Fig. 3.4 in the next chapter). They contain similarity scores proportional to:

$$s(i, j) = s(j, i) = \log \frac{q_{ij}}{p_i * p_j} \tag{2.1}$$

with:

q_{ij} the probability of finding amino acid *i* and amino acid *j* aligned in a reliable alignment of related protein sequences.

p_i the proportion of amino acid *i* in proteins and thus the probability of finding *i* at a randomly chosen position of a protein.

$p_i * p_j$ (by consequence) the probability of finding amino acid *i* and amino acid *j* aligned in a random alignment.

The simplest method for handling gaps is to penalize every gap symbol. This is, however, unsatisfactory because deletions/insertions of more than one base/amino acid at a time are common. Therefore a gap of *n* positions is often penalized using $a + b * n$, where *a* is the *gap penalty* and *b* the *gap length penalty* (the approach followed by GCG, STADEN, NCBI BLAST, FASTA and CLUSTAL), or by the functionally equivalent $a + b * (n - 1)$, where *a* is the *gap opening penalty* and *b* the *gap extension penalty* (the approach followed by EMBOSS and WU-BLAST). Note that the choice of an appropriate gap penalty is a difficult bioinformatic problem that has not yet been solved appropriately and that it depends, among other things, on the scoring matrix used (if the values in the matrix are higher, so must be the gap penalty). An example of how a pairwise alignment is scored using the BLOSUM50 matrix and a gap penalty in conjunction with a gap length penalty is shown in Fig. 2.4.

```
AFEIWG---M

AY-LWHCTRM
```

Fig. 2.4   An example of how an alignment is scored. Using the BLOSUM50 table and a gap penalty
of 10 + 2n one obtains a similarity score of 3, as 5 (pair A-A) + 4 (pair F-Y) − 12 (one
position gap) + 2 (pair I-L) + 15 (pair W-W) − 2 (pair G-H) − 16 (three positions gap) + 7
(pair M-M).

Note that the similarity score is always defined symmetrically; it does not matter
which of the two sequences is on top. Note also that the similarity score can
eventually be negative, but this would mean that the alignment is of poor quality,
e.g. an alignment between proteins where hydrophobic amino acids have been
matched with hydrophilic amino acids.

Sometimes you can make a rapid and reliable alignment manually because the
sequences are so conserved or because a lot of experimental data are available that
show which bases/amino acids must correspond. If this is not the case, computational approaches must be employed to search for the *optimal alignment*, that is
the alignment yielding the highest possible similarity score for a particular scoring
scheme. Note, however, that nothing guarantees that the optimal alignment is also
the "true" alignment (in the biological sense of the term)! Indeed, when altering
the scoring scheme, different optimal alignments can be found (see Chapter 3).

To find the optimal alignment it is not necessary to write down explicitly all the
alternative alignments. A method exists that guarantees to find the best possible
alignment while minimizing the amount of calculation needed. This is the so-
called **dynamic programming** and *backtracking algorithm*. An explanation of how
it works is provided by Box 3.1 of Chapter 3. This algorithm yields the best **global
alignment** and was first proposed by Needleman and Wunsch (1970). Gotoh (1982)
proposed a trick to minimize the amount of computation needed for handling gap
penalties of type $a + b * n$. It can happen that two sequences are only recognizably
homologous in a limited region. To handle this situation, Smith and Waterman
(1981) proposed a modified version of the algorithm that searches the best **local
alignment**, allowing for stretches of unaligned sequences.

An obvious procedure for searching a database is to make, one by one, the optimal
alignments between your sequence (the query sequence) and all the sequences
in the database. The Smith–Waterman (SW) will usually be preferred over the
Needleman–Wunsch algorithm, because sequences in the databases are often of
varying lengths. One can then recover the sequences that yield a similarity score
above a certain threshold or, better, the sequences that satisfy some statistical
criterion that shows that the alignment is better than what you could obtain by
chance (see the explanation of the **BLAST**, and **FASTA** algorithms below for examples

of such criteria). This type of analysis can, however, take a lot of computer time and the time needed increases linearly with the size of the database. A search of a sequence of common size (1000 bases or 300 amino acids) against a general database can easily take several hours or several days on a standard computer! In order to obtain the result in a reasonable amount of time, such a search is usually performed on a cluster of computers, whereby the individual computers each search a part of the database. An example of this is the public MPsrch server (formerly known as Blitz) of EMBL–EBI, which in its present version uses a Compaq cluster and allows only protein searches. To save computation time, the algorithm can be hard-coded in a dedicated processor; a commercially available example of such a device is the BioXL/H manufactured by Biocceleration Ltd. (Beith-Yitschak, Israel).

2.4.2 Basic Local Alignment Search Tool (BLAST)

Since one cannot always afford a fancy machine or a long search time on a standard computer, bioinformaticians have developed so-called "heuristic" algorithms, which allow searching a database in considerably less time, at the price, however, of not having the absolute guarantee to find all the sequences that yield the highest optimal alignment scores. The most popular one is BLAST. The earlier version 1 of BLAST developed at NCBI produced alignments without gaps (Altschul *et al.,* 1990). After Warren Gish of the BLAST team moved to Washington University (Saint Louis, Missouri, USA), a new version at NCBI called interchangeably BLAST, BLAST 2, "GAPPED BLAST" or NCBI BLAST (Altschul *et al.,* 1997) has been developed in parallel with WU-BLAST at Washington University. Compared to NCBI BLAST, WU-BLAST has more optional parameters that allow trading off speed against sensitivity. Here, we will restrict our explanation to the algorithm used by NCBI BLAST:

- BLAST starts by searching for "words" (oligonucleotides/oligopeptides) shared by the query sequence and a database sequence. For nucleic acids the default word length is 11, for proteins 3. For proteins, BLAST performs "neighboring" (Fig. 2.5a): it searches not only *identical* oligopeptides but also *similar* oligopeptides.
- When a common "word" has been found, BLAST tries to extend the "hit" by adding base/amino acid pairs (Fig. 2.5b). The growing alignment is scored, using for nucleic acids a simple match/mismatch scoring scheme (by default +1/−3) and for proteins a scoring matrix (by default the BLOSUM62 table). If the score drops more than a certain amount below the already found highest score ("X" in Fig. 2.5b), BLAST gives up extending further and considers that one must take the alignment up to the position corresponding to that high score. The procedure is repeated in the other direction. So one obtains a local alignment without gaps, called "High Scoring segment Pair" (HSP). BLAST retains an HSP if its $E()$ is below a chosen threshold. Because an HSP will often include several "hits," protein BLAST saves time by only

(a)   Neighborhood word list:

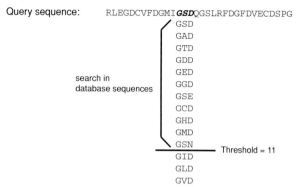

Query sequence: RLEGDCVFDGMI*GSD*QGSLRFDGFDVECDSPG

GSD
GAD
GTD
GDD
GED
search in   GGD
database sequences   GSE
CCD
GHD
GMD
GSN ——— Threshold = 11
GID
GLD
GVD

(b)   Extending a word hit into an HSP:

Query sequence: EGDCVFDGMI*GSD*QGSL
E   C+ +G   *G+D* GS+
Database sequence: EAGCLQNGQR*GTD*VGSV

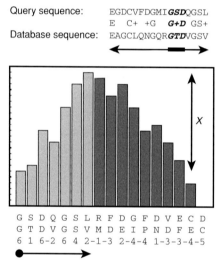

G S D Q G S L R F D G F D V E C D
G T D V G S V M D E I P N D F E C
6 1 6-2 6 4 2-1-3 2-4-4 1-3-3-4-5

Fig. 2.5   (a) Neighborhood list for the word "GSD" in the query sequence. Only words yielding a similarity greater than the threshold (11) according to the BLOSUM62 matrix will be considered as a hit. (b) Extending a word hit into a High Scoring segment Pair (HSP). The procedure of scoring a growing alignment based on the BLOSUM62 matrix is shown for the sequence to the right of the word. If the score drops more than a certain amount below the highest score found so far ("*X*" in Fig. 2.5b), BLAST aborts further extending and settles with the alignment up to the position corresponding to the highest score.

starting the extension procedure if there is a second "hit" in the proximity of the first one. This procedure is repeated for every sequence in the database.
- If the score of an HSP is above a certain threshold, BLAST makes an SW alignment, but saves time by searching only the squares of the dynamic programming matrix in the neighborhood of the HSP (it does not allow the score to drop more than a certain amount). The HSP is retained if its $E()$ is below a chosen threshold.

■ It is possible to find several HSPs for one database sequence. If BLAST does not make an alignment with gaps or if all the HSPs are not included in the gapped alignment, these HSPs are reported separately.

BLAST computes $E()$ (or **E-value**), which is the number of sequences yielding a similarity score of at least S that one would expect to find by chance or, in other words, an estimation of the number of *false positives*:

$$E() = m * n * K * e^{-\lambda * S} \tag{2.2}$$

where m is the length of the query sequence, n the total length of the database, and K and λ the so-called "Karlin and Altschul" parameters, which depend on the scoring scheme and the base/amino acid composition of the database. Stephen Altschul from the NCBI derived a rigorous statistical formula that allows computation of K and λ in case of an alignment without gaps (HSP), assuming that the scores follow an ***extreme value distribution*** (Karlin & Altschul, 1990). He also derived a formula in case one finds several HSPs for one database sequence ("Sum statistics", see Karlin & Altschul, 1993). There is no rigorous analytical formula in case of alignments with gaps. The experts, however, believe that the given formula is still approximately valid and BLAST uses a K and λ obtained by simulation (by searching a random sequence against a random database). The probability P of finding an alignment with a score of at least S by chance alone is then $1 - e^{-E}$.

BLAST computes also a "bit score," which does not depend directly on the scoring scheme used and thus reflects better the significance of the alignment:

$$s = \frac{\lambda * s - \ln K}{\ln 2} \tag{2.3}$$

The current BLAST package contains an everything-in-one program BLASTALL, which offers five types of database searches:

BLASTN   compares a nucleic acid sequence with all the sequences of a nucleic acid database and their complements.

BLASTP   compares a protein sequence with all the sequences in a protein database.

BLASTX   compares a nucleic acid sequence, translated in the six translation frames, with all the sequences of a protein database.

TBLASTN   compares a protein sequence with all the sequences of a nucleic acid database, translated on-the-fly in the six translation frames.

TBLASTX   compares a nucleic acid sequence, translated in the six translation frames, with all the sequences of a nucleic acid database, translated on-the-fly in the six translation frames. TBLASTX does not make gapped alignments.

There is also a program BLASTPGP (for protein searches only), which computes $E()$ using a more accurate formula taking into account the composition of the query and database sequence (Schaffer *et al.*, 2001) and also offers the following types of searches:

> Position Specific Iterated BLAST (PSI-BLAST) simply searches a protein against a protein database first, then picks out the best query sequence/database sequence alignments, merges them into a multiple sequence alignment, transforms the multiple sequence alignment into an amino acid frequency profile, and searches the profile against the database. The procedure can be repeated until the profile converges; that means until the sequences found by the profile are the same as the sequences used to make it.
>
> Pattern-Hit Initiated BLAST (PHI-BLAST) searches both a pattern (defined in PROSITE format) and a protein sequence against a protein database and finds sequences that match the pattern and show, in the same region, a significant local similarity. It is possible to perform a PHI-BLAST as a first round and go on with PSI-BLAST.

For those who want to learn how to use BLAST in an expert manner, we can recommend the book by I. Korf *et al.* (2003).

2.4.3 FASTA

FASTA (Pearson & Lipman, 1988) is older than BLAST and results from the work of William "Bill" Pearson at the University of Virginia (Charlottesville, USA). On the FASTA distribution site, you can still find version 2 in parallel with version 3, because the non-database searching programs were not ported when the statistics were improved (Pearson, 1998). Briefly, the algorithm goes as follows:

- FASTA searches all the "words" or *k*-tuples that are common between the query sequence and a database sequence, similar to BLAST but without "neighboring." By default, *k* is 6 for nucleic acids and 2 for proteins, but you can decrease *k* to obtain a greater sensitivity at the price of a significant increase in computing time.
- FASTA tries to assemble the "words" into long contiguous segments. These segments are scored using a simple match/mismatch scoring scheme for nucleic acids (by default +5/−4) and a scoring matrix for proteins (by default the BLOSUM50 table); the 10 best ones are retained. The highest score of all is called *init1*.
- FASTA tries to assemble segments that are compatible with the same alignment (that is, which do not overlap). A score is computed by adding the scores of the segments and by subtracting joining penalties. The combination that yields the highest score (*initn*) is retained. This procedure is repeated for every sequence in the database.
- For those database sequences that yield an initn above a certain chosen threshold, FASTA makes the SW alignment (score *opt*), but saves time by searching only the squares of the dynamic programming matrix in a band around the already-found segments.

▪ FASTA computes a *Z-score* and an *E*() value and lists the database sequences that result in an *E*() below a chosen threshold.

The *Z*-score is a similarity score corrected for the effect of the different lengths of the database sequences:

$$Z = \frac{s - \rho^* \ln(n) - \mu}{\sigma_s} \qquad (2.4)$$

where the mean score for unrelated sequences of length n, $\mu + \rho * \ln(n)$, is estimated by linear regression using as dataset a collection of sequences found during the database search that yielded very low opt scores. You can also force FASTA to use randomly shuffled copies of the database sequences instead. This is particularly useful if you want to assess the false positive rate in a specialized database or a set of sequences that you defined yourself. In such cases, even sequences with the lowest similarity to the query (low opt scores) can still represent true homologous, and as such, they would not be appropriate to compute a mean score for "unrelated" sequences.

The probability *P* that an unrelated sequence yields a score at least equal to the observed score is estimated, assuming that the *Z*-scores follow an extreme value distribution. Contrary to BLAST, which treats the complete database as one long sequence, FASTA considers the comparisons with the sequences of the database as statistically independent events and computes *E*() by multiplying *P* with the number of sequences in the database. FASTA also computes a *bit score*, similar to that of BLAST, but using a *K* and λ derived from the database sequences. Note that it is possible to find twice the same alignment with different *Z*-score, *E*() value and bit score if the same subsequence is found as part of a shorter and a longer database sequence.

The FASTA package contains programs that perform the following searches:

FASTA   compares a nucleic acid sequence and its complement with all the sequences of a nucleic acid database, or compares a protein sequence with all the sequences of a protein database.

FASTX   compares a nucleic acid sequence and its complement with all the sequences of a protein database, aligning translated codons with amino acids and allowing for gaps between codons (Pearson *et al.*, 1997)

FASTY   the same as fastx, but allowing also gaps within a codon.

TFASTA   compares a protein sequence with all the sequences of a nucleic acid database, translated on-the-fly in the six translation frames.

TFASTX   compares a protein sequence with all the sequences of a nucleic acid database and their complements, aligning translated codons with amino acids and allowing for gaps between codons.

TFASTY   the same as tfastx, but allows also gaps within a codon.

SSEARCH   the same as `fastA`, but uses the slow and precise SW algorithm.

2.4.4 Other tools and some general considerations

Among other database search tools, SSAHA (available as part of the Ensembl interface, see Ning *et al.*, 2001) and BLAT (available as part of the UCSC Genome Browser interface, see Kent, 2002) are worthwhile mentioning here. They are very fast because they do not search the sequences themselves but use an index with oligonucleotides, which is made each time the database is updated. They are suited for searching a genome to see if it contains a sequence stretch very similar to a query sequence.

Most database search tools compute an $E()$ value, an estimation of the number of false positives expected by chance. For the inexperienced user, who can be bewildered by the multiplicity of scores and statistics in the program outputs, this $E()$-value is the most useful. An often-cited rule of thumb is: if $E() < 0.1$, you can accept the sequence hit as a homolog with reasonable confidence; if $0.1 < E() < 10$, you are in the so-called "twilight zone," which means, that the hit could be a homolog, but you should not take it for granted without additional evidence; if $E() > 10$, you are in the so-called "midnight zone," which means, do not accept the result, because even if the hit is a true homolog, it has diverged beyond recognition and the alignment presented is likely to be incorrect. Beware of over-interpreting the exact value of $E()$, which is calculated based on assumptions that do not necessarily hold in reality (see, for example, an article by Brenner *et al.*, 1998, arguing that the BLAST $E()$ might be underestimated). In any case, very low $E()$ values indicate that the hit almost certainly represents a true homolog.

Protein similarity searches are inherently more sensitive than nucleic acid searches. The reason is a simple effect of the sequence "alphabet." Indeed, since there are only four bases in nucleic acids, one expects to find 25% of identities by chance alone. So, when sequences start diverging by substitutions, nucleic acids reach a point much earlier where you cannot infer homology from sequence similarity. If you are interested in a (potentially) coding DNA, it is therefore preferable to perform a protein similarity search. In this case, one can take advantage of the fact that BLAST and FASTA packages allow "translated" searches (see above).

It has been shown that, for *most* protein families, BLAST (even the old "ungapped" BLAST) is outperforming FASTA, and it performs nearly as well as the SW algorithm. The reason is that BLAST can rather easily detect substitutions in proteins thanks to the "neighboring." The scoring matrices precisely reflect the fact that one often finds similar amino acids rather than identical amino acids at corresponding positions. Furthermore, the HSPs found by BLAST often correspond to the segments forming the conserved "core" of the protein. In coding DNA, deletions/insertions that do not encompass a multiple of three bases are rare because they cause a shift in reading frame, and furthermore deletions/insertions in the "core" of a protein are rare because they change the relative position of the segments. Non-coding

DNA has, however, the tendency to rapidly accumulate deletions/insertions. In this case, FASTA is known to perform better than the BLASTN option of BLAST, because it can assemble smaller segments. Furthermore, the default settings in BLAST can be adapted to search rapidly for closely related sequences, while BLAST sensitivity can be increased by decreasing the word size (and in the case of WU-BLAST by enabling "neighboring" for nucleic acids). Keep in mind, however, that nucleic acid comparisons do not probe evolution very far.

Finding *all the sequences* of a family is something other than finding *a few related sequences*. To achieve a maximum sensitivity, it is not enough to perform an SW search. Note that, for protein searches, the usage of a high number BLOSUM/low number PAM substitution matrix gives the best probability for finding sequence fragments that are very closely related but very short, because one attaches a greater importance to conserved positions. Using a low number BLOSUM/high number PAM matrix yields the best probability of finding sequence fragments that are very distantly related, because one attaches a greater importance to amino acids aligned with a "similar" amino acid. This results from the way these tables were derived: BLOSUM62 is made from local alignments with at least 62% of the columns fully conserved, PAM250 should reflect evolution at a distance of 250 substitutions per 100 amino acids (see also Chapter 9). So, if you want to find just a few similar sequences, it is usually sufficient to run FASTA or BLAST once with the default parameter settings. But, if you want to be certain to have found as many as possible potentially homologous sequences, it will be necessary to perform several database searches (with BLAST, FASTA and/or eventually SW) using different scoring schemes. What one search does not find, another might. An alternative and complementary strategy to find all the sequences of a family is to perform new searches with, as a query, some of the sequences you found already. PSI-BLAST (see above) and SCAN-PS are also potentially useful; the latter works in a way similar to PSI-BLAST but uses a rigorous SW algorithm. These programs find a self-consistent set of proteins, which can be well suited for phylogenetic analysis, as well as some remote relatives that a simple run might not find.

The number of false positives, reflected by the $E()$ value, will increase with database size. By searching a small set instead of a complete database (e.g. only the bacterial sequences or the *E. coli* sequences), you can reduce the "noise" and save computation time. In this context, remember the utility of the specialized databases (see Section 2.2.3). Note also that the NCBI BLAST server allows the user to perform a BLAST search against a set of sequences resulting from an Entrez search (McGinnis & Madden, 2004).

Proteins typically are composed of several domains. If your query protein shares a domain with hundreds of proteins in the database and shares another domain with only a few other sequences, the second domain might be missed. Similarly, eukaryotic DNA sequences often contain repetitive elements such as Alu repeats that can produce a long but uninteresting list of hits at the top of the output. Therefore, if your BLAST or FASTA query repeatedly results in alignments covering the same range of your query sequence, it is a good idea to cut out that range (or replace it by XXX) and then run the search again. In addition, regions with biased composition like poly-A tails or histidine-rich regions yield a lot of uninteresting "hits." At the NCBI, some "filters" have been developed that automatically remove such regions of low compositional complexity: DUST for DNA (Hancock & Armstrong, 1994) and SEG for protein (Wootton & Federhen, 1993). They can be obtained as standalone programs and they are also implemented in BLAST. Note that, in the simple NCBI BLAST (BLASTALL), the filters are enabled by default, while in PSI/PHI-BLAST (BLASTPGP) and WU-BLAST they are disabled by default, but this can always be changed to suit your needs.

PRACTICE

Marc Van Ranst and Philippe Lemey

2.5 Database searching using ENTREZ

ENTREZ is an easy-to-use integrated text-based cross-database search and retrieval system developed by NCBI. It links a large number of databases including nucleotide and protein sequences, PubMed, 3-D protein structures, complete genomes, taxonomy, and others. ENTREZ has a comprehensive *Help Manual*, which you can consult at *http://www.ncbi.nlm.nih.gov/books/bv.fcgi?rid=helpentrez*.

On the ENTREZ homepage (*http://www.ncbi.nlm.nih. gov/gquery/gquery.fcgi*), you can specify a search term in the input box. The example in Fig. 2.6 shows the result for the search term "human papillomavirus type 13." You can perform a concurrent search across all ENTREZ databases by typing your search text into the box at the top of the page and then clicking "Go," or you can search a single database by entering your search term and clicking the database's icon instead of "Go."

At the time of writing, you could find eight articles in the scientific literature on human papillomavirus type 13, and the full text of all eight are freely available on PubMed Central. Figure 2.7 shows the first four most recent articles on human papillomavirus type 13. When you click for example on the fourth item on the list, it will link you to the abstract of this article.

When you would use the same search term "human papillomavirus type 13" to search the nucleotide database, ENTREZ finds five nucleotide sequences in the CoreNucleotide records (the main database), 0 in the Expressed Sequence Tags (EST) database, and 0 in the Genome Survey Sequence (GSS) database (Fig. 2.8). When you click on the blue hyperlink of the CoreNucleotide records, ENTREZ will present the records where it found the text string "human papillomavirus type 13" (Fig. 2.9). You will notice that the first record is the complete sequence of *Caenorhabditis elegans*. Somewhere in the annotation of the *C. elegans* genome, the search engine encountered the text string "human papillomavirus type 13." The second and fourth records contain the complete genome of human papillomavirus type 13, and the third record contains the complete sequence of a related virus, the pygmy chimpanzee papillomavirus. Thus, although this ENTREZ nucleotide will, in most instances, allow you to find the information that you wanted, it will not necessarily be in the first record in the list.

Clicking on a blue accession number hyperlink in the list will bring you to the GenBank file of the sequence (Fig. 2.10). A detailed explanation of the Genbank flat file format is provided at *http://www.ncbi.nlm.nih.gov/Sitemap/samplerecord.html*. The Genbank format is well suited to incorporate a single sequence and

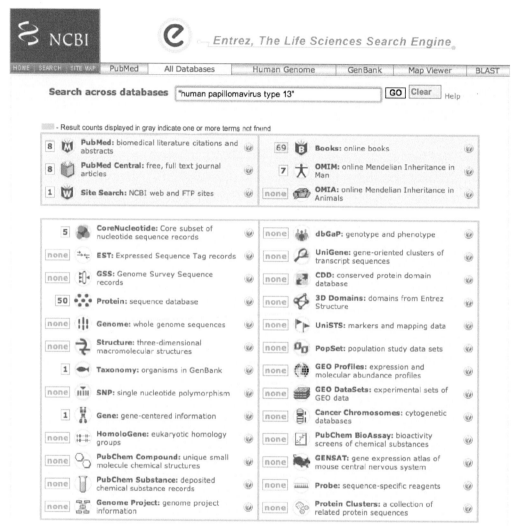

Fig. 2.6   An Entrez search across all databases using the search term "human papillomavirus type 13."

its annotation. For phylogenetic analysis, however, multiple sequences need to be formatted in a single file. Unfortunately, different programs have been using different input file formats resulting in about 18 formats that have been commonly used in molecular biology (an overview of most sequence formats can be found at *http://emboss.sourceforge.net/docs/themes/SequenceFormats.html*). Therefore, much of the effort involved in carrying out a comprehensive sequence analysis is devoted to preparing and/or converting files. The most popular sequence formats are illustrated in Box 2.4. Although file formats can be converted manually

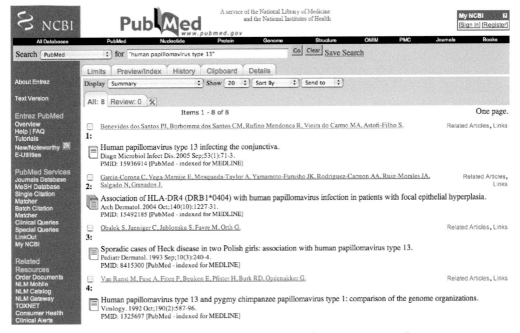

Fig. 2.7   PubMed results for the search term "human papillomavirus type 13."

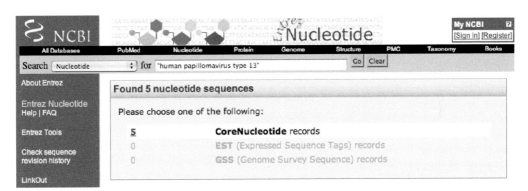

Fig. 2.8   ENTREZ nucleotide database results for "human papillomavirus type 13."

in any text editor, there exist sequence conversion programs, like READSEQ written by Don Gilbert, to perform this task in an automated fashion. An online version of this tool is available at *http://iubio.bio.indiana.edu/cgi-bin/readseq.cgi*. Fortunately, many alignment and phylogenetic programs have implemented sequence conversion code to import and/or export different file formats, and newly developed programs tend to use PHYLIP, NEXUS or FASTA as standard input format.

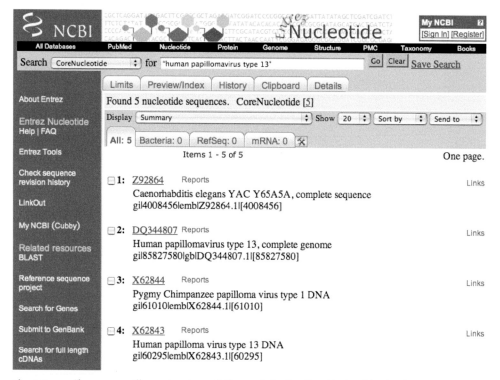

Fig. 2.9   "human papillomavirus type 13" records in the CoreNucleotide database.

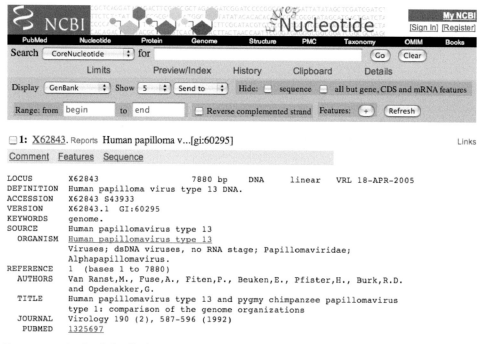

Fig. 2.10   Genbank flat file for accession number X62843.

Box 2.4 Common formats of sequence data input files

FASTA

FASTA is probably the simplest text-based format for representing either nucleic acid sequences or protein sequences. The format begins with single-line sequence description preceded by ">." The rest of the line can be used for description, but it is recommended to restrict the total number of characters to 80. On the next line, the actual sequence is represented in the standard IUB/IUPAC amino acid and nucleic acid codes. For aligned sequences, this may contain gap "-" symbols. FASTA format is sometimes referred to as Pearson/Fasta format. Many programs like CLUSTALW (see Chapter 3), PAUP* (use the "tonexus" command) (see Chapter 8), HYPHY (see Chapter 14), RDP (see Chapter 16), and DAMBE (see Chapter 20) can read or import **FASTA** format. To indicate this type of format, the extensions ".fas" or ".fasta" are generally used.

```
>Taxon1
AATTCCCCAGCTTTCCACCAAGCTC
>Taxon2
AATTCCACAGCTTTCCACCAAGCTC
>Taxon3
AACTCCAGCACATTCCACCAAGCTC
>Taxon4
AACTCCACAACATTCCACCAAGCTC
```

NEXUS is a modular format for systematic data and can contain both sequence data and phylogenetic trees in "block" units (Maddison et al., 1997). An example of a NEXUS file containing a data block is shown below. For a description of the tree format, see the practice section in Chapter 5. Data and trees are public blocks containing information used by several programs. Private blocks containing relevant information for specific programs are being used by PAUP* (see Chapter 8), MRBAYES (see Chapter 7), FIGTREE (see Chapter 5), SPLITSTREE (Chapter 21), etc. Box 8.4 in Chapter 8 further outlines the NEXUS format and explains how to convert data sets from PHYLIP to NEXUS format. Note that comments can be added in the NEXUS file using "[]"; the example below uses two lines of comments to indicate sequence positions in the alignment. NEXUS formatted files usually have the ".nex," ".nexus," or ".nxs" extension.

```
#NEXUS

Begin DATA;
   Dimensions ntax = 4 nchar = 25;
   Format datatype = NUCLEOTIDE gap = -;
   Matrix
   [     1          11         21      ]
   [     |          |          |       ]
   Taxon1 AATTCCCCAGCTTTCCACCAAGCTC
   Taxon2 AATTCCACAGCTTTCCACCAAGCTC
   Taxon3 AACTCCAGCACATTCCACCAAGCTC
   Taxon4 AACTCCACAACATTCCACCAAGCTC
;
End;
```

PHYLIP

PHYLIP is the standard input file format for programs in the PHYLIP package (see Chapters 4 and 5) and has subsequently been implemented as input file format for many other programs, like TREE-PUZZLE, PHYML and IQPNNI (see Chapter 6). The first line of the input file contains the number of taxa and the number of characters (in this case alignment sites), separated by blanks. In many cases, the taxon name needs to be restricted to ten characters and filled out to the full ten characters by blanks if shorter. Special characters like parentheses ("(" and ")"), square brackets ("[" and "]"), colon (":"), semicolon (";") and comma (",") should be avoided; this is true for most sequence formats. The name should be on the same line as the first character of the data for that taxon. PHYLIP files frequently have the ".phy" extension.

```
4 25
Taxon1    AATTCCCCAGCTTTCCACCAAGCTC
Taxon2    AATTCCACAGCTTTCCACCAAGCTC
Taxon3    AACTCCAGCACATTCCACCAAGCTC
Taxon4    AACTCCACAACATTCCACCAAGCTC
```

The PHYLIP format can be "sequential" or "interleaved." In the sequential format, the character data can run on to a new line at any time:

```
4 65
Taxon1    AATTCCCCAGCTTTCCACCAAGCTCTGCAAGATCCCAGAGTCAAGGGCCTG-
TATTTTCCTGCTGG
Taxon2    AATTCCACAGCTTTCCACCAAGCTCTGCAAGATCCCAGAGTCAGGGGCCTG-
TATTTTCCTGCTGG
Taxon3    AACTCCAGCACATTCCACCAAGCTCTGCTAGATCC---
AGTGAGGGGCCTATACGTTCCTGCTGG
Taxon4    AACTCCACAACATTCCACCAAGCTCTGCTAGATCCCAGAGTGAGGGGCCTTTAT-
TATCCTGCTGG
```

The PHYLIP interleaved format has the first part of each of the sequences (50 sites in the example below), then some lines giving the next part of each sequence, and so on.

```
4 65
Taxon1    AATTCCCCAG CTTTCCACCA AGCTCTGCAA GATCCCAGAG TCAAGGGCCT
Taxon2    AATTCCACAG CTTTCCACCA AGCTCTGCAA GATCCCAGAG TCAGGGGCCT
Taxon3    AACTCCAGCA CATTCCACCA AGCTCTGCTA GATCC---AG TGAGGGGCCT
Taxon4    AACTCCACAA CATTCCACCA AGCTCTGCTA GATCCCAGAG TGAGGGGCCT

GTATTTTCCT GCTGG
GTATTTTCCT GCTGG
ATACGTTCCT GCTGG
TTATTATCCT GCTGG
```

To make the alignment easier to read, there is a blank every ten characters in the example above (any such blanks are allowed). In some cases, the interleaved format can or needs to be specified using "I" on the first line (e.g. "4 65 I"). Note that NEXUS formatted alignments can also be interleaved or sequential. Programs in the PAML package (see Chapter 11) use the PHYLIP sequential format, but allow for more characters in the taxon name (up to 30). PAML considers two consecutive spaces as the end of a species name, so that the species name does not have to have exactly 30 (or 10) characters.

CLUSTAL

CLUSTAL is not widely supported as input format in phylogenetic programs but, since it is the standard output format of a popular alignment software (see Chapter 3), we include it in our overview. The format is recognized by the word CLUSTAL at the beginning of the file. CLUSTAL is an interleaved format with blocks that repeat the taxa names, which should not contain spaces or exceed 30 characters and which are followed by white space. The sequence alignment output from CLUSTAL software is usually given the default extension ".aln." CLUSTAL also indicates conserved residues in the alignment using a "*" for each block.

```
CLUSTAL W (1.83.1) multiple sequence alignment


Taxon1          AATTCCCCAGCTTTCCACCAAGCTC
Taxon2          AATTCCACAGCTTTCCACCAAGCTC
Taxon3          AACTCCAGCACATTCCACCAAGCTC
Taxon4          AACTCCACAACATTCCACCAAGCTC
                ** ***    * *************
```

MEGA

In this format, the "#Mega" keyword indicates that the data file is prepared for analysis using MEGA (see Chapters 4 and 5). It must be present on the very first line in the data file. On the second line, the word "Title" must be written, which can be followed by some description of data on the same line. Comments may be written on one or more lines right after the title line and before the data. Each taxon label must be written on a new line staring with "#" (not including blanks or tabs). The sequences can be formatted in both sequential and interleaved format.

```
#Mega
Title: example data set

! commenting
#Taxon1
AATTCCCCAGCTTTCCACCAAGCTC
#Taxon2
AATTCCACAGCTTTCCACCAAGCTC
#Taxon3
AACTCCAGCACATTCCACCAAGCTC
#Taxon4
AACTCCACAACATTCCACCAAGCTC
```

Some advanced program for population genetic inference, like BEAST (see Chapter 18) and LAMARC (see Chapter 19) use an *XML* file format. However, those programs usually provide file conversion tools that accept FASTA or NEXUS format.

The human papilloma virus type 13 genome can be displayed and downloaded in FASTA format using the `Display (FASTA)` and `Show, Send to (File)` options. One can focus on a particular gene region using the `Range (from.. to..)` and `Refresh` Option.

Multiple sequences can be obtained by typing accession numbers separated by blanks, for example, search the CoreNucleotide database for "AY843504 AY843505 AY843506." Three entries for primate TRIM5α sequences will be retrieved, which can be saved in FASTA format (`Display FASTA` and `Show, Send to File`). Alternatively, the accession numbers can be typed in plain text file with a single accession number on each line, which can be uploaded for sequence retrieval (from CoreNucleotide) by the Batch Entrez system (*http://www.ncbi.nlm.nih.gov/entrez/batchentrez.cgi?db=Nucleotide*). Finally, if a set of DNA sequences has been collected to analyze the evolutionary relationships within a particular population (single or multiple species population), they can be submitted and stored as a "PopSet." Such collections are submitted to GenBank via `Sequin`, often as a sequence alignment. To illustrate this, go back to the ENTREZ homepage and specify the search term "Sawyer." At the time of writing, this resulted in 16 entries in the PopSet database. When you click on hyperlink of the PopSet section, the 16 journal article titles should appear with the first author represented as a hyperlink. Follow the link of "Sawyer" with title "Positive selection of primate TRIM5α identifies a critical species-specific retroviral restriction domain." This will bring you to the population set from this study (Fig. 2.11). By scrolling down, you will notice that this collection holds 17 TRIM5α sequences from different species. A multiple sequence alignment can be generated and the sequences can be saved to a single file in **FASTA** format (`Display FASTA` and `Show, Send to File`). This is a subset of the sequences that will be further analyzed in Chapters 3, 4, 5, and 11.

2.6 BLAST

On the BLAST homepage at NCBI (*http://www.ncbi.nlm.nih.gov/BLAST/*), a choice is offered between the different BLAST programs (Fig. 2.12) through different hyperlinks. In this exercise we will compare a nucleotide and protein BLAST search for one of the TRIM5α sequences.

Fig. 2.11   PopSet page for the TRIM5α collection.

As query for a nucleotide BLAST (BLASTN), you can use a sequence in FASTA format, an accession number or a GenBank identifier (gi) (Fig. 2.13). We will use the TRIM5α nucleotide sequences in FASTA format (with accession number AY843504.1). You must also specify the database you want to search. If the complete GenBank database needs to be searched, the choice should be "Nucleotide collection (nr/nt)" database. We will also select BLASTN under the Program Selection header, which optimizes the search for somewhat similar nucleotide sequences. The parameters of the algorithm can be adjusted by clicking on Algorithm parameters at the bottom of the page. Increase the Max target sequences to 20000 to make sure we will see all the BLASTN hits in the results page. The default threshold for the $E()$ value is 10 and the default word size is 11. Click on the Blast button to initiate the search.

At the time of writing, this search resulted in 2372 hits. A graphical overview with color-coded alignment scores is provided for the most similar sequences. For all hits, there are discontinuous stretches with high alignment scores. This is

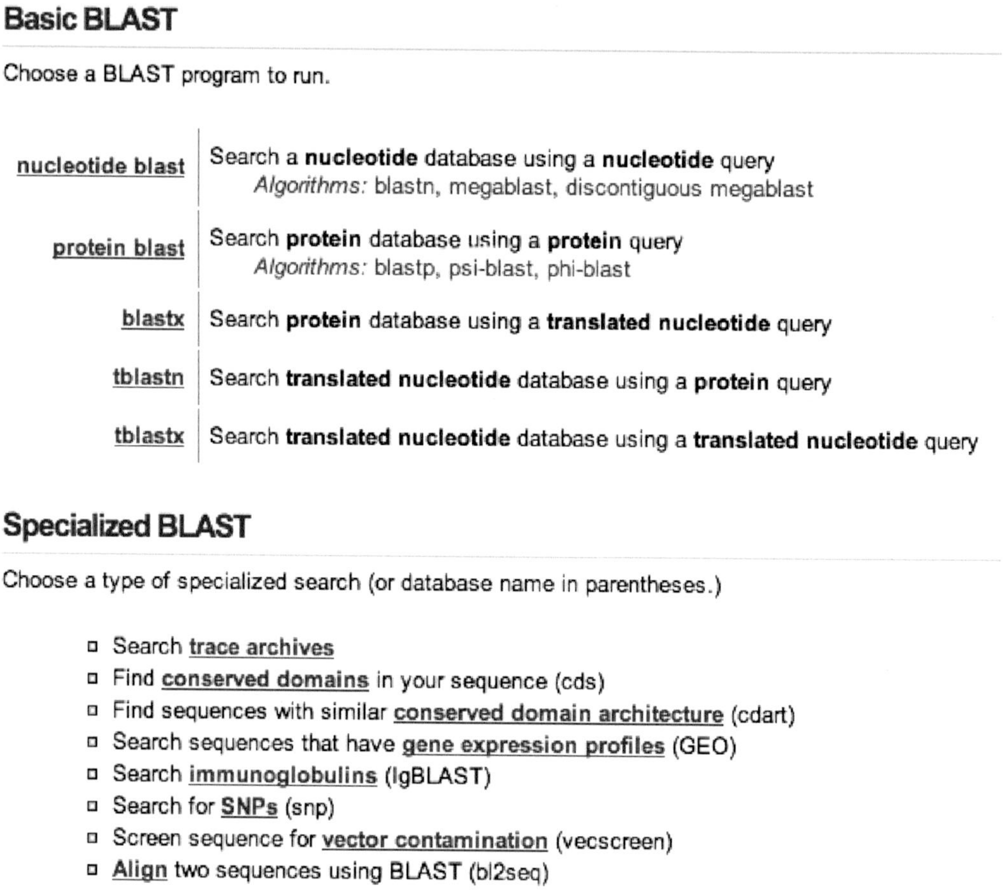

Basic BLAST

Choose a BLAST program to run.

nucleotide blast	Search a **nucleotide** database using a **nucleotide** query *Algorithms:* blastn, megablast, discontiguous megablast
protein blast	Search **protein** database using a **protein** query *Algorithms:* blastp, psi-blast, phi-blast
blastx	Search **protein** database using a **translated nucleotide** query
tblastn	Search **translated nucleotide** database using a **protein** query
tblastx	Search **translated nucleotide** database using a **translated nucleotide** query

Specialized BLAST

Choose a type of specialized search (or database name in parentheses.)

- ☐ Search <u>trace archives</u>
- ☐ Find <u>conserved domains</u> in your sequence (cds)
- ☐ Find sequences with similar <u>conserved domain architecture</u> (cdart)
- ☐ Search sequences that have <u>gene expression profiles</u> (GEO)
- ☐ Search <u>immunoglobulins</u> (IgBLAST)
- ☐ Search for <u>SNPs</u> (snp)
- ☐ Screen sequence for <u>vector contamination</u> (vecscreen)
- ☐ <u>Align</u> two sequences using BLAST (bl2seq)

Fig. 2.12   Different basic and specialized BLAST programs.

because the input sequence contained several stretches of "Ns" for regions that were not sequenced, which will not be taken into account in the word hits and BLAST extension with database sequences. The output of the BLASTN search also contains a table with sequences producing significant alignments (part of this table is shown in Fig. 2.14). The first three hits have the same Max score (1168) and maximum identity, one of which is, in fact, the query sequence. The first 11 hits are all TRIM5α sequences from African green monkeys. Scrolling down the list you will find many more TRIM5α sequences and eventually other isoforms or members of the tripartite motif containing protein family. At the bottom of the list are hits that resulted in an *E*() value of 6, which places these hits in the so-called "twilight zone" (see Section 2.4.4).

We also perform a protein BLAST search using the amino acid translation of the TRIM5α coding sequence (accession number AAV91975.1). For this search,

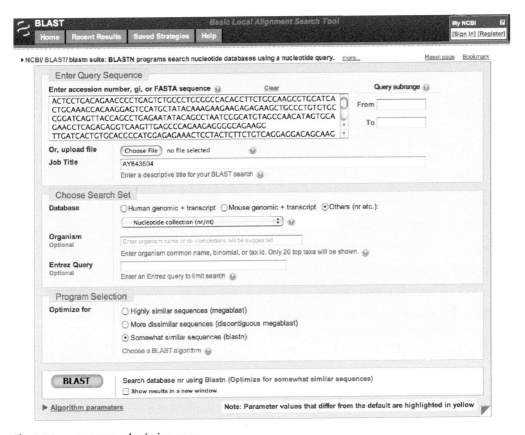

Fig. 2.13   BLASTN submission page.

Accession	Description	Max score	Total score	Query coverage	E value	Max ident	Links
AY625002.1	Cercopithecus aethiops cell-line CV-1 Trim5 alpha mRNA, com	1168	2698	69%	0.0	100%	
AB210051.1	Chlorocebus aethiops TRIM5 mRNA for tripartite motif protein	1168	2704	69%	0.0	100%	
AY843504.1	Cercopithecus aethiops TRIM5alpha (TRIM5) gene, complete c	1168	2707	69%	0.0	100%	
AY593973.2	Cercopithecus tantalus triparite motif protein TRIM5alpha (TR	1159	2698	69%	0.0	100%	
AY669399.1	Cercopithecus aethiops TRIM5-alpha (TRIM5) mRNA, complete	1153	2682	69%	0.0	100%	
AY710301.1	Cercopithecus sabaeus TRIM5 (TRIM5) gene, exon 8 and part	1153	1153	30%	0.0	99%	
DQ437606.1	Chlorocebus tantalus TRIM5 alpha (TRIM5) gene, exon 8 and	1144	1144	30%	0.0	99%	
AB210050.1	Chlorocebus aethiops TRIM5 mRNA for tripartite motif protein	1144	2679	69%	0.0	100%	
AY740613.1	Cercopithecus aethiops tantalus TRIM5 alpha (TRIM5) mRNA,	1141	2664	69%	0.0	100%	
AY625003.1	Cercopithecus aethiops cell-line Vero Trim5 alpha mRNA, com	1132	2662	69%	0.0	100%	
AY740612.1	Cercopithecus aethiops pygerythrus TRIM5 alpha (TRIM5) mR	1132	2653	69%	0.0	100%	
AY740619.1	Erythrocebus patas TRIM5 alpha (TRIM5) mRNA, complete cd	1113	2592	69%	0.0	99%	
DQ437602.1	Macaca nemestrina TRIM5 alpha (TRIM5) gene, exon 8 and pa	823	1090	30%	0.0	98%	
EF113923.1	Cercocebus torquatus atys tripartite motif-containing 5 alpha	820	2595	69%	0.0	100%	
EF113921.1	Cercocebus torquatus atys tripartite motif-containing 5 alpha	820	2566	69%	0.0	100%	
EF113916.1	Macaca mulatta tripartite motif-containing 5 alpha isoform (TF	820	2593	69%	0.0	100%	
EF113914.1	Macaca mulatta tripartite motif-containing 5 alpha isoform (TF	820	2577	69%	0.0	100%	
DQ437604.1	Cercocebus torquatus atys TRIM5 alpha (TRIM5) gene, exon 8	820	1086	30%	0.0	98%	
DQ842021.1	Macaca mulatta tripartite motif-containing 5 alpha isoform mR	820	2584	69%	0.0	100%	UG

Fig. 2.14   Part of the BLASTN output table for the TRIM5α search.

we specify blastp under Program Selection and set the Max target sequences to 20000. The default word size is now 3. During the search, a page will be displayed that shows four putative conserved domains detected in the query. The BLAST search now provides 7799 hits with the query sequence (AAV91975.1) heading the list.

>☐pdb|2IWG|B 🆂 Chain B, Complex Between The Pryspry Domain Of Trim21 And Igg Fc

pdb|2IWG|E 🆂 Chain E, Complex Between The Pryspry Domain Of Trim21 And Igg Fc
Length=181

```
Score =  117 bits (293),  Expect = 2e-24, Method: Composition-based stats.
Identities = 76/221 (34%), Positives = 110/221 (49%), Gaps = 50/221 (22%)

Query  299  YWVDVTLAPNNISH-AVIAEDKRQVSYQNPQIMYQAPGSSFGSLTNFNYCTGVLGSQSIT  357
            + V +TL P+  +    +++ED+RQV   + Q                         QSI
Sbjct  1    HMVHITLDPDTANPWLILSEDRRQVRLGDTQ---------------------QSIP  35

Query  358  SRKLTNFNYCTGVLGSQSITSGKHYWEVDVSKKSAWILGVCAGFQPDATYNIEQNENY--  415
               + F+    VLG+Q  SGKHYWEVDV+ K AW LGVC          ++ +  ++
Sbjct  36   GNE-ERFDSYPMVLGAQHFHSGKHYWEVDVTGKEAWDLGVCRD-------SVRRKGHFLL  87

Query  416  QPKYGYWVIGLQEGDKYSVFQDSSSHTPFAPFIVPLSVIICPDRVGVFVDYEACTVSFFN  475
              K G+W I L    KY         +    PL + + P +VG+F+DYEA  VSF+N
Sbjct  88   SSKSGFWTIWLWNKQKYEAGTYPQT---------PLHLQVPPCQVGIFLDYEAGMVSFYN  138

Query  476  ITNHGFLIYKFSQCSFSKPVFPYLNP-----RKCTVPMTLC  511
            IT+HG LIY FS+C+F+ P+ P+ +P     K T P+TLC
Sbjct  139  ITDHGSLIYSFSECAFTGPLRPFFSPGFNDGGKNTAPLTLC  179
```

Fig. 2.15   Pairwise alignment of the TRIM5α query sequence with the subject sequence in the GenBank database for which a protein structure has been determined.

Using "TRIM5" as query would not result in any hits in the NCBI Structure database. However, we might find homologous proteins with known structures in the database using our sequence similarity search. Scroll down the BLASTP hit list and look for an "S" in a red square next to the $E()$ column. The first hit in the list is "pdb|2IWG|B." When you click on the hyperlink in the Score column (117), the pairwise alignment of this subject (sbjct) sequence with your query sequence will appear (Fig. 2.15). The alignment of the Pryspry Domain Of Trim21 and the query has 76 identities in 221 aligned residues (34%). Clicking on the "S" in a red square will bring you to the structure database. There is another hit in the list (pdb | 2FBE | A) with Spry domain structures ($E()$ value 1e-09). These structures can be useful for homology modeling of this TRIM5α domain. Note that there are also structures in the list for other conserved domains in this protein.

2.7 FASTA

As a comparison, we now perform a fastA search with the same TRIM5α protein query at *http://fasta.bioch.virginia.edu/fasta_www2/fasta_list2.shtml*. Click on the Protein-protein FastA link and paste in the TRIM5α protein sequence in FASTA format (AAV91975.1) (Fig. 2.16). To have similar settings as the previous BLAST search, select the NCBI NR non-redundant database under

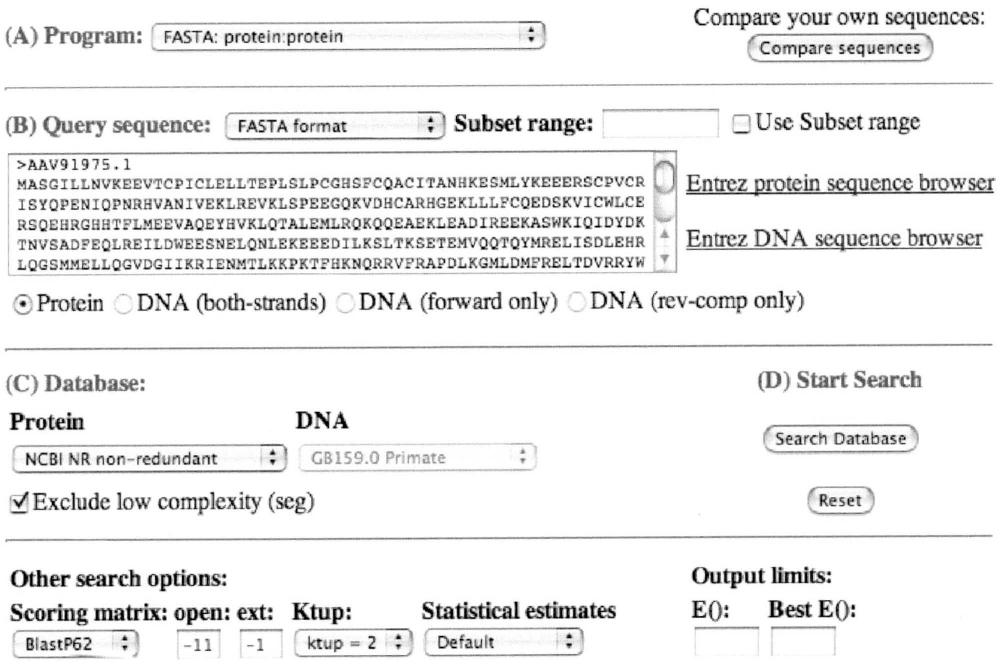

(A) Program: [FASTA: protein:protein   :]

Compare your own sequences:
[Compare sequences]

(B) Query sequence: [FASTA format   :] **Subset range:** [   ] ☐ Use Subset range

```
>AAV91975.1
MASGILLNVKEEVTCPICLELLTEPLSLPCGHSFCQACITANHKESMLYKEEERSCPVCR
ISYQPENIQPNRHVANIVEKLREVKLSPEEGQKVDHCARHGEKLLLFCQEDSKVICWLCE
RSQEHRGHHTFLMEEVAQEYHVKLQTALEMLRQKQQEAEKLEADIREEKASWKIQIDYDK
TNVSADFEQLREILDWEESNELQNLEKEEEDILKSLTKSETEMVQQTQYMRELISDLEHR
LQGSMMELLQGVDGIIKRIENMTLKKPKTFHKNQRRVFRAPDLKGMLDMFRELTDVRRYW
```

Entrez protein sequence browser

Entrez DNA sequence browser

⦿ Protein ◯ DNA (both-strands) ◯ DNA (forward only) ◯ DNA (rev-comp only)

(C) Database:

Protein   **DNA**

[NCBI NR non-redundant   :] [GB159.0 Primate   :]

☑ Exclude low complexity (seg)

(D) Start Search

[Search Database]

[Reset]

Other search options:

Scoring matrix: open: ext: Ktup: Statistical estimates

[BlastP62 :] [-11] [-1] [ktup = 2 :] [Default   :]

Output limits:

E():   Best E():

[   ] [   ]

Fig. 2.16   Protein–protein FASTA submission page.

(C) Database and BlastP62 as Scoring matrix under Other search options. The fastA search retrieves about 5000 records, with again the query sequence (AAV91975.1) heading the list of entries with best scores. The first part of the output file contains a histogram showing the distribution of the observed ("=") and expected ("*") Z-scores. (See Section 2.4.3 for an explanation of Z-score.) The theoretic curve ("*"), following the extreme value distribution, is very similar to observed results ("=") and provides a qualitative assessment of how well the statistical theory fits the similarity scores calculated by the program. Below the histogram, FASTA displays a listing of the best scores (with 10 as E() value cut-off), followed by the alignments of the regions of best overlap between the query and search sequences. Clicking on the align link in the score list will bring you to the relevant alignment, which, in turn, provides a link to the Entrez database entry for the hit.

Multiple sequence alignment

THEORY

Des Higgins and Philippe Lemey

3.1 Introduction

From a biological perspective, a sequence alignment is a hypothesis about *homology* of multiple residues in protein or nucleotide sequences. Therefore, aligned residues are assumed to have diverged from a common ancestral state. An example of a multiple sequence alignment is shown in Fig. 3.1. This is a set of amino acid sequences of globins that have been aligned so that *homologous* residues are arranged in columns "as much as possible." The sequences are of different lengths, implying that gaps (shown as hyphens in the figure) must be used in some positions to achieve the alignment. The gaps represent a deletion, an insertion in the sequences that do not have a gap, or a combination of insertions and deletions. The generation of alignments, either manually or using an automatic computer program, is one of the most common tasks in computational sequence analysis because they are required for many other analyses such as structure prediction or to demonstrate sequence similarity within a family of sequences. Of course, one of the most common reasons for generating alignments is that they are an essential prerequisite for phylogenetic analyses. Rates or patterns of change in sequences cannot be analysed unless the sequences can be aligned.

3.2 The problem of repeats

It can be difficult to find the optimal alignment for several reasons. First, there may be repeats in one or all the members of the sequence family; this problem is shown

The Phylogenetic Handbook: a Practical Approach to Phylogenetic Analysis and Hypothesis Testing,
Philippe Lemey, Marco Salemi, and Anne-Mieke Vandamme (eds.). Published by Cambridge
University Press. © Cambridge University Press 2009.

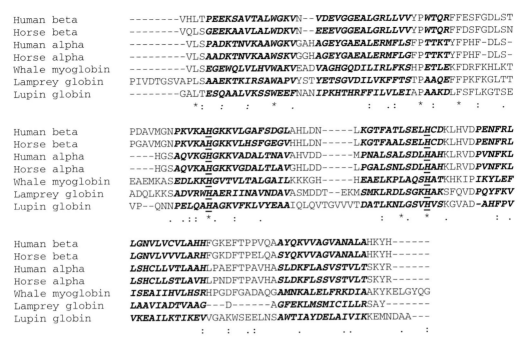

Human beta	--------VHLT**PEEKSAVTALWGKV**N--**VDEVGGEALGRLLVV**YP**WTQR**FFESFGDLST
Horse beta	--------VQLS**GEEKAAVLALWDKV**N--**EEEVGGEALGRLLVV**YP**WTQR**FFDSFGDLSN
Human alpha	---------VLS**PADKTNVKAAWGKV**GAH**AGEYGAEALERMFLS**F**P**TTKT**Y**FPHF-DLS-
Horse alpha	---------VLS**AADKTNVKAAWSKV**GGH**AGEYGAEALERMFLG**F**P**TTKT**Y**FPHF-DLS-
Whale myoglobin	---------VLS**EGEWQLVLHVWAKV**EAD**VAGHGQDILIRLFKS**H**P**ETLE**K**FDRFKHLKT
Lamprey globin	PIVDTGSVAPLS**AAEKTKIRSAWAP**VYST**YETSGVDILVKFFTS**T**P**AAQE**FFPKFKGLTT
Lupin globin	--------GALT**ESQAALVKSSWEEF**NAN**IPKHTHRFFILVLEI**A**P**AAKD**LFSFLKGTSE
	*: : : * . : .: * : * : .

Human beta	PDAVMGN**PKVKAHGKKVLGAF**S**DGL**AHLDN-----L**KGTFATLSELH**C**D**KLHVD**PENFRL**
Horse beta	PGAVMGN**PKVKAHGKKVLHSFGEGV**HHLDN-----L**KGTFAALSELH**C**D**KLHVD**PENFRL**
Human alpha	----HGS**AQVKGHGKKVADALTNAV**AHVDD-----M**PNALSALSDLH**AHKLRVD**PVNFKL**
Horse alpha	----HGS**AQVKAHGKKVGDALTLAV**GHLDD-----L**PGALSNLSDLH**AHKLRVD**PVNFKL**
Whale myoglobin	EAEMKAS**EDLKKHGVTVLTALGAIL**KKKGH-----H**EAELKPLAQSH**ATKHKIP**IKYLEF**
Lamprey globin	ADQLKKS**ADVRWHAERIINAVNDAV**ASMDDT--EKM**SMKLRDLSGKH**AKSFQVD**PQYFKV**
Lupin globin	VP--QNN**PELQAHAGKVFKLVYEAA**IQLQVTGVVVT**DATLKNLGSVH**VSKGVAD-**AHFPV**
	. .:: *. : . : *. * . : .

Human beta	**LGNVLVCVLAHH**FGKEFTPPVQA**AYQKVVAGVANALA**HKYH------
Horse beta	**LGNVLVVVLARH**FGKDFTPELQA**SYQKVVAGVANALA**HKYH------
Human alpha	**LSHCLLVTLAAH**LPAEFTPAVHA**SLDKFLASVSTVLT**SKYR------
Horse alpha	**LSHCLLSTLAVH**LPNDFTPAVHA**SLDKFLSSVSTVLT**SKYR------
Whale myoglobin	**ISEAIIHVLHSR**HPGDFGADAQG**AMNKALELFRKDIA**AKYKELGYQG
Lamprey globin	**LAAVIADTVAAG**---D------A**GFEKLMSMICILLR**SAY-------
Lupin globin	**VKEAILKTIKEV**VGAKWSEELNS**AWTIAYDELAIVIK**KEMNDAA---
	: : .: . . .. . :

Fig. 3.1   Multiple alignment of seven amino acid sequences. Identical amino acid positions are marked with asterisks (*) and biochemically conserved positions are marked with colons and periods (less conserved). The lupin sequence is a leghemoglobin and the lamprey sequence is a cyanohemoglobin. The whale sequence is from the sperm whale. The approximate positions of the alpha helices are typed in italics and bold font. The positions of two important histidine residues are underscored, and are responsible for binding the prosthetic heme and oxygen.

in the simple diagram in Fig. 3.2. It is not clear which example of a repeat unit should line up between the different members of the family. If there are large-scale repeats such as with duplications of entire protein domains, then the problem can be partly solved by excising the domains or repeat units and conducting a phylogenetic analysis of the repeats. This is only a partial solution because a single domain in one ancestral protein can give rise to two equidistant repeat units in one descendant protein and three in another; therefore, it will not be obvious how the repeat units should line up with one another. With small-grain repeats, such as those involving single nucleotides or with microsatellite sequences, the problem is even worse.

As shown in Fig. 3.2b, it is not obvious in the highlighted box if one alignment of the sequences is better than any other, except maybe cosmetically. One can only conclude that there are some repeated C residues. Fortunately, these alignment details often make no difference in a phylogenetic context and, in some cases, these

(a)

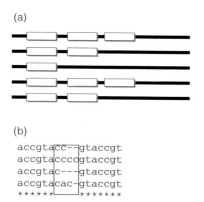

(b)

```
accgtacc--gtaccgt
accgtacccdgtaccgt
accgtac---gtaccgt
accgtacac-gtaccgt
******     *******
```

Fig. 3.2   (a) A simple diagram showing the distribution of a repeated domain (marked with a rectangle) in a set of sequences. When there are different numbers of repeats in the sequences, it can be difficult to know exactly how to arrange the alignment. Some domains will be equally similar to several domains in other sequences. (b) A simple example of some nucleotide sequences in which the central boxed alignment is completely ambiguous.

regions are simply ignored in phylogenetic inference. All computer programs will experience difficulties to unambiguously disentangle these repeats and, frequently, there is no ideal alignment. These small-scale repeats are especially abundant in some nucleotide sequences, where they can accumulate in particular positions due to mistakes during DNA replication. They tend to be localized to non-protein coding regions or to hypervariable regions; these will be almost impossible to align unambiguously and should be treated with caution. In amino acid sequences, such localized repeats of residues are unusual, whereas large-scale repeats of entire protein domains are very common.

3.3 The problem of substitutions

If the sequences in a data set accumulate few substitutions, they will remain similar and they will be relatively easy to align. The many columns of unchanged residues should make the alignment straightforward. In these cases, both manual and computerized alignment will be clear and unambiguous; in reality, sequences can and do change considerably.

For two amino acid sequences, it becomes increasingly difficult to find a good alignment once the sequence identity drops below approximately 25% (see also Chapter 9). Sequence identity, a commonly used and simple measure of sequence similarity, is the number of identical residues in an alignment divided by the total number of aligned positions, excluding any positions with a gap. Of course, a nonsensical alignment with a gap every two or three residues, for example, can result in a misleading measure of sequence similarity. In general, however,

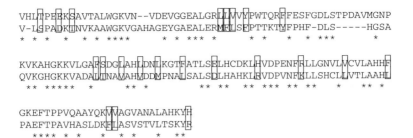

```
VHLTPEEKSAVTALWGKVN--VDEVGGEALGRLLVVYPWTQRFFESFGDLSTPDAVMGNP
V-LSPADKTNVKAAWGKVGAHAGEYGAEALERMFLSFPTTKTYFPHF-DLS-----HGSA
 *    *   * *  * ****   *  * *** *    * *    *  * ***    *

KVKAHGKKVLGAFSDGLAHLDNLKGTFATLSELHCDKLHVDPENFRLLGNVLVCVLAHHF
QVKGHGKKVADALTNAVAHVDDMPNALSALSDLHAHKLRVDPVNFKLLSHCLLVTLAAHL
 ** * *****  *    ** *     ** **  ** *** ** **   *     ** *

GKEFTPPVQAAYQKVVAGVANALAHKYH
PAEFTPAVHASLDKFLASVSTVLTSKYR
 **** * *    *  *   *     **
```

Fig. 3.3   An alignment of the amino acid sequences of human alpha globin (below) and human beta globin (above). The boxed pairs of residues are all between pairs of similar amino acids (i.e. biochemically similar side chains). The most conserved pairs are S and T (small polar); F and Y (aromatic); D and E (acidic, negatively charged); any two of H, K, and R (positively charged); and any two of F, I, L, M, or V (hydrophobic). Many of the other unmarked pairs also represent some conservation of biochemical property.

the measure is rather robust. Consider any pair of aligned homologous protein sequences with some positions where one or both residues have changed during evolution. When examining the nature of these changes, two patterns emerge (e.g. Fig. 3.3). First, the identities and differences are not evenly distributed along the sequence. Particular blocks of alignment with between 5 and 20 residues will have more identities and similar amino acids (i.e. biochemically similar; discussed in the next section) than elsewhere. These blocks will typically correspond to the conserved secondary-structure elements of α-helices and β-strands in the proteins, and are more functionally constrained than the connecting loops of irregular structure. The same pattern is observed when examining the gap locations in the sequences. This pattern of blocks is more obvious in multiple alignments of protein sequences (see Fig. 3.1). One benefit of this block-like similarity is that there will be regions of alignment which are clear and unambiguous and which most computer programs can find easily, even between distantly related sequences. However, it also means that there will be sequence regions that are more difficult to align and, if the sequences are sufficiently dissimilar, impossible to align unambiguously.

The second observed pattern is that most of the pairs of aligned, but non-identical residues, are biochemically similar (i.e. their side chains are similar). Similar amino acids such as leucine and isoleucine or serine and threonine tend to replace each other more often than dissimilar ones (see Fig. 3.3). This conservation of biochemical character is especially true in the conserved secondary-structure regions and at important sites such as active sites or ligand-binding domains. By far the most important biochemical property to be conserved is polarity/hydrophobicity. The amino acid size also matters but the exact conservation pattern will depend on the specific role of the amino acid in the protein. These patterns of conservation can significantly aid sequence alignment.

To carry out an automatic alignment, it is necessary to quantify biochemical similarity. Previous attempts focused on chemical properties and/or the genetic code; however, for both database searching and multiple alignment, the most popular amino acid scoring schemes are based on empirical studies of aligned proteins. Until the early 1990s, the most powerful method resulted from the work of Margaret Dayhoff and colleagues, who produced the famous PAM series of weight matrices (Dayhoff *et al.*, 1978) (see also Chapter 9). Protein substitution matrices are 20 × 20 tables that provide scores for all possible pairs of aligned amino acids. The higher the score, the more weight is attached to the pair of aligned residues. The PAM matrices were derived from the original empirical data using a sophisticated evolutionary model, which allowed for the use of different scoring schemes depending on how similar the sequences were. Although this was a powerful capability, most biologists used the default table offered by whatever software they were using. Currently, the PAM matrices have largely been superseded by the BLOSUM matrices of Jorja and Steven Henikoff (1992), which are also available as a series, depending on the similarity of the sequences to be aligned. The most commonly used BLOSUM62 matrix, shown in Fig. 3.4, illustrates how different pairs of identical residues get different scores. Less weight is assigned to residues which change readily during evolution, such as alanine or serine, and more weight is assigned to those that change less frequently, such as tryptophan. The remaining scores are either positive or negative, depending on whether a particular pair of residues are more or less likely to be observed in an alignment.

Given the table in Fig. 3.4, the alignment problem is finding the arrangement of amino acids resulting in the highest score summed over all positions. This is the basis of almost all commonly used alignment computer programs. One notable exception is software based on the so-called **hidden Markov models** (**HMMs**), which use probabilities rather than scores; however, these methods use a concept related to amino acid similarity.

For nucleotide sequences, it is much harder to distinguish conservative from non-conservative substitutions. In addition, two random sequences of equal base compositions will be 25% identical. As a consequence, it may be difficult to make a sensible alignment if the sequences have diverged significantly, especially if the nucleotide compositions are biased and/or if there are many repeats. This is one reason why, if there is a choice, *it is important to align protein coding sequences at the amino acid level.*

3.4 The problem of gaps

Insertions and deletions also accumulate as sequences diverge from each other. In proteins, these are concentrated between the main secondary structure elements,

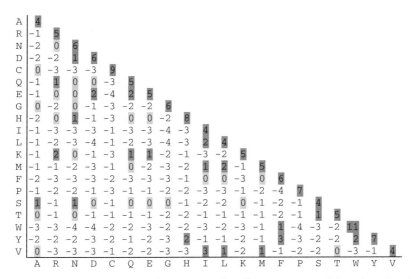

Fig. 3.4   The BLOSUM62 matrix. Similarity scores for all 20 × 20 possible pairs of amino acids, including identical pairs (the diagonal elements). Negative numbers represent pairs that are not commonly observed in real alignments; positive numbers represent commonly found pairs (shaded). These scores allow us to quantify the similarity of sequences in an alignment, when combined with gap scores (i.e. gap penalties).

just as for substitutions. Normally, no distinction is made between insertions and deletions, which are sometimes collectively referred to as *indels*. In alignments, they show up as gaps inserted in sequences in order to maximize an alignment score.

If there were no *indels*, the optimal alignment for sequences could be found by simply sliding one over the other. To choose the alignment with the highest score, the weight matrix scores are counted for each pair of aligned residues. With the presence of *indels*, a proper alignment algorithm is required to find the arrangement of gaps that maximizes the score. This is usually accomplished using a technique called **dynamic programming** (Needleman & Wunsch, 1970; Gotoh, 1982), which is an efficient method for finding the best alignment and its score. However, there is an additional complication: if gaps of any size can be placed at any position, it is possible to generate alignments with more gaps than residues. To prevent excessive use of gaps, *indels* are usually penalized (a penalty is subtracted from the alignment score) using so-called gap penalties (GPs). The most common formula for calculating GPs follows.

$$GP = g + e(l - 1) \tag{3.1}$$

where *l* is the length of the gap, *g* is a gap opening penalty (charged once per gap) and *e* is a gap-extension penalty (charged once per hyphen in a gap). These penalties are often referred to as **affine gap penalties**. The formula is quite flexible in that

```
VHLTPEEKSAVTALWGKVNVDEVGGEAL...
V----------------NEEEVGGEAL...
```

Fig. 3.5   A simple example to illustrate the problem of end gaps. The second sequence is missing a section from its N-terminus, and the end V aligns with the N-terminal V of the first sequence. This is clearly a nonsensical alignment, but it gets exactly the same score as the one with the V moved across to join the rest of the second sequence – unless end gaps are free (there is no GP for end gaps).

it allows the number and the lengths of gaps to be controlled separately by setting different values for g and e. However, there is no particular mathematical, statistical or biological justification for this formula. It is widely used because it works often well and it is straightforward to implement it in computer programs. In practice, the values of g and e are chosen arbitrarily, and there is no reason to believe that gaps simply evolve as the formula suggests. Significantly, the alignment with the highest alignment score may or may not be the correct alignment in a biological sense. Finally, it is common to make end gaps free (un-penalized), which takes into account that, for various biological and experimental reasons, many sequences are missing sections from the ends. Not making end gaps free risks getting nonsensical alignments, such as the one shown in Fig. 3.5.

3.5 Pairwise sequence alignment

3.5.1 Dot-matrix sequence comparison

Probably the oldest way to compare two nucleotide or amino acid sequences is a *dot matrix representation* or *dot plot*. In a *dot plot* two sequences are compared visually by creating a matrix where the columns are the character positions in sequence 1 and the rows are the character positions in sequence 2 (see Fig. 3.6). A dot is placed in location (i, j) if the ith character of sequence 1 matches the jth character of sequence 2. For nucleotide sequences, long protein sequences, or very divergent proteins, better resolution can be obtained by employing a sliding window approach to compare blocks of N residues (usually 5 to 20 amino acids or nucleotides). Two blocks are said to match if a mismatch level is not exceeded. For example, the user can look for DNA or amino acid stretches of length 10 (window size $= 10$) with the mismatch limit equal to 3, which means that a dot will be placed in the graph only if at least 7 out of 10 residues in the two 10-nucleotide stretches compared have identical residues. This procedure filters a considerable amount of noise that can be attributed to random matches of single residues. A high similarity region will show as a diagonal in the *dot plot*. This method can be used to explore similarity quickly in large genomic fragments, and thus evaluate homology among sequences, and can reveal features like internal repetitive elements when proteins

Sequence 1

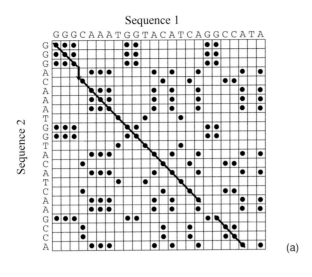

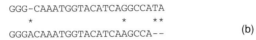

```
GGG-CAAATGGTACATCAGGCCATA
     *        *   **
GGGACAAATGGTACATCAAGCCA--
```
(b)

Fig. 3.6   (a). DNA dot plot representation with stringency 1 and window size 1 (see text). The nucleotides at each site of Sequence 1 and Sequence 2 are reported on the x (5′ → 3′ from left to right) and y axes (5′ → 3′ from top to bottom), respectively. Each nucleotide site of a sequence is compared with another sequence, and a dot is placed in the graph in case of identical residues. The path drawn by the diagonal line represents regions where the two sequences are identical; the vertical arrow indicates a deletion in Sequence 1 or an insertion in Sequence 2. (b) Sequence alignment resulting from following the path indicated in the dot plot representation.

are compared to themselves. However, their use has become somewhat obselete and we mainly discuss the method for historical and educational purposes.

3.5.2 Dynamic programming

For two sequences, dynamic programming can find the best alignment by scoring all possible pairs of aligned residues and penalizing gaps. This is a relatively rapid procedure on modern computers, requiring time and memory proportional to the product of the sequence lengths (N). This is usually referred to as an "N^2 algorithm" or as having complexity O(N^2). In practice, unless the two sequences are enormous, pairwise alignment through dynamic programming can be achieved in a matter of seconds. Dynamic programming is based on *Bellman's principle of optimality*, which states that any subsolution of an optimal solution is itself an optimal solution. In Box 3.1, we use an amino acid example to illustrate how this principle can be used to find the best pairwise alignment.

Box 3.1 Dynamic programming

The dynamic programming algorithm guarantees us to find the optimal scoring alignment of two sequences without enumerating all possible solutions. The solution to the problem can be considered as the optimal scoring path in a matrix. The scoring system in our example is constituted by the BLOSUM62 (see Fig. 3.4) substitution matrix and a simple linear gap penalty g taking the value of -8. We will introduce affine gap penalties in a later example. A matrix for sequence X (GRQTAGL) and sequence Y (GTAYDL) filled with the relevant BLOSUM62 substitution scores and gap penalties is shown below:

Sequence X

	i	*1*	*2*	*3*	*4*	*5*	*6*	*7*	*8*
j		*	G	R	Q	T	A	G	L
1	*	0	-8	-8	-8	-8	-8	-8	-8
2	G	-8	6	-2	-2	-2	0	6	-4
3	T	-8	-2	-1	-1	5	0	-2	-1
4	A	-8	0	5	-1	0	4	0	-1
5	Y	-8	-3	-2	-1	-2	-2	-3	-1
6	D	-8	-1	-2	0	-1	-2	-1	-4
7	L	-8	-4	-2	-2	-1	-1	-4	4

(Sequence Y labels the rows.)

To allow for end gaps, a boundary condition is imposed through the use of a complete gap column ($i = 1$, denoted by *) and a complete gap row ($j = 1$). In a first step of the algorithm, we need to find optimal alignments for smaller subsequences. This can be achieved by finding the best scoring subpath for each element (i, j) in the matrix (F). For each element, we need to consider three options: a substitution event (X_i, Y_i) moving from $(i-1, j-1)$ to (i, j), an insertion in X (or deletion in Y) moving from $(i-1, j)$ to (i, j), and a deletion in X (or insertion in Y) moving from $(i, j-1)$ to (i, j). Hence, the best subpath up to (i, j), will be determined by the score:

$$F(i, j) = \max \begin{cases} F(i-1, j-1) + s(X_i, Y_i), \\ F(i-1, j) - g, \\ F(i, j-1) - g \end{cases} \qquad (3.2)$$

where $s(X_i, Y_i)$ is the score defined by the BLOSUM62 substitution matrix. We can apply this equation to fill the matrix recursively: column-wise from top left to bottom right.

Importantly, we also will use a *pointer* to keep track of the movement in the matrix that resulted in the best score $F(i, j)$. For the first gap column ($i = 1$), only sequential gap events can be considered, reducing (3.2) to $F(i, j) = F(i, j-1) - g$. The same is true for the gap row ($j = 1$). The first element for which (1.2) needs to be fully evaluated is ($i = 2, j = 2$). For this element the best score is 6 as a result of:

$$F(2,2) = \max \begin{cases} F(1,1) + s(X_2, Y_2) = 6, \\ F(1,2) - 8 = -16, \\ F(2,1) - 8 = -16 \end{cases}$$

In this case, a pointer indicates a substitution event (X_i, Y_i) moving from $(i-1, j-1)$ to (i, j). We continue by finding the best score for element ($i = 2, j = 3$), given by:

$$F(2,3) = \max \begin{cases} F(1,2) + s(X_2, Y_3) = -10, \\ F(1,3) - 8 = -16, \\ F(2,2) - 8 = -2 \end{cases}$$

In which case a pointer is kept to indicate a deletion in X (or insertion in Y) moving from $(i, j-1)$ to (i, j). Repeating this procedure results in the matrix:

Note that for some elements, e.g. ($i = 4, j = 3$) two pointers are kept indicating paths that resulted in an equal score. Now that the matrix has been completed, we need to identify the best alignment based on all the pointers we have kept (represented by the arrows). We start from the score in the final element and follow the path indicated by the arrows, a procedure known as *traceback*. It is exactly here that we rely on Bellman's

Box 3.1 (*cont.*)

optimality principle: if the final element reports the score for the best path, then the sub-solution leading to this element is also lying on the optimal path. So, if we are able to identify the optimal subpath leading to this element, than the preceding element is guaranteed to lie on the optimal path allowing us to continue the *traceback* procedure from this sub-solution. If two paths can be followed (for example, from $(i-4, j=3)$ to $(i=3, j=3)$ or to $(i=3, j=2)$), than an arbitrary choice must be made (indicated by the dashed circles). Each step in the *traceback* procedure can be associated with the assembly of two symbols to construct the alignment: adding symbol X_i and Y_j if the step was to $(i-1, j-1)$, X_i and a gap if the step was to $(i-1, j)$, or a gap and Y_j if the step was to $(i, j-1)$. Hence, the alignment is constructed from right to left. Depending on the choice made at $(i=4, j=3)$ in our example, this assembly results in the alignment:

```
GRQTAGL   or   GRQTAGL
GT-AYDL        G-TAYDL
```

It is not surprising that these alignments have an equal score since both proposed amino acid substitutions (R↔T) and (Q↔T) are equally penalized by BLOSUM62 (Fig. 3.4).

Since gaps are often observed as blocks in sequence alignments, a simple linear gap penalty will generally be biologically unreasonable in real-world data. Therefore, affine gap penalties (see 3.1) have been the preferred choice for penalizing gaps in many algorithms. Here, we will extend our example using a gap-opening penalty of −8 and a gap extension penalty of −2. In this case, the matrix recursively completed with scores for the optimal subpath to each element is given by:

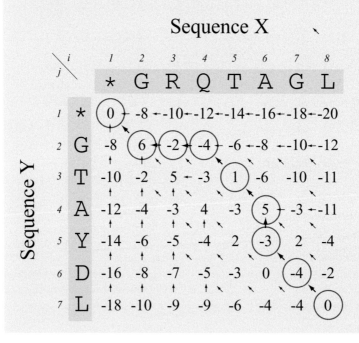

Note that the use of affine gap penalties significantly reduces the cost for sequential gaps (e.g. for column $i = 1$). Affine gap penalties introduce an additional complexity of having to keep track of multiple values. To evaluate the cost for a subpath that involves inserting a gap, we need to consider that the previous step may also have required the insertion of a gap. In this case, a gap extension needs to be penalized; in the other case, we need to penalize a gap opening. Interestingly, this scoring scheme results in a different optimal path, with which only a single alignment can be associated:

```
GRQTA-GL
G--TAYDL
```

This example illustrates that different scoring schemes, for which the gap penalties values are usually chosen arbitrarily, can result in different alignments. Therefore, it is generally recommended to explore how changing default values in alignment programs influences the quality of the sequence alignments.

3.6 Multiple alignment algorithms

The dynamic programming approach described in Box 3.1 can easily be generalized to more than two sequences. In this case, the alignment maximizing the similarity of the sequences in each column (using an amino acid weight matrix as usual) is found while allowing some minimal number and lengths of gaps. The task is most commonly expressed as finding the alignment that gives the best score for the following formula, called the *weighted sum of pairs* or WSP *objective function*:

$$\sum_i \sum_j W_{ij} D_{ij} \qquad (3.3)$$

For any multiple alignment, a score between each pair of sequences (D_{ij}) is calculated. Then, the WSP function is simply the sum of all of these scores, one for each possible pair of sequences. There is an extra weight term W_{ij} for each pair, that is, by default, always equal to one, but enables weighting some pairs more than others. This can be extremely useful to give more weight to pairs that are more reliable or more important than others. Alternatively, it can be used to give less weight to sequences with many close relatives because these are overrepresented in the data set. Although the weighting can be inspired by evolutionary relationships, the WSP fails to fully exploit phylogeny and does not incorporate an evolutionary model (Edgar & Batzoglou, 2006). Dynamic programming can be used to find a multiple alignment that gives the best possible score for the WSP function. Unfortunately, the time and memory required grows exponentially with the number of sequences as the complexity is $O(N^M)$, where M is the number of sequences and N is the sequence length. This quickly becomes impossible to compute for more than four sequences of even modest lengths.

One elegant solution to the computational complexity came from the MSA program of Lipman *et al.* (1989). It used a so-called ***branch-and-bound*** technique to eliminate many unnecessary calculations making it possible to compute the WSP function for five to eight sequences. The precise number of sequences depended on the length and similarity; the more similar the sequences, the faster the calculation and the larger the number of sequences that could be aligned. Although this is an important program, its use is still limited by the small number of sequences that can be managed. In tests with BALIBASE (see below for more information on testing multiple alignment methods), this program performs extremely well although it is not able to handle all test cases (any with more than eight sequences). The FASTMSA program, a highly optimized version of MSA, is faster and uses less memory, but it is still limited to small data sets.

In the following paragraphs, we discuss several heuristics or approximate alternatives for multiple sequence alignment and some of their most popular implementations. A comprehensive listing of multiple sequence programs and their availability can be found in Table 3.1. An updated version of this table with updated links to download pages and servers running web-based applications will be maintained on the website accompanying this book (*www.thephylogenetichandbook.org*).

3.6.1 Progressive alignment

Multiple alignments are the necessary prerequisite for phylogenetic analysis. Conversely, if the phylogenetic relationships in a set of sequences were known, this information could be useful to generate an alignment. Indeed, this mutual relationship was the basis of an early multiple alignment method (Sankoff, 1985) which simultaneously generated the tree and alignment; unfortunately, the particular method is too complex for routine use. One very simple shortcut is to make a quick and approximate tree of the sequences and use this to make a multiple alignment, an approach first suggested by Hogeweg and Hesper (1984). The method is heuristic in a mathematical sense insofar as it makes no guarantees to produce an alignment with the best score according to the optimality criterion. Nonetheless, this method is extremely important because it is the way the vast bulk of automatic alignments are generated. As judged by test cases, it performs very well, although not quite as well as those methods that use the WSP objective function. This lack of sensitivity is only visible with the most difficult test cases; for similar sequences, progressive alignment is perfectly adequate. The sheer speed of the method, however, and its great simplicity, make it extremely attractive for routine work.

A phylogenetic tree showing the relatedness of the sequences in Fig. 3.1, is shown in Fig. 3.7. A similar tree can be generated quickly by making all possible pairwise alignments between all the sequences and calculating the observed distance

Table 3.1 Multiple alignment programs

Software	Reference and URL*	Description
ABA	Raphael et al., 2004 *http://nbcr.sdsc.edu/euler/aba_v1.0/*	A-Bruijn Aligner (ABA) is a method that represents an alignment as a directed graph, which can contain cycles accommodating repeated and/or shuffled elements
ALIGN-M	Van Walle, Lasters, & Wyns, 2004 *http://bioinformatics.vub.ac.be/software/software.html*	ALIGN-M will not align parts with low confidence, thereby focusing on specificity which is important for sequences with low similarity
ALLALL	Korostensky & Gonnet, 1999 *http://www.cbrg.ethz.ch/services/AllvsAll*	Starting from a set of related peptides, ALLALL determines the relationship of each peptide sequence versus the others and uses these results to construct multiple alignments
AMAP	Schwartz & Pachter, 2007 *http://bio.math.berkeley.edu/amap/* *http://baboon.math.berkeley.edu/mavid/*	AMAP uses a sequence annealing algorithm, which is an incremental method for building multiple sequence alignments one match at a time
BALIPHY	Suchard & Redelings, 2006 *http://www.biomath.ucla.edu/msuchard/baliphy/index.php*	BALI-PHY performs joint Bayesian estimation of alignment and phylogeny. The program considers near-optimal alignments when estimating the phylogeny
CHAOS-DIALIGN	Brudno & Morgenstern, 2002 *http://dialign.gobics.de/* *http://dialign.gobics.de/chaos-dialign-submission*	The local alignment tool CHAOS rapidly identifies chains of pairwise similarities, which are subsequently used as anchor points to speed up the multiple-alignment program
CLUSTALW	Thompson, Higgins, & Gibson, 1994 *ftp://ftp-igbmc.u-strasbg.fr/pub/ClustalW/* *http://www.ebi.ac.uk/clustalw*	CLUSTALW is the prototypical progressive alignment program and the most widely used multiple alignment tool
COBALT	Papadopoulos & Agarwala, 2007 *ftp://ftp.ncbi.nlm.nih.gov/pub/agarwala/cobalt*	COBALT can combine a collection of pairwise constraints derived from database searches, sequence similarity, and user input, and incorporates them into a progressive multiple alignment
CONTRAALIGN	Do, Woods, & Batzoglou, 2006 *http://contra.stanford.edu/contralign/* *http://contra.stanford.edu/contralign/server.html*	CONTRAALIGN uses discriminative learning techniques, introduced in machine learning literature, to perform multiple alignment

(cont.)

Table 3.1 (*cont.*)

Software	Reference and URL*	Description
DCA	Stoye, 1998; Stoye, Moulton, & Dress, 1997 *http://bibiserv.techfak.uni-bielefeld.de/dca/* *http://bibiserv.techfak.uni-bielefeld.de/dca/submission.html*	The DCA (divide-and-conquer) algorithm is a heuristic approach to sum-of-pairs (SP) optimal alignment
DIALIGN	Morgenstern, 2004 *http://bibiserv.techfak.uni-bielefeld.de/dialign* *http://bibiserv.techfak.uni-bielefeld.de/dialign/submission.html*	DIALIGN does not use gap penalties and constructs pairwise and multiple alignments by comparing whole segments of the sequences, an efficient approach when sequences share only local similarities
DIALIGN-T	Subramanian et al., 2005 *http://dialign-t.gobics.de/* *http://dialign-t.gobics.de/submission?type = protein*	DIALIGN-T is a reimplementation of the segment-based ("local") alignment approach of DIALIGN with better heuristics
EXPRESSO (3DCOFFEE)	Armougom et al., 2006; O'Sullivan et al., 2004 *http://www.tcoffee.org/*	EXPRESSO is a webserver that runs BLAST to retrieve PDB structures of close homologues of the input sequences, which are used as templates to guide the alignment
ITERALIGN	Brocchieri & Karlin, 1998 *http://giotto.stanford.edu/~luciano/iteralign.html.* Currently in the process of moving ITERALIGN to a faster more reliable server	A symmetric-iterative method for multiple alignment that combines motif finding and dynamic programming procedures. The method produces alignment blocks that accommodate indels and are separated by variable-length unaligned segments
KALIGN	Lassmann & Sonnhammer, 2005 *http://msa.cgb.ki.se*	KALIGN employs the Wu-Manber string-matching algorithm, to improve both the accuracy and speed of multiple sequence alignment
MACAW	Schuler, Altschul, & Lipman, 1991 Can be downloaded at: *http://genamics.com/software/downloads/macawnta.exe*	The Multiple Alignment Construction & Analysis Workbench (MACAW) is an interactive program that allows the user to construct multiple alignments by locating, analyzing, editing, and combining "blocks" of aligned sequence segments
MAFFT	Katoh et al., 2005; Katoh et al., 2002 *http://align.bmr.kyushu-u.ac.jp/mafft/software/* *http://align.bmr.kyushu-u.ac.jp/mafft/online/server/*	MAFFT offers a range of multiple alignment methods including iterative refinement and consistency-based scoring approaches, e.g. L-INS-i (accurate; recommended for <200 sequences), FFT-NS-2 (fast; recommended for >2000 sequences)

Map	The Map program computes a global alignment of sequences using an iterative pairwise method: two sequences are aligned by computing a best overlapping alignment without penalizing terminal gaps and without heavily penalizing long internal gaps in short sequences	Huang, 1994 *http://genome.cs.mtu.edu/map.html*
Match-Box	Match-Box software performs protein sequence multiple alignment that does not require a gap penalty and is particularly suitable for finding and aligning conserved structural motifs	Depiereux & Feytmans, 1992 *http://www.fundp.ac.be/sciences/biologie/bms/matchbox_submit.shtml*
Mavid/Amap	Mavid is a progressive-alignment approach incorporating maximum-likelihood inference of ancestral sequences, automatic guide-tree construction, protein-based anchoring of ab-initio gene predictions, and constraints derived from a global homology map of the sequences	Bray & Pachter, 2004 *http://baboon.math.berkeley.edu/mavid/*
Msa	Msa uses the *branch-and-bound* technique to eliminate many unnecessary calculations making it possible to compute the WSP function for a limited number of sequences	Lipman, Altschul, & Kececioglu, 1989 *http://www.ncbi.nlm.nih.gov/CBBresearch/Schaffer/msa.html*
Multalin	Multalin is a conventional dynamic-programming method of pairwise alignment and hierarchical clustering of the sequences using the matrix of the pairwise alignment scores	Corpet, 1988 download: *ftp://ftp.toulouse.inra.fr/pub/multalin/* *http://bioinfo.genopole-toulouse.prd.fr/multalin/*
Multi-Lagan	Multi-Lagan has been developed to generate multiple alignments of long genomic sequences at any evolutionary distance	Brudno et al., 2003 *http://lagan.stanford.edu/lagan_web/index.shtml*
Mummals	Mummals uses probabilistic consistency, like ProbCons, but incorporates complex pairwise alignment HMMs and better estimation of HMM parameters	Pei & Grishin, 2006 *http://prodata.swmed.edu/mummals/mummals.php*
Murlet	Murlet is an alignment tool for structural RNA sequences using a variant of the Sankoff algorithm and an efficient scoring system that reduces time and space requirements	Kiryu et al., 2007 *http://www.ncrna.org/papers/Murlet/*
Musca	An algorithm for constrained alignment of multiple sequences that identifies two relatively simpler sub-problems whose solutions are used to obtain the alignment of the sequences	Parida, Floratos, & Rigoutsos, 1998 *http://cbcsrv.watson.ibm.com/Tmsa.html*

(cont.)

Table 3.1 (cont.)

Software	Reference and URL*	Description
MUSCLE	Edgar, 2004a; Edgar, 2004b *http://www.drive5.com/muscle/* *http://phylogenomics.berkeley.edu/cgi-bin/muscle/input_muscle.py*	MUSCLE is multiple alignment software that includes fast distance estimation using Kmer counting, progressive alignment using the log-expectation score, and refinement using tree-dependent restricted partitioning
PCMA	Pei, Sadreyev, & Grishin, 2003 *ftp://iole.swmed.edu/pub/PCMA/*	PCMA performs fast progressive alignment of highly similar sequences into relatively divergent groups, which are then aligned based on profile-profile comparison and consistency
PILEUP	Based on Feng & Doolittle, 1987 implemented in GCG: *http://www.gcg.com*	PILEUP creates a multiple sequence alignment from a group of related sequences using a simplification of the progressive alignment method of Feng & Doolittle (1987)
PIMA	Smith & Smith, 1992 Download: *http://genamics.com/software/downloads/* *http://searchlauncher.bcm.tmc.edu/multi-align/Options/pima.html*	PIMA performs multiple alignment using an extension of the covering pattern construction algorithm. It performs all pairwise comparisons and clusters the resulting scores. Each cluster is multiply aligned using a pattern-based alignment algorithm
POA	Lee, Grasso, & Sharlow, 2002 *http://bioinfo.mbi.ucla.edu/poa2/*	POA uses a graph representation of a multiple sequence alignment (MSA), called a POA-MSA, that can itself be aligned directly by pairwise dynamic programming, eliminating the need to reduce the MSA to a profile
PRALINE	Simossis & Heringa, 2005 *http://zeus.cs.vu.nl/programs/pralinewww/*	PRALINE is a multiple sequence alignment program that implements global or local preprocessing, predicted secondary structure information and iteration capabilities
PRIME	Yamada, Gotoh, & Yamana, 2006 *http://prime.cbrc.jp/*	PRIME is a group-to-group sequence alignment algorithm with piecewise linear gap cost, instead of traditional affine gap cost
PROBALIGN	Roshan & Livesay, 2006 *http://www.cs.njit.edu/usman/probalign*	PROBALIGN combines amino acid posterior probability estimation using partition function methods and computation of maximal expected accuracy alignment
PROBCONS	Do et al., 2005 *http://probcons.stanford.edu/*	PROBCONS uses a combination of probabilistic modeling and consistency-based alignment techniques and, according to the original authors, performs very well in terms of accuracy

ProDA	Phuong et al., 2006 *http://proda.stanford.edu/*	ProDA has been developed to identify and align all homologous regions appearing, not necessarily colinear, in a set of sequences
Promals	Pei & Grishin, 2007 *http://prodata.swmed.edu/promals/promals.php*	Promals uses probabilistic consistency-based scoring applied to progressive alignment in combination with profile information from database searches and secondary structure prediction
Prrn/Prrp	Based on Gotoh, 1996 *http://prrn.hgc.jp/* *http://prrn.ims.u-tokyo.ac.jp/*	Prrn/Prrp programs are prototypical iterative refinement tools using a doubly nested randomized iterative (DNR) method to make alignment, phylogenetic tree and pair weights mutually consistent
PSAlign	Sze, Lu, & Yang, 2006 *http://faculty.cs.tamu.edu/shsze/psalign/*	PSAlign uses consistency-based pairwise alignment, like the first stage of the programs TCoffee or ProbCons, but replaces the second heuristic progressive step by an exact preserving alignment step
Qoma	Zhang & Kahveci, 2006 *http://www.cise.ufl.edu/~tamer/other_files/msa.html*	Quasi-Optimal Multiple Alignment uses a graph-based method to represent an initial alignment and performs local improvements in the SP score using a greedy algorithm
Refiner	Chakrabarti et al., 2006 *ftp://ftp.ncbi.nih.gov/pub/REFINER*	A method to refine multiple sequence alignments by iterative realignment of its individual sequences with a predetermined conserved core model of a protein family
Saga	Notredame & Higgins, 1996 *http://www.tcoffee.org/Projects_home_page/saga_home_page.html*	Saga is based on the WSP objective function but uses a genetic algorithm to find the best alignment. This stochastic optimisation technique grows a population of alignments and evolves it over time using a process of selection and crossing
Sam	Hughey & Krogh, 1996; Krogh et al., 1994 *http://www.cse.ucsc.edu/research/compbio/sam.html*	The Sequence Alignment and Modeling system (Sam) is a collection of software tools for creating, refining, and using linear hidden Markov models for biological sequence analysis

(cont.)

Table 3.1 (cont.)

Software	Reference and URL*	Description
SATCHMO	Edgar & Sjolander, 2003 *http://www.drive5.com/lobster/index.htm* *http://phylogenomics.berkeley.edu/cgi-bin/satchmo/input_satchmo.py*	SATCHMO simultaneously constructs a tree and multiple sequence alignments for each internal node of the tree. These alignments of subsets predict which positions are alignable and which are not
SPEM	Zhou & Zhou, 2005 *http://sparks.informatics.iupui.edu/* *http://sparks.informatics.iupui.edu/Softwares-Services_files/spem.htm*	SPEM builds profiles and uses secondary structure information for pairwise alignment. These pairwise alignments are refined using consistency-based scoring and form the basis of a progressive multiple alignment
T-COFFEE	Notredame, Higgins, & Heringa, 2000 *http://www.tcoffee.org/Projects_home-page/t_coffee_home-page.html* *http://www.tcoffee.org/*	Tree-based Consistency objective function for alignment evaluation is the prototypical consistency-based alignment method capable of combining a collection of multiple/pairwise, global/local alignments into a single one
TRACKER/Anchored DIALIGN	Morgenstern et al., 2006 *http://dialign.gobics.de/anchor/index.php* *http://dialign.gobics.de/anchor/submission.php*	Tracker is a semi-automatic version of the alignment program DIALIGN that can take predefined constraints into account and uses these as 'anchor points' in the alignment

When two URLs are provided, the first one represents the link for download or the link to the home page, whereas the second one is a link to a webserver.

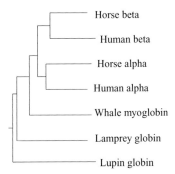

Fig. 3.7   A rooted tree showing the possible phylogenetic relationships between the seven globin sequences in Fig. 3.1. Branch lengths are drawn to scale.

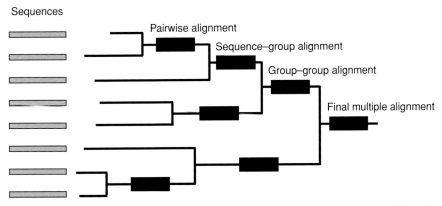

Fig. 3.8   Schematic representation of progressive alignment. This procedure begins with the aligning the two most closely related sequences (pairwise alignment) and subsequently adds the next closest sequence or sequence group to this initial pair (sequence–group or group–group alignment). This process continues in an iterative fashion along the guide tree, adjusting the positioning of indels in all grouped sequences.

(proportion of residues that differ between the two sequences) in each case. Such distances are used to make the tree with one of the widely available distance methods such as the **neighbor-joining** (**NJ**) method of Saitou and Nei (1987) (see Chapter 5). NJ trees, for example, can be calculated quickly for as many as a few hundred sequences. Next, the alignment is gradually built up by following the branching order in the tree (Fig. 3.8). The two closest sequences are aligned first using dynamic programming with GPs and a weight matrix (Fig. 3.8). For further alignment, the two sequences are treated as one, such that any gaps created between the two cannot be moved. Again, the two closest remaining sequences or pre-aligned groups of sequences are aligned to each other (Fig. 3.8). Two unaligned sequences or two sub-alignments can be aligned or a sequence can added to a sub-alignment, depending on which is more similar. The process is repeated until all the sequences are aligned. Once the initial tree is generated, the multiple alignment can be carried out with only *N*-1 separate alignments for *N* sequences. This process is fast enough to align many hundreds of sequences.

CLUSTAL

The most commonly used software for progressive alignment is CLUSTALW (Thompson *et al.*, 1994) and CLUSTALX (Thompson *et al.*, 1997). These programs are freely available as source code and/or executables for most computer platforms (Table 3.1). They can also be run using servers on the Internet at a number of locations. These programs are identical to each other in terms of alignment method but offer either a simple text-based interface (CLUSTALW) suitable for high-throughput tasks or a graphical interface (CLUSTALX). In the discussion that follows, we will refer only to CLUSTALW but all of it applies equally to CLUSTALX. CLUSTALW will take a set of input sequences and carry out the entire progressive alignment procedure automatically. The sequences are aligned in pairs in order to generate a distance matrix that can be used to make a simple initial tree of the sequences. This guide tree is stored in a file and is generated using the Neighbor-Joining method of Saitou and Nei (1987). This produces an unrooted tree (see Chapter 1), which is used to guide the multiple alignment. Finally, the multiple alignment is carried out using the progressive approach, as described above.

CLUSTALW has specific features which help it make more accurate alignments. First, sequences are down-weighted according to how closely related they are to other sequences (as judged by the guide tree). This is useful because it prevents large groups of similar sequences from dominating an alignment. Secondly, the weight matrix used for protein alignments varies, depending on how closely related the next two sequences or sets of sequences are. One weight matrix can be used for closely related sequences that gives high scores to identities and low scores otherwise. For very distantly related sequences, the reverse is true; it is necessary to give high scores to conservative amino acid matches and lower scores to identities. CLUSTALW uses a series of four matrices chosen from either the BLOSUM, or PAM series (Henikoff & Henikoff, 1992 and Dayhoff *et al.*, 1978, respectively). During alignment, the program attempts to vary GPs in a sequence- and position-specific manner, which helps to align sequences of different lengths and different similarities. Position-specific GPs are used in an attempt to concentrate gaps in the loops between the secondary structure elements, which can be set either manually using a mask or automatically. In the latter case, GPs are lowered in runs of hydrophilic residues (likely loops) or at positions where there are already many gaps. They are also lowered near some residues, such as Glycine, which are known to be common near gaps from empirical analysis (Pascarella & Argos, 1992). GPs are raised adjacent to existing gaps and near certain residues. These and other parameters can be set by the user before each alignment. Although CLUSTALW is still the most popular alignment tool, several more recent methods have now been shown to perform better in terms of accuracy and/or speed.

3.6.2 Consistency-based scoring

Progressive alignment is fast and simple, but it does have one obvious drawback: a local-minimum problem. Any alignment errors (i.e. misaligned residues or whole domains) that occur during the early alignment steps cannot be corrected later as more data are added. This may be due to an incorrect tree topology in the guide tree, but it is more likely due to simple errors in the early alignments. The latter occurs when the alignment with the best score is not the best one biologically or not the best if one considers all of the sequences in a data set. An effective way to overcome this problem is to use "consistency-based" scoring (Kececioglu, 1992), which is based on the WSP function and has been successfully used as an improvement over progressive alignment (Notredame *et al.*, 1998). Consistency-based alignment techniques use intermediate sequence information to improve the quality of pairwise comparisons and search for the multiple sequence alignment that maximizes the agreement with a set of pairwise alignments computed for the input sequences (such a procedure acknowledges that the pairwise alignments A–B and B–C may imply an A–C alignment that is different from the alignment directly computed for A and C) (Do *et al.*, 2004; Edgar & Batzoglou, 2006).

T-COFFEE

Tree-based consistency objective function for alignment evaluation (T-COFFEE) is the prototypical consistency-based alignment method (Notredame *et al.*, 2000). Although T-COFFEE is relatively slow, it generally results in more accurate alignments than CLUSTALW, PRRP (Gotoh, 1996), and DIALIGN (Morgenstern *et al.*, 1996; Morgenstern, 2004) when tested on BALIBASE (see below). Indeed, this increase in accuracy is most pronounced for difficult test cases and is found across all of the BALIBASE test sets. The method is based on finding the multiple alignment that is most consistent with a set of pairwise alignments between the sequences. These pairwise alignments can be derived from a mixture of sources such as different alignment programs or from different types of data such as structure superpositions and sequence alignments. These are processed to find those aligned pairs of residues in the initial data set which are most consistent across different alignments. This information is used to compile data on which residues are most likely to align in which sequences. Finally, this information is used to build up the multiple alignment using progressive alignment, a fast and simple procedure that requires no other parameters (GPs or weight matrices). The main disadvantage of T-COFFEE over CLUSTALW is that the former is much more computationally demanding and cannot handle large alignment problems as can the latter.

3.6.3 Iterative refinement methods

The pioneering work on alignment algorithms by Gotoh eventually resulted in the PRRN and PRRP programs (now unified to "ordinary" or "serial" PRRN) (Gotoh 1995, 1996 and 1999). PRRN uses an iterative scheme to gradually work toward the optimal alignment. At each iteration cycle, the sequences are randomly split into two groups. Within each group, the sequences are kept in fixed alignment, and the two groups are then aligned to each other using dynamic programming. This cycle is repeated until the score converges. The alignments produced by PRRN are excellent, as judged by performance using BALIBASE and other test cases (Notredame *et al.*, 1998; Thompson *et al.*, 1999). The program is, however, relatively slow with more than 20 sequences and not widely adopted.

Recently, more efficient implementations of iterative refinement methods have been developed under the name of MAFFT and MUSCLE (Katoh *et al.*, 2002, 2005; Edgar, 2004a,b). These have the same basic strategy of building a progressive alignment, to which horizontal refinement subsequently is applied. MAFFT applies the "fast Fourier Transform" to rapidly detect homologous regions and also uses an improved scoring system. MUSCLE allows fast distance estimation using "*k*-mer counting," progressive alignment using a novel profile function, and refinement using "tree-dependent restricted partitioning." Both programs are fine-tuned towards high-throughput applications, offering significant improvements in scalability while still performing relatively accurately. It should be noted that recent versions of MAFFT (v.5 or higher) implement variants of the algorithm with an objective function combining the WSP score and COFFEE-like score, which could be classified under consistency-based alignment.

3.6.4 Genetic algorithms

The SAGA program (Notredame & Higgins, 1996) is based on the WSP objective function but uses a *genetic algorithm* instead of dynamic programming to find the best alignment. This stochastic optimization technique grows a population of alignments and evolves it over time using a process of selection and crossing to find the best alignment. Comparisons with the performance of MSA, which implements the *branch-and-bound* technique, suggest that SAGA can find the optimal alignment in terms of the WSP function. The advantage of SAGA, however, is its ability to deliver good alignments for more than eight sequences; the disadvantage is that it is still relatively slow, perhaps taking many hours to compute to a good alignment for 20 or 30 sequences. The program must also be run several times because the stochastic nature of the algorithm does not guarantee the same results for different runs. Despite these reservations, SAGA has proven useful because it allows the testing of alternative WSP scoring functions.

3.6.5 Hidden Markov models

An interesting approach to alignment is the so-called hidden Markov models or HMMs, which are based on probabilities of residue substitution and gap insertion and deletion (Krogh *et al.*, 1994). HMMs have been shown to be extremely useful in a wide variety of situations in computational molecular biology, such as locating introns and exons or predicting promoters in DNA sequences. HMMs are also very useful for summarizing the diversity of information in an existing alignment of sequences and predicting whether new sequences belong to the family (e.g. Eddy, 1998; Bateman *et al.*, 2000). Packages are available that simultaneously generate such an HMM and find the alignment from unaligned sequences. These methods used to be inaccurate (e.g. see results in Notredame *et al.*, 1998); however, some progress has been made and the SAM method (Hughey & Krogh, 1996) is now roughly comparable to CLUSTALW in accuracy, although not as easy or as fast to use.

3.6.6 Other algorithms

The DIVIDE AND CONQUER, DCA (Stoye *et al.*, 1997), program also computes alignments according to the WSP scoring function. The algorithm finds sections of alignment, which when joined together head to tail, will give the best alignment. Each section is found using the MSA program and so, ultimately, it is limited in the number of sequences it can handle, even if more than MSA.

Consistency-based progressive alignment methodology has been combined with probabilistic HMMs to account for suboptimal alignments (Do *et al.*, 2004). The PROBCONS program implements this approach called posterior probability-based scoring (Durbin *et al.*, 1998), with additional features such as unsupervised expectation-maximization parameter training. This makes the approach highly accurate, but unfortunately, also computationally expensive, limiting practical application to data sets of less than 100 sequences. More complex pairwise alignment HMMs that incorporate local structural information and better estimation of HMM parameters have recently been implemented in MUMMALS (Pei & Grishin, 2006).

All methods described above try to globally align sequences, which "forces" the alignment to span the entire length of all query sequences. Recently developed methods, like ALIGN-M (Van Walle *et al.*, 2004), DIALIGN (Morgenstern *et al.*, 1998; Morgenstern, 1999; Subramanian *et al.*, 2005), POA (Lee *et al.*, 2002; Grasso & Lee, 2004) and SATCHMO (Edgar & Sjölander, 2003), have relaxed the constraint of global alignability to make possible the alignment of sequences highly variable in length or sequences with different domain structure. The DIALIGN method of Morgenstern (1999) is based on finding sections of local multiple alignment in a set of sequences; that is, sections similar across all sequences, but relatively short. DIALIGN will therefore be useful if the similarity is localized to isolated domains. In practice, the algorithm performed well in two of the BALIBASE test sets: those with long insertions and deletions; otherwise, CLUSTALW easily outperforms DIALIGN. Although DIALIGN allows unalignable regions, the alignable domains must still appear in the same order in the sequences being compared. To overcome this problem, PRODA has recently been developed to identify and align all homologous regions appearing, not necessarily colinear, in a set of sequences (Phuong et al., 2006).

3.7 Testing multiple alignment methods

The most common way to test an alignment method is by assessing its performance on test cases. Exaggerated claims have been made about how effective some methods are or why others should be avoided. The user-friendliness of the software and its availability are important secondary considerations, but ultimately, it is the quality of the alignments that matters most. In the past, many new alignment programs have been shown to perform better than older programs, using one or two carefully chosen test cases. The superiority of the new method is then often verbally described by authors who selected particular parameter values (i.e. GPs and weight matrix) that favor the performance of their program. Of course, authors do not choose test cases for which their programs perform badly and, to be fair, some underappreciated programs perform well on some test cases and are still potentially useful.

One solution is to use benchmark data sets, generated by experts, with extensive reference to secondary and tertiary structural information. For structural RNA sequences, the huge alignments of ribosomal RNA sequences can be used as tests; these comparisons, however, are difficult and not necessarily generalizable. For proteins, an excellent collection of test cases is called BALIBASE (Thompson et al., 1999), to which we have referred to in the description of some programs. BALIBASE is a collection of 141 alignments representing five different types of alignment situations: (1) equidistant (small sets of phylogenetically equidistant sequences); (2) orphan (as for Type 1 but with one distant member of the family); (3) two families (two sets of related sequences, distantly related to each other); (4) long insertions (one or more sequences have a long insertion); (5) long deletions. These test cases are made from sets of sequences for which there is extensive tertiary-structure information. The sequences and structures were carefully compared and multiple alignments were produced by manual editing.

The use of these test cases involves several assumptions. Although contradictory scenarios can be envisaged, the first assumption is that an alignment maximizing structural *similarity* is somehow "correct." A more serious assumption is that the reference alignment has been made correctly. Although parts of alignments will be difficult to judge, even with tertiary information, this can be largely circumvented by selecting and focusing only on those regions of core alignment from each test case that are clearly and unambiguously aligned. More recently, several new benchmarks have been proposed, including OXBENCH (Raghava *et al.*, 2003), PREFAB (Edgar, 2004a), SABmark (Van Walle *et al.*, 2005), IRMBASE (Subramanian *et al.*, 2005), and an extended version of BALIBASE (BaliBase 3, available at *http://www-bio3d-igbmc.u-strasbg.fr/balibase/*). These new test cases have generally been constructed by automatic means implying that the overall quality and accuracy of an individual alignment cannot be guaranteed. However, averaging accuracy scores over a large set of such alignments can still result in a meaningful ranking of multiple alignment tools.

3.8 Which program to choose?

Validation of multiple alignment programs using benchmark data sets provides useful information about their biological accuracy. Although this generally is of major concern, it is not the only criterion for choosing a particular software tool. Execution time and memory usage, in particular, can be limiting factors in practice. This is the main drawback of programs that achieve high accuracy, like T-COFFEE and PROBCONS. It is interesting to note that CLUSTALW is still relatively memory efficient compared with modern programs.

As a guideline to choosing alignment programs, we follow here the recommendations made by Edgar and Batzoglou (2006), which are summarized in Table 3.2. Because multiple alignment is an ongoing and active research field, these recommendations can evolve relatively rapidly. If accuracy is the only concern, which is generally the case for limited size data sets, it is recommended to apply different alignment methods and compare the results using the ALTAVIST web server (Morgenstern, 2003).

Incorporation of structural information in protein sequence alignment can provide significant improvement in alignment accuracy. This has inspired the development of programs that use such information in an automated fashion, like 3DCOFFEE (O'Sullivan *et al.*, 2004). The 3DCOFFEE algorithm has been implemented in a webserver called EXPRESSO that only requires input sequences (Armougom *et al.*, 2006). EXPRESSO automatically runs BLAST to identify close homologs of the sequences within the PDB database (see Chapter 2). These PDB structures are used as templates to guide the alignment of the original sequences using structure-based sequence alignment methods. Homology

Table 3.2   Typical alignment tasks and recommended procedures

Input data	Recommendations
2–100 sequences of typical protein length (maximum around 10 000 residues) that are approximately globally alignable	Use PROBCONS, T-COFFEE, and MAFFT or MUSCLE, compare the results using ALTAVIST. Regions of agreement are more likely to be correct. For sequences with low percent identity, PROBCONS is generally the most accurate, but incorporating structure information (where available) via 3DCOFFEE (a variant of T-COFFEE) can be extremely helpful
100–500 sequences that are approximately globally alignable	Use MUSCLE or one of the MAFFT scripts with default options. Comparison using ALTAVIST is possible, but the results are hard to interpret with larger numbers of sequences unless they are highly similar
>500 sequences that are approximately globally alignable	Use MUSCLE with a faster option (we recommend maxiters-2) or one of the faster MAFFT scripts
Large numbers of alignments, high-throughput pipeline	Use MUSCLE with faster options (e.g. maxiters-1 or maxiters-2) or one of the faster MAFFT scripts
2–100 sequences with conserved core regions surrounded by variable regions that are not alignable	Use DIALIGN
2–100 sequences with one or more common domains that may be shuffled, repeated or absent	Use PRODA
A small number of unusually long sequences (say, >20 000 residues)	Use CLUSTALW. Other programs may run out of memory, causing an abort (e.g. a segmentation fault)

This table is published in *Current Opinion in Structural Biology*, **16**, Edgar R.C. & Batzoglou S., Multiple sequence alignment, 368–373, Copyright Elsevier (2006).

information is also exploited by PRALINE (Simossis & Heringa, 2005), which uses PSI-BLAST (see Chapter 2) to retrieve homologs and build profiles for these sequences. SPEM also builds profiles, and additionally, uses secondary structure information (Zhou & Zhou, 2005). COBALT uses pairwise constraints derived from database searches, in particular from the Conserved Domain Database and PROSITE protein motif database, or from user input, and incorporates these into a progressive multiple alignment (Papadopoulos & Agarwala, 2007). Finally, PROMALS uses probabilistic consistency-based scoring applied to progressive alignment in combination with profile information from database searches and secondary structure prediction (Pei & Grishin, 2007). This approach has been shown to be a particular improvement for highly divergent homologs.

All these programs try to exploit information that does not merely come from the input sequences, a research direction that deserves more attention in the future. In addition, parameter selection needs to be further explored in alignment tools, as benchmarking results may be sensitive to parameter choices. Finally, the rapid development of new alignment algorithms remains unmatched by independent large-scale comparisons of the software implementations, making it difficult to make justified recommendations. A recent comparison of frequently used programs, including CLUSTALW, DIALIGN, T-COFFEE, POA, MUSCLE, MAFFT, PROBCONS, DIALIGN-T and KALIGN, indicated that the iterative approach available in MAFFT, and PROBCONS were consistently the most accurate, with MAFFT being the faster of the two (Nuin *et al.*, 2006).

3.9 Nucleotide sequences vs. amino acid sequences

Nucleotide sequences may be coding or non-coding. In the former case, they may code for structural or catalytic RNA species but more commonly for proteins. In the case of protein-coding genes, the alignment can be accomplished based on the nucleotide or amino acid sequences. This choice may be biased by the type of analysis to be carried out after alignment (Chapter 9); for example, silent changes in closely related sequences may be counted. In this case, an amino acid alignment will not be of much use for later analysis. By contrast, if the sequences are only distantly related, an analysis of amino acid or of nucleotide differences can be performed. Regardless of the end analysis desired, amino acid alignments are easier to carry out and less ambiguous than nucleotide alignments, which is also true for sequence database searching (Chapter 2). Another disadvantage of nucleotide alignment is that most programs do not recognize a codon as a unit of sequence and can break up the reading frame during alignment. Particular two-sequence alignment and database search programs can be exceptions (e.g. SEARCHWISE and PAIRWISE by Birney *et al.*, 1996). A typical approach is to carry out the alignment at the amino acid level and to then use this to generate a corresponding nucleotide sequence alignment. Different computer programs are available to perfom such an analysis; for example, PROTAL2DNA by Catherine Letondal: *http:// bioweb.pasteur.fr/seqanal/interfaces/protal2dna.html*, REVTRANS (Wernersen & Pedersen, 2003), TRANSALIGN (Bininda-Emonds, 2005) or DAMBE (Xia & Xie 2001; see Chapter 20).

If the sequences are not protein coding, then the only choice is to carry out a nucleotide alignment. If the sequences code for structural RNA (e.g. small sub-unit ribosomal RNA [SSU rRNA]), these will be constrained by function to conserve primary and secondary structure over at least some of their lengths. Typically, there are regions of clear nucleotide identity interspersed by regions that are free to change rapidly. The performance of automatic software depends on the circumstances, but specific algorithms are being developed (e.g. MURLET;

Kiryu *et al.*, 2007). With rRNA, the most commonly used programs manage to align the large conserved core sections, which are conserved across very broad phylogenetic distances. These core alignment blocks, however, will be interspersed by the highly variable expansion segments; most programs have difficulties with them. Consideration of the secondary structures, perhaps using a dedicated RNA editor, can help but it may still be difficult to find an unambiguous alignment. Excluding these regions from further analysis should be seriously considered; if the alignment is arbitrary, further analysis would be meaningless anyway. Fortunately, there are usually enough clearly conserved blocks of alignment to make a phylogenetic analysis possible.

If the nucleotide sequences are non-coding (e.g. SINES or introns), then alignment may be difficult once the sequences diverge beyond a certain level. Sequences that are highly unconstrained can accumulate indels and substitutions in all positions; these rapidly become unalignable. There is no algorithmic solution and the situation will be especially difficult if the sequences have small repeats (see Section 3.2). Even if a "somewhat optimal" alignment is obtained using a particular alignment score or parsimony criterion, there may be no reason to believe it is biologically reasonable. Such scores are based on arbitrary assumptions about the details of the evolutionary processes that have shaped the sequences. Even if the details of the assumptions are justifiable, the alignment may be so hidden that it has become unrecoverable. Caution should be taken if one alignment cannot be justified over an alternative one.

3.10 Visualizing alignments and manual editing

Although there is a continuous effort to improve biological accuracy of multiple alignment tools, manually refined alignments remain superior to purely automated methods. This will be particularly the case when manual editing takes advantage of additional knowledge (other than sequence data), which is difficult to take into account automatically (although see 3DCOFFEE, PRALINE and SPEM in Section 3.8). Manual editing is necessary to correct obvious alignment errors and to remove sections of dubious quality. Analogous to automatic alignment, manual editing should be performed on the amino acid level for protein coding sequences. In this way, information about protein domain structure, secondary structure, or amino acid physicochemical properties can be taken into consideration. In Chapter 9, some general recommendations for manual adjustment of a protein alignment are presented.

After refinement, regions that are still aligned unambiguously need to be deleted and a decision needs to be made on how to handle gaps. Not necessarily all positions with gaps need to be discarded (often referred to as "gap stripping") because they

Table 3.3 Multiple alignment editors

Editor	Reference/author and URL	Operation System	Software features
BASE-BY-BASE	Brodie *et al.*, 2004 *http://athena.bioc.uvic.ca/* (look for **BASE BY BASE** under Workbench)	Java Web Start Application: W, L, M	**BbB** is a user-friendly alignment viewer with extensive annotation capabilities, including primer display. Although visualization is its main feature, **BbB** also includes editing and alignment facilities for nucleotide and protein sequences, as well as some additional sequence analysis tools
BIOEDIT	Hall T., 1999 *http://www.mbio.ncsu.edu/BioEdit/* *bioedit.html*	W	**BIOEDIT** is a feature-rich manual alignment editor that includes many standard analysis tools and the possibility to configure several external tools to be run through the **BIOEDIT** interface (e.g. **TREEVIEW**, **BLAST** and **CLUSTALW**). The program allows easy toggling between nucleotide and amino acid translations during editing
GDE (and **MACGDE**)	De Oliveira *et al.*, 2003; Smith *et al.*, 1994 *http://www.bioafrica.net/GDElinux/index.html*	L, M	**GDE** is as a multiple sequence alignment editor that can call external command-line programs in a highly customizable fashion (e.g. **READSEQ**, **BLAST**, **CLUSTALW**, **PHYLIP** programs, etc.). Therefore, it is ideally suited for flexibly pipelining high-throughput analyses
GENEDOC	Nicholas, Nicholas, & Deerfield, 1997 *http://www.nrbsc.org/gfx/genedoc/index.html*	W	**GENEDOC** is full-featured software for visualizing, editing and analyzing multiple sequence alignments. The program has extensive shading features (incl. structural and biochemical), allows scoring while arranging, and can "re-gap" a DNA alignment when changes were made to the protein alignment
GENEIOUS	Biomatters *http://www.geneious.com/*	W, L, M	**GENEIOUS** is well-maintained, feature-rich commercial software for visualization and analysis of sequences, including an interface that integrates sequences, alignments, 3-D structures and tree data. The program can perform nucleotide alignment via amino acid translation and back and has user-friendly visualization and editing facilities

(cont.)

Table 3.3 (cont.)

Editor	Reference/author and URL	Operation System	Software features
JALVIEW	Clamp et al., 2004 http://www.jalview.org/	Java Web Start Application: W, L, M	In addition to visualizing and editing protein and nucleotide sequences, JALVIEW can realign selected sequences using MUSCLE, MAFFT or CLUSTALW, and infer distance-based trees
MACCLADE	Maddison & Maddison, 1989 http://macclade.org/macclade.html	M	MACCLADE is commercial software for exploring phylogenetic trees and character data, editing data with its specialized spreadsheet editor, and interpreting character evolution using parsimony. Sequence alignment can be done both manually, with extensive selecting and editing functions, and automatically. Amino acid translations can be shown alongside the nucleotide data
SE-AL	Rambaut A. http://tree.bio.ed.ac.uk/software/seal/	M	SE-AL is a user-friendly multiple alignment editor particularly useful for manipulating protein coding DNA/RNA sequences since it has the ability to translate in real time DNA to amino acids whilst editing the alignment, meaning that changes made in the amino acid translation are in fact made to the underlying DNA sequences
SEAVIEW	Galtier et al., 1996 http://pbil.univ-lyon1.fr/software/seaview.html	W, L, M	SEAVIEW is a graphical multiple sequence alignment editor capable of reading and writing various alignment formats (NEXUS, MSF, CLUSTAL, FASTA, PHYLIP, MASE). It allows the user to manually edit the alignment, and also to run DOT-PLOT or CLUSTALW/MUSCLE programs to locally improve the alignment

W, Windows; M, MacOs; L, Linux.

can still contain useful information. If the gaps have been inserted unambiguously and the alignment columns are not overly gapped, preferably less than 50%, they can be kept in the alignment. By unambiguous, we mean that it is obvious that a gap needs to be inserted in a particular column rather than the preceding or the following one to recover the true positional homology. The way the information in gapped columns is used in phylogenetic analysis depends on the inference method. Distance-based methods (Chapter 5) either ignore all sites that include gaps or missing data (complete deletion) or compute a distance for each pair of sequences ignoring only gaps in the two sequences being compared (pairwise deletion). In likelihood analysis (Chapters 6 and 7), gaps are generally treated as unknown characters. So, if a sequence has a gap at a particular site, then that site simply has no information about the phylogenetic relationships of that sequence, but it can still contribute information about the phylogenetic relationships of other sequences.

There are several software programs available for manual editing, visualizing and presenting alignments. A detailed software review is available online at the webpage of the Pasteur Institute (*http://bioweb.pasteur.fr/cgi-bin/seqanal/review-edital.pl*). In Table 3.3, we present a selection of some useful, user-friendly alignment editors with a brief description of their specific features, their availability, and the operating system on which they can be run. Although editors are useful tools in sequence analysis, funding and/or publication is not always obvious. As a consequence, researchers may sometimes need to resort to well-maintained and supported commercial software packages.

PRACTICE

Des Higgins and Philippe Lemey

In this practical exercise, we will align TRIM5α gene sequences from different primate species. TRIM5α is a retroviral restriction factor that protects most Old World monkey cells against HIV infection. This data set was originally analyzed by Sawyer *et al.* (2005), and is also used in the practical exercise on molecular clock analysis (Chapter 11). We will employ progressive alignment (CLUSTALX), consistency-based scoring (T-COFFEE) and iterative refinement (MUSCLE) to create different protein alignments and compare the results using the ALTAVIST web server. The exercise is designed to cover a program with a graphical user interface, a webserver and a command-line program. We will align the sequences at the amino acid level, compare different alignment algorithms, generate a corresponding nucleotide alignment and manually refine the result. Both the amino acid and nucleotide sequences ("primatesAA.fasta" and "primatesNuc.fasta," respectively) are available for download at *www.thephylogenetichandbook.org*.

3.11 CLUSTAL alignment

3.11.1 File formats and availability

As discussed in the previous chapter, the most common file formats are Genbank, EMBL, SWISS-PROT, FASTA, PHYLIP, NEXUS, and Clustal. The database formats Genbank, EMBL, and SWISS-PROT are typically used for a single sequence and most of each entry is devoted to information about the sequence. These are generally not used for multiple alignment. Nonetheless, CLUSTAL can read these formats and write out alignments (including gaps, "-") in different formats, like PHYLIP and Clustal, used exclusively for multiple alignments. In fact, CLUSTAL programs can be used as alignment converters, being able to read sequence files in the following formats: NBRF/PIR, EMBL/SWISS-PROT, FASTA, GDE, Clustal, GCG/MSF and NBRF/PIR; and write alignment files in all of the following formats: NBRF/PIR, GDE, Clustal, GCG/MSF and PHYLIP.

CLUSTALW and CLUSTALX are both freely available and can be downloaded from the EMBL/EBI file server (*ftp://ftp.ebi.ac.uk/pub/software/*) or from ICGEB in Strasbourg, France (*ftp://ftp-igbmc.u-strasbg.fr/pub/*ClustalW/ and *ftp://ftp-igbmc.u-strasbg.fr/pub/*ClustalX/). These sites are also accessible through the phylogeny programs website maintained by Joe Felsenstein (*http://evolution.genetics.washington.edu/phylip/software.html*). The programs are available for PCs running MSDOS or Windows, Macintosh computers, VAX VMS and for Unix/Linux. In each case, CLUSTALX (X stands for X windows) provides a graphical user interface with colorful display of alignments. CLUSTALW has an older, text-based interface and is less attractive for occasional use. Nonetheless, it does have extensive command line facilities making it extremely useful for embedding in scripts for high-throughput use. The CLUSTAL algorithm is also implemented in a number of commercial packages.

CLUSTAL is also directly accessible from several servers across the Internet, which is especially attractive for occasional users, but it is not without drawbacks. First, users generally do not have access to the full set of features that CLUSTAL provides. Second, it can be complicated to send and retrieve large numbers of sequences or alignments. Third, it can take a long time to carry out large alignments with little progress indication; there may even be a limit on the number of sequences that can be aligned. Nonetheless, excellent CLUSTAL servers can be recommended at EBI (*http://www.ebi.ac.uk/clustalw/*) and BCM search launcher at *http://searchlauncher.bcm.tmc.edu/*.

3.11.2 Aligning the primate Trim5α amino acid sequences

Download the primatesAA.fasta file from the website accompanying the phylogenetic handbook (*http://www.thephylogenetichandbook.org*), which contains 22 primate Trim5α amino acid sequences in fasta format. Open CLUSTALX and open the sequence file using File → Load Sequences. The graphical display allows the user to slide over the unaligned protein sequences. Select Do complete Alignment from the Alignment menu. CLUSTALX performs the progressive alignment (progress can be followed up in the lower left corner), and creates an output guide tree file and an output alignment file in the default Clustal format. It is, however, possible to choose a different format in the Output Format Options from the Alignment menu, such as, for example, the PHYLIP format which, in contrast to CLUSTAL, can be read by many of the phylogeny packages. The file with ".dnd" extension contains the guide tree generated by CLUSTALX in order to carry out the multiple alignment (see Section 3.6.1). Note that the guide tree is a very rough phylogenetic tree and it should not be used to draw conclusions about the evolutionary relationships of the taxa under investigation! CLUSTALX also allows the user to change the alignment parameters (from Alignment Parameters in the Alignment menu). Unfortunately, there are no general rules for choosing the best set of parameters, like the gap-open penalty or the gap-extension penalty. If an alignment shows, for example, too many large gaps, the user can try to increase the gap-opening penalty and redo the alignment. CLUSTALX indicates the degree of conservation at the bottom of the aligned sequences, which can be used to evaluate a given alignment. By selecting Calculate Low-Scoring Segment and Show Low-Scoring Segments from the Quality menu, it is also possible to visualize particularly unreliable parts of the alignment, which may be better to exclude from the subsequent analyses. As noted in Section 3.9, the user should keep in mind that alignment scores are based on arbitrary assumptions about the details of the evolutionary process and are not always biologically plausible. Save the amino acid alignment using File → Save Sequences as and selecting FASTA output format, which automatically saves the file with fasta extension to the directory that contains the input file (and may thus overwrite the input file).

3.12 T-COFFEE **alignment**

Although standalone **T-COFFEE** software is also available for different operating systems (Windows, Unix/linux, and MacosX), we will perform the consistency-based alignment using the **T-COFFEE** webserver in this exercise (available at *http://www.tcoffee.org/*). We select the regular submission form that only requires us to upload the "primatesAA.fasta" file or paste the sequences in the submission window. An email notification with a link to the results page will be sent if the email address is filled in, but this is not required. As noted above, **T-COFFEE** computations are more time-consuming than **CLUSTAL** progressive alignment. A job submitted on the webserver should take less than 2 minutes to complete. When the alignment procedure is completed, a new page will appear with links to the output files. Save the alignment in fasta format to your desktop (e.g. as "primatesAA_tcoffee.fas"). The score_pdf file contains a nicely quality-colored alignment in pdf format; the same output is presented in an html file.

3.13 MUSCLE **alignment**

To compute an iteratively refined alignment, we will use the standalone **MUSCLE** software, available for Windows, Unix/linux, and MacosX at *http://www.drive5.com/*. **MUSCLE** is a command-line program and thus requires the use of a terminal (Unix/Linux/MacosX) or a DOS-Window on a Windows operating system. Copy the input file ("primatesAA.fasta") to the **MUSCLE** folder, open a terminal/DOS-Window and go to the **MUSCLE** folder (using "cd"). To execute the program type muscle −in primatesAA.fasta −out primatesAA_muscle.fasta. In a DOS-Window the executable with extension needs to be specified: muscle.exe −in primatesAA.fasta −out primatesAA_muscle.fasta. The program completes the alignment procedure in a few seconds and writes out the outfile in fasta format ("primatesAA_muscle.fasta").

3.14 Comparing alignments using the ALTAVIST web tool

To evaluate results obtained using the different alignment algorithms, we use a simple and user-friendly WWW-based software program called ALTAVIST (*http://bibiserv.techfak.uni-bielefeld.de/altavist/*). ALTAVIST compares two alternative multiple alignments and uses different color-codes to indicate local agreement and conflict. The regions where both alignments coincide are generally considered to be more reliable than regions where they disagree. This is similar to the reasoning that clusters present in phylogenetic trees, reconstructed by various algorithms, are more reliable than clusters that are not consistently present. A color printout of the result pages obtained for the following alignment comparisons will be useful for further manual editing in Section 3.16.

Go to the ALTAVIST webserver and click on OPTION 2: Enter two different pre-calculated alignments of a multiple sequence set. Upload or paste the alignments generated by CLUSTALX and T-COFFEE, enter the title "ClustalX" and "T-Coffee," respectively and click submit. In the results page, the two alignments are shown with colored residues (several parts of this alignment are shown in Fig. 3.9). When all residues in a single column are shown in the same color, which is the case for the first 46 columns, these residues are lined up in the same way in both alignments. If a residue has a different color, e.g. the arginine "R" for "Howler" in column 52, it is aligned differently in the second alignment (in column 47 instead of 52). The same is true for the glutamic acid residues "E" in the second block (column 89). With respect to column 89 and 90, there is no obvious reason to prefer the first or second alignment since the total alignment score for these columns is the same and phylogenetic inference would not be influenced.

Different colors are used to distinguish groups of residues where the alignment coincides *within* groups but not *between* different groups. An example of this can be observed in the sixth alignment block (column 342–343; Fig. 3.9): the "LT" residues in AGM, Tant_cDNA and AGM_cDNA are aligned by CLUSTALX as well as T-Coffee, but not with the same residues from the other taxa. Also the "PS" residues in Rhes_cDNA and baboon are in common columns in the two alignments, but not aligned with the same residues (e.g. not with the "LT" residues from AGM, Tant_cDNA, and AGM_cDNA in the second alignment), explaining why they have different color-code. In the second alignment, all residues have the same color as in the first alignment allowing straightforward comparison of the two alignments. The color-coded comparison reveals that large blocks of alignment discrepancies are concentrated close to gapped regions. So, if gapped regions are to be stripped from this alignment, it is recommended also to delete the neighboring ambiguously aligned columns. The same comparison for the CLUSTALX and MUSCLE alignments

```
Human        M..SMLDK-GE..PE-G..PKPQIIYGARGTRYQTFV-------------------N..QPDAMCNI
Chimp        M..SMLDK-GE..PE-G..PKPQIIYGARGTRYQTFM-------------------N..QPDAMCNI
Gorilla      M..SMLDK-GE..PE-G..PKPQIIYGAQGTRYQTFM-------------------N..QPDATCNI
Orangutan    M..STLDK-GE..PE-G..PEPQIIYGAQGTTYQTYV-------------------N..QPDAMYNI
Gibbon       M..SMPDE-GE..PEEG..PEPQIIFEAQGTISQTFV-------------------N..QPDAMYNI
AGM          M..SMLYKEEE..PEEG..QNPQIMYQAPGSSFGSLTNFNYCTGVLGSQSITSRKLTN..QPDATYNI
Tant_cDNA    M..SMLYKEEE..PEEG..QNPQIMYQAPGSSFGSLTNFNYCTGVLGSQSITSRKLTN..QPDATYNI
AGM_cDNA     M..SMLYKEEE..PEEG..RNPQIMYQSPGSLFGSLTNFSYCTGVPGSQSITSGKLTN..QPDATYNI
Rhes_cDNA    M..SMLYKEGE..PEEG..RNPQIMYQAPGTLFTFPS-------------------LTN..QSDAMYNI
Baboon       M..SMLYKEGE..PEEG..RNPQITYQAPGTLFSFPS-------------------LTN..QPDAMYNI
Patas        M..SMLYKEEE..PEEG..RNPQIMYWAQGKLFQSLK-------------------N..QPDAMYDV
Colobus      M..SMLYKEGE..PEEG..PNPQIMYRAQGTLFQSLK-------------------N..QPDAMYNI
DLangur      M..SMLYKEGE..PEEG..PNPQIMCRARGTLFQSLK-------------------N..QPDAMYNI
PMarmoset    M..STLHQ-GE..PEEG..Q-VPI-HQPLV-------------------------K..KCNAKWNV
Tamarin      M..STPHQ-GE..PEEG..Q-FQI-HQPSV-------------------------K..KCNAKWNV
Owl          M..SMPHQ-GE..PEEG..Q-KRI-YQPFL-------------------------K..KRTASCSV
Squirrel     M..SMLHQ-GE..PEER..Q-KPI-RHLLV-------------------------K..KCTANQSV
Titi         M..STLHQ-GE..PEEG..Q-EWI-HQSSG-------------------------R..KCAANRNG
Saki         M..SMLHQ-GE..PEEG..Q-ERI-HQSFG-------------------------K..KCTANRNG
Spider       M..STLHQ-GE..PEEG..Q-EQI-HQPSV-------------------------K..KCTAN--V
Woolly       M..STLHQ-GE..PEEG..Q-KQR-HRPSV-------------------------K..KCTAN--V
Howler       M..S-----RE..PEEG..Q-EQIHHHPSM-------------------------E..KCIGN--F
             |  |        |   |  |                                       |  |    |
             1  47       52  88 91 325                                  463 402  409
```

Fig. 3.9   Parts of the TRIM5α CLUSTALX alignment coloured by ALTAVIST. The colours are based on the comparison between CLUSTALX and T-COFFEE. The different parts of the alignment shown are interspersed with two dots; the position of the start and end of the column of each part are indicated at the bottom.

reveals very similar discrepancies. Not surprisingly, there are fewer differences between the T-COFFEE and MUSCLE alignments. We will use this information when manually editing the sequence alignment (Section 3.16).

3.15 From protein to nucleotide alignment

All three programs discussed above frequently break up the coding reading frame when aligning the TRIM5α nucleotide sequences. This would invalidate further codon-based analysis, e.g. inference of positively selected sites like originally performed on this data set (Sawyer *et al.*, 2005; see also Chapter 14), and extensive manual editing would be required to restore the reading frame. To avoid this, the protein alignment can be used to generate a corresponding nucleotide sequence alignment. This procedure also takes advantage of the fact that protein alignments are less ambiguous and faster to compute.

We will create a TRIM5α nucleotide alignment using the PROTAL2DNA web-application available at *http://bioweb.pasteur.fr/seqanal/interfaces/protal2dna.html*. Go to WWW-submission form and enter your email address. Upload the CLUSTALX protein alignment and the unaligned nucleotide sequences ("primates-Nuc.fasta," available at *www.phylogenetichandbook.org*). Because the order in which sequences appear in both files may be different, select the option identify corresponding DNA sequences by same ID or name (-i). Select the Pearson/Fasta Output Alignment format and run the application.

Save the output file "protal2DNA.out" to your computer and change the name to "primatesNuc_pro2DNA.fasta."

The complete procedure of aligning protein sequences and generating the corresponding nucleotide alignment can also be performed in an automated fashion using programs like RevTrans, transAlign and Dambe. We will briefly discuss how to do this using Dambe, a feature-rich PC program for data analysis in molecular biology (*http://dambe.bio.uottawa.ca/software.asp*), assuming a fully working version of the program is installed on your PC.

(i) Copy the "primatesNuc.fasta" file to the Dambe folder, run DAMBE.exe, and open the file by choosing Open standard sequence file from the File menu. The window "Open" appears: set the Files of type option to Pearson/Fasta, browse to the Dambe folder and open "primatesNuc.phy."

(ii) In the "Sequence Info" window, select Protein-coding Nuc. Seqo and keep the standard genetic code. Click the Go! button.

(iii) Select Work on Amino Acid Sequence from the Sequences menu and confirm that all sequences have a complete start codon.

(v) Select Align Sequences Using ClustalW from the Alignment menu, keep default settings and Click the Go! button.

(vi) When the protein alignment is completed select Align Nuc. Seq. to aligned AA seq. in buffer from the Alignment menu. Have a look at the requirements for the back-translation and proceed by opening the original "primatesNuc.fasta" file. The corresponding nucleotide alignment can be saved in different output formats by selecting Save or Convert Sequence Format from the File menu. Note that Dambe has several automated features to delete gapped columns or other parts of the alignment using the Get Rid of Segment option in the Sequences menu.

3.16 Editing and viewing multiple alignments

As recommended in Section 3.10, we will edit the alignment at the amino acid level. However, if we wish to perform further analysis on the nucleotide sequences, it would be preferable to edit the nucleotide alignment by making changes in the appropriate amino acid translation mode. Two editors allowing the user to flexibly toggle between nucleotides and amino acids while making changes to the alignment are BioEdit for Windows and Se-Al for MacOS. We will now briefly discuss how to delete ambiguously aligned regions in the coding gene alignment based on the AltAVist comparisons and provide some basic instructions for using either BioEdit or Se-Al.

Open the coding gene alignment "primatesNuc_pro2DNA.fasta" obtained by the PROTAL2DNA web-application (BioEdit & Se-AL: File → Open; Note that shortcut keys are available for both programs). Changing the background colour can make the view more clear (BIOEDIT: `View → View Mode → Inverse Background Colored`; SE-AL: `Alignment → Use Block Colours`). Switch to amino acid translation (BIOEDIT: `Ctrl+A` to select all sequences, `Sequence → Toggle Translation or Ctrl+G`; SE-AL: `Alignment → Alignment Type → Amino Acid` or `command+T`).

The first alignment ambiguity indicated by the ALTAVIST comparisons is situated in columns 47 to 52, where different algorithms did not agree how to align the arginine "R" for "Howler" (Fig. 3.9). The most conservative option here would be to delete these six columns all together. In both programs, these columns can be selected by click-dragging the mouse over the relevant alignment region. In SE-AL, the columns can simply be deleted using `Edit → Cut or command-X` (Fig. 3.10). In BIOEDIT, we need to switch to the `Edit` mode in the alignment window menu (Fig. 3.10), and revert (`Ctrl+G`) to the nucleotide sequences. No direct editing on the amino acid translation is allowed, but our selection remains in the nucleotide view and can be deleted using `Edit → Cut or Ctrl-X`. Note that SE-AL can only "undo" the last action, whereas BioEdit does not have this limitation. The next alignment differences were indicated in columns 89 and 90. However, as discussed above, the differences are, in fact, equivalent for phylogenetic inference and no editing is required for this purpose. Based on the ambiguities indicated by ALTAVIST (both CLUSTALX vs. T-COFFEE and CLUSTALX vs. MUSCLE) and the presence of highly gapped columns, the alignment regions that should preferably be deleted are 47–52, 326–362, 403–408 and 425–511 (numbering according to the protein translation of "primatesNuc_pro2DNA.fasta"). The first three of these regions are shown in Fig. 3.9. When these regions are being deleted from the beginning of the aligment the numbering will change to 47–52, 320–356, 360–365 and 376–462. The edited alignments can now be saved to your computer in different formats (BIOEDIT: `File → Save As ...` or `File → Export → Sequence Alignment`; SE-AL: `File → Export or command+E`). A file with the modified alignment has been made available to check whether the correct editing was performed.

In this exercise, editing was restricted to deleting unreliable regions. This will be the case in many real-world alignment situations. However, considering only a single alignment, it can be necessary to correct obvious alignment errors. Both programs have extensive facilities to perform these refinements using select and slide, grab and drag, copy, cut, and paste functions.

3.17 Databases of alignments

In the future, it will become more clear how many different protein and RNA sequence families actually exist, and large alignments of these will be available

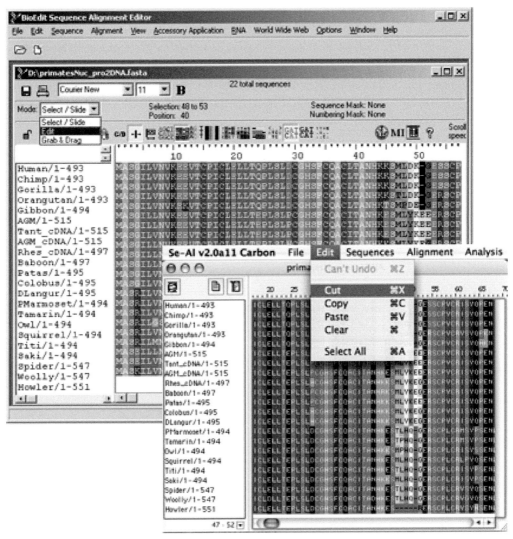

Fig. 3.10   Screen shots of BioEdit and Se-Al. In both programs, the protein coding alignment is loaded, translated to amino acids, and the columns 47 to 52 are selected. In BioEdit, the Mode is being switched to Edit; in Se-Al, the columns are being deleted.

through computer networks. In the meantime, huge alignments for the biggest and most complicated families (rRNA) already exist, which are maintained and updated regularly (*http://www.psb.ugent.be/rRNA/*). Also databases for specific organisms with alignments are being maintained (e.g. HIV: *http://www.hiv.lanl.gov/*). In this context, it is interesting to note that programs like **Clustal** have utilities to align sequences or alignments against pre-built alignments. There are numerous databases of protein alignments, each with different advantages and disadvantages, different file formats, etc. The situation is settling down with the INTERPRO project (*http://www.ebi.ac.uk/interpro/index.html*), which aims to link some of the biggest and most important protein family databases. INTERPRO is an invaluable resource for obtaining detailed information on a family of interest and for downloading sample alignments. These alignments can be used as the basis for further phylogenetic work or simply for educational or informational purposes.

Section III

Phylogenetic inference

Genetic distances and nucleotide substitution models

THEORY

Korbinian Strimmer and Arndt von Haeseler

4.1 Introduction

One of the first steps in the analysis of aligned nucleotide or amino acid sequences typically is the computation of the matrix of *genetic distances* (or *evolutionary distances*) between all pairs of sequences. In the present chapter we discuss two questions that arise in this context. First, what is a reasonable definition of a genetic distance, and second, how to estimate it using statistical models of the substitution process.

It is well known that a variety of evolutionary forces act on DNA sequences (see Chapter 1). As a result, sequences change in the course of time. Therefore, any two sequences derived from a common ancestor that evolve independently of each other eventually diverge (see Fig. 4.1). A measure of this divergence is called a genetic distance. Not surprisingly, this quantity plays an important role in many aspects of sequence analysis. First, by definition it provides a measure of the similarity between sequences. Second, if a *molecular clock* is assumed (see Chapter 11), then the genetic distance is linearly proportional to the time elapsed. Third, for sequences related by an evolutionary tree, the branch lengths represent the distance between the nodes (sequences) in the tree. Therefore, if the exact amount of sequence divergence between all pairs of sequences from a set of n sequences is known, the genetic distance provides a basis to infer the evolutionary tree relating the sequences. In particular, if sequences actually evolved according to a tree and if

The Phylogenetic Handbook: a Practical Approach to Phylogenetic Analysis and Hypothesis Testing,
Philippe Lemey, Marco Salemi, and Anne-Mieke Vandamme (eds.). Published by Cambridge
University Press. © Cambridge University Press 2009.

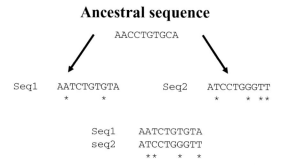

Ancestral sequence

```
                    AACCTGTGCA

Seq1    AATCTGTGTA          Seq2    ATCCTGGGTT
         *        *                  *     * **


         Seq1    AATCTGTGTA
         seq2    ATCCTGGGTT
                 **    *    *
```

Fig. 4.1   Two sequences derived from the same common ancestral sequence mutate and diverge.

the correct genetic distances between all pairs of sequences are available, then it is computationally straightforward to reconstruct this tree (see next chapter).

The substitution of nucleotides or amino acids in a sequence is usually modeled as a random event. Consequently, an important prerequisite for computing genetic distances is the prior specification of a ***model of substitution,*** which provides a statistical description of this stochastic process. Once a mathematical model of substitution is assumed, then straightforward procedures exist to infer genetic distances from the data.

In this chapter we describe the mathematical framework to model the process of nucleotide substitution. We discuss the most widely used classes of models, and provide an overview of how genetic distances are estimated using these models, focusing especially on those designed for the analysis of nucleotide sequences.

4.2 Observed and expected distances

The simplest approach to measure the divergence between two strands of aligned DNA sequences is to count the number of sites where they differ. The proportion of different homologous sites is called ***observed distance***, sometimes also called ***p-distance***, and it is expressed as the number of nucleotide differences per site.

The p-distance is a very intuitive measure. Unfortunately, it suffers from a severe shortcoming: if the degree of divergence is high, p-distances are generally not very informative with regard to the number of substitutions that actually occurred. This is due to the following effect. Assume that two or more mutations take place consecutively at the same site in the sequence, for example, suppose an A is being replaced by a C, and then by a G. As result, even though two replacements have occurred, only one difference is observed (A to G). Moreover, in case of a back-mutation (A to C to A) we would not even detect a single replacement. As a consequence, the observed distance p underestimates the true

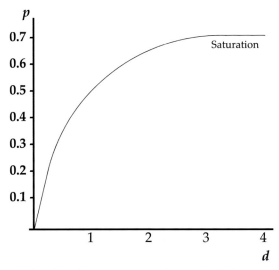

Fig. 4.2   Relationships between expected ***genetic distance*** d and observed ***p-distance***.

genetic distance d, i.e. the actual number of substitutions per site that occurred. Figure 4.2 illustrates the general relationship between d and p. As evolutionary time goes by, multiple substitutions per site will accumulate and, ultimately, sequences will become random or **saturated** (see Chapter 20). The precise shape of this curve depends on the details of the **substitution model** used. We will calculate this function later.

Since the genetic distance cannot be observed directly, statistical techniques are necessary to infer this quantity from the data. For example, using the relationship between d and p given in Fig. 4.2, it is possible to map an observed distance p to the corresponding genetic distance d. This transformation is generally non-linear. On the other hand, d can also be inferred directly from the sequences using ***maximum likelihood*** methods.

In the next sections we will give an intuitive description of the substitution process as a stochastic process. Later we will emphasize the "mathematical" mechanics of nucleotide substitution and also outline how ***maximum likelihood estimators (MLEs)*** are derived.

4.3 Number of mutations in a given time interval *(optional)*

To count the number of mutations $X(t)$ that occurred during the time t, we introduce the so-called *Poisson process* which is well suited to model processes like radioactive decay, phone calls, spread of epidemics, population growth, and so on. The structure of all these phenomena is as follows: at any point in time an event, i.e. a mutation, can take place. That is to say, per unit of time a mutation occurs with intensity or rate μ. The number of events that can take place is an integer number.

Let $P_n(t)$ denote the probability that exactly n mutations occurred during the time t:

$$P_n(t) = P(X(t) = n) \tag{4.1}$$

If t is changed, this probability will change.

Let us consider a time interval δt. It is reasonable to assume that the occurrence of a new mutation in this interval is independent of the number of mutations that happened so far. When δt is small compared to the rate μ, $\mu\delta t$ equals the probability that exactly one mutation happens during δt. The probability of no mutation during δt is obviously $1 - \mu\delta t$. In other words, we are assuming that, at the time $t + \delta t$, the number of mutations either remains unchanged or increases by one. More formally

$$P_0(t + \delta t) = P_0(t) \cdot (1 - \mu\delta t) \tag{4.2}$$

That is the probability of no mutation up to time $t + \delta t$ is equal to the probability of no mutation up to time t multiplied by the probability that no mutation took place during the interval $(t, t + \delta t)$. If we observe exactly n mutations during this period, two possible scenarios have to be considered. In the first scenario, $n - 1$ mutations occurred up to time t and exactly one mutation occurred during δt, with the probability of observing n mutations given by $P_{n-1}(t) \cdot \mu\delta t$. In the second scenario, n mutations already occurred at time t and no further mutation takes place during δt, with the probability of observing n mutations given by $P_n(t) \cdot (1 - \mu\delta t)$. Thus, the total probability of observing n mutations at time $t + \delta t$ is given by the sum of the probabilities of the two possible scenarios:

$$P_n(t + \delta t) = P_{n-1}(t) \cdot \mu\delta t + P_n(t) \cdot (1 - \mu\delta t) \tag{4.3}$$

Equations (4.2) and (4.3) can be rewritten as:

$$[P_0(t + \delta t) - P_0(t)]/\delta t = -\mu P_0(t) \tag{4.4a}$$

$$[P_n(t + \delta t) - P_n(t)]/\delta t = \mu[P_{n-1}(t) - P_n(t)] \tag{4.4b}$$

When δt tends to zero, the left part of (4.4a, b) can be rewritten (ignoring certain regularity conditions) as the first derivative of $P(t)$ with respect to t

$$P_0'(t) = -\mu \cdot P_0(t) \tag{4.5a}$$

$$P_n'(t) = \mu \cdot [P_{n-1}(t) - P_n(t)] \tag{4.5b}$$

These are typical differential equations which can be solved to compute the probability $P_0(t)$ that no mutation has occurred at time t. In fact, we are looking for a function of $P_0(t)$ such that its derivative equals $P_0(t)$ itself multiplied by the rate μ. An obvious solution is the exponential function:

$$P_0(t) = \exp(-\mu t) \tag{4.6}$$

That is, with probability $\exp(-\mu t)$ no mutation occurred in the time interval $(0, t)$. Alternatively, we could say that probability that the first mutation occurred at time $x \geq t$ is given by:

$$F(x) = 1 - \exp(-\mu t) \tag{4.7}$$

This is exactly the density function of the *exponential distribution* with parameter μ. In other words, the time to the first mutation is exponentially distributed: the longer the time, the higher the probability that a mutation occurs. Incidentally, the times between any two mutations are also exponentially distributed with parameter μ. This is the result of our underlying assumption that the mutation process "does not know" how many mutations already occurred.

Let us now compute the probability that a single mutation occurred at time t: $P_1(t)$. Recalling (4.5b), we have that:

$$P_1'(t) = \mu \cdot [P_0(t) - P_1(t)] \tag{4.8}$$

From elementary calculus, we remember the well-known rule of products to compute the derivative of a function $f(t)$, when $f(t)$ is of the form $f(t) = h(t)g(t)$:

$$f''(t) = g'(t)h(t) + g(t)h'(t) \tag{4.9}$$

Comparing (4.9) with (4.8), we get the idea that $P_1(t)$ can be written as the product of two functions, i.e. $P_1(t) = h(t) g(t)$ where $h(t) = P_0(t) = \exp(-\mu t)$ and $g(t) = \mu t$. Thus $P_1(t) = (\mu t) \exp(-\mu t)$. If we compute the derivative, we reproduce (4.8). Induction leads to (4.10):

$$P_n(t) = [(\mu t)^n \exp(-\mu t)] / n! \tag{4.10}$$

This formula describes the *Poisson distribution*, that is, the number of mutations up to time t is Poisson distributed with parameter μt. On average, we expect μt mutations with variance μt. It is important to note that the parameters μ, nucleotide substitutions per site per unit time, and t, the time, are confounded, meaning that we cannot estimate them separately but only through their product μt (number of mutations per site up to time t). We will show in the practical part of the chapter an example from literature on how to use (4.10).

$$
Q = \begin{pmatrix}
-\mu(a\pi_C+b\pi_G+c\pi_T) & a\mu\pi_C & b\mu\pi_G & c\mu\pi_T \\
g\mu\pi_A & -\mu(g\pi_A+d\pi_G+e\pi_T) & d\mu\pi_G & e\mu\pi_T \\
h\mu\pi_A & i\mu\pi_C & -\mu(h\pi_A+j\pi_C+f\pi_T) & f\mu\pi_T \\
j\mu\pi_A & k\mu\pi_C & l\mu\pi_G & -\mu(i\pi_A+k\pi_C+l\pi_G)
\end{pmatrix}
$$

$$\begin{matrix} \mathbf{A} & \mathbf{C} & \mathbf{G} & \mathbf{T} \end{matrix}$$

Fig. 4.3   Instantaneous rate matrix **Q**. Each entry in the matrix represents the instantaneous substitution rate form nucleotide i to nucleotide j (rows, and columns, follow the order **A, C, G, T**). m is the mean instantaneous substitution rate; $a, b, c, d, e, f, g, h, i, j, k, l$, are relative rate parameters describing the relative rate of each nucleotide substitution to any other. π_A, π_C, π_T, π_G, are frequency parameters corresponding to the nucleotide frequencies (Yang, 1994). Diagonal elements are chosen so that the sum of each row is equal to zero.

4.4 Nucleotide substitutions as a *homogeneous Markov process*

The nucleotide substitution process of DNA sequences outlined in the previous section (i.e. the Poisson process) can be generalized to a so-called *Markov process* which uses a **Q matrix** that specifies the relative rates of change of each nucleotide along the sequence (see next section for the mathematical details). The most general form of the **Q** matrix is shown in Fig. 4.3. Rows follow the order A, C, G, and T, so that, for example, the second term of the first row is the instantaneous rate of change from base A to base C. This rate is given by the product of μ, the mean instantaneous substitution rate, times the frequency of base A, times a, a relative rate parameter describing, in this case, how often the substitution A to C occurs during evolution with respect to the other possible substitutions. In other words, each non-diagonal entry in the matrix represents the flow from nucleotide i to j, while the diagonal elements are chosen in order to make the sum of each row equal to zero since they represent the total flow that leaves nucleotide i.

Nucleotide substitution models like the ones summarized by the **Q** matrix in Fig. 4.3 belong to a general class of models known as *time-homogeneous time-continuous stationary **Markov models***. When applied to modeling nucleotide substitutions, they all share the following set of underlying assumptions:

(1) At any given site in a sequence, the rate of change from base i to base j is independent from the base that occupied that site prior i (*Markov property*).
(2) Substitution rates do not change over time (**homogeneity**).
(3) The relative frequencies of A, C, G, and T ($\pi_A, \pi_C, \pi_G, \pi_T$) are at equilibrium (**stationarity**).

These assumptions are not necessarily biologically plausible. They are the consequence of modeling substitutions as a stochastic process. Within this general framework, we can still develop several sub-models. In this book, however, we will examine only the so-called time-reversible models, i.e. those ones assuming for any

$$Q = \begin{pmatrix} -\tfrac{3}{4}\mu & \tfrac{1}{4}\mu & \tfrac{1}{4}\mu & \tfrac{1}{4}\mu \\ \tfrac{1}{4}\mu & -\tfrac{3}{4}\mu & \tfrac{1}{4}\mu & \tfrac{1}{4}\mu \\ \tfrac{1}{4}\mu & \tfrac{1}{4}\mu & -\tfrac{3}{4}\mu & \tfrac{1}{4}\mu \\ \tfrac{1}{4}\mu & \tfrac{1}{4}\mu & \tfrac{1}{4}\mu & -\tfrac{3}{4}\mu \end{pmatrix}$$

Fig. 4.4   Instantaneous rate matrix **Q** for the Jukes and Cantor model (JC69).

two nucleotides that the rate of change from i to j is always the same than from j to i ($a = g$, $b = h$, $c = i$, $d = j$, $e = k$, $f = g$ in the **Q** matrix). As soon as the **Q** matrix, and thus the evolutionary model, is specified, it is possible to calculate the probabilities of change from any base to any other during the evolutionary time t, $\mathbf{P}(t)$, by computing the matrix exponential

$$\mathbf{P}(t) = \exp(\mathbf{Q}t) \tag{4.11}$$

(for an intuitive explanation of why, consider in analogy the result that led us to (4.6)). When the probabilities $\mathbf{P}(t)$ are known, this equation can also be used to compute the expected genetic distance between two sequences according to the evolutionary models specified by the **Q** matrix. In the next section we will show how to calculate $\mathbf{P}(t)$ and the expected genetic distance in case of the simple Jukes and Cantor model of evolution (Jukes & Cantor, 1969), whereas for more complex models only the main results will be discussed.

4.4.1 The Jukes and Cantor (JC69) model

The simplest possible nucleotide substitution model, introduced by Jukes and Cantor in 1969 (JC69), specifies that the equilibrium frequencies of the four nucleotides are 25% each, and that during evolution any nucleotide has the same probability to be replaced by any other. These assumptions correspond to a **Q** matrix with $\pi_A = \pi_C = \pi_G = \pi_T = 1/4$, and $a = b = c = g = e = f = 1$ (see Fig. 4.4). The matrix fully specifies the rates of change between pairs of nucleotides in the JC69 model. In order to obtain an analytical expression for p we need to know how to compute $P_{ii}(t)$, the probability of a nucleotide to remain the same during the evolutionary time t, and $P_{ij}(t)$, the probability of replacement. This can be done by solving the exponential $\mathbf{P}(t) = \exp(\mathbf{Q}t)$ (4.11), with **Q** as the instantaneous rate matrix for the JC69 model. The detailed solution requires the use of matrix algebra (see next section for the relevant mathematics), but the result is quite straightforward:

$$P_{ii}(t) = 1/4 + 3/4 \exp(-\mu t) \tag{4.12a}$$

$$P_{ij}(t) = 1/4 - 1/4 \exp(-\mu t) \tag{4.12b}$$

From these equations, we obtain for two sequences that diverged t time units ago:

$$p = 3/4[1 - \exp(-2\mu t)] \tag{4.13}$$

and solving for μt we get:

$$\mu t = -1/2 \log(1 - 4/3 p) \tag{4.14}$$

Thus the right-hand side gives the number of substitutions occurring in both of the lines leading to the shared ancestral sequence. The interpretation of the above formula is very simple. Under the JC69 model $3/4\mu t$ is the number of substitutions that actually occurred per site (see **Q** matrix in Fig. 4.4). Therefore, $d = 2(3/4\,\mu t)$ is the genetic distance between two sequences sharing a common ancestor. On the other hand, p is interpreted as the observed distance or p-distance, i.e. the observed proportion of different nucleotides between the two sequences (see Section 4.4). Substituting μt with $2/3d$ in (4.14) and re-arranging a bit, we finally obtain the Jukes and Cantor correction formula for the genetic distance d between two sequences:

$$d = -3/4 \ln(1 - 4/3 p) \tag{4.15a}$$

It can also be demonstrated that the variance $V(d)$ will be given by:

$$V(d) = 9p(1 - p)/(3 - 4p)^2 n \tag{4.15b}$$

(Kimura & Ohta, 1972). More complex nucleotide substitution models can be implemented depending on which parameters of the **Q** matrix we decide to estimate (see Section 4.6 below). In the practical part of this chapter we will see how to calculate pairwise genetic distances for the example data sets according to different models. Chapter 10 will discuss a statistical test that can help select the best-fitting nucleotide substitution model for a given data set.

4.5 Derivation of Markov Process *(optional)*

In this section we show how the stochastic process for nucleotide substitution can be derived from first principles such as detailed balance and the Chapman–Kolmogorov equations. To model the substitution process on the DNA level, it is commonly assumed that a replacement of one nucleotide by another occurs randomly and independently, and that nucleotide frequencies π_i in the data do not change over time and from sequence to sequence in an alignment. Under these assumptions the mutation process can be modeled by a *time-homogeneous stationary* **Markov process**.

In this model, essentially each site in the DNA sequence is treated as a random variable with a discrete number n of possible states. For nucleotides there are four states ($n = 4$), which correspond to the four nucleotide bases A, C, G, and T. The **Markov process** specifies the transition probabilities from one state to the other, i.e. it gives the probability of the replacement of nucleotide i by nucleotide j after a certain period of time t. These probabilities are collected in the transition probability matrix $\mathbf{P}(t)$. Its components $P_{ij}(t)$ satisfy the conditions:

$$\sum_{j=1}^{n} P_{ij}(t) = 1 \tag{4.16}$$

and

$$P_{ij}(t) > 0 \quad \text{for } t > 0 \tag{4.17}$$

Moreover, it also fulfills the requirement that

$$\mathbf{P}(t + s) = \mathbf{P}(t) + \mathbf{P}(s) \tag{4.18}$$

known as the Chapman–Kolmogorov equation, and the initial condition

$$P_{ij}(0) = 1, \quad \text{for } i = j \tag{4.19a}$$

$$P_{ij}(0) = 0, \quad \text{for } i \neq j \tag{4.19b}$$

For simplicity it is also often assumed that the substitution process is reversible, i.e. that

$$\pi_i \, P_{ij}(t) = \pi_j \, P_{ji}(t) \tag{4.20}$$

holds. This additional condition on the substitution process, known as detailed balance, implies that the substitution process has no preferred direction. For small t the transition probability matrix $\mathbf{P}(t)$ can be linearly approximated (Taylor expansion) by:

$$\mathbf{P}(t) \approx \mathbf{P}(0) + t\mathbf{Q} \tag{4.21}$$

where $\mathbf{Q}$ is called rate matrix. It provides an infinitesimal description of the substitution process. In order not to violate (4.16) the rate matrix $\mathbf{Q}$ satisfies

$$\sum_{i=1}^{n} Q_{ij} = 0 \tag{4.22}$$

which can be achieved by defining

$$Q_{ii} = -\sum_{i \neq j}^{n} Q_{ij} \tag{4.23}$$

Note that $Q_{ij} > 0$, since we can interpret them as the flow from nucleotide i to j, $Q_{ii} < 0$ is then the total flow that leaves nucleotide i, hence it is less than zero. In contrast to $\mathbf{P}$, the rate matrix $\mathbf{Q}$ does not comprise probabilities. Rather, it describes the amount of change of the substitution probabilities per unit time. As can be seen from (4.21) the rate matrix is the first derivative of $\mathbf{P}(t)$, which is constant for all t in a time-homogeneous Markov process. The total number of substitutions per unit time, i.e. the total rate μ, is

$$\mu = -\sum_{i=1}^{n} \pi_i Q_{ii} \tag{4.24}$$

so that the number of substitutions during time t equals $d = \mu t$. Note that, in this equation, μ and t are confounded. As a result, the rate matrix can be arbitrarily scaled, i.e. all entries can be multiplied with the same factor without changing the overall substitution pattern, only the unit in which time t is measured will be affected. For a reversible process $\mathbf{P}$, the rate matrix $\mathbf{Q}$ can be decomposed into rate parameters R_{ij} and nucleotide frequencies π_i.

$$Q_{ij} = R_{ij}, \quad \pi_j, \text{ for } i \neq j \tag{4.25}$$

The matrix $\mathbf{R} = R_{ij}$ is symmetric, $R_{ij} = R_{ji}$, and has vanishing diagonal entries, $R_{ii} = 0$.

From the Chapman–Kolmogorov (4.18) we get the forward and backward differential equations:

$$\frac{d}{dt}\mathbf{P}(t) = \mathbf{P}(t)\mathbf{Q} = \mathbf{Q}\mathbf{P}(t) \tag{4.26}$$

which can be solved under the initial condition (4.19a,b) to give

$$\mathbf{P}(t) = \exp(t\mathbf{Q}). \tag{4.27}$$

For a reversible rate matrix $\mathbf{Q}$ (4.20) this quantity can be computed by spectral decomposition (Bailey, 1964)

$$P_{ij}(t) = \sum_{m=1}^{n} \exp(\lambda_m t) U_{mi} U_{jm}^{-1} \tag{4.28}$$

where the λ_i are the eigenvalues of $\mathbf{Q}$, $\mathbf{U} = (U_{ij})$ is the matrix with the corresponding eigenvectors, and $\mathbf{U}^{-1}$ is the inverse of $\mathbf{U}$.

Choosing a model of nucleotide substitution in the framework of a reversible rate matrix amounts to specifying explicit values for the matrix $\mathbf{R}$ and for the frequencies π_i. Assuming n different states, the model has $n-1$ independent frequency parameters π_i (as $\sum \pi_i = 1$) and $[n(n-1)/2]-1$ independent rate parameters (as the scaling of the rate matrix is irrelevant, and $R_{ij} = R_{ji}$ and $R_{ii} = 0$). Thus, in the case of nucleotides ($n = 4$) the substitution process is governed by 3 independent frequency parameters π_i and 5 independent rate parameters R_{ij}.

4.5.1 Inferring the expected distances

Once the rate matrix $\mathbf{Q}$ or, equivalently, the parameters π_i and R_{ij}, are fixed, the substitution model provides the basis to statistically infer the genetic distance d between two DNA sequences. Two different techniques exist, both of which are widely used. The first approach relies on computing the exact relationship between d and p for the given model (see Fig. 4.2). The probability that a substitution is observed after time t is

$$p = 1 - \sum_{i=1}^{n} \pi_i P_{ii}(t) \tag{4.29}$$

With the definition of μ (equation 4.24) and $t = d/\mu$ we obtain

$$p = 1 - \sum_{i=1}^{n} \pi_i P_{ii}\left(-\frac{d}{\sum_{i=1}^{n} \pi_i Q_{ii}}\right) \tag{4.30}$$

This equation can then be used to construct a *method of moments estimator* of the expected distance by solving for d and estimating p (observed proportion of different sites) from the data. This formula is a generalization of (4.13).

Another way to infer the expected distance between two sequences is to use a maximum-likelihood approach. This requires the introduction of a **likelihood function** $L(d)$ (see Chapter 6 for more details). The likelihood is the probability to observe the two sequences given the distance d. It is defined as

$$L(d) = \prod_{s=1}^{l} \pi_{x_{A(s)}} P_{x_{A(s)} x_{B(s)}}\left(\frac{d}{\mu}\right) \tag{4.31}$$

where $x_{A(s)}$ *is the state at site* $s = 1, \ldots, l$ *in sequence A and* $P_{x_{A(s)} x_{B(s)}}(\frac{d}{\mu})$ *is the transition probability.* A value for d that maximizes $L(d)$ is called a **maximum likelihood estimate (MLE)** of the genetic distance. To find this estimate, numerical optimization routines are employed, as analytical results are generally not available. Estimates of error of the inferred genetic distance can be computed for both the methods of moments estimator (4.30) and the likelihood estimator (4.31) using standard statistical techniques. The so-called "delta" method can be employed to compute the variance of an estimate obtained from (4.30), and the *Fisher information criterion* is helpful to estimate the asymptotic variance of maximum likelihood estimates. For details we refer to standard statistics textbooks.

4.6 Nucleotide substitution models

If all of the eight free parameters of a reversible nucleotide rate matrix $\mathbf{Q}$ are specified, the *general time reversible* model (GTR) is derived (see Fig. 4.5). However,

$$Q = \begin{pmatrix} -\mu(a\pi_C + b\pi_G + c\pi_T) & a\mu\pi_C & b\mu\pi_G & c\mu\pi_T \\ a\mu\pi_A & -\mu(a\pi_A + d\pi_G + e\pi_T) & d\mu\pi_G & e\mu\pi_T \\ b\mu\pi_A & d\mu\pi_C & -\mu(b\pi_A + d\pi_C + f\pi_T) & f\mu\pi_T \\ c\mu\pi_A & e\mu\pi_C & f\mu\pi_G & -\mu(c\pi_A + e\pi_C + f\pi_G) \end{pmatrix}$$

$$\begin{array}{cccc} \mathbf{A} & \mathbf{C} & \mathbf{G} & \mathbf{T} \end{array}$$

Fig. 4.5   **Q** matrix of the general time reversible (GTR) model of nucleotide substitutions.

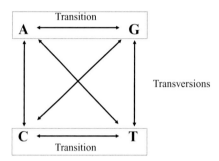

Fig. 4.6   The six possible substitution patterns for nucleotide data.

it is often desirable to reduce the number of free parameters, in particular when parameters are unknown (and hence need to be estimated from the data). This can be achieved by introducing constraints reflecting some (approximate) symmetries of the underlying substitution process. For example, nucleotide exchanges all fall into two major groups (see Fig. 4.6). Substitutions where a purine is exchanged by a pyrimidine or vice versa (A↔C, A↔T, C↔G, G↔T) are called *transversions* (*Tv*), all other substitutions are *transitions* (*Ts*). Additionally, one may wish to distinguish between substitutions among purine and pyrimidines, i.e. purine transitions (A↔G) Ts$_R$, and pyrimidine transitions (C↔T) Ts$_Y$. When these constraints are imposed, only two independent rate parameters (out of five) remain, the ratio κ of the Ts and Tv rates and the ratio γ of the two types of transition rates. This defines the Tamura–Nei (TN93) model (Tamura & Nei, 1993) which can be written as

$$R_{ij}^{TN} = \kappa \left(\frac{2\gamma}{\gamma + 1} \right) \quad \text{for Ts}_Y \tag{4.32a}$$

$$R_{ij}^{TN} = \kappa \left(\frac{2}{\gamma + 1} \right) \quad \text{for Ts}_R \tag{4.32b}$$

$$R_{ij}^{TN} = 1 \quad \text{for Tv} \tag{4.32c}$$

If $\gamma = 1$ and therefore the purine and pyrimidine transitions have the same rate, this model reduces to the HKY85 model (Hasegawa *et al.*, 1985)

$$R_{ij}^{HKY} = \kappa \quad \text{for Ts} \tag{4.33a}$$

$$R_{ij}^{HKY} = 1 \quad \text{for Tv} \tag{4.33b}$$

If the base frequencies are uniform ($p_i = 1/4$), the HKY85 model further reduces to the Kimura 2-parameter (K80) model (Kimura, 1980). For $\kappa = 1$, the HKY85 model is called F81 model (Felsenstein, 1981) and the K80 model degenerates to the Jukes and Cantor (JC69) model. The F84 model (Thorne *et al.*, 1992; Felsenstein, 1993) is also a special case of the TN93 model. It is similar to the HKY85 model but uses a slightly different parameterization. A single parameter τ generates the κ and γ parameters of the TN93 model (4.32a,b,c) in the following fashion. First, the quantity

$$\rho = \frac{\pi_R \pi_Y [\pi_R \pi_Y \tau - (\pi_A \pi_G + \pi_C \pi_T)]}{(\pi_A \pi_G \pi_Y + \pi_C \pi_T \pi_R)} \tag{4.34}$$

is computed which then determines both

$$\kappa = 1 + \frac{1}{2} \rho \left(\frac{1}{\pi_R} + \frac{1}{\pi_Y} \right) \tag{4.35}$$

and

$$\gamma = \frac{\pi_Y + \rho}{\pi_Y} \frac{\pi_R}{\pi_R + \rho} \tag{4.36}$$

of the TN93 model, where π_A, π_C, etc. are the base frequencies, π_R and π_Y are the frequency of purines and pyrimidines.

The hierarchy of the substitution models discussed above is shown in Fig. 4.7.

4.6.1 Rate heterogeneity among sites

It is a well-known phenomenon that the rate of nucleotide substitution can vary substantially for different positions in a sequence. For example, in protein coding genes third codon positions mutate usually faster than first positions, which, in turn, mutate faster than second positions. Such a pattern of evolution is commonly explained by the presence of different evolutionary forces for the sites in question. In the previous sections we have ignored this problem and silently assumed rate homogeneity over sites, but rate heterogeneity can play a crucial part in the inference of genetic distances. To account for the site-dependent rate variation, a plausible

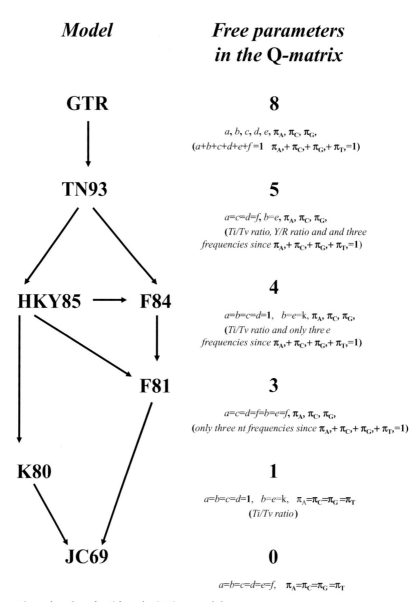

Model   *Free parameters in the Q-matrix*

GTR   **8**

$a, b, c, d, e, \pi_A, \pi_C, \pi_G,$
$(a+b+c+d+e+f=1 \quad \pi_A+\pi_C+\pi_G+\pi_T=1)$

TN93   **5**

$a=c=d=f, b=e, \pi_A, \pi_C, \pi_G,$
(Ti/Tv ratio, Y/R ratio and and three
frequencies since $\pi_A+\pi_C+\pi_G+\pi_T=1)$

HKY85 ⟶ **F84**   **4**

$a=b=c=d=1, \quad b=e=k, \pi_A, \pi_C, \pi_G,$
(Ti/Tv ratio and only three
frequencies since $\pi_A+\pi_C+\pi_G+\pi_T=1)$

F81   **3**

$a=c=d=f=b=e=f, \pi_A, \pi_C, \pi_G,$
(only three nt frequencies since $\pi_A+\pi_C+\pi_G+\pi_T=1)$

K80   **1**

$a=b=c=d=1, \quad b=e=k, \quad \pi_A=\pi_C=\pi_G=\pi_T$
(Ti/Tv ratio)

JC69   **0**

$a=b=c=d=e=f, \quad \pi_A=\pi_C=\pi_G=\pi_T$

Fig. 4.7   Hierarchy of nucleotide substitution models.

model for distribution of rates over sites is required. The common approach is to use a gamma (Γ) distribution with expectation 1.0 and variance $1/\alpha$.

$$Pdf(r) = \alpha^\alpha r^{\alpha-1}/\exp(\alpha r)\Gamma(\alpha) \tag{4.37}$$

By adjusting the shape parameter α, the Γ-distribution accommodates for varying degree of rate heterogeneity (see Fig. 4.8). For $\alpha > 1$, the distribution is bell-shaped

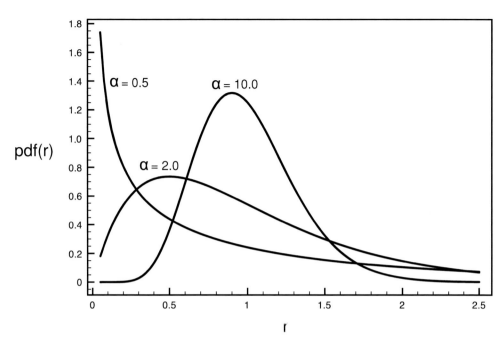

Fig. 4.8   Different shapes of the Γ-distribution depending on the α-shape parameter.

and models weak rate heterogeneity over sites. The relative rates drawn from this distribution are all close to 1.0. For $\pi < 1$, the Γ-distribution takes on its characteristic L-shape, which describes situations of strong rate heterogeneity, i.e. some positions have very large substitution rates but most other sites are practically invariable.

Rather than using the continuous Γ-distribution it is computationally more efficient to assume a discrete Γ-distribution with a finite number c of equally probable rates $q_1, q_2, \ldots, q_c$. Usually, 4–8 discrete categories are enough to obtain a good approximation of the continuous function (Yang, 1994b). A further generalization is provided by the approach of Kosakovsky et al. (2005) who propose a two stage hierarchical Beta–Gamma model for fitting the rate distribution across sites.

PRACTICE

Marco Salemi

4.7 Software packages

A large number of software packages are available to compute genetic distances from DNA sequences. An exhaustive list is maintained by Joe Felsenstein at *http://evolution.genetics.washington.edu/PHYLIP/software.html*. Among others, the programs PHYLIP, PAUP* (see Chapter 8), TREE-PUZZLE (Schmidt *et al.*, 2002; see Chapter 6), MEGA4 (Kumar *et al.*, 1993), DAMBE (Xia, 2000; see Chapter 20), and PAML (Yang, 2000; see Chapter 11) provide the possibility to infer genetic distances and will be discussed in this book.

Phylogeny Inference Package (PHYLIP), was one of the first freeware phylogeny software to be developed (Felsenstein, 1993). It is a package consisting of several programs for calculating genetic distances and inferring phylogenetic trees according to different algorithms. Pre-compiled executables files are available for Windows3.x/95/98, Windows Vista, pre-386 and 386 DOS, Macintosh (non-PowerMac), and MacOSX. A complete description of the package including the instructions for installation on different machines can be found at *http://evolution.gs.washington.edu/phylip.html*. The PHYLIP software modules that will be discussed throughout the book are briefly summarized in Box 4.1.

TREE-PUZZLE was originally developed to reconstruct phylogenetic trees from molecular sequence using maximum likelihood with a fast tree-search algorithm called *quartet puzzling* (Strimmer & von Haeseler, 1995; see Chapter 6). The program also computes pairwise maximum likelihood distances according to a number of models of nucleotide substitution. Versions of TREE-PUZZLE for UNIX, MacOSX, and Windows95/98/NT can be freely downloaded from the TREE-PUZZLE web page at *http://www.TREE-PUZZLE.de/*. The quartet-puzzling algorithm to infer phylogenetic trees will be described in detail in Chapter 6. In what follows, it is shown how to compute genetic distances according to different evolutionary models using TREE-PUZZLE.

Installation of these programs should make the PHYLIP folder and the TREE-PUZZLE folder visible on your local computer. These folders contain several files, including executable applications, documentation, and source codes. PHYLIP version 3.66 has three subdirectories: doc, exe, src; the executables are in the exe folder. The doc directory contains an extensive documentation, whereas the source codes are in src. In TREE-PUZZLE version 5.3 the program executable can be found in the src folder within the TREE-PUZZLE folder. Any of the software modules within PHYLIP and TREE-PUZZLE works in the same basic way: they need a

Box 4.1 The PHYLIP software package

PHYLIP contains several executables for analyzing molecular as well as morphological data and drawing phylogenetic trees. However, only some of the software modules included in the package will be discussed in this book; that is, those dealing with DNA and amino acid sequences analysis. These modules are summarized herein. Information about the other modules can be found in the PHYLIP documentation available with the program at *http://evolution.genetics.washington.edu/PHYLIP/software.html.*

PHYLIP executable	input data	type of analysis	book chapters
DNAdist.exe	aligned DNA sequences	calculates genetic distances using different nucleotide substitution models	5
ProtDist.exe	aligned protein sequences	calculates genetic distances using different amino acid substitution matrices	See also 9
Neighbor.exe	genetic distances	calculates NJ or UPGMA trees	5
Fitch.exe	genetic distances	calculates Fitch–Margoliash trees	5
Kitch.exe	genetic distances	calculates Fitch–Margoliash trees assuming a molecular clock	5
DNAML.exe	aligned DNA sequences	calculates maximum likelihood trees	See also 6
ProtPars.exe	aligned protein sequences	calculates maximum parsimony trees	See also 8
SeqBoot.exe	aligned DNA or protein sequences	generates bootstrap replicates of aligned DNA or protein sequences	5
Consense.exe	phylogenetic trees (usually obtained from bootstrap replicates)	generates a consensus tree	5

file containing the input data, for example, aligned DNA sequences in PHYLIP format (see Box 2.2 in Chapter 2 for different file formats), to be placed in the same directory where the program resides; it produces one or more output files in text format (usually called outfile and outtree), containing the analysis result. By default, any application reads the data from a file named infile (no extension type!) if such a file is present in the same directory, otherwise the user is asked to enter the name of the input file. Other details about PHYLIP modules are summarized in Box 4.1.

Molecular Evolutionary Genetics Analysis (MEGA4) is a sophisticated program originally developed to carry out a number of distance and parsimony based analysis of both nucleotide and amino acid sequences (Kumar *et al.*, 2004). One of the advantages of MEGA4 is the possibility to calculate standard errors of distance estimates either using analytical formulas derived for a specific evolutionary model or using **bootstrap** analysis. The latest version of the program also includes an excellent data editor that allows multiple sequences to be aligned using a native implementation of the Clustal algorithm (see Chapter 3) and to manually edit aligned and unaligned sequences. The software is freeware and can be downloaded from *http://www.megasoftware.net/overview.html*. The website also contains a detailed overview of the program capabilities, installation instructions, and extensive on-line documentation. MEGA4 only works under Windows, but it can be run in Mac by using a PC emulator (which, unfortunately, makes the program very slow) or under the Windows partition installed in the new Macs with Intel processors.

The aligned sequences in PHYLIP format (required for the analysis with PHYLIP or TREE-PUZZLE) or FASTA format (required for the analysis in MEGA4) can be downloaded from *www.thephylogenetichandbook.org*.

4.8 Observed vs. estimated genetic distances: the JC69 model

In what follows, we will use as an example the primate Trim5α sequences also used in Chapter 3 and Chapter 11 ('`Primates.phy`'). Figure 4.9a shows a matrix with pairwise *p*-distances, i.e. number of different sites between two sequences divided by the sequence length, for the primates data. The matrix is written in lower-triangular form. Comparison of Human and Chimp sequences reveals 13 point mutations over 1500 nucleotides giving an observed distance $p = 13/1500 = 0.008667$. The estimated genetic distance according to the JC69 model, obtained by substituting the observed distance in (4.15a) (see Section 4.4.1), is 0.008717. The two measures agree up to the fourth digit. This is not surprising because, as we have seen in the first part of the chapter, the relationship between observed and estimated genetic distance is approximately linear when the evolutionary rate is low and sequences share a relatively recent common ancestor. If mutations occur according to a Poisson process (Section 4.3), we expect to see only a few nucleotide substitutions between the two sequences, and no more than one substitution per site. Simply counting the observed differences between two aligned sequences is going to give an accurate measure of the genetic distance. Human and Chimp split about five million year ago (MYA) and the evolutionary rate (μ) of cellular genes is approximately 10^{-9} nucleotide substitutions per site per year (Britten, 1986). According to (4.10) the number of mutations between the Human and Chimp lineage up to time $t = 5 \times 10^6$ is Poisson distributed with parameter

$\mu = 10^{-9}$. On average, we expect to see $2\mu t = 0.01$ mutations per nucleotide site along the two phylogenetic lineages since their divergence from the common ancestor with variance 0.01. For two sequences 1500 nucleotides long we should observe a mean of 15 mutations with a 95% confidence interval of 15 ± 7.59. Indeed, the observed genetic distance between Human and Chimp in our example falls within the expected interval.

As discussed in Section 4.2, however, using the p-distance to compare more distantly related species is likely to lead to systematic underestimation of the genetic distance. By comparing the p-distance matrix in Fig. 4.9a with the JC69 distance matrix in Fig. 4.9b, we can see that the larger the observed distance the larger the discrepancy with the JC69 distance. In our alignment there are 234 mutations between Human and Squirrel (p-distance $= 0.156$). **Assuming that the JC69 model correctly describes the evolutionary process** (which, as we will see, is actually not true) the p-distance underestimates the actual genetic distance by about 11% ($d = 0.1749$). JC69 distances, as well as distances according to more complex evolutionary models, can be easily calculated with the **DNADIST** program of the **PHYLIP** software package.

Place the file `Primates.phy`, containing the primate nucleotide alignment in **PHYLIP** format, in the directory **PHYLIP\EXE**, or in the same directory as where the **PHYLIP** software module **DNADIST** is on your computer. Rename the file infile and start **DNADIST** by double clicking on its icon. A new window will appear with the following menu:

```
Nucleic acid sequence Distance Matrix program, version 3.66

Settings for this run:
  D   Distance (F84, Kimura, Jukes-Cantor, LogDet)?   F84
  G   Gamma distributed rates across sites?           No
  T   Transition/transversion ratio?                  2.0
  C   One category of substitution rates?             Yes
  W   Use weights for sites?                          No
  F   Use empirical base frequencies?                 Yes
  L   Form of distance matrix?                         Square
  M   Analyze multiple data sets?                      No
  I   Input sequences interleaved?                     Yes
  0   Terminal type (IBM PC, VT52, ANSI)?             IBM PC
  1   Print out the data at start of run?              No
  2   Print indications of progress of run?            Yes

Y   to accept these or type letter for one to change
```

Type D followed by the `enter` key and again until the model selected is Jukes–Cantor. In the new menu the option T and option F will no longer be present,

(a)

```
[Human]
[Chimp]      0.0087
[Gorilla]    0.0133 0.0087
[Orangutan]  0.0267 0.0233 0.0253
[Gibbon]     0.0380 0.0347 0.0353 0.0340
[Rhes_cDNA]  0.0680 0.0653 0.0673 0.0687 0.0773
[Baboon]     0.0653 0.0627 0.0647 0.0660 0.0747 0.0087
[AGM_cDNA]   0.0660 0.0633 0.0640 0.0660 0.0753 0.0153 0.0127
[Tant_cDNA]  0.0660 0.0633 0.0640 0.0653 0.0760 0.0153 0.0127 0.0027
[Patas]      0.0667 0.0633 0.0653 0.0653 0.0753 0.0227 0.0200 0.0180 0.0180
[Colobus]    0.0540 0.0513 0.0533 0.0553 0.0640 0.0253 0.0240 0.0240 0.0240 0.0267
[DLangur]    0.0560 0.0533 0.0553 0.0587 0.0660 0.0260 0.0247 0.0247 0.0247 0.0273 0.0087
[PMarmoset]  0.1400 0.1380 0.1420 0.1440 0.1480 0.1587 0.1573 0.1533 0.1533 0.1593 0.1467 0.1500
[Tamarin]    0.1407 0.1393 0.1433 0.1440 0.1487 0.1613 0.1600 0.1553 0.1553 0.1613 0.1487 0.1520 0.0347
[Squirrel]   0.1560 0.1520 0.1560 0.1560 0.1593 0.1740 0.1713 0.1707 0.1707 0.1753 0.1640 0.1627 0.0727 0.0713
[Titi]       0.1453 0.1440 0.1467 0.1473 0.1493 0.1620 0.1613 0.1573 0.1573 0.1587 0.1500 0.1500 0.0600 0.0600 0.0740
[Saki]       0.1367 0.1360 0.1387 0.1407 0.1440 0.1560 0.1547 0.1527 0.1533 0.1580 0.1440 0.1487 0.0567 0.0573 0.0727 0.0393
[Howler]     0.1533 0.1487 0.1533 0.1527 0.1573 0.1727 0.1700 0.1700 0.1720 0.1720 0.1567 0.1647 0.0800 0.0800 0.0907 0.0740 0.0787
[Spider]     0.1420 0.1400 0.1420 0.1400 0.1467 0.1567 0.1553 0.1520 0.1527 0.1567 0.1447 0.1480 0.0633 0.0633 0.0740 0.0607 0.0633 0.0547
[Woolly]     0.1453 0.1447 0.1467 0.1440 0.1540 0.1607 0.1593 0.1547 0.1540 0.1627 0.1500 0.1520 0.0727 0.0727 0.0800 0.0653 0.0687 0.0627 0.0300
```

(b)

```
[Human]
[Chimp]      0.0087
[Gorilla]    0.0135 0.0087
[Orangutan]  0.0272 0.0237 0.0258
[Gibbon]     0.0390 0.0355 0.0362 0.0348
[Rhes_cDNA]  0.0713 0.0684 0.0705 0.0720 0.0816
[Baboon]     0.0684 0.0654 0.0676 0.0691 0.0787 0.0087
[AGM_cDNA]   0.0691 0.0662 0.0669 0.0691 0.0794 0.0155 0.0128
[Tant_cDNA]  0.0691 0.0662 0.0669 0.0684 0.0801 0.0155 0.0128 0.0027
[Patas]      0.0698 0.0662 0.0684 0.0684 0.0794 0.0230 0.0203 0.0182 0.0182
[Colobus]    0.0560 0.0532 0.0553 0.0575 0.0669 0.0258 0.0244 0.0244 0.0244 0.0272
[DLangur]    0.0582 0.0553 0.0575 0.0611 0.0691 0.0265 0.0251 0.0251 0.0251 0.0278 0.0087
[PMarmoset]  0.1550 0.1525 0.1574 0.1599 0.1649 0.1783 0.1766 0.1715 0.1715 0.1791 0.1632 0.1674
[Tamarin]    0.1558 0.1541 0.1591 0.1599 0.1657 0.1817 0.1800 0.1741 0.1741 0.1817 0.1657 0.1699 0.0355
[Squirrel]   0.1749 0.1699 0.1749 0.1749 0.1791 0.1980 0.1945 0.1936 0.1936 0.1997 0.1834 0.1851 0.0764 0.0750
[Titi]       0.1615 0.1599 0.1632 0.1640 0.1665 0.1825 0.1817 0.1766 0.1783 0.1859 0.1674 0.1674 0.0625 0.0625 0.0779
[Saki]       0.1509 0.1501 0.1533 0.1558 0.1599 0.1749 0.1732 0.1707 0.1715 0.1774 0.1599 0.1640 0.0589 0.0596 0.0764 0.0404
[Howler]     0.1715 0.1667 0.1715 0.1707 0.1766 0.1971 0.1962 0.1962 0.1928 0.1954 0.1825 0.1851 0.0846 0.0846 0.0966 0.0779 0.0831
[Spider]     0.1574 0.1550 0.1574 0.1550 0.1632 0.1757 0.1741 0.1707 0.1707 0.1757 0.1649 0.1607 0.0662 0.0669 0.0779 0.0633 0.0662 0.0568
[Woolly]     0.1615 0.1607 0.1632 0.1599 0.1724 0.1808 0.1791 0.1732 0.1724 0.1834 0.1699 0.1674 0.0764 0.0742 0.0846 0.0684 0.0720 0.0654 0.0306
```

Fig. 4.9 (a) Pairwise *p*-distance and (b) Jukes-and-Cantor matrix for the primate TRIM5α sequences.

since under the JC69 model all nucleotide substitutions are equally likely and base frequency is assumed to be 0.25 for each base (see Section 4.4.1). Type y followed by the enter key to carry out the computation of genetic distances. The result is stored in a file called outfile, which can be opened with any text editor. The format of the output matrix, square or lower triangular, can be chosen before starting the computation by selecting option L. Of course, each pairwise distance can be obtained by replacing p in (4.15a) with the observed distance given in Fig. 4.9. That is exactly what the program DNADIST with the current settings has done: first it calculates p-distances, and then it uses the JC69 formula to convert them in genetic distances.

4.9 Kimura 2-parameter (K80) and F84 *genetic distances*

The K80 model relaxes one of the main assumptions of the JC69 model allowing for a different instantaneous substitution rate between transitions and transversions ($a = c = d = f = 1$ and $b = e = \kappa$ in the **Q** matrix) (Kimura, 1980). Similarly to what has been done in Section 4.4, by solving the exponential $\mathbf{P}(t) = \exp(\mathbf{Q}t)$ for $\mathbf{P}(t)$ the K80 correction formula for the expected genetic distance between two DNA sequences is obtained:

$$d = 1/2\ln(1/(1 - 2P - Q)) + 1/4\ln(1/(1 - 2Q)) \qquad (4.38a)$$

where P and Q are the proportion of the transitional and transversional differences between the two sequences, respectively. The variance of the K80 distances is calculated by:

$$V(d) = 1/n[(A^2 P + B^2 Q - (AP + BQ)^2] \qquad (4.38b)$$

with A = 1/(1–2P-Q) and B = 1/2[(1/1–2P–Q) + (1/1–2Q)].

K80-distances can be obtained with DNAdist by choosing Kimura 2-parameter within the D option. The user can input an empirical *transition/transversion ratio* (*Ti/Tv*) by selecting option T from the main menu. *Ti/Tv* is the probability of *any* transition (over a single unit of time) divided by the probability of *any* transversion (over a single unit of time), which can be obtained by dividing the sum of the probabilities of transitions (four terms) by the sum of the probabilities of transversions (eight terms). The default value for *Ti/Tv* in DNAdist is 2.0. Considering that there are twice more possible transversions than transitions, the default value of *Ti/Tv* = 2.0 in DNADIST assumes that, during evolution, transitional changes are about four times more likely than transversional ones. When an empirical *Ti/Tv* value for the set of organisms under investigation is not known from the literature, it is good practice to estimate it directly from the data. A general strategy to estimate the *Ti/Tv* ratio of aligned DNA sequences will be discussed in Chapter 6. Note that some programs use the transition/transversion *rate* ratio (κ) instead of the expected *Ti/Tv* ratio, which is the instantaneous rate of transitions divided by the instantaneous rate of transversions and does not involve the equilibrium base frequencies. Depending on the equilibrium base frequencies, this rate ratio will be about twice the *Ti/Tv* ratio.

The genetic distance estimated with the K80 model (*Ti/Tv* = 2.0) between Human and Chimp (0.008722), is still not significantly different from the *p*-distance for the same reasons discussed above. The K80 distance between Squirrel and Human is 0.180, which is slightly larger than the one estimated by the JC69. However, even small changes in the distance can influence the topology of phylogenetic trees inferred with distance-based methods. The K80 model still relies on very restricted assumptions such as that of equal frequency of the four bases at equilibrium. The HKY85 (Hishino *et al.*, 1985) and F84 (Felsenstein, 1984; Kishino & Hasegawa, 1989) models relax that assumption allowing for unequal frequencies; their **Q** matrices are slightly different, but both models essentially share the same set of assumptions: a bias in the rate of transitional with respect to the rate of transversional substitutions and unequal base frequencies (which are usually set to the empirical frequencies). F84 is the default model in PHYLIP version 3.66. Since the F84 model assumes unequal base frequencies, **DNADIST** empirically estimates the frequencies for each sequence (option F) and it uses the average value over all sequences to compute pairwise distances. When no is selected in option F, the program asks the user to input the base frequencies in order A, C, G, T/U separated by blank spaces. As with the K80 model, F84 scores transitional and transversional substitutions differently and it is possible to input an empirical *Ti/Tv* ratio with the option T.

4.10 More complex models

The TN93 model (Tamura & Nei, 1993), an extension of the F84 model, allows different nucleotide substitution rates for purine (A↔G) and pyrimidine (C↔T) transitions ($b \neq e$ in the correspondent **Q** matrix). TN93 genetic distances can be computed with TREE-PUZZLE by selecting from the menu: `Pairwise distances only (no tree)` in option k, and TN (Tamura & Nei, 1993) in option m. The menu allows the user to input empirical *Ti/Tv* bias and pyrimidine/purine transition bias, otherwise the program will estimate those parameters from the data set (see Chapter 6 for the details). Genetic distances can also be obtained according to simpler models. For example, by selecting HKY in option m, TREE-PUZZLE computes HKY85 distances. Since the JC69 model is a further simplification of the HKY85 model where equilibrium nucleotide frequencies are equal and there is no nucleotide substitution bias (see above), JC69 distances can be calculated with TREE-PUZZLE by selecting the HKY85 model and setting nucleotide frequencies equal to 0.25 each (option f in the menu) and the Ti/Tv ratio equal to 0.5 (option t). Note that since there are twice more transversions than transitions (see Fig. 4.6) the Ti/Tv ratio needs to be set to 0.5 and not to 1 in order to reduce the HKY85 model to the JC69! The distance matrix in square format is written to the `outdist` file and can be opened with any text editor. The program also outputs an `oufile` with several statistics about the data set (the file is mostly self-explanatory, but see also Section 4.13 and Chapter 6).

4.10.1 Modeling rate heterogeneity among sites

The JC69 model assumes that all sites in a sequence change at a uniform rate over time. More complex models allow particular substitutions, for example, transitions, to occur at different rate than others, for example, transversions, but any particular substitution rate between nucleotide i and nucleotide j is the same among different sites. Section 4.6.1 pointed out that such an assumption is not realistic, and it is especially violated in coding regions where different codon positions usually evolve at different rates. Replacements at the second codon position are always ***non-synonymous***, i.e. they change the encoded amino acid (see Chapter 1), whereas, because of the degeneracy of the genetic code, 65% of the possible replacements at the third codon position are ***synonymous***, i.e. no change in the encoded amino acid. Finally only 4% of the possible replacements at the first codon position are synonymous. Since mutations in a protein sequence will most of the time reduce the ability of the protein to perform its biological function, they are rapidly removed from the population by purifying selection (see Chapter 1). As a consequence, over time, mutations will accumulate more rapidly at the third rather than at the second or the first codon position. It has been shown, for example, that in each coding region of the human T-cell lymphotropic viruses (HTLVs), a group of human oncogenic retroviruses, the third codon positions evolve about eight times faster than the first and 16 times faster than the second positions (Salemi *et al.*, 2000). It is possible to model rate heterogeneity over sites by selecting the option:

```
B One category of substitution rates? Yes
```

in the main menu of DNADIST (which toggles this option to No) and to choose up to nine different categories of substitution rates. The program then asks for input of the relative substitution rate for each category as a non-negative real number. For example, consider how to estimate the genetic distances for the primates data set using the JC69 model, but assuming that mutations at the third position accumulate ten times faster than at the first and 20 times faster than at the second codon position. Since only the relative rates are considered, one possibility is to set the rate at the first codon position equal to 1, the rate at the second to 0.5, and the rate at the third to 10. It is also necessary to assign each site in the aligned data set to one of the three rate categories. PHYLIP assigns rates to sites by reading an additional input file with default name "categories" containing a string of digits (a new line or a blank can occur after any character in this string) representing the rate category of each site in the alignment. For example, to perform the calculation with the primates data set, we need to prepare a text file called categories (with no extension) containing the following string:

```
12312312311231231231[ ...]
```

Each number in the line above represents a nucleotide position in the aligned data set: for example, the first four numbers, 1231, refer to the first four positions in the alignment and they assign the first position to rate category 1, the second position to rate category 2, the third position to rate category 3, the fourth position to rate category 1 again, and so forth. In the primates data set, sequences are in the correct reading frame, starting at the first codon position and ending at a third codon position, and there are 1500 positions. Thus the `categories` file has to be prepared in the following way:

   123123123 (and so forth up to 1500 digits)

An appropriately edited file (`Primates_cdp_categories.phy`) can be found at *www.thephylogenetichandbook.org*. After renaming this file as "`categories`," the following exercise can be carried out:

  (i) Place the input files (`primates.phy` and `categories`) in the PHYLIP folder and run DNADIST
 (ii) Select option C and type 3 to choose three different rate `categories`
(iii) At the prompt of the program asking to specify the relative rate for each category type: 1 0.5 10 and press enter
 (iv) Choose the desired evolutionary model as usual and run the calculation.

If there is no information about the distribution and the extent of the relative substitution rates across sites, rate heterogeneity can be modeled using a G-distribution (Yang, 1994b), a negative binomial distribution (Xia, 2000) or a two-stage hierarchical Beta–Gamma model (Kosakovsky *et al.*, 2005). As discussed in Section 4.6.1, a single parameter α describes the shape of the Γ-distribution (Fig. 4.8): L-shaped for $\alpha < 1$ (strong rate heterogeneity), or bell-shaped for $\alpha > 1$ (weak rate heterogeneity). Which value of α is the most appropriate for a given data set, however, is usually not known. The next few chapters will discuss how to estimate *a* with different approaches and how to estimate genetic distances with Γ-distributed rates across sites (in particular, see Chapter 6). However, it is important to keep in mind that, even though different sites across the genome do change at different rates (Li, 1997), the use of a *discrete* Γ-distribution to model rate heterogeneity over sites has no biological justification. It merely reflects our ignorance about the underlying distribution of rates. It is widely used because it allows both low and high rate heterogeneity among sites to be modeled easily and flexibly by varying the *a* parameter.

Chapter 10 will show how to compare different evolutionary models. In such a way it is also possible to test whether a nucleotide substitution model implementing Γ-distributed rates across sites usually fits the data significantly better than a model assuming uniform rates. Genetic distances with Γ-models can be estimated by selecting the option G Gamma distributed rates across sites? in DNAdist. For example, to estimate F84+Γ distances for the primates data set, just run DNAdist as before, type G followed by the enter key (the menu will change to display G Gamma distributed rates across sites? Yes), and type y followed again by the enter key. Notice that, before running the analysis, the program will ask to enter the coefficient of variation CV, which is required for the specific computational implementation of G-models in PHYLIP. The relationship between α and CV is CV $= 1/\sqrt{a}$. Therefore, to use $a = 0.5$ we digit the value 1.414 and press the enter key. As usual, the calculated distances will be written to outfile.

To illustrate the effect of model complexity and rate heterogeneity among sites on distance estimation, the genetic distances for Human–Squirrel are shown for different substitution models in Fig. 4.10. When correcting for "multiple hits," increasingly complex models have only a marginal effect on evolutionary distances for this data set, whereas modeling rate heterogeneity (shown using circles) has a profound effect on distance estimation. The practice section of the next chapter demonstrates that this can also have an important impact on phylogenetic inference.

4.11 Estimating standard errors using MEGA4

The program MEGA4 can estimate genetic distance estimates using most of the nucleotide substitution models (with and without Γ-distribution) discussed above. In addition, it is possible to use MEGA4 to calculate the standard errors of the estimated distances either analytically or by bootstrapping (a statistical technique that is often used to assess the robustness of phylogenetic inference, see Chapter 5). Standard errors for JC69 or K80 models are calculated in MEGA4 by employing the same variance formulas given in (4.15b) and (4.37), and can be useful to perform statistical tests comparing different distance estimates (for example, to decide whether or not two sets of distances are significantly different). The idea behind bootstrapping, on the other hand, is to generate a large number of random replicates (usually 1000–10 000) by randomly resampling with replacement from

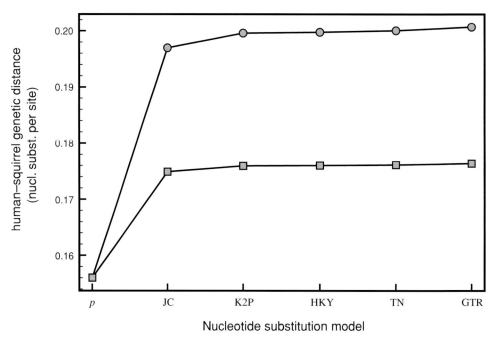

Fig. 4.10   Pairwise genetic distance for Human–Squirrel TRIM5α sequences using different evolutionary models: p = p-distance, JC = Jukes and Cantor, K2P = Kimura 2-parameter, HKY = Hasegawa–Kishino–Yano (85), TN = Tamura–Nei, GTR = General Time-Reversible. Distances computed using gamma distributed rate among sites (a = 1.0) are indicated using circles.

the original data points (for example, each column in a given alignment) so that each replicate contains exactly the same number of data points as the original set (for example, the same number of columns, i.e. sites, of the original alignment). As a consequence, some of the original data points may be absent in particular replicates, whereas others may be present more than once. The variance of each parameter (for example, the genetic distance between two sequences) is then calculated using the distribution of the parameter values obtained for each random replicate. The underlying idea of this technique is to evaluate how robust the parameter values are with respect to small changes in the original data.

To perform analyses in MEGA4, the sequence data must first be imported in the program. Sequences in FASTA format with the extension .fas under Windows should be associated automatically with MEGA4 as soon as the program is installed. By double clicking on the file "Primates.fas" the nucleotide sequences appear in the Alignment Explorer window of MEGA4 where they can be aligned, translated to amino acid and edited in different ways. By closing the Alignment Explorer window the program asks whether the user wants to save the data in MEGA format. After choosing yes, you can save the file in MEGA on your computer. The user is now asked whether to open the data file in MEGA4. By selecting yes again, the sequences are displayed in a new window called `Sequence Data Explorer`. To perform specific analyses, the user needs to close the `Sequence Data Explorer` window and select the appropriate menu from the main MEGA4 window. As an example, we will obtain JC69 genetic distances with Γ-distributed rates across sites (parameter $\alpha = 0.5$) and standard errors calculated by 1000 bootstrap replicates:

(1) From the `Distances` menu, select `Choose model...`
(2) >Click on the green square to the right of the row saying `->Model` and select `Nucleotide>Jukes and Cantor`
(3) Click on the green square to the right of the row `->Rates among sites` and select `Different(Gamma Distributed)`
(4) Click on the button to the right of the row `->Gamma Parameter`, select 0.5, click the Ok button at the bottom of the window.
(5) From the `Distances` menu select `Compute Pairwise`
(6) Click on the green square to the right of the row `->Compute` and select `Distance & Std. Err.`
(7) Click the button to the right of the row `Std. Err. Computation by`
(8) In the new window select `Bootstrap` and `1000 Replications`.
(9) Select the `Option Summary` tab and run the analysis by clicking the `Compute` button at the bottom of the window.

Estimated distances and standard errors appear in a new window with a square matrix providing pairwise distances in the lower triangular part (in gray) and standard errors in the upper triangular part (in blue). The matrix can be printed and/or exported in text format from the `File` menu of the `Pairwise Distances` window.

4.12 The problem of substitution saturation

It can be demonstrated that two randomly chosen, aligned DNA sequences of the same length and similar base composition would have, on average, 25% identical residues. Moreover, if gaps are allowed, as much as 50% of the residues can be identical. This is the reason why the curve in Fig. 4.2, showing the relationship between p-distance and genetic distance, reaches a plateau for p between 0.5 and 0.75 (i.e. for similarity scores between 50% and 25%). Beyond that point, it is not possible anymore to infer the expected genetic distance from the observed one, and the sequences are said to be saturated or to have reached **substitution saturation**. The similarities between them are likely to be the result of chance alone rather than **homology** (common ancestry). **In other words, when full saturation is reached, the phylogenetic signal is lost, i.e. the sequences are no longer informative about the underlying evolutionary processes.** In such a situation any estimate of genetic distances or phylogenetic trees, no matter which method is used (parsimony, distance or even maximum likelihood methods!), **is going to be meaningless** since gene sequences will tend to cluster according to the degree of similarity in their base (or amino acid) composition, irrespectively of their true genealogy. The problem of saturation is often overlooked in phylogeny reconstruction, when in fact it is rather crucial (see Chapter 20). For example, in coding regions, third codon positions, which usually evolve much faster than first and second (see above), are likely to be saturated especially when distantly related taxa are compared. One way to avoid this problem is to exclude third positions from the analysis, or to analyze the corresponding amino acid sequence (see Chapter 9).

The program DAMBE implements different methods to check for saturation in a data set of aligned nucleotide sequences (see Chapter 20). Here, a graphical exploration tool is introduced. The method takes advantage of the empirical observation that, in most data sets, transitional substitutions happen more frequently than transversional ones. Therefore, by plotting the observed number of transitions and transversions against the corrected genetic distance for the $n(n-1)/2$ pairwise comparison in an alignment of n taxa, transitions and transversions should both increase linearly with the genetic distance, with transitions being higher than transversions. However, as the genetic distance (the evolutionary time) increases, i.e. more divergent sequences are compared, saturation is reached and transversions will eventually outnumber transitions. This is because by chance alone there are eight possible transversions but only four transitions (see Fig. 4.6). In coding sequences, saturation will be more pronounced in the rapidly evolving third codon position.

Figure 4.11 (a) and (b) show the result for the Primates data set (Primates.fas file) analyzed using DAMBE. To analyze first and second codon positions (Fig. 4.11a) or third codon position (Fig. 4.11b) separately, select the item `Work on codon position 1 and 2` or `Work on codon position 3` from the `Sequences` menu before starting any analysis (the original sequences can be restored by choosing `Restore sequences` from the same menu). The transition and transversion vs. divergence plot can be obtained by selecting the item from the `Graphics` menu in DAMBE.

The plots show that both transitions and transversions grow approximately linear with the genetic distance indicating no saturation in the Primates data set. Figure 4.11c, on the other hand, is an example of substitution saturation at the third codon position in envelope gp120 HIV-1 sequences aligned with simian immunodeficiency virus (SIVcpz) isolated from Chimpanzees (the data set was previously used in Salemi et al., 2001). Saturation becomes evident for sequence pairs with F84 distances greater than 0.70. Therefore, in spite of SIVcpz and HIV-1 belonging to the same phylogenetic lineage and sharing a common ancestor

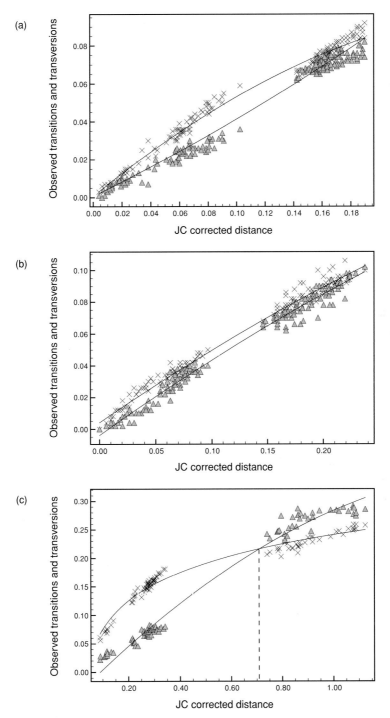

Fig. 4.11   Plotting the observed transitions and transversions against a corrected genetic distance using DAMBE. (a) First and second codon position of the primate data set. (b) Third codon position of the primate data set. (c) Third codon position of an HIV/SIV envelope data set.

within the last 300 years (Salemi *et al.*, 2001) any phylogenetic inference based on the signal present at the third codon position has to be considered unreliable. Substitution saturation in this case is due to the extremely fast evolutionary rate of HIV-1, around 10^{-3} nucleotide substitutions per site per year. In addition to the graphical tool introduced here, DAMBE also implements different statistical tests to assess substitution saturation (see Chapter 20).

4.13 Choosing among different evolutionary models

After a few exercises, it should become clear that genetic distances inferred according to different evolutionary models can lead to rather different results. Tree-building algorithms such as **UPGMA** and ***Neighbor-Joining*** (see next chapter) are based on pairwise distances among taxa: unreliable estimates will lead to inaccurate branch lengths and, in some cases, to the wrong tree topology. When confronted with model choice, the most complex model with the largest number of parameters is not necessarily the most appropriate. A model with fewer parameters will produce estimates with smaller variances. Since we always analyze a finite data sample, our parameter estimates will be associated with sampling errors. Although for sequences of at least 1000 nucleotides, these may be reasonably small, it has been shown that models with more parameters still produce a larger error than simpler ones (Tajima & Nei, 1984; Gojobori *et al.*, 1992; Zharkikh, 1994). When a simple evolutionary model (for example, JC or F84) fits the data not significantly worse than a more complex model, the former should be preferred (see Chapter 10). Finally, complex models can be computationally daunting for analyzing large data sets, even using relatively fast computers.

Another basic assumption of *Time-homogeneous time-continuous stationary Markov models*, the class of nucleotide substitution models discussed in the present chapter (Section 4.4), is that the base composition among the sequences being analyzed is at equilibrium, i.e. each sequence in the data set is supposed to have similar base composition, which does not change over time. Such an assumption usually holds when closely related species are compared, but it may be violated for very divergent *taxa* risking flawed estimates of genetic distances. TREE-PUZZLE implements by default a chi-square test comparing whether the nucleotide composition of each sequence is significantly different with respect to the frequency distribution assumed in the model selected for the analysis. The result is written in the outfile at the end of any computation. As an example, Fig. 4.12 shows the results of the chi-square test for a set of mtDNA sequences from different organisms. There are significant differences in base composition for 8 of the 17 taxa included in the data: a severe violation of the equal frequency assumption in the homogeneous Markov process. In this case, a more reliable estimate of the genetic distance can be

```
SEQUENCE COMPOSITION (SEQUENCES IN INPUT ORDER)

                  5% chi-square test        p-value
      Lungfish Au        passed               6.20%
      Lungfish SA        failed               0.62%
      Lungfish Af        failed               1.60%
      Frog               passed              58.01%
      Turtle             passed              44.25%
      Sphenodon          passed              59.78%
      Lizard             passed              38.67%
      Crocodile          failed               2.51%
      Bird               failed               0.00%
      Human              failed               0.85%
      Seal               passed              68.93%
      Cow                passed              59.11%
      Whale              passed              97.83%
      Mouse              failed               1.43%
      Rat                passed              39.69%
      Platypus           failed               3.46%
      Opossum            failed               0.01%


The chi-square test compares the nucleotide composition of each
sequence to the frequency distribution assumed in the maximum
likelihood model.
```

Fig. 4.12   Comparing nucleotide composition using a chi-squared test for a mitochondrial DNA data set as outputted by TREE-PUZZLE.

obtained with the LogDet method, which has been developed to deal specifically with this kind of problem (Steel, 1994; Lockart et al., 1994). The method estimates the distance d between two aligned sequences by calculating:

$$d = -\ln[\det F] \tag{4.39}$$

Det F is the determinant of a 4×4 matrix where each entry represents the proportion of sites having any possible nucleotide pair within the two aligned sequences. A mathematical justification for (4.39) is beyond the scope of this book. An intuitive introduction to the LogDet method can be found in Pages and Holmes (1998), whereas a more detailed discussion is given by Swofford et al. (1996). LogDet distances can be calculated using the program PAUP* and their application to the mtDNA data set is discussed in the practical part of Chapter 8. Chapter 10 will focus on statistical tests for selecting the best evolutionary model for the data under investigation.

Phylogenetic inference based on distance methods

THEORY

Yves Van de Peer

5.1 Introduction

In addition to **maximum parsimony** (**MP**) and **likelihood** methods (see Chapters 6, 7 and 8), pairwise distance methods form the third large group of methods to infer evolutionary trees from sequence data (Fig. 5.1). In principle, distance methods try to fit a tree to a matrix of pairwise **genetic distances** (Felsenstein, 1988). For every two sequences, the distance is a single value based on the fraction of positions in which the two sequences differ, defined as **p-distance** (see Chapter 4). The p-distance is an underestimation of the true genetic distance because some of the nucleotide positions may have experienced multiple substitution events. Indeed, because mutations are continuously fixed in the genes, there has been an increasing chance of multiple substitutions occurring at the same sequence position as evolutionary time elapses. Therefore, in distance-based methods, one tries to estimate the number of substitutions that have actually occurred by applying a specific **evolutionary model** that makes particular assumptions about the nature of evolutionary changes (see Chapter 4). When all the pairwise distances have been computed for a set of sequences, a tree topology can then be inferred by a variety of methods (Fig. 5.2).

Correct estimation of the genetic distance is crucial and, in most cases, more important than the choice of method to infer the tree topology. Using an unrealistic evolutionary model can cause serious artifacts in tree topology, as previously shown

The Phylogenetic Handbook: a Practical Approach to Phylogenetic Analysis and Hypothesis Testing,
Philippe Lemey, Marco Salemi, and Anne-Mieke Vandamme (eds.). Published by Cambridge
University Press. © Cambridge University Press 2009.

	Character-based methods	Non-character-based methods
Methods based on an explicit model of evolution	Maximum likelihood methods	Pairwise distance methods
Methods not based on an explicit model of evolution	Maximum parsimony methods	

Fig. 5.1   Pairwise distance methods are non-character-based methods that make use of an explicit substitution model.

Step 1
Estimation of evolutionary distances

```
3 T T C A A T C A G G C  C C G A
  | |   | |            |  |
1 T C A A G T C A G G T  T C G A
    |     | |        |  |
2 T C C A G T T A G A C  T C G A
  |   | |   | |   | |
3 T T C A A T C A G G C  C C G A
```

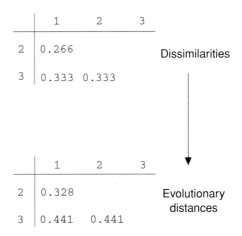

	1	2	3
2	0.266		
3	0.333	0.333	

Dissimilarities

Convert dissimilarity into evolutionary distance by correcting for multiple events per site, e.g. Jukes & Cantor (1969):

$$d_{AB} = -\frac{3}{4}\ln\left(1 - \frac{4}{3}\,0.266\right) = 0.328$$

	1	2	3
2	0.328		
3	0.441	0.441	

Evolutionary distances

Step 2
Infer tree topology on the basis of estimated evolutionary distances

Fig. 5.2   Distance methods proceed in two steps. First, the evolutionary distance is computed for every sequence pair. Usually, this information is stored in a matrix of pairwise distances. Second, a tree topology is inferred on the basis of the specific relationships between the distance values.

in numerous studies (e.g. Olsen, 1987; Lockhart *et al.*, 1994; Van de Peer *et al.*, 1996; see also Chapter 10). However, because the exact historical record of events that occurred in the evolution of sequences is not known, the best method for estimating the genetic distance is not necessarily self-evident.

Substitution models are discussed in Chapters 4 and 10. The latter discusses how to select the best-fitting evolutionary model for a given data set of aligned nucleotide or amino acid sequences in order to get accurate estimates of genetic distances. In the following sections, it is assumed that genetic distances were estimated using an appropriate evolutionary model, and some of the methods used for inferring tree topologies on the basis of these distances are briefly outlined. However, by no means should this be considered a complete discussion of distance methods; additional discussions are in Felsenstein (1982), Swofford *et al.* (1996), Li (1997), and Page & Holmes (1998).

5.2 Tree-inference methods based on genetic distances

The main distance-based tree-building methods are cluster analysis and minimum evolution. Both rely on a different set of assumptions, and their success or failure in retrieving the correct phylogenetic tree depends on how well any particular data set meets such assumptions.

5.2.1 Cluster analysis (UPGMA and WPGMA)

Clustering methods are tree-building methods that were originally developed to construct taxonomic *phenograms* (Sokal & Michener, 1958; Sneath & Sokal, 1973); that is, trees based on overall phenotypic similarity. Later, these methods were applied to phylogenetics to construct *ultrametric trees*. *Ultrametricity* is satisfied when, for any three taxa, A, B, and C,

$$d_{AC} \leq \max(d_{AB}, d_{BC}). \tag{5.1}$$

In practice, (5.1) is satisfied when two of the three distances under consideration are equal and as large (or larger) as the third one. Ultrametric trees are *rooted* trees in which all the end nodes are equidistant from the root of the tree, which is only possible by assuming a *molecular clock* (see Chapter 11). Clustering methods such as the *unweighted-pair group method with arithmetic means* (*UPGMA*) or the **weighted-pair group method with arithmetic means** (*WPGMA*) use a sequential clustering algorithm. A tree is built in a stepwise manner, by grouping sequences or groups of sequences – usually referred to as *operational taxonomic units* (*OTUs*) – that are most similar to each other; that is, for which the genetic distance is the smallest. When two OTUs are grouped, they are treated as a new single OTU (Box 5.1). From the new group of OTUs, the pair for which the similarity is highest is again identified, and so on, until only two OTUs are left. The method applied in Box 5.1 is actually the WPGMA, in which the averaging of the distances is not based on the total number of OTUs in the respective clusters. For example, when OTUs A, B (which have been grouped before), and C are grouped into a new node "*u*," then the distance from node "*u*" to any other node "*k*" (e.g. grouping D and E) is computed as follows:

$$d_{uk} = \frac{d_{(A,B)k} + d_{Ck}}{2} \tag{5.2}$$

Box 5.1 Cluster analysis (Sneath & Sokal, 1973)

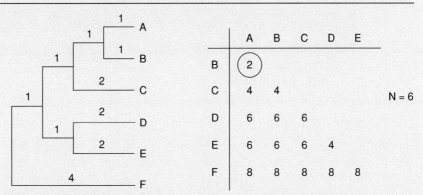

Cluster analysis proceeds as follows:

(1) Group together (cluster) these OTUs for which the distance is minimal; in this case group together A and B. The depth of the divergence is the distance between A and B divided by 2.

(2) Compute the distance from cluster (A, B) to each other OTU

$d_{(AB)C} = (d_{AC} + d_{BC})/2 = 4$
$d_{(AB)D} = (d_{AD} + d_{BD})/2 = 6$
$d_{(AB)E} = (d_{AE} + d_{BE})/2 = 6$
$d_{(AB)F} = (d_{AF} + d_{BF})/2 = 8$

	(AB)	C	D	E
C	4			
D	6	6		
E	6	6	4	
F	8	8	8	8

Repeat steps 1 and 2 until all OTUs are clustered (repeat until $N = 2$)

$N = N - 1 = 5$

(1) Group together (cluster) these OTUs for which the distance is minimal, e.g. group together D and E. Alternatively, (AB) could be grouped with C.

Box 5.1 *(cont.)*

(2) Compute the distance from cluster (D, E) to each other OTU (cluster)

$$d_{(DE)(AB)} = (d_{D(AB)} + d_{E(AB)})/2 = 6$$
$$d_{(DE)C} = (d_{DC} + d_{EC})/2 = 6$$
$$d_{(DE)F} = (d_{DF} + d_{EF})/2 = 8$$

	(AB)	C	(DE)
C	4		
(DE)	6	6	
F	8	8	8

$N = N - 1 = 4$

(1) Group together these OTUs for which the distance is minimal, e.g. group (A, B) and C

(2) Compute the distance from cluster (A, B, C) to each other OTU (cluster)

$$d_{(ABC)(DE)} = (d_{(AB)(DE)} + d_{C(DE)})/2 = 6$$
$$d_{(ABC)F} = (d_{(AB)F} + d_{CF})/2 = 8$$

	(ABC)	(DE)
(DE)	6	
F	8	8

$N = N - 1 = 3$

(1) Group together these OTUs for which the distance is minimal, e.g. group (A, B, C) and (D, E)

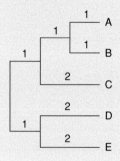

(2) Compute the distance from cluster (A, B, C, D, E) to OTU F

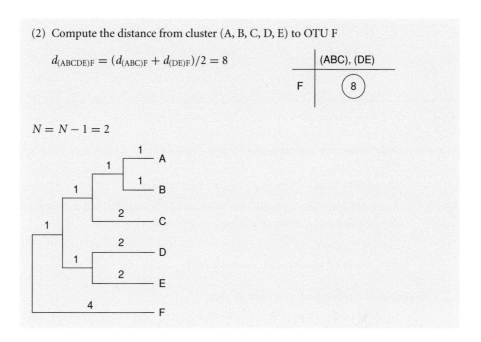

$$d_{(ABCDE)F} = (d_{(ABC)F} + d_{(DE)F})/2 = 8$$

$$N = N - 1 = 2$$

Conversely, in UPGMA, the averaging of the distances is based on the number of OTUs in the different clusters; therefore, the distance between "u" and "k" is computed as follows:

$$d_{uk} = \frac{\left(N_{AB} d_{(A,B)k} + N_C d_{Ck} \right)}{(N_{AB} + N_C)} \tag{5.3}$$

where N_{AB} equals the number of OTUs in cluster AB (i.e. 2) and N_C equals the number of OTUs in cluster C (i.e. 1). When the data are **ultrametric**, UPGMA and WPGMA have the same result. However, when the data are not ultrametric, they can differ in their inferences.

Until about 15 years ago, clustering was often used to infer evolutionary trees based on sequence data, but this is no longer the case. Many computer-simulation studies have shown that clustering methods such as UPGMA are extremely sensitive to unequal rates in different lineages (e.g. Sourdis & Krimbas, 1987; Huelsenbeck & Hillis, 1993). To overcome this problem, some have proposed methods that convert non-ultrametric distances into ultrametric distances. Usually referred to as *transformed distance methods*, these methods correct for unequal rates among different lineages by comparing the sequences under study to a reference sequence or an **outgroup** (Farris, 1977; Klotz *et al.*, 1979; Li, 1981). Once the distances are made ultrametric, a tree is constructed by clustering, as explained previously. Nevertheless, because there are now better and more effective methods to cope with non-ultrametricity and non-clock-like behavior, there is little reason left to

$$d_{AB} + d_{CD} \leq \min \ (d_{AC} + d_{BD}, d_{AD} + d_{BC})$$

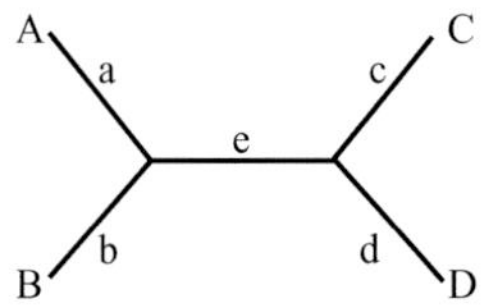

$$d_{AB} = a + b \qquad d_{AC} = a + e + c \qquad d_{AD} = a + e + d$$
$$d_{CD} = c + d \qquad d_{BD} = b + e + d \qquad d_{BC} = b + e + a$$

$$(a + b + c + d) \leq \min [\ (a + b + c + d + 2e) , (a + b + c + d + 2e) \ ]$$

Fig. 5.3   Four-point condition. Letters on the branches of the unrooted tree represent branch lengths. The function min[ ] returns the minimum among a set of values.

use cluster analysis or transformed distance methods to infer distance trees for nucleotide or amino acid sequence data.

5.2.2 Minimum evolution and neighbor-joining

Because of the serious limitations of ordinary clustering methods, algorithms were developed that reconstruct so-called **additive distance** trees. Additive distances satisfy the following condition, known as the **four-point metric condition** (Buneman, 1971): for any four taxa, A, B, C, and D,

$$d_{AB} + d_{CD} \leq \max(d_{AC} + d_{BD}, d_{AD} + d_{BC}) \qquad (5.4)$$

Only **additive distances** can be fitted precisely into an *unrooted* tree such that the genetic distance between a pair of OTUs equals the sum of the lengths of the branches connecting them, rather than an average, as in the case of cluster analysis. Why (5.4) needs to be satisfied is explained by the example shown in Fig. 5.3. When A, B, C, and D are related by a tree in which the sum of branch lengths connecting two terminal taxa is equal to the genetic distance between them, such as the tree in Fig. 5.3, $d_{AB} + d_{CD}$ is always smaller or equal than the minimum between $d_{AC} + d_{BD}$ and $d_{AD} + d_{BC}$ (see Fig. 5.3). The equality only occurs when the four sequences are related by a star-like tree; that is, only when the internal branch length of the tree in Fig. 5.3 is $e = 0$ (see Fig. 5.3). If (5.4) is not satisfied, A, B, C, and D cannot be represented by an additive distance tree because, to maintain the additivity of the genetic distances, one or more branch lengths of any tree relating them should be negative, which would be biologically meaningless. Real data sets often fail to satisy the four-point condition; this problem is the origin of the discrepancy between *actual* distances (i.e. those estimated from pairwise comparisons among nucleotide or amino acid sequences) and tree distances (i.e. those actually fitted into a tree) (see Section 5.2.3).

If the genetic distances for a certain data set are ultrametric, then both the ultrametric tree and the additive tree will be the same if the additive tree is rooted at the same point as the *ultrametric tree*. However, if the genetic distances are not ultrametric due to non-clock-like behavior of the sequences, additive trees will almost always be a better fit to the distances than ultrametric trees. However, because of the finite amount of data available when working with real sequences, stochastic errors usually cause deviation of the estimated genetic distances from perfect tree additivity. Therefore, some systematic error is introduced and, as a result, the estimated tree topology may be incorrect.

Minimum evolution (*ME*) is a distance method for constructing additive trees that was first described by Kidd & Sgaramella-Zonta (1971); Rzhetsky & Nei (1992) described a method with only a minor difference. In ME, the tree that minimizes the lengths of the tree, which is the sum of the lengths of the branches, is regarded as the best estimate of the phylogeny:

$$S = \sum_{i=1}^{2n-3} v_i \tag{5.5}$$

where n is the number of taxa in the tree and v_i is the ith branch (remember that there are $2n$–3 branches in an unrooted tree of n taxa). For each tree topology, it is possible to estimate the length of each branch from the estimated pairwise distances between all OTUs. In this respect, the method can be compared with the maximum parsimony (MP) approach (see Chapter 8), but in ME, the length of the tree is inferred from the genetic distances rather than from counting individual nucleotide substitutions over the tree (Rzhetsky & Nei, 1992, 1993; Kumar, 1996). The minimum tree is not necessarily the "true" tree. Nei *et al.* (1998) have shown that, particularly when few nucleotides or amino acids are used, the "true" tree may be larger than the minimum tree found by the optimization principle used in ME and MP. A drawback of the ME method is that, in principle, all different tree topologies have to be investigated to find the minimum tree. However, this is impossible in practice because of the explosive increase in the number of tree topologies as the number of OTUs increases (Felsenstein, 1978); an exhaustive search can no longer be applied when more than ten sequences are being used (see Chapter 1).

A good heuristic method for estimating the ME tree is the **neighbor-joining (NJ) method**, developed by Saitou & Nei (1987) and modified by Studier & Keppler (1988). Because NJ is conceptually related to clustering, but without assuming a clock-like behavior, it combines computational speed with uniqueness of results.

NJ is today the method most commonly used to construct distance trees. Box 5.2 is an example of a tree constructed with the NJ method. The method adopts the ME criterion and combines a pair of sequences by minimizing the S value (see 5.5) in each step of finding a pair of neighboring OTUs. Because the S value is not minimized globally (Saitou & Nei, 1987; Studier & Keppler, 1988),

Box 5.2 The neighbor-joining method (Saitou & Nei, 1987; modified from Studier & Keppler, 1988)

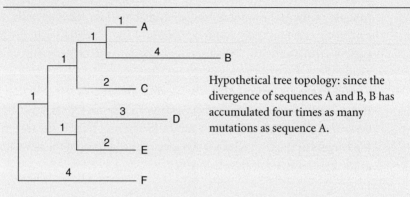

Hypothetical tree topology: since the divergence of sequences A and B, B has accumulated four times as many mutations as sequence A.

Suppose the following matrix of pairwise evolutionary distances:

	A	B	C	D	E
B	5				
C	(4)	7			
D	7	10	7		
E	6	9	6	5	
F	8	11	8	9	8

Clustering methods (discussed in Box 5.1) would erroneously group sequences A and C, since they assume clock-like behavior. Although sequences A and C look more similar, sequences A and B are more closely related.

Neighbor-joining proceeds as follows:

(1) Compute the net divergence r for every endnode ($N = 6$)

$$r_A = 5 + 4 + 7 + 6 + 8 = 30 \qquad r_D = 38$$
$$r_B = 5 + 7 + 10 + 9 + 11 = 42 \qquad r_E = 34$$
$$r_C = 32 \qquad\qquad\qquad\qquad r_F = 44$$

(2) Create a rate-corrected distance matrix; the elements are defined by $M_i = d_{ij} - (r_i + r_j)/(N - 2)$

$$M_{AB} = d_{AB} - (r_A + r_B)/(N - 2) = 5 - (30 + 42)/4 = -13$$
$$M_{AC} = \ldots.$$

$\ldots$

	A	B	C	D	E
B	⊙−13				
C	−11.5	−11.5			
D	−10	−10	−10.5		
E	−10	−10	−10.5	⊙−13	
F	−10.5	−10.5	−11	−11.5	−11.5

(3) Define a new node that groups OTUs i and j for which M_i is minimal For example, sequences A and B are neighbors and form a new node U (but, alternatively, OTUs D and E could have been joined; see further)

(4) Compute the branch lengths from node U to A and B

$$S_{AU} = d_{AB}/2 + (r_A - r_B)/2(N - 2) = 1$$
$$S_{BU} = d_{AB} - S_{AU} = 4$$

or alternatively

$$S_{BU} = d_{AB}/2 + (r_B - r_A)/2(N - 2) = 4$$
$$S_{AU} - d_{AB} - S_{BU} = 1$$

(5) Compute new distances from node U to each other terminal node

$$d_{CU} = (d_{AC} + d_{BC} - d_{AB})/2 = 3$$
$$d_{DU} = (d_{AD} + d_{BD} - d_{AB})/2 = 6$$
$$d_{EU} = (d_{AE} + d_{BE} - d_{AB})/2 = 5$$
$$d_{FU} = (d_{AF} + d_{BF} - d_{AB})/2 = 7$$

	U	C	D	E
C	3			
D	6	7		
E	5	6	5	
F	7	8	9	8

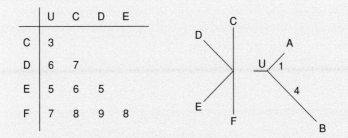

(6) $N = N - 1$; repeat step 1 through 5

Box 5.2 *(cont.)*

(1) Compute the net divergence r for every endnode ($N = 5$)

$$r_B = 21 \quad r_E = 24$$
$$r_C = 24 \quad r_F = 32$$
$$r_D = 27$$

(2) Compute the modified distances:

	U	C	D	E
C	(−12)			
D	−10	−11		
E	−10	−10	(−12)	
F	−10.7	−10.7	−10.7	−10.7

(3) Define a new node: e.g. U and C are neighbors and form a new node V; alternatively, D and E could be joined

(1) Compute the net divergence r for every endnode ($N = 4$)

$$r_V = 15 \quad r_E = 17$$
$$r_D = 19 \quad r_F = 23$$

(2) Compute the modified distances

	V	D	E
D	−12		
E	−12	(−13)	
F	(−13)	−12	−12

(3) Define a new node: e.g. D and E are neighbors and form a new node W; alternatively, F and V could be joined

(4) Compute the branch lengths from node V to C and U

$$S_{UV} = d_{CU}/2 + (r_U - r_C)/2(N - 2) = 1$$
$$S_{CV} = d_{CU} - S_{UV} = 2$$

(5) Compute distances from V to each other terminal node

$$d_{DV} = (d_{DU} + d_{CB} - d_{CU})/2 = 5$$
$$d_{EV} = (d_{EU} + d_{CB} - d_{CU})/2 = 4$$
$$d_{FV} = (d_{FU} + d_{CF} - d_{CU})/2 = 6$$

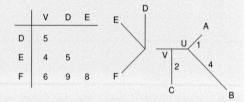

	V	D	E
D	5		
E	4	5	
F	6	9	8

(6) $N = N - 1$; repeat step 1 through 5

(4) Compute the branch lengths from node W to E and D

$$S_{DW} = d_{DE}/2 + (r_D - r_E)/2(N - 2) = 3$$
$$S_{DW} = d_{DE} - S_{DW} = 2$$

(5) Compute distances from W to each other terminal node

$$d_{VW} = (d_{DV} + d_{EV} - d_{DE})/2 = 2$$
$$d_{FW} = (d_{Dr} + d_{gr} - d_{DE})/2 = 6$$

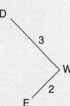

	W	V
V	2	
F	6	6

(6) $N = N - 1$; repeat step 1 through 5

(1) Compute the net divergence r for every endnode ($N = 3$)

$$r_V = 8 \quad r_F = 17 \quad r_W = 8$$

(2) Compute the modified distances

	W	V
V	(−14)	
F	(−14)	(−14)

(3) Define a new node: e.g. V and F are neighbors and form a new node X; Alternatively, W and V could be joined, or W and F could be joined

(4) Compute the branch lengths from X to V and F

$$S_{VX} = d_{FV}/2 + (r_V - r_F)/2(N-2) = 1$$
$$S_{FX} = d_{FV} - S_{VX} = 5$$

(5) Compute distances from X to each other terminal node

$$d_{WX} = (d_{FW} + d_{VW} - d_{FV})/2 = 1$$

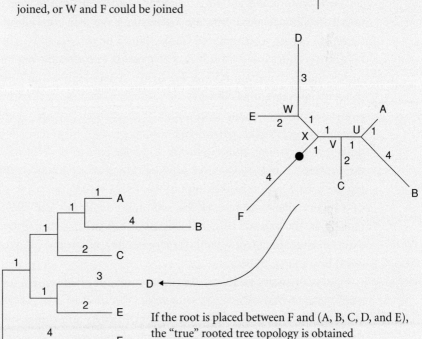

If the root is placed between F and (A, B, C, D, and E), the "true" rooted tree topology is obtained

the NJ tree may not be the same as the ME tree if pairwise distances are not additive (Kumar, 1996). However, NJ trees have proven to be the same or similar to the ME tree (Saitou & Imanishi, 1989; Rzhetsky & Nei, 1992, 1993; Russo et al., 1996; Nei et al., 1998). Several methods have been proposed to find ME trees, starting from an NJ tree but evaluating alternative topologies close to the NJ tree by conducting local rearrangements (e.g. Rzhetsky & Nei, 1992). Nevertheless, it is questionable whether this approach is really worth considering (Saitou & Imanishi, 1989; Kumar, 1996), and it has been suggested that combining NJ and bootstrap analysis (Felsenstein, 1985) might be the best way to evaluate trees using distance methods (Nei *et al.*, 1998).

Recently, alternative versions of the NJ algorithm have been proposed, including **BIONJ** (Gascuel, 1997), **generalized neighbor-joining** (Pearson *et al.*, 1999), **weighted neighbor-joining** or **weighbor** (Bruno *et al.*, 2000), **neighbor-joining maximum-likelihood** (NJML; Ota & Li, 2000), **QuickJoin** (Mailund & Pedersen, 2004), **multi-neighbor-joining** (Silva *et al.*, 2005) and **relaxed neighbor-joining** (Evans *et al.*, 2006). BIONJ and weighbor both consider that long genetic distances present a higher variance than short ones when distances from a newly defined node to all other nodes are estimated (see Box 5.2). This should result in higher accuracy when distantly related sequences are included in the analysis. Furthermore, the weighted neighbor-joining method of Bruno *et al.* (2000) uses a likelihood-based criterion rather than the ME criterion of Saitou & Nei (1987) to decide which pair of OTUs should be joined. NJML divides an initial neighbor-joining tree into subtrees at internal branches having bootstrap values higher than a threshold (Ota & Li, 2000). A topology search is then conducted using the *maximum-likelihood* method only re-evaluating branches with a bootstrap value lower than the threshold. The generalized neighbor-joining method of Pearson *et al.* (1999) keeps track of multiple, partial, and potentially good solutions during its execution, thus exploring a greater part of the tree space. As a result, the program is able to discover topologically distinct solutions that are close to the ME tree. Multi-neighbor-joining also keeps various partial solutions resulting in a higher chance to recover the minimum evolution tree (Silva *et al.*, 2005). QuickJoin and relaxed neighbor-joining use heuristics to improve the speed of execution, making them suitable for large-scale applications (Mailund & Pedersen, 2004; Evans *et al.*, 2006).

Figure 5.4 shows two trees based on evolutionary distances inferred from 20 small subunit ribosomal RNA sequences (Van de Peer *et al.*, 2000a). The tree in Fig. 5.4a was constructed by clustering (UPGMA) and shows some unexpected results. For example, the sea anemone, *Anemonia sulcata*, clusters with the fungi rather than the other animals, as would have been expected. Furthermore, neither the basidiomycetes nor the ascomycetes form a clear-cut *monophyletic* grouping. In contrast, on the NJ tree all animals form a highly supported monophyletic grouping, and the same is true for basidiomycetes and ascomycetes. The NJ tree also shows why clustering could not resolve the right relationships. Clustering methods are sensitive to unequal rates of evolution in different lineages; as is clearly seen, the branch length of *Anemonia sulcata* differs greatly from that of the

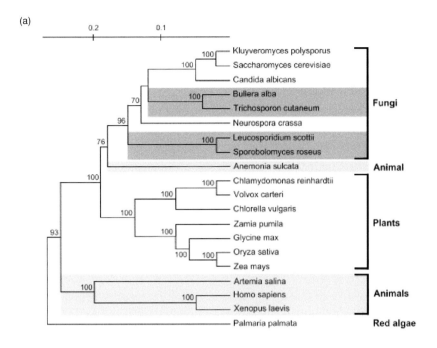

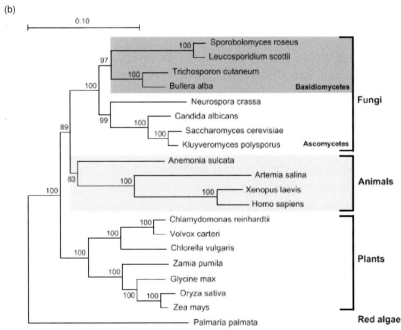

Fig. 5.4   Phylogenetic trees based on the comparison of 20 small subunit ribosomal RNA sequences. Animals are indicated by light gray shading; dark gray shading indicates the basidiomycetes. The scales on top measure evolutionary distance in substitutions per nucleotide. The red alga *Palmaria palmata* was used to root the tree. (a) Ultrametric tree obtained by clustering. (b) Neighbor-joining tree.

other animals. Also, different basidiomycetes have evolved at different rates and, as a result, they are split into two groups in the tree obtained by clustering (see Fig. 5.4a).

5.2.3 Other distance methods

It is possible for every tree topology to estimate the length of all branches from the estimated pairwise distances between all OTUs (e.g. Fitch & Margoliash, 1967; Rzhetsky & Nei, 1993). However, when summing the branch lengths between sequences, there is usually some discrepancy between the distance obtained (referred to as the *tree* distance or **patristic distance**) and the distance as estimated directly from the sequences themselves (the observed or actual distances) due to deviation from tree additivity (see Section 5.2.2). Whereas ME methods try to find the tree for which the sum of the lengths of branches is minimal, other distance methods have been developed to construct additive trees depending on goodness of fit measures between the actual distances and the tree distances. The best tree, then, is that tree that minimizes the discrepancy between the two distance measures. When the criterion for evaluation is based on a *least-squares fit*, the goodness of fit F is given by the following:

$$F = \sum_{i,j} w_{ij}(D_{ij} - d_{ij})^2 \tag{5.6}$$

where D_{ij} is the observed distance between i and j, d_{ij} is the tree distance between i and j, and w_{ij} is different for different methods. For example, in the Fitch and Margoliash method (1967), w_{ij} equals $1/D_{ij}^2$; in the Cavalli-Sforza and Edwards approach (1967), w_{ij} equals 1. Other values for w_{ij} are also possible (Swofford *et al.*, 1996) and using different values can influence which tree is regarded as the best. To find the tree for which the discrepancy between actual and tree distances is minimal, one has in principle to investigate all different tree topologies. However, as with ME, distance methods that are based on the evaluation of an explicit criterion, such as goodness of fit between observed and tree distances, suffer from the explosive increase in the number of different tree topologies as more OTUs are examined. Therefore, heuristic approaches, such as **stepwise addition** of sequences and local and global rearrangements, must be applied when trees are constructed on the basis of ten or more sequences (e.g. Felsenstein, 1993).

5.3 Evaluating the reliability of inferred trees

The two techniques used most often to evaluate the reliability of the inferred tree or, more precisely, the reliability of specific clades in the tree are bootstrap analysis (Box 5.3) and **jackknifing** (see Section 5.3.2).

5.3.1 Bootstrap analysis

Bootstrap analysis is a widely used sampling technique for estimating the statistical error in situations in which the underlying *sampling distribution* is either unknown or difficult to derive analytically (Efron & Gong, 1983). The bootstrap method offers a useful way to approximate the underlying distribution by resampling from the original data set. Felsenstein (1985) first applied this technique to the estimation of confidence intervals for phylogenies inferred from sequence data. First, the sequence data are bootstrapped, which means that a new alignment is obtained from the original by randomly choosing columns from it with replacements. Each column in the alignment can be selected more than once or not at all until a new set of sequences, a *bootstrap replicate*, the same length as the original one has been constructed. Therefore, in this resampling process, some characters will not be included at all in a given bootstrap replicate and others will be included once, twice, or more. Second, for each reproduced (i.e. artificial) data set, a tree is constructed, and the proportion of each clade among all the bootstrap replicates is computed. This proportion is taken as the statistical confidence supporting the monophyly of the subset.

Two approaches can be used to show bootstrap values on phylogenetic trees. The first summarizes the results of bootstrapping in a *majority-rule consensus* tree (see Box 5.3, Option 1), as done, for example, in the PHYLIP software package (Felsenstein, 1993). The second approach superimposes the bootstrap values on the tree obtained from the original sequence alignment (see Box 5.3, Option 2). In this case, all bootstrap trees are compared with the tree based on the original alignment and the number of times a cluster (as defined in the original tree) is also found in the bootstrap trees is recorded. Although in terms of general statistics the theoretical foundation of the bootstrap has been well established, the statistical properties of the bootstrap estimation applied to sequence data and evolutionary relationships are less well understood; several studies have reported on this problem (Zharkikh & Li, 1992a, b; Felsenstein & Kishino, 1993; Hillis & Bull, 1993). Bootstrapping itself is a neutral process that only reflects the phylogenetic signal (or noise) in the data as detected by the tree-construction method used. If the tree-construction method makes a bad estimate of the phylogeny due to systematic errors (caused by incorrect assumptions in the tree-construction method), or if the sequence data are not representative of the underlying distribution, the resulting confidence intervals obtained by the bootstrap are not meaningful. Furthermore, if the original sequence data are biased, the bootstrap estimates will be too. For example, if two sequences are clustered together because they both share an unusually high GC content, their artificial clustering will be supported by bootstrap analysis at a high confidence level. Another example is the artificial grouping of sequences with an increased *evolutionary rate*. Due to the systematic underestimation of

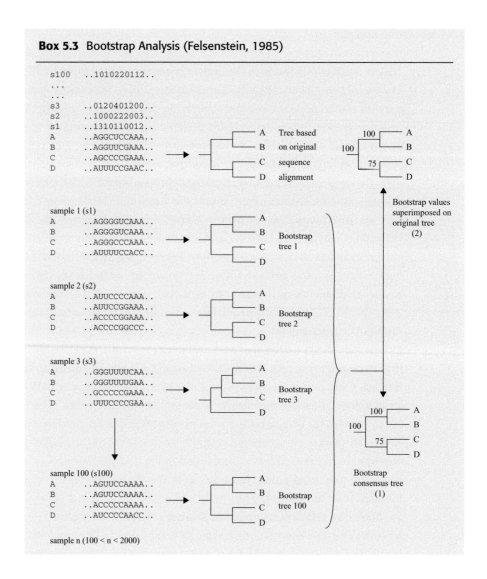

Box 5.3 Bootstrap Analysis (Felsenstein, 1985)

the genetic distances when applying an unrealistically simple substitution model, distant species either will be clustered together or drawn toward the root of the tree. When the bootstrap trees are inferred on the basis of the same incorrect evolutionary model, the early divergence of long branches or the artificial clustering of long branches (the so-called ***long-branch attraction***) will be supported at a high bootstrap level. Therefore, when there is evidence of these types of artifacts, bootstrap results should be interpreted with caution.

In conclusion, bootstrap analysis is a simple and effective technique to test the relative stability of groups within a phylogenetic tree. The major advantage of the bootstrap technique is that it can be applied to basically all tree-construction methods, although it must be remembered that applying the bootstrap method multiplies the computer time needed by the number of bootstrap samples requested. Between 200 and 2000 resamplings are usually recommended (Hedges, 1992; Zharkikh & Li, 1992a). Overall, under normal circumstances, considerable confidence can be given to branches or groups supported by more than 70% or 75%; conversely, branches supported by less than 70% should be treated with caution (Zharkikh & Li, 1992a; see also Van de Peer *et al.*, 2000b for a discussion about the effect of species sampling on bootstrap values).

5.3.2 Jackknifing

An alternative resampling technique often used to evaluate the reliability of specific clades in the tree is the so-called ***delete-half jackknifing*** or jackknife. Jackknife randomly purges half of the sites from the original sequences so that the new sequences will be half as long as the original. This resampling procedure typically will be repeated many times to generate numerous new samples. Each new sample (i.e. new set of sequences) – no matter whether from bootstrapping or jackknifing – will then be subjected to regular phylogenetic reconstruction. The frequencies of subtrees are counted from reconstructed trees. If a subtree appears in all reconstructed trees, then the jackknifing *value* is 100%; that is, the strongest possible support for the subtree. As for bootstrapping, branches supported by a jackknifing *value* less than 70% should be treated with caution.

5.4 Conclusions

Pairwise distance methods are tree-construction methods that proceed in two steps. First, for all pairs of sequences, the genetic distance is estimated (Swofford *et al.*, 1996) from the observed sequence dissimilarity (*p*-distance) by applying a correction for multiple substitutions. The genetic distance thus reflects the expected mean number of changes per site that have occurred, since two sequences diverged from their common ancestor. Second, a phylogenetic tree is constructed by considering the relationship between these distance values. Because distance methods strongly reduce the phylogenetic information of the sequences (to basically one value per sequence pair), they are often regarded as inferior to character-based methods (see Chapters 6, 7 and 8). However, as shown in many studies, this is not necessarily so, provided that the genetic distances were estimated accurately (see Chapter 10). Moreover, contrary to maximum parsimony, distance methods have the advantage – which they share with maximum-likelihood methods – that an appropriate substitution model can be applied to correct for multiple mutations. Popular distance methods such as the NJ and the Fitch and Margoliash methods have long proven to be quite efficient in finding the "true" tree topologies or those that are close (Saitou & Imanishi, 1989; Huelsenbeck & Hillis, 1993; Charleston *et al.*, 1994; Kuhner & Felsenstein, 1994; Nei *et al.*, 1998). NJ has the advantage of being very fast, which allows the construction of large trees including hundreds of sequences; this significant difference in speed of execution compared to other distance methods has undoubtedly accounted for the popularity of the method (Kuhner & Felsenstein, 1994; Van de Peer & De Wachter, 1994).

Distance methods are implemented in many different software packages, including PHYLIP (Felsenstein, 1993), MEGA4 (Kumar *et al.*, 1993), TREECON (Van de Peer & Dewachter, 1994), PAUP* (Swofford, 2002), DAMBE (Xia, 2000), and many more.

PRACTICE

Marco Salemi

5.5 Programs to display and manipulate phylogenetic trees

In the following sections, we will discuss two applications that are useful for displaying, editing, and manipulating phylogenetic trees: TREEVIEW and FIGTREE. TREEVIEW 1.6.6 (*http://taxonomy.zoology.gla.ac.uk/rod/treeview.html*) is a user-friendly and freely available program for high-quality display of phylogenetic trees, such as the ones reconstructed using PHYLIP, MEGA, TREE-PUZZLE (Chapter 6) or PAUP* (Chapter 8). The program also implements some tree manipulation functionalities, for example, defining outgroups and re-rooting trees. Program versions available for MacOsX and PC use almost identical interfaces; a manual and installation instructions are available on the website, but note that a printer driver needs to be installed to run the program under Windows. FIGTREE is a freeware application for visualization and sophisticated editing of phylogenetic trees. Trees can be exported in PDF format for publication quality figures or saved in nexus format with editing information included as a FIGTREE block. The program is written in JAVA and both MacOSX and Windows executables can be downloaded from *http://tree.bio.ed.ac.uk/software/figtree/*. The program MEGA, discussed in the previous chapter, also contains a built-in module for displaying and manipulation of phylogenetic trees.

Phylogenetic trees are almost always saved in one of two formats: NEWICK or NEXUS. The NEWICK standard for a computer-readable tree format makes use of the correspondence between trees and nested parentheses; an example for a four-taxon tree is shown in Fig. 5.5. In this notation, a tree is basically a string of balanced pairs of parenthesis with every two balanced parentheses representing an internal node. Branch lengths for terminal branches and internal nodes are written after a colon. The NEXUS format incorporates NEWICK formatting along with other commands and usually has a separate taxa-definition block (see Box 8.4 for more details on the NEXUS alignment format). The NEXUS equivalent for the tree in Fig. 5.5 with branch lengths is:

```
#NEXUS
Begin trees;
   Translate
        1 A,
        2 B,
        3 C,
```

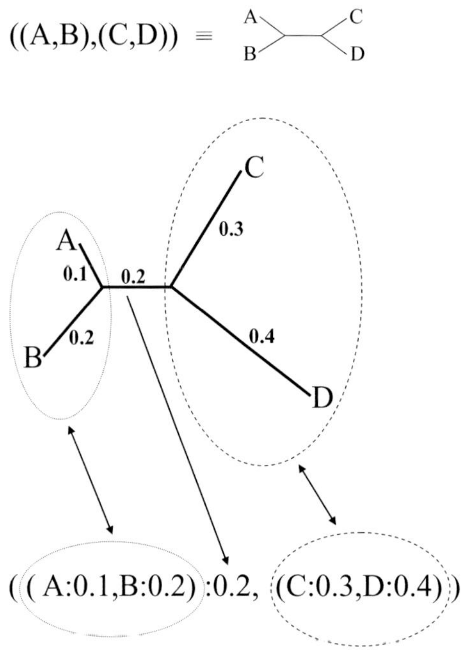

$$((A,B),(C,D)) \equiv$$

$$(\;(\;A:0.1,B:0.2\;):0.2,\;(C:0.3,D:0.4)\;)$$

Fig. 5.5   NEWICK representation of phylogenetic trees. A hypothetical unrooted tree of four taxa (A, B, C, and D) with numbers along the branches indicating estimated genetic distances and its description in NEWICK format (see text for more details).

```
    4 D,
    ;
tree PAUP_1 = [&U] ((1:0.1,2:0.2):0.2,(3:0.3,4:0.4));
End;
```

5.6 Distance-based phylogenetic inference in PHYLIP

The PHYLIP software package implements four different distance-based tree-building methods: the Neighbour-Joining (NJ) and the UPGMA methods, carried out by the program NEIGHBOR.EXE; the *Fitch–Margoliash* method, carried out by the program FITCH.EXE; and the Fitch and Margoliash method assuming a molecular clock, carried out by the program KITCH.EXE. The UPGMA method and the algorithm implemented in the program KITCH.EXE both assume ultra-metricity of the sequences in the data set, i.e. that the sequences are contemporaneous and accumulate mutations over time at a more or less constant rate. As discussed above, ultrametric methods tend to produce less reliable phylogenetic trees when mutations occur at significantly different rates in different lineages.

Non-ultrametric methods, such as NJ or Fitch–Margoliash, do not assume a molecular clock, and they are better in recovering the correct phylogeny in situations when different lineages exhibit a strong heterogeneity in evolutionary rates. Chapter 11 will discuss how to test the molecular clock hypothesis for contemporaneously and serially sampled sequences. However, the statistical evaluation of the molecular clock and the substitution model best fitting the data (see previous chapter and Chapter 10) require the knowledge of the tree topology relating the operational taxonomic units (OTUs, i.e. the taxa) under investigation. Therefore, the first step in phylogenetic studies usually consists of constructing trees using a simple evolutionary model, like the JC69 or the Kimura 2-parameter model (Kimura, 1980), and tree-building algorithms not assuming a molecular clock. The reliability of each clade in the tree is then tested with bootstrap analysis or Jackknifing (see Section 5.3). In this way it is possible to infer one or more trees that, although not necessarily the true phylogeny, are reasonable hypotheses for the data. Such "approximate" trees are usually appropriate to test a variety of evolutionary hypotheses using maximum likelihoods methods, including the model of nucleotide substitution and the molecular clock (see Chapters 10 and 11). Moreover, when a "reasonable" tree topology is known, the free parameters of any model, for example, the *transition/transversion ratio* or the shape parameter α of the Γ-distribution (see previous chapter), can be estimated from the data set by maximum likelihood methods. These topics will be covered in Chapters 6 and 10.

In what follows, we will use the Windows versions of Phylip, Mega, TreeView and FigTree. However, the exercises could also be carried out under MacOSX using the Mac executables of each application (Phylip, TreeView and FigTree) or using a virtual PC emulator (Mega). The data sets required for the exercises can be downloaded from *www.thephylogenetichandbook.org*.

5.7 Inferring a Neighbor-Joining tree for the primates data set

To infer a NJ tree using the program Neighbor.exe from the Phylip package for the primates alignment (`primates.phy` file: alignment in sequential Phylip format), an input file with pairwise evolutionary distances is required. Therefore, before starting the neighbor-joining program, first calculate the distance matrix using the program DNAdist.exe, as explained in the previous chapter, employing the F84 model and an empirical transition/transversion ratio of 2. The matrix in the `outfile` is already in the appropriate format, and it can be used directly as input file for Neighbor.exe. Rename the `outfile` file to `infile` and run Neighbor.exe by double-clicking the application's icon in the `exe` folder within the PHYLIP folder; the following menu will appear:

```
Neighbor-Joining/UPGMA method version 3.66

Settings for this run:
N Neighbor-joining or UPGMA tree?        Neighbor-joining
O                      Outgroup root?    No, use as
                                         outgroup species 1
L       Lower-triangular data matrix?    No
R       Upper-triangular data matrix?    No
S                       Subreplicates?   No
J  Randomize input order of species?     No. Use input order
M           Analyze multiple data sets?  No
0    Terminal type (IBM PC, ANSI, none)? IBM PC
1     Print out the data at start of run No
2  Print indications of progress of run  Yes
3                         Print out tree Yes
4       Write out trees onto tree file?  Yes
Y  to accept these or type the letter for one to change
```

Option N allows the user to choose between NJ and UPGMA as tree-building algorithm. Option O asks for an outgroup. Since NJ trees do not assume a molecular clock, the choice of an outgroup merely influences the way the tree is drawn; for now, leave option O unchanged. Options L and R allow the user to use as input a pairwise distance matrix written in lower-triangular or upper-triangular format. The rest of the menu is self-explanatory: enter Y to start the computation. Outputs are written into `outfile` and `outtree`, respectively. The content of the `outfile` can be explored using a text editor and contains a description of the tree topology and a table with the branch lengths. The `outtree` contains a description of the tree in the so-called NEWICK format (see Fig. 5.5). Run `TreeView.exe` and select Open from the File menu. Choose `All Files` in `Files of type` and open the `outtree` just created in the `exe` sub folder of the PHYLIP folder. The tree in Fig. 5.6a will appear. At the top of the `TreeView` window a bar with four buttons (see Fig. 5.6a) indicates the kind of tree graph being displayed. The highlighted button indicates that the tree is shown as a *cladogram*, which only displays the phylogenetic relationships among the taxa in the data set. In this kind of graph branch lengths are not drawn proportionally to evolutionary distances so that only the topology of the tree matters. To visualize the *phylogram* (i.e. the tree with branch lengths drawn proportionally to the number of nucleotide substitutions per site along each lineage) click the last button to the right of the bar at the top of the window. The tree given in Fig. 5.6b will appear within the TREEVIEW window. This time, branch lengths are drawn proportionally to genetic distances and the

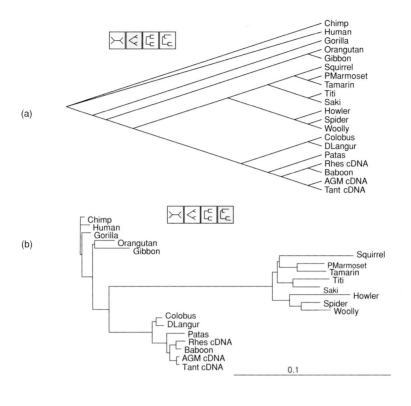

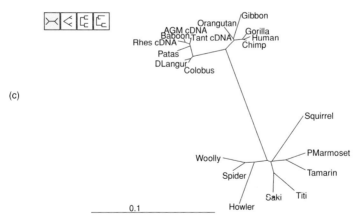

Fig. 5.6   Different TREEVIEW displays for the Neighbor-joining tree of the Primates data set. Genetic distances are calculated with the F84 model with an empirical transition/transversion ratio of 2. (a) Slanted cladogram. (b) Rooted Phylogram. The scale on bottom represents genetic distances in substitutions per nucleotide. (c) Unrooted phylogram. The scale on top represents genetic distances in substitutions per nucleotide. Note that, even if the trees in panel (a) and (b) appear to be rooted, the NJ method actually infers unrooted tree topologies. Therefore, the position of the root is meaningless and the trees shown in A and B should be considered equivalent to the one given in panel (c) (see text for more details).

bar at the bottom of the tree indicates a length corresponding to 0.1 nucleotide substitutions per site.

It is important to keep in mind that a NJ tree is unrooted, and the fact that the Chimp and Human sequences appear to be the first to branch off does not mean that these are in fact the oldest lineages! The unrooted tree can be displayed by clicking on the first button to the left of the bar at the top of the `TreeView` window (see Fig. 5.6c). The unrooted phylogram clearly shows three main monophyletic clades (i.e. group of taxa sharing a common ancestor). The first one, at the bottom of the phylogram (Fig. 5.6c) includes DNA sequences from the so-called New World monkeys: woolly monkey (Woolly), spider monkey (Spider), Bolivian red howler (Howler), white-faced saki (Saki), Bolivian grey titi (Titi), tamarin, pygmy marmoset (PMarmoset), squirrel monkey (Squirrel). A second monophyletic clade on the top left of the phylogram includes sequences from Old World monkeys: African green monkey (AGM), baboon (Baboon), tantalus monkey (Tant), rhesus monkey (Rhes), patas monkey (Patas), dour langur (DLangur), kikuyu colobus (Colobus). The Old World monkeys clade appears to be more closely related to the third monophyletic clade shown on the top right of the tree in Fig. 5.6c, which includes DNA sequences from the Hominids group: Gibbon, Orangutan, Gorilla, Chimp, and Human. If we assume a roughly constant evolutionary rate among these three major clades, the branch lengths in Fig. 5.6b and c suggest that Hominids and Old World monkeys have diverged more recently and the root of the tree should be placed on the lineage leading to the New World monkeys. This observation is in agreement with the estimated divergence time between Old World monkeys and Hominids dating back to about 23 million years ago, and the estimated split between New World monkeys and Old World monkeys dating back to about 33 millions years ago (Sawyer *et al.*, 2005). This confirms that we can place the root of the tree on the branch connecting the New World monkeys and the Old World monkeys/Hominids clades. This will correspond to the root halfway between the two most divergent taxa in the tree. This rooting technique, called ***midpoint rooting*** is useful to display a tree with a meaningful evolutionary direction when an outgroup (see Chapter 1 and below for more details) is not available in the data set under investigation. To obtain the midpoint rooted tree, open the `outtree` file with the program FIGTREE and choose the option `Midpoint Root` from the `Tree` menu. The program creates a midpoint rooted tree that can be displayed in the window by clicking the `Next` button on the top left of the `FigTree` window. To increase the taxa font size click on the little triangle icon on the `Tip Labels` bar (on the left of the window, see Fig. 5.7) and increase the `Font Size` to 12 by using the up and down buttons on the right of the `Font Size` display box. The FIGTREE application also implements other user-friendly tools for editing phylogenetic trees. For example, by selecting the `Appearance` bar on the right

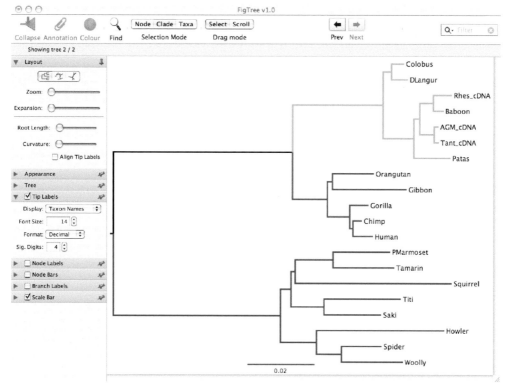

Fig. 5.7   Editing phylogenetic trees with FIGTREE. FIGTREE screenshot (MacosX) of the NJ phylogenetic tree for the primates data set shown in Fig. 5.6. The FIGTREE command panels on top and left allow various manipulation of the tree (see text). The tree was midpoint rooted and branches colored according to specific lineages: New World monkeys, Old World monkeys, and Hominids.

of the window we can increase the Line Weight (for example to 2 or 3), which results in thicker branches. The Scale Bar option allows increasing the Font Size and the Line Weight of the scale bar at the bottom of the tree. We can also display branch lengths by checking the box on the left of the Branch Labels bar. For publication purposes it is sometimes useful to display different colors for the branches of a tree. For example, we may want to color the Hominids clade in green, the Old World monkey one in red, and the New World monkey one in blue. Select the Clade button on the top of the FIGTREE window. Click on the internal branch leading to the Hominids clades: that branch and all its descending branches within the clade will be highlighted. Click on the Color icon displayed on the top of the FIGTREE window and select the color red. Repeat the same procedure for the other two main clades choosing different colors. The edited tree can be saved in NEWICK format by selecting Save as . . . from the File menu, which contains additional information such as the color of the branch lengths in a FIGTREE block. The tree can be re-opened later, but only `FigTree` is capable of displaying the editing information in the FIGTREE block. The tree can also be exported as PDF file by selecting `Export PDF...` from the `File` menu. An example of the primates tree midpoint rooted and edited with FIGTREE is displayed in Fig. 5.7.

5.7.1 Outgroup rooting

Midpoint rooting only works if we have access to independent information (like in the case discussed above), and/or when we can safely assume that the evolutionary rates along different branches of the tree are not dramatically different. When rates are dramatically different, a long branch may represent faster accumulation of mutations rather than an older lineage, and placing the root at the midpoint of the tree may be misleading. Chapter 11 will discuss in detail the molecular clock (constancy of evolutionary rates) hypothesis and the so-called *local clock* and *relaxed clock* models that can be used to investigate the presence of different evolutionary rates along different branches of a tree.

An alternative rooting technique consists of including an outgroup in the data set and placing the root at the midpoint of the branch that connects the outgroup with the rest (ingroup) of the taxa. As an example, we will use the `mtDNA.phy` data set including mitochondrial DNA sequences from birds, reptiles, several mammals, and three sequences from lungfish. After obtaining the NJ tree with F84-corrected distances as above, open the tree in TREEVIEW. The unrooted phylogram is shown in Fig. 5.8a. It can be seen that birds and crocodiles share a common ancestor, they are called *sister taxa*. A monophyletic clade can also be distinguished for all the mammalian taxa (platypus, opossum, mouse, rat, human, cow, whale, seal). From systematic studies based on morphological characters it is known that lungfish belongs to a clearly distinct phylogenetic lineage with respect to the amniote vertebrates (such as reptiles, birds, and mammals). Thus, the three lungfish sequences (LngfishAf, LngfishSA, LngfishAu) can be chosen as outgroups and the tree can be rooted to indicate the evolutionary direction.

(i) choose `Define outgroup...` from the `Tree` menu and select the three lungfish sequences (LngFishAu, LngFishSA, LngFishAf)
(ii) choose `Root with outgroup...` from the `Tree` menu

The rooted phylogram displayed by TREEVIEW is shown in Fig. 5.8b.

Choosing an outgroup in order to estimate the root of a tree can be a tricky task. The chosen outgroup must belong to a clearly distinct lineage with respect to the ingroup sequences, i.e. the sequences in the data set under investigation, but it does not have to be so divergent that it cannot be aligned unambiguously

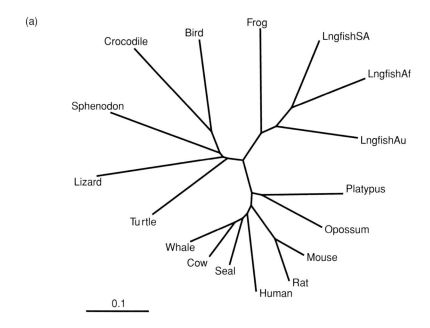

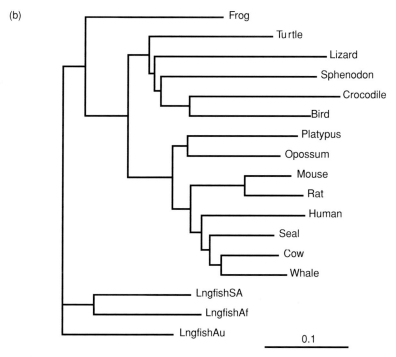

Fig. 5.8   Neighbor-joining tree of the mtDNA data set. Genetic distances were calculated with the F84 model and a transition/transversion ratio of 2. The scale at the bottom represents genetic distances in nucleotide substitutions per site. (a) Unrooted phylogram. (b) Rooted phylogram using the Lungfish sequences as outgroup.

against them. Therefore, before aligning outgroup and ingroup sequences and performing phylogenetic analyses, it can be useful to evaluate the similarity between the potential outgroup and some of the ingroup sequences using Dot Plots (see Section 3.10 in Chapter 3). If the Dot Plot does not show a clear diagonal, it is better to choose a different outgroup or to align outgroup and ingroup sequences only in the genome region where a clear diagonal is visible in the Dot Plot.

5.8 Inferring a *Fitch–Margoliash* tree for the mtDNA data set

The Fitch–Margoliash tree is calculated with the program Fitch.exe by employing the same distance matrix used for estimating the NJ tree in Section 5.7. The only option to be changed is option G (select Yes), which slows down a little the computation but increases the probability of finding a tree minimizing the difference between estimated pairwise distances and **patristic distances** (see Section 5.2.3). Again, the tree written to the outtree file can be displayed and edited with the TREEVIEW or FIGTREE program. The phylogram rooted with DNA sequences from lungfish is shown in Fig. 5.9.

5.9 Bootstrap analysis using PHYLIP

The mtDNA data set discussed in this book was originally obtained to support the common origin of birds and crocodiles versus an alternative hypothesis proposing mammals as the lineage most closely related to birds (Hedges, 1994). Both the NJ and the Fitch–Margoliash tree show that birds and crocodiles cluster together. However, the two trees differ in clustering of turtles and lizards (compare Figs. 5.8 and 5.9). It is not unusual to obtain slightly different tree topologies using different tree-building algorithms. To evaluate the hypothesis of crocodile and bird monophyly appropriately, the reliability of the clustering in the phylogenetic trees estimated above must be assessed. As discussed in Section 5.3, one of the most widely used methods to evaluate the reliability of specific branches in a tree is bootstrap analysis. Bootstrap analysis can be carried out using PHYLIP as following:

(i) Run the program Seqboot.exe using as infile the aligned mtDNA sequences; the following menu will appear:

```
Bootstrapping algorithm, version 3.66

Settings for this run:
D     Sequence, Morph, Rest., Gene Freqs?    Molecular
                                             sequences
J Bootstrap, Jackknife, Permute, Rewrite?    Bootstrap
%    Regular or altered sampling fraction?   regular
```

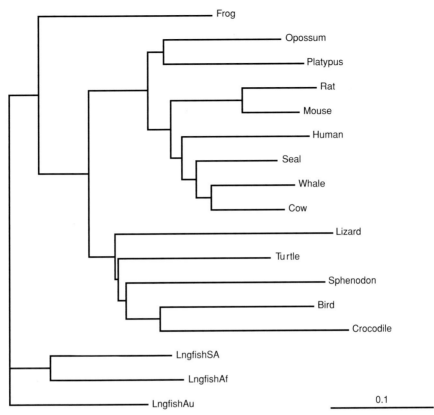

Fig. 5.9   Fitch–Margoliash tree of the mtDNA data set. Genetic distances were calculated with the F84 model and a transition/transversion ratio of 2. The scale at the bottom represents genetic distances in nucleotide substitutions per site. The phylogram was rooted using the Lungfish sequences as outgroup.

```
B      Block size for block-bootstrapping?   1 (regular
                                             bootstrap)
R             How many replicates?           100
W         Read weights of characters?        No
C          Read categories of sites?         No
S    Write out data sets or just weights?    Data sets
I          Input sequences interleaved?      Yes
0    Terminal type (IBM PC, ANSI, none)?     IBM PC
1      Print out the data at start of run    No
2    Print indications of progress of run    Yes
Y to accept these or type the letter for one to change
```

Option J selects the kind of analysis (bootstrap by default). Option R determines the number of replicates. Type R and enter 1000. When entering Y, the program asks for a random number seed. The number should be odd and it is used to feed a random number generator, which is required to generate replicates of the data set with random resampling of the alignment columns (see Section 5.3). Type 5 and the program will generate the replicates and write them to the `outfile`:

(ii) Rename the `outfile` to `infile` and run `DNAdist.exe`. After selecting option M the following option should appear in the menu:

```
Multiple data sets or multiple weights? (type D or W)
```

Type D followed by the `enter` key, then type 1000 and press the `enter` key again. Type Y to accept the options followed by the `enter` key. `DNAdist.exe` will compute 1000 distance matrices from the 1000 replicates in the original mtDNA alignment and write them to the `outfile`.

(iii) Rename the `outfile` to `infile`. It is now possible to calculate, using the new `infile`, 1000 NJ or Fitch–Margoliash trees with `Neighbor.exe` or `Fitch.exe` by selecting option M from their menus. As usual, the trees will be written in the `outtree` file. Since the computation of 1000 Fitch–Margoliash trees can be very slow, especially if the option G is selected, enter 200 in option M so that only the first 200 replicates in the `infile` will be analyzed by `Fitch.exe` (keep in mind, however, that for publication purposes 1000 replicates is more appropriate). Enter again a random number and type Y.

(iv) The collection of trees in the `outtree` from the previous step is the input data to be used with the program `Consense.exe`. Rename `outtree` to `intree`, run `Consense.exe`, and enter Y. Note that there is no indication of the progress of the calculation given by `Consense.exe`, which depending on the size of the data set can take a few seconds to a few minutes. The consensus tree (see Section 5.3), written in the `outtree` in the usual NEWICK format (Fig. 5.5), can be viewed with TREEVIEW. Detailed information about the bootstrap analysis are also contained in the `outfile`.

The bootstrap values can be viewed in TREEVIEW by selecting `Show internal edges labels` from the Tree menu. Figure 5.10a shows the NJ bootstrap consensus tree (using 1000 bootstrap replicates) as it would be displayed by TREEVIEW. In 998 out of our 1000 replicates, birds and crocodile cluster together (bootstrap value 99.8%). A slightly different value (e.g. 99.4%) could have been obtained when a different set of 1000 bootstrap replicates were analyzed, e.g. by feeding a different number to the random number generator. This is an excellent support that strengthens our confidence in the monophyletic origin of these two species. The difference in clustering between the NJ and Fitch–Margoliash tree (Lizard vs. Turtle monophyletic with Crocodile–Bird–Sphenodon) is only poorly supported. A similar weak support would be obtained for the Fitch–Margoliash bootstrap analysis indicating that, in both cases, there is considerable uncertainty about the evolutionary relationships of these taxa. Note that the program by default draws rooted trees using an arbitrarily chosen outgroup and that branch lengths in this consensus tree, represented in cladogram style, are meaningless. It is important

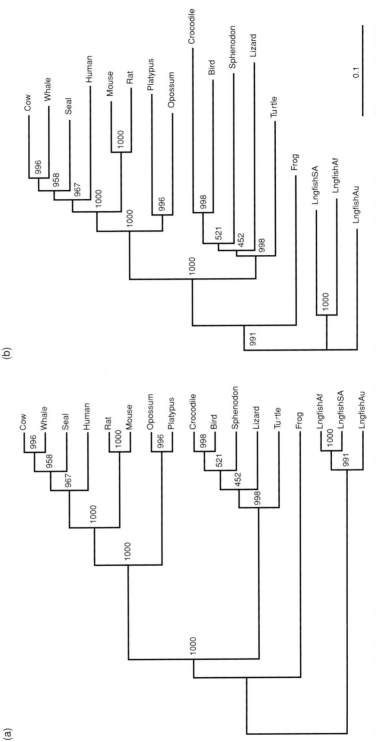

Fig. 5.10 (a) Neighbor-joining consensus tree for 1000 bootstrap replicates of the mtDNA data set as displayed in TreeView. (b) Inferred neighbor-joining tree for the mtDNA data set with bootstrap values. In both cases, the bootstrap values are shown to the right of the node representing the most recent common ancestor of the clade they support.

to distinguish between the consensus tree and the "real" tree constructed using the NJ method. The two trees do not necessarily have the same topology (see Box 5.4). The bootstrap consensus tree is inherently more unresolved since it is a consensus of different clustering hypothesis. When the trees are different, the topology of the consensus tree has to be disregarded since it is not based on the real data set but on bootstrap re-samplings of the original data. For publication the bootstrap values are usually displayed on a tree inferred from the original alignment, which usually includes branch lengths proportional to genetic distances. For example, Fig. 5.10b shows the NJ tree of the mtDNA data set (the same from Fig. 5.8b) where the bootstrap values from the consensus tree in Fig. 5.10a have been added. In this case, the topologies are the same because all the clusters in the inferred tree are also in the bootstrap consensus tree. The tree from TREEVIEW can be exported as a graphic file by selecting `Save as graphic . . .` from the `File` menu. This file can then be imported in most graphics programs and bootstrap values can be added on the branches like in Fig. 5.10b. The consensus tree can also be viewed and edited using FIGTREE.

5.10 Impact of genetic distances on tree topology: an example using MEGA4

In the previous chapter we have introduced several nucleotide substitution models based on the **Markov process** and we have shown how genetic distance estimates are affected by the underlying assumptions of each model. Since pairwise genetic distances are used in turn by distance-based algorithms to infer topology and branch lengths of phylogenetic trees, it is not surprising that distances estimated according to different models might produce different trees.

As pointed out at the beginning of this chapter, it can be demonstrated that the NJ method is able to infer the true phylogenetic history (the true tree) of a set of taxa when the estimated genetic distances represent, in fact, the true genetic distances, i.e. the actual number of nucleotide substitutions per site between each pair of lineages. Such a condition is rarely met by real data sets. As we have seen in the previous chapter, Markov models, from the simple Jukes and Cantor to the complex General Time Reversible, are at best a reasonable approximation of the evolutionary process and more realistic models can be implemented at the cost of longer computation time and larger variance of the estimates. Even assuming that the selected substitution model accurately represents the evolutionary process for a given set of taxa, random fluctuations of the Poisson process will affect our estimates to some degree. Therefore, it is crucial not only to choose an appropriate model of nucleotide substitution that is complex enough to describe the underlying evolutionary process without increasing excessively the variance of the estimates,

---

Box 5.4 Bootstrap analysis with PHYLIP Consense.exe outfile

Part of the output file produced by the program Consense.exe for the 1000 bootstrap replicates of the *mtDNA* data set discussed in Section 5.6 is shown below.

```
Majority-rule and strict consensus tree program, version 3.573c


Species in order:


Opossum
Platypus
Rat
Mouse
Whale
Cow
Seal
Human
Lizard
Bird
Crocodile
Turtle
Sphenodon
Frog
LngfishSA
LngfishAf
LngfishAu


Sets included in the consensus tree


Set (species in order)    How many times out of 1000.00


..........  ...****          1000.00
..**......  .......          1000.00
.........**  *******         1000.00
....**....  .......           999.00
..........  ....**.           999.00
.........**  ***....          999.00
..******..  .......           998.00
..........  ....***           997.00
.........*  *......           994.00
....***...  .......           903.00
..********  *******           879.00
....****..  .......           757.00
.........**  *.*....           669.00
.........*  *.*....           666.00
```

Box 5.4 *(cont.)*

```
Sets NOT included in consensus tree:

Set (species in order)        How many times out of 1000.00
.........*    ***....            245.00
..*****...    .......            170.00
.........**   *......            159.00
.........*.   ..*....            112.00
.*******..    .......             88.00
.........*    **.....             77.00
....**.*..    .......             75.00
..**...*..    .......             66.00
.........**   **.....             46.00
.*......**    *******             33.00
.........*.   .*.....             24.00
......**..    .......             12.00
..**..*...    .......              9.00
..........    *.*....              5.00
..........    ...*..*              3.00
..****....    .......              3.00
..****.*..    .......              3.00
..........    .**....              2.00
..**..**..    .......              2.00
.........*.   *.....&              1.00
....******    ****"**              1.00
.......***    *******              1.00
.....**...    .......              1.00
.........*    *******              1.00
..........    .....**              1.00
```

The outfile first lists the *taxa* included in the tree (Species in order). The second section of the file, Sets included in the consensus tree, lists the clades which are present in more than 50% of the *bootstrap replicates* and are therefore included in the *consensus tree*. The clades are indicated as follows: each " . " represents a *taxon* in the same order as it appears in the list and *taxa* belonging to the same clade are represented by "*". For example, "..** .... 1000.00" means that the clade joining rat and mouse (the third and fourth species in the list) is present in all 1000 trees estimated from the *bootstrap replicates*. It can happen that a particular clade present in the original tree is not included in the *consensus tree* because a different topology with other clades was better supported by the bootstrap test. In this case, the *bootstrap value* of that particular clade can be found in the third section, Sets NOT included in consensus tree.

but also to realize that, except in a few cases, a phylogenetic tree is at best an educated guess that needs to be carefully evaluated and interpreted.

Chapter 10 will describe a general statistical framework that can be used to select the best fitting nucleotide substitution model for any given data set. In what follows we will show, using the primates data set as an example, how different evolutionary assumptions can affect the tree estimated with a tree-based algorithm like NJ, and can occasionally lead to dramatically "wrong" tree topologies. We will perform tree inference using the MEGA4 program and the file `primates.meg`. The file contains aligned DNA sequences in mega format (see Chapter 4) and can be downloaded from *www.thephylogenetichandbook.org*.

Using MEGA4 it is possible to estimate a NJ tree and perform the bootstrap test in an automated fashion. The program will display the tree in a new window and superimpose bootstrap support values along each branch of the tree. To estimate a NJ tree using Kimura-2P (K2P) corrected distances and perform bootstrap analysis on 1000 replicates, open the `primates.meg` file in MEGA4 and select the submenu `Bootstrap Test of Phylogeny > Neighbor-Joining ...` from the `Phylogeny` menu in the MEGA4 main window. The `Analysis Preferences` window will appear. Click on the green square to the right of the `Gaps/missing data` row and select `pairwise deletion` (specifying that for each pair of sequences only gaps in the two sequences being compared should be ignored). Similarly, select in the `Model` row `Nucleotide > Kimura-2-parameter`. To set the number of bootstrap replicates, click on the `Test of Phylogeny` tab on the top of the window and enter 1000 in the `Replications` cell. Select again the `Option Summary` tab and click on the `compute` button at the bottom of the window. After a few seconds (or a few minutes, depending on the speed of your computer processor) the NJ tree with bootstrap values will appear in the `Tree Explorer` window. By default, the tree is midpoint rooted and should look like the one given in Fig. 5.11a. If the location of the root needs to be placed on any other branch of the tree, this can be done by selecting the top button on the left side of the window (the button is indicated by an icon representing a phylogenetic tree with a green triangle on its left), placing the mouse on the branch chosen as the new root and clicking on it: a re-rooted tree will be displayed in the same window. To go back to the midpoint-rooted tree, simply select `Root on Midpoint` from the `View` menu.

In a similar way, NJ trees can be obtained using different nucleotide substitution models, with and without Γ-distributed rates across sites and with a different shape parameter α (see previous chapter) by selecting the appropriate options in the `Analysis Preferences` window. Figure 5.11a, b, c, d show the NJ trees (with 1000 bootstrap replicates each) obtained with the K2P model without Γ-distribution (5.11a), and with Γ-distribution using different α values: $\alpha = 0.5$

(a) K2P

(b) K2P + Γ (α = 0.5)

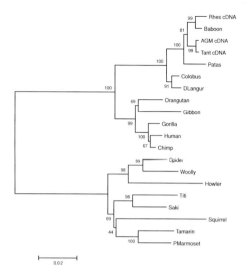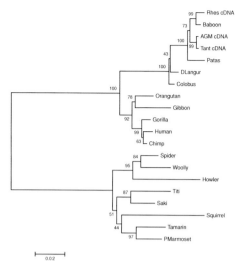

(c) K2P + Γ (α = 0.25)

(d) K2P + Γ (α = 0.10)

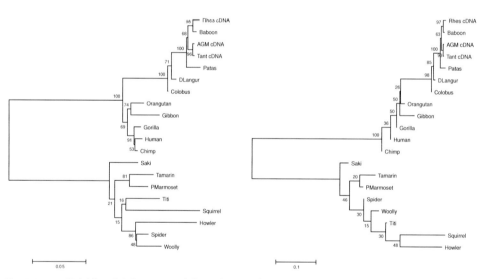

Fig. 5.11   Neighbor-joining trees of the Primates data set inferred by using different nucleotide substitution models. The scale at the bottom measures genetic distances in nucleotide substitutions per site. All phylograms were midpoint rooted and bootstrap values are now shown to the left of the node representing the most recent common ancestor of the clade they support. (a) NJ tree with Kimura 2-parameter estimated distances. (b) NJ tree with Kimura 2-parameter estimated distances using Γ-distributed rates along sites with α-parameter of 0.5. (c) NJ tree with Kimura 2-parameter estimated distances using Γ-distributed rates along sites with α-parameter of 0.25. (d) NJ tree with Kimura 2-parameter estimated distances using Γ-distributed rates along sites with α-parameter = 0.1. (e) NJ tree with Tamura–Nei estimated distances. (f) NJ tree with Tamura–Nei estimated distances using Γ-distributed rates along sites with α-parameter of 0.5. (g) NJ tree with Tamura–Nei estimated distances using Γ-distributed rates along sites with α-parameter of 0.25. (h) NJ tree with Tamura–Nei estimated distances using Γ-distributed rates along sites with α-parameter of 0.1.

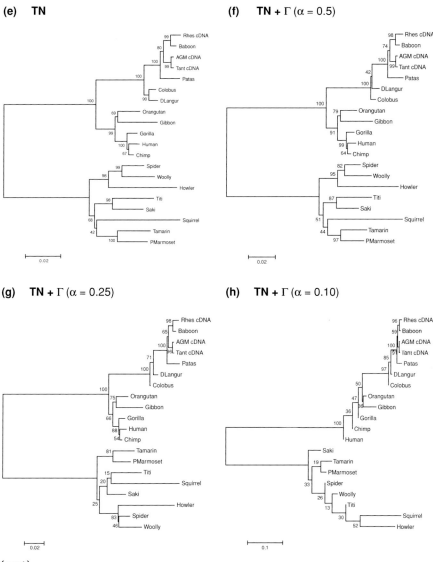

Fig. 5.11   *(cont.)*

(Fig. 5.11b), $\alpha = 0.25$ (Fig. 5.11c), and $\alpha = 0.1$ (Fig. 5.11d). Figure 5.11e, f, g, h show trees obtained with the Tamura–Nei model (TN, see previous chapter) without (Fig. 5.11e), and with (Fig. 5.11f, g, h) different Γ-distributions. The K2P and the TN tree without Γ-distribution and with Γ-distribution but moderate rate heterogeneity ($\alpha = 0.5$) appear very similar except for the position of colobus and langur taxa that are monophyletic in the trees in Fig. 5.11a and 5.11e but **paraphyletic** (with the colobus sequence branching off in trees in Fig. 5.11b and

5.11f). The K2P and TN trees obtained using $\alpha = 0.25$ (strong rate heterogeneity) also show a discordant position for the white-faced saki sequence (Fig. 5.11c and 5.11g). The K2P and the TN trees with $\alpha = 0.1$ show again a similar (but not identical!) topology. However, such trees are in obvious contradiction to all the known data about Hominids evolutionary relationships. In fact, according to the K2P tree (Fig. 5.11d) the chimp lineage is the oldest lineage at the base of the Hominids clade, while according to the TN tree (Fig. 5.11h), it is the human lineage that branches off at the root of the Hominids clade. In general, a different α parameter of the Γ-distribution tends to have a greater impact on the tree topology than a different evolutionary model. However, the example shows (hopefully, in a convincing way) that the nucleotide substitution model and the parameters of the model (especially α) can significantly impact the phylogenetic reconstruction and, therefore, need to be chosen carefully. The following chapter and Chapter 10 will discuss how maximum likelihood methods can be used to estimate substitution model parameters and to test different evolutionary hypotheses within a rigorous statistical framework.

5.11 Other programs

Several other programs are freely available to compute phylogenetic trees with distance-based methods. Some of them, like PAUP*, will be discussed in the next chapters. A comprehensive list of tree visualization and editing software is provided at *http://bioinfo.unice.fr/biodiv/Tree_editors.html*. A rather complete list of the phylogeny software available is maintained by Joe Felsenstein at *http://evolution.genetics.washington.edu/PHYLIP/software.html* and most packages can be downloaded following the links of his web page. They always contain complete documentation on how to install and run them properly. However, the programmer usually assumes the user has a background in phylogeny and molecular evolution, and for those programs not discussed in this book it may be necessary to read the original papers in order to understand the details or the meaning of the implemented computation.

Phylogenetic inference using maximum likelihood methods

THEORY

Heiko A. Schmidt and Arndt von Haeseler

6.1 Introduction

The concept of *likelihood* refers to situations that typically arise in natural sciences in which given some data **D**, a decision must be made about an adequate explanation of the data. Thus, a specific model and a hypothesis are formulated in which the model as such is generally not in question. In the phylogenetic framework, one part of the model is that sequences actually evolve according to a tree. The possible hypotheses include the different tree structures, the branch lengths, the parameters of the *model of sequence evolution*, and so on. By assigning values to these elements, it is possible to compute the probability of the data under these parameters and to make statements about their plausibility. If the hypothesis varies, the result is that some hypotheses produce the data with higher probability than others. Coin-tossing is a standard example. After flipping a coin $n = 100$ times, $h = 21$ heads and $t = 79$ tails were observed. Thus, $\mathbf{D} = (21, 79)$ constitutes a sufficient summary of the data. The model then states that, with some probability, $\theta \in [0, 1]$ heads appear when the coin is flipped. Moreover, it is assumed that the outcome of each coin toss is independent of the others, that θ does not change during the experiment, and that the experiment has only two outcomes (head or tail). The model is now fully specified. Because both, heads and tails, were obtained, θ must be larger than zero and smaller than 1. Moreover, any probability textbook explains

The Phylogenetic Handbook: a Practical Approach to Phylogenetic Analysis and Hypothesis Testing,
Philippe Lemey, Marco Salemi, and Anne-Mieke Vandamme (eds.). Published by Cambridge
University Press. © Cambridge University Press 2009.

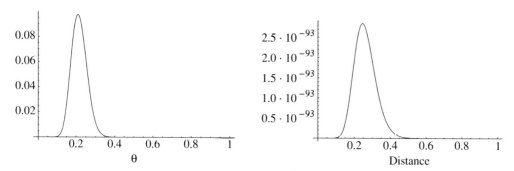

Fig. 6.1   Left: likelihood function of a coin-tossing experiment showing 21 heads and 79 tails. Right: likelihood function of the Jukes–Cantor model of sequence evolution for a sequence with length 100 and 21 observed differences.

that the probability to observe exactly $H = h$ heads in n tosses can be calculated according to the binomial distribution:

$$\Pr[H = h] = \binom{n}{h}\theta^h(1 - \theta)^{n-h} \tag{6.1}$$

Equation (6.1) can be read in two ways. First, if θ is known, then the probability of $h = 0, \ldots, n$ heads in n tosses can be computed. Second, (6.1) can be seen as a function of θ, where n and h are given; this defines the so-called **likelihood function**

$$L(\theta) = \Pr[H = h] = \binom{n}{h}\theta^h(1 - \theta)^{n-h} \tag{6.2}$$

From Fig. 6.1, which illustrates the likelihood function for the coin-tossing example, it can be seen that some hypotheses (i.e. choices of θ) generate the observed data with a higher probability than others. In particular, (6.2) becomes maximal if $\theta = \frac{21}{100}$. This value also can be computed analytically. For ease of computation, first compute the logarithm of the likelihood function, which results in sums rather than products:

$$\log[L(\theta)] = \log\binom{n}{h} + h\log\theta + (n - h)\log(1 - \theta) \tag{6.3}$$

The problem is now to find the value of θ ($0 < \theta < 1$) maximizing the function. From elementary calculus, it is known that relative extrema of a function, $f(x)$, occur at critical points of f, i.e. values x_0 for which either $f'(x_0) = 0$ or $f'(x_0)$ is undefined. Differentiation of (6.3) with respect to θ yields:

$$L'(\theta) = \frac{\partial \log[L(\theta)]}{\partial\theta} = \frac{h}{\theta} - \frac{n - h}{1 - \theta} \tag{6.4}$$

This derivative is equal to zero if $\theta_0 = \frac{h}{n}$, positive, i.e. $L'(\theta) > 0$, for $0 < \theta < \theta_0$, and negative for $\theta_0 < \theta < 1$, so that $\log[L(\theta)]$ attains its maximum at $\theta_0 = \frac{h}{n}$. We say $\hat{\theta} = \frac{h}{n}$ is the *maximum likelihood estimate (MLE)* of the probability of observing a head in a single coin toss (the hat ^ notation indicates an estimate rather than the unknown value of θ). In other words, when the value of θ is selected that maximizes (6.3), the observed data are produced with the highest likelihood, which is precisely the *maximum likelihood (ML) principle*. However, the resulting likelihoods are usually small (e.g. $L(21/100) \approx 0.0975$); conversely, the likelihoods of competing hypotheses can be compared by computing the odds ratio. Note that the hypothesis that the coin is fair ($\theta = 1/2$) results in a likelihood of $L(1/2) \approx 1.61 \cdot 10^{-9}$; thus, the MLE of $\hat{\theta} = 0.21$ is $6 \cdot 10^7$ times more likely to produce the data than $\theta = 0.5$! This comparison of odds ratios leads to the statistical test procedure discussed in more detail in Chapters 8, 10–12, and 14.

In evolution, point mutations are considered chance events, just like tossing a coin. Therefore, at least in principle, the probability of finding a mutation along one branch in a **phylogenetic tree** can be calculated by using the same maximum-likelihood framework discussed previously. The main idea behind phylogeny inference with maximum-likelihood is to determine the tree topology, branch lengths, and parameters of the evolutionary model (e.g. *transition/transversion* ratio, base frequencies, rate variation among sites) (see Chapter 4) that maximize the probability of observing the sequences at hand. In other words, the likelihood function is the conditional probability of the data (i.e. sequences) given a hypothesis (i.e. a model of substitution with a set of parameters θ and the tree τ, including branch lengths):

$$L(\tau, \theta) = \Pr(\text{Data}|\tau, \theta)$$

$$= \Pr(\text{aligned sequences}|\text{tree, model of evolution}) \quad (6.5)$$

The MLEs of τ and θ (named $\hat{\tau}$ and $\hat{\theta}$) are those making the likelihood function as large as possible:

$$\hat{\tau}, \hat{\theta} = \operatorname*{argmax}_{\tau, \theta} L(\tau, \theta) \quad (6.6)$$

Before proceeding to the next section, some cautionary notes are necessary. First, the likelihood function must not be confused with a probability. It is defined in terms of a probability, but it is the probability of the observed event, not of the unknown parameters. The parameters have no probability because they do not depend on chance. Second, the probability of getting the observed data has nothing to do with the probability that the underlying model is correct. For example, if the model states that the sequences evolve according to a tree, although they have recombined, then the final result will still be a single tree that gives rise to the maximum-likelihood value (see also Chapter 15). The probability of the data being given the MLE of the parameters does not provide any hints that the model assumptions are in fact true. One can only compare the maximum-likelihood values with other likelihoods for model parameters that are elements of the model. To determine whether the hypothesis of tree-like evolution is reasonable, the types of relationship allowed among sequences must be enlarged; this is discussed in Chapter 21.

6.2 The formal framework

Before entering the general discussion about maximum-likelihood tree reconstruction, the simplest example (i.e. reconstructing a maximum-likelihood tree for two sequences) is considered. A tree with two **taxa** has only one branch connecting the two sequences; the sole purpose of the exercise is reconstructing the branch length that produces the data with maximal probability.

6.2.1 The simple case: maximum-likelihood tree for two sequences

In what follows, it is assumed that the sequences are evolving according to the Jukes and Cantor model (see Chapter 4). Each position evolves independently from the remaining sites and with the same **evolutionary rate**. The alignment has length l for the two sequences $S_i = (s_i^1, \ldots, s_i^l)$, $(i = 1, 2)$, where s_i^j is the nucleotide, the amino acid, or any other letter from a finite alphabet at sequence position j in sequence i. The likelihood function is, then, according to (4.31) (Chapter 4):

$$L(d) = \prod_{j=1}^{l} \pi_{s_1^j} P_{s_1^j s_2^j}\left(-\frac{4d}{3}\right) \tag{6.7}$$

where d, the number of substitutions per site, is the parameter of interest and $P_{xy}(t)$ is the probability of observing nucleotide y if nucleotide x was originally present, and $\pi_{s_1^j}$ is the probability of character s_1^j in the equilibrium distribution. From (4.12a) and (4.12b), the following is obtained:

$$P_{xy}\left(-\frac{4}{3}d\right) = \begin{cases} \frac{1}{4}\left(1 + 3\exp\left[-\frac{4}{3}d\right]\right) \equiv \tilde{P}_{xx}(d), & \text{if } x = y \\ \frac{1}{4}\left(1 - \exp\left[-\frac{4}{3}d\right]\right) \equiv \tilde{P}_{xy}(d), & \text{if } x \neq y \end{cases} \tag{6.8}$$

To infer d, the relevant statistic is the number of identical pairs of nucleotides (l_0) and the number of different pairs (l_1), where $l_0 + l_1 = l$. Therefore, the alignment is summarized as $\mathbf{D} = (l_0, l_1)$ and the score is computed as:

$$\log[L(d)] = C + l_0 \log\left[\tilde{P}_{xx}(d)\right] + l_1 \log\left[\tilde{P}_{xy}(d)\right] \tag{6.9}$$

which is maximal if

$$d = -\frac{3}{4}\log\left[1 - \frac{4}{3} \cdot \frac{l_1}{l_1 + l_0}\right]. \tag{6.10}$$

```
L20571    ...AAAGTAATGAAGAAGAACAACAGGAAGTCATGGAGCTTATACATA...
AF10138   ...ATGGAGAAGAAGAAG--------AGACTCTGGCTAAGTTATTGT...
X52154    ...ATGGAGAAGAAGAAG--------AGAGACTGGAACAGCTTATCC...
U09127    ...ATGGGGATAGAGAGGAATTATCCTTGCTGGTGGACATGGGGGATT...
U27426    ...AGGGGGATACAGATGAATTGGCAACACTTGTGGAAATGGGGAACT...
U27445    ...AAGGGGATACGGACGAATTGGCAACACTTCTGGAGATGGGGAACT...
U067158   ...AGGGGGACACTGAGGAATTATCAACAATGGTGGATATGGGGCGTC...
U09126    ...GAGGGGATACAGAGGAATTGGAAACAATGGTGGATATGGGGCATC...
U27399    ...AGGGAGATGAGGAGGAATTGTCAGCATTTGTGGGGATGGGGCACC...
U43386    ...AGGGAGATGCAGAGGAATTATCAGCATTTATGGAAATGGGGCATC...
L02317    ...AAGGAGATCAGGAAGAATTATCAGCACTTGTGGAGATGGGGCACC...
AF025763  ...AAGGGGATCAGGAAGAATTGTCAGCACTTGTGGAGATGGGGCATG...
U08443    ...AAGGAGATGAGGAAGCATTGTCAGCACTTATGGAGAGGGGGCACC...
AF042106  ...AAGGGGATCAGGAAGAATTATCGGCACTTGTGGACATGGGGCACC...
```

Fig. 6.2   Part of the mtDNA sequence alignment used as a relevant example throughout the book.

This result is not influenced by the constant C, which only changes the height of the maximum but not its "location." Please note that, the MLE of the number of substitutions per site equals the method-of-moments estimate (see (4.15a)). Therefore, the maximum-likelihood tree relating the sequences S_1 and S_2 is a straight line of length d, with the sequences as endpoints.

This example was analytically solvable because it is the simplest model of sequence evolution and, more importantly, because only two sequences – which can only be related by *one* tree – were considered. The following sections set up the formal framework to study more sequences.

6.2.2 The complex case

When the data set consists of $n > 2$ aligned sequences, rather than computing the probability $P_{xy}(t)$ of observing two nucleotides x and y at a given site in two sequences, the probability of finding a certain column or pattern of nucleotides in the data set is computed. Let D_j denote the nucleotide pattern at site $j \in \{1, \ldots, l\}$ in the alignment (Fig. 6.2). The unknown probability obviously depends on the model of sequence evolution, M, and the tree, τ relating the n sequences with the number of substitutions along each branch of the tree (i.e. the branch lengths). In theory, each site could be assigned its own model of sequence evolution according to the general time reversible model (see Chapter 4) and its own set of branch lengths. Then, however, the goal to reconstruct a tree from an alignment becomes almost computationally intractable and, hence, several simplifications are needed. First, it is assumed that each site s in the alignment evolves according to the same model M; for example, the Tamura–Nei (TN) model (see (4.32a, b, c)) (i.e. γ, κ, and π are assumed the same for each site in the alignment). The assumption also implies that all sites evolve at the same rate μ (see (4.24)). To overcome this simplification, the rate at a site is modified by a rate-specific factor, $\rho_j > 0$.

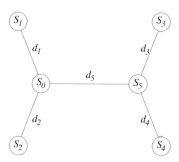

Fig. 6.3   Four-sequence tree, with branch lengths d_1, d_2, d_3, and d_4 leading to sequences S_1, S_2, S_3, and S_4 and branch length d_5 connecting the "ancestral" sequences S_0 and S_5.

Thus, the ingredients for the probability of a certain site pattern are available, and

$$\Pr\left[D_j|\tau, M, \rho_j\right], \quad j = 1, \dots, l \tag{6.11}$$

specifies the probability to observe pattern D_j. If it is also assumed that each sequence site evolves independently (i.e. according to τ and M, with a site specific rate ρ_j), then the probability of observing the alignment (data) $\mathbf{D} = (D_1, \dots, D_l)$ equals the product of the probabilities at each site, as follows:

$$L\left(\tau, M, \rho|\mathbf{D}\right) \equiv \Pr\left[\mathbf{D}|\tau, M, \rho\right] = \prod_{j=1}^{l} \Pr\left[D_j|\tau, M, \rho_j\right] \tag{6.12}$$

When the data are fixed, (6.12) is again a likelihood function (like (6.2) and (6.5)), which allows for the two ways of looking at it (see the previous section). First, for a fixed choice of τ, M, and the site rate vector ρ, the probability to observe the alignment $\mathbf{D}$ can be computed with (6.11). Second, for a given alignment $\mathbf{D}$, (6.12) can be used to find the **MLEs**.

   In what follows, the two issues are treated separately. However, to simplify the matter, it is assumed that the site-specific rate factor ρ_j is drawn from a Γ-**distribution** with expectation 1 and variance $\frac{1}{\alpha}$ (Uzzel & Corbin, 1971; Wakeley, 1993), where α defines the shape of the distribution (see also Section 4.6.1).

6.3 Computing the probability of an alignment for a fixed tree

Consider the tree τ with its branch lengths (i.e. number of substitutions per site), the model of sequence evolution M with its parameters (e.g. transition/transversion ratio, stationary base composition), and the site-specific rate factor $\rho_j = 1$ for each site j. The goal is to compute the probability of observing one of the 4^n possible patterns in an alignment of n sequences. The tree displayed in Fig. 6.3 illustrates the principle for four sequences ($n = 4$). Because the model M is a submodel of the GTR class – that is, a time-reversible model (see Chapter 4) – we can assign any point as a root to the tree for the computation of its likelihood (Pulley Principle, Felsenstein, 1981). Here, we will assume that evolution started from sequence S_0 and then proceeded along the branches of tree τ with branch lengths d_1, d_2, d_3, d_4, and d_5. To compute $\Pr[\,D_j,\,\tau,\,M,\,1\,]$ for a specific site j, where $D_j = (s_1^j, s_2^j, s_3^j, s_4^j)$ are the nucleotides observed, it is necessary to know the ancestral states s_0^j and s_5^j. The conditional probability of the data, given the ancestral states, then will be as follows:

$$
\Pr\left[D_j,\,\tau,\,M,\,1 \middle| s_0^j,\,s_5^j \right] \\
= P_{s_0^j s_1^j}(d_1) \cdot P_{s_0^j s_2^j}(d_2) \cdot P_{s_0^j s_5^j}(d_5) \cdot P_{s_5^j s_3^j}(d_3) \cdot P_{s_5^j s_4^j}(d_4) \tag{6.13}
$$

The computation follows immediately from the considerations in Chapter 4. However, in almost any realistic situation, the ancestral sequences are not available. Therefore, one sums over all possible combinations of ancestral states of nucleotides gaining a so-called *maximum average likelihood* (Steel & Penny, 2000). As discussed in Section 4.4, nucleotide substitution models assume **stationarity**; that is, the relative frequencies of A, C, G, and T ($\pi_A, \pi_C, \pi_G, \pi_T$) are at equilibrium. Thus, the probability for nucleotide s_0^j will equal its stationary frequency $\pi(s_0^j)$, from which it follows that

$$
\Pr\left[D_j,\,\tau,\,M,\,1 \right] \\
= \sum_{s_0^j} \sum_{s_5^j} \pi(s_0^j) \cdot P_{s_0^j s_1^j}(d_1) \cdot P_{s_0^j s_2^j}(d_2) \cdot P_{s_0^j s_5^j}(d_5) \cdot P_{s_5^j s_3^j}(d_3) \cdot P_{s_5^j s_4^j}(d_4)
$$

$$\tag{6.14}$$

Although this equation looks like one needs to compute exponentially many summands, the sum can be efficiently assessed by evaluating the likelihoods moving from the end nodes of the tree to the root (Felsenstein, 1981). In each step, starting from the leaves of the tree, the computations for two nodes are joined and replaced by the joint value at the ancestral node (see Section 6.3.1 for details). This process bears some similarity to the computation of the minimal number of substitutions on a given tree in the **maximum parsimony** framework (Fitch, 1971) (see Chapter 8). However, contrary to maximum parsimony, the distance (i.e. number of substitutions) between the two nodes is considered. Under the maximum parsimony framework, if two sequences share the same nucleotide, then the most recent common ancestor also carries this nucleotide (see Chapter 8). In the maximum-likelihood framework, this nucleotide is shared by the ancestor only with a certain probability, which gets smaller if the sequences are only very remotely related.

6.3.1 Felsenstein's pruning algorithm

Equation (6.14) shows how to compute the likelihood of a tree for a given position in a sequence alignment. To generalize this equation for more than four sequences, it is necessary to sum all the possible assignments of nucleotides at the $n - 2$ inner nodes of the tree. Unfortunately, this straightforward computation is not feasible, but the amount of computation can be reduced considerably by noticing the following recursive relationship in a tree. Let $D_j = (s_1^j, s_2^j, s_3^j, \ldots, s_n^j)$ be a pattern at a site j, with tree τ and a model M fixed. Nucleotides at inner nodes of the tree are abbreviated as x_i with $i = n + 1, \ldots, 2n - 2$. For an inner node i with offspring o_1 and o_2, the vector $(\mathbf{L}_j^i = L_j^i(A), L_j^i(C), L_j^i(G), L_j^i(T))$ is defined recursively as

$$L_j^i(s) = \left[\sum_{x \in \{A,C,G,T\}} P_{sx}(d_{o_1}) L_j^{o_1}(x) \right] \cdot \left[\sum_{x \in \{A,C,G,T\}} P_{sx}(d_{o_2}) L_j^{o_2}(x) \right]$$

$$s \in \{A, C, G, T\} \tag{6.15}$$

and for the leaves

$$L_j^i(s) = \begin{cases} 1, & \text{if } s = s_i^j \\ 0, & \text{otherwise} \end{cases} \tag{6.16}$$

where d_{o_1} and d_{o_2} are the number of substitutions connecting node i and its descendants in the tree (Fig. 6.4). Without loss of generality, it is assumed that the node $2n - 2$ has three offspring: o_1, o_2, and o_3, respectively. For this node, (6.15) is modified accordingly. This equation allows an efficient computation of the likelihood for each alignment position (Fig. 6.4) by realizing that

$$\Pr\left[D_j, \tau, M, 1\right] = \sum_{s \in \{A,C,G,T\}} \pi_s L_j^{2n-2}(s) \tag{6.17}$$

Equation (6.17) then can be used to compute the likelihood of the full alignment with the aid of (6.12). In practice, the calculation of products is avoided, moving instead to log-likelihoods; that is, (6.12) becomes

$$\log\left[L\left(\tau, M, 1\right)\right] = \log\left[\prod_{j=1}^{l} \Pr\left[D_j, \tau, M, 1\right]\right]$$

$$= \sum_{j=1}^{l} \log\left[\Pr\left[D_j, \tau, M, 1\right]\right] \tag{6.18}$$

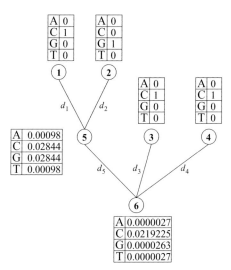

Fig. 6.4   Likelihood computation on a four-taxa tree for an alignment site pattern $D_j = (C, G, C, C)$ with all branch lengths $d_1, \ldots d_5$ set to 0.1. According to (6.8) the probability to observe no mutation after $d = 0.1$ is 0.9058 and for a specific nucleotide pair 0.0314. The values $L^i_j(s)$ at the leaves are computed with (6.16), those at the internal nodes with (6.15). For example, to obtain $L^5_j(C)$ at node five, (6.15) reduces to $P_{CC}(d_1) \cdot P_{CG}(d_2) = 0.9058 \cdot 0.0314$. The position likelihood according to (6.17) is 0.0054886, the according log-likelihood is −5.2051.

6.4 Finding a maximum-likelihood tree

Equations (6.15) through (6.18) show how to compute the probability of an alignment, if everything were known. In practice, however, branch lengths of the tree are unknown. Branch lengths are computed numerically by maximizing (6.18); that is, by finding those branch lengths for tree τ maximizing the log-likelihood function, which is accomplished by applying numerical routines like Newton–Raphson or Brent's method (Press *et al.*, 1992). Such a computation is usually time-consuming and typically the result depends on the numerical method.

Nevertheless, maximizing the likelihood for a single tree is not the biggest challenge in phylogenetic reconstruction; the daunting task is to actually find the tree among all possible tree structures that maximizes the global likelihood. Unfortunately, for any method that has an explicit optimality criterion (e.g. maximum parsimony, distance methods, and maximum-likelihood), no efficient algorithms are known that guarantee the localization of the best tree(s) in the huge space of all possible tree topologies. The naïve approach to simply compute the maximum-likelihood value for each tree topology is prohibited by the huge number of tree structures, even for moderately sized data sets. The number of (unrooted) binary tree topologies increases tremendously with the number of taxa (n), which can be computed according to

$$t_n = \frac{(2n-5)!}{2^{n-3}(n-3)!} = \prod_{i=1}^{n}(2i-5) \tag{6.19}$$

When computing the maximum-likelihood tree, the model parameters and branch lengths have to be computed for each tree, and then the tree that yields the highest likelihood is selected. Because of the numerous tree topologies, testing all possible trees is impossible, and it is also computationally not feasible to estimate the model parameters for each tree. Thus, various heuristics are used to suggest reasonable trees, including ***stepwise addition*** (e.g. used in Felsenstein's PHYLIP package: program DNAML, Felsenstein, 1993) and ***star decomposition*** (MOLPHY, Adachi & Hasegawa, 1996) as well as the ***neighbor-joining*** (NJ) algorithm (Saitou & Nei, 1987). Stepwise addition and NJ are discussed in Chapter 8 and Chapter 5, respectively. However, to make this chapter self-consistent we briefly summarize the various heuristics. In our parlance, we are looking for the tree with the highest likelihood. However, the tree rearrangement operations themselves are independent of the objective function.

6.4.1 Early heuristics

Stepwise addition was probably among the first heuristics to search for a maximum-likelihood of a tree. The procedure starts from the unrooted tree topology for three taxa randomly selected from the list of n taxa. Then one reconstructs the corresponding maximum likelihood tree. To extend this tree we randomly pick one of the remaining $n-3$ taxa. This taxon is then inserted into each branch of the best tree. The branch, where the insertion leads to the highest likelihood, will be called insertion branch. Thus, we have a local decision criterion that selects the tree with the highest likelihood from a list of $2k-3$ trees, if k taxa are already in the sub-tree. The resulting tree will then be used to repeat the procedure. After $n-3$ steps, a maximum-likelihood tree is obtained, that is at least locally optimal. That means given the insertion order of the taxa and given the local decision criterion no better tree is possible.

However, we have only computed the maximum-likelihood for $\sum_{i=3}^{n}(2i-5) = (n-2)^2$ trees. Thus, it is possible that another insertion order of the taxa will provide trees with a higher likelihood. To reduce the risk of getting stuck in such local optima, tree-rearrangement operations acting on the full tree were suggested.

6.4.2 Full-tree rearrangement

Full-tree rearrangement operations change the structure of a given tree with n leaves. They employ the following principle. From a starting tree a number of trees (the neighborhood of the starting tree) are generated according to specified rules.

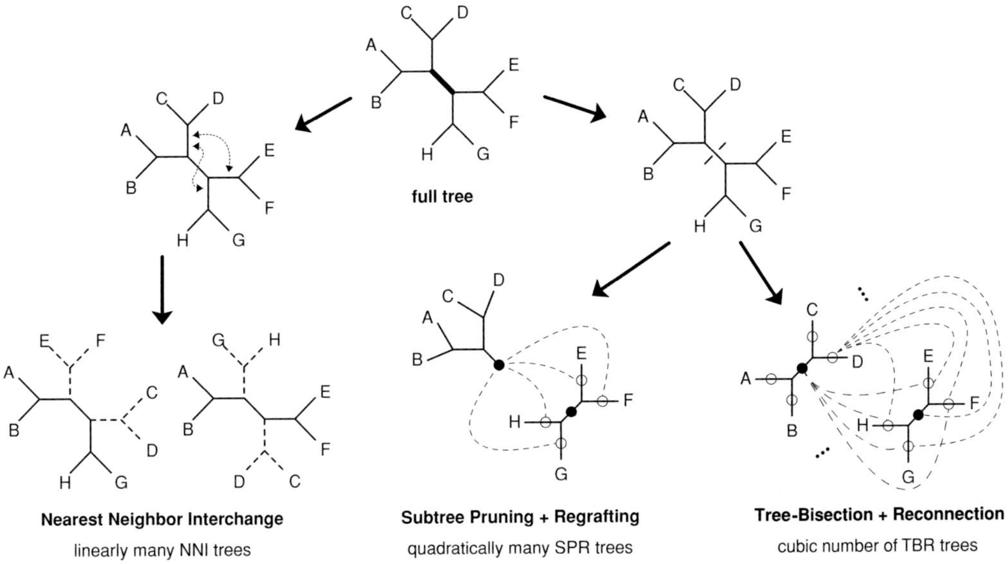

Fig. 6.5   The three basic tree rearrangement operations (NNI, SPR, and TBR) on the thick branch in the full tree. In SPR and TBR all pairs of "circled" branches among the two subtrees will be connected (dashed lines), except the two filled circles to each other, since this yields the full tree again.

For each resulting tree, the maximum-likelihood value is computed. The tree with the highest likelihood is then used to repeat the procedure. The rearrangement typically stops if no better tree is found. This tree is then said to be a locally optimal tree. The chance of actually having determined the globally optimal tree, however, depends on the data and the size of neighborhood.

Three full-tree rearrangement operations are currently popular: ***Nearest neighbor interchange*** (***NNI***), ***sub-tree pruning and regrafting*** (***SPR***) and ***tree-bisection and reconnection*** (***TBR***), confer Fig. 6.5 and see Chapter 8 for more details. Depending on the operation, the size of the neighborhood grows linearly (NNI), quadratically (SPR), or cubically (TBR) with the number of taxa in the full tree.

Different approaches are applied to limit the increase of computation time of the SPR or TBR, while still taking advantage of their extended neighborhood. We will briefly describe some programs and search schemes. Some of these packages have implemented different variants and extensions of the insertion or rearrangement operations. We will not explain them in full detail, but rather refer to the corresponding publications.

6.4.3 DNAml and fastDNAml

The DNAml program (PHYLIP package, Felsenstein, 1993) and its descendant, fastDNAml (Olsen *et al.*, 1994; Stewart *et al.*, 2001), search by stepwise addition.

Although not turned on by default, the programs allow to apply SPR rearrangements after all sequences have been added to the tree.

Moreover, FASTDNAML provides tools to do full tree rearrangements after each insertion step. The user may choose either NNI or SPR, and can also restrict the SPR neighborhood by setting a maximal number of branches to be crossed between pruning and inserting point of the subtree.

6.4.4 PHYML and PHYML-SPR

PHYML (Guindon & Gascuel, 2003) reduces the running time by a mixed strategy. It uses a fast distance based method, BioNJ (Gascuel, 1997), to quickly compute a full initial tree. Then they apply *fastNNI* operations to optimize that tree. During fastNNI all possible NNI trees are evaluated (optimizing only the branch crossed by the NNI) and ranked according to their ML value. Those NNIs which increase the ML value most, but do not interfere with each other, are simultaneously applied to the current tree. Simultaneously applying different NNIs saves time and makes it possible to walk quickly through tree space. On the new current tree fastNNI is repeated until no ML improvement is possible.

Due to their limited range of topological changes NNIs are prone to get stuck in local optima. Hence, a new SPR-based version, PHYML-SPR (Hordijk & Gascuel, 2006), has been devised taking advantage of the larger neighborhood induced by SPR. To compensate for the increased computing time, PHYML-SPR evaluates the SPR neighborhood of the current tree by fast measures like distance-based approaches to determine a ranked list of most promising SPR tree candidates (Hordijk & Gascuel, 2006; for more details). Their likelihood is then assessed by only optimizing the branch lengths on the path from the pruning to the insertion point. If a better tree is found, it takes the status of new current tree.

A fixed number of best candidate trees according to their likelihood are then optimized by adjusting all branch lengths. If now a tree has a higher likelihood than the current one, this tree replaces the old one.

PHYML-SPR allows to alternate SPR and fastNNI-based iterations. Iteration continues until no better tree is found.

6.4.5 IQPNNI

IQPNNI (Vinh & von Haeseler, 2004) uses BioNJ (Gascuel, 1997) to compute the starting tree and fastNNI for likelihood optimization. IQPNNI, however, applies a different strategy to reduce the risk of getting stuck in local optima. When the current tree cannot be improved anymore, IQPNNI randomly removes taxa from the current tree and re-inserts them using a fast quartet-based approach. The new tree is again optimized with fastNNI. If the new tree is better, it then becomes the new starting tree, otherwise the original current tree is kept.

This procedure is either repeated for a user-specified number of iterations or IQPNNI applies a built-in stopping rule, that uses a statistical criterion, to abandon further search (Vinh & von Haeseler, 2004).

Furthermore, IQPNNI provides ML tree reconstruction for various codon models like Goldman & Yang (1994) or Yang & Nielsen (1998). Refer to Chapter 14 for details on such complex models.

6.4.6 RAxML

The RAxML program (Stamatakis, 2006) builds the starting tree based on maximum parsimony (Chapter 8) and optimizes with a variant of SPR called *lazy subtree rearrangement* (LSR, Stamatakis *et al.*, 2005). LSR combines two tricks to reduce the computational demand of SPR operations. First, it assigns a maximal distance between pruning and insertion point for the SPR operations to restrict the size of the neighborhood. The maximal SPR distance (< 25 branches) is determined at the start of the program. Second, LSR optimizes only the branch that originates at the pruning point and the three newly created at the insertion point. The LSRs are repeated many times always using the currently best tree. For the 20 best trees, found during the LSR, the final ML-value is re-optimized by adjusting all branch lengths. The LSR and re-optimization is repeated until no better tree is found.

6.4.7 Simulated annealing

Simulated annealing (Kirkpatrick *et al.*, 1983) is an attempt to find the maximum of complex functions (possibly with multiple peaks), where standard (hill climbing) approaches may get trapped in local optima. One starts with an initial tree, then samples the tree-space by accepting with a reasonable probability a tree with a lower likelihood (down-hill move). Trees with higher likelihood (up-hill moves) are always accepted. This is conceptually related to **Markov chain Monte Carlo** (see Chapters 7 and 18). However, as the process continues the down-hill probability is decreased. This decrease is modeled by a so-called cooling schedule. The term "annealing" is borrowed from crystal formation. Initially (high temperature) there is a lot of movement (almost every tree is accepted), then as the temperature is lowered the movements get smaller and smaller. If the decrease in temperature is modeled adequately, then the process will eventually find the ML tree. However, to model the decrease in temperature is not trivial.

First introduced in a parsimony context (Lundy, 1985; Dress & Krüger, 1987), simulated annealing to reconstruct ML trees is applied by SSA (Salter & Pearl, 2001) and RAxML-SA (Stamatakis, 2005). Furthermore, Fleissner *et al.* (2005) use simulated annealing to construct alignments and trees simultaneously.

6.4.8 Genetic algorithms

Genetic algorithms (GA) are an alternative search technique to solve complex optimization problems. They borrow the nomenclature and the optimization decision from evolutionary biology. In fact, GA are a special category of evolutionary algorithms (Bäck & Schwefel, 1993).

The basic ingredients of GA are a population of individuals (in our case a collection of trees) a fitness function (maximum likelihood function according to (6.17)) that determines the offspring number. According to the principles of evolution a tree can mutate (change in branch lengths, NNI, SPR, TBR operations), even trees can exchange sub-trees (recombination). For the mutated tree, the fitness function is computed. The individuals of the next generation are then randomly selected from the mutant trees and the current non-mutated trees according to their fitness (selection step). Typically, one also keeps track of the fittest individual (the tree with the best likelihood). After several generations, evolution stops and the best tree is output.

After having been introduced to phylogenetics in the mid-1990s (e.g. Matsuda, 1995), GARLI (Zwickl, 2006), METAPIGA (Lemmon & Milinkovitch, 2004), and GAML (Lewis, 1998) are examples for applications of GA in phylogenetic inference.

6.5 Branch support

As should be clear by now, none of the above methods guarantee to detect the optimal tree. Hence, biologists usually apply a plethora of methods, and if those reconstruct similar trees one tends to have more confidence in the result.

Typically, tree reconstruction methods are searching for the best tree, leaving the user with a single tree and ML value, but without any estimate of the reliability of its sub-trees.

Several measures are used to assess the certainty of a tree or its branches. The ML values from competing hypotheses can be used in a *likelihood ratio test* (*LRT*, see Chapters 10, 11, and 14) or other tests (Chapter 12).

The support of branches are often assessed by employing statistical principles. The most widely used approach to assess branch support seems to be *bootstrapping* (Efron, 1979; Felsenstein, 1985), where pseudo-samples are created by randomly drawing with replacement l columns from the original l-column alignment, i.e. a column from the data alignment can occur more than once or not at all in a pseudo-sample. From each pseudo-sample a tree is reconstructed and a consensus tree is constructed, incorporating those branches that occur in the majority of the reconstructed trees. These percentages are used as indicator for the reliability of branches. See Chapter 5 for details on branch support analysis using the bootstrap.

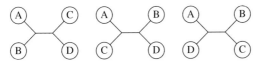

Fig. 6.6   The three different informative tree topologies for the quartet $q = (A, B, C, D)$.

Very similar to the bootstrap is jackknifing, where only a certain percentage of the columns are drawn without replacement (Quenouille, 1956).

Finally, the trees sampled in a Bayesian MCMC analysis are usually summarized in a consensus tree and the tree sample can be used to derive approximate ***posterior probabilities*** for each split or clade (Ronquist & Huelsenbeck, 2003; see also next chapter).

Another method to measure branch support is the ***quartet puzzling*** method implemented in the TREE-PUZZLE software, that will be explained in the following section. Although TREE-PUZZLE is nowadays not faster than most of the above mentioned ML methods, it is usually faster than running at least 100 bootstraps with an ML method and certainly faster than a Bayesian MCMC analysis.

6.6 The quartet puzzling algorithm

Quartet puzzling (Strimmer & von Haeseler, 1996) utilizes quartets, i.e. groups of four sequences. Quartets are the smallest set of taxa for which more than one unrooted tree topology exists. The three different quartet tree topologies are shown in Fig. 6.6. Quartet-based methods use the advantage that the quartet trees can be quickly evaluated with maximum-likelihood. However, there exist $\binom{n}{4} = \frac{n!}{4!(n-4)!}$ possible quartets in a set of n taxa.

Quartet puzzling is performed in four steps.

6.6.1 Parameter estimation

First TREE-PUZZLE estimates the parameters for the evolutionary model. To this end:

(i) The pairwise distance matrix D is estimated for all pairs of sequences in the input alignment and a Neighbor Joining tree is constructed from D.
(ii) Then, maximum-likelihood branch lengths are computed for the NJ topology and parameters of the sequence evolution are estimated.
(iii) Based on these estimates, a new D and NJ tree are computed and Step (ii) is repeated.

Steps (ii) and (iii) are repeated until the estimates of the model parameters are stable.

6.6.2 ML step

To produce the set of tree topologies, the likelihoods of all $3 \times \binom{n}{4}$ quartet tree topologies are evaluated. Then, for each quartet and each topology the corresponding highest likelihood is stored. The algorithm takes into account that two topologies may have similar likelihoods (partly resolved quartet) or that even no topology (unresolved quartet) gains sufficient support (Strimmer *et al.*, 1997).

6.6.3 Puzzling step

Based on the set of supported quartet topologies, trees are constructed by adding taxa in random order. Each taxon is inserted into that branch least contradicted by the set of relevant quartet trees.

This step is repeated many times with different input orders, producing a large set of intermediate trees.

6.6.4 Consensus step

The set of intermediate trees is subsequently summarized by a majority rule consensus tree, the so-called quartet puzzling tree, where the percent occurrences for each branch are considered **puzzle support values**.

6.7 Likelihood-mapping analysis

The chapter so far has discussed the problem of reconstructing a phylogenetic tree and assessing the reliability of its branches. A maximum-likelihood approach may also be used to study the amount of *evolutionary information* contained in a data set. The analysis is based on the maximum-likelihood values for the three possible four taxa trees. If L_1, L_2, and L_3 are the likelihoods of trees T_1, T_2, and T_3, then one computes the posterior probabilities of each tree T_i as $p_i = \frac{L_i}{L_1 + L_2 + L_3}$. Since the p_i terms sum to 1, the probabilities p_1, p_2, and p_3 can be reported simultaneously as a point P lying inside an equilateral triangle, each corner of the triangle representing one of the three possible tree topologies (Fig. 6.7a). If P is close to one corner – for example, the corner T_1 – the tree T_1 receives the highest support. In a maximum-likelihood analysis, the tree T_i, which satisfies $p_i = \max\{p_1, p_2, p_3\}$, is selected as the MLE. However, this decision is questionable if P is close to the center of the triangle. In that case, the three likelihoods are of similar magnitude; in such situations, a more realistic representation of the data is a star-like tree rather than an artificially *strictly bifurcating tree* (see Section 1.7 in Chapter 1).

Therefore, the ***likelihood-mapping method*** (Strimmer & von Haeseler, 1997) partitions the area of the equilateral triangle into seven regions (Fig. 6.7b). The three trapezoids at the corners represent the areas supporting strictly bifurcating trees (i.e. Areas 1, 2, and 3 in Fig. 6.7b). The three rectangles on the sides represent

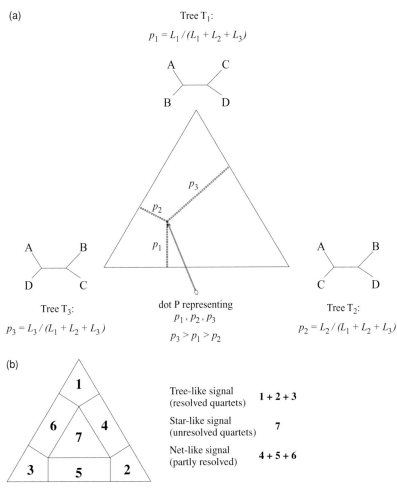

Fig. 6.7   Likelihood mapping. (a) The three posterior probabilities p_1, p_2, and p_3 for the three possible unrooted trees of four taxa are reported as a point (P) inside an equilateral triangle, where each corner represents a specific tree topology with likelihood L_1, L_2, and L_3, respectively. (b) Seven main areas in the triangle supporting different evolutionary information.

regions where the decision between two trees is not obvious (i.e. Areas 4, 5, and 6 in Fig. 6.7b for trees 1 and 2, 2 and 3, and 3 and 1). The center of the triangle represents sets of points P where all three trees are equally supported (i.e. Area 7 in Fig. 6.7b). Given a set of n aligned sequences, the likelihood-mapping analysis works as follows. The three likelihoods for the three tree topologies of each possible quartet (or of a random sample of the quartets) are reported as a dot in an equilateral triangle like the one in Fig. 6.7a. The distribution of points in the seven areas of the triangle (see Fig. 6.7b) gives an impression of the tree-likeness of the data. Note that, because the method evaluates quartets computed from n sequences, which one of the three topologies is supported by any corner of the triangle is not relevant. Only the percentage of points belonging to the areas 1, 2, and 3 is relevant to get an impression about the amount of tree-likeness in the data. To summarize the three corners (Areas $1 + 2 + 3$; see Fig. 6.7b) represent fully resolved tree topologies; Area 7 represents star-like phylogenies (Fig. 6.7b); the three Areas $4 + 5 + 6$ (see Fig. 6.7b) represent network-like phylogeny, where the data support conflicting tree topologies (see also Chapter 21).

From a biological standpoint, a likelihood mapping analysis showing more than 20%–30% of points in the star-like or network-like area suggests that the data are not reliable for phylogenetic inference. The reasons why an alignment may not be suitable for tree reconstruction are multiple, e.g. noisy data, alignment errors, recombination, etc. In the latter case, methods that explore and display conflicting trees, such as **bootscanning** (see Chapter 16), *split decomposition* or Neighbor-Net (see Chapter 21 for network analysis) may give additional information. A more detailed study on quartet mapping is given in Nieselt-Struwe & von Haeseler (2001).

PRACTICE

Heiko A. Schmidt and Arndt von Haeseler

6.8 Software packages

A number of software packages are available to compute maximum-likelihood trees from DNA or amino acid sequences. A detailed list can be found at Joe Felsenstein's website, *http://evolution.genetics.washington.edu/PHYLIP/software.html.*

Because program packages are emerging at a rapid pace, the reader is advised to visit this website for updates.

6.9 An illustrative example of an ML tree reconstruction

In what follows, the `hivALN.phy` file (available at *http://www.thephylogenetichandbook.org*) will be analyzed with the latest version 3.2 of IQPNNI (Vinh & von Haeseler, 2004) to infer a maximum likelihood tree and TREE-PUZZLE 5.3 (Schmidt *et al.*, 2002) to compute support values for the branches of the ML tree.

6.9.1 Reconstructing an ML tree with IQPNNI

Place the `hivALN.phy` file in the same folder as the IQPNNI executable and start `iqpnni`. The following text appears:

```
WELCOME TO IQPNNI 3.2 (sequential version)

Please enter a file name for the sequence data:
```

Type the filename hivALN.phy **and press enter.**

```
GENERAL OPTIONS
  o              Display as outgroup? L20571
  n              Number of iterations? 200
  s                 Stopping rule? No, stop after 200 iterations


IQP OPTIONS
  p  Probability of deleting a sequence? 0.3
  k              Number representatives? 4


SUBSTITUTION PROCESS
  d       Type of sequence input data? Nucleotides
  m              Model of substitution? HKY85 (Hasegawa et al. 1985)
  t          Ts/Tv ratio (0.5 for JC69)? Estimate from data
  f               Base frequencies? Estimate from data
```

```
RATE HETEROGENEITY
    r           Model of rate heterogeneity? Uniform rate

quit [q], confirm [y], or change [menu] settings:
```

Each option can be selected to change the setting by typing the corresponding letter. For example, if the user types m then each keystroke will change the model of sequence evolution. After a number of strokes the default HKY85 reappears. The letter n allows to change the number of iterations. By default the number of iterations is set to twice the number of sequences. The user is advised to set the limit as high as possible (at least 200) or to use the stopping rule.

For example, a typical run to infer a tree based on DNA sequences would start with the following setting:

```
GENERAL OPTIONS
    o                 Display as outgroup? L20571
    n              Number of iterations? 200
    s                    Stopping rule? No, stop after 200 iterations

IQP OPTIONS
    p   Probability of deleting a sequence? 0.3
    k             Number representatives? 4

SUBSTITUTION PROCESS
    d          Type of sequence input data? Nucleotides
    m              Model of substitution? HKY85 (Hasegawa et al. 1985)
    t          Ts/Tv ratio (0.5 for JC69)? Estimate from data
    f               Base frequencies? Estimate from data

RATE HETEROGENEITY
    r           Model of rate heterogeneity? Gamma distributed rates
    i        Proportion of invariable sites? No
    a   Gamma distribution parameter alpha? Estimate from data
    c      Number of Gamma rate categories? 4

quit [q], confirm [y], or change [menu] settings:
```

Entering y starts the program. As model of sequence evolution HKY is selected (see Sections 4.6 and 4.9 in Chapter 4). Rate heterogeneity is modeled with a Γ-distribution. The shape parameter α of the Γ-distribution is estimated with the aid of four discrete categories (see Section 4.6.1). An appropriate model can also be selected based on one of the approaches discussed in Chapter 10.

IQPNNI also infers trees from amino acid sequence alignments and it is possible to compute trees from coding DNA sequences using different codon models.

During the optimization messages like

```
(1) Optimizing gamma shape parameter ...
Gamma distribution shape = 0.57224
Optimizing transition/transversion ratio ...
Transition/transversion ratio: 1.7074
LogL = -17408.85673


(2) Optimizing gamma shape parameter ...
Gamma distribution shape = 0.54188
Optimizing transition/transversion ratio ...
Transition/transversion ratio: 1.7951
LogL = -17396.4208
```

will appear.

After optimizing the model parameters the IQPNNI continues with the tree search as described. Whenever a better tree is found IQPNNI outputs

```
Doing Nearest Neighbour Interchange... 1 s
We have constructed the initial tree !!!
The currently best log likelihood = -17395.94491
29 Iterations / time elapsed = 0h:0m:8s (will finish in 0h:0m:47s)
GOOD NEWS: BETTER TREE FOUND: THE CURRENTLY BEST LOG LIKELIHOOD = -
17392.7264
76 Iterations / time elapsed = 0h:0m:19s (will finish in 0h:0m:31s)
120 Iterations / time elapsed = 0h:0m:30s (will finish in 0h:0m:20s)
179 Iterations / time elapsed = 0h:0m:41s (will finish in 0h:0m:4s)
200 Iterations / time elapsed = 0h:0m:45s
```

Looking at the output above highlights that IQPNNI could well escape the local optimum it was stuck for the first 28 iterations. Strategies only based on BioNJ and fastNNI like the original PHYML strategy, which is roughly equivalent to the reconstruction of the initial tree, would have finished on that local optimum.

Then IQPNNI optimizes the model parameters as described above and re-estimates the likelihood of the final tree:

```
Optimizing the final tree topology as well as branch lengths...
Final best log likelihood: -17392.72496
Constructing the majority rule consensus tree...
Estimating site rates by empirical Bayesian...
The results were written to following files:
    1. hivALN.phy.iqpnni
    2. hivALN.phy.iqpnni.treefile
    3. hivALN.phy.iqpnni.treels
    4. hivALN.phy.iqpnni.rate
Total Runtime: 0h:0m:46s
Finished!!!
```

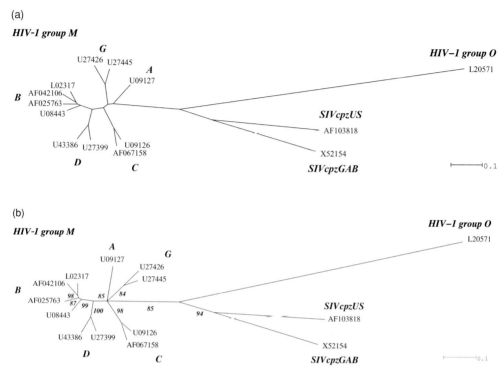

Fig. 6.8   ML-tree reconstructed with IQPNNI (a) and the quartet puzzling consensus tree (b) for the HIV/SIV data set. The major groups of HIV-1 and the Group M subtypes are indicated in bold, the puzzle support values are given at the branches.

The outfile of IQPNNI, a text file here called hivALN.phy.iqpnni, summarizes the results of the phylogenetic analyses. hivALN.phy.iqpnni.treefile contains the ML tree in NEWICK format. The tree can be displayed using the TREE-VIEW and FIGTREE program. The reconstructed tree is shown in Fig. 6.8a.

Valuable information can be obtained from the stopping rule even if the analysis is not finished by the stopping rule: If more than three times a better tree has been found during the iterations, IQPNNI estimates how many additional iterations are necessary to be sure according to a 5% confidence level that the current search will not produce a tree with a better likelihood. If the rule suggests more iterations than executed, one is advised to re-run the analysis. IQPNNI also offers the option to continue the current analysis (see the IQPNNI manual for details).

Repeating the example will produce slightly different output, that means, the best tree might be found in different iterations. This is due to the stochastic component in the IQPNNI procedure (Section 6.4.5) when randomly removing and then re-inserting taxa into the tree.

6.9.2 Getting a tree with branch support values using quartet puzzling

Most tree reconstruction methods output only a tree topology but do not provide information about the significance of the branching pattern.

Here, we will analyze the above data with TREE-PUZZLE to get support values for the IQPNNI tree.

To do this, place the `hivALN.phy` file (available at *http://www. thephylogenetichandbook.org*) in the same folder as the TREE-PUZZLE binary and run the executable. The following text appears:

```
WELCOME TO TREE-PUZZLE 5.3.


Please enter a file name for the sequence data:
```

Type the filename `hivALN.phy` and press enter.

```
Input data set (hivALN.phy) contains 14 sequences of length 2352
     1. L20571
     2. AF103818
     3. X52154
     4. U09127
     5. U27426
     6. U27445
     7. AF067158
     8. U09126
     9. U27399
    10. U43386
    11. L02317
    12. AF025763
    13. U08443
    14. AF042106
(consists very likely of nucleotides)


GENERAL OPTIONS
   b                    Type of analysis? Tree reconstruction
   k                Tree search procedure? Quartet puzzling
   v         Quartet evaluation criterion? Approximate maximum likeli-
hood (ML)
   u             List unresolved quartets? No
   n             Number of puzzling steps? 1000
   j             List puzzling step trees? No
   9   List puzzling trees/splits (NEXUS)? No
   o                  Display as outgroup? L20571 (1)
   z      Compute clocklike branch lengths? No
   e                  Parameter estimates? Approximate (faster)
   x              Parameter estimation uses? Neighbor-joining tree
```

```
SUBSTITUTION PROCESS
d            Type of sequence input data? Auto: Nucleotides
h             Codon positions selected? Use all positions
m              Model of substitution? HKY (Hasegawa et al. 1985)
t    Transition/transversion parameter? Estimate from data set
f              Nucleotide frequencies? Estimate from data set
RATE HETEROGENEITY
w         Model of rate heterogeneity? Uniform rate

Quit [q], confirm [y], or change [menu] settings:
```

Each option can be selected and changed by typing the corresponding
letter. For example, if the user types b repeatedly, then the option will
change from Tree reconstruction to Likelihood mapping to Tree
reconstruction again. The letter k cycles from Quartet puzzling to
Evaluate user defined trees to Consensus of user defined
trees to Pairwise distances only (no tree). A typical run to infer
a tree based on DNA sequences starts with the following setting:

```
GENERAL OPTIONS
b                     Type of analysis? Tree reconstruction
k              Tree search procedure? Quartet puzzling
v         Quartet evaluation criterion? Approximate maximum likeli-
hood (ML)
u             List unresolved quartets? No
n             Number of puzzling steps? 10000
j             List puzzling step trees? No
9    List puzzling trees/splits (NEXUS)? No
o                  Display as outgroup? L20571 (1)
z     Compute clocklike branch lengths? No
e                  Parameter estimates? Approximate (faster)
x           Parameter estimation uses? Neighbor-joining tree
SUBSTITUTION PROCESS
d            Type of sequence input data? Auto: Nucleotides
h             Codon positions selected? Use all positions
m              Model of substitution? HKY (Hasegawa et al. 1985)
t    Transition/transversion parameter? Estimate from data set
f              Nucleotide frequencies? Estimate from data set
RATE HETEROGENEITY
w         Model of rate heterogeneity? Gamma distributed rates
a   Gamma distribution parameter alpha? Estimate from data set
c     Number of Gamma rate categories? 8

Quit [q], confirm [y], or change [menu] settings:
```

By entering y, TREE-PUZZLE computes a quartet puzzling tree based on 10 000 intermediate trees, using approximate likelihoods to estimate the quartet trees. The model parameter estimates are also approximated and are based on an NJ tree computed at the beginning of the optimization routine. In the example, the HKY model is selected (see Sections 4.6 and 4.9). Again, Γ-distributed rate heterogeneity among sites is modeled, where the shape parameter α is estimated with the aid of eight discrete categories (see Section 4.6.1). If these settings are confirmed (type "y"), the following output will appear on the screen:

```
Optimizing missing substitution process parameters
Optimizing missing rate heterogeneity parameters
Optimizing missing substitution process parameters
Optimizing missing rate heterogeneity parameters
Optimizing missing substitution process parameters
Optimizing missing rate heterogeneity parameters
Writing parameters to file hivALN.phy.puzzle
Writing pairwise distances to file hivALN.phy.dist
Computing quartet maximum likelihood trees
Computing quartet puzzling trees
Computing maximum likelihood branch lengths (without clock)

All results written to disk:
        Puzzle report file:       hivALN.phy.puzzle
        Likelihood distances:     hivALN.phy.dist
        Phylip tree file:         hivALN.phy.tree

The parameter estimation took 11.00 seconds (= 0.18 min-
utes = 0.00 hours)
The ML step took               8.00 seconds (= 0.13 minutes = 0.00 hours)
The puzzling step took         2.00 seconds (= 0.03 minutes = 0.00 hours)
The computation took 23.00 seconds (= 0.38 minutes = 0.01 hours)
    including input 272.00 seconds (= 4.53 minutes = 0.08 hours)
```

The puzzle report file `hivALN.phy.puzzle`, the most important file, summarizes all results of the phylogenetic analyses. Because the content of the report file is self-explanatory, it is not discussed here. `hivALN.phy.dist` contains the matrix of pairwise distances based on the model parameters. `hivALN.phy.tree` contains the quartet puzzling tree (Fig. 6.8b) in NEWICK notation that can be displayed with the TREEVIEW or FIGTREE program (see Chapter 5).

TREE-PUZZLE can output all the different tree topologies computed during puzzling steps (option j). In the HIV example, 685 different intermediate trees were found, in which the most frequent tree occurred at about 6.6%. Therefore, the quartet puzzling algorithm can be used to generate a collection of plausible candidate trees, and this collection can subsequently be employed to search for the most likely tree. It is known that the consensus tree does not necessarily coincide with the maximum-likelihood tree, especially when the consensus is not fully resolved (Cao *et al.*, 1998). Thus, to get the maximum-likelihood tree option j should be changed to unique topologies, which will be output in the .ptorder file. A typical line of this file looks like the following:

```
[1.  657 6.57 14 68510000](L20571,((AF10138,X52154),
 (U09127,(((U27426,U27445),  (U067158,U09126)),
 ((U27399,U43386),(((L02317,AF042106),AF025763),
 U08443))))));
```

The first column is a simple numbering scheme, in which each tree is numbered according to its frequency (i.e. second column and third column). Column four (14) gives the first time among 10 000 (column 6) puzzling steps, when the tree was found. Column five shows how many different trees were found (685).

If one enables option 9 before running the analysis, TREE-PUZZLE also outputs all splits and trees in NEXUS format to the file hivALN.phy.nex which can be analyzed with the SPLITSTREE software (see Chapter 21).

Similar to bootstrap analysis (Chapter 5), the resulting intermediate trees and, hence, the resulting support values might differ slightly due to the randomization of the insertion order in the puzzling step (Section 6.6). Consequently, the number of unique intermediate tree topologies and their percentages will also vary slightly. Please note, the intermediate tree found most often does not necessarily coincide with the maximum-likelihood tree, but often the ML tree is among the intermediate trees.

To compute the maximum-likelihood tree among all intermediate trees, run TREE-PUZZLE again with the following settings:

```
GENERAL OPTIONS
b                       Type of analysis? Tree reconstruction
k                 Tree search procedure? Evaluate user defined trees
z       Compute clocklike branch lengths? No
e                   Parameter estimates? Approximate (faster)
x             Parameter estimation uses? Neighbor-joining tree

SUBSTITUTION PROCESS
d        Type of sequence input data? Auto: Nucleotides
h           Codon positions selected? Use all positions
m              Model of substitution? HKY (Hasegawa et al., 1985)
t  Transition/transversion parameter? Estimate from data set
f              Nucleotide frequencies? Estimate from data set
```

```
RATE HETEROGENEITY
w        Model of rate heterogeneity? Gamma distributed rates
a Gamma distribution parameter alpha? Estimate from data set
c    Number of Gamma rate categories? 8

Quit [q], confirm [y], or change [menu] settings: y
```

Now TREE-PUZZLE will compute the maximum-likelihood values for all intermediate trees using the model parameter estimates from the iterative procedure.
The computation takes time, but it provides more insight about the data and likelihood surface of the trees. The resulting `.puzzle` file shows the results of the ML
analysis for the different topologies at the end of the file. In addition, the likelihoods
of the trees are compared by various tests (see Chapter 12 for details).

6.9.3 Likelihood-mapping analysis of the HIV data set

Option b in the GENERAL OPTIONS menu of TREE-PUZZLE switches from `Tree
reconstruction` to `Likelihood mapping`. When the number of quartets
for the data is below 10 000, the program computes the posterior probabilities of
the quartet trees (see Section 6.7) for all quartets; otherwise, only the posterior
probabilities for 10 000 random quartets are computed. In the latter case, the
user can decide how many quartets to evaluate by selecting from the GENERAL
OPTIONS menu option n, number of quartets, and typing the number. Typing
0 forces TREE-PUZZLE to analyze all the possible quartets; however, this is time-
consuming for large data sets. A random selection serves the same purpose.

Box 6.1 shows part of the outfile from the likelihood-mapping analysis of the
1001 possible quartets for the HIV data set. Some 92.72% of all quartets are tree-
like, i.e. they are located close to the corners of the triangle. Only 4.4% of all
quartets lie in the rectangles and 2.9% in the central triangle: they represent the
unresolved part of the data. Because most of the quartets are tree-like and only a
fraction of about 7% do not support a unique phylogeny, an overall phylogenetic
tree with a good resolution is expected. Nevertheless, the percentages of up to 4.4%
unresolved and 2.9% partly resolved quartets still implies a considerable amount
of noisy or conflicting signal in the data set.

Please note that, if the likelihood mapping is repeated on a random subset, the
resulting percentages may naturally differ slightly.

6.10 Conclusions

The HIV quartet puzzling consensus tree (Fig. 6.8b) calculated in the previous
section is well resolved for the most part and all resolved clusters coincide with

Box 6.1 Likelihood-mapping analysis with TREE-PUZZLE

TREE-PUZZLE writes results of the likelihood-mapping analysis at the end of the report file `hivALN.phy.puzzle` in the `LIKELIHOOD MAPPING STATISTICS` section. The results are given for the whole data set and also for each sequence to help with identifying outliers. For example, for the HIV data set (alignment: `hivALN.phy`; substitution model: HKY with Γ-distributed rate heterogeneity, parameters estimated via maximum likelihood):

```
LIKELIHOOD MAPPING STATISTICS
```

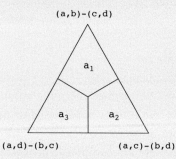
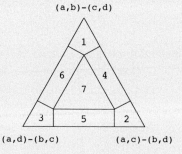

```
[...]

Quartet resolution per sequence:
```

	name	#quartets	resolved	partly	unresolved
1	L20571	286	261 (91.26)	12 (4.20)	13 (4.55)
2	AF103818	286	271 (94.76)	14 (4.90)	1 (0.35)
3	X52154	286	260 (90.91)	13 (4.55)	13 (4.55)
4	U09127	286	255 (89.16)	17 (5.94)	14 (4.90)
5	U27426	286	260 (90.91)	15 (5.24)	11 (3.85)
6	U27445	286	273 (95.45)	11 (3.85)	2 (0.70)
7	AF067158	286	261 (91.26)	18 (6.29)	7 (2.45)
8	U09126	286	263 (91.96)	13 (4.55)	10 (3.50)
9	U27399	286	270 (94.41)	9 (3.15)	7 (2.45)
10	U43386	286	264 (92.31)	8 (2.80)	14 (4.90)
11	L02317	286	268 (93.71)	12 (4.20)	6 (2.10)
12	AF025763	286	268 (93.71)	13 (4.55)	5 (1.75)
13	U08443	286	267 (93.36)	12 (4.20)	7 (2.45)
14	AF042106	286	271 (94.76)	9 (3.15)	6 (2.10)

```
                    1001  928 ( 92.71)  44 ( 4.40)  29 ( 2.90)


Overall quartet resolution:

Number of resolved        quartets (regions 1+2+3):  928 (= 92.71%)
Number of partly resolved quartets (regions 4+5+6):   44 (=  4.40%)
Number of unresolved      quartets (region 7):        29 (=  2.90%)
```

TREE-PUZZLE also outputs a drawing of the likelihood-mapping triangles in encapsulated Postscript format (`hivALN.phy.eps`) that can be printed or imported in a graphics application to be edited.

clusters in the IQPNNI ML tree (Fig. 6.8a). However, the quartet puzzling tree contains a **polytomy** (see Section 1.7 in Chapter 1) joining Subtypes A, C, G, and the B/D clade (Fig. 6.8). This is not surprising in light of the noise in the data revealed by likelihood mapping. Such a lack of resolution may suggest a *star-like radiation* at the origin of the HIV-1 Group M subtypes, but it could also imply the presence of *inter-subtype recombination* in HIV-1. The latter speculation from the first edition of this book (Strimmer & von Haeseler, 2003) was recently substantiated by the findings of (Abecasis *et al.*, 2007) that subtype G might be a recombinant instead of a pure subtype as previously thought.

Such issues are analyzed further in Chapters 15, 16, and 21, which discuss general methods for detecting and investigating recombination and conflicting phylogenetic signals in molecular data.

Bayesian phylogenetic analysis using MrBayes

THEORY

Fredrik Ronquist, Paul van der Mark, and John P. Huelsenbeck

7.1 Introduction

What is the probability that Sweden will win next year's world championships in ice hockey? If you're a hockey fan, you probably already have a good idea, but even if you couldn't care less about the game, a quick perusal of the world championship medalists for the last 15 years (Table 7.1) would allow you to make an educated guess. Clearly, Sweden is one of only a small number of teams that compete successfully for the medals. Let's assume that all seven medalists the last 15 years have the same chance of winning, and that the probability of an outsider winning is negligible. Then the odds of Sweden winning would be 1:7 or 0.14. We can also calculate the frequency of Swedish victories in the past. Two gold medals in 15 years would give us the number 2:15 or 0.13, very close to the previous estimate. The exact probability is difficult to determine but most people would probably agree that it is likely to be in the vicinity of these estimates.

You can use this information to make sensible decisions. If somebody offered you to bet on Sweden winning the world championships at the odds 1:10, for instance, you might not be interested because the return on the bet would be close to your estimate of the probability. However, if you were offered the odds 1:100, you might be tempted to go for it, wouldn't you?

As the available information changes, you are likely to change your assessment of the probabilities. Let's assume, for instance, that the Swedish team made it to

The Phylogenetic Handbook: a Practical Approach to Phylogenetic Analysis and Hypothesis Testing, Philippe Lemey, Marco Salemi, and Anne-Mieke Vandamme (eds.). Published by Cambridge University Press. © Cambridge University Press 2009.

Table 7.1   Medalists in the ice hockey world championships 1993–2007

Year	Gold	Silver	Bronze
1993	Russia	Sweden	Czech Republic
1994	Canada	Finland	Sweden
1995	Finland	Sweden	Canada
1996	Czech Republic	Canada	United States
1997	Canada	Sweden	Czech Republic
1998	Sweden	Finland	Czech Republic
1999	Czech Republic	Finland	Sweden
2000	Czech Republic	Slovakia	Finland
2001	Czech Republic	Finland	Sweden
2002	Slovakia	Russia	Sweden
2003	Canada	Sweden	Slovakia
2004	Canada	Sweden	United States
2005	Czech Republic	Canada	Russia
2006	Sweden	Czech Republic	Finland
2007	Canada	Finland	Russia

the finals. Now you would probably consider the chance of a Swedish victory to be much higher than your initial guess, perhaps close to 0.5. If Sweden lost in the semifinals, however, the chance of a Swedish victory would be gone; the probability would be 0.

This way of reasoning about probabilities and updating them as new information becomes available is intuitively appealing to most people and it is clearly related to rational behavior. It also happens to exemplify the Bayesian approach to science. Bayesian inference is just a mathematical formalization of a decision process that most of us use without reflecting on it; it is nothing more than a probability analysis. In that sense, Bayesian inference is much simpler than classical statistical methods, which rely on sampling theory, asymptotic behavior, statistical significance, and other esoteric concepts.

The first mathematical formulation of the Bayesian approach is attributed to Thomas Bayes (c. 1702–1761), a British mathematician and Presbyterian minister. He studied logic and theology at the University of Edinburgh; as a Non-Conformist, Oxford and Cambridge were closed to him. The only scientific work he published during his lifetime was a defense of Isaac Newton's calculus against a contemporaneous critic (*Introduction to the Doctrine of Fluxions*, published anonymously in 1736), which apparently got him elected as a Fellow of the Royal Society in 1742. However, it is his solution to a problem in so-called inverse probability that made him famous. It was published posthumously in 1764 by his friend Richard Price in the *Essay Towards Solving a Problem in the Doctrine of Chances*.

Assume we have an urn with a large number of balls, some of which are white and some of which are black. Given that we know the proportion of white balls, what is the probability of drawing, say, five white and five black balls in ten draws? This is a problem in forward probability. Thomas Bayes solved an example of the converse of such problems. Given a particular sample of white and black balls, what can we say about the proportion of white balls in the urn? This is the type of question we need to answer in Bayesian inference.

Let's assume that the proportion of white balls in the urn is p. The probability of drawing a white ball is then p and the probability of drawing a black ball is $1 - p$. The probability of obtaining, say, two white balls and one black ball in three draws would be

$$\Pr(2\text{white}, 1\text{black}|\,p) = p \times p \times (1 - p) \times \binom{3}{2} \tag{7.1}$$

The vertical bar indicates a condition; in this case we are interested in the probability of a particular outcome given (or conditional) on a particular value of p. It is easy to forget the last factor (3 choose 2), which is the number of ways in which we can obtain the given outcome. Two white balls and one black ball can be the result of drawing the black ball in the first, second or third draw. That is, there are three ways of obtaining the outcome of interest, 3 choose 2 (or 3 choose 1 if we focus on the choice of the black ball; the result is the same). Generally, the probability of obtaining a white balls and b black balls is determined by the function

$$f(a, b|\,p) = p^a (1 - p)^b \binom{a + b}{a} \tag{7.2}$$

which is the **probability mass function** (Box 7.1) of the so-called binomial distribution. This is the solution to the problem in forward probability, when we know the value of p. Bayesians often, somewhat inappropriately, refer to the forward probability function as the **likelihood function.**

But given that we have a sample of a white balls and b black balls, what is the probability of a particular value of p? This is the reverse probability problem, where we are trying to find the function $f(p|a, b)$ instead of the function $f(a, b|\,p)$. It turns out that it is impossible to derive this function without specifying our prior beliefs about the value of p. This is done in the form of a probability distribution on the possible values of p (Box 7.1), the **prior probability distribution** or just **prior** in everyday Bayesian jargon. If there is no previous information about the value of p, we might associate all possible values with the same probability, a so-called uniform probability distribution (Box 7.1).

Box 7.1 Probability distributions

A function describing the probability of a discrete random variable is called a ***probability mass function***. For instance, this is the probability mass function for throwing a dice, an example of a *discrete uniform distribution*:

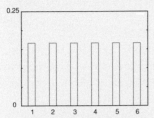

For a continuous variable, the equivalent function is a ***probability density function***. The value of this function is not a probability, so it can sometimes be larger than one. Probabilities are obtained by integrating the density function over a specified interval, giving the probability of obtaining a value in that interval. For instance, a *continuous uniform distribution* on the interval (0,2) has this probability density function:

Most prior probability distributions used in Bayesian phylogenetics are uniform, exponential, gamma, beta or Dirichlet distributions. Uniform distributions are often used to express the lack of prior information for parameters that have a uniform effect on the likelihood in the absence of data. For instance, the discrete uniform distribution is typically used for the topology parameter. In contrast, the likelihood is a negative exponential function of the branch lengths, and therefore the *exponential distribution* is a better choice for a *vague* prior on branch lengths. The exponential distribution has the density function $f(x) = \lambda e^{-\lambda x}$, where λ is known as the *rate* parameter. The expectation (mean) of the exponential distribution is $1/\lambda$.

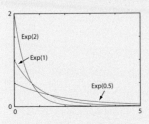

The *gamma* distribution has two parameters, the shape parameter α and the scale parameter β. At small values of α, the distribution is L-shaped and the variance is large;

Box 7.1 (*cont.*)

at high values it is similar to a normal distribution and the variance is low. If there is considerable uncertainty concerning the shape of the prior probability distribution, the gamma may be a good choice; an example is the rate variation across sites. In these cases, the value of α can be associated with a uniform or an exponential prior (also known as a *hyperprior* since it is a prior on a parameter of a prior), so that the MCMC procedure can explore different shapes of the gamma distribution and weight each according to its posterior probability. The sum of exponentially distributed variables is also a gamma distribution. Therefore, the gamma is an appropriate choice for the prior on the tree height of clock trees, which is the sum of several presumably exponentially distributed branch lengths.

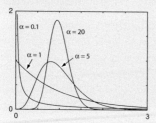

The *beta* and *Dirichlet* distributions are used for parameters describing proportions of a whole, so called simplex parameters. Examples include the stationary state frequencies that appear in the instantaneous rate matrix of the substitution model. The exchangeability or rate parameters of the substitution model can also be understood as proportions of the total exchange rate (given the stationary state frequencies). Another example is the proportion of invariable and variable sites in the invariable sites model. The beta distribution, denoted Beta(α_1, α_2), describes the probability on two proportions, which are associated with the weight parameters $\alpha_1 > 0$ and $\alpha_2 > 0$. The Dirichlet distribution is equivalent except that there are more than two proportions and associated weight parameters.

A Beta(1, 1) distribution, also known as a flat beta, is equivalent to a uniform distribution on the interval (0,1). When $\alpha_1 = \alpha_2 > 1$, the distribution is symmetric and emphasizes equal proportions, the more so the higher the weights. When $\alpha_1 = \alpha_2 < 1$, the distribution puts more probability on extreme proportions than on equal proportions. Finally, if the weights are different, the beta is skewed towards the proportion defined by the weights; the expectation of the beta is $\alpha/(\alpha + \beta)$ and the mode is $(\alpha - 1)/(\alpha + \beta - 2)$ for $\alpha > 1$ and $\beta > 1$.

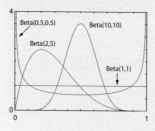

Box 7.1 (*cont.*)

Assume that we toss a coin to determine the probability p of obtaining heads. If we associate p and $1 - p$ with a flat beta prior, we can show that the posterior is a beta distribution where $\alpha_1 - 1$ is the number of heads and $\alpha_2 - 1$ is the number of tails. Thus, the weights roughly correspond to counts. If we started with a flat Dirichlet distribution and analyzed a set of DNA sequences with the composition 40 A, 50 C, 30 G, and 60 T, we might expect a posterior for the stationary state frequencies around Dirichlet(41, 51, 31, 61) if it were not for the other parameters in the model and the blurring effect resulting from looking back in time. Wikipedia (http://www.wikipedia.org) is an excellent source for additional information on common statistical distributions.

Thomas Bayes realized that the probability of a particular value of p, given some sample (a, b) of white and black balls, can be obtained using the probability function

$$f(p|a, b) = \frac{f(p)\,f(a, b|p)}{f(a, b)} \tag{7.3}$$

This is known as Bayes' theorem or Bayes' rule. The function $f(p|a, b)$ is called the ***posterior probability distribution***, or simply the ***posterior***, because it specifies the probability of all values of p after the prior has been updated with the available data.

We saw above how we can calculate $f(a, b|p)$, and how we can specify $f(p)$. How do we calculate the probability $f(a, b)$? This is the unconditional probability of obtaining the outcome (a, b) so it must take all possible values of p into account. The solution is to integrate over all possible values of p, weighting each value according to its prior probability:

$$f(a, b) = \int_0^1 f(p)\,f(a, b|p)\,\mathrm{d}p \tag{7.4}$$

We can now see that the denominator is a normalizing constant. It simply ensures that the posterior probability distribution integrates to 1, the basic requirement of a proper probability distribution.

A Bayesian problem that occupied several early workers was an analog to the following. Given a particular sample of balls, what is the probability that p is larger than a specified value? To solve it analytically, they needed to deal with complex integrals. Bayes made some progress in his *Essay*; more important contributions were made later by Laplace, who, among other things, used Bayesian reasoning and novel integration methods to show beyond any reasonable doubt that the probability of a newborn being a boy is higher than 0.5. However, the analytical complexity of most Bayesian problems remained a serious problem for a long time and it is only in the last few decades that the approach has become popular due to the combination of efficient numerical methods and the widespread availability of fast computers.

7.2 Bayesian phylogenetic inference

How does Bayesian reasoning apply to phylogenetic inference? Assume we are interested in the relationships between man, gorilla, and chimpanzee. In the standard case, we need an additional species to root the tree, and the orangutan would be appropriate here. There are three possible ways of arranging these species in a phylogenetic tree: the chimpanzee is our closest relative, the gorilla is our closest relative, or the chimpanzee and the gorilla are each other's closest relatives (Fig. 7.1).

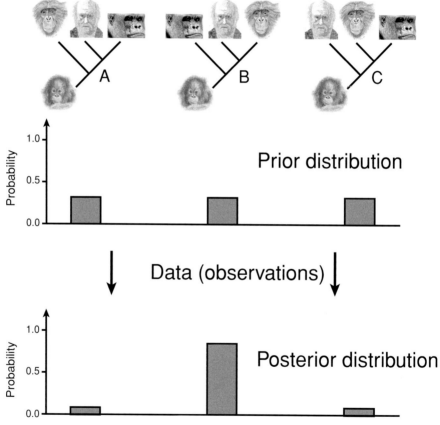

Fig. 7.1   A Bayesian phylogenetic analysis. We start the analysis by specifying our prior beliefs about the tree. In the absence of background knowledge, we might associate the same probability to each tree topology. We then collect data and use a stochastic evolutionary model and Bayes' theorem to update the prior to a posterior probability distribution. If the data are informative, most of the posterior probability will be focused on one tree (or a small subset of trees in a large tree space).

Before the analysis starts, we need to specify our prior beliefs about the relationships. In the absence of background data, a simple solution would be to assign equal probability to the possible trees. Since there are three trees, the probability of each would be one-third. Such a prior probability distribution is known as a *vague* or *uninformative prior* because it is appropriate for the situation when we do not have any prior knowledge or do not want to build our analysis on any previous results.

To update the prior we need some data, typically in the form of a molecular sequence alignment, and a stochastic model of the process generating the data on the tree. In principle, Bayes' rule is then used to obtain the posterior probability distribution (Fig. 7.1), which is the result of the analysis. The posterior specifies the probability of each tree given the model, the prior, and the data. When the data are informative, most of the posterior probability is typically concentrated on one tree (or a small subset of trees in a large tree space).

If the analysis is performed correctly, there is nothing controversial about the posterior probabilities. Nevertheless, the interpretation of them is often subject to considerable discussion, particularly in the light of alternative models and priors.

To describe the analysis mathematically, designate the matrix of aligned sequences X. The vector of model parameters is contained in θ (we do not distinguish in our notation between vector parameters and scalar parameters). In the ideal case, this vector would only include a topology parameter τ, which could take on the three possible values discussed above. However, this is not sufficient to calculate the probability of the data. Minimally, we also need branch lengths on the tree; collect these in the vector v. Typically, there are also some **substitution model** parameters to be considered but, for now, let us use the Jukes Cantor substitution model (see below), which does not have any free parameters. Thus, in our case, $\theta = (\tau, v)$.

Bayes' theorem allows us to derive the posterior distribution as

$$f(\theta|X) = \frac{f(\theta)\,f(X|\theta)}{f(X)} \tag{7.5}$$

The denominator is an integral over the parameter values, which evaluates to a summation over discrete topologies and a multidimensional integration over possible branch length values:

$$f(X) = \int f(\theta)\,f(X|\theta)\,\mathrm{d}\theta \tag{7.6}$$

$$= \sum_{\tau} \int_{v} f(v)\,f(X|\tau, v)\,\mathrm{d}v \tag{7.7}$$

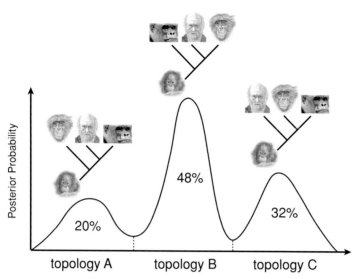

Fig. 7.2   Posterior probability distribution for our phylogenetic analysis. The x-axis is an imaginary
one-dimensional representation of the parameter space. It falls into three different regions
corresponding to the three different topologies. Within each region, a point along the axis
corresponds to a particular set of branch lengths on that topology. It is difficult to arrange
the space such that optimal branch length combinations for different topologies are close
to each other. Therefore, the posterior distribution is multimodal. The area under the curve
falling in each tree topology region is the posterior probability of that tree topology.

Even though our model is as simple as phylogenetic models come, it is impossible
to portray its parameter space accurately in one dimension. However, imagine for a
while that we could do just that. Then the parameter axis might have three distinct
regions corresponding to the three different tree topologies (Fig. 7.2). Within each
region, the different points on the axis would represent different branch length
values. The one-dimensional parameter axis allows us to obtain a picture of the
posterior probability function or surface. It would presumably have three distinct
peaks, each corresponding to an optimal combination of topology and branch
lengths.

To calculate the posterior probability of the topologies, we integrate out the
model parameters that are not of interest, the branch lengths in our case. This
corresponds to determining the area under the curve in each of the three topology
regions. A Bayesian would say that we are marginalizing or deriving the *marginal
probability distribution* on topologies.

Why is it called marginalizing? Imagine that we represent the parameter space
in a two-dimensional table instead of along a single axis (Fig. 7.3). The columns in
this table might represent different topologies and the rows different branch length
values. Since the branch lengths are continuous parameters, there would actually

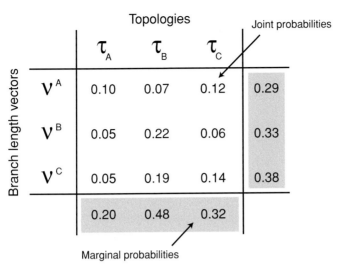

Fig. 7.3   A two-dimensional table representation of parameter space. The columns represent different tree topologies, the rows represent different branch length bins. Each cell in the table represents the joint probability of a particular combination of branch lengths and topology. If we summarize the probabilities along the margins of the table, we get the marginal probabilities for the topologies (bottom row) and for the branch length bins (last column).

be an infinite number of rows, but imagine that we sorted the possible branch length values into discrete bins, so that we get a finite number of rows. For instance, if we considered only short and long branches, one bin would have all branches long, another would have the terminal branches long and the interior branch short, etc.

Now, assume that we can derive the posterior probability that falls in each of the cells in the table. These are *joint probabilities* because they represent the joint probability of a particular topology and a particular set of branch lengths. If we summarized all joint probabilities along one axis of the table, we would obtain the marginal probabilities for the corresponding parameter. To obtain the marginal probabilities for the topologies, for instance, we would summarize the entries in each column. It is traditional to write the sums in the margin of the table, hence the term marginal probability (Fig. 7.3).

It would also be possible to summarize the probabilities in each row of the table. This would give us the marginal probabilities for the branch length combinations (Fig. 7.3). Typically, this distribution is of no particular interest but the possibility of calculating it illustrates an important property of Bayesian inference: there is no sharp distinction between different types of model parameters. Once the posterior probability distribution is obtained, we can derive any marginal distribution of interest. There is no need to decide on the parameters of interest before performing the analysis.

7.3 Markov chain Monte Carlo sampling

In most cases, including virtually all phylogenetic problems, it is impossible to derive the posterior probability distribution analytically. Even worse, we can't even estimate it by drawing random samples from it. The reason is that most of the posterior probability is likely to be concentrated in a small part of a vast parameter space. Even with a massive sampling effort, it is highly unlikely that we would obtain enough samples from the interesting region(s) of the posterior. This argument is particularly easy to appreciate in the phylogenetic context because of the large number of tree topologies that are possible even for small numbers of taxa. Already at nine taxa, you are more likely to be hit by lightning (odds 3:100 000) than to find the best tree by picking one randomly (odds 1:135, 135). At slightly more than 50 taxa, the number of topologies outnumber the number of atoms in the known universe – and this is still considered a small phylogenetic problem.

The solution is to estimate the posterior probability distribution using **Markov chain Monte Carlo sampling**, or **MCMC** for short. **Markov chains** have the property that they converge towards an equilibrium state regardless of starting point. We just need to set up a Markov chain that converges onto our posterior probability distribution, which turns out to be surprisingly easy. It can be achieved using several different methods, the most flexible of which is known as the *Metropolis algorithm*, originally described by a group of famous physicists involved in the Manhattan project (Metropolis *et al.*, 1953). Hastings (1970) later introduced a simple but important extension, and the sampler is often referred to as the **Metropolis–Hastings** method.

The central idea is to make small random changes to some current parameter values, and then accept or reject those changes according to the appropriate probabilities. We start the chain at an arbitrary point θ in the landscape (Fig. 7.4). In the next generation of the chain, we consider a new point θ^* drawn from a proposal distribution $f(\theta^*|\theta)$. We then calculate the ratio of the posterior probabilities at the two points. There are two possibilities. Either the new point is uphill, in which case we always accept it as the starting point for the next cycle in the chain, or it is downhill, in which case we accept it with a probability that is proportional to the height ratio. In reality, it is slightly more complicated because we need to take asymmetries in the proposal distribution into account as well. Formally, we accept

Markov chain Monte Carlo steps

1. Start at an arbitrary point (θ)
2. Make a small random move (to θ^*)
3. Calculate height ratio (r) of new state (to θ^*) to old state (θ)
   - (a) $r > 1$: new state accepted
   - (b) $r < 1$: new state accepted with probability r
       if new state rejected, stay in old state
4. Go to step 2

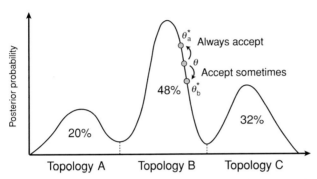

Fig. 7.4   The Markov chain Monte Carlo (MCMC) procedure is used to generate a valid sample from the posterior. One first sets up a Markov chain that has the posterior as its stationary distribution. The chain is then started at a random point and run until it converges onto this distribution. In each step (generation) of the chain, a small change is made to the current values of the model parameters (step 2). The ratio r of the posterior probability of the new and current states is then calculated. If $r > 1$, we are moving uphill and the move is always accepted (3a). If $r < 1$, we are moving downhill and accept the new state with probability r (3b).

or reject the proposed value with the probability

$$r = \min\left(1, \frac{f(\theta^*|X)}{f(\theta|X)} \times \frac{f(\theta|\theta^*)}{f(\theta^*|\theta)}\right) \tag{7.8}$$

$$= \min\left(1, \frac{f(\theta^*)f(X|\theta^*)/f(X)}{f(\theta)f(X|\theta)/f(X)} \times \frac{f(\theta|\theta^*)}{f(\theta^*|\theta)}\right) \tag{7.9}$$

$$= \min\left(1, \frac{f(\theta^*)}{f(\theta)} \times \frac{f(X|\theta^*)}{f(X|\theta)} \times \frac{f(\theta|\theta^*)}{f(\theta^*|\theta)}\right) \tag{7.10}$$

The three ratios in the last equation are referred to as the *prior ratio*, the *likelihood ratio*, and the *proposal ratio* (or *Hastings ratio*), respectively. The first two ratios correspond to the ratio of the numerators in Bayes' theorem; note that the complex integral in the denominator of Bayes' theorem, $f(X)$, cancels out in the second step because it is the same for both the current and the proposed states. Because of this, r is easy to compute.

The Metropolis sampler works because the relative equilibrium frequencies of the two states θ and θ^* is determined by the ratio of the rates at which the chain moves back and forth between them. Equation (7.10) ensures that this ratio is the same as the ratio of their posterior probabilities. This means that, if the Markov chain is allowed to run for a sufficient number of generations, the amount of time it spends sampling a particular parameter value or parameter interval is proportional to the posterior probability of that value or interval. For instance, if the posterior probability of a topology is 0.68, then the chain should spend 68% of its time sampling that topology at **stationarity**. Similarly, if the posterior probability of a branch length being in the interval $(0.02, 0.04)$ is 0.11, then 11% of the chain samples at stationarity should be in that interval.

For a large and parameter-rich model, a mixture of different Metropolis samplers is typically used. Each sampler targets one parameter or a set of related parameters (Box 7.2). One can either cycle through the samplers systematically or choose among them randomly according to some proposal probabilities (MRBAYES does the latter).

Box 7.2 Proposal mechanisms

Four types of proposal mechanisms are commonly used to change continuous variables. The simplest is the *sliding window* proposal. A continuous uniform distribution of width w is centered on the current value x, and the new value x^* is drawn from this distribution. The "window" width w is a tuning parameter. A larger value of w results in more radical proposals and lower acceptance rates, while a smaller value leads to more modest changes and higher acceptance rates.

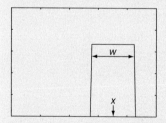

The *normal* proposal is similar to the sliding window except that it uses a normal distribution centered on the current value x. The variance σ^2 of the normal distribution determines how drastic the new proposals are and how often they will be accepted.

Box 7.2 (*cont.*)

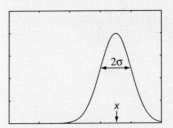

Both the sliding window and normal proposals can be problematic when the effect on the likelihood varies over the parameter range. For instance, changing a branch length from 0.01 to 0.03 is likely to have a dramatic effect on the posterior but changing it from 0.51 to 0.53 will hardly be noticeable. In such situations, the *multiplier* proposal is appropriate. It is equivalent to a sliding window with width λ on the log scale of the parameter. A random number u is drawn from a uniform distribution on the interval $(-0.5, 0.5)$ and the proposed value is $x^* = mx$, where $m = e^{\lambda u}$. If the value of λ takes the form $2 \ln a$, one will pick multipliers m in the interval $(1/a, a)$.

The *beta* and *Dirichlet* proposals are used for simplex parameters. They pick new values from a beta or Dirichlet distribution centered on the current values of the simplex. Assume that the current values are (x_1, x_2). We then multiply them with a value α, which is a tuning parameter, and pick new values from the distribution Beta($\alpha x_1, \alpha x_2$). The higher the value of α, the closer the proposed values will be to the current values.

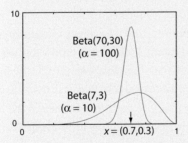

More complex moves are needed to change topology. A common type uses stochastic branch rearrangements (see Chapter 8). For instance, the extending **subtree pruning and regrafting** (extending SPR) move chooses a subtree at random and then moves its attachment point, one branch at a time, until a random number u drawn from a uniform on $(0, 1)$ becomes higher than a specified extension probability p. The extension probability p is a tuning parameter; the higher the value, the more drastic rearrangements will be proposed.

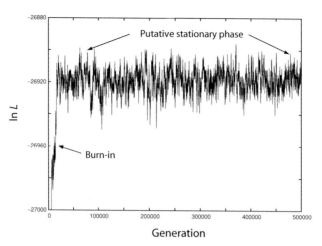

Fig. 7.5   The likelihood values typically increase very rapidly during the initial phase of the run because the starting point is far away from the regions in parameter space with high posterior probability. This initial phase of the Markov chain is known as the burn in. The burn-in samples are typically discarded because they are so heavily influenced by the starting point. As the chain converges onto the target distribution, the likelihood values tend to reach a plateau. This phase of the chain is sampled with some thinning, primarily to save disk space.

7.4 Burn-in, mixing and convergence

If the chain is started from a random tree and arbitrarily chosen branch lengths, chances are that the initial likelihood is low. As the chain moves towards the regions in the posterior with high probability mass, the likelihood typically increases very rapidly; in fact, it almost always changes so rapidly that it is necessary to measure it on a log scale (Fig. 7.5). This early phase of the run is known as the ***burn-in***, and the burn-in samples are often discarded because they are so heavily influenced by the starting point.

As the chain approaches its stationary distribution, the likelihood values tend to reach a plateau. This is the first sign that the chain may have converged onto the target distribution. Therefore, the plot of the likelihood values against the generation of the chain, known as the ***trace plot*** (Fig. 7.5), is important in monitoring the performance of an MCMC run. However, it is extremely important to confirm convergence using other diagnostic tools because it is not sufficient for the chain to reach the region of high probability in the posterior, it must also cover this region adequately. The speed with which the chain covers the interesting regions of the posterior is known as its ***mixing behavior***. The better the mixing, the faster the chain will generate an adequate sample of the posterior.

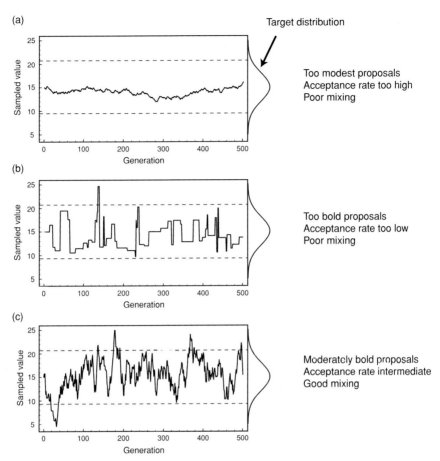

Fig. 7.6   The time it takes for a Markov chain to obtain an adequate sample of the posterior depends critically on its mixing behavior, which can be controlled to some extent by the proposal tuning parameters. If the proposed values are very close to the current ones, all proposed changes are accepted but it takes a long time for the chain to cover the posterior; mixing is poor. If the proposed values tend to be dramatically different from the current ones, most proposals are rejected and the chain will remain on the same value for a long time, again leading to poor mixing. The best mixing is obtained at intermediate values of the tuning parameters, associated with moderate acceptance rates.

The mixing behavior of a Metropolis sampler can be adjusted using its tuning parameter(s). Assume, for instance, that we are sampling from a normal distribution using a sliding window proposal (Fig. 7.6). The sliding window proposal has one tuning parameter, the width of the window. If the width is too small, then the proposed value will be very similar to the current one (Fig. 7.6a). The posterior probabilities will also be very similar, so the proposal will tend to be accepted. But each proposal will only move the chain a tiny distance in parameter space, so it will take the chain a long time to cover the entire region of interest; mixing is poor.

A window that is too wide also results in poor mixing. Under these conditions, the proposed state is almost always very different from the current state. If we have reached a region of high posterior probability density, then the proposed state is also likely to have much lower probability than the current state. The new state will therefore often be rejected, and the chain remains in the same spot for a long time (Fig. 7.6b), resulting in poor mixing. The most efficient sampling of the target distribution is obtained at intermediate acceptance rates, associated with intermediate values of the tuning parameter (Fig. 7.6c).

Extreme acceptance rates thus indicate that sampling efficiency can be improved by adjusting proposal tuning parameters. Studies of several types of complex but unimodal posterior distributions indicate that the optimal acceptance rate is 0.44 for one-dimensional and 0.23 for multi-dimensional proposals (Roberts *et al.*, 1997; Roberts & Rosenthal, 1998, 2001). However, multimodal posteriors are likely to have even lower optimal acceptance rates. Adjusting the tuning parameter values to reach a target acceptance rate can be done manually or automatically using adaptive tuning methods (Roberts & Rosenthal, 2006). Bear in mind, however, that some samplers used in Bayesian MCMC phylogenetics have acceptance rates that will remain low, no matter how much you tweak the tuning parameters. In particular, this is true for many tree topology update mechanisms.

Convergence diagnostics help determine the quality of a sample from the posterior. There are essentially three different types of diagnostics that are currently in use: (1) examining autocorrelation times, **effective sample sizes**, and other measures of the behavior of single chains; (2) comparing samples from successive time segments of a single chain; and (3) comparing samples from different runs. The last approach is arguably the most powerful way of detecting convergence problems. The drawback is that it wastes computational power by generating several independent sets of burn-in samples that must be discarded.

In Bayesian MCMC sampling of phylogenetic problems, the tree topology is typically the most difficult parameter to sample from. Therefore, it makes sense to focus our attention on this parameter when monitoring convergence. If we start several parallel MCMC runs from different, randomly chosen trees, they will initially sample from very different regions of tree space. As they approach stationarity, however, the tree samples will become more and more similar. Thus, an intuitively appealing convergence diagnostic is to compare the variance among and within tree samples from different runs.

Perhaps the most obvious way of achieving this is to compare the frequencies of the sampled trees. However, this is not practical unless most of the posterior probability falls on a small number of trees. In large phylogenetic problems, there is often an inordinate number of trees with similar probabilities and it may be extremely difficult to estimate the probability of each accurately.

The approach that we and others have taken to solve this problem is to focus on *split* (clade) *frequencies* instead. A split is a partition of the tips of the tree into two non-overlapping sets; each branch in a tree corresponds to exactly one such split. For instance, the split ((human, chimp),(gorilla, orangutan)) corresponds to the branch uniting the human and the chimp in a tree rooted on the orangutan. Typically, a fair number of splits are present in high frequency among the sampled trees. In a way, the dominant splits (present in, say, more than 10% of the trees) represent an efficient diagnostic summary of the tree sample as a whole. If two tree samples are similar, the split frequencies should be similar as well. To arrive at an overall measure of the similarity of two or more tree samples, we simply calculate the average standard deviation of the split frequencies. As the tree samples become more similar, this value should approach zero.

Most other parameters in phylogenetic models are continuous scalar parameters. An appropriate convergence diagnostic for these is the ***Potential Scale Reduction Factor (PSRF)*** originally proposed by Gelman and Rubin (1992). The PSRF compares the variance among runs with the variance within runs. If chains are started from over-dispersed starting points, the variance among runs will initially be higher than the variance within runs. As the chains converge, however, the variances will become more similar and the PSRF will approach 1.0.

7.5 Metropolis coupling

For some phylogenetic problems, it may be difficult or impossible to achieve convergence within a reasonable number of generations using the standard approach. Often, this seems to be due to the existence of isolated peaks in tree space (also known as tree islands) with deep valleys in-between. In these situations, individual chains may get stuck on different peaks and have difficulties moving to other peaks of similar probability mass. As a consequence, tree samples from independent runs tend to be different. A topology convergence diagnostic, such as the standard deviation of split frequencies, will indicate that there is a problem. But are there methods that can help us circumvent it?

A general technique that can improve mixing, and hence convergence, in these cases is ***Metropolis Coupling***, also known as MCMCMC or $(MC)^3$ (Geyer, 1991). The idea is to introduce a series of Markov chains that sample from a *heated* posterior probability distribution (Fig. 7.7). The heating is achieved by raising the posterior probability to a power smaller than 1. The effect is to flatten out the posterior probability surface, very much like melting a landscape of wax.

Because the surface is flattened, a Markov chain will move more readily between the peaks. Of course, the heated chains have a target distribution that is different from the one we are interested in, sampled by the ***cold chain***, but we can use them

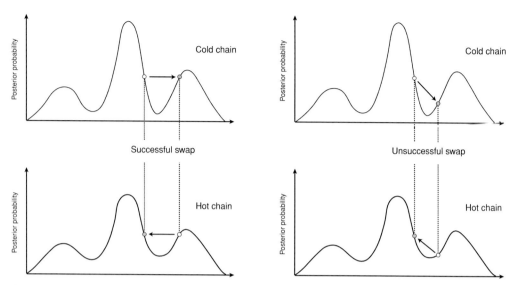

Fig. 7.7   Metropolis Coupling uses one or more *heated* chains to accelerate mixing in the so-called *cold* chain sampling from the posterior distribution. The heated chains are flattened out versions of the posterior, obtained by raising the posterior probability to a power smaller than one. The heated chains can move more readily between peaks in the landscape because the valleys between peaks are shallower. At regular intervals, one attempts to swap the states between chains. If a swap is accepted, the cold chain can jump between isolated peaks in the posterior in a single step, accelerating its mixing over complex posterior distributions.

to generate proposals for the cold chain. With regular intervals, we attempt to swap the states between two randomly picked chains. If the cold chain is one of them, and the swap is accepted, the cold chain can jump considerable distances in parameter space in a single step. In the ideal case, the swap takes the cold chain from one tree island to another. At the end of the run, we simply discard all of the samples from the heated chains and keep only the samples from the cold chain.

In practice, an incremental heating scheme is often used where chain i has its posterior probability raised by the temperature factor

$$T = \frac{1}{1 + \lambda i} \tag{7.11}$$

where $i \in \{0, 1, \ldots, k\}$ for k heated chains, with $i = 0$ for the cold chain, and λ is the temperature factor. The higher the value of λ, the larger the temperature difference between adjacent chains in the incrementally heated sequence.

If we apply too much heat, then the chains moving in the heated landscapes will walk all over the place and are less likely to be on an interesting peak when we try to swap states with the cold chain. Most of the swaps will therefore be rejected and the heating does not accelerate mixing in the cold chain. On the other hand, if we do not heat enough, then the chains will be very similar, and the heated chain will not mix more rapidly than the cold chain. As with the proposal tuning parameters, an intermediate value of the heating parameter λ works best.

7.6 Summarizing the results

The stationary phase of the chain is typically sampled with some thinning, for instance every 50th or 100th generation. This is done primarily to save disk space, since an MCMC run can easily generate millions of samples. Once an adequate sample is obtained, it is usually trivial to compute an estimate of the marginal posterior distribution for the parameter(s) of interest. For instance, this can take the form of a frequency histogram of the sampled values. When it is difficult to visualize this distribution or when space does not permit it, various summary statistics are used instead.

Most phylogenetic model parameters are continuous variables and their estimated posterior distribution is summarized using statistics such as the mean, the median, and the variance. Bayesian statisticians typically also give the 95% **credibility interval**, which is obtained by simply removing the lowest 2.5% and the highest 2.5% of the sampled values. The credibility interval is somewhat similar to a confidence interval but the interpretation is different. A 95% credibility interval actually contains the true value with probability 0.95 (given the model, prior, and data) unlike the confidence interval, which has a more complex interpretation.

The posterior distribution on topologies and branch lengths is more difficult to summarize efficiently. If there are few topologies with high posterior probability, one can produce a list of the best topologies and their probabilities, or simply give the topology with the maximum posterior probability. However, most posteriors contain too many topologies with reasonably high probabilities, and one is forced to use other methods.

One way to illustrate the topological variance in the posterior is to list the topologies in order of decreasing probabilities and then calculate the cumulative probabilities so that we can give the estimated number of topologies in various **credible sets**. Assume, for instance, that the five best topologies have the estimated probabilities (0.35, 0.25, 0.20, 0.15, 0.03), giving the cumulative probabilities (0.35, 0.60, 0.80, 0.95, 0.98). Then the 50% credible set has two topologies in it, the 90% and the 95% credible sets both have four trees in them, etc. We simply pass down the list and count the number of topologies we need to include before the target probability is met or superseded. When these credible sets are large, however, it is difficult to estimate their sizes precisely.

The most common approach to summarizing topology posteriors is to give the frequencies of the most common splits, since there are much fewer splits than topologies. Furthermore, all splits occurring in at least 50% of the sampled trees are guaranteed to be compatible and can be visualized in the same tree, a *majority rule consensus tree*. However, although the split frequencies are convenient, they do have limitations. For instance, assume that the splits ((A,B),(C,D,E)) and ((A,B,C),(D,E)) were both encountered in 70% of the sampled trees. This could mean that 30% of the sampled trees contained neither split or, at the other extreme, that all sampled trees contained at least one of them. The split frequencies themselves only allow us to approximately reconstruct the underlying set of topologies.

The sampled branch lengths are even more difficult to summarize adequately. Perhaps the best way would be to display the distribution of sampled branch length values separately for each topology. However, if there are many sampled topologies, there may not be enough branch length samples for each. A reasonable approach, taken by MrBayes, is then to pool the branch length samples that correspond to the same split. These pooled branch lengths can also be displayed on the consensus tree. However, one should bear in mind that the pooled distributions may be multimodal since the sampled values in most cases come from different topologies, and a simple summary statistic like the mean might be misleading.

A special difficulty appears with branch lengths in clock trees. Clock trees are rooted trees in which branch lengths are proportional to time units (see Chapter 11). Even if computed from a sample of clock trees, a majority rule consensus tree with mean pooled branch lengths is not necessarily itself a clock tree. This problem is easily circumvented by instead using mean pooled node depths instead of branch lengths (for Bayesian inference of clock trees, see also Chapter 18).

7.7 An introduction to phylogenetic models

A phylogenetic model can be divided into two distinct parts: a tree model and a substitution model. The tree model we have discussed so far is the one most commonly used in phylogenetic inference today (sometimes referred to as the different-rates or *unrooted model*, see Chapter 11). Branch lengths are measured in amounts of expected evolutionary change per site, and we do not assume any correlation between branch lengths and time units. Under time-reversible substitution models, the likelihood is unaffected by the position of the root, that is, the tree is unrooted. For presentation purposes, unrooted trees are typically rooted between a specially designated reference sequence or group of reference sequences, the *outgroup*, and the rest of the sequences.

Alternatives to the standard tree model include the strict and *relaxed clock* tree models. Both of these are based on trees, whose branch lengths are strictly proportional to time. In strict clock models, the evolutionary rate is assumed to be constant so that the amount of evolutionary change on a branch is directly proportional to its time duration, whereas relaxed clock models include a model component that accommodates some variation in the rate of evolution across the tree. Various prior probability models can be attached to clock trees. Common examples include the uniform model, the birth-death process, and the coalescent process (for the latter two, see Chapter 18).

The substitution process is typically modeled using Markov chains of the same type used in MCMC sampling. For instance, they have the same tendency towards an equilibrium state. The different substitution models are most easily described in terms of their instantaneous rate matrices, or Q *matrices*. For instance, the general time-reversible model (GTR) is described by the rate matrix

$$
Q = \begin{bmatrix}
- & \pi_C r_{AC} & \pi_G r_{AG} & \pi_T r_{AT} \\
\pi_A r_{AC} & - & \pi_G r_{CG} & \pi_T r_{CT} \\
\pi_A r_{AG} & \pi_C r_{CG} & - & \pi_T r_{GT} \\
\pi_A r_{AT} & \pi_C r_{CT} & \pi_G r_{GT} & -
\end{bmatrix}
$$

Each row in this matrix gives the instantaneous rate of going from a particular state, and each column represents the rate of going to a particular state; the states are listed in alphabetical sequence A, C, G, T. For instance, the second entry in the first row represents the rate of going from A to C. Each rate is composed of two factors; for instance, the rate of going from A to C is a product of π_C and r_{AC}. The rates along the diagonal are commonly omitted since their expressions are slightly more complicated. However, they are easily calculated since the rates in each row always sum to zero. For instance, the instantaneous rate of going from A to A (first entry in the first row) is $-\pi_C r_{AC} - \pi_G r_{AG} - \pi_T r_{AT}$.

It turns out that, if we run this particular Markov chain for a long time, it will move towards an equilibrium, where the frequency of a state i is determined exactly by the factor π_i given that $\sum \pi_i = 1$. Thus, the first rate factor corresponds to the stationary state frequency of the receiving state. The second factor, r_{ij}, is a parameter that determines the intensity of the exchange between pairs of states, controlling for the stationary state frequencies. For instance, at equilibrium we will have π_A sites in state A and π_C sites in state C. The total instantaneous rate of going from A to C over the sequence is then π_A times the instantaneous rate of the transition from A to C, which is $\pi_C r_{AC}$, resulting in a total rate of A to C changes over the sequence of $\pi_A \pi_C r_{AC}$. This is the same as the total rate of the reverse changes over the sequence, which is $\pi_C \pi_A r_{AC}$. Thus, there is no net change of the state proportions, which is the definition of an equilibrium, and the factor r_{AC} determines how intense the exchange between A and C is compared with the exchange between other pairs of states.

Many of the commonly used substitution models are special cases or extensions of the GTR model. For instance, the Jukes Cantor model has all rates equal, and the Felsenstein 81 (F81) model has all exchangeability parameters (r_{ij}) equal. The *covarion* and *covariotide* models have an independent on–off switch for each site, leading to a composite instantaneous rate matrix including four smaller rate matrices: two matrices describing the switching process, one being a zero-rate matrix, and the last describing the normal substitution process in the on state.

In addition to modeling the substitution process at each site, phylogenetic models typically also accommodate rate variation across sites. The standard approach is to assume that rates vary according to a gamma distribution (Box 7.1) with mean 1. This results in a distribution with a single parameter, typically designated α, describing the shape of the rate variation (see Fig. 4.8 in Chapter 4). Small values of α correspond to large amounts of rate variation; as α approaches infinity, the model approaches rate constancy across sites. It is computationally expensive to let the MCMC chain integrate over a continuous gamma distribution of site rates, or to numerically integrate out the gamma distribution in each step of the chain. The standard solution is to integrate out the gamma using a discrete approximation with a small number of rate categories, typically four to eight, which is a reasonable compromise. An alternative is to use MCMC sampling over discrete rate categories.

Many other models of rate variation are also possible. A commonly considered model assumes that there is a proportion of invariable sites, which do not change at all over the course of evolution. This is often combined with an assumption of gamma-distributed rate variation in the variable sites.

It is beyond the scope of this chapter to give a more detailed discussion of phylogenetic models but we present an overview of the models implemented in MRBAYES 3.2, with the command options needed to invoke them (Fig. 7.8). The MRBAYES manual provides more details and references to the different models. A simulation-based presentation of Markov substitution models is given in (Huelsenbeck & Ronquist, 2005) and further details can be found in Chapter 4 and Chapter 10.

7.8 Bayesian model choice and model averaging

So far, our notation has implicitly assumed that Bayes' theorem is conditioned on a particular model. To make it explicit, we could write Bayes' theorem:

$$f(\theta|X, M) = \frac{f(\theta|M)\, f(X|\theta, M)}{f(X|M)} \tag{7.12}$$

It is now clear that the normalizing constant, $f(X|M)$, is the probability of the data given the chosen model after we have integrated out all parameters. This

(a)   *Models supported by MrBayes 3 (simplified)*

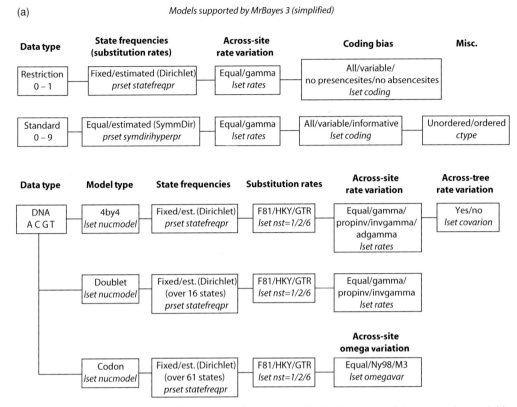

Fig. 7.8   Schematic overview of the models implemented in MrBayes 3. Each box gives the available settings in normal font and then the program commands and command options needed to invoke those settings in italics.

quantity, known as the *model likelihood*, is used for Bayesian model comparison. Assume we are choosing between two models, M_0 and M_1, and that we assign them the prior probabilities $f(M_0)$ and $f(M_1)$. We could then calculate the ratio of their posterior probabilities (the posterior odds) as

$$\frac{f(M_0|X)}{f(M_1|X)} = \frac{f(M_0) f(X|M_0)}{f(M_1) f(X|M_1)} = \frac{f(M_0)}{f(M_1)} \times \frac{f(X|M_0)}{f(X|M_1)} \tag{7.13}$$

Thus, the posterior odds is obtained as the prior odds, $f(M_0)/f(M_1)$, times a factor known as the ***Bayes factor***, $B_{01} = f(X|M_0)/f(X|M_1)$, which is the ratio of the model likelihoods. Rather than trying to specify the prior model odds, it is common to focus entirely on the Bayes factor. One way to understand the Bayes factor is that it determines how much the prior model odds are changed by the data when calculating the posterior odds. The Bayes factor is also the same as

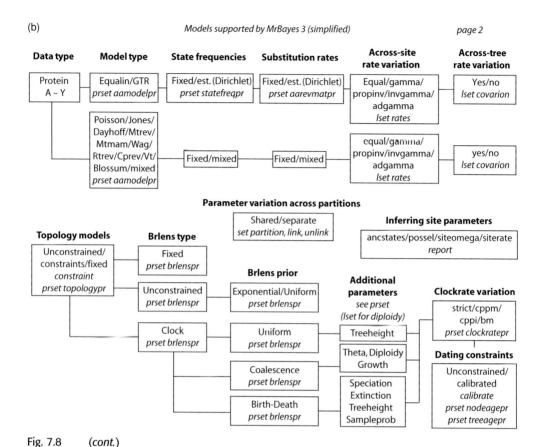

Fig. 7.8   *(cont.)*

the posterior odds when the prior odds are 1, that is, when we assign equal prior probabilities to the compared models.

Bayes factor comparisons are truly flexible. Unlike **likelihood ratio tests**, there is no requirement for the models to be nested. Furthermore, there is no need to correct for the number of parameters in the model, in contrast to comparisons based on the **Akaike Information Criterion** (Akaike, 1974) or the confusingly named **Bayesian Information Criterion** (Schwarz, 1978). Although it is true that a more parameter-rich model always has a higher **maximum likelihood** than a nested submodel, its model likelihood need not be higher. The reason is that a more parameter-rich model also has a larger parameter space and therefore a lower prior probability density. This can lead to a lower model likelihood unless it is compensated for by a sufficiently large increase in the likelihood values in the peak region.

The interpretation of a Bayes factor comparison is up to the investigator but some guidelines were suggested by Kass and Raftery (1995) (Table 7.2).

Table 7.2   Critical values for Bayes factor comparisons

$2 \ln B_{01}$	B_{01}	Evidence against M_1
0 to 2	1 to 3	Not worth more than a bare mention
2 to 6	3 to 20	Positive
6 to 10	20 to 150	Strong
>10	>150	Very strong

From Kass & Raftery (1995).

The easiest way of estimating the model likelihoods needed in the calculation of Bayes factors is to use the *harmonic mean* of the likelihood values from the stationary phase of an MCMC run (Newton & Raftery, 1994). Unfortunately, this estimator is unstable because it is occasionally influenced by samples with very small likelihood and therefore a large effect on the final result. A stable estimator can be obtained by mixing in a small proportion of samples from the prior (Newton & Raftery, 1994). Even better accuracy, at the expense of computational complexity, can be obtained by using thermodynamic integration methods (Lartillot & Philippe, 2006). Because of the instability of the harmonic mean estimator, it is good practice to compare several independent runs and only rely on this estimator when the runs give consistent results.

An alternative to running a full analysis on each model and then choosing among them using the estimated model likelihoods and Bayes' factors is to let a single Bayesian analysis explore the models in a predefined model space (using ***reversible-jump MCMC***). In this case, all parameter estimates will be based on an average across models, each model weighted according to its posterior probability. For instance, MRBAYES 3 uses this approach to explore a range of common fixed-rate matrices for amino acid data (see practice in Chapter 9 for an exercise).

Different topologies can also be considered different models and, in that sense, all Markov chains that integrate over the topology parameter also average across models. Thus, we can use the posterior sample of topologies from a single run to compare posterior probabilities of topology hypotheses.

For instance, assume that we want to test the hypothesis that group A is monophyletic against the hypothesis that it is not, and that 80% of the sampled trees have A monophyletic. Then the posterior model odds for A being monophyletic would be $0.80/0.20 = 4.0$. To obtain the Bayes factor, one would have to multiply this with the inverse of the prior model odds (see 7.13). If the prior assigned equal prior probability to all possible topologies, then the prior model odds would be determined by the number of trees consistent with each of the two hypotheses, a ratio that is easy to calculate. If one class of trees is empty, a conservative estimate of the Bayes factor would be obtained by adding one tree of this class to the sample.

7.9 Prior probability distributions

We will end with a few cautionary notes about priors. Beginners often worry excessively about the influence of the priors on their results and the subjectivity that this introduces into the inference procedure. In most cases however, the exact form of the priors (within rather generous bounds) has negligible influence on the posterior distribution. If this is a concern, it can always be confirmed by varying the prior assumptions.

The default priors used in MrBayes are designed to be vague or uninformative probability distributions on the model parameters. When the data contain little information about some parameters, one would therefore expect the corresponding posterior probability distributions to be diffuse. As long as we can sample adequately from these distributions, which can be a problem if there are many of them (Nylander *et al.*, 2004), the results for other parameters should not suffer. We also know from simulations that the Bayesian approach does well even when the model is moderately overparameterized (Huelsenbeck & Rannala, 2004). Thus, the Bayesian approach typically handles weak data quite well.

However, the parameter space of phylogenetic models is vast and occasionally there are large regions with inferior but not extremely low likelihoods that attract the chain when the data are weak. The characteristic symptom is that the sample from the posterior is concentrated on parameter values that the investigator considers unlikely or unreasonable, for instance in comparison with the maximum likelihood estimates. We have seen a few examples involving models of rate variation applied to very small numbers of variable sites. In these cases, one can either choose to analyze the data under a simpler model (probably the best option in most cases) or include background information into the priors to emphasize the likely regions of parameter space.

PRACTICE

Fredrik Ronquist, Paul van der Mark, and John P. Huelsenbeck

7.10 Introduction to MrBayes

The rest of this chapter is devoted to two tutorials that will get you started using MrBayes 3 (Huelsenbeck & Ronquist, 2001; Ronquist & Huelsenbeck, 2003). For more help using the program, visit its website at http://www.mrbayes.net. The program has a command–line interface and should run on a variety of computer platforms, including clusters of Macintosh and UNIX computers. Note that the computer should be reasonably fast and should have a lot of RAM memory (depending on the size of the data matrix, the program may require hundreds of megabytes of memory). The program is optimized for speed and not for minimizing memory requirements.

Throughout the tutorial text, we will use typewriter font for what you see on the screen and what is in the input file. What you should type is given in **bold font**.

7.10.1 Acquiring and installing the program

MrBayes 3 is distributed without charge by download from the MrBayes website (*http://mrbayes.net*). If someone has given you a copy of MrBayes 3, we strongly suggest that you download the most recent version from this site. The site also gives information about the MrBayes users' email list and describes how you can report bugs or contribute to the project.

MrBayes 3 is a plain-vanilla program that uses a command–line interface and therefore behaves virtually the same on all platforms – Macintosh, Windows, and Unix. There is a separate download package for each platform. The Macintosh and Windows versions are ready to use after unzipping. If you decide to run the program under Unix/Linux, or in the Unix environment on a Mac OS X computer, then you will need to compile the program from the source code first. The MrBayes website provides detailed instructions on how to do this.

In addition to the normal serial version, MrBayes 3 is also available in a parallel version that uses MPI to distribute chains across two or more available processors. You can use this version to run MrBayes on a computer cluster or on a single machine with several processors or processor cores available. See the MrBayes website for detailed instructions.

All three packages of MrBayes come with example data files. These are intended to show various types of analyses you can perform with the program, and you can use them as templates for your own analyses. Two of the files, primates.nex and cynmix.nex, will be used in the tutorials that follow.

7.10.2 Getting started

Start MRBAYES by double-clicking the application icon (or typing **./mb** in the Unix environment) and you will see the information below:

```
                     MrBayes 3.2.0

               (Bayesian Analysis of Phylogeny)


                            by


   John P. Huelsenbeck, Fredrik Ronquist, and Paul van der Mark


             Section of Ecology, Behavior and Evolution
                  Division of Biological Sciences
                University of California, San Diego
                      johnh@biomail.ucsd.edu


                  School of Computational Science
                      Florida State University
                       ronquist@scs.fsu.edu
                       paulvdm@scs.fsu.edu


           Distributed under the GNU General Public License


        Type ''help'' or ''help <command>'' for information
                 on the commands that are available.


MrBayes >
```

The order of the authors is randomized each time you start the program, so don't be surprised if the order differs from the one above. Note the MrBayes > prompt at the bottom, which tells you that MRBAYES is ready for your commands.

7.10.3 Changing the size of the MRBAYES window

Some MRBAYES commands will output a lot of information and write fairly long lines, so you may want to change the size of the MRBAYES window to make it easier to read the output. On Macintosh and Unix machines, you should be able to increase the window size simply by dragging the margins. On a Windows machine, you cannot increase the size of the window beyond the preset value by simply dragging the margins, but (on Windows XP, 2000 and NT) you can change both the size of the screen buffer and the console window by right-clicking on the blue title bar of the MRBAYES window and then selecting Properties in the menu that appears. Make sure the Layout tab is selected in the window that appears, and then set the "Screen Buffer Size" and "Window Size" to the desired values.

7.10.4 Getting help

At the MrBayes > prompt, type **help** to see a list of the commands available. Most commands allow you to set values (options) for different parameters. If you type **help** <**command**>, where <command> is any of the listed commands, you will see the help information for that command as well as a description of the available options. For most commands, you will also see a list of the current settings at the end. Try, for instance, **help lset** or **help mcmc**. The **lset** settings table looks like this:

```
Parameter     Options                                  Current Setting
---------------------------------------------------------------------
Nucmodel      4by4/Doublet/Codon                       4by4
Nst           1/2/6                                    1
Code          Universal/Vertmt/Mycoplasma/
              Yeast/Ciliates/Metmt                     Universal
Ploidy        Haploid/Diploid                          Diploid
Rates         Equal/Gamma/Propinv/Invgamma/Adgamma     Equal
Ngammacat     <number>                                 4
Usegibbs      Yes/No                                   No
Gibbsfreq     <number>                                 100
Nbetacat      <number>                                 5
Omegavar      Equal/Ny98/M3                            Equal
Covarion      No/Yes                                   No
Coding        All/Variable/Noabsencesites/
              Nopresencesites                          All
Parsmodel     No/Yes                                   No
---------------------------------------------------------------------
```

Note that MRBAYES 3 supports abbreviation of commands and options, so in many cases it is sufficient to type the first few letters of a command or option instead of the full name.

A complete list of commands and options is given in the command reference, which can be downloaded from the program website (*http://www.mrbayes.net*). You can also produce an ASCII text version of the command reference at any time by giving the command **manual** to MRBAYES. Further help is available in a set of hyperlinked html pages produced by Jeff Bates and available on the MRBAYES website. Finally, you can get in touch with other MRBAYES users and developers through the mrbayes-users' email list (subscription information on the MRBAYES website).

7.11 A simple analysis

This section is a tutorial based on the `primates.nex` data file. It will guide you through a basic Bayesian MCMC analysis of phylogeny, explaining the most important features of the program. There are two versions of the tutorial. You will first find a Quick-Start version for impatient users who want to get an analysis started immediately. The rest of the section contains a much more detailed description of the same analysis.

7.11.1 Quick start version

There are four steps to a typical Bayesian phylogenetic analysis using MRBAYES:

(i) Read the NEXUS data file.
(ii) Set the evolutionary model.
(iii) Run the analysis.
(iv) Summarize the samples.

In more detail, each of these steps is performed as described in the following paragraphs.

(1) At the `MrBayes >` prompt, type **execute primates.nex**. This will bring the data into the program. When you only give the data file name (primates.nex), the MRBAYES program assumes that the file is in the current directory. If this is not the case, you have to use the full or relative path to your data file, for example, **execute ../taxa/primates.nex**. If you are running your own data file for this tutorial, beware that it may contain some MRBAYES commands that can change the behavior of the program; delete those commands or put them in square brackets to follow this tutorial.

(2) At the `MrBayes >` prompt, type **lset nst=6 rates=invgamma**. This sets the evolutionary model to the GTR model with gamma-distributed rate variation across sites and a proportion of invariable sites. If your data are not DNA or RNA, if you want to invoke a different model, or if you want to use non-default priors, refer to the manual available from the program website.

(3.1) At the `MrBayes >` prompt, type **mcmc ngen = 10 000 samplefreq = 10**. This will ensure that you get at least a thousand samples (10 000/10) from the posterior probability distribution. For larger data sets you probably want to run the analysis longer and sample less frequently (the default sampling frequency is every hundredth generation and the default number of generations is one million). During the run, MRBAYES prints samples of substitution model parameters to one or more files ending with the suffix ".p" and tree samples to one or more files ending with the suffix ".t". You can find the predicted remaining time to completion of the analysis in the last column printed to screen.

(**3.2**) If the standard deviation of split frequencies is below 0.05 (or 0.01 for more precise results) after 10 000 generations, stop the run by answering **no** when the program asks Continue the analysis? (yes/no). Otherwise, keep adding generations until the value falls below 0.05 (or 0.01).

(**4.1**) Summarize the parameter values by typing **sump burnin = 250** (or whatever value corresponds to 25% of your samples). The program will output a table with summaries of the samples of the substitution model parameters, including the mean, mode, and 95% credibility interval of each parameter. Make sure that the potential scale reduction factor (PSRF) is reasonably close to 1.0 for all parameters (ideally below 1.02); if not, you need to run the analysis longer.

(**4.2**) Summarize the trees by typing **sumt burnin = 250** (or whatever value corresponds to 25% of your samples). The program will output a *cladogram* with the posterior probabilities for each split and a *phylogram* with mean branch lengths. The trees will also be printed to a file that can be read by tree drawing programs such as TREEVIEW (see Chapter 5), MACCLADE, MESQUITE, and FIGTREE (see Chapter 5).

It does not have to be more complicated than this; however, as you get more proficient you will probably want to know more about what is happening behind the scenes. The rest of this section explains each of the steps in more detail and introduces you to all the implicit assumptions you are making and the machinery that MRBAYES uses in order to perform your analysis.

7.11.2 Getting data into MRBAYES

To get data into MRBAYES, you need a so-called NEXUS file that contains aligned nucleotide or amino acid sequences, morphological ("standard") data, restriction site (binary) data, or any mix of these four data types. The NEXUS data file is often generated by another program, such as MACCLADE or MESQUITE. Note, however, that MRBAYES version 3 does not support the full NEXUS standard, so you may have to do a little editing of the file for MRBAYES to process it properly. In particular, MRBAYES uses a fixed set of symbols for each data type and does not support user-defined symbols. The supported symbols are A, C, G, T, R, Y, M, K, S, W, H, B, V, D, N for DNA data; A, C, G, U, R, Y, M, K, S, W, H, B, V, D, N for RNA data; A, R, N, D, C, Q, E, G, H, I, L, K, M, F, P, S, T, W, Y, V, X for protein data; 0, 1 for restriction (binary) data; and 0, 1, 2, 3, 4, 5, 6, 5, 7, 8, 9 for standard (morphology) data. In addition to the standard one-letter ambiguity symbols for DNA and RNA listed above, ambiguity can also be expressed using the NEXUS parenthesis or curly braces notation. For instance, a taxon polymorphic for states 2 and 3 can be coded as (23), (2,3), 23, or 2,3 and a taxon with either amino acid A or F can be coded as (AF), (A,F), AF or A,F. Like most other statistical phylogenetics programs, MRBAYES effectively treats polymorphism and uncertainty the same way (as uncertainty), so it does not matter whether you use parentheses or curly braces. If you have symbols in your matrix other than the ones supported by MRBAYES, you will need to replace them before processing the data block in MRBAYES. You will also need to remove the "Equate" and "Symbols" statements in the "Format" line if they are included. Unlike the NEXUS standard, MRBAYES supports data blocks that contain mixed data types as described below.

To put the data into MRBAYES, type **execute <filename>** at the MrBayes > prompt, where **<filename>** is the name of the input file. To process our example file, type **execute primates.nex** or simply **exe primates.nex** to save some typing. Note that the input file must be located in the same folder (directory) where you started the MRBAYES application (or else you will have to give the path to the file) and the name of the input file should not have blank spaces. If everything proceeds normally, MRBAYES will acknowledge that it has read the data in the DATA block of the NEXUS file by outputting some information about the file read in.

7.11.3 Specifying a model

All of the commands are entered at the MrBayes > prompt. At a minimum two commands, `lset` and `prset`, are required to specify the evolutionary model that will be used in the analysis. Usually, it is also a good idea to check the model settings prior to the analysis using the `showmodel` command. In general, `lset` is used to define the structure of the model and `prset` is used to define the prior probability distributions on the parameters of the model. In the following, we will specify a GTR + I + Γ model (a General Time Reversible model with a proportion of invariable sites and a gamma-shaped distribution of rate variation across sites) for the evolution of the mitochondrial sequences and we will check all of the relevant priors. If you are unfamiliar with stochastic models of molecular evolution, we suggest that you consult Chapters 4 and 10 in this book or a general text, such as Felsenstein (2003).

In general, a good start is to type **help lset**. Ignore the help information for now and concentrate on the table at the bottom of the output, which specifies the current settings. It should look like this:

```
Model settings for partition 1:
```

Parameter	Options	Current Setting
Nucmodel	4by4/Doublet/Codon	4by4
Nst	1/2/6	1
Code	Universal/Vertmt/Mycoplasma/	
	Yeast/Ciliates/Metmt	Universal
Ploidy	Haploid/Diploid	Diploid

```
Rates         Equal/Gamma/Propinv/Invgamma/Adgamma    Equal
Ngammacat     <number>                                4
Usegibbs      Yes/No                                  No
Gibbsfreq     <number>                                100
Nbetacat      <number>                                5
Omegavar      Equal/Ny98/M3                           Equal
Covarion      No/Yes                                  No
Coding        All/Variable/Noabsencesites/
              Nopresencesites                         All
Parsmodel     No/Yes                                  No
```

First, note that the table is headed by Model settings for partition
1. By default, MrBayes divides the data into one partition for each type of data
you have in your DATA block. If you have only one type of data, all data will be
in a single partition by default. How to change the partitioning of the data will be
explained in the second tutorial.

The Nucmodel setting allows you to specify the general type of DNA model.
The Doublet option is for the analysis of paired stem regions of ribosomal DNA
and the Codon option is for analyzing the DNA sequence in terms of its codons.
We will analyze the data using a standard nucleotide substitution model, in which
case the default 4by4 option is appropriate, so we will leave Nucmodel at its
default setting.

The general structure of the substitution model is determined by the Nst setting.
By default, all substitutions have the same rate (Nst=1), corresponding to the F81
model (or the JC model if the stationary state frequencies are forced to be equal
using the prset command, see below). We want the GTR model (Nst=6) instead
of the F81 model so we type **lset nst=6**. MrBayes should acknowledge that it has
changed the model settings.

The Code setting is only relevant if the Nucmodel is set to Codon. The Ploidy
setting is also irrelevant for us. However, we need to change the Rates setting from
the default Equal (no rate variation across sites) to Invgamma (gamma-shaped
rate variation with a proportion of invariable sites). Do this by typing **lset rates =
invgamma**. Again, MrBayes will acknowledge that it has changed the settings. We
could have changed both lset settings at once if we had typed **lset nst = 6 rates =
invgamma** in a single line.

We will leave the Ngammacat setting (the number of discrete categories used
to approximate the gamma distribution) at the default of four. In most cases, four
rate categories are sufficient. It is possible to increase the accuracy of the likelihood
calculations by increasing the number of rate categories. However, the time it will
take to complete the analysis will increase in direct proportion to the number of

rate categories you use, and the effects on the results will be negligible in most cases.

The default behavior for the discrete gamma model of rate variation across sites is to sum site probabilities across rate categories. To sample those probabilities using a Gibbs sampler, we can set the Usegibbs setting to Yes. The Gibbs sampling approach is much faster and requires less memory, but it has some implications you have to be aware of. This option and the Gibbsfreq option are discussed in more detail in the MrBayes manual.

Of the remaining settings, it is only Covarion and Parsmodel that are relevant for single nucleotide models. We will use neither the parsimony model nor the covariotide model for our data, so we will leave these settings at their default values. If you type **help lset** now to verify that the model is correctly set, the table should look like this:

```
Model settings for partition 1:

    Parameter    Options                                 Current Setting
    --------------------------   -----------------------------------------

    Nucmodel     4by4/Doublet/Codon                      4by4
    Nst          1/2/6                                   6
    Code         Universal/Vertmt/Mycoplasma/
                 Yeast/Ciliates/Metmt                    Universal
    Ploidy       Haploid/Diploid                         Diploid
    Rates        Equal/Gamma/Propinv/Invgamma/Adgamma    Invgamma
    Ngammacat    <number>                                4
    Usegibbs     Yes/No                                  No
    Gibbsfreq    <number>                                100
    Nbetacat     <number>                                5
    Omegavar     Equal/Ny98/M3                           Equal
    Covarion     No/Yes                                  No
    Coding       All/Variable/Noabsencesites/
                 Nopresencesites                         All
    Parsmodel    No/Yes                                  No
    --------------------------------------------------------------------
```

7.11.4 Setting the priors

We now need to set the priors for our model. There are six types of parameters in the model: the topology, the branch lengths, the four stationary frequencies of the nucleotides, the six different nucleotide substitution rates, the proportion of invariable sites, and the shape parameter of the gamma distribution of rate variation. The default priors in MrBayes work well for most analyses, and we will not change any of them for now. By typing **help prset** you can obtain a list of the default settings for the parameters in your model. The table at the end of the help information reads:

```
Model settings for partition 1:

   Parameter        Options                         Current Setting
   -----------------------------------------------------------------------
   Tratiopr         Beta/Fixed                      Beta(1.0,1.0)
   Revmatpr         Dirichlet/Fixed                 Dirichlet
                                                    (1.0,1.0,1.0,1.0,1.0,1.0)
   Aamodelpr        Fixed/Mixed                     Fixed(Poisson)
   Aarevmatpr       Dirichlet/Fixed                 Dirichlet(1.0,1.0,...)
   Omegapr          Dirichlet/Fixed                 Dirichlet(1.0,1.0)
   Ny98omega1pr     Beta/Fixed                      Beta(1.0,1.0)
   Ny98omega3pr     Uniform/Exponential/Fixed       Exponential(1.0)
   M3omegapr        Exponential/Fixed               Exponential
   Codoncatfreqs    Dirichlet/Fixed                 Dirichlet(1.0,1.0,1.0)
   Statefreqpr      Dirichlet/Fixed                 Dirichlet(1.0,1.0,1.0,1.0)
   Shapepr          Uniform/Exponential/Fixed       Uniform(0.0,200.0)
   Ratecorrpr       Uniform/Fixed                   Uniform(-1.0,1.0)
   Pinvarpr         Uniform/Fixed                   Uniform(0.0,1.0)
   Covswitchpr      Uniform/Exponential/Fixed       Uniform(0.0,100.0)
   Symdirihyperpr   Uniform/Exponential/Fixed       Fixed(Infinity)
   Topologypr       Uniform/Constraints             Uniform
   Brlenspr         Unconstrained/Clock             Unconstrained:Exp(10.0)
   Treeheightpr     Exponential/Gamma               Exponential(1.0)
   Speciationpr     Uniform/Exponential/Fixed       Uniform(0.0,10.0)
   Extinctionpr     Uniform/Exponential/Fixed       Uniform(0.0,10.0)
   Sampleprob       <number>                        1.00
   Thetapr          Uniform/Exponential/Fixed       Uniform(0.0,10.0)
   Nodeagepr        Unconstrained/Calibrated        Unconstrained
   Treeagepr        Fixed/Uniform/
                    Offsetexponential               Fixed(1.00)
   Clockratepr      Strict/Cpp/Bm                   Strict
   Cppratepr        Fixed/Exponential               Exponential(0.10)
   Psigammapr       Fixed/Exponential/Uniform       Fixed(1.00)
   Nupr             Fixed/Exponential/Uniform       Fixed(1.00)
   Ratepr           Fixed/Variable=Dirichlet        Fixed
   -----------------------------------------------------------------------
```

We need to focus on `Revmatpr` (for the six substitution rates of the GTR rate matrix); `Statefreqpr` (for the stationary nucleotide frequencies of the GTR rate matrix); `Shapepr` (for the shape parameter of the gamma distribution of rate variation); `Pinvarpr` (for the proportion of invariable sites); `Topologypr` (for the topology); and `Brlenspr` (for the branch lengths).

The default prior probability density is a flat Dirichlet (all values are 1.0) for both `Revmatpr` and `Statefreqpr`. This is appropriate if we want to estimate these parameters from the data assuming no prior knowledge about their values. It is possible to fix the rates and nucleotide frequencies, but this is generally not recommended. However, it is occasionally necessary to fix the nucleotide frequencies to be equal, for instance, in specifying the JC and SYM models. This would be achieved by typing **prset statefreqpr = fixed(equal)**.

If we wanted to specify a prior that puts more emphasis on equal nucleotide frequencies than the default flat Dirichlet prior, we could, for instance, use **prset statefreqpr = Dirichlet(10,10,10,10)** or, for even more emphasis on equal frequencies, **prset statefreqpr = Dirichlet(100,100,100,100)**. The sum of the numbers in the Dirichlet distribution determines how focused the distribution is, and the balance between the numbers determines the expected proportion of each nucleotide (in the order A, C, G, and T). Usually, there is a connection between the parameters in the Dirichlet distribution and the observations. For example, you can think of a Dirichlet (150,100,90,140) distribution as one arising from observing 150 As, 100 Cs, 90 Gs, and 140 Ts in some set of reference sequences. If your set of sequences is independent of those reference sequences, but this reference set is clearly relevant to the analysis of your sequences, it might be reasonable to use those numbers as a prior in your analysis.

In our analysis, we will be cautious and leave the prior on state frequencies at its default setting. If you have changed the setting according to the suggestions above, you need to change it back by typing **prset statefreqpr = Dirichlet(1,1,1,1)** or **prs st = Dir(1,1,1,1)** if you want to save some typing. Similarly, we will leave the prior on the substitution rates at the default flat Dirichlet(1,1,1,1,1,1) distribution.

The `Shapepr` parameter determines the prior for the α (shape) parameter of the gamma distribution of rate variation. We will leave it at its default setting, a uniform distribution spanning a wide range of α values. The prior for the proportion of invariable sites is set with `Pinvarpr`. The default setting is a uniform distribution between 0 and 1, an appropriate setting if we don't want to assume any prior knowledge about the proportion of invariable sites.

For topology, the default `Uniform` setting for the `Topologypr` parameter puts equal probability on all distinct, fully resolved topologies. The alternative is to constrain some nodes in the tree to always be present, but we will not attempt that in this analysis.

The `Brlenspr` parameter can either be set to unconstrained or clock-constrained. For trees without a molecular clock (unconstrained) the branch length prior can be set either to exponential or uniform. The default exponential prior with parameter 10.0 should work well for most analyses. It has an expectation of $1/10 = 0.1$, but allows a wide range of branch length values (theoretically from 0 to infinity). Because the likelihood values vary much more rapidly for short branches than for long branches, an exponential prior on branch lengths is closer to being uninformative than a uniform prior.

7.11.5 Checking the model

To check the model before we start the analysis, type **showmodel**. This will give an overview of the model settings. In our case, the output will be as follows:

```
Model settings:

    Datatype  = DNA
        Nucmodel  = 4by4
        Nst       = 6
                    Substitution rates, expressed as proportions
                    of the rate sum, have a Dirichlet prior
                    (1.00,1.00,1.00,1.00,1.00,1.00)
        Covarion  = No
        # States  = 4
                    State frequencies have a Dirichlet prior
                    (1.00,1.00,1.00,1.00)
        Rates     = Invgamma
                    Gamma shape parameter is uniformly dist-
                    ributed on the interval (0.00,200.00).
                    Proportion of invariable sites is uniformly dist-
                    ributed on the interval (0.00,1.00).
                    Gamma distribution is approximated using 4 categories.
                    Likelihood summarized over all rate categories
                       in each generation.

Active parameters:

Parameters
------------------
    Revmat            1
    Statefreq         2
    Shape             3
    Pinvar            4
    Topology          5
    Brlens            6
------------------


1 --   Parameter  = Revmat
            Type       = Rates of reversible rate matrix
            Prior      = Dirichlet(1.00,1.00,1.00,1.00,1.00,1.00)
```

```
2 --   Parameter  = Pi
              Type        = Stationary state frequencies
              Prior       = Dirichlet

3 --   Parameter  = Alpha
              Type        = Shape of scaled gamma distribution of site rates
              Prior       = Uniform(0.00,200.00)

4 --   Parameter  = Pinvar
              Type        = Proportion of invariable sites
              Prior       = Uniform(0.00,1.00)

5 --   Parameter  = Tau
              Type        = Topology
              Prior       = All topologies equally probable a priori
              Subparam.   = V

6 --   Parameter  = V
              Type        = Branch lengths
              Prior       = Unconstrained:Exponential(10.0)
```

Note that we have six types of parameters in our model. All of these parameters will be estimated during the analysis (to fix them to some estimated values, use the prset command and specify a fixed prior). To see more information about each parameter, including its starting value, use the showparams command. The startvals command allows one to set the starting values of each chain separately.

7.11.6 Setting up the analysis

The analysis is started by issuing the mcmc command. However, before doing this, we recommend that you review the run settings by typing **help mcmc**. In our case, we will get the following table at the bottom of the output:

```
Parameter        Options                 Current Setting
-------------------------------------------------------------
Seed             <number>                144979379
Swapseed         <number>                1587146502
Ngen             <number>                10000
Nruns            <number>                2
Nchains          <number>                4
Temp             <number>                0.200000
Reweight         <number>,<number>        0.00 v 0.00  ^
Swapfreq         <number>                1
Nswaps           <number>                1
```

```
Samplefreq       <number>                10
Printfreq        <number>                100
Printall         Yes/No                  Yes
Printmax         <number>                8
Mcmcdiagn        Yes/No                  Yes
Diagnfreq        <number>                1000
Diagnstat        Avgstddev/Maxstddev     Avgstddev
Minpartfreq      <number>                0.20
Allchains        Yes/No                  No
Allcomps         Yes/No                  No
Relburnin        Yes/No                  Yes
Burnin           <number>                0
Burninfrac       <number>                0.25
Stoprule         Yes/No                  No
Stopval          <number>                0.05
Savetrees        Yes/No                  No
Checkpoint       Yes/No                  Yes
Checkfreq        <number>                100000
Filename         <name>                  primates.nex.<p/t>
Startparams      Current/Reset           Current
Starttree        Current/Random          Current
Nperts           <number>                0
Data             Yes/No                  Yes
Ordertaxa        Yes/No                  No
Append           Yes/No                  No
Autotune         Yes/No                  Yes
Tunefreq         <number>                100
Scientific       Yes/No                  Yes
---------------------------------------------------------
```

The Seed is simply the seed for the random number generator, and Swapseed is the seed for the separate random number generator used to generate the chain swapping sequence (see below). Unless they are set to user-specified values, these seeds are generated from the system clock, so your values are likely to be different from the ones in the screen dump above. The Ngen setting is the number of generations for which the analysis will be run. It is useful to run a small number of generations first to make sure the analysis is correctly set up and to get an idea of how long it will take to complete a longer analysis. We will start with 10 000 generations. To change the Ngen setting without starting the analysis we use the mcmcp command, which is equivalent to mcmc except that it does not start the analysis. Type **mcmcp ngen = 10 000** to set the number of generations to 10 000. You can type **help mcmc** to confirm that the setting was changed appropriately.

By default, MrBayes will run two simultaneous, completely independent, analyses starting from different random trees (Nruns = 2). Running more than one analysis simultaneously allows MrBayes to calculate convergence diagnostics on the fly, which is very helpful in determining when you have a good sample from the posterior probability distribution. The idea is to start each run from different randomly chosen trees. In the early phases of the run, the two runs will sample very different trees, but when they have reached convergence (when they produce a good sample from the posterior probability distribution), the two tree samples should be very similar.

To make sure that MrBayes compares tree samples from the different runs, check that Mcmcdiagn is set to yes and that Diagnfreq is set to some reasonable value, such as every 1000th generation. MrBayes will now calculate various run diagnostics every Diagnfreq generation and print them to a file with the name <Filename>.mcmc. The most important diagnostic, a measure of the similarity of the tree samples in the different runs, will also be printed to screen every Diagnfreq generation. Every time the diagnostics are calculated, either a fixed number of samples (burnin) or a percentage of samples (burninfrac) from the beginning of the chain is discarded. The relburnin setting determines whether a fixed burnin (relburnin = no) or a burnin percentage (relburnin = yes) is used. By default, MrBayes will discard the first 25% samples from the cold chain (relburnin = yes and burninfrac = 0.25).

By default, MrBayes uses Metropolis coupling to improve the MCMC sampling of the target distribution. The Swapfreq, Nswaps, Nchains, and Temp settings together control the Metropolis coupling behavior. When Nchains is set to 1, no heating is used. When Nchains is set to a value n larger than 1, then $n - 1$ heated chains are used. By default, Nchains is set to 4, meaning that MrBayes will use three heated chains and one "cold" chain. In our experience, heating is essential for some data sets but it is not needed for others. Adding more than three heated chains may be helpful in analyzing large and difficult data sets. The time complexity of the analysis is directly proportional to the number of chains used (unless MrBayes runs out of physical RAM memory, in which case the analysis will suddenly become much slower), but the cold and heated chains can be distributed among processors in a cluster of computers and among cores in multicore processors using the MPI version of the program, greatly speeding up the calculations.

MrBayes uses an incremental heating scheme, in which chain i is heated by raising its posterior probability to the power $1/(1 + i\lambda)$, where λ is the temperature controlled by the Temp parameter (see Section 7.5). Every Swapfreq generation, two chains are picked at random and an attempt is made to swap their states. For many analyses, the default settings should work nicely. If you are running many more than three heated chains, however, you may want to increase the number of swaps (`Nswaps`) that is tried each time the chain stops for swapping. If the frequency of swapping between chains that are adjacent in temperature is low, you may want to decrease the `Temp` parameter.

The `Samplefreq` setting determines how often the chain is sampled. By default, the chain is sampled every 100th generation, and this works well for most analyses. However, our analysis is so small that we are likely to get convergence quickly. Therefore, it makes sense to sample the chain more frequently, say every 10th generation (this will ensure that we get at least 1000 samples when the number of generations is set to 10 000). To change the sampling frequency, type **mcmcp samplefreq = 10**.

When the chain is sampled, the current values of the model parameters are printed to file. The substitution model parameters are printed to a `.p` file (in our case, there will be one file for each independent analysis, and they will be called `primates.nex.run1.p` and `primates.nex.run2.p`). The `.p` files are tab delimited text files that can be imported into most statistics and graphing programs (including Tracer, see Chapter 18). The topology and branch lengths are printed to a `.t` file (in our case, there will be two files called `primates.nex.run1.t` and `primates.nex.run2.t`). The `.t` files are NEXUS tree files that can be imported into programs like PAUP*, TreeView and FigTree. The root of the `.p` and `.t` file names can be altered using the Filename setting.

The `Printfreq` parameter controls the frequency with which the state of the chains is printed to screen. You can leave `Printfreq` at the default value (print to screen every 100th generation).

The default behavior of MrBayes is to save trees with branch lengths to the `.t` file. Since this is what we want, we leave this setting as it is. If you are running a large analysis (many taxa) and are not interested in branch lengths, you can save a considerable amount of disk space by not saving branch lengths.

When you set up your model and analysis (the number of runs and heated chains), MrBayes creates starting values for the model parameters. A different random tree with predefined branch lengths is generated for each chain and most substitution model parameters are set to predefined values. For instance, stationary state frequencies start out being equal and unrooted trees have all branch lengths set to 0.1. The starting values can be changed by using the **Startvals** command. For instance, user-defined trees can be read into MrBayes by executing a NEXUS file with a "trees" block and then assigned to different chains using the `Startvals` command. After a completed analysis, MrBayes keeps the parameter values of the last generation and will use those as the starting values for the next analysis unless the values are reset using `mcmc starttrees = random startvals = reset`.

Since version 3.2, MrBayes prints all parameter values of all chains (cold and heated) to a checkpoint file every Checkfreq generations, by default every 100 000 generations. The checkpoint file has the suffix .ckp. If you run an analysis and it is stopped prematurely, you can restart it from the last checkpoint by using mcmc append = yes. MrBayes will start the new analysis from the checkpoint; it will even read in all the old trees and include them in the convergence diagnostic if needed. At the end of the new run, you will obtain parameter and tree files that are indistinguishable from those you would have obtained from an uninterrupted analysis. Our data set is so small that we are likely to get an adequate sample from the posterior before the first checkpoint.

7.11.7 Running the analysis

Finally, we are ready to start the analysis. Type **mcmc**. MrBayes will first print information about the model and then list the proposal mechanisms that will be used in sampling from the posterior distribution. In our case, the proposals are the following:

```
The MCMC sampler will use the following moves:
   With prob.  Chain will change
       3.45 %   param. 1 (Revmat) with Dirichlet proposal
       3.45 %   param. 2 (Pi) with Dirichlet proposal
       3.45 %   param. 3 (Alpha) with Multiplier
       3.45 %   param. 4 (Pinvar) with Sliding window
      17.24 %   param. 5 (Tau) and 6 (V) with Extending subtree swapper
      34.48 %   param. 5 (Tau) and 6 (V) with Extending TBR
      17.24 %   param. 5 (Tau) and 6 (V) with Parsimony-based SPR
      17.24 %   param. 6 (V) with Random brlen hit with multiplier
```

The exact set of proposals and their relative probabilities may differ depending on the exact version of the program that you are using. Note that MrBayes will spend most of its effort changing the topology (Tau) and branch length (V) parameters. In our experience, topology and branch lengths are the most difficult parameters to integrate over and we therefore let MrBayes spend a large proportion of its time proposing new values for those parameters. The proposal probabilities and tuning parameters can be changed with the Propset command, but be warned that inappropriate changes of these settings may destroy any hopes of achieving convergence.

After the initial log likelihoods, MrBayes will print the state of the chains every 100th generation, like this:

```
Chain results:


  1 -- [-5723.498] (-5729.634) (-5727.207) (-5731.104) * [-5721.779] (-5731.701) (-5737.807) (-5730.336)

 100 -- (-5726.662) (-5728.374) (-5733.144) [-5722.257] * [-5721.199] (-5726.193) (-5732.098) (-5732.563) -- 0:03:18

 200 -- [-5729.666] (-5721.116) (-5731.222) (-5731.546) * (-5726.632) [-5731.803] (-5738.420) (-5729.889) -- 0:02:27

 300 -- [-5727.654] (-5725.420) (-5736.655) (-5725.982) * (-5722.774) [-5743.637] (-5729.989) [-5729.954] -- 0:02:09

 400 -- [-5728.809] (-5722.467) (-5742.752) (-5729.874) * (-5723.731) (-5739.025) [-5719.889] (-5731.096) -- 0:02:24

 500 -- [-5728.286] (-5723.060) (-5738.274) (-5726.420) * [-5724.408] (-5733.188) (-5719.771) (-5725.882) -- 0:02:13

 600 -- [-5719.082] (-5728.268) (-5728.040) (-5731.023) * (-5727.788) (-5733.390) [-5723.994] (-5721.954) -- 0:02:05

 700 -- [-5717.720] (-5725.982) (-5728.786) (-5732.380) * (-5722.842) (-5727.218) [-5720.717] (-5729.936) -- 0:01:59

 800 -- (-5725.531) (-5729.259) (-5743.762) [-5731.019] * (-5729.238) [-5731.272] (-5722.135) (-5727.906) -- 0:02:06

 900 -- [-5721.976] (-5725.464) (-5731.774) (-5725.830) * (-5727.845) [-5723.992] (-5731.020) (-5728.988) -- 0:02:01

1000 -- (-5724.549) [-5723.807] (-5726.810) (-5727.921) * (-5729.302) [-5730.518] (-5733.236) (-5727.348) -- 0:02:06


Average standard deviation of split frequencies: 0.000000


1100 -- [-5724.473] (-5726.013) (-5723.995) (-5724.521) * (-5734.206) (-5720.464) [-5727.936] (-5723.821) -- 0:02:01

...

9000 -- (-5741.070) (-5728.937) (-5738.787) [-5719.056] * (-5731.562) [-5722.514] (-5721.184) (-5731.386) -- 0:00:13


Average standard deviation of split frequencies: 0.000116


9100 -- (-5747.669) [-5726.528] (-5738.190) (-5725.938) * (-5723.844) (-5726.963) [-5723.221] (-5724.665) -- 0:00:11

9200 -- (-5738.994) (-5725.611) (-5734.902) [-5723.275] * [-5718.420] (-5724.197) (-5730.129) (-5724.800) -- 0:00:10

9300 -- (-5740.946) (-5728.599) [-5729.193] (-5731.202) * (-5722.247) [-5723.141] (-5729.026) (-5727.039) -- 0:00:09

9400 -- (-5735.178) (-5726.517) [-5726.557] (-5728.377) * (-5721.659) (-5723.202) (-5734.709) [-5726.191] -- 0:00:07

9500 -- (-5731.041) (-5730.340) [-5721.900] (-5730.002) * (-5724.353) [-5727.075] (-5735.553) (-5725.420) -- 0:00:06

9600 -- [-5726.318] (-5737.300) (-5725.160) (-5731.890) * (-5721.767) [-5730.250] (-5742.843) (-5725.866) -- 0:00:05

9700 -- [-5726.573] (-5735.158) (-5728.509) (-5724.753) * (-5722.873) [-5729.740] (-5744.456) (-5723.282) -- 0:00:03

9800 -- (-5728.167) (-5736.140) (-5729.682) [-5725.419] * (-5723.056) (-5726.630) (-5729.571) [-5720.712] -- 0:00:02

9900 -- (-5738.486) (-5737.588) [-5732.250] (-5728.228) * (-5726.533) (-5733.696) (-5724.557) [-5722.960] -- 0:00:01

10000 -- (-5729.797) (-5725.507) [-5727.468] (-5720.465) * (-5729.313) (-5735.121) (-5722.913) [-5726.844] -- 0:00:00


Average standard deviation of split frequencies: 0.000105


Continue with analysis? (yes/no):
```

If you have the terminal window wide enough, each generation of the chain will print on a single line.

The first column lists the generation number. The following four columns with negative numbers each correspond to one chain in the first run. Each column corresponds to one physical location in computer memory, and the chains actually shift positions in the columns as the run proceeds. The numbers are the log likelihood values of the chains. The chain that is currently the cold chain has its value surrounded by square brackets, whereas the heated chains have their values surrounded by parentheses. When two chains successfully change states, they trade column positions (places in computer memory). If the Metropolis coupling works well, the cold chain should move around among the columns; this means that the cold chain successfully swaps states with the heated chains. If the cold chain gets stuck in one of the columns, then the heated chains are not successfully contributing states to the cold chain, and the Metropolis coupling is inefficient. The analysis may then have to be run longer or the temperature difference between chains may have to be lowered.

The star column separates the two different runs. The last column gives the time left to completion of the specified number of generations. This analysis approximately takes 1 second per 100 generations. Because different moves are used in each generation, the exact time varies somewhat for each set of 100 generations, and the predicted time to completion will be unstable in the beginning of the run. After a while, the predictions will become more accurate and the time will decrease predictably.

7.11.8 When to stop the analysis

At the end of the run, MRBAYES asks whether or not you want to continue with the analysis. Before answering that question, examine the average standard deviation of split frequencies. As the two runs converge onto the stationary distribution, we expect the average standard deviation of split frequencies to approach zero, reflecting the fact that the two tree samples become increasingly similar. In our case, the average standard deviation is zero after 1000 generations, reflecting the fact that both runs sampled the most probable tree in the first few samples. As the runs pick up some of the less probable trees, the standard deviation first increases slightly and then decreases to end up at a very low value. In larger phylogenetic problems, the standard deviation is typically moderately large initially and then increases for some time before it starts to decrease. Your values can differ slightly because of stochastic effects. Given the extremely low value of the average standard deviation at the end of the run, there appears to be no need to continue the analysis beyond 10 000 generations so, when MRBAYES asks "`Continue with analysis? (yes/no):`", stop the analysis by typing "**no**."

Although we recommend using a convergence diagnostic, such as the standard deviation of split frequencies, there are also simpler but less powerful methods of determining when to stop the analysis. The simplest technique is to examine the log likelihood values (or, more exactly, the log probability of the data given the parameter values) of the cold chain, that is, the values printed to screen within square brackets. In the beginning of the run, the values typically increase rapidly (the absolute values decrease, since these are negative numbers). This is the "burn-in" phase and the corresponding samples typically are discarded. Once the likelihood of the cold chain stops increasing and starts to randomly fluctuate within a more or less stable range, the run may have reached stationarity, that is, it may be producing a good sample from the posterior probability distribution. At stationarity, we also expect different, independent runs to sample similar likelihood values. Trends in likelihood values can be deceiving though; you're more likely to detect problems with convergence by comparing split frequencies than by looking at likelihood trends.

When you stop the analysis, MRBAYES will print several types of information useful in optimizing the analysis. This is primarily of interest if you have difficulties in obtaining convergence, which is unlikely to happen with this analysis. We give a few tips on how to improve convergence at the end of the chapter.

7.11.9 Summarizing samples of substitution model parameters

During the run, samples of the substitution model parameters have been written to the `.p` files every `samplefreq` generation. These files are tab-delimited text files that look something like this:

```
[ID: 9409050143]
   Gen      LnL        TL      r(A<->C)  ...  pi(G)     pi(T)     alpha     pinvar
   1       -5723.498   3.357   0.067486  ...  0.098794  0.247609  0.580820  0.124136
   10      -5727.478   3.110   0.030604  ...  0.072965  0.263017  0.385311  0.045880
   ....
   9990    -5727.775   2.687   0.052292  ...  0.086991  0.224332  0.951843  0.228343
   10000   -5720.465   3.290   0.038259  ...  0.076770  0.240826  0.444826  0.087738
```

The first number, in square brackets, is a randomly generated ID number that lets you identify the analysis from which the samples come. The next line contains the column headers, and is followed by the sampled values. From left to right, the columns contain: (1) the generation number (Gen); (2) the log likelihood of the cold chain (LnL); (3) the total tree length (the sum of all branch lengths, TL); (4) the six GTR rate parameters (r(A<->C), r(A<->G) etc.); (5) the four stationary nucleotide frequencies (pi(A), pi(C) etc.); (6) the shape parameter of the gamma distribution of rate variation (alpha); and (7) the proportion of invariable sites (pinvar). If you use a different model for your data set, the .p files will, of course, be different.

MRBAYES provides the sump command to summarize the sampled parameter values. Before using it, we need to decide on the burn-in. Since the convergence diagnostic we used previously to determine when to stop the analysis discarded the first 25% of the samples and indicated that convergence had been reached after 10 000 generations, it makes sense to discard 25% of the samples obtained during the first 10 000 generations. Since we sampled every 10th generation, there are 1000 samples (1001 to be exact, since the first generation is always sampled) and 25% translates to 250 samples. Thus, summarize the information in the .p file by typing "**sump burnin** = **250.**" By default, sump will summarize the information in the .p file generated most recently, but the filename can be changed if necessary.

The sump command will first generate a plot of the generation versus the log probability of the data (the log likelihood values). If we are at stationarity, this plot should look like "white noise," that is, there should be no tendency of increase or decrease over time. The plot should look something like this:

```
+------------------------------------------------------------+ -5718.96
|                      2                           12        |
|           2                                                |
|               1 2                            2             |
|          1      22      1*1    2 22        2   1     2      |
|    2     2 2           1 2 2      2     2        1          |
|      11       1     2        1  2   2 2        2 2    2 |
|   1  2  1       1    12     1 1     *   2      2            |
| 1 11 2   *     2       1         1      2  2     1       *| |
| *2       2     1    22            2  1         211    2?   |
|      2  1                       1  1    11 1    22  1       |
| *      1      2          2       1       1 2         1*  |
|                         2         1        1     1         |
|      1         111   2    1        1 1                     |
|  22                 1                              1 |     |
|                                            1     1        |
+------+-----+-----+-----+-----+-----+-----+-----+-----+-----+ -5729.82
^                                                          ^
2500                                                     10000
```

If you see an obvious trend in your plot, either increasing or decreasing, you probably need to run the analysis longer to get an adequate sample from the posterior probability distribution.

At the bottom of the sump output, there is a table summarizing the samples of the parameter values:

```
Model parameter summaries over the runs sampled in files
   ''primates.nex.run1.p'' and ''primates.nex.run2.p'':
   (Summaries are based on a total of 1502 samples from 2 runs)
   (Each run produced 1001 samples of which 751 samples were included)
```

			95 % Cred. Interval			
Parameter	Mean	Variance	Lower	Upper	Median	PSRF *
TL	2.954334	0.069985	2.513000	3.558000	2.941000	1.242

r(A<->C)	0.044996	0.000060	0.030878	0.059621	0.044567	1.016
r(A<->G)	0.470234	0.002062	0.386927	0.557040	0.468758	1.025
r(A<->T)	0.038107	0.000073	0.023568	0.056342	0.037172	1.022
r(C<->G)	0.030216	0.000189	0.007858	0.058238	0.028350	1.001
r(C<->T)	0.396938	0.001675	0.317253	0.476998	0.394980	1.052
r(G<->T)	0.019509	0.000158	0.001717	0.047406	0.018132	1.003
pi(A)	0.355551	0.000150	0.332409	0.382524	0.357231	1.010
pi(C)	0.320464	0.000131	0.298068	0.343881	0.320658	0.999
pi(G)	0.081290	0.000043	0.067120	0.095940	0.080521	1.004
pi(T)	0.242695	0.000101	0.220020	0.261507	0.243742	1.030
alpha	0.608305	0.042592	0.370790	1.142317	0.546609	1.021
pinvar	0.135134	0.007374	0.008146	0.303390	0.126146	0.999

```
----------------------------------------------------------------
* Convergence diagnostic (PSRF = Potential scale reduction factor [Gelman

  and Rubin, 1992], uncorrected) should approach 1 as runs converge. The

  values may be unreliable if you have a small number of samples. PSRF should

  only be used as a rough guide to convergence since all the assumptions

  that allow one to interpret it as a scale reduction factor are not met in

  the phylogenetic context.
```

For each parameter, the table lists the mean and variance of the sampled values, the lower and upper boundaries of the 95% credibility interval, and the median of the sampled values. The parameters are the same as those listed in the .p files: the total tree length (TL), the six reversible substitution rates (r(A<->C), r(A<->G), etc.), the four stationary state frequencies (pi(A), pi(C), etc.), the shape of the gamma distribution of rate variation across sites (alpha), and the proportion of invariable sites (pinvar). Note that the six rate parameters of the GTR model are given as proportions of the rate sum (the Dirichlet parameterization). This parameterization has some advantages in the Bayesian context; in particular, it allows convenient formulation of priors. If you want to scale the rates relative to the G–T rate, just divide all rate proportions by the G–T rate proportion.

The last column in the table contains a convergence diagnostic, the Potential Scale Reduction Factor (PSRF). If we have a good sample from the posterior probability distribution, these values should be close to 1.0. If you have a small number of samples, there may be some spread in these values, indicating that you may need to sample the analysis more often or run it longer. In our case, we can probably obtain more accurate estimates of some parameters easily by running the analysis slightly longer.

7.11.10 Summarizing samples of trees and branch lengths

Trees and branch lengths are printed to the .t files. These files are NEXUS-formatted tree files with a structure like this:

```
#NEXUS
  [ID: 9409050143]
  [Param: tree]
  begin trees;
translate
     1 Tarsius_syrichta,
     2 Lemur_catta,
     3 Homo_sapiens,
     4 Pan,
     5 Gorilla,
     6 Pongo,
     7 Hylobates,
     8 Macaca_fuscata,
     9 M_mulatta,
    10 M_fascicularis,
    11 M_sylvanus,
    12 Saimiri_sciureus;
tree rep.1 = ((12:0.486148,(((((3:0.042011,4:0.065025):0.034344,5:0.051939...
    ...
tree rep.10000 = (((((10:0.087647,(8:0.013447,9:0.021186):0.030524):0.0568...
  end;
```

To summarize the tree and branch length information, type "**sumt burnin =
250.**" The sumt and sump commands each have separate burn-in settings, so it
is necessary to give the burn-in here again. Most MRBAYES settings are persistent
and need not be repeated every time a command is executed, but the settings are
typically not shared across commands. To make sure the settings for a particular
command are correct, you can always use help <command> before issuing the
command.

The sumt command will output, among other things, summary statistics for
the taxon bipartitions, a tree with clade credibility (posterior probability) values,
and a phylogram (if branch lengths have been saved). The summary statistics (see
below) describe each split (clade) in the tree sample (dots for the taxa that are on
one side of the split and stars for the taxa on the other side; for instance, the sixth
split (ID 6) is the terminal branch leading to taxon 2 since it has a star in the second
position and a dot in all other positions). Then it gives the number of times the
split was sampled (\#obs), the probability of the split (Probab.), the standard
deviation of the split frequency (Stdev(s)) across runs, the mean (Mean(v))
and variance (Var(v)) of the branch length, the Potential Scale Reduction Factor
(PSRF), and finally the number of runs in which the split was sampled (Nruns).
In our analysis, there is overwhelming support for a single tree, so almost all splits
in this tree have a posterior probability of 1.0.

```
Summary statistics for taxon bipartitions:

ID --
Partition          #obs   Probab.  Stdev(s)  Mean(v)   Var(v)    PSRF   Nruns
-----------------------------------------------------------------------------
----
 1 -- .......**...  1502  1.000000 0.000000  0.035937  0.000083  1.000    2
 2 -- .........*..  1502  1.000000 0.000000  0.056738  0.000148  1.006    2
 3 -- ........*...  1502  1.000000 0.000000  0.022145  0.000037  1.077    2
 4 -- ..........*.  1502  1.000000 0.000000  0.072380  0.000338  1.007    2
 5 -- .......*....  1502  1.000000 0.000000  0.017306  0.000037  1.036    2
 6 -- .*..........  1502  1.000000 0.000000  0.345552  0.003943  1.066    2
 7 -- .**********   1502  1.000000 0.000000  0.496361  0.006726  1.152    2
 8 -- ..*********   1502  1.000000 0.000000  0.273113  0.003798  1.021    2
 9 -- .......***..  1502  1.000000 0.000000  0.045900  0.000315  1.002    2
10 -- .......****.  1502  1.000000 0.000000  0.258660  0.002329  1.041    2
11 -- ..*.........  1502  1.000000 0.000000  0.049774  0.000110  1.014    2
12 -- ...*........  1502  1.000000 0.000000  0.062863  0.000147  1.000    2
13 -- .....*......  1502  1.000000 0.000000  0.146137  0.000665  1.060    2
14 -- ...........*  1502  1.000000 0.000000  0.430463  0.004978  1.045    2
15 -- ......*.....  1502  1.000000 0.000000  0.173405  0.000940  1.053    2
16 -- ..***.......  1502  1.000000 0.000000  0.080733  0.000375  1.023    2
17 -- ..****......  1502  1.000000 0.000000  0.055286  0.000409  1.064    2
18 -- ..*****.....  1502  1.000000 0.000000  0.116993  0.001254  1.046    2
19 -- ....*.......  1502  1.000000 0.000000  0.059082  0.000219  1.014    2
20 -- ..*********.  1501  0.999334 0.000942  0.124653  0.001793  1.141    2
21 -- ..**........  1500  0.998668 0.000000  0.030905  0.000135  1.030    2
-----------------------------------------------------------------------------
```

The clade credibility tree (upper tree) gives the probability of each split or clade in the tree, and the phylogram (lower tree) gives the branch lengths measured in expected substitutions per site:

```
Clade credibility values:

/----------------------------------------------------- Tarsius_syrichta (1)
|
|----------------------------------------------------- Lemur_catta (2)
|
|                                      /-------- Homo_sapiens (3)
|                           /--100--+
|                           |        \-------- Pan (4)
|                  /--100--+
|                  |        \--------------- Gorilla (5)
|         /---100--+
|         |        \---------------------- Pongo (6)
+
|              /--100--+
```

```
   |                    |      \------------------------------- Hylobates (7)
   |                    |
   |                    |                       /-------- Macaca_fuscata (8)
   |         /--100--+         /--100--+
   |         |          |                       |     \-------- M_mulatta (9)
   |         |          |           /--100--+
   |         |          |                       |     \--------------- M_fascicularis (10)
   \--100--+         \-------100------+
             |                         \------------       ------- M_sylvanus (11)
             |
             \--------------------------------------------------- Saimiri_sciureus (12)
```

Phylogram (based on average branch lengths):

```
/--------------------------------------- Tarsius_syrichta (1)
|
|------------------------- Lemur_catta (2)
|
|                                               /---- Homo_sapiens (3)
|                                           /-+
|                                           | \----- Pan (4)
|                                      /------+
|                                      |     \---- Gorilla (5)
|                                /---+
+                                |   \----------- Pongo (6)
|                            /--------+
|                            |        \------------ Hylobates (7)
|                            |
|                            |                   /-- Macaca_fuscata (8)
|               /---------+              /-+
|               |            |           | \-- M_mulatta (9)
|               |            |         /---+
|               |            |           |  \---- M_fascicularis (10)
\-------------------+        \-------------------+
                |                             \------ M_sylvanus (11)
                |
                \------------------------------- Saimiri_sciureus (12)
```

|-------------| 0.200 expected changes per site

In the background, the sumt command creates three additional files. The first is a .parts file, which contains the list of taxon bipartitions, their posterior probability (the proportion of sampled trees containing them), and the branch lengths associated with them (if branch lengths have been saved). The branch length values are based only on those trees containing the relevant bipartition.

The second generated file has the suffix `.con` and includes two consensus trees. The first one has both the posterior probability of clades (as interior node labels) and the branch lengths (if they have been saved) in its description. A graphical representation of this tree can be generated in the program TREEVIEW by Rod Page or FIGTREE by Andrew Rambaut (see Chapter 5 and Chapter 18). The second tree only contains the branch lengths and it can be imported into a wide range of tree-drawing programs such as MACCLADE and MESQUITE. The third file generated by the `sumt` command is the `.trprobs` file, which contains the trees that were found during the MCMC search, sorted by posterior probability.

7.12 Analyzing a partitioned data set

MRBAYES handles a wide variety of data types and models, as well as any mix of these models. In this example we will look at how to set up a simple analysis of a combined data set, consisting of data from four genes and morphology for 30 taxa of gall wasps and outgroups. A similar approach can be used, for example, to set up a partitioned analysis of molecular data coming from different genes. The data set for this tutorial is found in the file `cynmix.nex`.

7.12.1 Getting mixed data into MRBAYES

First, open up the NEXUS data file in a text editor. The DATA block of the NEXUS file should look familiar but there are some differences compared to the `pri-mates.nex` file in the format statement:

```
Format datatype = mixed(Standard:1-166,DNA:167-3246)
interleave=yes gap=- missing=?;
```

First, the datatype is specified as `datatype = mixed(Standard:1-166 DNA:167-3246)`. This means that the matrix contains standard (morphology) characters in columns 1-166 and DNA characters in the remaining columns. The mixed datatype is an extension to the NEXUS standard. This extension was originated by MRBAYES 3 and may not be compatible with other phylogenetics programs.

Second, the matrix is interleaved. It is often convenient to specify mixed data in interleaved format, with each block consisting of a natural subset of the matrix, such as the morphological data or one of the gene regions.

7.12.2 Dividing the data into partitions

By default, MRBAYES partitions the data according to data type. There are only two data types in the matrix, so this model will include only a morphology (standard) and a DNA partition. To divide the DNA partition into gene regions, it is convenient to first specify character sets. In principle, this can be done from the command line, but it is more convenient to do it in a MRBAYES block in the data file. With the MRBAYES distribution, we added a file `cynmix-run.nex` with a complete MRBAYES block. For this section, we are going to create a command block from scratch, but you can consult the `cynmix-run.nex` for reference.

In your favorite text editor, create a new file called `cynmix-command.nex` in the same directory as the `cynmix.nex` file and add the following new MRBAYES block (note that each line must be terminated by a semicolon):

```
#NEXUS

begin mrbayes;
execute cynmix.nex;
charset morphology = 1-166;
charset COI = 167-1244;
charset EF1a = 1245-1611;
charset LWRh = 1612-2092;
charset 28S = 2093-3246;
```

The first line is required to comply with the NEXUS standard. With the `exe-cute` command, we load the data from the `cynmix.nex` file and the `charset` command simply associates a name with a set of characters. For instance, the character set `COI` is defined above to include characters 167 to 1244. The next step is to define a partition of the data according to genes and morphology. This is accomplished with the line (add it after the lines above):

```
partition favored = 5: morphology, COI, EF1a, LWRh, 28S;
```

The elements of the `partition` command are: (1) the name of the partitioning scheme (`favored`); (2) an equal sign (=); (3) the number of character divisions in the scheme (5); (4) a colon (:); and (5) a list of the characters in each division, separated by commas. The list of characters can simply be an enumeration of the character numbers (the above line is equivalent to `partition favored = 5: 1-166, 167-1244, 1245-1611, 1612-2092, 2093-3246;`) but it is often more convenient to use predefined character sets as we did above. The final step is to tell MRBAYES that we want to work with this partitioning of the data instead of with the default partitioning. We do this using the `set` command:

```
set partition = favored;
```

Finally, we need to add an `end` statement to close the MRBAYES block. The entire file should now look like this:

```
#NEXUS

begin mrbayes;
  execute cynmix.nex;
  charset morphology = 1-166;
  charset COI = 167-1244;
  charset EF1a = 1245-1611;
  charset LWRh = 1612-2092;
  charset 28S = 2093-3246;
  partition favored = 5: morphology, COI, EF1a, LWRh, 28S;
  set partition = favored;
end;
```

When we read this block into MRBAYES, we will get a partitioned model with the first character division being morphology, the second division being the COI gene, etc. Save the data file, exit your text editor, and finally launch MRBAYES and type **execute cynmix-command.nex** to read in your data and set up the partitioning scheme.

7.12.3 Specifying a partitioned model

Before starting to specify the partitioned model, it is useful to examine the default model. Type "**showmodel**" and you should get this table as part of the output:

```
Active parameters:

                  Partition(s)
    Parameters    1  2  3  4  5
    ------------------------------
    Statefreq     1  2  2  2  2
    Topology      3  3  3  3  3
    Brlens        4  4  4  4  4
    ------------------------------
```

There is a lot of other useful information in the output of showmodel but this table is the key to the partitioned model. We can see that there are five partitions in the model and four active (free) parameters. There are two stationary state frequency parameters, one for the morphological data (parameter 1) and one for the DNA data (parameter 2). Then there is also a topology parameter (3) and a set of branch length parameters (4). Both the topology and branch lengths are the same for all partitions.

Now, assume we want a separate GTR + Γ + I model for each gene partition. All the parameters should be estimated separately for the individual genes. Assume further that we want the overall evolutionary rate to be (potentially) different across partitions, and that we want to assume gamma-shaped rate variation for the morphological data. We can obtain this model by using `lset` and `prset` with the `applyto` mechanism, which allows us to apply the settings to specific partitions. For instance, to apply a GTR + Γ + I model to the molecular partitions, we type **lset applyto = (2,3,4,5) nst = 6 rates = invgamma**. This will produce the following table when **showmodel** is invoked:

```
Active parameters:


                   Partition(s)
     Parameters    1  2  3  4  5

     ------------------------------

     Revmat        .  1  1  1  1
     Statefreq     2  3  3  3  3
     Shape         .  4  4  4  4
     Pinvar        .  5  5  5  5
     Topology      6  6  6  6  6
     Brlens        7  7  7  7  7

     ------------------------------
```

As you can see, all molecular partitions now evolve under the correct model but all parameters (`statefreq`, `revmat`, `shape`, `pinvar`) are shared across partitions. To unlink them such that each partition has its own set of parameters, type: **unlink statefreq = (all) revmat = (all) shape = (all) pinvar = (all)**. Gamma-shaped rate variation for the morphological data is enforced with **lset applyto = (1) rates = gamma**. The trickiest part is to allow the overall rate to be different across partitions. This is achieved using the `ratepr` parameter of the `prset` command. By default, `ratepr` is set to `fixed`, meaning that all partitions have the same overall rate. By changing this to variable, the rates are allowed to vary under a flat Dirichlet prior. To allow all our partitions to evolve under different rates, type **prset applyto = (all) ratepr = variable**.

The model is now essentially complete but there is one final thing to consider. Typically, morphological data matrices do not include all types of characters. Specifically, morphological data matrices do not usually include any constant (invariable) characters. Sometimes, *autapomorphies* are not included either, and the matrix is restricted to parsimony-informative characters. For MRBAYES to calculate the probability of the data correctly, we need to inform it of this ascertainment (coding) bias. By default, MRBAYES assumes that standard data sets include all variable characters but no constant characters. If necessary, one can change this setting using `lset coding`. We will leave the `coding` setting at the default, though. Now, **showmodel** should produce this table:

```
Active parameters:

                      Partition(s)
        Parameters      1   2   3   4   5
        ------------------------------------

        Revmat          .   1   2   3   4
        Statefreq       5   6   7   8   9
        Shape          10  11  12  13  14
        Pinvar          .  15  16  17  18
        Ratemultiplier 19  19  19  19  19
        Topology       20  20  20  20  20
        Brlens         21  21  21  21  21
        ------------------------------------
```

7.12.4 Running the analysis

When the model has been completely specified, we can proceed with the analysis essentially as described above in the tutorial for the `primates.nex` data set. However, in the case of the `cynmix.nex` data set, the analysis will have to be run longer before it converges.

When looking at the parameter samples from a partitioned analysis, it is useful to know that the names of the parameters are followed by the character division (partition) number in curly braces. For instance, `pi(A){3}` is the stationary frequency of nucleotide A in character division 3, which is the EF1a division in the above analysis.

In this section we have used a separate NEXUS file for the MRBAYES block. Although one can add this command block to the data file itself, there are several advantages to keeping the commands and the data blocks separate. For example, one can create a set of different analyses with different parameters in separate "command" files and submit all those files to a job scheduling system on a computer cluster.

7.12.5 Some practical advice

As you continue exploring Bayesian phylogenetic inference, you may find the following tips helpful:

(1) If you are anxious to get results quickly, you can try running without Metropolis coupling (heated chains). This will save a large amount of computational time at the risk of having to start over if you have difficulties getting convergence. Turn off heating by setting the `mcmc` option `nchains` to 1 and switch it on by setting `nchains` to a value larger than 1.

(2) If you are using heated chains, make sure that the acceptance rate of swaps between adjacent chains are in the approximate range of 10% to 70% (the

acceptance rates are printed to the .mcmc file and to screen at the end of the run). If the acceptance rates are lower than 10%, decrease the temperature constant (mcmc temp=<value>); if the acceptance rates are higher than 70%, increase it.

(3) If you run multiple simultaneous analyses or use Metropolis coupling and have access to a machine with several processors or processor cores, or if you have access to a computer cluster, you can speed up your analyses considerably by running MRBAYES in parallel under MPI. See the MRBAYES website for more information about this.

(4) If you are using automatic optimization of proposal tuning parameters, and your runs are reasonably long so that MRBAYES has sufficient time to find the best settings, you should not have to adjust proposal tuning parameters manually. However, if you have difficulties getting convergence, you can try selecting a different mix of topology moves than the one used by default. For instance, the random SPR move tends to do well on some data sets, but it is switched off by default because, in general, it is less efficient than the default moves. You can add and remove topology moves by adjusting their relative proposal probabilities using the propset command. Use showmoves allavailable = yes first to see a list of all the available moves.

For more information and tips, turn to the MRBAYES website (*http://mrbayes. net*) and the MRBAYES users' email list.

Phylogeny inference based on parsimony and other methods using PAUP*

THEORY

David L. Swofford and Jack Sullivan

8.1 Introduction

Methods for inferring evolutionary trees can be divided into two broad categories: those that operate on a matrix of discrete characters that assigns one or more attributes or character states to each taxon (i.e. sequence or gene-family member); and those that operate on a matrix of pairwise distances between taxa, with each distance representing an estimate of the amount of divergence between two taxa since they last shared a common ancestor (see Chapter 1). The most commonly employed discrete-character methods used in molecular phylogenetics are *parsimony* and *maximum likelihood* methods. For molecular data, the character-state matrix is typically an aligned set of DNA or protein sequences, in which the states are the nucleotides A, C, G, and T (i.e. DNA sequences) or symbols representing the 20 common amino acids (i.e. protein sequences); however, other forms of discrete data such as restriction-site presence/absence and gene-order information also may be used.

Parsimony, maximum likelihood, and some distance methods are examples of a broader class of phylogenetic methods that rely on the use of *optimality criteria*. Methods in this class all operate by explicitly defining an *objective function* that returns a score for any input tree topology. This tree score thus allows any two or more trees to be ranked according to the chosen optimality criterion. Ordinarily, phylogenetic inference under **criterion-based methods** couples the selection of

The Phylogenetic Handbook: a Practical Approach to Phylogenetic Analysis and Hypothesis Testing, Philippe Lemey, Marco Salemi, and Anne-Mieke Vandamme (eds.). Published by Cambridge University Press. © Cambridge University Press 2009.

a suitable optimality criterion with a search for an optimal tree topology under that criterion. Because the number of tree topologies grows exponentially with the number of taxa (see Table 1.4), criterion-based methods are necessarily slower than algorithmic approaches, such as **UPGMA** or **neighbor-joining** (**NJ**), which simply cluster taxa according to a prescribed set of rules and operations (see Chapter 5). However, we believe that criterion-based methods have a strong advantage in that the basis for preferring one tree over another is made mathematically precise, unlike the case for algorithmic methods. For example, NJ was originally described (Saitou & Nei, 1987) as a method for approximating a tree that minimizes the sum of least-squares branch lengths – the **minimum-evolution** criterion (see Chapter 5). However, it rarely achieves this goal for data sets of non-trivial size, and rearrangements of the NJ tree that yield a lower minimum-evolution score can usually be found. This result makes it difficult to defend the presentation of an NJ tree as the most reasonable estimate of a phylogeny. With criterion-based methods, it can at least be said that a given tree topology was the best that could be found according to the criterion. If others want to dispute that tree, they are free to criticize the criterion or search for better trees according to the criterion, but it is clear why the tree was chosen. It is true that **bootstrapping** or **jackknifing methods** (see Chapter 5) can be used to quantify uncertainty regarding the groupings implied by an NJ tree, but fast approximations are also available for criterion-based methods.

Although it contains additional capabilities, the PAUP* program (Swofford, 2002) is primarily a program for estimating phylogenies using criterion-based methods. It includes support for parsimony, maximum likelihood (nucleotide data), and distance methods. This chapter discusses the use of PAUP* as a tool for phylogenetic inference. First, some theoretical background is provided for parsimony analysis, which – unlike distance and maximum likelihood methods (see Chapters 5 and 6) – is not treated elsewhere in this book. Next, strategies are described for searching for optimal trees that are appropriate for any of the existing optimality criteria. Finally, in the Practice section, many of the capabilities of PAUP* are illustrated using the real-data examples common to other chapters in this book.

8.2 Parsimony analysis – background

Although the first widely used methods for inferring phylogenies were pairwise distance methods, parsimony analysis has been the predominant approach used to construct phylogenetic trees from the early 1970s until relatively recently; despite some limitations, it remains an important and useful technique. The basic idea underlying parsimony analysis is simple: one seeks the tree, or collection of trees, that minimizes the amount of evolutionary change (i.e. transformations of one character state into another) required to explain the data (Kluge & Farris, 1969; Farris, 1970; Fitch, 1971).

The goal of minimizing evolutionary change is often defended on philosophical grounds. One line of argument is the notion that, when two hypotheses provide equally valid explanations for a phenomenon, the simpler one should always be preferred. This position is often referred to as "Ockham's Razor": shave away all that is unnecessary. To use simplicity as a justification for parsimony methods in phylogenetics, one must demonstrate a direct relationship between the number of character-state changes required by a tree topology and the complexity of the corresponding hypotheses. The connection is usually made by asserting that each instance of *homoplasy* (i.e. sharing of identical character states that cannot be explained by inheritance from the common ancestor of a group of taxa) constitutes an *ad hoc* hypothesis, and that the number of such ad hoc hypotheses should be minimized. Related arguments have focused on the concepts of falsifiability and corroboration most strongly associated with the writings of Karl Popper, suggesting that parsimony is the only method consistent with a hypothetico-deductive framework for hypothesis testing. However, as argued recently by de Queiroz and Poe (2001), careful interpretation of Popper's work does not lead unambiguously to parsimony as the method of choice. Furthermore, the linkage between parsimony and simplicity is tenuous, as highlighted by the recent work of Tuffley and Steel (1997). It demonstrates that parsimony and *likelihood* become equivalent under an extremely parameter-rich likelihood model that assigns a separate parameter for each character (site) on every branch of the tree, which is hardly a "simple" model. So, despite more than 20 years of ongoing debate between those who advocate the exclusive use of parsimony methods and those who favor maximum likelihood and related model-based statistical approaches, the issue remains unsettled and the camps highly polarized. Although we place ourselves firmly on the statistical side, we believe that parsimony methods will remain part of a complete phylogenetic-analysis toolkit for some time because they are fast and have been demonstrated to be quite effective in many situations (e.g. Hillis, 1996). In this sense, parsimony represents a useful "fallback" method when model-based methods cannot be used due to computational limitations.

Although parsimony methods are most effective when rates of evolution are slow (i.e. the expected amount of change is low), it is often asserted incorrectly that this is an "assumption" of parsimony methods. In fact, parsimony can perform extremely well under high rates of change as long as the lengths of the branches on the true underlying tree do not exhibit certain kinds of pathological inequalities (Hillis *et al.*, 1994) (see also *long-branch attraction*, Section 5.3). Nonetheless, it is difficult to state what the assumptions of parsimony analysis actually are. Conditions can be specified in which parsimony does very well in recovering the true evolutionary tree; however, alternative conditions can be found where it fails horribly. For example, in the so-called *Felsenstein zone*, standard parsimony methods converge to the wrong tree with increasing certainty as more data are accumulated. This is because a greater proportion of identical character states will be shared by chance between unrelated taxa than are shared by related taxa due to common ancestry (Felsenstein, 1978; Swofford *et al.*, 1996). Because parsimony requires no explicit assumptions other than the standard one of independence among characters (and some might even argue with that), all one can say is that it assumes that the conditions that would cause it to fail do not apply to the current analysis. Fortunately, these conditions are now relatively well understood. The combined use of model-based methods – less susceptible to long-branch attraction (see Section 5.3) and related artifacts but limited by computational burden – with faster parsimony methods that permit greater exploration of alternative tree topologies provides a mechanism for making inferences maintaining some degree of protection against the circumstances that could cause parsimony estimates alone to be misleading.

8.3 Parsimony analysis – methodology

The problem of finding optimal trees under the parsimony criterion can be separated into two subproblems: (1) determining the amount of character change, or tree length, required by any given tree; and (2) searching over all possible tree topologies for the trees that minimize this length. The first problem is easy and fast, whereas the second is slow due to the extremely large number of possible tree topologies for anything more than a small number of taxa.

8.3.1 Calculating the length of a given tree under the parsimony criterion

For n taxa, an unrooted binary tree (i.e. a fully bifurcating tree) contains n terminal nodes representing those sequences, $n - 2$ internal nodes, and $2n - 3$ branches (edges) that join pairs of nodes. Let τ represent some particular tree topology, which could be, for example, an arbitrarily chosen tree from the space of all possible trees. The length of this tree is given by

$$L(\tau) = \sum_{j=1}^{N} l_j \tag{8.1}$$

where N is the number of sites (characters) in the alignment and l_j is the length for a single site j. This length l_j is the amount of character change implied by a most parsimonious reconstruction that assigns a character state x_{ij} to each node i for each site j. For terminal nodes, the character state assignment is fixed by the input data. Thus, for binary trees:

$$l_j = \sum_{k=1}^{2N-3} c_{a(k),b(k)} \tag{8.2}$$

where $a(k)$ and $b(k)$ are the states assigned to the nodes at either end of branch k, and c_{xy} is the cost associated with the change from state x to state y. In the simplest form of parsimony (Fitch, 1971), this cost is simply 1 if x and y are different or 0 if they are identical. However, other cost schemes may be chosen. For example, one common scheme for nucleotide data is to assign a greater cost to *transversions* than to *transitions* (see Chapter 1), reflecting the observation that the latter occur more frequently in many genes and, therefore, are accorded less weight. The cost scheme can be represented as a *cost matrix*, or *step matrix*, that assigns a cost for the change between each pair of character states. In general, the cost matrix is symmetric (e.g. $c_{AG} = c_{GA}$), with the consequence that the length of the tree is the same regardless of the position of the root. If the cost matrix contains one or more elements for which $c_{xy} \neq c_{yx}$, then different rootings of the tree may imply different lengths, and the search among trees must be done over rooted trees rather than unrooted trees.

Direct algorithms for the determination of the l_j are available and are described briefly in this section. First, however, it will be instructive to examine the calculation of tree length using the brute-force approach of evaluating all possible character-state reconstructions. In general, with an alphabet size of r (e.g. $r = 4$ states for nucleotides, $r = 20$ states for amino acids) and T taxa, the number of these reconstructions for each site is equal to r^{T-2}. Consider the following example:

$$j$$

W ... ACA**G**GAT ...
X ... ACA**C**GCT ...
Y ... GTA**A**GGT ...
Z ... GCA**C**GAC ...

Suppose that the tree ((W, Y),(X, Z)) is being evaluated (see Fig. 5.5 about the NEWICK representation of phylogenetic trees), that the lengths for the first $j - 1$ sites have been calculated, and that the length of site j is to be determined next. Because there are four sequences, the number of reconstructions to be evaluated is $4^{(4-2)} = 16$. The lengths implied by each of these reconstructions under two different cost schemes are shown in Fig. 8.1. With equal costs, the minimum length is two steps, and this length is achievable in three ways (i.e. internal nodes assignment "A–C," "C–C," and "G–C"). If a similar analysis for the other two trees is conducted, both of the trees ((W, X),(Y, Z)) and ((W, Z),(Y, X)) are also found to have lengths of two steps. Thus, this character does not discriminate among

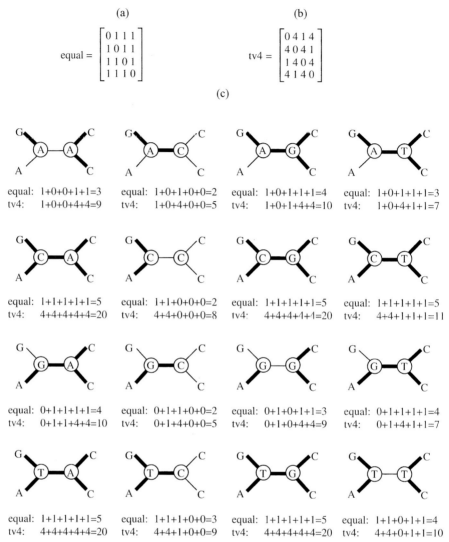

(a)

$$\text{equal} = \begin{bmatrix} 0 & 1 & 1 & 1 \\ 1 & 0 & 1 & 1 \\ 1 & 1 & 0 & 1 \\ 1 & 1 & 1 & 0 \end{bmatrix}$$

(b)

$$\text{tv4} = \begin{bmatrix} 0 & 4 & 1 & 4 \\ 4 & 0 & 4 & 1 \\ 1 & 4 & 0 & 4 \\ 4 & 1 & 4 & 0 \end{bmatrix}$$

(c)

equal: 1+0+0+1+1=3
tv4:   1+0+0+4+4=9

equal: 1+0+1+0+0=2
tv4:   1+0+4+0+0=5

equal: 1+0+1+1+1=4
tv4:   1+0+1+4+4=10

equal: 1+0+1+1+1=3
tv4:   1+0+4+1+1=7

equal: 1+1+1+1+1=5
tv4:   4+4+4+4+4=20

equal: 1+1+0+0+0=2
tv4:   4+4+0+0+0=8

equal: 1+1+1+1+1=5
tv4:   4+4+4+4+4=20

equal: 1+1+1+1+1=5
tv4:   4+4+1+1+1=11

equal: 0+1+1+1+1=4
tv4:   0+1+1+4+4=10

equal: 0+1+1+0+0=2
tv4:   0+1+4+0+0=5

equal: 0+1+0+1+1=3
tv4:   0+1+0+4+4=9

equal: 0+1+1+1+1=4
tv4:   0+1+4+1+1=7

equal: 1+1+1+1+1=5
tv4:   4+4+4+4+4=20

equal: 1+1+1+0+0=3
tv4:   4+4+1+0+0=9

equal: 1+1+1+1+1=5
tv4:   4+4+4+4+4=20

equal: 1+1+0+1+1=4
tv4:   4+4+0+1+1=10

Fig. 8.1   Determination of the length of a tree by brute-force consideration of all possible state assignments to the internal nodes. Calculations are for one site of one of the possible trees for the four taxa W, X, Y, and Z (character states: A, C, G, T). (a) Cost matrix that assigns equal cost to all changes from one nucleotide to another. (b) Cost matrix that assigns four times as much weight to transversions as to transitions (rows and columns are ordered A, C, G, T). (c) All 16 possible combinations of state assignments to the two internal nodes and the lengths under each cost scheme. Minimum lengths are two steps for equal costs and five steps for 4 : 1 tv : ti weighting.

the three tree topologies and is said to be *parsimony-uninformative* under this cost scheme. With 4 : 1 transversion : transition weighting, the minimum length is five steps, achieved by two reconstructions (i.e. internal node assignments "A–C" and "G–C"). However, similar evaluation of the other two trees (not shown) finds a minimum of eight steps on both trees (i.e. two transversions are required rather than one transition plus one transversion). Under these unequal costs, the character becomes informative in the sense that some trees have lower lengths than others, which demonstrates that the use of unequal costs may provide more information for phylogeny reconstruction than equal-cost schemes.

The method used in this example, in principle, could be applied to every site in the data set, summing these lengths to obtain the total tree length for each possible tree, and then choosing the tree that minimizes the total length. Obviously, for real applications, a better way is needed for determining the minimum lengths that does not require evaluation of all r^{n-2} reconstructions. A straightforward **dynamic programming** algorithm (Sankoff & Rousseau, 1975) provides such a method for general cost schemes; the methods of Farris (1970), Fitch (1971), and others handle special cases of this general system with simpler calculations. Sankoff's algorithm is illustrated using the example in Box 8.1; the original papers may be consulted for a full description of the algorithm. Dynamic programming operates by solving a set of subproblems and by assembling those solutions in a way that guarantees optimality for the full problem. In this instance, the best length that can be achieved for each subtree – given each of the possible state assignments to each node – is determined, moving from the tips toward the root of the tree. Upon arriving at the root, an optimal solution for the full tree is guaranteed. A nice feature of this algorithm is that, at each stage, only one pair of vectors of conditional subtree lengths above each node needs to be referenced; knowing how those subtree lengths were obtained is unnecessary.

For simple cost schemes, the full dynamic programming algorithm described in Box 8.1 is not necessary. Farris (1970) and Fitch (1971) described algorithms for characters with ordered and unordered states, respectively, which can be proven to yield optimal solutions (Hartigan, 1973; Swofford & Maddison, 1987); these two algorithms are equivalent for binary (i.e. two-state) characters. Fitch's algorithm, illustrated in Box 8.2, is relevant for sequence data when the cost of a change from any state to any other state is assumed to be equal.

8.4 Searching for optimal trees

Having specified a means for calculating the score of a tree under our chosen criterion, the more difficult task of searching for an optimal tree can be confronted. The methods described in the following sections can be used for parsimony,

Box 8.1 Calculation of the minimum tree length under general cost schemes using Sankoff's algorithm

For symmetric cost matrixes, we can root an unrooted tree arbitrarily to determine the minimum tree length. Then, for each node i (labeled in boldface italics), we compute a conditional-length vector $\mathbf{S}_{ij}$ containing the minimum possible length above i, given each of the possible state assignments to this node for character j (for simplicity, we drop the second subscript because only one character is considered in turn). Thus, s_{ik} is the minimum possible length of the subtree descending from node i if it is assigned state k. For the tip sequences, this length is initialized to 0 for the state(s) actually observed in the data or to infinity otherwise. The algorithm proceeds by working from the tips toward the root, filling in the vector at each node based on the values assigned to the node's children (i.e. immediate descendants).

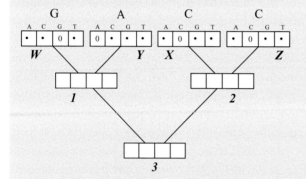

We now visit node 1. For each element k of this vector, we consider the costs associated with each of the four possible assignments to each of the child nodes W and Y, and the cost needed to reach these states from state k, which is obtained from the cost matrix (**C** in this example, we use the cost matrix from Fig. 8.1B, which represents a 4 : 1 tv : ti weighting). This calculation is trivial for nodes ancestral to two terminal nodes because only one state needs to be considered for each child. Thus, if we assign state A to node 1, the minimum length of the subtree above node 1 given this assignment is the cost of a change from A to G in the left branch, plus the cost of a (non-) change from A to A in the right branch: $s_{1A} = c_{AG} + c_{AA} = 1 + 0 = 1$. Similarly, s_{1C} is the sum of c_{CG} (left branch) and c_{CA} (right branch), or 8. Continuing in this manner, we obtain

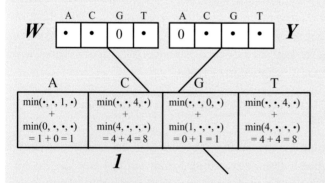

for the subtree of node 1.

The calculation for node 2 proceeds analogously:

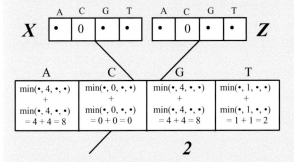

The calculation for the root node (node 3) is somewhat more complicated: for each state k at this node, we must explicitly consider each of the four state assignments to each of the child nodes 1 and 2. For example, when calculating the length conditional on the assignment of state A to node 3, for the left branch we consider in turn all four of the assignments to node 1. If node 1 is assigned state A as well, the length would be the sum of 1 (for the length above node 1) plus 0 (for the non-change from state A to state A). If we instead choose state C for node 1, the length contributed by the left branch would be 8 (for the length above node 1) plus 4 (for the change from A to C). The same procedure is used to determine the conditional lengths for the right branch. By summing these two values for each state k, we obtain the entire conditional-length vector for node 3:

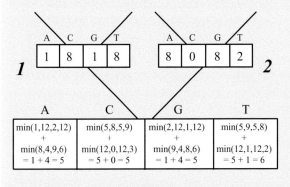

Since we are now at the root of the tree, the conditional-length vector s_3 provides the minimum possible lengths for the full tree given each of the four possible state assignments to the root, and the minimum of these values is the tree length we seek. Observe that this length, 5, is the same value we obtained using the brute-force enumeration shown in Fig. 8.1. As an exercise, the reader may wish to verify that other rootings yield the same length.

This algorithm provides a means of calculating the length required by any character on any tree under any cost scheme. We can obtain the length of a given tree by repeating the procedure outlined here for each character and summing over characters. In principle, we can then find the most parsimonious tree by generating and evaluating all possible trees, although this exhaustive-search strategy would only be feasible for a relatively small number of sequences (i.e. 11 in the current version of PAUP*).

Box 8.2 Calculation of the minimum tree length using Fitch's algorithm for equal costs

As for the general case, we can root an unrooted tree arbitrarily to determine the minimum tree length. We will assign a **state set** X_i to each node i; this set represents the set of states that can be assigned to each node so that the minimum possible length of the subtree above that node can be achieved. Also, for each node, we will store an accumulated length s_i, which represents the minimum possible (unconditional) length in the subtree descending from node i.

The state sets for the terminal nodes are initialized to the states observed in the data, and the accumulated lengths are initialized to zero. For the example of Fig. 8.1, this yields:

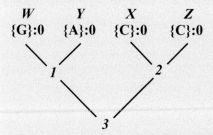

For binary trees, the state-set calculation for each internal node i follows two simple rules, based on the state sets and accumulated lengths of the two descendant child nodes, denoted $L(i)$ and $R(i)$:

(1) Form the intersection of the two child state sets: $X_{L(i)} \cap X_{R(i)}$. If this intersection is non-empty, let X_i equal this intersection. Set the accumulated length for this node to be the sum of the accumulated lengths for the two child nodes: $s_i = s_{L(i)} + s_{R(i)}$.
(2) If the intersection was empty (i.e. $X_{L(i)}$ and $X_{R(i)}$ are disjoint), let X_i equal the union of the two child state sets: $X_{L(i)} \cup X_{R(i)}$. Set the accumulated length for this node to be the sum of the accumulated lengths for the two child nodes *plus one* : $s_i = s_{L(i)} + s_{R(i)} + 1$.

Proceeding to node 1 in our example, we find that the intersection of $\{G\}$ and $\{A\}$ is the empty set (ϕ); therefore, by Rule 2, we let $X_1 = \{A\} \cup G = \{A, G\}$ and $s_1 = 0 + 0 + 1 = 1$:

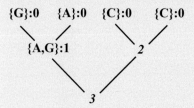

For node 2, the intersection of {C} and {C} is (obviously) {C}; therefore, by Rule 1, we obtain:

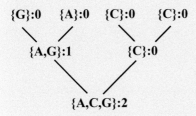

{G}:0   {A}:0   {C}:0   {C}:0

{A,G}:1   {C}:0

3

Finally, for node 3, the intersection of {A, G} and {C} $= \phi$, so the state set at the root is equal to the union of these sets, and the corresponding accumulated length equals $1 + 0 + 1 = 2$:

{G}:0   {A}:0   {C}:0   {C}:0

{A,G}:1   {C}:0

{A,C,G}:2

Thus, the length required by the tree is equal to two steps, as we obtained previously using the brute-force approach. As an exercise, the reader may wish to verify that other rootings of the tree yield the same length.

least-squares distance criteria, and maximum likelihood. Regrettably, this search is complicated by the huge number of possible trees for anything more than a small number of taxa.

8.4.1 Exact methods

For 11 or fewer taxa, a brute-force **exhaustive search** is feasible; an algorithm is needed that can guarantee generation of all possible trees for evaluation using the previously described methods. The procedure outlined in Box 8.3 is used in PAUP*. This algorithm recursively adds the tth taxon in a stepwise fashion to all possible trees containing the first $t - 1$ taxa until all n taxa have been joined. The algorithm is easily modified for rooted trees by including one additional artificial taxon that locates the root of each tree. In this case, the first three trees generated represent each of the three possible rootings of an unrooted three-taxon tree, and the algorithm proceeds as in the unrooted case. Thus, the number of rooted trees for n taxa is equal to the number of unrooted trees for $n + 1$ taxa.

Box 8.3 Generation of all possible trees

The approach for generation of all possible unrooted trees is straightforward. Suppose that we have a data set containing six sequences (i.e. taxa). We begin with the only tree for the first three taxa in the data set, and connect the fourth taxon to each of the three branches on this tree.

This generates all three of the possible unrooted trees for the first four taxa. Now, we connect the fifth taxon, E, to each branch on each of these three trees. For example, we can join taxon E to the tree on the right above:

By connecting taxon E to the other two trees in a similar manner, we generate all 15 of the possible trees for the first five taxa. Finally, we connect the sixth taxon, F, to all locations on each of these 15 trees (7 branches per tree), yielding a total of 105 trees. Thus, the full-search tree can be represented as follows (only the paths toward the full 6-taxon trees are shown):

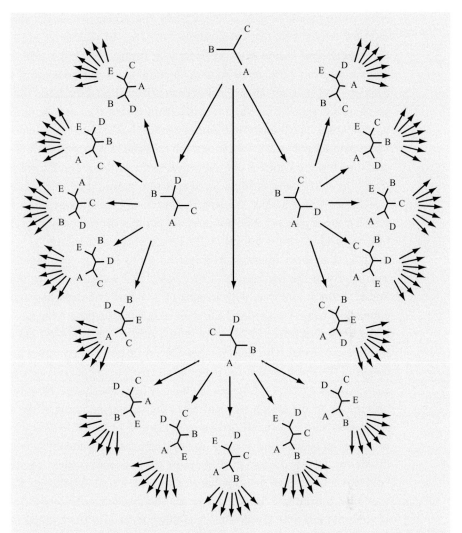

The lengths (or likelihoods or distance scores) of each of these 105 trees can now be evaluated, and the set of optimal trees identified.

It is clear from the description of the algorithm for generating all possible trees in Box 8.3 that the number of possible trees grows by a factor that increases by two with each additional taxon, as expressed in the following relationship:

$$B(t) = \prod_{i=3}^{t} (2i - 5) \tag{8.3}$$

where $B(t)$ is the number of unrooted trees for t taxa. For example, the number of unrooted trees for 7 taxa is $1 \times 3 \times 5 \times 7 \times 9 = 945$ and the number of unrooted trees for 20 taxa is over 2×10^{20}. Clearly, the exhaustive-search method can be used for only a relatively small number of taxa. An alternative exact procedure, the **branch-and-bound method** (Hendy & Penny, 1982), is useful for data sets containing from 12 to 25 or so taxa, depending on the "messiness" of the data. This method operates by implicitly evaluating all possible trees, but cutting off paths of the search tree when it is determined that they cannot possibly lead to optimal trees. The branch-and-bound method is illustrated for a hypothetical six-taxon data set in Fig. 8.2. We present the example as if parsimony is the optimality criterion, but this choice is not important to the method. The algorithm effectively traces the same route through the search tree as would be used in an exhaustive search (see Box 8.3), but the length of each tree encountered at a node of the search tree is evaluated even if it does not contain the full set of taxa. Throughout this traversal, an upper bound on the length of the optimal tree(s) is maintained; initially, this upper bound can simply be set to infinity. The traversal starts by moving down the left branch of the search tree successively connecting taxa D and E to the initial tree, with lengths of 221 and 234 steps, respectively. Then, connecting taxon F provides the first set of full-tree lengths. After this connection, it is known that a tree of 241 steps exists, although it is not yet known whether this tree is optimal. This number therefore, is taken as a new upper bound on the length of the optimal tree (i.e. the optimal tree length cannot be longer than 241 steps because a tree at this length has already been identified). Now, the algorithm backtracks on the search tree and takes the second path out of the 221-step, 4-taxon tree. The 5-taxon tree containing taxon E obtained by following this path requires 268 steps. Thus, there is no point in evaluating the seven trees produced by connecting taxon F to this tree because they cannot possibly require fewer than 268 steps, and a tree of 241 steps has already been found. By cutting off paths in this way, large portions of the search tree may be avoided and a considerable amount of computation time saved. The algorithm proceeds to traverse the remainder of the search tree, cutting off paths where possible and storing optimal trees when they are found. In this example, a new optimal tree is found at a length of 229 steps, allowing the upper bound on the tree length to be further reduced. Then, when the 233-step tree containing the first five taxa is encountered, the seven trees that would be derived from it can be immediately rejected because they would also require at least 233 steps. The algorithm terminates when the root of the search tree has been visited for the last time, at which time all optimal trees will have been identified.

Several refinements to the branch-and-bound method improve its performance considerably. Briefly, these include (1) using a heuristic method such as **stepwise addition** (discussed later in this section) or NJ (see Chapter 5) to find a tree whose length provides a smaller initial upper bound, which allows earlier termination of

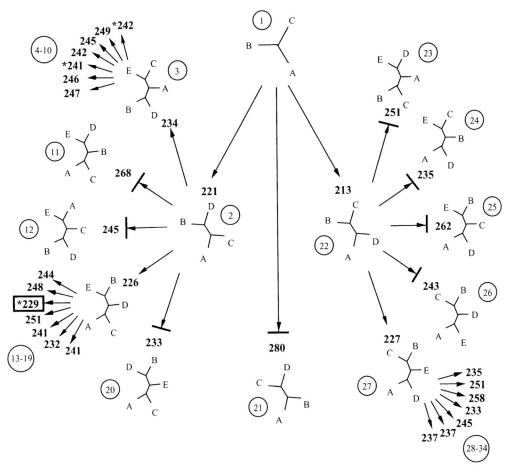

Fig. 8.2   The branch-and-bound algorithm for the exact solution of the problem of finding an optimal parsimony tree. The search tree is the same as shown in Box 8.3, with tree lengths for a hypothetical data set shown in boldface type. If a tree lying at a node of this search tree (thus joining fewer taxa) has a length that exceeds the current lower bound on the optimal tree length, this path of the search tree is terminated (indicated by a cross-bar), and the algorithm backtracks and takes the next available path. When a tip of the search tree is reached (i.e. when we arrive at a tree containing the full set of taxa), the tree is either optimal (and therefore retained) or suboptimal (and rejected). When all paths leading from the initial three-taxon tree have been explored, the algorithm terminates, and all most parsimonious trees will have been identified. Asterisks indicate points at which the current lower bound is reduced. See text for additional explanation; circled numbers represent the order in which phylogenetic trees are visited in the search tree.

search paths in the early stages of the algorithm; (2) ordering the sequential addition of taxa in a way that promotes earlier cutoff of paths (rather than just adding them in order of their appearance in the data matrix); and (3) using techniques such as pairwise character incompatibility to improve the *lower* bound on the minimum length of trees that can be obtained by continuing outward traversal of the search tree (allowing earlier cutoffs). All of these refinements are implemented in PAUP* (see the program documentation for further information).

The branch-and-bound strategy may be used for any optimality criterion whose objective function is guaranteed to be non-decreasing as additional taxa are connected to the tree. Obviously, this is true for parsimony; increasing the variability of the data by adding additional taxa cannot possibly lead to a decrease in tree length. It is also true for maximum likelihood and some distance criteria, including least-squares methods that score trees by minimizing the discrepancy between observed and path-length distances (see Chapter 5). However, it does not work for the minimum-evolution distance criterion. In minimum evolution, one objective function is optimized for the computation of branch lengths (i.e. least-squares fit), but a different one is used to score the trees (i.e. sum of branch lengths). Unfortunately, the use of these two objective functions makes it possible for the minimum-evolution score to decrease when a new taxon is joined to the tree, invalidating the use of the branch-and-bound method in this case.

8.4.2 Approximate methods

When data sets become too large for the exact searching methods described in the previous section to be used, it becomes necessary to resort to the use of heuristics: approximate methods that attempt to find optimal solutions but provide no guarantee of success. In PAUP* and several other programs, a two-phase system is used to conduct approximate searches. In the simplest case, an initial starting tree is generated using a "greedy" algorithm that builds a tree sequentially according to some set of rules. In general, this tree will not be optimal, because decisions are made in the early stages without regard for their long-term implications. After this starting tree is obtained, it is submitted to a round of perturbations in which neighboring trees in the perturbation scheme are evaluated. If some perturbation yields a better tree according to the optimality criterion, it becomes the new "best" tree and it is, in turn, submitted to a new round of perturbations. This process continues until no better tree can be found in a full round of the perturbation process.

One of the earliest and still most widely used greedy algorithms for obtaining a starting tree is *stepwise addition* (e.g. Farris, 1970), which follows the same kind of search tree as the branch-and-bound method described previously. However, unlike the exact exhaustive enumeration of branch-and-bound methods, stepwise addition commits to a path out of each node on the search tree that looks most promising at the moment, which is not necessarily the path leading to a global optimum. In the example in Fig. 8.2, Tree 22 is shorter than Trees 2 or 21; thus, only trees derivable from Tree 22 remain as candidates. Following this path ultimately leads to selection of a tree of 233 steps (Fig. 8.3), which is only a local rather than a global optimum. The path leading to the optimal 229-step tree was rejected because it appeared less promising at the 4-taxon stage. This tendency to become "stuck" in local optima is a common property of greedy heuristics, and they are often called *local-search methods* for that reason.

Because stepwise addition rarely identifies a globally optimal tree topology for real data and any non-trivial number of taxa, other methods must be used to improve the solution. One such class of methods involves tree-rearrangement perturbations known as *branch-swapping*. These methods all involve cutting off one or more pieces of a tree (subtrees) and reassembling them in a way that is locally different from the original tree. Three kinds of branch-swapping moves are used in PAUP*, as well as in other programs. The simplest type of rearrangement is a *nearest-neighbor interchange (NNI)*, illustrated in Fig. 8.4. For any binary tree containing T terminal taxa, there are $T - 3$ internal branches. Each branch is visited, and the two topologically distinct rearrangements that can be obtained by swapping a subtree connected to one end of the branch with a subtree connected to the other end of the branch are evaluated. This procedure generates a relatively small number of perturbations whose lengths or scores can be compared to the original tree. A more extensive rearrangement scheme is *subtree pruning and regrafting (SPR)*, illustrated in Fig. 8.5, which involves clipping off all possible subtrees from the main tree and reinserting them at all possible locations, but avoiding pruning and grafting operations that would generate the same tree redundantly. The most extensive rearrangement strategy available in PAUP* is *tree bisection and reconnection (TBR)*, illustrated in Fig. 8.6. TBR rearrangements involve cutting a tree into two subtrees by cutting one branch, and then reconnecting the two subtrees by creating a new branch that joins a branch on one subtree to a branch on the other. All possible pairs of branches are tried, again avoiding redundancies.

Note that the set of possible NNIs for a tree is a subset of the possible SPR rearrangements, and that the set of possible SPR rearrangements is, in turn, a subset of the possible TBR rearrangements. For TBR rearrangements, a "reconnection distance" can be defined by numbering the branches from zero starting at the cut branch (Fig. 8.6c and 8.6e). The reconnection distance is then equal to the sum of numbers of the two branches that are reconnected. The reconnection distances then have the following properties: (1) NNIs are the subset of TBRs that have a reconnection distance of 1; (2) SPRs are the subset of TBRs so that exactly one of the

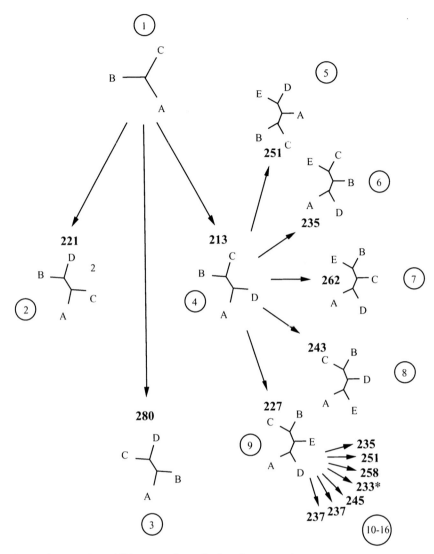

Fig. 8.3   A greedy stepwise-addition search applied to the example in Fig. 8.2. The best four-taxon tree is determined by evaluating the lengths of the three trees obtained by joining Taxon D to Tree 1 containing only the first three taxa. Taxa E and F are then connected to the five and seven possible locations, respectively, on Trees 4 and 9, with only the shortest trees found during each step being used for the next step. In this example, the 233-step tree obtained is not a global optimum (see Fig. 8.2). Circled numbers indicate the order in which phylogenetic trees are evaluated in the stepwise-addition search.

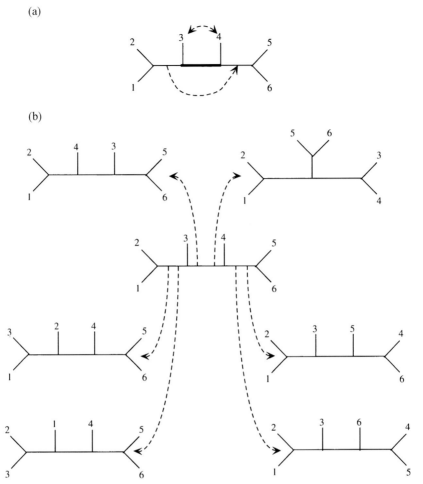

Fig. 8.4   Nearest-neighbor interchange (NNI) rearrangements. (a) An NNI around the central branch. (b) All six of the possible NNIs for the tree in (a).

two reconnected branches is numbered zero; and (3) TBRs that are neither NNIs nor SPRs are those for which both reconnected branches have non-zero numbers. The reconnection distance can be used to limit the scope of TBR rearrangements tried during the branch-swapping procedure.

The default strategy used for each of these rearrangement methods is to visit branches of the "current" tree in some arbitrary and predefined order. At each branch, all of the non-redundant branch swaps are tried and the score of each resulting tree is obtained (e.g. using the methods described in Box. 8.1 for parsimony). If a rearrangement is successful in finding a shorter tree, the previous tree is discarded and the rearrangement process is restarted on this new tree. If

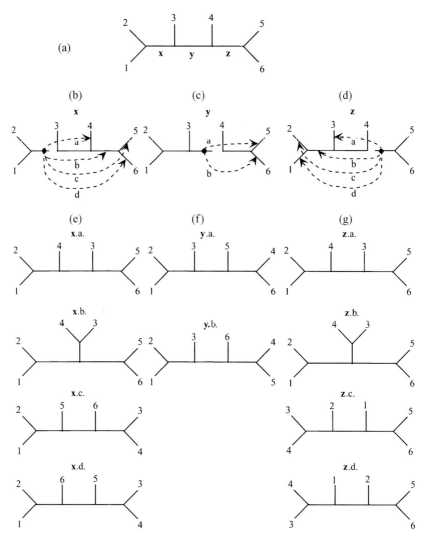

Fig. 8.5   Subtree pruning-regrafting (SPR) rearrangements. (a) A tree to be rearranged. (b), (c), (d)
SPRs resulting from pruning of branches **x**, **y**, and **z**, respectively. In addition to these
rearrangements, all terminal taxa (i.e. leaves) would be pruned and reinserted elsewhere
on the tree. (e), (f), (g) Trees resulting from regrafting of branches **x**, **y**, and **z**, respectively,
to other parts of the tree.

all possible rearrangements have been tried without success in finding a better
tree, the swapping process terminates. Optionally, when trees are found that are
equal in score to the current tree (e.g. equally parsimonious trees or trees that have
identical likelihoods within round-off error), they are appended to a list of optimal
trees. In this case, when the arrangement of one tree finishes, the next tree in the
list is obtained and input to the branch-swapping algorithm. If a rearrangement

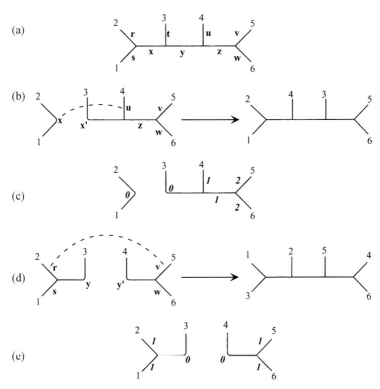

Fig. 8.6   Tree bisection–reconnection (TBR) rearrangements. (a) A tree to be rearranged. (b) Bisection of branch **x** and reconnection to branch **u**; other TBRs would connect **x** to **z**, **v**, and **w**,
respectively. (c) Branch-numbering for reconnection distances involving branch **x** (see text).
(d) Bisection of branch **y** and reconnection of branch **r** to **v**; other TBRs would connect **r** to
w, **r** to **y'**, **s** to **v**, **s** to **w**, **s** to **y'**, **y** to **v**, and **y** to **w**, respectively. (e) Branch-numbering for
reconnection distances involving branch **y** (see text). All other branches, both internal and
external, also would be cut in a full round of TBR swapping.

of this next tree yields a better tree than any found so far, all trees in the current
list are discarded and the entire process is restarted using the newly discovered
tree. The algorithm then terminates when every possible rearrangement has been
tried on each of the stored trees. In addition to identifying multiple and equally
good trees, this strategy often identifies better trees than would be found if only
a single tree were stored at any one time. This can happen when all of the trees
within one rearrangement of the current tree are no better than the current tree;
however, some of the adjacent trees can, in turn, be rearranged to yield trees that are
better.

   Although they are often quite effective, **hill-climbing algorithms**, such as the
branch-swapping methods implemented in PAUP*, are susceptible to the problem
of entrapment in local optima. By only accepting proposed rearrangements that

are equal to, or better than, the current best tree, these algorithms eventually reach the peak of the slope on which they start; however, the peak may not represent a global optimum. One generally successful method for improving the chances of obtaining a globally optimal solution is to begin the search from a variety of starting points in the hope that at least one of them will climb the right hill. An option available in PAUP* is to start the search from randomly chosen tree topologies. However, in practice, these randomly chosen trees usually fit the data so poorly that they end up merely climbing a foothill on a rugged landscape, and usually fail to find trees that are as good as those resulting from using a starting tree obtained by stepwise addition. An alternative method takes advantage of the fact that, for data sets of non-trivial size and complexity, varying the sequence in which taxa are added during stepwise addition may produce different tree topologies that each fit the data reasonably well. Starting branch-swapping searches from a variety of random-addition-sequence replicates thereby provides a mechanism for performing multiple searches that each begins at a relatively high point on some hill, increasing the probability that the overall search will find an optimal tree.

Random-addition-sequence searches are also useful in identifying multiple "islands" of trees (Maddison, 1991) that may exist. Each island represents all of the trees that can be obtained by a sequence of rearrangements, starting from any tree in the island, keeping and rearranging all optimal trees that are discovered. If two optimal trees exist so that it is impossible to reach one tree by a sequence of rearrangements starting from the other without passing through trees that are suboptimal, these trees are on different islands. Because trees from different islands tend to be topologically dissimilar, it is important to detect multiple islands when they exist.

The methods described previously are generally effective for data sets containing up to 100 or so taxa. However, for larger data sets, they are not as efficient as some newer methods that use a variety of stochastic-search and related algorithms that are better able to avoid entrapment in local optima (Lewis, 1998; Goloboff, 1999; Nixon, 1999; Moilanen, 2001). Nixon's (1999) "parsimony ratchet" can be implemented in PAUP* using the PAUPRAT program (Sikes & Lewis, 2001). In Section 6.4 of Chapter 6, different heuristic approaches in a maximum likelihood framework are discussed.

PRACTICE

David L. Swofford and Jack Sullivan

The basic capabilities of PAUP* (Swofford, 2002; *http://paup.csit.fsu.edu*) are illustrated by analyzing the three different data sets. Because PAUP* does not currently support model-based analyses of amino-acid sequence data, only parsimony analysis and basic tree searching are described using a glycerol-3-phosphate dehydrogenase data set. Distance methods and a variety of additional capabilities are demonstrated using an amniote mitochondrial DNA data set. Finally, the flexibility of PAUP* for evaluating and comparing maximum likelihood models is described through the use of an HIV data set. These data sets are also used in other chapters in this book and are available at *www.thephylogenetichandbook.org*. By performing the three different analyses, we illustrate the convenience and power of being able to switch easily between different optimality criteria in a single program. Versions of PAUP* are available with a full graphical user interface (Macintosh, system 9 or earlier), a partial graphic user interface (Microsoft Windows), and a command–line-only interface (UNIX/Linux, MacOSX, and Microsoft Windows console). Because the command–line interface is available on all platforms, it is used exclusively for the analyses that follow. For each example, it is assumed that PAUP* has been successfully installed and is invoked by typing paup at the operating-system prompt, or double-clicking the executable application in the PAUP* folder (Windows versions). Instructions on how to purchase and install PAUP* for different platforms can be found at *http://paup.csit. fsu.edu*. At the time of writing, great effort is being invested into a major update of the program.

PAUP* uses the NEXUS format for input data files and command scripts. The details of this format, a command-reference manual, and quick-start tutorial describing data-file formats are available at *http://paup.csit.fsu.edu/downl.html*. Box 8.4 covers these topics and outlines how to convert the example data sets from PHYLIP to NEXUS format.

Box 8.4 The PAUP* program

The PAUP* (Phylogenetic Analysis Using Parsimony* and other methods) program is distributed by Sinauer Associates and purchasing information can be found at *http://paup.csit.fsu.edu*. The PAUP* installer creates a PAUP folder on the local computer containing the executable application plus a number of other files, including sample data files in NEXUS format and an extensive documentation in pdf format. Versions of the program are available for Macintosh, MS Windows, and UNIX/Linux. The Mac version is controlled using either a full graphical interface (MacOS 9 or earlier) or a command–line interface; only the command–line interface is available for the other platforms. At the time of writing, however, graphical interfaces are being developed for all platforms. Because it is intuitively easy to switch to the Mac graphic interface for the user familiar with the PAUP* commands, this book focuses on how to use the program through the command–line interface. What follows is a brief introduction to the NEXUS format and the main PAUP* commands. More details can be found in the quick-start tutorial available at *http://paup.csit.fsu.edu/downl.html* and in the pdf documentation in the PAUP folder.

The NEXUS format
PAUP* reads input files in NEXUS format. A NEXUS file must begin with the string
#NEXUS
on the first line. The actual sequences plus some additional information about the data set are then written in the next lines within a so-called data block. A data block begins with the line
Begin data;
and it must end with the line
End;
Here is an example of a NEXUS file containing four sequences ten nucleotides long:
(Dimensions ntax = 4 nchar = 10):
#NEXUS
        Begin data;
        Dimensions ntax = 4 nchar = 10;
        Format datatype = nucleotide gap = — missing = ? interleave;
        Matrix
L20571   ATGACAG-AA
AF10138 ATGAAAG-AT
X52154   AT?AAAGTAT
U09127   ATGA??GTAT
;
End;

The sequences are DNA with gaps indicated by a – symbol, and missing characters indicated by a ? symbol (format datatype = nucleotide gap = — missing = ? matchchar = . Interleave). The aligned sequences are reported on the line after the keyword Matrix. Each sequence starts on a new line with its name followed by a few blank spaces and then the nucleotide (or amino-acid) sequence itself. A semicolon (;) must be placed in the line after the last sequence (to indicate the end of the Matrix block).

Running PAUP* (MS Windows version)
By double-clicking the PAUP executable inside the PAUP folder (the version available at the time of this writing is called win-paup4b10.exe), an Open window appears asking to select a file to be edited or executed. By choosing edit, the file is opened in a text-editor window, so that it is possible, for example, to convert its format to NEXUS by manual editing. If the file is already in NEXUS format, it can be executed by selecting Execute in the Open window. The PAUP* Display window appears showing information about the data in the NEXUS file (e.g. number of sequences, number of characters) which are now ready to be analyzed by entering PAUP* commands in the command–line at the bottom of the PAUP* Display window. By clicking on cancel, an empty PAUP* Display window appears. To quit

the program, just click the x in the upper right corner of the PAUP* Display window, or choose exit from the File menu.

During a phylogenetic analysis with PAUP*, it is good practice to keep a record of all the different steps, displayed in the program Display window, in a log file. To do that at the beginning of a new analysis, choose Log output to disk from the File menu. The log file can be viewed later with any text editor, including PAUP* working in edit mode (described previously). The program can convert aligned sequences from PHYLIP and other formats to NEXUS. In the following example, the file hivALN.phy (available at *www.thephylogenetichandbook.org*) is converted from PHYLIP to NEXUS format:

(1) Place the file hivALN.phy in the PAUP folder.
(2) Run PAUP* by double-clicking the executable icon and click on cancel.
(3) Type in the command–line the following:
(4) toNexus fromFile = hivALN.phy toFile = hivALN.nex;
(5) Click execute below the command–line or press Enter.

A file called hivALN.nex will appear in the PAUP folder.

The command toNexus converts files to the NEXUS format. The option fromFile = indicates the input file with the sequences in format other than NEXUS. toFile = is used to name the new NEXUS file being created. Note that PAUP* is case-insensitive, which means that, for example, the commands toNexus and TONEXUS are exactly the same. In what follows, uppercase and lowercase in commands or options are only used for clarity reasons. The file gdp.phy, containing the protein example data set (available at *www.thephylogenetichandbook.org*) in PHYLIP interleaved format (see Box 2.4), is translated into NEXUS with the following command block:

```
toNexus fromFile = gdp.phy toFile = gdp.nex Datatype = protein
Interleaved = yes;
```

Note that it is necessary to specify protein for the option Datatype and to set Interleaved to yes (the starting default option is no).

Commands and options
As discussed in this chapter, PAUP* can perform distance-, parsimony-, and maximum-likelihood-based phylogenetic analyses of molecular sequence data. Such analyses are carried out by entering commands on the command–line. For most commands, one or more options may be supplied. Each option has a default value, but the user can assign an alternative value among those allowed for that particular option by typing option = value. Any command and its options with assigned values can be typed in the command–line at the bottom of the PAUP* Display window and executed by clicking execute below the command–line or pressing Enter. A semicolon (;) is used to indicate the end of a particular command. If desired, multiple commands can be entered, each separated from the next by a semicolon, in which case the commands are executed in the order they appear in the command–line. The complete list of available options for a given command can be obtained by entering
command–name ?;
in the PAUP* command–line, where *command-name* represents the name of a valid PAUP* command. In this way, it is also possible for the user to check the current value of each option. Try, for example, to type and execute Set ? (when only one command is executed, the semicolon can be omitted). The following list will be displayed in the PAUP* window:

```
Usage: Set [options ...];
Available options:
Keyword  ── Option type ─────────── Current default-
Criterion    Parsimony|Likelihood|Distance Parsimony
MaxTrees     <integer-value> 100
```
This list is an abridged version of the possible options and their current values for the `Set` command. `Set` is used to choose whether to perform a parsimony-, maximum-likelihood-, or distance-based analysis by assigning the appropriate value – selected among the `Option` type in the list – to the `Criterion` option. By default, PAUP* performs parsimony-based analysis. By typing `set Criterion=Likelihood` or `set Criterion=Distance` in the command line, the analysis criterion switches to maximum likelihood or distance, respectively.

MaxTrees is another option that can be used in conjunction with the `Set` command. `MaxTrees` controls the maximum number of trees that will be saved by PAUP* when a heuristic tree search with parsimony or maximum likelihood is performed (see Sections 8.4.2 and 8.6). The default value is 100, but the user can enter a different integer number (e.g. 1000) by typing `set MaxTrees = 1000`.

8.5 Analyzing data with PAUP* through the command–line interface

A typical phylogenetic analysis with PAUP* might follow these steps:

(1) Choose a phylogenetic optimality criterion: parsimony (default when the program starts), maximum likelihood (Set criterion=likelihood) or distance `Set criterion=distance`).

(2) Choose the appropriate settings for the selected parsimony, likelihood, or distance analysis with the `PSet`, `LSet`, or `DSet` command, respectively.

(3) Compute the tree topology with the HSearch (all optimality criteria) or NJ (distance) command (see Section 8.7).

(4) Evaluate the reliability of the inferred tree(s) with bootstrapping (`Bootstrap` command). Also, for maximum likelihood, the zero-branch-length test is available (see Section 8.8).

(5) Examine the tree in the PAUP* `Display` window with the `DescribeTrees` command; or save it in a tree file with the `SaveTrees` command by typing `savetrees file=tree – file – name`).

A note of caution: By default, the `SaveTrees` command saves trees in NEXUS format without including branch lengths. To include branch lengths, type:

```
savetrees file=tree-file-name brlens=yes;
```

Trees in NEXUS format can be opened with TREEVIEW of FIGTREE (see Chapter 5) or reimported in PAUP* (after executing the original sequence data file) with the following command:

```
gettrees file=tree-file-name;
```

Complete details can be found in the pdf manual in the PAUP folder.

8.6 Basic parsimony analysis and tree-searching

Assuming that the file gpd.nex is located in the current directory, start PAUP* and execute the file (execute gpd.nex). The number of taxa in this data set (i.e. 12) is effectively the maximum for which an exhaustive search is feasible. This can be accomplished using the command:

```
alltrees fd=histogram;
```

The fd=histogram option requests output of the frequency distribution of tree lengths in the form of a histogram with the default number (i.e. 20) of class intervals. The resulting output follows:

```
Exhaustive search completed:
  Number of trees evaluated = 654729075
  Score of best tree found = 812
  Score of worst tree found = 1207
  Number of trees retained = 1
  Time used = 00:59:02 (CPU time = 00:57:27.8)


Frequency distribution of tree scores:

                mean = 1108.333294 sd = 48.441980
                g1 = -0.817646    g2 = 0.531509


  812.00000 +-----------------------------------
  831.75000 | (337)
  851.50000 | (4337)
  871.25000 | (26850)
  891.00000 | (96474)
  910.75000 | (305244)
  930.50000 | (761800)
  950.25000 |# (1797170)
  970.00000 |## (3373829)
  989.75000 |#### (6937378)
 1009.50000 |####### (12789821)
 1029.25000 |############ (21003098)
 1049.00000 |################# (32167531)
 1068.75000 |#################### (49814274)
 1088.50000 |###################### (70216478)
 1108.25000 |####################### (89088850)
 1128.00000 |########################### (98980066)
 1147.75000 |############################## (114383861)
 1167.50000 |############################# (103567496)
 1187.25000 |######################### (46015542)
 1207.00000 |## (3398639)
            +-----------------------------------
```

Thus, it is known with certainty that the shortest tree requires 812 steps and that there is only one tree of this length. Furthermore, although there are many trees slightly longer than the shortest tree, these "good" trees represent only a small fraction of the more than 6.5×10^8 possible trees for 12 taxa. This search requires significant computation time (i.e. approximately 1 hour on a moderately fast Intel Core 2 Duo Powerbook). If the distribution of tree lengths is not of interest, the branch-and-bound algorithm may be used to find an exact solution more quickly by typing:

```
bandb;
```

This search takes less than 1 second and finds the same MP tree. If the data set had been much larger, finding exact solutions using the exhaustive-search or branch-and-bound methods would not have been feasible, in which case the HSearch command could be used to perform a heuristic search. For example,

```
hsearch/addseq=random nreps=100;
```

would request a search from 100 starting trees generated by stepwise addition with random-addition sequences, followed by TBR (the default) branch-swapping on each starting tree. In this small example, the majority of the random-addition-sequence replicates finds the best tree obtained using the exact methods, but a small percentage of random-addition-sequence replicates finds a tree that requires one additional evolutionary change. These two trees are different islands.

To examine the trees, the ShowTrees or DescribeTrees commands are available. ShowTrees simply outputs a diagram representing each tree topology, but provides no other information:

```
showtrees all;
```

The best tree is shown in the output as follows:

```
Tree number 1 (rooted using default outgroup)
/------------------------------------ gpd1yeast
|
|           /--------------------------- gpdadrome
|           |
|           |                    /-------- gpdhuman
|           |            /-----+
+---------+           |            \------- gpdarabit
|           |     /------+
|           |     |            \------------- gpdamouse
|           \-------+
|                    \--------------------- gpdacaeel
|
```

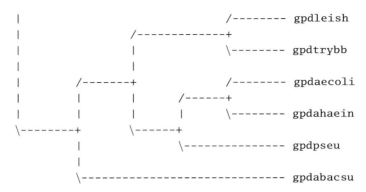

Remember that, in general, parsimony analysis does not determine the location of the root. By default, PAUP* assumes that the first taxon in the data matrix is the only *outgroup* taxon and roots the tree at the adjacent internal node leading to a basal *polytomy* as indicated previously. In this example, the root should probably be located between the cluster of bacterial (gpdaecoli, gpdahaein, gpdabacsu and gpdpseu) and Trypanosomatidae (gpdleish, gpdtrybb) sequences and the cluster of other eukaryote homologues. This can be achieved by specifying the taxa of one cluster as the outgroup, for example:

```
outgroup gpdleish gpdtrybb gpdaecoli gpdahaein gpdabacsu
gpdpseu;
```

or, alternatively, by

```
outgroup 7-12;
```

The tree may now be output according to the new outgroup specification by reissuing the ShowTrees command. Because this will not display the ingroup as *monophyletic*, it is advisable to also specify:

```
Set outroot=mono;
```

Instead of ShowTrees, the DescribeTrees command can be used to request output of the tree in *phylogram* format, where the branches of the tree are drawn proportionally to the number of changes (see Chapter 5) assigned under the parsimony criterion. Other information is also available through additional DescribeTrees options. For example, the command

```
describetrees 1/plot=phylogram diagnose;
```

outputs a drawing of the best tree in phylogram format (now reflecting the new rooting), plus a table showing the number of changes required by each character on the first tree, the minimum and maximum possible number of changes on any tree, and several goodness-of-fit measures derived from the following values:

```
Tree number 1 (rooted using user-specified outgroup)

Tree length = 812
Consistency index (CI) = 0.8300
Homoplasy index (HI) = 0.1700
CI excluding uninformative characters = 0.8219
HI excluding uninformative characters = 0.1781
Retention index (RI) = 0.7459
Rescaled consistency index (RC) = 0.6191
```

```
         /------------------------------- gpd1yeast
         |
         |              /------------ gpdadrome
         22            |
         ||            |                     /-- gpdhuman
         ||            |                  /-13
         |\------------16              |  \---- gpdarabit
         |             |      /------14
         |             |      |       \-- gpdamouse
         |             \----15
         |                   \---------- gpdacaeel
         |
         |                   /--------- gpdleish
         |         /---------17
         |         |         \--------- gpdtrybb
         |         |
         |  /---20              /----- gpdaecoli
         |  |   |           /--------18
         |  |   |           |         \------ gpdahaein
         \-21  \---------19
         |              \---------- gpdpseu
         |
         \----------------- gpdabacsu
```

```
Character diagnostics:
                 Min  Tree  Max                                   G-
Character Range steps steps steps   CI    RI    RC    HI    fit
-----------------------------------------------------------------
1              2    2    2    6  1.000 1.000 1.000 0.000 1.000
3              5    5    5    6  1.000 1.000 1.000 0.000 1.000
.              .    .    .    .    .     .     .     .     .
.              .    .    .    .    .     .     .     .     .
.              .    .    .    .    .     .     .     .     .
233            6    6    6    7  1.000 1.000 1.000 0.000 1.000
234            1    1    2    2  0.500 0.000 0.000 0.500 0.750
(24 constant characters not shown)
```

To assess whether this result is strongly supported, bootstrap or jackknife analysis (see Chapter 5) can be performed using the `Bootstrap` or `Jackknife` commands. For example, the command

`bootstrap nreps=1000;`

requests a bootstrap analysis with 1000 pseudoreplicate samples using the current optimality criterion and associated settings (`hsearch/addseq=random nreps=100`).

Results of the bootstrap analysis can be summarized using a ***majority-rule consensus tree*** that indicates the frequency in which each bipartition, or division of the taxa into two groups, was present in the trees obtained for each bootstrap replicate (see Box 5.3). Interpretation of *bootstrap values* is complicated (e.g. Hillis & Bull, 1993) (see also Chapter 5), but clearly clades with close to 100% support can be considered very robust. When the bootstrap analysis is completed, the majority-rule consensus with bootstrap values is shown:

```
Bootstrap 50% majority-rule consensus tree
```

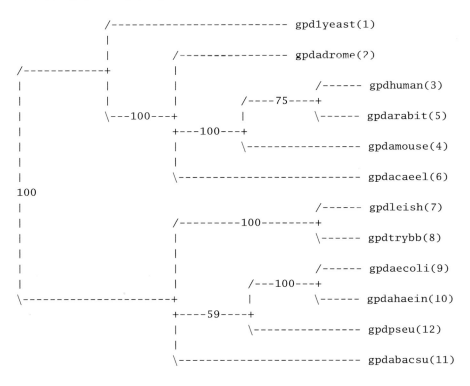

In addition to the consensus tree, the output includes a table showing the frequency of all groups that were found at a frequency of 5% or higher, so that support for groups not included in the consensus also can be evaluated as follows:

```
Bipartitions found in one or more trees and frequency of occur-
rence (bootstrap support values):
1 1
123456789012     Freq         %
-----------------------------------
........**..   1000.00  100.0%
..***.......   1000.00  100.0%
......******   1000.00  100.0%
......**....    999.67  100.0%
.*****......    997.07   99.7%
..*.*.......    747.89   74.8%
........**.*    589.53   59.0%
.*...*......    498.14   49.8%
......****.*    476.07   47.6%
..****......    407.86   40.8%
........****    313.48   31.3%
........***.    230.47   23.0%
..**........    177.42   17.7%
......*****.    135.95   13.6%
.****.......     91.80    9.2%
..........**     77.38    7.7%
...**.......     74.70    7.5%
......****..     67.33    6.7%
......**...*     59.53    6.0%
```

Note: the format of this output follows the same convention of the output file from the program Consense.exe of the PHYLIP package; see Box 5.4). Another measure of nodal support is the **decay index**, or **Bremer support** (Bremer, 1988; Donoghue *et al.*, 1992). If the shortest tree inconsistent with the monophyly of some group is three steps longer than the most parsimonious tree (on which the group is monophyletic), the decay index is equal to 3. These values can be calculated in two ways. For groups that decay quickly, all trees within one step, two steps, and so on, of the shortest tree can be stored and a strict consensus tree calculated for each length. The following commands can be used to find groups that occur in all trees within 1, 2, and 3 steps, respectively, of the shortest tree:

```
hsearch keep=813; contree;
hsearch keep=814; contree;
hsearch keep=815; contree;
```

For example, a group consisting of human (gpdhuman) and rabbit (gpdarabit) is found in all trees of length 813 and shorter, but this group disappears from the consensus of all trees of length 814 and shorter, indicating that there are some trees at 814 steps that do not contain this 2-taxon clade. The decay index is therefore, $814 - 812 = 2$. Evaluating the meaning of decay indexes is probably even more difficult than bootstrap values, but any group for which the decay index is less than 4 should probably be viewed with suspicion (DeBry, 2001).

An alternative method for calculating decay indexes is to use the *converse constraints* feature of PAUP*. In general, a *constraint tree* is a user-defined tree that limits the search space to those trees that are either compatible or incompatible with the constraint tree. For monophyly constraints, a tree T is compatible with the constraint tree if it is equal to the constraint tree or if the constraint tree can be obtained by collapsing one or more branches of T. For calculating decay indexes, the monophyly of the group of interest is indicated and a search is performed using converse constraints. For the example discussed previously, the constraint tree can be defined using the command

```
constraints humanrabbit=((3,5));
```

The tree resulting from this definition can then be viewed using the show-constr command:

```
Constraint-tree "humanrabbit":
/----------------------------------- gpd1yeast
|
+----------------------------------- gpdadrome
|
|              /--------------------- gpdhuman
+------------+
|              \--------------------- gpdarabit
|
+----------------------------------- gpdamouse
|
+----------------------------------- gpdacaeel
|
+----------------------------------- gpdleish
|
+----------------------------------- gpdtrybb
|
+----------------------------------- gpdaecoli
|
+----------------------------------- gpdahaein
|
+----------------------------------- gpdabacsu
|
\----------------------------------- gpdpseu
```

Next, a constrained search is conducted as follows:

```
HSearch enforce constraints=humanrabbit converse addseq=random
nreps=100;
```

It is especially desirable to use multiple random-addition-sequence starting points when searching under converse constraints to improve the chances of identifying the shortest trees that are incompatible with the constraint tree. In this case, five trees of length 814 are found in the constrained search, yielding a decay index of 2, as was determined previously. This second approach is much more useful for determining larger decay values, where saving all trees within *n* steps of the shortest tree becomes more unwieldy as *n* increases. As is often but not always the case, the modest decay index in the example mirrors a modest bootstrap value for the same group.

8.7 Analysis using distance methods

The same search strategies discussed previously may be used for searching trees under distance optimality criteria. Because distance methods were covered extensively in Chapter 5, what follows only illustrates how to perform some commonly used analyses with PAUP*, using the vertebrate mtDNA data set as an example.

The DSet command is used to set the type of distance to be computed:

```
dset distance=jc;
```

This will set the distances to Jukes–Cantor distances (see Chapter 4); other distance options available in PAUP* (as well as the current settings) can be viewed by typing Dset?. To calculate a standard NJ tree (Saitou & Nei, 1987) according to the current distance transformation, use the command

```
nj;
```

Before proceeding with further analyses, note that the vertebrate mtDNA data set exhibits heterogeneous base frequencies among taxa, as can be seen from the output of the command

```
basefreq;
```

In the presence of such heterogeneity, an appropriate distance transformation is the LogDet distance (see Section 4.13), which can be selected using the command

```
dset distance=logdet;
```

Criterion-based searches are performed using the same commands as were previously described for parsimony analysis. However, because the default optimality criterion is parsimony, the following command must be issued to force use of distance-based optimality criteria.

```
set criterion=distance;
```

By default, distance analyses use an NJ method to obtain a starting tree (`addseq` `=nj`). This can be overridden to use random addition sequences, as follows:

```
hsearch/start=stepwise addseq=random nreps=100;
```

The resulting tree can then be output as a phylogram:

```
describe/plot=phylogram;
```

```
Minimum evolution score = 2.47415

/-------------------- LngfishAu
|
|                     /--------- LngfishSA
|     /-----------18
|     |               \--------- LngfishAf
|     |
|     |               /--------- Frog
|     |               |
|     |               |    /------------ Turtle
\--32 |               |    |
|                     20   /---------------- Crocodile
|                     |\---19
|                     |    \----------- Bird
|           /---22
\---31 |    |/-------------- Sphenodon
      |     21
      |     \---------------- Lizard
      |
      |               /--------- Human
      |               |
      |               /25  /------- Seal
      \----30         | |  |
            |         |\-24 /------- Cow
            |         |  |  \-23
            |    /--27        \------- Whale
            |    |  |
            |    |  |         /----- Mouse
            |    |  \-----26
            \--29            \------ Rat
              |
              |  /------------ Platypus
              \-28
                 \---------- Opossum
```

This tree places turtles with archosaurs. Constraints may be used to find the best tree consistent with the conventional placement of turtle (i.e. basal reptile):

```
constraints turtlebasal = ((Crocodile, Bird, Sphenodon, Lizard),
Turtle);
```

If you previously enforced converse constraints, this option will persist in the current PAUP* session unless you specify `converse = no`:

```
hsearch enforce converse=no constraints=turtlebasal;
```

The score of the resulting tree is 2.47570, which is barely worse than the best unconstrained tree. The low support for the placement of the turtle can be confirmed using bootstrap analysis:

```
bootstrap nreps=1000 / enforce=no;
```

The bootstrap support results, in fact, in only about 51% (this value can slightly vary because of the stochastic nature of bootstrap resampling).

Minimum evolution (see Chapter 5) is the default method used for distance analyses in PAUP*. This method uses least-squares fit to determine the branch lengths, but then uses the sum of these branch lengths as the objective function for computing the score of each tree. The unweighted least-squares and Fitch–Margoliash (weighted least-squares) variants are also available by replacing n with 0 or 2, respectively, in the following command

```
dset objective=ls power=n;
```

prior to beginning a distance-criterion search.

One particularly problematic aspect of distance analysis involves the handling of negative branch lengths, which often occur in unconstrained optimization of the least-squares fit between path-length and observed distances. Because negative branch lengths have no obvious biological meaning, they probably should be avoided. A constrained optimization is available in PAUP* to perform the least-squares optimization under the restriction of branch-length non-negativity, which can be requested as follows:

```
dset negbrlen=prohibit;
```

We believe that this option provides the most appropriate way to deal with the negative-branch length problem, but analyses run significantly more slowly when this option is specified. The default setting in PAUP* is to allow negative branch lengths when performing the least-squares fit calculations, but to set negative values to zero before computing the score of the tree (`negbrlen = setzero`). Thus, if the least-squares fit does not require negative lengths, the tree scores obtained in this way are identical, whereas trees for which the fit is improved by allowing negative branch lengths are still penalized accordingly. Negative branch lengths also may be set to their absolute value (i.e., `negbrlen = setabsval`) (Kidd & Sgaramella-Zonta, 1971) or simply used "as is" (i.e. `negbrlen = allow`) (Rzhetsky & Nei, 1992).

It is also possible to input distances directly into **PAUP*** using a NEXUS format Distances block. This feature permits mixing the tree-searching capabilities of **PAUP*** with distance transformations that are not available within the program (e.g. model-based protein distances, but their implementation is in development at the time of writing). Examples for using user-supplied distance matrixes are given in the command-reference document in the PAUP folder.

8.8 Analysis using maximum likelihood methods

Maximum likelihood methods are discussed in Chapter 6, including the ***quartet-puzzling*** algorithm implemented in **TREE-PUZZLE** as a heuristic strategy for a maximum likelihood tree search. **PAUP*** is also able to carry out a number of maximum-likelihood-based analyses on nucleotide sequence data. In particular, the program can perform exhaustive, branch-and-bound, and various types of heuristic searches, described previously, on aligned nucleotide sequences. The following example assumes that the HIV data set in NEXUS format (`hivALN.nex`) (see Box 8.4) has already been executed in **PAUP***.

The first task in any maximum likelihood analysis is to choose an appropriate model. The default model in **PAUP*** is the HKY model (see Chapter 4) with a transition : transversion ratio of 2; however, this model should never be used uncritically. Although automated model-selection techniques are available and useful (Posada & Crandall, 1998) (see Chapter 10), we find that manual model selection can be highly informative and may identify good models that might otherwise be missed. Unfortunately, we cannot recommend a precise solution that is best for all situations (if such a solution were available, then it could simply be automated). Instead, we outline the following general strategy, which basically follows the "top-down dynamical" approach of Chapter 10, but without enforcing any particular *a priori* specification of models to be compared.

(1) Start with some reasonable tree for the data. This tree need not be optimal in any sense, but it should at least be a "good" tree under some criterion.
(2) Set the likelihood model to the most complex one available in **PAUP*** (six-substitution-type general time-reversible model with some fraction of invariable sites and rates at variable sites following a gamma distribution ($= GTR + I + \Gamma$)).
(3) Estimate all model parameters (i.e. relative substitution rates, base frequencies, proportion of invariable sites, gamma shape parameter) and calculate the likelihood score of the tree using the `LScores` command.

(4) Examine the parameter estimates carefully in an attempt to identify possible model simplifications that will not greatly reduce the fit of the model. Evaluate the simplified model and decide whether it is acceptable. If the attempted simplification is rejected, return to the previous model.
(5) Repeat Step 4 until no further acceptable simplifications can be found.
(6) Proceed to search for an optimal tree using the model and parameter estimates selected previously.

Some of these steps deserve further elaboration. The reason that the tree used for Step 1 is not critical is that parameter estimates do not vary much from tree to tree, as long as the trees are reasonably good explanations of the data (i.e. those that are much better than randomly chosen trees and include clades that are well supported under any optimality criterion) (Yang, 1994; Sullivan *et al.*, 1996; unpublished observations). The determination of whether to reject a simplified model in Step 4 can be accomplished using ***likelihood-ratio tests*** (LRTs), the ***AIC*** or ***BIC*** criteria (see Chapter 10), or more subjective methods. Model-selection criteria are somewhat controversial. LRTs offer the appeal of statistical rigor, but the choice of an appropriate alpha level for the test (e.g. 0.05, 0.01, 0.001) is arbitrary. Furthermore, even if a simple model is strongly rejected in favor of a more complex model, the simpler model may still be better for tree inference because it requires the estimation of fewer parameters. As models become more complex, more parameters must be estimated from the same amount of data, so that the variance of the estimates increases. Conversely, use of an oversimplified model generally leads to biased estimates. We follow Burnham and Anderson (1998) and others in preferring models that optimize the trade-off between bias and variance as a function of the number of parameters in the model. Unfortunately, it seems impossible to avoid some degree of subjectivity when selecting models, but this should not be taken as a reason to avoid model-based methods entirely (Sullivan & Swofford, 2001).

We illustrate our model-selection strategy using the HIV data set. First, we obtain a starting tree, which can be, for example, a parsimony, an NJ, or a maximum likelihood tree obtained under a simple model such as Jukes–Cantor. In practice, the choice of this tree rarely makes a difference in the model selected (unpublished observations). In the absence of further information, the LogDet transformation represents a good all-purpose distance, and an NJ tree computed from the LogDet distance matrix can serve as the starting tree:

```
dset distance=logdet; nj;
```

Steps 2 and 3 are then accomplished using the LSet and/or LScores commands:

```
lset nst=6 rmatrix=estimate basefreq=esti-
mate rates=gamma shape=estimate pinvar=estimate;
lscores;
```

Or, equivalently:

```
lscores/nst=6 rmatrix=estimate basefreq=esti-
mate rates=gamma shape=estimate pinvar=estimate;
```

NST represents the number of substitution types. Instead of two (i.e. transitions vs. transversions), it is assumed that all six of the pairwise substitution rates are potentially different. Γ-distributed rates for variable sites (see Chapter 4) are specified via rates=gamma. Rather than taking default values, all parameters are estimated from the data. The resulting output follows. Note that PAUP* reports the tree score as the negative log likelihood rather than the log likelihood itself; therefore, smaller likelihood scores are better. This is done so that all optimizations performed by the program can be treated as minimization problems.

```
Likelihood scores of tree(s) in memory:
  Likelihood settings:
    Number of substitution types = 6
    Substitution rate-matrix parameters estimated via ML
    Nucleotide frequencies estimated via ML
  Among-site rate variation:
  Assumed proportion of invariable sites = estimated
  Distribution of rates at variable sites = gamma
  (discrete approximation)
    Shape parameter (alpha) = estimated
    Number of rate categories = 4
    Representation of average rate for each category = mean
  These settings correspond to the GTR+G+I model
  Number of distinct data patterns under this model = 1013
  Molecular clock not enforced
  Starting branch lengths obtained using
    Rogers-Swofford approximation method
  Branch-length optimization = one-dimensional
    Newton-Raphson with pass limit = 20, delta = 1e-06
  -ln L (unconstrained) = 12463.69362


Tree                    1
-------------------
-ln  L    17321.15334
Base frequencies:
    A          0.359090
    C          0.182946
    G          0.214583
    T          0.243381
```

```
Rate matrix R:
   AC          2.27643
   AG          4.89046
   AT          0.84752
   CG          1.14676
   CT          4.77744
   GT          1.00000
P_inv          0.159207
Shape          0.907088
```

Searching for simplifications, it is apparent that the substitution rates for the two transitions, r_{AG} and r_{CT}, are nearly equal (i.e. close to 4.8). Thus, an obvious first try at simplifying the model involves equating these two rates and reducing the number of parameters by one. This is accomplished by specifying a particular submodel of the GTR model using the RClass option, which allows entries in the rate matrix to be pooled into larger classes. The classification of substitution types follows the order r_{AC}, r_{AG}, r_{AT}, r_{CG}, r_{CT}, r_{GT}. All rates assigned the same alphabetic character are pooled into a single category. Thus, the following command pools the two transitions (r_{AG} and r_{CT}) into one class, but leaves each transversion distinct, and re-estimates all other parameters (all options from the previous command are still in effect):

```
lscores/rclass=(a b c d b e);
```

The resulting output shows that the tree score is only trivially worse than the full GTR+I+Γ model (17321.153 versus 17321.188):

```
Tree                   1
--------------------
-ln L      17321.18757
Base frequencies:
   A           0.359680
   C           0.182335
   G           0.215248
   T           0.242737
Rate matrix R:
   AC          2.28603
   AG          4.84836
   AT          0.84841
   CG          1.14921
   CT          4.84836
   GT          1.00000
P_inv          0.159112
Shape          0.907157
```

Thus, it is appropriate to accept the simpler model with the two transitions pooled into one category.

Often, models with either Γ-distributed rates or invariable sites alone fit the data nearly as well as those with both types of rate variation. To examine this possibility for the HIV data set, the reduced model is evaluated, but with all sites variable (pinv = 0) and following a gamma distribution (still the current setting):

```
lscores/pinv=0;
```

The likelihood score obtained for this model is 17325.566, or 4.378 units worse. An LRT can be performed by multiplying this score difference by 2 and comparing to a chi-squared distribution with 1 degree of freedom (see Chapter 10). The *P*-value for this test is 0.0031, suggesting that the invariable sites plus gamma model $(I + \Gamma)$ fits significantly better than the model alone. The AIC (see Chapter 10) also favors the $I + \Gamma$ rate-variation model.

Next, we evaluate the model with some sites invariable, and all variable sites evolving at the same rate:

```
lsc/rates=equal pinv=est;
```

The likelihood score for this model, 17455.811, is even worse, so it also is rejected in favor of the $I + \Gamma$ model.

After failing to simplify the model by eliminating among-site rate-variation parameters, attention can be returned to the substitution rate matrix. Inspection of this matrix reveals that the relative rate for the CG substitution ($r_{CG} = 1.149$) is not much different than $r_{GT} = 1$. Thus, these two substitutions also can be pooled and the resulting likelihood score calculated:

```
lsc/rclass=(a b c d b d) rates=gamma shape=estimate
pinv=estimate;
```

The resulting likelihood score and parameter estimates follow:

```
Tree                    1
--------------------
-ln L     17321.52578
Base frequencies:
  A           0.359704
  C           0.183294
  G           0.215397
  T           0.241606
Rate matrix R:
  AC          2.16054
  AG          4.58543
  AT          0.80136
```

```
        CG          1.00000
        CT          4.58543
        GT          1.00000
      P_inv         0.161272
      Shape         0.914014
```

The likelihood score (17321.526) is not much worse than the previous acceptable model (17321.188). The difference in scores is equal to 0.338, and an LRT using twice this value with 1 degree of freedom yields a *P*-value of 0.41. Thus, the simpler model is preferred. Although we do not show the results, further attempts to simplify the model by additional substitution-pooling (e.g. rclass=(a b c c b c)) or assuming equal base frequencies (i.e. basefreq=equal) yield likelihood scores that are significantly worse than this model. It is interesting that the model chosen here, to our knowledge, has not been "named" and is not always available as a model setting in phylogenetic programs.

As an aside, the RClass option may be used to select other specific models. For example, lset rclass=(a b a a c a); can be used to specify the Tamura–Nei model (see Chapter 4), which assumes one rate for transversions, a second rate for purine transitions, and a third rate for pyrimidine transitions. RClass is meaningful only when the substitution rate matrix is being estimated (rmatrix=estimate) and the number of substitution types is set to six (nst=6). Specific (relative) rates also may be specified using the RMatrix option. For example, the command

```
lset rmatrix=(1 5 1 1 2.5);
```

also specifies a Tamura–Nei model in which purine and pyrimidine transitions occur at a rate of 5 and 2.5 times (respectively) that of transversions. Note that these rates are assigned relative to the G–T rate, which is assigned a value of 1 and is not included in the list of RMatrix values.

Having chosen a model, we can now proceed to search for an optimal tree. To set the criterion to maximum likelihood, execute the following command:

```
set criterion=likelihood;
```

Because the criterion is now likelihood, the HSearch command will perform a maximum likelihood tree search using the default or current model settings. To reset the model chosen, the following command can be issued:

```
lset nst=6 rclass=(a b c d b d) rmatrix=estimate base-
freq=estimate rates=gamma shape=estimate pinv=estimate;
```

With these settings, Paup* will estimate all model parameters on each tree evaluated in the search. Whereas in general this is the ideal method, it is too slow for data sets containing more than a small number of taxa. A more computationally tractable approach uses a successive approximations strategy that alternates between parameter estimation and tree search. First, the parameter values are estimated on a starting tree (e.g. the same NJ tree used for the model-selection procedure). These values are then fixed at their estimated values for the duration of a heuristic search. If the search finds a new tree that is better than the tree for which the parameter values were estimated, then the parameter values are re-estimated on this new tree. The iteration continues until the same tree is found in two consecutive searches. To implement this iterative strategy for the HIV example, the parameter values must first be fixed to the values estimated during the model-selection procedure. An easy way to do this is as follows (assuming that the NJ tree is still in memory from before):

```
lscores;
lset rmatrix=previous basefreq=previous shape=previous
pinv=previous;
```

Each parameter set to "previous" in this command is assigned the value that was most recently estimated for that parameter (i.e. in the preceding LScores command when the parameters were set to "estimate"). A standard heuristic search using simple stepwise addition followed by TBR rearrangements (see Section 8.4.2) is performed using the command hseach. If the search strategy options have been changed due to previous settings in this session, set them back to default (AddSeq=simple NReps=10):

```
hsearch;
```

A tree of likelihood score 17316.909 is found, which is an improvement over the NJ starting tree. Consequently, the model parameters are re-estimated on this new tree:

```
lset rmatrix=estimate basefreq=estimate shape=estimate
pinv=estimate; Lscores;
```

With the newly optimized parameter values, the likelihood score of the tree found by the previous search improves only very slightly to 17316.896:

```
Tree                    1
--------------------
-ln L    17316.89554
Base frequencies:
    A          0.359776
    C          0.183578
    G          0.215216
    T          0.241429
```

```
Rate matrix R:
    AC          2.17187
    AG          4.58751
    AT          0.80055
    CG          1.00000
    CT          4.58751
    GT          1.00000
P_inv           0.156966
Shape           0.900917
```

The parameter estimates are very close to those obtained using the original NJ tree (discussed previously), which is why the likelihood score did not improve appreciably. Nonetheless, to make sure that the change in parameter values will not lead to a new optimal tree, the search should be repeated, fixing the parameter values to these new estimates:

```
lset rmatrix=previous basefreq=previous shape=previous
pinv=previous;
hsearch;
```

This search finds the same tree as the previous one did, so the iterations can stop. Analyses of both real and simulated data sets provide reassurance that this iterative, successive-approximations strategy can be safely used as a much faster approach in ML tree topology estimation (Sullivan et al., 2005).

Although it is not important for the relatively small HIV data set, it may be desirable for data sets containing larger numbers of taxa to reduce the scope of the rearrangements using the ReconLimit option. If ReconLimit is set to x, then any rearrangement for which the reconnection distance (see Section 8.4.2) exceeds x is not evaluated. For data sets of 30 or more taxa, use of values such as ReconLimit=12 can often substantially reduce the time required for a TBR search without greatly compromising its effectiveness.

As a starting tree for the heuristic search, it is possible to use an NJ tree (instead of a tree obtained by stepwise addition) with the command HSearch Start = NJ;. Although the program finds the same maximum likelihood tree for the HIV example data set, this will not necessarily be true for other data sets. In general, the use of different starting trees permits a more thorough exploration of the tree space, and it is always a good idea to compare results of alternative search strategies. Use of random-addition sequences (i.e. HSearch AddSeq=Random;) is especially desirable. The default number of random-addition-sequence replicates is rather small (10); the option NReps=n can be used to specify a different number of replicates n.

As for parsimony and distance methods, PAUP* can perform bootstrapping and jackknifing analyses under the likelihood criterion, although these analyses may be very time-consuming. In addition, a zero-branch-length test can be performed on the maximum likelihood tree by setting the ZeroLenTest option of the LSet command to full and then executing the DescribeTrees command:

```
lset zerolentest=full; describetrees;
```

If more than one tree is in memory, the result of the test is printed on the screen (and/or in the log file: see Box 8.4) for each tree. In the zero-branch-length test, the statistical support for each branch of a previously estimated maximum likelihood tree is obtained as follows. After collapsing the branch under consideration (by constraining its length to zero), the likelihood of the tree is recalculated and compared with the likelihood of the original tree. If the likelihood of the former is significantly worse than that of the latter according to an LRT, the branch is considered statistically supported. As a result of executing these commands, PAUP* prints out a list of all pairs of neighboring nodes in the tree (including terminal nodes), each pair with a corresponding *p-value* giving the support for the branch connecting the two nodes:

```
Branch lengths and linkages for tree #1 (unrooted)
```

Node	Connected to node	Branch length	Standard error	--- L.R. test --- lnL diff	P*
L20571(1)	26	0.95536	0.06838	399.948	<0.001
15	26	0.10463	0.02367	8.449	<0.001
AF103818 (2)	15	0.33927	0.02887	137.486	<0.001
X52154 (3)	15	0.33452	0.02858	125.842	<0.001
25	26	0.21406	0.02630	32.721	<0.001
U09127 (4)	25	0.07713	0.00837	46.112	<0.001
24	25	0.02213	0.00682	4.197	0.004
16	24	0.06190	0.00724	110.645	<0.001
U27426 (5)	16	0.06438	0.00663	132.276	<0.001
U27445 (6)	16	0.04666	0.00582	86.951	<0.001
23	24	0.01643	0.00438	14.143	<0.001
17	23	0.07067	0.00754	157.899	<0.001
AF067158 (7)	17	0.05477	0.00603	126.407	<0.001
U09126 (8)	17	0.04146	0.00536	73.886	<0.001
22	23	0.03517	0.00583	47.726	<0.001
18	22	0.04925	0.00614	99.613	<0.001
U27399 (9)	18	0.04582	0.00557	99.612	<0.001
U43386 (10)	18	0.05442	0.00601	132.606	<0.001
21	22	0.03817	0.00549	69.776	<0.001
20	21	0.01060	0.00319	11.560	<0.001
19	20	0.00639	0.00237	8.840	<0.001
L02317 (11)	19	0.03638	0.00456	136.741	<0.001

AF042106 (14)	19	0.04899	0.00529	199.615	<0.001
AF025763 (12)	20	0.04503	0.00512	190.207	<0.001
U08443 (13)	21	0.04405	0.00519	138.658	<0.001

| Sum | | 2.81762 | | | |

```
* Probability of obtaining a likelihood ratio as large or larger than the
observed ratio under the null
hypothesis that a branch has zero length (full reoptimization after forc-
ing a branch length to zero)
```

```
-Ln likelihood = 17316.89554
```

In this case, all the branches in the maximum likelihood tree seem to be robustly supported by this test. However, the test only assesses whether a given resolution of a trichotomy is a significant improvement relative to leaving the trichotomy unresolved. For example, it is possible for incompatible groups on different trees to both receive significant support using this test. The fact that not all of the clades received high bootstrap values in the parsimony and NJ trees, as well as in maximum likelihood bootstrap analyses (not shown), highlights the need for caution in interpreting the results of this test.

Phylogenetic analysis using protein sequences

THEORY

Fred R. Opperdoes

9.1 Introduction

In addition to nucleotide sequences, protein sequences are frequently used for the analysis of evolutionary relationships and, as we will learn below, they are often preferred for constructing phylogenetic trees. As DNA and RNA sequences are subject to gradual change over evolutionary time resulting from the incorporation of mutations, insertions or deletions, the translated products of genes will inevitably undergo similar evolutionary changes. Not all genetic mutations in the genes will lead to mutations in the corresponding proteins, but when amino acid mutations occur, they may affect the protein in three possible ways. They may be harmful (*deleterious*) to the protein, in which case **purifying selection** will eliminate them; they may be neutral, which means that they will have no effect on the function of the protein; or they may be beneficial. As neutral mutations are not penalized by natural selection, they can accumulate over time according to the process of **genetic drift** (see Chapter 13). Beneficial mutations usually become fixed more rapidly in the population due to **positive selection**. As a result, sequences of **orthologous** proteins from two species that have diverged from a common ancestor will differ only slightly when there has been little time for mutations to occur, whereas more differences will have accumulated when ample time has passed.

Different proteins may evolve at highly different rates. The so-called "housekeeping" proteins evolve rather slowly because they catalyze essential functions

The Phylogenetic Handbook: a Practical Approach to Phylogenetic Analysis and Hypothesis Testing,
Philippe Lemey, Marco Salemi, and Anne-Mieke Vandamme (eds.). Published by Cambridge
University Press. © Cambridge University Press 2009.

within cells. For example, the cytochrome c sequences of humans and great apes (gorilla, orangutan, and chimpanzee) are identical and differ from that of rhesus monkeys, spider monkeys, and Western Tarsiers by only 1%, 6%, and 10%, respectively. Bacterial and protist sequences still share between 30% and 50% identical residues with that of primates and it is therefore not surprising that cytochrome c sequences were used to infer one of the first phylogenetic trees for organisms that diverged several billion years ago (McLaughlin & Dayhoff, 1973). Similar patterns apply to other essential and widely shared house-keeping proteins. However, proteins that are less essential generally evolve more rapidly. Genetic drift may therefore quickly erase any sequence similarity between two proteins of identical function even if they shared a common ancestor at some time in the past. The early protein trees were produced using protein sequences obtained from the direct sequencing of peptides by Edmann degradation. Now, with large-scale automated DNA sequencing, protein sequences can be easily obtained from the translation of the open reading frames of the corresponding gene sequences. These are all widely available from the appropriate databases (see Chapter 2). Many programs used for the analysis of DNA sequences are also capable of handling protein sequences.

9.2 Protein evolution

9.2.1 Why analyze protein sequences?

The sequence of a coding gene contains all the necessary information to create functional proteins, and its nucleotides directly incorporate the mutations that result from replication errors, radiation damage, oxidative stress, or chemical modification. Therefore, it is often advocated that DNA rather than protein sequences should be used in molecular evolution studies. However, there are many reasons why it may be more appropriate to use protein sequences, as we will argue below. Although DNA contains all the necessary information to create a protein, it is generally not the DNA itself that is subject to natural selection. The catalysts of virtually all of the chemical transformations in the cell are proteins. Proteins are the fundamental building blocks of life and they are the units of life on which natural selection acts. While DNA sequences consist of only four bases A, G, C, and T, the functional properties of proteins are determined by a sequence of 20 possible amino acids, which leads to a much higher resolution at large evolutionary distances. The underlying DNA sequence, which is related to the amino acid sequence of the proteins via the genetic code, reflects this selection process in combination with species-specific pressures on the DNA sequence (see below). Thus, while there is sufficient room for the DNA sequence to change and respond to requirements of GC content, the protein sequence may remain almost unaltered and not reflect such adaptations.

Table 9.1 The Genetic Code

The relationship between the codons of nucleic acids, and the amino acids for which they code, is embodied in the Genetic Code (slight variations of it are found in protists, in mitochondria, and in chloroplasts). The 64 possible triplets of bases in a codon, and the amino acid coded for are shown.

First Position	Second Position				Third Position
	U(T)	C	A	G	
U(T)	Phe	Ser	Tyr	Cys	U(T)
	Phe	Ser	Tyr	Cys	C
	Leu	Ser	STOP	STOP	A
	Leu	Ser	STOP	Trp	G
C	Leu	Pro	His	Arg	U(T)
	Leu	Pro	His	Arg	C
	Leu	Pro	Gln	Arg	A
	Leu	Pro	Gln	Arg	G
A	Ile	Thr	Asn	Ser	U(T)
	Ile	Thr	Asn	Ser	C
	Ile	Thr	Lys	Arg	A
	Met	Thr	Lys	Arg	G
G	Val	Ala	Asp	Gly	U(T)
	Val	Ala	Asp	Gly	C
	Val	Ala	Glu	Gly	A
	Val	Ala	Glu	Gly	G

9.2.2 The genetic code and codon bias

The nucleotide sequence of a coding gene is read as a non-overlapping triplet code. Each triplet is interpreted via the so-called universal genetic code (Table 9.1). Four different nucleotides taken three at a time can result in 64 different possible triplet codes; more than enough to encode 20 amino acids. One codon, ATG, translated to methionine, serves as the initiation codon and thus represents the start of a protein; other amino acids may be encoded by 1 to 6 different triplet codes, and finally 3 of the 64 codes, called stop (or termination) codons, specify the "end of the peptide sequence." Codons have been degenerated with a *wobble* in the third position. For instance, the amino acids, leucine, serine, and arginine are all encoded by six triplets. Valine, proline, threonine, alanine, and glycine are each encoded by four triplets. Since the base composition of DNA may vary among organisms, not all organisms have the same codon preference. Yeasts, protists, and animals all have different preferences, and as a consequence, the same protein

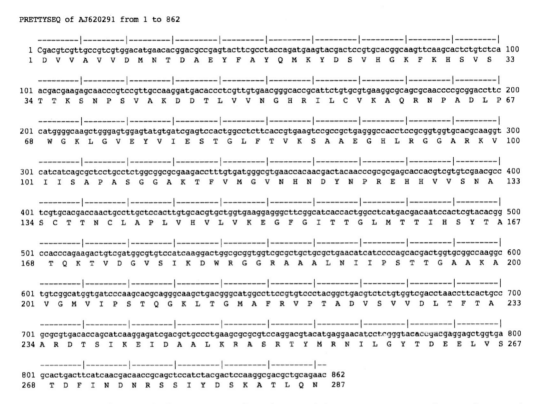

PRETTYSEQ of AJ620291 from 1 to 862

Fig. 9.1   Translation of the open reading frame of *Trypanosoma congolense* glycosomal glyceraldehyde-3-phosphate dehydrogenase gene fragment using the PRETTYSEQ program of the EMBOSS suite.

sequence in each of these organisms is the result of different DNA sequences and these differences are related to codon bias and not merely to stochastic evolution. Where multiple codons specify the same amino acid, the different codons are used with unequal frequency, depending on the nature of the gene and its level of expression. This frequency distribution varies widely between species and is referred to as *codon usage bias*. Another complication is that, although the genetic code in Table 9.1 is called *universal*, this is not entirely true. Slightly different codes are being used by different genetic systems, and this should be kept in mind when one works with proteins derived from either mitochondria or protists. Some protists use the codons TAA and TGA to encode glutamine, rather than "STOP," and in mitochondria, the codon TGA encodes tryptophane, rather than "STOP." The translation of a nucleotide sequence into its corresponding protein sequence via the use of the genetic code will result in a sequence of one-letter codes (Fig. 9.1). Table 9.2 lists the one-letter and three-letter code for the different amino acids.

Table 9.2 Single and 3-letter codes for amino acids. All proteins are polymers of the 20 naturally occurring L-amino acids

Alanine	Ala	A
Cysteine	Cys	C
Aspartic AciD	Asp	D
Glutamic Acid	Glu	E
Phenylalanine	Phe	F
Glycine	Gly	G
Histidine	His	H
Isoleucine	Ile	I
Lysine	Lys	K
Leucine	Leu	L
Methionine	Met	M
AsparagiNe	Asn	N
Proline	Pro	P
Glutamine	Gln	Q
ARginine	Arg	R
Serine	Ser	S
Threonine	Thr	T
Valine	Val	V
Tryptophan	Trp	W
TYrosine	Tyr	Y

In addition, B may be used for Asx (aspartate or asparagine) and X for Glx (glutamate or glutamine). J, O and U are not used. The one-letter code is invariably used when comparing and aligning sequences of proteins. Most are easily remembered by their initial letters. Note that cysteine and methionine are the only two sulphur-containing amino acids.

9.2.3 Look-back time

Homologous sequences diverge over time and tend to incorporate mutations more or less at random. This will make two sequences more different from each other as time elapses. The chance that a certain position in the protein sequence incorporates a second mutation, so obscuring the first mutational event, or even a back mutation resulting in no observable difference, will increase with the total number of mutations that have been incorporated and thus will also increase over time. Therefore, one will underestimate the total number of mutations that two sequences have incorporated after their separation, and the relative level of underestimation will increase with time. In protein-coding nucleotide sequences the first and second position of each codon are less prone to nucleotide substitutions, because this will almost always lead to a change in amino acid in the corresponding position of the protein. In most cases, natural selection will act against such mutations (purifying selection). The third position, also called the *wobble* position (see above), may be frequently substituted without directly affecting the corresponding amino acid. When comparing protein-coding sequences that have diverged for possibly hundreds of millions of years, it is very likely that the wobble bases in the codons will have become randomized. When comparing distantly related protein-coding sequences, it has been standard practice in phylogenetic analyses to exclude the wobble bases by removing every third nucleotide from the protein-coding sequence. Since this operation is very similar to the translation of DNA triplets into their corresponding amino acids and because it may be cumbersome to remove the wobble bases from the DNA sequences, it is recommended to simply translate an open reading frame in a nucleotide sequence into its corresponding protein sequence. From the arguments above, it can be concluded that the phylogenetic information is more rapidly erased in DNA sequences compared to protein sequences. This is illustrated in a different way below.

DNA is composed of only four different units: A, G, C, and T. If gaps were not allowed, on average 25% of residues in two randomly chosen aligned sequences would be identical. However, as soon as gaps are allowed, as much as 50% of residues in two randomly chosen aligned sequences can be identical. Such a situation may obscure any genuine relationship that may exist between two gene sequences, especially when comparing distantly related or rapidly evolving gene sequences. In contrast, the alignment of proteins with their 20 amino acids is less cumbersome. On average, 5% of residues in two randomly chosen and aligned sequences would be identical. Even after the introduction of gaps, still only 10%–15% of residues in two such sequences would be identical. Therefore, as a result of translating gene sequences into their corresponding protein sequences the latter are much more easy to align. Thus, translation of DNA into amino acids significantly improves the signal to noise ratio, which can sharpen up phylogenetic information considerably.

As a result of these different levels of resolution imposed by DNA and protein sequences, the use of gene sequences for the creation of phylogenetic trees of distantly related taxa is not recommended. In such situations only the use of translated protein sequences may lead to more reliable results. In the case of more closely related taxa, however, **substitution saturation** of *wobble* bases will not pose a problem and gene sequences may contain more information than protein sequences. A full discussion on nucleotide substitution saturation is provided in Chapter 20.

The difference in sensitivity in comparing homologous nucleotide or amino acid sequences becomes particularly clear when one carries out BLAST searches of a nucleotide sequence against all the nucleotide sequences present in a database such as Genbank or EMBL (Chapter 2). For the identification of closely related or identical sequences, the BLASTN algorithm, which directly compares two nucleotide sequences, provides excellent results. However, more distantly related sequences cannot be easily identified and are only found by using their corresponding protein sequences as a query against a protein sequence database such as SwissProt or Uniprot in a BLASTP search. An alternative method that also uses the protein sequence as an intermediate is the TBLASTX algorithm. This algorithm takes the translation of the open reading frame of the query sequence and compares this with all the nucleotide sequences in the database after translation of the six open reading frames into the corresponding protein sequences (Fig. 9.2).

9.2.4 Nature of sequence divergence in proteins (the PAM unit)

Non-synonymous substitutions in the DNA of protein-encoding genes will alter the corresponding protein sequence. With time, an increasing amount of mutations will be incorporated and, as a consequence, the two descendants of one ancestral protein will diverge. The observed sequence difference of two proteins that experienced mutations is, however, not linear with time but takes the course of a negative exponential (Fig. 9.3). This can be attributed to an increasing probability of reverse changes (*back mutations*) and *multiple hits.* Consequently, the observed percentage of difference between two homologous protein sequences is not proportional to the actual evolutionary difference and the level of underestimation will be more serious at larger evolutionary distances. The same scenario is true for nucleotide sequences, requiring appropriate nucleotide **substitution models** to correct for multiple hits (Chapter 4). A measure that is proportional to the true evolutionary distance between two proteins is the **PAM** value, which was introduced by Dayhoff in 1978. PAM is the number of Accepted Point Mutations per 100 amino acids. Examples of some PAM values and their corresponding observed distances are given in Table 9.3. Two homologous proteins that have had a common ancestor and have a PAM distance of 250–300 (80%–85% difference), or more, cannot reliably be distinguished from two randomly chosen and aligned proteins of similar length for the reason explained above. Therefore, phylogenetic analyses using protein sequences must generally be limited to proteins that differ by less than 80%–85%.

Proteins with functions that are non-essential to the survival of the organism may in general evolve at a relatively high rate. As a result, evolutionary information can be quickly erased and such proteins can therefore only be used for phylogenetic studies of closely related organisms. However, house-keeping proteins, such as histones, enzymes of the replication machinery, enzymes of core metabolism and proteins of the cytoskeleton, evolve slowly and incorporate between 1 to 10 mutations per 100 residues and per 100 million years (Table 9.4). Therefore, it takes a considerable time before these proteins have incorporated so many substitutions as to erase all evolutionary information. Because of this slow rate of evolution, house-keeping proteins are excellent markers to trace evolutionary relationships over long periods of time. For instance, the slow mutation rate of

Nucleotide sequence against nucleic acid database
Query= LmjF30.2080 |||hypothetical protein, conserved|Leishmania major
         (1656 letters)
Database: EMBL (release + updates)
66277815 sequences; 69403745503 total letters

	Score (bits)	E Value
Sequences producing significant alignments:		
em:CT005267[Leishmania major] Leishmani ...	3283	0.0e+00
em:CV663049[Leishmania donovani chagas ...	430	2.5e-116
em:DX961126[Trypanosoma cruzi] CHORI10 ...	54	6.1e-03
em:DX948996[Trypanosoma cruzi] CHORI10 ...	48	3.8e-01
em:AL939105[Streptomyces coelicolor] St ...	46	1.5e+00
em:BX628111[Anopheles gambiae (African ...	46	1.5e+00
em:AV628110[Chlamydomonas reinhardtii] ...	46	1.5e+00
em:AY686591[Escherichia coli] Escherich ...	46	1.5e+00
em:AI226793[Mus musculus (house mouse) ...	46	1.5e+00
em:BM016710[Homo sapiens (human)] 6036 ...	46	1.5e+00
em:CR529048[Anopheles gambiae (African ...	46	1.5e+00
em:CP000386[Rubrobacter xylanophilus DS ...	46	1.5e+00
em:AZ220839[Trypanosoma brucei] Sheare ...	46	1.5e+00
em:AC074258[Trypanosoma brucei] Trypano ...	46	1.5e+00
em:BC032244[Homo sapiens (human)] Homo ...	44	5.9e+00

Protein sequence against protein database
Query= LmjF30.2080 |||hypothetical protein, conserved|Leishmania major
         (551 letters)
Database: UniProt
4103579 sequences; 1350324263 total letters

	Score (bits)	E Value
Sequences producing significant alignments:		
trembl:Q4Q785_LEIMA Hypothetical protein...	1021	1.1e-296
trembl:Q4DCK6_TRYCR Hypothetical protein...	473	1.2e-131
trembl:Q584X8_9TRYP Hypothetical protein...	469	1.3e-130
trembl:Q4DB07_TRYCR Hypothetical protein...	466	1.1e-129
trembl:Q0ID02_SYNS3 R3H domain protein ...	203	2.2e-50
trembl:Q31QL2_SYNP7 Single-stranded nuc ...	191	8.7e-47
trembl:Q3AUV7_SYNS9 ATPase. ...	189	4.3e-46
trembl:Q2JIJ9_SYNJB R3H domain protein ...	189	4.3e-46
trembl:A0H4H4_9CHLR Single-stranded nuc ...	189	4.3e-46
trembl:Q05QB0_9SYNE Hypothetical protei ...	188	5.7e-46
trembl:Q3J6X9_NITOC Single-stranded nuc ...	188	5.7e-46

Fig. 9.2   Comparison of the sensitivities of Blast searches of a nucleic acid against a nucleic acid database and of a protein sequence against a protein database.

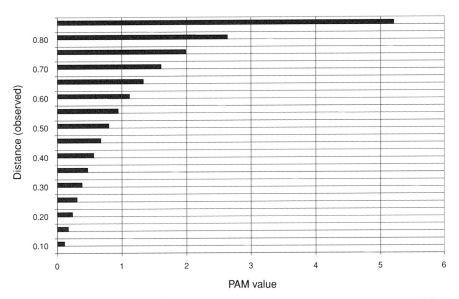

Fig. 9.3   Relationship between observed percentage of distance between two sequences and their PAM value. When the PAM distance value between two distantly related proteins nears the value 250, it becomes difficult to tell whether the two proteins are homologous, or whether they are two randomly taken proteins that can be aligned by chance. In that case we speak of the "twilight zone." (Modified from Doolittle, 1987.)

glyceraldehyde-3-phosphate dehydrogenase provides a theoretical look-back window larger than the duration of live on earth. In conclusion, proteins can be excellent tools to study the evolutionary relationships of both closely as well as distantly related taxa, provided an appropriate set of sequences is being selected.

9.2.5 Introns and non-coding DNA

When confronted with a DNA sequence, a biologist needs to figure out which parts code for a protein sequence. This can be particularly problematic for a eukaryotic genome that contains much more DNA than is needed to encode proteins; the sequence of a random piece of DNA is likely to encode no protein whatsoever. Eukaryotic genes, in general, have been fragmented into exons and interspersing introns. Due to differences in evolutionary pressure on exons and introns, the rate of incorporation of base substitutions in these two elements of eukaryotic genes may be dramatically different. Therefore, a study of the evolution of a protein using its DNA sequence should only include coding sequences. This requires that, in every DNA sequence, all the introns are edited out. Many of these problems can simply be avoided by the sequencing of the corresponding mRNA (via its cDNA) rather than the DNA itself, because the separate exons have been joined in one continuous stretch of RNA that contains much less extraneous material. Millions of (partial)

Table 9.3 Correspondence of observed differences
between proteins and their evolutionary distance

Observed percentage difference	Evolutionary distance in PAM
1	1
5	5
10	11
15	17
20	23
25	30
30	38
35	47
40	56
45	67
50	80
55	94
60	112
65	133
70	159
75	195
80	246
85	328 <- Twilight zone

As the evolutionary distance increases, the probability of superimposed mutation becomes greater resulting in a lower observed
percent difference. (Adapted from Table 23 in Dayhoff, 1978.)

sequences are now available in the so-called EST (*expressed sequence tag*) databases
being compiled for many organisms. For most protein-coding genes, not only the
nucleotide sequence, but also the translated protein sequence is available in the
nucleic acid database or in a secondary protein database (see below).

9.2.6 Choosing DNA or protein?

If possible, it is recommended to analyze a data set both ways (DNA and protein);
however, with the limitation that, for very distantly related taxa, nucleic acid
sequences have probably lost most, if not all, phylogenetic signal. Keep in mind
that, for a group of taxa that are relatively closely related (like different isolates of the
same or similar viruses), DNA-based analysis is probably a good recommendation,
since there may be fewer problems like differences in codon bias or saturation
of the third position of codons. It is nevertheless strongly recommended to carry
out an analysis in parallel on the protein sequence data. Moreover, where there is
ambiguity in the alignment of gene sequences, it is recommended translating the

Table 9.4 The highly different rates at which proteins evolve

Type of protein	Rate of Change (PAMs/100 myrs)	Theoretical Lookback Time (myrs)
Pseudogenes	400	45
Fibrinopeptides	90	200
Ig lambda chain C region	27	670
Somatotropin	25	800
Ribonucleases	21	850
Haemoglobin alpha chain	12	1500
Acid proteases	8	2300
Cytochrome c	4	5000
Adenylate kinase	3.2	6000
Glyceraldehyde-P dehydrogenase	2	9000
Glutamate dehydrogenase	0.9	18000

Useful lookback time = 360 PAMs

(Adapted from Table 1 in Dayhoff, 1978)

sequences first to their corresponding protein sequences, then aligning the protein sequences, and determining the position of gaps in the DNA sequences according to the more reliable protein alignment (see Chapter 3).

9.3 Construction of phylogenetic trees

9.3.1 Preparation of the data set

Retrieval of protein sequences

To reconstruct a phylogenetic tree using protein sequences, one needs to start from a set of homologous sequences, i.e. sequences that all have evolved from the same common ancestral sequence. Protein sequences can be obtained from primary databases, such as GenBank, EMBL, and DDBJ, which, in addition to nucleotide sequences, also store the translation products from gene sequences. However, the preferred source is secondary databases dedicated to protein sequences such as SwissProt, UniProt, and Genpept. SwissProt (*http://www.expasy.ch/*), a curated database, is the most reliable source for the collection of such sequences. In this database each sequence has been checked by experts and is annotated extensively. Moreover, homologous sequences from different species have highly similar accession codes, facilitating the recognition and simultaneous retrieval of all related sequences. A disadvantage of this database is that it is always one step behind other databases because of the necessary human intervention. The Protein Identification Resource (PIR; *http://pir.georgetown.edu/*) database contains many more protein sequences, but these have not all been checked, and the database is redundant. This is also the case with the GenPept (translated Genbank, *http://www.ncbi.nlm.nih.gov/*), and the UniProt/TrEMBL (translated EMBL sequences, *http://www.ebi.ac.uk/uniprot/*) databases. However, if you do not want to miss any sequence homologous to your protein, even if it is an incomplete sequence or a pseudogene, these are the databases to include in your search. The Enzyme database (*http://www.expasy.ch/enzyme/*) is also a useful tool for simultaneously retrieving all homologous sequences of one specific enzyme. These databases can all be accessed via the internet and allow the sequences to be downloaded, not only as the complete entry in a database-specific format, but also in the FASTA format, which can be handled by most sequence analysis programs without further modification. More information on sequence databases can be found in Chapter 2.

Alignment of two-protein sequences

Comparative analysis of protein sequences is not always straightforward because homologous proteins are almost never identical, but only similar. In addition to substitutions, there will be insertions and deletions in one sequence relative to the other. Moreover, phylogenetic analyses should be carried out on homologous residues only (i.e. those residues in each of the sequences that originate from a common ancestral residue). Thus it is essential that two or more sequences are properly aligned. Several excellent programs have been developed for the alignment of protein sequences and are available in the public domain such as CLUSTALW for use with command-line terminals or CLUSTALX for use with a graphical user interface (Thompson *et al.*, 1994). The algorithms used for the alignment of proteins are the same as those described for nucleic acids (see Chapter 3). To align two homologous sequences, a procedure similar to the dot-matrix method can be used (Fig. 9.4). In this method the two sequences to be aligned are written out as column and row headings of a matrix. Dots are put in the matrix wherever the residues in the two sequences are identical. If the two sequences are identical, there will be dots in all the diagonal elements of the matrix. If the two sequences are different, but can be aligned without gaps, there will be dots in most of the diagonal elements. If a gap occurred in one of the sequences, the alignment diagonal will be shifted vertically or horizontally. Horizontal or vertical shifts from the diagonal due to the presence of gaps are given a penalty. Alignment programs to create pairwise and multiple alignments can be downloaded or accessed directly via the internet (see the practical part of this chapter and Chapter 3).

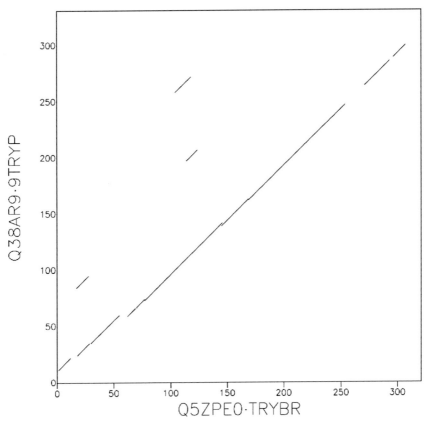

Fig. 9.4   A dot matrix of two homologous glyceraldehyde-3-phosphate dehydrogenase sequences that are 56% identical. The DOTMATCHER program of the EMBOSS suite was used to generate the plot.

    The choice of the value for each positive score relative to that of the gap penalty strongly influences the quality of the resulting alignment. Too low a gap penalty value will lead to a situation where residues in highly dissimilar regions will not be aligned with each other, but rather with gaps. Too high a gap penalty will lead to alignment of non-homologous residues or regions. Most programs allow the user to select an appropriate weight matrix for the scoring of either identities or similarities of amino acids, to adjust the gap penalty value, and to change the size of the window that scans the diagonal. Figure 9.4 shows such a dot matrix for two homologous trypanosome glyceraldehyde-3-phosphate dehydrogenases (GAPDH). It is obvious that there is usually no problem in aligning two sequences as long as they are of similar length and have more than 50% identity.

Many different weight matrices have been developed for sequence alignment programs to reflect some theoretical or observed rules of mutation. Empirical weight matrices are generally based on the observed substitution rates of amino acids in large collections of homologous sequences in databases such as SwissProt. The choice of the weight matrix is left to the scientific judgment of the user, and will depend on the data set being analyzed. The PAM250 (Dayhoff, 1978) and **BLOSUM62** (Henikoff & Henikoff, 1992) are *log odds matrices*, where the probabilities have been determined from a limited (PAM) and a very large (BLOSUM) database of aligned homologous protein sequences. These two scoring matrices are implemented in the CLUSTALW and CLUSTALX programs. The Jones–Taylor–Thornton (JTT) matrix (Jones *et al.*, 1992) is an update to the PAM matrix. The Gonnet matrix (Gonnet *et al.*, 1992) is a scoring matrix based on the alignment of the entire Swiss-Prot database where 1.7×10^6 matches were used from sequences differing by 6.4 to 100 PAM. It is implemented in the Darwin program by the same authors. The Gonnet, PAM, and BLOSUM matrices are now widely used in alignment programs. The Whelan and Goldman (WAG) matrix is an improvement on the JTT matrix based on *likelihood* methods (Whelan & Goldman, 2001). Müller and Vingron (2000) have also presented a protein substitution matrix (VT) based on a method that approximates *maximum likelihood.* In some cases, there are specific protein matrices available for the type of data, e.g. mtREV for mitochondrial DNA (Adachi & Hasegawa, 1996), or the type of organism, e.g. HIV (Nickle *et al.*, 2007).

Several algorithms for the pairwise alignment of protein sequences have been developed, including the well-known Pearson–Lipman algorithm (Pearson & Lipman, 1988), used in Pearson's FASTA program, the Needleman–Wunsch (1970), the Smith–Waterman algorithm (1981), and the BLAST algorithms (Altschul *et al.*, 1990). They are implemented in sequence-comparison programs for the search of homologous sequences in large sequence databases and in multiple alignment programs.

Multiple sequence alignment

For the construction of reliable phylogenetic trees, the quality of a multiple protein sequence alignment is of the utmost importance. There are many programs available to perform the multiple alignment task (see Chapter 3). Most programs quickly align pairs of sequences and roughly determine the degrees of identity between each pair. The sequences are then aligned more precisely in a progressive way, using an appropriate amino acid replacement matrix, starting from the two most related sequences (e.g. CLUSTALX). If possible, it is advisable to modify the alignment parameters such as amino acid replacement matrices and gap penalties to obtain the best possible alignment and to keep in mind that most multiple sequence alignment programs perform better with sequences of similar length.

A very convenient and rapid way to create an alignment of homologous protein sequences starting from a single protein sequence is to submit the query sequence to the EBI BLAST server (http://www.ebi.ac.uk/blast2/). This allows the search of an appropriate protein database such as SwissProt or Uniprot for all homologous sequences. The best hits then are automatically aligned using the CLUSTALDB utility, and the resulting multiple alignment can then be imported in most alignment programs or editors. Special subsets of protein databases can be chosen such as the non-redundant Uniref90 or Uniref50 subsets of the UniProt database. In these subsets no pair of sequences in the representative set has >90% or >50% mutual sequence identity, respectively. This allows the creation of a set of homologous sequences with a wide representation of phylogenetic diversity (sometimes over the various kingdoms of life) with exclusion of many very similar or identical sequences.

Manual adjustment of a protein alignment

An automatically produced multiple sequence alignment often needs manual adjustment to improve its quality, especially at the position of gaps. Such improvement can be obtained by using all the knowledge that is available about a specific type of protein. Information about active site residues and elements of secondary structure, such as α-helices, β-strands and loops, may be of great help here. While manually adjusting multiple alignments one should have knowledge of the physico-chemical properties of the 20 amino acids and keep in mind a number of rules of thumb for the mutability of the various amino acids. In a folded protein, the residues D, R, E, N, and K (Table 9.2) are preferably mutated to residues of similar properties. Since they are polar or charged, they are found mainly on the surface of the folded protein. Moreover, since they play a less important role in protein folding, they mutate rather easily. Hydrophobic residues (F, A, M, I, L, Y, V, and W) are replaced preferentially by other hydrophobic ones. These residues are mainly positioned internally and determine the folding of the protein. As a consequence, they mutate rather slowly. The residues C, H, Q, S, and T are generally indifferent and may be replaced with any other type of residue. The residues (D, R, E, N, K, C, H, Q, S, and T), when conserved throughout the alignment, are very likely residues that are involved in the active site of an enzyme. So the multiple alignment should be adjusted in such a way as to maintain the positional homology of these residues. Periodicity of charged residues may provide information about the presence of elements of secondary structure such as α-helices and β-strands. α-Helices have a repetition of 3.6 residues per turn. Stretches of more than 12 amino acids with a charged amino acid every third or fourth position in the sequence may be indicative of the presence of an amphipathic α-helix with charged residues at one side and apolar residues at the other side. Short stretches with a repetition of charged amino acids

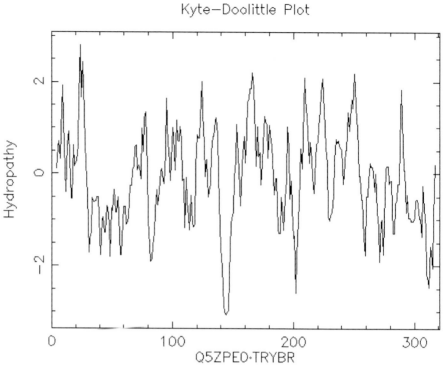

Kyte–Doolittle Plot

Q5ZPEO·TRYBR

Fig. 9.5   Hydrophobicity (or hydropathy) profile according to Kyte and Doolittle (1982) of *Try-panosoma brucei* glyceraldehyde-3-phosphate dehydrogenase. The program PEPWINDOW of the EMBOSS suit was used.

every second residue may be indicative of a β-strand structure. Gaps are almost never found in elements of secondary structure, but only in regions with loops. Moreover, the residues P and G interfere with secondary structure elements and thus have a preference for loop regions. Since loops easily acquire or lose residues you should always try to align gaps together with P and G residues. However, because of the inherent difficulties of unambiguously aligning sequence residues associated with gap regions, it is strongly advised removing all columns with gaps before submitting the alignment to a tree-building program. Hydrophobicity (or hydropathy) profiles according to Kyte and Doolittle (1982) of two homologous proteins are, in general, strikingly similar (Fig. 9.5) and may also be of help in adjusting the alignment manually. Another very useful tool for the manual alignment of proteins, in case there is a crystal structure available for at least one of the enzymes in the alignment, is the feature table of the SwissProt entries. This contains information about the active site residues and secondary structure elements of the protein.

9.3.2 Tree-building

Methods for inferring a protein phylogeny

The goal of many evolutionary analyses is to reconstruct an evolutionary tree describing the relationships of the various taxa with respect to each other. Two classes of methods exist for the inference of evolutionary trees: (i) *distance-based* methods, using a matrix that contains pairwise distance values between all sequences in the alignment, such as **UPGMA, Neighbor-joining (NJ)**, and Fitch and Margoliash (see Chapter 5), and (ii) *character-based* methods that carry out calculations on the individual residues of the sequences such as *maximum likelihood*, **maximum parsimony** and *Bayesian inference* (see Chapters 6–8). In general, distance methods are fast, while character-based methods are computationally expensive and produce results much more slowly.

Distance-based tree-building algorithms for proteins are identical to the ones described for nucleotide sequences (Chapter 5). Unweighted Pair Group with Arithmetic Mean (UPGMA) is a sequential clustering algorithm that starts from a matrix of **genetic distances.** The tree reconstruction method is sensitive to unequal rates of evolution and should only be used when the evolutionary rate can be assumed to be constant. This problem is overcome by the Fitch and Margoliash (Fitch, 1981) and NJ methods (Saitou & Nei, 1987; Felsenstein, 1993). Distance methods can take advantage of an evolutionary model to transform observed distances into evolutionary distances. This can be achieved by an empirical substitution matrix, like PAM or BLOSUM. An alternative is to use the approximate formula developed by Kimura (1983) that estimates true genetic distance d from the observed distance p:

$$d = -\ln(1 - p - 0.2p^2)$$

where p is the proportion of observed distances between any two homologous sequences. The calculation of evolutionary distances with this formula is very fast, applicable to very large data sets, and implemented in the CLUSTALW, CLUSTALX and PROTDIST (PHYLIP package, Felsenstein, 1993) programs. The formula is based on statistical considerations (see Kimura, 1983) and is usually reliable for $p \leq 0.7$ (Nei, 1985). Because this approach does not consider the specific type of amino acid replacement, some information is ignored. When p exceeds 0.85, d rapidly grows to infinite values in agreement with the fact that any phylogenetic signal gets lost at such distances. Protein distance estimates can be obtained with the CLUSTAL programs, with PROTDIST and with TREEPUZZLE (Strimmer & Von Haeseler, 1997). The TREEPUZZLE program implements a large set of substitution matrices (see Chapter 6), and Chapter 10 discusses how to select the best-fit model for your protein sequences. Ideally, distance methods should result in branch lengths that are proportional to the (corrected) distances between the taxa.

The maximum parsimony method uses the sequence information itself rather than the deduced distance information. It calculates for all possible trees the tree that represents the minimum number of substitutions summed over all informative sites. Maximum parsimony does not assume an explicit evolutionary model, and therefore appropriate branch lengths cannot be estimated for divergent taxa. The method frequently finds many equally parsimonious trees. Examples of programs available for maximum parsimony analysis are PROTPARS (Felsenstein, 1993) and PAUP* (by David Swofford, 1998) (see Chapter 8).

Maximum likelihood is a method that evaluates a hypothesis about the evolutionary history in terms of the probability that the proposed model, and the hypothesized history would give rise to the observed data set (see Chapter 6). In this probabilistic framework, a history with a higher probability of producing the observed data is preferred to a history with a lower probability. The method searches for the tree that yields maximum probability of observing the data. Due to the fact that the information about all characters present in the data set is used for the analysis, and likelihood computation involves averaging over all possible character states at internal nodes in a tree, maximum likelihood analyses are very CPU intensive and take a long time to complete. Popular programs to analyze protein sequences using maximum likelihood are PROTML from the MOLPHY package (Adachi & Hasegawa, 1996), PHYML (Guindon & Gascuel, 2003) and TREEPUZZLE (Strimmer & von Haeseler, 1997). The latter two programs apply heuristic methods that are much faster than PROTML, but they do not necessarily guarantee to find the best tree (see Chapter 6). Bayesian inference of protein phylogenies, which combines both likelihood and *prior* (see Chapter 7), can be performed using MRBAYES (Huelsenbeck & Ronquist, 2001; Ronquist & Huelsenbeck, 2003). An example of this is presented in the practice section of this chapter.

Rooting a tree

Most methods for phylogeny inference yield trees that are unrooted. Thus from a tree itself it is impossible to tell which of the OTUs branched off before all the others. To root a tree one should include an *outgroup* in the data set. An outgroup is an OTU for which external information (e.g. paleontological information) indicates that this taxon branched off before all other taxa.

An interesting way to root a tree is to exploit an event of gene duplication, which took place before speciation and that has led to the formation of *paralogous* isoenzymes. An example of this is the gene duplication leading to the various vacuolar ATPases, present in archaebacteria, eubacteria, and eukaryotes, which provided a root to the "tree of life" (Gogarten *et al.*, 1989).

Reliability of tree topologies

Tree topologies may strongly depend on several issues: the alignment accuracy, the inference method (distance, parsimony, maximum likelihood and Bayesian methods); the number of OTUs included in the alignment; for some algorithms, the order of the OTUs in the alignment; the selection of an appropriate outgroup and finally, the presence of widely varying branch lengths, etc. None of the methods guarantees to find the one tree with the correct ("true") topology. Most do not even guarantee to find the best tree according to their optimality criterion. So, as to have an idea about the reliability of the topology of the resulting tree, the following strategies can be recommended:

(1) Apply different tree-building methods to the data set.
(2) Vary the parameters used by the different programs, such as the seed value and jumble factor for the order of OTU addition in PHYLIP programs. Use different starting trees in heuristic searches (e.g. random-addition-sequence replicates, see Chapter 8).
(3) Apply the most appropriate model for amino acid substitution (see Chapter 10). When in doubt, apply various substitution models.
(4) Add or remove one or more OTUs and see how this influences the tree topology. Do not include mosaic (recombinant) sequences (see Chapter 15).
(5) Include an appropriate outgroup that may serve as a root for your tree.
(6) Be aware of taxa with very long branches that may be subject to the so-called **long-branch attraction** effect, which may lead to an anomalous positioning of that taxon.
(7) Apply "**bootstrap**" or "**jackknife**" analyses to your data set and prepare a consensus tree of 100–1000 replicates depending on size of the data set and on computer power (see Chapter 5). Alternatively, compute **posterior probabilities** (Chapter 7).
(8) Keep in mind that, in the case of bootstrap analysis, only nodes that occur in more than 95% of the cases are considered to be highly reliable.

Finally, a tree should be considered robust to inference methods and reliable when widely different methods infer similar or identical tree topologies and such topologies are generally supported by good bootstrap values (i.e. more than 95%).

PRACTICE

Fred R. Opperdoes and Philippe Lemey

9.4 A phylogenetic analysis of the Leishmanial glyceraldehyde-3-phosphate dehydrogenase gene carried out via the Internet

The practical exercise on amino acid phylogenetic inference described here can be carried out entirely via the World Wide Web. The analysis deals with an open reading frame (ORF) coding for the *Leishmania mexicana* NAD-dependent glyceraldehyde-3-phosphate dehydrogenase (G3PD, EC.1.2.1.12), an enzyme that is associated with microbodies in the haemoflagellated protozoan parasite *L. mexicana* (Hannaert *et al.*, 1992), and for which a crystal structure is available in the PDB database (1GYP). The exercise consists of the following steps:

(1) Consulting the Enzyme database and collecting all available protein sequences of a specific enzyme.
(2) Searching the Protein database for the presence of a 3-D structure.
(3) Scanning a DNA sequence for the presence of open reading frames and translating an open reading frame into its corresponding protein sequence.
(4) Performing a BLASTP homology search against SwissProt using a protein sequence.
(5) Creating a multiple alignment of homologous protein sequences.
(6) Converting the file format from Clustal to Phylip.
(7) Constructing a maximum likelihood tree using PHYML.
(8) Visualizing the inferred tree.

Before starting the analysis project, we collect some general information about the enzyme we are going to study. First, we connect to the Enzyme, or EC database, available at the Expasy server in Switzerland (*http://www.expasy.ch/enzyme/*), which contains general information about enzymes, their official names and their EC numbers, the reactions they catalyze and the pathways they are involved in. It also provides access to all protein sequences available in the SwissProt database. Once you are in the Enzyme database, select the enzyme by its EC number, 1.2.1.12, and get the requested information. Familiarize yourself with the information available in the database. Try also to find publications in the PubMed database (*http://www.ncbi.nlm.nih.gov/entrez/query.fcgi*) that deal with this *Leishmania* enzyme. Now retrieve the SwissProt entry referenced as G3PG_LEIME (and not G3PC_LEIME) and study its content (Fig. 9.6a). The sequence can be saved in SwissProt format (g3pg_leime.sw) or in FASTA format as g3pg_leime.fasta (see the link in the lower right corner in Fig. 9.6b); we will further use the latter format in this exercise.

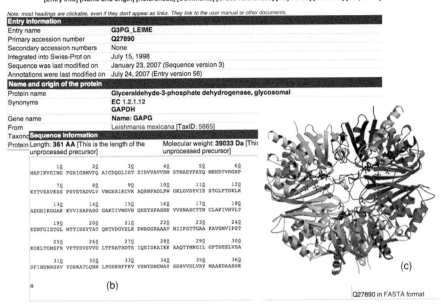

UniProtKB/Swiss-Prot entry Q27890

Fig. 9.6   Retrieving information about *Leishmania mexicana* NAD-dependent glyceraldehyde-3-phosphate dehydrogenase. (a) First part of the SwissProt entry referenced as G3PG_LEIME. (b) Sequence information panel of the SwissProt entry. (c) *Leishmania* G3PD three-dimensional structure (1GYP) available in the Brookhaven Protein Database.

Using the information available in the various databases, you probably noticed that the crystal structure of the *Leishmania* G3PD has been solved (Kim *et al.*, 1995). This three-dimensional structure can be found in the Brookhaven Protein Database or PDB (*http://www.rcsb.org/pdb/*) under the accession code 1GYP; have a look at this information (Fig. 9.6c). Note that the secondary structure information is also available in the SwissProt entry (see above).

The gene for the *Leishmania* G3PD and its flanking nucleotides are available from the EMBL database at the EBI (European Bioinformatic Institute, Hinxton, UK, *http://www.ebi.ac.uk*) under the accession number EMBL:X65226. Go to the simple sequence retrieval page (*http://www.ebi.ac.uk/cgi-bin/emblfetch*), enter the accession number and select FASTA format before retrieving and saving the sequence on your computer. To scan the DNA sequence for the presence of open reading frames, submit the entire nucleotide sequence to the EMBOSS TRANSEQ utility (*http://www.ebi.ac.uk/emboss/transeq/*). The nucleotide sequence contains two almost identical open reading frames, from position 266 to 1351 in the second reading frame and from position 2101 to 3186 in the first reading frame, which encode proteins of 361 amino acids each that end with -SKM, a typical microbody targeting sequence. Translate the second open reading frame ranging from position 2101 to 3186 with a length of 1086 nucleotides (in the first reading frame); after a few seconds the server returns the following output:

```
> X65226_1 L.mexicana gap gene for glycosomal glyceraldehyde-3-phosphate dehydrogenase
MAPIKVGINGFGRIGRMVFQAICDQGLIGTEIDVVAVVDMSTNAEYFAYQMKHDTVHGRPKYTV
EAVKSSPSVETADVLVVNGHRIKCVKAQRNPADLPWGKLGVDYVIESTGLFTDKLKAEGHIKGG
AKKVVISAPASGGAKTIVMGVNQHEYSPASHHVVSNASCTTNCLAPIVHVLTKENFGIETGLMTT
IHSYTATQKTVDGVSLKDWRGGRAAAVNIIPSTTGAAKAVGMVIPSTKGKLTGMSFRVPTPDVSV
VDLTFRATRDTSIQEIDKAIKKAAQTYMKGILGFTDEELVSADFINDNRSSVYDSKATLQNNLPG
EKRFFKVVSWYDNEWAYSHRVVDLVRYMAAKDAASSKM*
```

To retrieve homologs of this sequence, a BLASTP search with a protein sequence as query against a protein database has a much better signal to noise ratio than a corresponding search using the nucleotide sequence. Therefore, we now use the *Leishmania* protein as query sequence in a BLASTP search in the hope to find all glyceraldehyde-3-phosphate dehydrogenase homologs in the protein databank. Although we could use the translated GenPep database of Genbank, or its EMBL equivalent UniProt, we prefer to search the non-redundant SwissProt database. Although this is a derived database, it is annotated extensively and has been checked thoroughly by protein scientists for redundancy and correctness of the included sequence information. For our search, we will use the BLAST server at the EBI (*http://www.ebi.ac.uk/blast2/*), which has a nice facility to create multiple alignments based on the query results. Paste your sequence into the sequence submission window and then select the options PROGRAM "Blastp" and DATABASE "UniProtKB/Swiss-Prot" protein database and use "BLOSUM62" as substitution matrix. The BLAST output should be returned within a few seconds. Because of the high signal to noise ratio of a protein search, only glyceraldehyde-3-phosphate dehydrogenases are reported at the top of the output.

Although the query sequence in the BLAST search was obtained from *Leishmania*, a eukaryotic protist, several bacterial sequences score equally well or better than many of the eukaryotic G3PD sequences. We will proceed with creating a multiple alignment for the first 50 sequences with the best scores in order to study the evolutionary relationship of the Leishmanial enzyme with the other glyceraldehyde-3-phosphate dehydrogenases. Note that the 50 best scoring hits might change as the database is updated. The BLASTP server allows the rapid creation of multiple alignments using the CLUSTAL program. Click on the button DBCLUSTAL, select the first 50 sequences and run the program. After a minute or so the alignment appears in nice colours. Note that the alignment should probably be edited for proper phylogenetic analysis (see Chapter 3), but this is beyond the scope of this exploratory exercise. In the results page, click on the link "Multiple Sequence Alignment" to obtain the alignment in Clustal format. To convert the file format, copy and paste the alignment in the submission box of the online READSEQ conversion tool (*http://iubio.bio.indiana.edu/cgi-bin/readseq.cgi*). Select "Phylip|Phylip4" as output sequence format and click "submit". Save this alignment in Phylip interleaved format as g3pd.phy.

To reconstruct a maximum likelihood tree, we will use PHYML available online at *http://atgc.lirmm.fr/phyml/*. The speed at which your job will be processed will depend on the server load, which is indicated at the top of the submission page. Upload the input file (g3pd.phy) by clicking on "Choose File". Select "File" instead of "Example file" and select "Amino-Acids" as "Data Type". Keep the default WAG substitution model, four rate categories for the gamma distribution parameter and all other default settings. Provide the remaining information including your name, country of residence, email and file format of the data file (important to interpret line breaks) before clicking "Execute & email results".

Once the PHYML job is processed on the server, it takes about 50 minutes to complete the maximum likelihood heuristic search. The results are returned by email and include a link to an ATV (A TREE VIEWER) applet to visualize phylogenetic trees. Following this link, click on "View tree" to start the applet and display the tree. By selecting "root/reroot", the tree can be rerooted by clicking on any node. Alternatively, you can paste the NEWICK tree string at the end of the email into the PHYFI online application available at *http://cgi-www.daimi.au.dk/cgi-chili/phyfi/go*. In addition to visualizing the tree with this application, the image can be downloaded in several formats. Another web server that can be used for similar purposes is the phylogenetic tree printer program PHYLODENDRON (*http://www.es.embnet.org/Doc/phylodendron/treeprint-form.html*). This program gives you various possibilities for presenting a tree and allows you to print trees in various formats. Select "tree diag" or "phenogram" for a tree where branch lengths are shown and select an appropriate output format. The PDF format also allows you to include hyperlinks to a database with the sequence information. In the case of SwissProt sequences, the "Base URL" shown in the web page should be changed to: "*http://www.expasy.org/uniprot/*". More advanced tree editing and visualizing can be performed using TREEVIEW and FIGTREE (see Chapter 5).

From the topology of the midpoint rooted tree (Fig. 9.7), it is immediately obvious that the G3PD of the protist *L. mexicana* (G3PG_LEIME) clusters with that of *Crithidia fasciculata* (G3PG_CRIFA) and then with G3PD of *Trypanosoma cruzi* (G3PG_TRYCR) and *Trypanosoma brucei brucei* (G3PG_TRYBB), all members

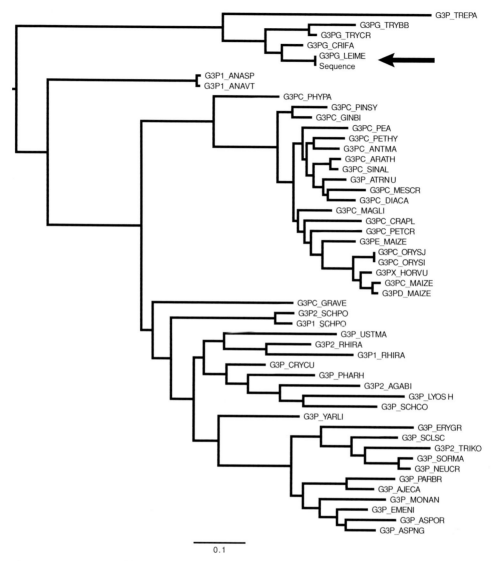

Fig. 9.7   Maximum likelihood tree of the *Leishmania mexicana* glyceraldehyde-3-phosphate dehydrogenase and 50 homologous sequences available in the SwissProt database. The horizontal bar represents 0.1 substitutions per amino acid site. The arrow indicates the *Leishmania
mexicana* sequence. Because the query was the same as the first hit obtained by BLASTP,
"Sequence" is clustered as an identical sequence with "G3PG_LEIME."

of the family Trypanosomatidae. Interestingly, these sequences are more related to
bacterial homologs, like G3P_TREPA from *Treponema pallidum*, than to other
eukaryotic G3PDs. This observation has previously been interpreted as an event of
horizontal gene transfer (Hannaert *et al.*, 2003).

9.5 A phylogenetic analysis of trypanosomatid glyceraldehyde-3-phosphate dehydrogenase protein sequences using Bayesian inference

In the second exercise, we will create a trypanosomatid phylogeny based on glyceraldehyde-3-phosphate dehydrogenase protein sequences using the program MrBayes (Huelsenbeck & Ronquist, 2001), which performs Bayesian *Markov chain Monte Carlo (MCMC)* analysis to infer a *posterior distribution* of trees (see Chapter 7). Trypanosomatidae are a family of hemoflagellate parasitic protists, which comprise trypanosomes and *Leishmania*, both parasites of man, *Crithidia* and *Herpetomonas*, parasites of insects and *Phytomonas* which is a parasite of plants. Trypanosomatidae are exclusively parasitic, whereas Bodonidae are free-living. Together, they form the order of the Kinetoplastida. The Kinetoplastida belong to the Euglenozoa, to which also *Euglena* – a green alga – belongs. The members of the Trypanosomatidae are supposed to have diverged several hundreds of million years ago (Fernandes *et al.*, 1993). The separation between Euglenids and Kinetoplastida must have taken place between 500 million and 1 billion years ago.

We start by collecting the protein sequences required for this exercise. A drawback of the SwissProt database is that it does not contain all available protein sequences, but only those that have been verified by human intervention. As a consequence, the SwissProt database contains a limited amount of trypanosomatid sequences for glyceraldehyde-3-phosphate dehydrogenase. We start with a glyceraldehyde-3-phosphate dehydrogenase sequence (LmjF30.2980) from the trypanosomatid *Leishmania major* responsible for the disfiguring disease "oriental sore" in the Middle East. This protein sequence is available in the genome database GeneDB (*http://www.genedb.org*; enter "glyceraldehyde" in the "search for gene" box, LmjF30.2980 is among the search results). To collect other sequences of this Trypanosomatid enzyme we have to resort to the UniProt database at the Expasy server (*http://www.expasy.ch*). Retrieve the following sequences and save them in FASTA format in a single file and align them with Clustal (see above or Chapter 3):

Accession Number	Organism	Entry name
Q4Q6Z4	*Leishmania major*	Q4Q6Z4_LEIMA
O96423	*Crithidia fasciculate*	G3PG_CRIFA
O96424	*Herpetomonas pessoai*	O96424_9TRYP
O96426	*Phytomonas* sp.	O96426_9TRYP
P22512	*Trypanosoma brucei brucei*	G3PG_TRYBB
A1YE37	*Bodo saltans* (bodonid)	A1YE37_9EUGL
A1YE38	*Parabodo caudatus* (bodonid)	A1YE38_9EUGL
O97105	*Trypanosoma rangeli*	O97105_TRYRA
Q5ZPF6	*Trypanosoma cruzi*	Q5ZPF6_TRYCR
Q43311	*Euglena gracilis*	Q43311_EUGGR
Q26753	*Trypanoplasma borreli* (bodonid)	Q26753_TRYBO
A1XIQ6	*Sergeia podlipaevi* (trypanosomatid)	A1XIQ6_9TRYP
Q76EI0	*Symbiodinium* sp. (dinoflagellate)	Q76EI0_9DINO

These protein sequences can also be obtained by accessing the BLAST2 server at EBI (*http://www.ebi.ac.uk/blast2/*). Paste the sequence LmjF30.2980 (available at *www.thephylogenetichandbook.org*) in the sequence window and select the "Uniprot Clusters 90%" database. This database only contains protein sequences that share less than 90% pairwise identity and thus is ideally suited to create a data set with a wide range of distantly related taxa with exclusion of duplicates or very similar sequences. Default settings can be used for all other options. Note that database updates can change the results of this BLAST search and a slightly different set of sequences than the one listed above might be retrieved.

Run the BLASTP program, select the 13 best scoring sequences in the output and create a multiple alignment in Clustal format as described in the first example. The resulting alignment should be trimmed and edited to remove all ambiguously aligned regions. This task can be performed using a dedicated alignment editor (see Chapter 3), or a regular text editor available on different platforms. For amino acid sequences that are shorter than most other sequences, missing characters can be coded using 'X' (see below). In order to use the edited alignment in the program MRBAYES, it should be converted into NEXUS format (see Chapter 2) and the "nchar" value should match exactly the number of character used per sequence (in this case 313). Instead of interactively setting up the MCMC analysis, which is fully explained in the practice section of Chapter 7, we will add a MRBAYES block at the end of the file:

```
#NEXUS

begin data;
    dimensions ntax = 14 nchar = 313;
    format interleave datatype = protein missing = X gap = -;

    matrix
    LmjF30.2980    INGFGRIGRMVLQAICDQGLIGNEIDVVAVVD
    Q4Q6Z4_LEIMA   INGFGRIGRMVLQAICDQGLIGNEIDVVAVVD
    G3PG_CRIFA     INGFGRIGRMVFQSMCEDNVLGTELDVVAVVD
    O96424_9TRYP   INGFGRIGRMVFQAMCEQGVLGKDFDVVAVVD
    O96426_9TRYP   INGFGRIGRNVFQAICEGNHLGTDIDVVAVAD
    G3PG_TRYBB     INGFGRIGRMVFQALCDDGLLGNEIDVVAVVD
    A1YE37_9EUGL   INGFGRIGRMVFQAIADQGLLGKEIEVVAVVD
    A1YE38_9EUGL   INGFGRIGRMVFQAICDQGLLGTEIDLVAVVD
    O97105_TRYRA   INGFGRIGRMVFQAMCEADLLGTEIEVVAVVD
    Q5ZPF6_TRYCR   IRPFGRIGRMVFQALCEDGLLGTEIDVVAVVD
    Q43311_EUGGR   INGFGRIGRMVFQALCDQGLLGTTFDVVGVVD
    Q26753_TRYBO   INGFGRIGRMVLQAICDQGLLGTEIDVVAVVV
```

```
A1XIQ6_9TRYP   XXXXXXXXXXXXXXXXXXXXXXXXXXXXXDVVAVVD
Q76EI0_9DINO   INGFGRIGRMVFQAICDQNLLGTKLDVVGVVD

[only a part of the alignment is shown for clarity]
;
end;

begin mrbayes;
  mcmcp filename=g3pdh.mb;
  prset aamodelpr=mixed;
  lset rates=equal;
  prset brlenspr=unconstrained:exponential(10.00);
  mcmcp ngen=200000 samplefreq=100 printfreq=100;
  mcmcp nruns=2 diagnfreq=1000 burninfrac=0.250000;
  mcmcp stoprule=yes stopval=0.004000;
  mcmcp nchains=4 swapfreq=1 temp=0.200000;
[ EDIT FILE AND REMOVE OUTCOMMENT BRACKETS AS NEEDED ]
[   mcmc; ]
[   sump burnin=XXX; ]
[   sumt contype=halfcompat burnin=XXX; ]
end;
```

The "prset" parameter sets the rate matrix for amino acid substitutions. This can be fixed to one of the available amino acid models or, as specified here using aamodelpr=mixed, the analysis can average over ten available models. This setting instructs the **Markov chain** to sample each model according to its probability and avoids having to select a particular model *a priori*. Furthermore, the command block tells MrBayes to run the MCMC for 200 000 generations, to sample and print parameters and trees every hundred generations, to stop the analysis as soon as the chain has converged to a standard deviation of split frequencies below 0.004, and to discard the first 25% of the sampled trees as a "***burn-in***." The standard deviation of split frequencies is a measure of the variance in clade support between two independent runs (MrBayes will run by default two independent runs). The lower the standard deviation, the more similar the tree sample of the independent runs and thus the more confidence we have the runs with different starting state have converged. An elaborate discussion of MCMC settings in Bayesian phylogenetics can be found in Chapter 7.

The edited alignment in NEXUS format is now saved to disk as a "text only" file under the name "g3pdh.mb" and is ready to be run by MrBayes (the file g3pdh.mb is also available at *www.thephylogenetichandbook.org*). The program reads the file, sets the options according to the settings specified in the MrBayes block and then waits for user input. At this stage it is still possible to change the settings,

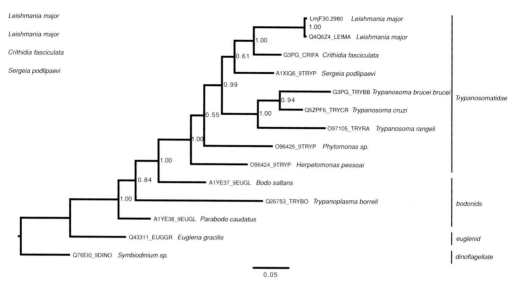

Fig. 9.8   Majority-rule consensus tree for the Bayesian MCMC analysis of the trypanosomatid glyceraldehyde-3-phosphate dehydrogenase sequences. The numbers at the nodes represent approximate posterior probabilities for the clades. The horizontal bar represents 0.05 substitutions per amino acid site.

if required. In our case, we should type "outgroup Q76EI0_9DINO" to root the tree of the Euglenozoa, which comprise the Kinetoplastida and the Euglenida, with a dinoflagellate. The analysis can be started using the "mcmc" command and stops when either 200 000 trees have been sampled or the trees have converged to a standard deviation of split frequencies less than 0.004, whichever comes first. At this stage, the command sump burnin=XXX (where XXX is the first number of trees to discard; in this case 100 is a reasonable number) summarizes the amino acid model sampling and calculates the posterior probability for each model. The sampled model is reported as an index: poisson(0), jones(1), dayhoff(2), mtrev(3), mtmam(4), wag(5), rtrev(6), cprev(7), vt(8), or blosum(9). In our case, the Wagner model is the only evolutionary model sampled after burn-in (see amino acid model probabilities), providing strong evidence that this model fits the data significantly better than the other models. The sumt burnin=XXX command instructs MRBAYES to summarize the sampled trees by creating a consensus tree. This *majority-rule consensus tree* can be visualized using applications like TREEVIEW and FIGTREE (see Chapter 5) or using the web applications described in the first exercise.

The consensus tree is shown in Fig. 9.8. The bodonids (Q26753, A1YE37, A1YE38) and the euglenid (Q43311) do not belong to the Trypanosomatidae, but together they belong to the Euglenozoa, to which also the Trypanosomatidae belong. The analysis shows that all Trypanosomatidae (Q4Q6Z4, G3PG_CRIFA, A1XIQ6, G3PG_TRYBB, Q5ZPF6, O97105, O96426, O96424) form a monophyletic group. Within this group, the genus Trypanosoma (G3PG_TRYBB, Q5ZPF6, O97105) forms a distinct monophyletic cluster.

Section IV

Testing models and trees

Selecting models of evolution

THEORY

David Posada

10.1 Models of evolution and phylogeny reconstruction

Phylogenetic reconstruction is a problem of statistical inference. Since statistical inferences cannot be drawn in the absence of probabilities, the use of a model of nucleotide substitution or amino acid replacement – a *model of evolution* – becomes indispensable when using DNA or protein sequences to estimate phylogenetic relationships among taxa. Models of evolution are sets of assumptions about the process of nucleotide or amino acid substitution (see Chapters 4 and 9). They describe the different probabilities of change from one nucleotide or amino acid to another along a phylogenetic tree, allowing us to choose among different phylogenetic hypotheses to explain the data at hand. Comprehensive reviews of models of evolution are offered elsewhere (Swofford *et al.*, 1996; Liò & Goldman, 1998).

As discussed in the previous chapters, phylogenetic methods are based on a number of assumptions about the evolutionary process. Such assumptions can be implicit, like in *parsimony* methods (see Chapter 8), or explicit, like in distance or *maximum likelihood* methods (see Chapters 5 and 6, respectively). The advantage of making a model explicit is that the parameters of the model can be estimated. Distance methods can only estimate the number of substitutions per site. However, maximum likelihood methods can estimate all the relevant parameters of the model of evolution. Parameters estimated via maximum likelihood have desirable statistical properties: as sample sizes get large, they converge to the true value and

The Phylogenetic Handbook: a Practical Approach to Phylogenetic Analysis and Hypothesis Testing,
Philippe Lemey, Marco Salemi, and Anne-Mieke Vandamme (eds.). Published by Cambridge
University Press. © Cambridge University Press 2009.

have the smallest possible variance among all estimates with the same expected value.

It is well known that the use of one model of evolution or another may change the results of a phylogenetic analysis (Sullivan & Joyce, 2005). When the model assumed is wrong, branch lengths, *transition/transversion* ratio, and divergence may be underestimated, while the strength of rate variation among sites may be overestimated. Simple models tend to suggest that a clade is significantly supported when it cannot be, and tests of evolutionary hypotheses (e.g. of the *molecular clock*, see Chapter 11) can become conservative. In general, phylogenetic methods may be less accurate (recover an incorrect tree more often), or may be inconsistent (converge to an incorrect tree with increased amounts of data) when the assumed model of evolution is wrong. Cases where the use of wrong models increases phylogenetic performance are the exception, and they rather represent a bias towards the true tree due to violated assumptions. Indeed, models are not important just because of their consequences in the phylogenetic analysis, but because the characterization of the evolutionary process at the molecular level is itself relevant.

Models of evolution make assumptions to make complex problems computationally tractable. A model becomes a powerful tool when, despite its simplified assumptions, it can fit the data and make accurate predictions about the problem at hand. The performance of a method is maximized when its assumptions are satisfied, and some indication of the fit of the data to the phylogenetic model is necessary. If the model used may influence the results of the analysis, it becomes crucial to decide which is the most appropriate model to work with.

Before proceeding further, a word of caution should be said when selecting best-fit models for heterogeneous data, for example, when joining different genes for the phylogenetic analysis, or coding and non-coding regions. Since different genomic regions are subjected to different selective pressures and evolutionary constraints, a single model of evolution may not fit well with all the data. Nowadays, some options exist for a combined analysis in which each data partition (e.g. different genes) has its own model (Nylander *et al.*, 2004). In addition, *mixture models* consider the possibility of the model varying in different parts of the alignment (Pagel & Meade, 2004).

10.2 Model fit

In general, models that are more complex will fit the data better than simpler ones just because they have more parameters. An *a priori* attractive procedure to select a model of evolution would be the arbitrary use of the most complex, parameter-rich model available. However, when using complex models a large number of parameters need to be estimated, and this has several disadvantages. First, the analysis becomes computationally difficult, and requires a large amount of time. Second, as more parameters need to be estimated from the same amount of data, more error is included in each estimate. Ideally, it would be advisable to incorporate as much complexity as needed, i.e. to choose a model that is intricate enough to explain the data, but not that complicated that requires impractical long computations or large data sets to obtain accurate estimates.

The best-fit model of evolution for a particular data set can be selected using sound statistical techniques. During the last few years different approaches have been proposed, to select the best-fit model of evolution within a collection of candidate models, like *hierarchical **likelihood ratio tests** (hLRTs), information criteria, Bayesian,* or *performance-based approaches.* In addition, although not considered here, the overall adequacy of a particular model can also be evaluated using different procedures (Goldman, 1993; Bollback, 2002).

Regardless of the model selection strategy chosen, the fit of a model can be measured through the **likelihood function**. The **likelihood** is proportional to the probability of the data (D) given a model of evolution (M), a vector of K model parameters (θ), a tree topology (τ), and a vector of branch lengths (v):

$$L = P(D|M, \theta, \tau, v) \tag{10.1}$$

When the goal is to compute the likelihood of a model, the parameter values and the tree affect the calculations, but they are not really what we want to infer (they are **nuisance parameters**). A standard strategy to "remove" nuisance parameters is to utilize their **maximum likelihood estimates** (**MLEs**), which are the values that make the likelihood function as large as possible:

$$\hat{\theta}, \hat{\tau}, \hat{v} = \max_{\theta,\tau,v} L(\theta, \tau, v) \tag{10.2}$$

Note that, to facilitate the computation, we usually work with the maximized log likelihood:

$$\ell = \ln P(D|M, \hat{\theta}, \hat{\tau}, \hat{v}) \tag{10.3}$$

Alternatively, in a Bayesian setting we can integrate the nuisance parameters out and obtain the *marginal probability* of the data given only the model ($P(D|M)$, also called *model likelihoods*), typically using computationally intensive techniques like **Markov chain Monte Carlo** (**MCMC**). Integrating out the tree, branch lengths, and model parameters to obtain $P(D|M)$ is represented by:

$$P(D|M) = \int\int\int P(D|M, \theta, \tau, v) P(\theta, \tau, v|M)\, d\theta\, d\tau\, dv \tag{10.4}$$

10.3 Hierarchical likelihood ratio tests (hLRTs)

A standard way of comparing the fit of two models is to contrast their log likelihoods using the likelihood ratio test (LRT) statistic:

$$LRT = 2(\ell_1 - \ell_0) \tag{10.5}$$

where ℓ_1 is the maximum log likelihood under the more parameter-rich, complex model (alternative hypothesis) and ℓ_0 is the maximum log likelihood under the less parameter-rich simple model (null hypothesis). The value of this statistic is always equal to, or greater than, zero, even if the simple model is closest to the true model, simply because the superfluous parameters in the complex model provide a better explanation of the stochastic variation in the data than the simpler model. When the models compared are nested (i.e. the null hypothesis is a special case of the alternative hypothesis) and the null hypothesis is correct, this statistic is asymptotically distributed as a χ^2 distribution with a number of degrees of freedom equal to the difference in number of free parameters between the two models. When the value of the LRT is significantly large, the conclusion is that the inclusion of additional parameters in the alternative model increases the likelihood of the data significantly, and consequently the use of the more complex model is favored. On the other hand, a small difference in the log likelihoods indicates that the alternative hypothesis does not explain the data significantly better than the null hypothesis.

That two models are nested means that one model (null model or constrained model) is equivalent to a restriction of the possible values that one or more parameters can take in the other model (alternative, unconstrained or full model). For example, the Jukes–Cantor model (JC) (1969) and the Felsenstein (F81) (1981) models are nested. This is because the JC model is a special case of the F81, where the base frequencies are set to be equal (0.25), while in the F81 model these frequencies can be different. When comparing two different nested models through an LRT, we are testing hypotheses about the data. The hypotheses tested are those represented by the difference in the assumptions among the models compared. Several hypotheses can be tested hierarchically to select the best-fit model for the data set at hand among a set of possible models. Are the base frequencies equal? Is there a transition/transversion bias? Are all transition rates equal? Are there invariable sites? Is there rate homogeneity among sites? And so on. For example, testing the equal base frequencies hypothesis can be done with a LRT comparing JC vs. F81, as these models only differ in the fact that F81 allows for unequal base frequencies (alternative hypothesis), while JC assumes equal base frequencies (null hypothesis). Indeed, the same hypothesis could also have been evaluated by comparing JC + Γ vs. F81 + Γ, or K80 + I vs. HKY + I, or SYM vs. GTR. An example of such a hierarchical LRT procedure (hLRT) for 24 models is shown in Fig. 10.1. The hLRTs can be easily accomplished by using the program MODELTEST (Posada & Crandall, 1998) for a set of 56 candidate models (see the practice section in this chapter).

10.3.1 Potential problems with the hLRTs

We should be aware that there are some potential problems derived from the use of pairwise LRTs for model selection (Posada & Buckley, 2004). The χ^2 distribution approximation for the *LRT* statistic may not be appropriate when the null model is equivalent to fixing some parameter at the boundary of its possible values (Whelan & Goldman, 1999). An example of this situation is the invariable sites test. In this case, the alternative hypothesis postulates that the proportion of invariable sites could range from 0 to 1. The null hypothesis (no invariable sites) is a special case of the alternative hypothesis, with the proportion of invariable sites fixed to 0, which is at the boundary of the range of the parameter in the alternative model. In this case, the use of a mixed χ^2 distribution (50% χ_0^2 and 50% χ_1^2) is appropriate. However, even after using the most appropriate χ^2 distribution, obtaining *correct P*-values for the LRT statistics can be difficult, because LRTs implicitly assume that at least one of the models compared is correct. Moreover, when the two competing hypotheses are not nested the χ^2 approximation may perform poorly when the data include very short sequences relative to the number of parameters to be estimated. In these cases, the null distribution of the LRT statistic can be approximated by *Monte Carlo simulation.*

In addition, which model comparison is used to compare which hypothesis depends on the starting model of the hierarchy, and on the order in which different hypotheses are performed. For example, it could be possible to start with the simple JC or with the most-complex GTR + I + Γ. In the same way, a test for equal base frequencies could be performed first followed by a test for rate heterogeneity among sites, or vice versa. Many hierarchies of LRTs are possible, and they can result in different models being selected (Posada & Crandall, 2001; Posada & Buckley, 2004), and in some cases they can even lead to the estimation of different trees (Pol, 2004).

10.4 Information criteria

A different approach for model selection is the simultaneous comparison of all competing models. The idea again is to include as much complexity in the model as needed. To do that, the likelihood of each model is penalized by a function of the number of free parameters in the model (K); the more parameters, the bigger the penalty. The **Akaike Information Criterion** *or* **AIC** (Akaike, 1974) is an asymptotically unbiased estimator of the Kullback–Leibler information quantity

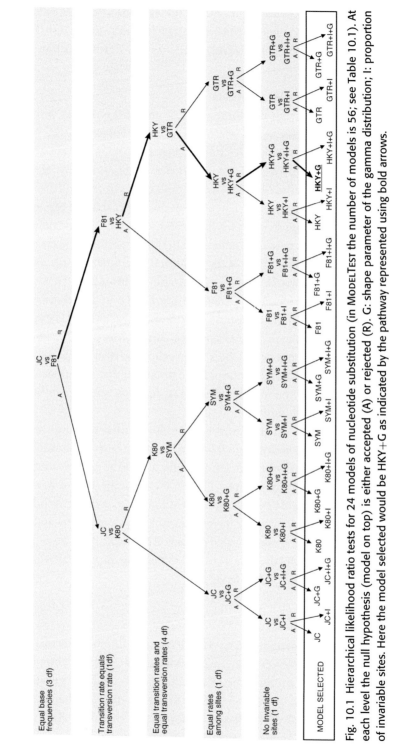

Fig. 10.1 Hierarchical likelihood ratio tests for 24 models of nucleotide substitution (in MODELTEST the number of models is 56; see Table 10.1). At each level the null hypothesis (model on top) is either accepted (A) or rejected (R). G: shape parameter of the gamma distribution; I: proportion of invariable sites. Here the model selected would be HKY+G as indicated by the pathway represented using bold arrows.

(Kullback & Leibler, 1951), which measures the expected distance between the true model and the estimated model:

$$AIC = -2\ell + 2K \tag{10.6}$$

We can think of the AIC as the amount of information lost when we use, say HKY85, to approximate the real process of molecular evolution. Hence, the model with the smallest AIC is preferred. If branch lengths are estimated *de novo* for every model, as is usually the case, K will include the number of branches (twice the number of taxa minus three). An advantage of the AIC is that it can be used to compare both nested and non-nested models. When sample size (n) is small compared with the number of parameters ($n/K < 40$) a corrected version of the AIC is recommended:

$$AIC_c = AIC + \frac{2K(K+1)}{n - K - 1} \tag{10.7}$$

Note that sample size is usually approximated by the total number of characters in the alignment, although what is the sample size of an alignment is still an open question. The AIC and AIC_c calculations are implemented in the programs MODELTEST and PROTTEST (Abascal *et al.*, 2005) for DNA and protein sequences, respectively.

10.5 Bayesian approaches

Model selection can be implemented in a Bayesian setting using *Bayes factors*, *posterior probabilities* or the *Bayesian Information Criterion*. Bayes factors are similar to the LRTs in that they compare the evidence (here, the model likelihoods) for two competing models:

$$B_{ij} = \frac{P(D|M_i)}{P(D|M_j)} \tag{10.8}$$

In this case, evidence for M_i is considered very strong if $B_{ij} > 150$, strong if $12 < B_{ij} < 150$, positive if $3 < B_{ij} < 12$, barely worth mentioning if $1 < B_{ij} < 3$, and negative (supports M_j) if $B_{ij} < 1$. Bayes factors for models of molecular evolution can be calculated using *reversible jump MCMC* (Huelsenbeck *et al.*, 2004) (see also Chapter 7).

In addition, when multiple models are considered, it is possible to choose the model with the highest *posterior probability* (Raftery, 1996). For R candidate models, the posterior probability of the *i*th model is:

$$P(M_i|D) = \frac{P(D|M_i)P(M_i)}{\sum_{r=1}^{R} P(D|M_r)P(M_r)} \tag{10.9}$$

where $P(M)$ are the model *prior probabilities*.

Both Bayes factors and model posterior probabilities can be difficult to compute. The Bayesian Information Criterion (BIC) (Schwarz, 1978) provides an approximate solution to the natural log of the Bayes factor:

$$BIC = -2\ell + K \log n \tag{10.10}$$

The smaller the BIC, the better the fit of the model to the data. Given equal priors for all competing models, choosing the model with the smallest BIC is equivalent to selecting the model with the maximum posterior probability. Because with standard alignments the natural log of n is usually >2, the BIC tends to choose simpler models than the AIC. Like the AIC, the BIC can be used to compare nested and non-nested models. The BIC calculation is implemented in the programs MODELTEST and PROTTEST.

10.6 Performance-based selection

Arguing that there is no guarantee that the best-fit models will produce the best estimates of phylogeny, Minin *et al.* (2003) developed a novel approach that selects models on the basis of their phylogenetic performance, measured as the expected error on branch lengths estimates weighted by their BIC. Under this *decision theoretic* framework (DT) the best model is the one with that minimizes the risk function:

$$C_i \approx \sum_{j=1}^{R} \left\| \hat{\mathbf{B}}_i - \hat{\mathbf{B}}_j \right\| \frac{e^{-BIC_i/2}}{\sum_{j=1}^{R} e^{-BIC_i/2}} \tag{10.11}$$

where

$$\left\| \hat{\mathbf{B}}_i - \hat{\mathbf{B}}_j \right\|^2 = \sum_{l=1}^{2t-3} (\hat{B}_{il} - \hat{B}_{jl})^2 \tag{10.12}$$

and where t is the number of taxa.

Indeed, simulations suggested that models selected with this criterion result in slightly more accurate branch length estimates than those obtained under models selected by the hLRTs (Minin *et al.*, 2003; Abdo *et al.*, 2005).

10.7 Model selection uncertainty

One big advantage of the AIC, Bayesian, and DT methods over the hLRTs is that they can rank models, allowing us to assess how confident we are in the model selected. Indeed, models could be ranked according to posterior probabilities, and credible intervals could easily be constructed by summing these probabilities. For other relative measures like the AIC or BIC, we could present their *differences* (Δ). For example, for the ith model, the AIC (or BIC) difference is:

$$\Delta AIC_i = AIC_i - \min AIC \qquad (10.13)$$

where min *AIC* is the smallest **AIC** value among all candidate models.

Very conveniently, we can use these *differences* to obtain the relative weight (w_i) of each model:

$$w_i = \frac{\exp(-1/2\Delta_i)}{\sum_{r=1}^{R} \exp(-1/2\Delta_r)} \qquad (10.14)$$

Note that the weights for every model add to 1, so it is easy to establish a 95% confidence set of models for the best models by summing the weights from largest to smallest from largest to smallest until the sum is just 0.95 (or similar).

10.8 Model averaging

Very interestingly, the model weights (or the posterior probabilities) allow us to obtain a *model-averaged* estimate (also called a *multimodel* estimate) of any parameter (Raftery, 1996; Wasserman, 2000; Burnham & Anderson, 2003; Hoeting *et al.*, 1999; Madigan & Raftery, 1994). For example, a model-averaged estimate of the relative substitution rate between adenine and cytosine (φ_{A-C}) using the model weights (w) for R candidate models would be:

$$\hat{\varphi}_{A-C} = \frac{\sum_{i=1}^{R} w_i \, I_{\varphi_{A-C}}(M_i)\varphi_{A-C_i}}{w + (\varphi_{A-C})}, \qquad (10.15)$$

where

$$w + (\varphi_{A-C}) = \sum_{i=1}^{R} w_i \, I_{\varphi_{A-C}}(M_i), \qquad (10.16)$$

and

$$I_{\varphi_{A-C}}(M_i) = \begin{cases} 1 & \text{if } \varphi_{A-C} \text{ is in model } M_i, \\ 0 & \text{otherwise} \end{cases} \qquad (10.17)$$

Remarkably, it is possible to construct a model-averaged estimate of the phylogeny itself. Also, note that some parameters do not have the same interpretation across different models. For example, the shape of the gamma distribution (α, commonly used to describe among-site rate variation) in a $+\Gamma$ model is not the same parameter as α in a $+I+\Gamma$ model.

Furthermore, if we sum up the weights of the models that contain a given parameter we will get an estimate of the *relative importance* of that parameter (ranging from 0 to 1). For example, the relative importance of the relative substitution rate between adenine and cytosine is simply the $w + (\varphi_{A-C})$ coefficient above. Because we usually do not explore all the possible combinations of parameters in the set of candidate models, the relative importance of some parameters can be correlated.

PRACTICE

David Posada

10.9 The model selection procedure

This section will demonstrate how to use PAUP* (Swofford, 1998) (see Chapter 8) and MODELTEST (Posada & Crandall, 1998) for selecting models of nucleotide substitution, and PHYML (Guindon & Gascuel, 2003) and PROTTEST (Abascal et al., 2005) for selecting models of amino acid replacement, using hLRTs, AIC and BIC. Although Bayes factors and model posterior probabilities can be approximated using MRBAYES (Ronquist & Huelsenbeck, 2003), and the DT strategy can be accomplished with PAUP* and DT-SEL (Minin et al., 2003), these procedures are beyond the scope of this chapter (although see new versions of MODELTEST). The data files required for this section are available at *www.thephylogenetichandbook.org.*

By far the most complicated and time-consuming step for the hLRTs, AIC, and BIC strategies is the estimation of the likelihood scores for each model. For the hLRTs the hypotheses should be nested, and therefore the likelihood scores are estimated on the same fixed topology (the *base tree*). This tree is used just to estimate parameters and likelihood scores of the different models, and it is not our final estimate of the phylogenetic relationships among the taxa under investigation. It has been shown that, as far as this tree is a reasonable estimate of the phylogeny (i.e. a maximum parsimony or a neighbor-joining tree; never use a random tree!), the parameter estimates and the model selected should be appropriate (Posada & Crandall, 2001; Abdo et al., 2005). In the case of the AIC and BIC, the models compared do not have to be nested, and one could relax the fixed-topology restriction. That is, we could estimate the maximum likelihood tree for each model to obtain the likelihood scores. However, in principle the latter has little effect on the model selection procedure, but increases computation time a great deal (Abdo et al., 2005).

10.10 MODELTEST

MODELTEST[†] is a simple program written in ANSI C that has been compiled for Macintosh, Windows, and Linux. The MODELTEST package is available for free and can be downloaded from the software section *http://darwin.uvigo.es.* The input of MODELTEST is a text file containing the log likelihood scores that correspond to each one of the 56 nucleotide substitution models specified in Table 10.1. Such input file can be generated by executing in PAUP* a particular

---

† See also the new program MODELTEST at the same website.

Table 10.1 Set of candidate models evaluated in MODELTEST. ef: equal base frequencies; uf: base frequencies. For each *model* we evaluate *model, model+I, model+G* and *model +I+G*, with I = proportion of invariable sites (1 parameter) and G = shape parameter of the gamma distribution (1 parameter)

Model	Free parameters	Base frequencies	Substitution code[a]	Reference
JC	0	equal	000000	(Jukes & Cantor, 1969)
F81	3	unequal	000000	(Felsenstein, 1981)
K80	1	equal	010010	(Kimura, 1980)
HKY	4	unequal	010010	(Hasegawa *et al.*, 1985)
TrNef	2	equal	010020	(Tamura & Nei, 1993)
TrN	5	unequal	010020	(Tamura & Nei, 1993)
K81	2	equal	012210	(Kimura, 1981)
K81uf	5	unequal	012210	(Kimura, 1981)
TIMef	3	equal	012230	(Posada, 2003)
TIM	6	unequal	012230	(Posada, 2003)
TVMef	4	equal	012314	(Posada, 2003)
TVM	7	unequal	012314	(Posada, 2003)
SYM	5	equal	012345	(Zharkikh, 1994)
GTR	8	unequal	012345	(Rodríguez *et al.*, 1990)

[a] Numbers refer to how many distinct parameters are used to describe the relative rates A⇔C, A⇔G, A⇔T, C⇔G, C⇔T, and G⇔T. For example, 010020 implies three different parameters: $0 \approx A\Leftrightarrow C = A\Leftrightarrow T = C\Leftrightarrow G = G\Leftrightarrow T$, $1 \approx A\Leftrightarrow G$, and $2 \approx C\Leftrightarrow T$.

block of commands available in the modelblock file included in the MODELTEST package. The complete procedure requires three steps:

(i) Execute the alignment file (NEXUS format) in PAUP* (see Chapter 8, Practice).
(ii) Open the command file modelblock and execute it. PAUP* estimates a neighbor-joining tree and the likelihood and parameter values for several models. This can take from several minutes to several hours depending on the number of taxa and the computer speed. Once finished, a file called model.scores will appear in the same directory as the modelblock file. It is often a good practice to rename this file as *yourdata*.scores.
(iii) Run MODELTEST using the file *yourdata*.scores as the input file. The easiest way is to use the MODELTEST WEB SERVER (Posada, 2006) at http://darwin.uvigo.es (Fig. 10.2). Alternatively, one can run MODELTEST locally. The Mac version of the program has a command–line interface asking to select an input file and to choose a name for the output file. The PC version requires using the console (go to Start > Accessories > Console prompt) or an MS-DOS window. Here you would type: modeltest.exe < model.scores > outfile and press enter. The program will save the outfile with the results in the same directory. Note that model.scores has to be in the same folder

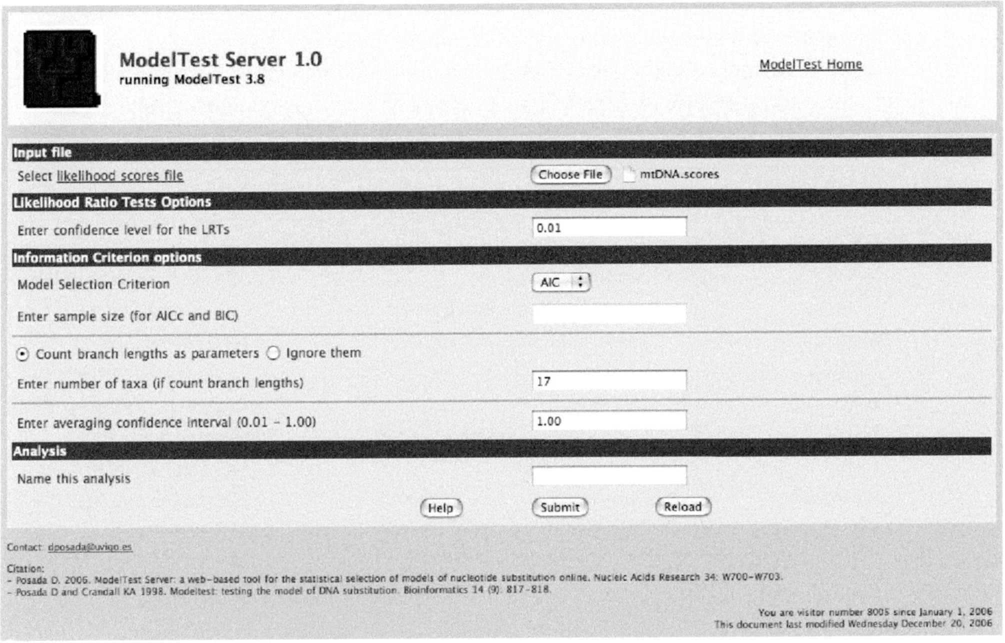

ModelTest Server 1.0
running ModelTest 3.8   ModelTest Home

Input file
Select likelihood scores file   Choose File   mtDNA.scores
Likelihood Ratio Tests Options
Enter confidence level for the LRTs   0.01
Information Criterion options
Model Selection Criterion   AIC
Enter sample size (for AICc and BIC)
⊙ Count branch lengths as parameters ○ Ignore them
Enter number of taxa (if count branch lengths)   17
Enter averaging confidence interval (0.01 – 1.00)   1.00
Analysis
Name this analysis
                          Help   Submit   Reload

Contact: dposada@uvigo.es

Citation:
- Posada D. 2006. ModelTest Server: a web-based tool for the statistical selection of models of nucleotide substitution online. Nucleic Acids Research 34: W700–W703.
- Posada D and Crandall KA 1998. Modeltest: testing the model of DNA substitution. Bioinformatics 14 (9): 817-818.

You are visitor number 8005 since January 1, 2006
This document last modified Wednesday December 20, 2006

Fig. 10.2   MODELTEST WEB SERVER at *http://darwin.uvigo.es*. Note that the specification to count branch lengths as parameters in the standalone MODELTEST software requires the option "-t17", where 17 is the number of taxa in this data set.

where modeltest.exe resides (such folder, called Modeltest, is created during the installation of the program).

The output of MODELTEST consists of a description of the hLRTs and AIC or BIC strategies. The first section of the output corresponds to the hLRTs strategy. Here, the particular LRTs performed and their associated *P*-values are listed, and the model selected with the corresponding parameter estimates (actually calculated by PAUP*) is described. In the second section of the output, the program will use the AIC or the BIC framework, depending on the option previously indicated by the user. Here, the model selected is described first. Then, three tables show the models ranked according to the AIC/BIC differences and AIC/BIC weights, the model-averaged parameter estimates and their relative parameter importance.

In addition, MODELTEST also provides a block of commands in NEXUS format, which can be executed in PAUP* an order to facilitate the implementation of the model selected. It is often convenient to paste this block of commands after the data block in the NEXUS file. This is very useful if the user wants to use this model for further analyses by PAUP*; for example, to perform a LRT of the molecular clock, or to estimate a tree using the best-fit model.

Table 10.2 Set of candidate models evaluated in PROTTEST. For each *model* (0 free parameters) we evaluate *model, model+F, model+I, model+G, model+F+I, model+F+G, model+I+G* and *model+F+I+G*, with F = amino acid frequencies (19 parameters), I = proportion of invariable sites (1 parameter) and G = shape parameter of the gamma distribution (1 parameter)

Model	Data type	Reference
JTT	nuclear	(Jones *et al.*, 1992)
MtREV	mitochondrial	(Adachi & Hasegawa, 1996)
MtMam	mammal mitochondria	(Cao *et al.*, 1998)
MtArt	arthropod mitochondria	(Abascal *et al.*, 2007)
Dayhoff	nuclear	(Dayhoff *et al.*, 1978)
WAG	nuclear	(Whelan & Goldman, 2001)
RtREV	retroviral polymerase	(Dimmic *et al.*, 2002)
CpREV	chloroplastic	(Adachi *et al.*, 2000)
Blosum62	nuclear	(Henikoff & Henikoff, 1992)
VT	nuclear	(Muller & Vingron, 2000)

10.11 PROTTEST

PROTTEST (Abascal *et al.*, 2005) is a program for the selection of models of amino acid replacement. It is written in Java, so it can run in any operating system with a Java Runtime Environment. PROTTEST performs similar calculations as MODELTEST, except for the hLRTs, which are not implemented because most of the amino acid models considered are not nested. Likelihood calculations are implemented using PHYML (Guindon & Gascuel, 2003), and some functions make use of the PAL library (Drummond & Strimmer, 2001).

PROTTEST provides a user-friendly interface where the user selects the different options for the analysis. In addition, it can be run at the command–line or through a web server (*http://darwin.uvigo.es*). Given an input file with an amino acid alignment in PHYLIP or NEXUS format, the program automatically calculates a BIONJ tree (Gascuel, 1997). Alternatively, a user-defined tree can be specified. Next, likelihoods and parameter estimates can be calculated using PHYML for every model upon this fixed topology, or a PHYML tree search can be implemented for every model. At the time of writing, PROTTEST selects among 80 empirical models of amino acid replacement (Table 10.2). These models are preferentially based on empirical amino acid replacement matrices for computational reasons. Such matrices have been estimated from large data sets consisting of diverse protein families.

```
-------------------------------------------------------------------
*                                                                 *
*              AKAIKE INFORMATION CRITERION (AIC)                 *
*                                                                 *
-------------------------------------------------------------------


  Model selected: GTR+I+G
    -lnL  =    21148.9277
    K     =    41
    AIC   =    42379.8555
    Base frequencies:
       freqA =        0.3771
       freqC =        0.2255
       freqG =        0.1759
       freqT =        0.2214
    Substitution model:
       Rate matrix
       R(a) [A-C] =         3.4141
       R(b) [A-G] =         5.0976
       R(c) [A-T] =         3.5059
       R(d) [C-G] =         0.4437
       R(e) [C-T] =        14.9945
       R(f) [G-T] =         1.0000
    Among-site rate variation
       Proportion of invariable sites (I) =       0.1595
       Variable sites (G)
         Gamma distribution shape parameter =      0.7279
```

Fig. 10.3   Partial output of MODELTEST WEB SERVER indicating the AIC model for the vertebrate mtDNA data set. Note that the maximum likelihood estimates shown by MODELTEST are actually obtained by PAUP*.

10.12 Selecting the best-fit model in the example data sets

10.12.1 Vertebrate mtDNA

The first data set is an alignment of mitochondrial sequences from several vertebrates including 17 taxa and 1998 nucleotides. The model selected by the hLRTs in MODELTEST 3.8 is TrN+I+G, while for the AIC (Fig. 10.3) and BIC (using the alignment length as sample size) was GTR+I+G. In both models base frequencies are unequal, there is a significant proportion of invariable sites (indicated by the +I), and there is rate heterogeneity among sites (indicated by +G). However, while the TrN model considers three types of substitutions (all transitions together and the two types of transversions), GTR considers that the six possible kinds of substitutions among the different bases occur at different rates (see Table 10.1). The hLRTs therefore selected a simpler model than the AIC or the BIC, which is often the case. In the output of MODELTEST, we can see that the TrN model cannot be rejected when it is tested against the TIM model, and therefore the full GTR scheme is never reached. However, note that TrN+I+G (-lnL = 21262.9570) is

Table 10.3 Parameter importances and model-averaged estimates for the vertebrate mtDNA data set

Parameter[a]	Importance	Model-averaged estimates
fA	1.0000	0.3771
fC	1.0000	0.2255
fG	1.0000	0.1759
fT	1.0000	0.2214
TiTv	0.0000	–
rAC	1.0000	3.4152
rAG	1.0000	5.0991
rAT	1.0000	3.5073
rCG	1.0000	0.4440
rCT	1.0000	15.0010
pinv(I)	0.0000	–
alpha(G)	0.0022	0.4738
pinv(IG)	0.9978	0.1595
alpha(IG)	0.9978	0.7280

[a] (I): averaged using only +I models; (G): averaged using only +G models; (IG): averaged using only +I+G models.

significantly worse than GTR+I+G (-lnL = 21148.9279) (LRT = 228.0582; d.f. = 3; P-value < 0.0001). In this case, the hLRTs seem to have been stuck in a local optimum (see Posada & Buckley, 2004).

According to the AIC or BIC weights (0.99 or 0.96, respectively), there is little uncertainty about the GTR+I+G model being the best model for this data set, so it would be very reasonable to use just this GTR+I+G for the next analyses. The relative parameter importance values (Table 10.3) indicate that all the parameters in this model are equally relevant. Given the large weights of the best model, the model-averaged estimates (Table 10.3) are indeed very similar to those obtained under the GTR+I+G model (Figure 10.3). Importantly, these parameter estimates characterize the molecular evolution of this gene. We can see that A is the most frequent nucleotide (0.3771), that the most common substitution is between C and T (15.0010), that around 16% of the sites have not changed (the proportion of invariable sites, p inv, is 0.1595), and that there is medium rate heterogeneity (the shape of the gamma distribution, α is 0.7280).

10.12.2 HIV-1 envelope gene

The second data set is the *env* gene from HIV and SIV strains. The alignment includes 14 taxa and 2202 nucleotides. In this case, the three criteria, hLRTs, AIC, and BIC select the same model, TVM+I+G. The TVM ("*transversional model*'; Posada, 2003) scheme considers one type of transition and four types of transversions. For this particular data set the hLRTs do not seem to have stuck in local optima, as GTR+I+G (-lnL = 15780.9895) is not significantly better than TVM+I+G (-lnL = 15780.9971). There is some uncertainty on the model selected, as the AIC and BIC weights for TVM+I+G are 0.70 and 0.56, respectively. For the AIC, the second best model is GTR+I+G with a weight of 0.26. For the BIC, the second best model is TVM+G, with a weight of 0.41. The relative parameter importances indicate that the A⇔G and C⇔T rate parameters do not need to be considered independently (importance = 0.2705). The model averaged estimates indicate that *A* is the most frequent nucleotide (0.3537), transitions are the most common type of substitutions, around 17% of the sites have not changed (p inv = 0.1721, and there is medium rate heterogeneity (α = 0.9436).

10.12.3 G3PDH protein

Finally, the third example data set is an amino acid alignment of the enzyme glycerol-3-phosphate dehydrogenase in bacteria, protozoa, and animals. It includes 12 taxa and 234 amino acid residues. The model selected by PROTTEST 1.3 for this data set after 2–3 hours running in a laptop was WAG+I+G (Whelan & Goldman, 2001), regardless of the criterion used (AIC, AICc, or BIC). From the output, we can see that there is no uncertainty in the selection; the WAG+I+G model has a weight of almost 1 under any criteria (look for AICw, AICcw, or BICw).

The WAG matrix was inferred from a large database of sequences comprising a broad range of nuclear protein families, and it is suited for distantly related amino acid sequences. Therefore, the model selected appears to be appropriate. Given its large weight, the relative importances of the different parameters are concentrated around +I+G. The model-averaged estimates indicate a very small proportion of invariables sites (p inv ~ 0.08 for the different criteria) and moderate rate variation among residues (α ~ 2). Note that this model does not include the +F (Cao *et al.*, 1994), which suggests that it is not worth including in the model the amino acid frequencies estimated (observed) from this particular alignment. Remember that each replacement matrix has attached its own set of amino acid frequencies estimated from a large protein database. Conveniently, PROTTEST has an option that allows the user to visualize at once the results for all criteria, including different ways of defining sample size for the AICc and the BIC (number of sites in the alignment, sum of site entropies, length × number of sequences scaled by the total entropy).

Molecular clock analysis

THEORY

Philippe Lemey and David Posada

11.1 Introduction

Between 1962 and 1965, before Kimura postulated the neutral theory of evolution (Kimura, 1968), Zuckerkandl and Pauling published two fundamental papers on the *evolutionary rate* of proteins (Zuckerkandl & Pauling, 1962; Zuckerkandl & Pauling, 1965). They noticed that the *genetic distance* of two sequences coding for the same protein, but isolated from different species, seems to increase linearly with the divergence time of the two species. Since several proteins showed a similar behavior, Zuckerkandl and Pauling hypothesized that the rate of evolution for any given protein is constant over time. This suggestion implies the existence of a sort of *molecular clock* ticking faster or slower for different genes but at a more or less constant rate for any given gene among different phylogenetic lineages (see Fig. 11.1). The clock hypothesis received an enormous popularity almost immediately for several reasons. If a molecular clock exists and the rate of evolution of a gene can be calculated, then this information can easily be used for dating the unknown divergence time between two species just by comparing their DNA or protein sequences. If, on the other hand, the information about the divergence time between two species (for example, estimated from fossil data) is known, then the rate of molecular evolution of a given gene can be inferred. An additional advantage of assuming a molecular clock is that it can render phylogenetic reconstruction much easier and more accurate (see Chapter 5).

The Phylogenetic Handbook: a Practical Approach to Phylogenetic Analysis and Hypothesis Testing, Philippe Lemey, Marco Salemi, and Anne-Mieke Vandamme (eds.). Published by Cambridge University Press. © Cambridge University Press 2009.

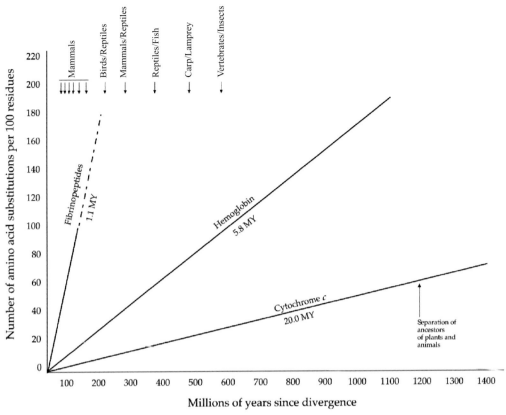

Fig. 11.1   Molecular clock ticking at different speed in different proteins. Fibrinopeptides are relatively unconstrained and have a high neutral substitution rate, whereas cytochrome c is more constrained and has a lower neutral substitution rate (after Hartl & Clark, 1997).

The molecular clock is not only a tool for estimating historical dates and rates of evolution; it also provides a formal description of the substitution process, which is central to our understanding of evolutionary processes in organisms. The molecular clock hypothesis is in perfect agreement with the neutral theory of evolution (Kimura, 1968, 1983). In fact, the existence of a clock seems to be a major support of the neutral theory against **positive selection** to explain the maintenance of polymorphisms (see Chapters 1 and 13). This implies that deviations from clock-like behavior may reveal adaptive evolution, relaxing functional constraints, or changes in **effective population size**. A detailed discussion of the molecular clock itself is beyond the scope of this book. Excellent reviews can be found in classical textbooks of molecular evolution (e.g. Hillis et al., 1996; Li, 1997; Page & Holmes, 1998). The next sections will focus on how to test the clock hypothesis for a group of taxa with known phylogenetic relationships and how sources of rate variation can be modeled.

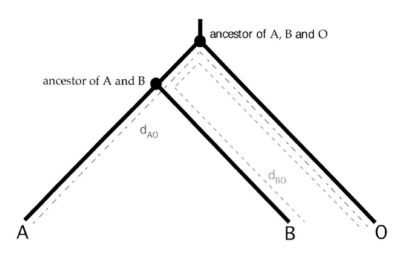

Under a perfect molecular clock
$$d_{AO} = d_{BO},$$
$$\text{then } d_{AO} - d_{BO} = 0$$

Fig. 11.2   The relative rate test. Under a molecular clock, the distance from A to O should be the same as the distance from B to O.

11.2 The relative rate test

According to the molecular clock hypothesis, two taxa sharing a common ancestor *t* years ago should have accumulated more or less the same number of substitutions during time *t*. In most cases, however, the ancestor is unknown and there is no possibility to test directly the constancy of the evolutionary rate. The problem can be solved by considering an **outgroup**, i.e. one or more distantly related species (Chapter 1). Figure 11.2 illustrates the idea for the triplet of taxa A–B–O. Under a perfect molecular clock, d_{AO}, the number of substitutions between taxon A and the outgroup, is expected to be equal to d_{BO}, the number of substitutions between taxon B and the outgroup. The **relative rate test** evaluates the molecular clock hypothesis comparing whether $d_{AO} - d_{BO}$ is significantly different from zero. When this is the case, the sign of the difference indicates which taxon is evolving faster or slower. The relative rate test assumes that the phylogenetic relationships among the taxa are known without error. This makes the test very problematic for taxa with still uncertain phylogeny, like the placental mammals. Note that, in such cases, it would not be a good idea to choose a very distant related species as an outgroup. A too distant outgroup implies a smaller impact on $d_{AO} - d_{BO}$. In addition, the more distant the outgroup, the higher the probability that multiple substitutions occurred at some sites. Therefore, the estimation of the genetic distance becomes less accurate, even if we employ a sophisticated *model of nucleotide substitution*. Furthermore, when many taxa are considered, the investigator is confronted with multiple non-independent tests. Extensions of the "triplet-based" relative test have been presented that compare the average or weighted-average distances of members of two clades to an outgroup (e.g. Robinson *et al.*, 1998; Takezaki *et al.*, 1995). Not only can independence be difficult to take into account, but also the power of this test to detect cases of rate variation can be relatively poor (Bromham *et al.*, 2000). A more powerful test of the molecular clock that fully takes into account all forms of non-independence is the *likelihood ratio test* (*LRT*).

11.3 Likelihood ratio test of the global molecular clock

The molecular clock LRT evaluates whether the evolutionary rate is homogeneous along all branches of a phylogenetic tree. The phylogeny of a group of taxa is known when the topology and the branch lengths of the phylogenetic tree relating them are known. Most of the tree-building algorithms, like *neighbor-joining* or the Fitch and Margoliash method, do not assume a molecular clock. Methods that do assume a clock, such as *UPGMA*, result in *ultrametric* tree topologies with terminal nodes all at equal distant from the root (see Fig. 11.3b). Of course, whatever the tree topology is, branch lengths can be estimated assuming a constant evolutionary rate along each branch. This requires the tree to be appropriately rooted. Clock-like phylogenetic trees are often rooted on the longest branch representing the oldest lineage (see Fig. 11.3b), which frequently satisfies the *midpoint-rooting* criterion (see Chapter 13). Rooting can also be performed using a suitable outgroup. Non-clock-like trees are unrooted (Fig. 11.3a); in them, a longer branch represents a lineage that might have evolved faster or that might be an older lineage.

Using *maximum likelihood* methods, branch lengths of a tree can be estimated by enforcing or not a molecular clock. In the maximum likelihood framework, the assumption of a molecular clock corresponds to the restriction that all branches have the same mean instantaneous substitution rate, μ, in the *Q matrix* (see Fig. 4.3 in Chapter 4). In the absence of a molecular clock (the *"different-rates"* or *unrooted model*), $2n-3$ branch lengths have to be inferred for a strictly bifurcating unrooted phylogenetic tree with n taxa (see Fig. 11.3a). If the molecular clock is enforced, the tree is rooted, and just $n-1$ branch lengths need to be estimated (Fig. 11.3b). This should appear obvious considering that, under a molecular clock, the tips of the tree should all be equally distant from the root (ultrametric tree, Fig. 11.3b). Thus, for any two taxa sharing a common ancestor, only the length of the branch from the ancestor to one of the taxa needs to be estimated, the other one having a tip at the same genetic distance.

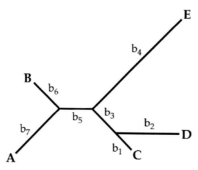

(a)

Non-clock-like phylogenetic tree
n taxa = 5

unrooted tree
2*n*-3 independent branches

All b_1, b_2, b_3, b_4, b_5, b_6 and b_7
need to be estimated

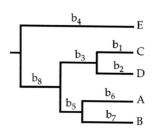

(b)

Clock-like phylogenetic tree
n taxa = 5

rooted tree
n-1 independent branches

Only b_1, b_3, b_4 and b_6,
for example, need to be estimated,
since under the molecular clock:

$$b_2 = b_1$$
$$b_5 = b_1 + b_3 - b_6$$
$$b_7 = b_6$$
$$b_8 = b_4 - b_5 - b_6$$

Fig. 11.3   Number of free parameters in clock and non-clock trees. (a) Under the unrooted model (= non-clock or different rates), all the branches need to be estimated ($2n - 3$). (b) Under the molecular clock, only $n - 1$ branches have to be estimated. The difference in the number of parameters between a non-clock and a clock model is $n - 2$, n is the number of taxa.

Statistically speaking, the molecular clock is the null hypothesis (the rate of evolution is equal for all branches of the tree) and represents a special case of the more general alternative hypothesis assuming a different rate for each branch (free-rates model). Thus, similar to the case of nested models of nucleotide substitution (see Chapter 10), the LRT can be used to evaluate whether or not the taxa have been evolving at the same rate (Felsenstein, 1988). In practice, a model of nucleotide (or amino acid) substitution is chosen and the branch lengths of a particular tree with and without enforcing the molecular clock are estimated. To assess the significance of the LRT, twice the log likelihood difference can be compared with a χ^2 distribution with $(2n - 3) - (n - 1) = n - 2$ degrees of freedom, since the only difference in parameter estimates is in the number of branch lengths that need to be estimated. It is important to note that the choice of the tree topology can be important for the results of the LRT. Ideally, a tree search in enforcing the molecular clock or not should result in the same tree topology. If this is not the case, a particular tree topology needs to be chosen for the LRT to be valid.

Frequently, researchers opt for the tree inferred using the more general unrooted model and root this tree topology to compute the likelihood under a clock model. Without an outgroup, the root that yields the highest likelihood value under a clock model would be the most appropriate choice. Conversely, a clock tree can also be unrooted to obtain the likelihood under the free-rates model. The former might result in a somewhat liberal LRT, while the latter is probably a more conservative LRT. In order to remove the influence of the outgroup in the outcome of this LRT (the outgroup can have a different rate of evolution from the ingroup; and we are interested in testing the clock hypothesis within the ingroup), it might be important to prune the outgroup from the tree and the alignment before the likelihood estimation.

It should be noted that particular precautions are needed before applying the LRT to test the molecular clock. We have already discussed the issue of selecting a phylogeny (see above). In addition, small data samples may lead to an inappropriate application of the χ^2 distribution to select a critical value; **_Monte Carlo simulation_** can be used in this case. Finally, recombination has been found to confound this test in such a way that the molecular clock is rejected when, in fact, all the lineages are evolving at the same rate (Schierup & Hein, 2000). A modified version of the relative ratio test has been proposed to overcome this difficulty (Posada, 2001).

11.4 Dated tips

Sequence data is often obtained by sampling individuals at different time points. During this sampling period, nucleotide substitutions may have accumulated if the sampling scheme extends into ancient times or when the organisms have evolved very fast during more recent times. The former case of **_measurably evolving populations_** (**_MEPs_**; Drummond _et al._, 2003) can be represented by well-characterized vertebrate subfossil material from which ancient DNA is reliably amplified (e.g. Barnes _et al._, 2002; Shapiro _et al._, 2004), while the latter is frequently observed in RNA virus evolution. Obviously, models to test the assumption of rate constancy for MEPs should take into account the date of sampling. Instead of enforcing an ultrametric tree topology, the tips of the tree are now constrained at distances from the root that are proportional to the sampling times (as is shown for three different sampling times in Fig. 11.4a). In this model, the evolutionary rate is an additional parameter that rescales the times of the internal nodes into units of expected number of substitutions per site (Rambaut, 2000). This implies that differences in sampling time are an important source of information to calibrate molecular clocks. The molecular clock model for serially sampled sequences was originally labeled as "the single rate dated tip" (SRDT) model – with the usual clock

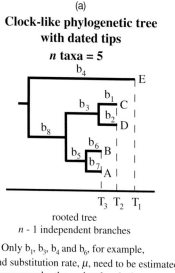

(a)
**Clock-like phylogenetic tree
with dated tips**

n taxa = 5

rooted tree
n - 1 independent branches

Only b_1, b_3, b_4 and b_6, for example,
and substitution rate, μ, need to be estimated
because under the molecular clock:

$$b_2 = b_1$$
$$b_5 = b_1 + b_3 - b_6 - \mu(T_3 - T_2)$$
$$b_7 = b_6$$
$$b_8 = b_4 - b_5 - b_6 - \mu(T_3 - T_1)$$

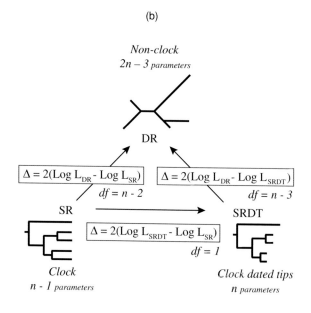

(b)

Non-clock
$2n - 3$ *parameters*

DR

$\Delta = 2(\text{Log } L_{DR} - \text{Log } L_{SR})$
$df = n - 2$

$\Delta = 2(\text{Log } L_{DR} - \text{Log } L_{SRDT})$
$df = n - 3$

SR

$\Delta = 2(\text{Log } L_{SRDT} - \text{Log } L_{SR})$
$df = 1$

SRDT

Clock
$n - 1$ *parameters*

Clock dated tips
n *parameters*

Fig. 11.4   Number of free parameters in a single rate dated tip (SRDT) model and likelihood ratio tests
for model comparisons. (a) Under the molecular clock applied to serial sampled data, n −
1 branches and an additional evolutionary rate parameter, μ, have to be estimated. (b) The
comparison of the single rate (SR) model and the different rates (DR) model constitutes
the molecular clock test for contemporaneously sampled sequences; the comparison of the
single rate dated tip (SRDT) model and the different rates (DR) model provides a molecular
clock test for serially sampled sequences. Comparing SR and SRDT allows to test whether
accommodating for sampling times significantly improves the fit of a clock model. The
degrees of freedom (*df*) are indicated for each LRT.

and non-clock model referred to as the single rate (SR) and different rates (DR)
model, respectively – and has n free parameters (n −1 branch lengths or internal
times and the substitution rate parameter). The LRT for testing the rate constancy
assumption in MEPs compares the likelihood scores of the SR and SRDT model
(Fig. 11.4b), with the critical value for twice the log likelihood difference being
evaluated against a χ^2 distribution with $(2n-3)-n = n-3$ degrees of freedom.
The SR model, which makes no accommodation for sampling times, is also a special case of the SRDT model (Fig. 11.4b). A LRT comparing these models provides
a way to statistically evaluate whether incorporating sampling times significantly
improves the clock model.

11.5 Relaxing the molecular clock

The initial findings of Zuckerkandl and Pauling raised much excitement about using the molecular clock to estimate evolutionary dates. However, empirical studies that employed rigorous statistical testing have revealed non-clock behavior in many gene sequences (e.g. Jenkins *et al.*, 2002), which sometimes caused molecular dating to be controversial. Bromham and Penny (2003) have provided a detailed discussion on the potential sources of rate variation, including **generation times**, replication and repair mechanisms and differences in natural selection (Bromham & Penny, 2003).

Unlike selecting models of nucleotide substitution (see Chapter 4), there is no hierarchy of models available to accommodate for increasing amount of rate variation. Pybus (2006) has compared this problem with a simple case of fitting a line through points representing a hypothetical data set (Fig. 11.5a) (Pybus, 2006). A global molecular clock, ticking at the same rate for all the taxa, is in many cases equivalent to an oversimplified and badly fitting model that assumes a straight line through the points (Fig. 11.5b). The non-clock model, represented by a model interconnecting the different points (Fig. 11.5c), has too many parameters to be informative about the underlying biological process (only branch lengths can be estimated in this case, not the separate contribution of evolutionary time and rate). Ideally, a model should be pursued that fits the data well using an intermediate amount of parameters (Fig. 11.5d). Such a model could bridge the gap between the two extreme clock and non-clock models.

Different attempts to develop models accommodating for a reasonable amount of rate variation have been undertaken. Rambaut and Bromham (1998) discussed a maximum likelihood model with two rates on a tree of four species (quartets) (Rambaut & Bromham, 1998). Yoder and Yang (2000) extended this idea and demonstrated that the clock hypothesis could be relaxed by allowing a constant rate of evolution within a particular clade but assuming different rates for different clades (a "*local molecular clock*" model: Yoder & Yang, 2000). Applied to the hypothetical data set, local molecular clocks could be represented by Fig. 11.5e. In practice, local clocks are being implemented by assigning separate rate parameters to different branches or collections of branches. A global molecular clock is a special case of a local molecular clock, which in turn is a special case of a free-rates model. This allows the models to be tested against each other using the LRT, but the comparison of local clock models with the global clock model is only valid when the null hypothesis is specified beforehand. It would not be appropriate to derive a hypothesis from the data itself (e.g. by inferring a tree under the unrooted model and judging which branches might be too long or too short in order to conform to an ultrametric tree structure). This is known as *post-hoc* hypothesis

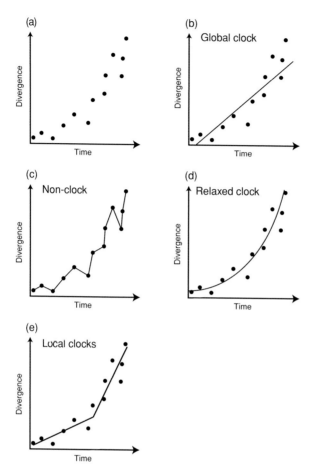

Fig. 11.5   Conceptual representation of model selection and the molecular clock. (a) Scatter plot of a hypothetical data set. (b) Fitting a simple linear regression model to the data. (c) Interconnecting all the data points which represents a complex highly parameterized model. (d) A model with an intermediate number of parameters represented by a curve. (e) A model representing the concept of a local clock. Although the variables of the hypothetical data set could be any two variables, we have chosen time and divergence as representative examples. In this case, a point in the graph is a comparison between two given sequences. The x-axis represents the time elapsed and the y-axis the genetic divergence since the two sequences diverged from their common ancestor. (Adapted from Pybus, 2006.)

testing, for which the usual χ^2 significance values result in too high a probability of incorrectly finding a difference (a similar problem of "data-dredging" is also discussed for selection analysis where different branches can be allowed to have their own non-synonymous/synonymous substitution rate ratio, see Chapter 14). An additional problem of local clock models arises when too many different rate parameters are specified, especially around the root, which can result in problems of identifiability (i.e. inability to separate the contribution of rate and time) (Yoder & Yang, 2000).

Thorne *et al.* (1998) presented a Bayesian treatment of models in which rates of evolution vary through time (Kishino *et al.*, 2001; Thorne *et al.*, 1998). Bayesian analysis requires specifying ***prior distributions*** (see Chapters 7 and 18); in this case, it is assumed that each branch has a rate that is drawn from a lognormal distribution, which is centered on the rate of the ancestral branch. Rates across branches are therefore "autocorrelated," with the strength of the correlations between ancestral and descendent branches depending either on the difference of the midpoints of the branches or the length of the ancestral branch. These models are inspired by the reasoning that closely related lineages have similar biological properties and therefore are expected to have similar evolutionary rates, and that significant changes in those rates are more plausible over longer time frames. ***Posterior distributions*** for the rates in these models are obtained using ***Markov chain Monte Carlo*** (***MCMC***) integration (see Chapters 7 and 18). Other variants of ***autocorrelated relaxed clock*** models have been proposed, but they all share the difficulty of specifying *a priori* the degree of autocorrelation among rates and specifying a particular prior for the rate at the root of the tree.

Recently, an alternative to the autocorrelated prior has been presented in which no correlation on adjacent branches is assumed. In this case, branch-specific rates are drawn independently and identically from an underlying rate distribution, like an exponential or lognormal distribution (Drummond *et al.*, 2006). These ***uncorrelated relaxed clock*** models have been implemented in a Bayesian framework for genealogy-based population genetics (see BEAST, Chapter 18). In contrast to previous relaxed clock models, which need to be applied on a fixed tree topology, the parameters of the uncorrelated models can be estimated by averaging over a set of plausible trees using MCMC (Drummond *et al.*, 2006). Therefore, these models are very useful for estimating evolutionary rates and divergence times in the face of phylogenetic uncertainty.

11.6 Discussion and future directions

The ability to deduce the time elapsed since two organisms diverged, based on the number of genetic differences between their gene sequences, has truly revolutionized evolutionary biology. A variety of statistical models and inference techniques have been developed to test the rate constancy assumption and to estimate historical dates and evolutionary rates. This has known many important research applications including pathogen epidemiology (e.g. Korber *et al.*, 2000; Lemey *et al.*, 2006), the study of the origin of the main types of animals (for review see Bromham & Penny, 2003), studies of historical changes in mammalian population sizes (Shapiro *et al.*, 2004), and even as evidence against deliberate virus transmission in an HIV outbreak (de Oliveira *et al.*, 2006). However, the assumption of a constant evolutionary rate often has been challenged by rigorous statistical testing of protein and DNA sequences. Therefore, it remained unclear how much confidence could be invested in molecular clock estimates. Nowadays, most popular phylogenetic inference techniques have abandoned the molecular clock and resorted to the unrooted model.

Much progress has been made to accommodate for sources of rate variation and to reintroduced time in phylogenetic inference (Drummond *et al.*, 2006). The uncorrelated relaxed clock method appropriately accounts for rate variation among lineages in the face of phylogenetic uncertainty and provides quantitative measures of rate variability and autocorrelation of rates along different lineages (Drummond *et al.*, 2006). Further developments also focus on the synonymous and non-synonymous component of the substitution process by implementing codon models in the relaxed clock framework (Seo *et al.*, 2004; Lemey *et al.*, 2007) (see Chapter 14 for codon models). This has the potential to provide further insights into the biological processes underlying evolutionary rate changes.

PRACTICE

Philippe Lemey and David Posada

11.7 Molecular clock analysis using PAML

Different software programs can be used to compute the likelihood under clock and non-clock models, including PAUP* (Swofford, 1998)(see Chapter 9), PAML (Yang, 1997), TREE-PUZZLE (Schmidt *et al.*, 2002) (Chapter 6), and HYPHY (Kosakovsky Pond *et al.*, 2005)(Chapter 14). In this practical section, we will use the PAML package that includes clock models for both contemporaneously and serially sampled sequences.

In this exercise, we will examine clock-like behavior in two nucleotide sequence alignments. The first data set ("primates.phy") contains the TRIM5α gene sequences for 21 different primate species. TRIM5α is present in most primates and encodes a retroviral restriction factor that protects Old World monkey cells against HIV infection (Stremlau *et al.*, 2004). Positions with gaps in the original alignment, analyzed by Sawyer *et al.* (2005), were removed for our analysis here. The second data set ("RSVA.phy") contains the partial attachment G protein coding sequences for 35 human respiratory syncytial subgroup A viruses (HRSV-A). RSV causes respiratory tract infection and is known to be responsible for the majority of infant hospitalizations due to respiratory illness. Sequences in this group A data set were sampled during 1956–2002, and they represent a subset of a larger molecular evolutionary HRSV-A study. Both data sets, in sequential PHYLIP format, and the associated rooted and unrooted maximum likelihood trees ("primates_unrooted.tre," "primates_rooted.tre," "RSVA_unrooted.tre," "RSVA_rooted.tre"), in NEWICK format, can be downloaded from the website accompanying this book (*www.thephylogenetichandbook.org*).

The PAML package includes different programs that can be run on all the popular computer systems (Windows, Unix, and MacosX). Here, we will use baseml (PAML v. 4) to perform maximum likelihood analysis of nucleotide sequences. Like most programs in the PAML package, baseml has a control file that specifies the names of the sequence data file, the tree structure file, the models and different options for the analysis; the default control file for baseml is called "baseml.ctl." A control file with analysis settings for the primate data set is shown in Box 11.1. According to these settings, a maximum likelihood analysis will be performed using the GTR model (model = 7) with gamma-distributed rate variation among sites (fix_alpha = 0, alpha = 0.5, ncatG = 6; with 0.5 being the initial value of the shape parameter that will be estimated for a descretized gamma distribution with 6 categories), and no molecular clock enforced (clock = 0).

Box 11.1

PAML (Phylogenetic Analysis by Maximum Likelihood) is a freeware software package for phylogenetic analysis of nucleotide and amino acid sequences using ***maximum likelihood***. Self-extracting archives for MacOS, Windows, and UNIX are available from *http://abacus.gene.ucl.ac.uk/software/paml.html*. The self-extracting archive creates a PAML directory containing several executable applications (extension.exe in Windows), the compiled files (extension .c, placed in the sub-directory `src`), an extensive documentation (in the doc sub-directory), and several files with example data sets. Each PAML executable also has a corresponding *control file*, with the same name but the extension .ctl, which needs to be edited with a text editor before running the module. For example, the program baseml.exe has a *control file* called `baseml.ctl`, which can be opened with any text editor. The control file for the non-clock analysis of the TRIM5α looks like the following:

```
      seqfile = primates.phy
     treefile = primates_unrooted.tre


      outfile = mlb      * main result file
        noisy = 9   * 0,1,2,3: how much rubbish on the screen
      verbose = 0   * 1: detailed output, 0: concise output
      runmode = 0   * 0: user tree; 1: semi-automatic; 2: automatic
                    * 3: StepwiseAddition; (4,5):PerturbationNNI


        model = 7   * 0:JC69, 1:K80, 2:F81, 3:F84, 4:HKY85
                    * 5:T92, 6:TN93, 7:REV, 8:UNREST, 9:REVu; 10:UNRESTu


        Mgene = 0   * 0:rates, 1:separate; 2:diff pi, 3:diff kapa, 4:all diff

*       ndata = 5
        clock = 0   * 0:no clock, 1:clock; 2:local clock; 3:CombinedAnalysis
    fix_kappa = 0   * 0: estimate kappa; 1: fix kappa at value below
        kappa = 5   * initial or fixed kappa

    fix_alpha = 0   * 0: estimate alpha; 1: fix alpha at value below
        alpha = 0.5 * initial or fixed alpha, 0:infinity (constant rate)
       Malpha = 0   * 1: different alpha's for genes, 0: one alpha
        ncatG = 6   * # of categories in the dG, AdG, or nparK models of
                    rates
        nparK = 0   * rate-class models. 1:rK, 2:rK fK, 3:rK MK(1/K), 4:rK MK

        nhomo = 0   * 0 1: homogeneous, 2: kappa for branches, 3: N1, 4: N2
        getSE = 0   * 0: don't want them, 1: want S.E.s of estimates
  RateAncestor = 1  * (0,1,2): rates (alpha>0) or ancestral states

   Small_Diff = 7e-6
    cleandata = 0   * remove sites with ambiguity data (1:yes, 0:no)?
*       icode = 0   * (with RateAncestor=1. try "GC" In data,model=4,Mgene=4)
```

```
*  fix_blength = -1 * 0: ignore, -1: random, 1: initial, 2: fixed
   method      = 0  * 0: simultaneous; 1: one branch at a time
```

Each executable has a similar *control file*. The software modules included in PAML usually require an alignment and a tree topology as input. The user has to edit the *control file* corresponding to the application he wants to employ. This editing consists of adding the name of the sequence input file (next to the = sign of the control variable: `seqfile= primates.phy` in the example above), adding the name of the file containing one or more *phylogenetic trees* for the data set under investigation (next to the = sign of the control variable: `treefile= primates_unrooted.tre` in the example above), and specifying a name for the output file where the results of the computation will be written (`outfile = mlb`). Other control variables are used to choose among different types of analysis. For example, `baseml.exe` can estimate *maximum likelihood* parameters of a number of *nucleotide substitution models* (see Chapter 4), given a set of aligned sequences and a tree. The control variable of `baseml.ctl` that needs to be edited in order to choose a model is, in fact, `model`. In the example above, by assigning 'model = 7' the GTR substitution model is chosen. Most of the other control variables are self-explanatory as well. After editing and saving the *control file*, the corresponding application can be executed by simply double-clicking on its icon both in Windows or typing 'baseml baseml.ctl' in a terminal window. A detailed documentation is included in the PAML package (doc sub-directory).

The GTR-γ model was chosen based on the ***Akaike Information Criterion*** (***AIC***, see Chapter 1) using the FINDMODEL website (*http://hcv.lanl.gov/content/hcv-db/findmodel/findmodel.html*). In the absence of a molecular clock, we have specified an unrooted tree topology (treefile = primates_unrooted.tre). Note that the sequence alignment is in PHYLIP format, but an additional space was inserted after each taxon name since at least two spaces are required to separate the name and the sequence in PAML.

11.8 Analysis of the primate sequences

`Baseml` can be executed from a command prompt by typing "baseml" (by default, the program will read in the "baseml.ctl" file in the folder) or "baseml baseml.ctl." If PAML is installed according to the author's instructions, the executables should be in the "bin" folder and the full path to this folder may have to be specified. After execution, the program will read the sequence and tree file specified in the control file and start the numerical optimization algorithm. The amount of information displayed during the run and logged in the output file, "mlb," can be controlled by the "noisy" and "verbose" settings, respectively. The only value of interest from the analysis using the non-clock model is the log likelihood value, which is shown at the end of the run or can be found after the "TREE" line in the main output file ("mlb"). For the non-clock model this should result in a value of -6408.906 (digits at decimal places can depend on the starting values of the parameters and the "Small_Diff" value).

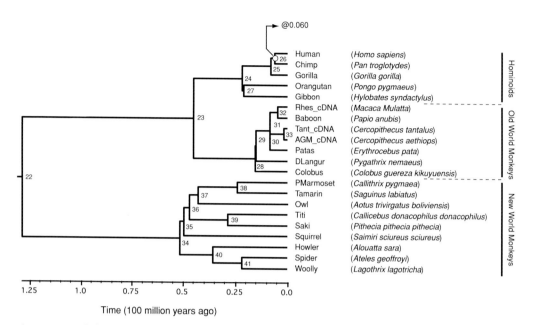

Fig. 11.6 Phylogenetic relationships of 21 primate species reconstructed using TRIM5α gene sequences. Species names are indicated between brackets. Branch lengths were estimated using a molecular clock. The time scale is based on the calibration of the molecular clock using a date of 6 MY for the human–chimp divergence. Because the relevant node was constrained to 0.06, the unit of time is 100 MY.

To perform the analyses using a strict molecular clock, we have to specify a rooted tree topology. The root of the primate tree is situated on the branch connecting the New World monkeys and Hominoids–Old World monkeys (Fig. 11.6). The rooting can be performed in phylogeny reconstruction programs like PAUP* or tree-editors like TREEVIEW (*http://taxonomy.zoology.gla.ac.uk/rod/treeview.html*) and TREEEDIT (*http://evolve.zoo.ox.ac.uk/software/TreeEdit/main.html*). Compared with the non-clock analysis, we now have to change the following settings in the control file: "treefile = primates_rooted.tre" and "clock = 1." Using a molecular clock, we can estimate absolute rates of evolution and divergence times if a fossil calibration point in the tree file is specified. As an example, we will calibrate the molecular clock using a divergence time of 6 million years (MY) for the human–chimpanzee common ancestor in the primate tree. Because calibration dates have to fall within the range of 0.00001-10 in PAML, we will set the date at 0.06 (so the time unit is 100 MY). This can be done by placing "@0.06" at the relevant node in the NEWICK formatted tree:

(((((Human,Chimp)"@0.060",Gorilla),(Orangutan,Gibbon)),(((((Rhes_cDNA,Baboon),(Tant_cDNA,AGM_cDNA)),Patas),DLangur),Colobus)),(((((PMarmoset,Tamarin),Owl),(Titi,Saki)),Squirrel),(Howler,(Spider,Woolly)))));

Multiple calibration points can be specified in this way. To check whether dates were added correctly to the NEWICK file, the tree can be displayed in TREEVIEW with the "Show Internal Edge Labels" option (the node label is indicated in Fig. 11.6). The "primates_rooted.tre" file contains the tree with the appropriate label. To obtain standard errors for our estimates we have to set "getSE = 1." After making the changes to the settings in the control file, baseml can be run as before. This should result in a log likelihood of −6444.633. The molecular clock LRT compares twice the log likelihood difference $(2^*[-6408.906 - (-6444.633)]) = 71.45$ with a χ^2 distribution with 19 $(n-2)$ degrees of freedom. The PAML package includes a program called chi2 to display critical values for the χ^2-distribution with different degrees of freedom. The 0.05 and 0.01 cut-off values for 19 degrees of freedom are 30.14 and 36.19, respectively. Given that our test statistic is much higher than these values, we can conclude that the molecular clock is rejected with $P < 0.01$. The "mlb" output file now contains a tree with branch lengths in units of time. This tree with timescale is shown in Fig. 11.6. The evolutionary rate, which corresponds to $8.092E10^{-8} \pm 1.985E10^{-8}$ substitutions per site per year, and dates for the nodes with standard errors are listed at the end of the "mlb" file. To determine which nodes the numbers refer to, a tree with node labels is written to the "rst" output file, which can be displayed with node numbers in TREEVIEW (node numbers are shown in Fig. 11.6). Caution needs to be taken in interpreting the divergence dates and their confidence intervals because the molecular clock assumption was clearly violated in this case.

11.9 Analysis of the viral sequences

The RSV data set is an example of sequences sampled at different time points. A scatter plot of genetic divergence as a function of sampling time suggests a roughly constant accumulation of nucleotide substitutions over time (Fig. 11.7). To analyze these sequences using the DR, SR, and SRDT models in PAML, different data and tree files are prepared (Table 11.1). To perform the analysis in the absence of a molecular clock assumption (DR model), the file "RSVA.phy" and the unrooted tree topology in the file "RSVA_unrooted.tre" are required. After setting the relevant options in the control file (seqfile = RSVA.phy, treefile = RSVA_unrooted.tre, model = 7; clock = 0, fix_alpha = 0, ncatG = 6), baseml can be run as before. This analysis results in a log likelihood value of −3035.051.

To test the molecular clock, we have to compare this value with the one obtained using the SRDT model, which assumes rate constancy and takes into account the sampling times. In analogy to fossil calibrations, sampling dates are encoded by the digits following the "@" symbol in sequence names. Both in the sequence files ("RSVA_dates.phy") and tree files ("RSVA_dates_rooted.tre"), the year of sampling

Table 11.1   Input files, molecular clock settings, number of parameters and log likelihood results for the analysis of the HRSVA data sets.

	Different rates (DR)	Single Rate Dated Tip (SRDT)	Single Rate (SR)
seqfile	RSVA.phy	RSVA_dates.phy	RSVA.phy
treefile	RSVA_unrooted.tre	RSVA_dates_rooted.tre	RSVA_rooted.tre
clock	0	1	1
Number of parameters	67	35	34
log likelihood	−3035.051	−3057.190	−3128.271

The number of parameters in the table includes the number of branch lengths that need to be estimated and the evolutionary rate parameter for the SRDT model, it does not include the substitution model parameters, which is the same number for all three models.

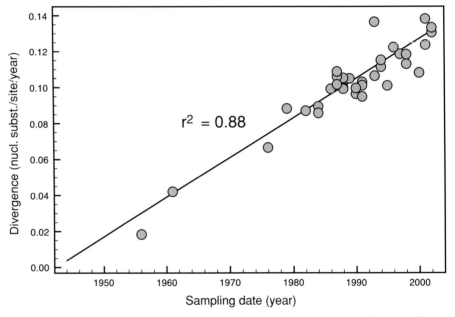

Fig. 11.7   Divergence as a function of sampling time for the RSVA data set. The coefficient of determination (r^2) value is indicated above the regression line.

is included in the sequence names. When baseml reads in sequences with the dates encoded, and a molecular clock is enforced (clock = 1), the SRDT model is automatically in effect. Since the SRDT model assumes rate constancy, a rooted tree topology is required. If no appropriate root can be assigned, it would be appropriate to compute the SRDT likelihood for every possible rooting. TREEEDIT has some functionality to generate all possible rooted topologies, and PAML programs will accept multiple trees in the tree file and can perform sequential maximum likelihood optimization on each one of them. Since this procedure is very time-consuming, we have saved the rooted tree topology that yielded the best likelihood in the file "RSVA_dates_rooted.tre." The SRDT analyses for this tree results in a log likelihood value of −3057.190. Twice the log likelihood difference ($2^*[-3035.051 - (-3057.190)] = 44.28$) can now be compared to with a χ^2 distribution with 32 ($n-3$) degrees of freedom. This LRT fails to reject the molecular clock at the 0.01 and 0.05 significance level ($P = 0.073$). The estimate for the substitution rates is 0.00206 ± 0.00020 substitutions per site per year, which is well in the range of rates expected for rapidly evolving RNA viruses. According to the estimated divergence dates in the output file, the most recent common ancestor (MRCA) for HRSVA is dated back to 1944.76 ± 2.85. Because dates encoded in the sequence names set the SRDT model by default, no dates should be specified to run the SR model. To apply the SR model, the file "RSVA.phy" and the rooted tree topology in the file "RSVA_rooted.tre" are required. After setting the relevant options in the control file (seqfile = RSVA.phy, treefile = RSVA_rooted.tre, model = 7; clock = 1, fix_alpha = 0, ncatG = 6), baseml can be run. The SR analysis results a log likelihood value of −3128.271. A LRT comparing SR and SRDT ($2^*[-3057.190 - (-3128.271)] = 142.162$, df = 1; χ^2 cut-off for 0.01 significance level = 6.635) indicates that accommodating sampling times provides a much better fit to the clock model. This is not surprising for a rapidly evolving virus and a range of sampling times of 1956–2002.

For comparison, we have also plotted evolutionary rates and divergence dates estimated using the Bayesian coalescent approach implemented in **BEAST** (Drummond *et al.*, 2002; Drummond & Rambaut, 2003) (Fig. 11.8). The XML input files are available to download; for information how to set up such files and run the program, we refer to Chapter 18. Very similar estimates are obtained using the *MCMC* methods that average over a set of plausible trees. Slightly higher standard deviations were inferred for the relaxed clock model. We would expect to see even larger standard deviations if the molecular hypothesis would be strongly rejected. It should be noted, however, that Bayesian and maximum likelihood analysis methods employ different methods to build confidence/credible intervals around the estimates. The error in the ***maximum likelihood estimates*** (**MLEs**) is deduced using a curvature method that relies on ***asymptomatic normality***. This method and others to obtain confidence intervals for MLEs are described in more detail in Chapter 14. The Bayesian credibility intervals are constructed using the marginal posterior density of a parameter. Mean values, standard deviations, and ***highest posterior density intervals*** (see Chapter 18) can be used to summarize the parameter values sampled during the MCMC procedure. The Bayesian relaxed clock inference also provides a *coefficient of variation* that quantifies the variability of the

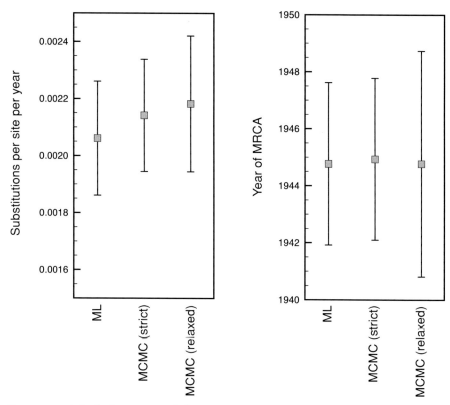

Fig. 11.8   Substitution rates and divergence dates estimated using maximum likelihood (ML) and Bayesian MCMC methods. The error bars represent standard deviations.

substitution rate among branches under a relaxed clock. A mean value of 0.24 suggests that rates on the branches of the HRSVA trees vary within about 24% of the mean rate, which is, apparently, still acceptable under the rate constancy assumption.

Testing tree topologies

THEORY

Heiko A. Schmidt

12.1 Introduction

Throughout the book a number of approaches have been exemplified to assess and compare various aspects of evolutionary trees and models.

To check the reliability of branches in a certain tree, one can use *(non-parametric) bootstrapping* or *jackknifing*, combining alignment subsampling and consensus trees to get support values on branches (Chapter 5). Furthermore, other methods that generate or sample sets of plausible trees can be used to get support values, like Bayesian *MCMC* sampling (Chapter 7) or *quartet puzzling* (Chapter 6).

Various approaches have been devised to determine a best-suited *evolutionary model* (Chapter 10). Such approaches are often based on the *maximum likelihood* values obtained for the models in question. Different measures are applied like *Akaike Information Criterion (AIC)*, *Bayesian Infomation Criterion (BIC)*, *Akaike Weights*, and other model selection techniques (refer to Johnson & Omland, 2004, for review) to correct for the additional parameters in the more complex models. Such techniques are also implemented in programs like MODELTEST to select the most useful model of evolution (see Posada & Buckley, 2004 and Chapter 10 for details).

In this chapter we will briefly review different techniques and tests to compare contradicting and, hence, non-nested tree topologies using their *likelihood* values. Since a large variation of testing approaches can be applied (see, e.g. Goldman

The Phylogenetic Handbook: a Practical Approach to Phylogenetic Analysis and Hypothesis Testing, Philippe Lemey, Marco Salemi, and Anne-Mieke Vandamme (eds.). Published by Cambridge University Press. © Cambridge University Press 2009.

et al., 2000), we will restrict ourselves to review a number of common tests for which easily accessible software implementations exist. We will briefly describe the different approaches, the hypotheses they test, and discuss possible problems and pitfalls.

12.2 Some definitions for distributions and testing

In the current context we are usually interested in whether the difference between two values, e.g. the likelihoods of two models or trees, are significantly different or could be explained by random effects.

To perform a test, we first have to state a *null hypothesis* H_0. The null hypothesis is the hypothesis of *no difference* and is usually the negation of the question we are interested in (Siegel & Castellan, 1988, p. 7). This null hypothesis has to be precise, since the test of significance is based on its rejection (Fisher, 1971). If the tested null hypothesis is rejected, the *alternative hypothesis* H_A is supported which typically reflects our question, like "are two likelihoods significantly different?"

There are two types of possible errors when testing the null hypothesis H_0. First, the null hypothesis is rejected when it is actually true (type I error). This result is also called a *false positive*. Second, an erroneous null hypothesis is failed to be rejected (type II error), also called *false negative*. The probability of a type I error is denoted by α. We set α to the largest probability of a type I error we are willing to accept; the *significance level*, typically $\alpha = 0.05$ (i.e. 5% error). This corresponds to a *confidence limit* of $(1 - \alpha) = 0.95$ or 95% (e.g. Siegel & Castellan, 1988; Zarkikh & Li, 1995).

Given a **sampling distribution** that reflects the probability of every possible sample value if drawn randomly, the null hypothesis (of no difference from expectation) can be tested. If the observed value μ_0 is in the region of rejection, i.e. outside the 95% *confidence interval* or *acceptance region*, the null hypothesis H_0 is rejected and the alternative hypothesis is supported. If that value falls inside the acceptance region, H_0 cannot be rejected at the chosen level of confidence (see Fig. 12.1a).

If we have prior knowledge about the direction of the effect of the alternative hypothesis, then a one-sided test is used. Note, that one-sided and two-sided tests do not differ in the size but in the location of the rejection region, i.e. in the one-sided test the region is entirely at one tail of the sample distribution (see Fig. 12.1b).

The significance level α has to be set in advance and determines the critical value below or above which the null hypothesis is rejected. The p-value, on the other hand, denotes the probability of obtaining a result equal or more extreme (with respect to the null hypothesis) than the observed value μ_0 and can only be

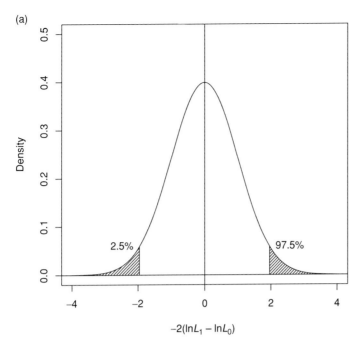

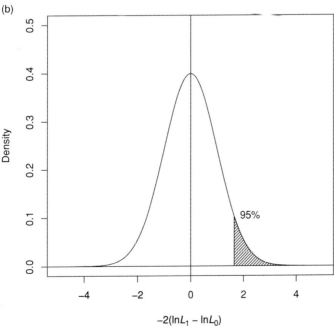

Fig. 12.1   Normal distributions for (a) a two-sided and (b) a one-sided test. An observed value μ_0 is (a) significantly different from the expectation μ if it is in one of the two shaded tails covering each 2.5% of the surface below the curve on either side or (b) significantly larger if it is above the upper 95% quantile. The null hypothesis cannot be rejected if μ_0 falls into the (unshaded) 95% confidence interval. These decisions depend on the significance level $\alpha = 0.05$.

determined after the test (Goodman, 1999). If the null hypothesis is rejected, the p-value is also necessarily less than α.

When testing tree topologies, there is a big difference whether the trees are selected *a priori* or *a posteriori*.

A priori means that the trees have been selected without knowledge about their support by the data or any optimizing analysis involved. Such trees might just be derived as logical alternative scenarios or, for example, from a **Markov chain** without prior knowledge about their likelihood values or probabilities. Hence, each of the trees of interest might be the one with the highest likelihood.

If the trees of interest are selected from an analysis to test whether, for example, the second, third, etc. tree is significantly worse than the best tree, that is the tree yielding the best maximum likelihood value called the *ML* tree (cf. Chapter 6), the trees are chosen *a posteriori*.

12.3 Likelihood ratio tests for nested models

If the two evolutionary models of interest are *nested*, meaning that the more parameter-rich model can be restricted to the simpler one by restricting its parameters, then **likelihood ratio tests** are straightforward to compare the likelihoods L_0 and L_1 of the two models based on a (single) tree. For convenience l_a denotes the log-likelihood $\ln L_a$ in the following. The **likelihood ratio** test (LRT) **statistic**

$$\Delta = -2\ln\frac{L_0}{L_1} = 2(l_1 - l_0) \tag{12.1}$$

follows approximately a χ^2 distribution for the respective degrees of freedom, that is, the number of additional parameters in the more parameter-rich model. L_1 is the likelihood of the alternative (more parameter-rich) model and L_0 that of the less parameter-rich null model. If their Δ value computed from (12.1) is located in the rejection area of the χ^2 distribution (the shaded area in Fig. 12.2) beyond the 95%-quantile (if a significance level of 5% is assumed), the null hypothesis is rejected and the alternative model is said to give a significantly higher likelihood L_1 compared to the null model. (Likelihood ratio tests of nested models are discussed in detail in Chapter 10.)

Although this methodology is straightforward for nested models, it is generally not applicable to compare different tree topologies. The problem with trees is that tree topologies cannot be interpreted as a single statistical parameter and, furthermore, it remains unclear how many parameters a tree represents with its possible groupings and branch lengths (Yang *et al.*, 1995; Huelsenbeck & Crandall, 1997).

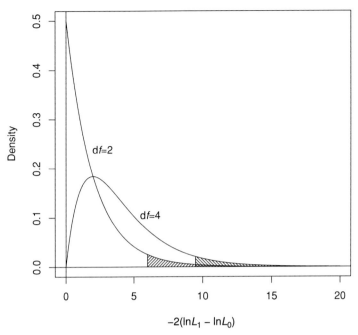

Fig. 12.2 χ^2 distribution for 2 and 4 degrees of freedom. An LRT is assumed significant, i.e. the more parameter-rich has a significantly higher likelihood L_1, if the $\wedge$ value is in the shaded 95% quantile.

If the tested models are not nested, the distribution of Δ is *Gaussian*, that means, according to a *normal distribution* (Cox, 1961, 1962). The shape of a normal distribution $\mathcal{N}(\mu, \sigma^2)$ is determined by its mean value μ and standard deviation σ.

Thus, the χ^2 distribution does not apply, and different steps must be taken to find the distribution that can be used to test the difference between two likelihoods.

12.4 How to get the distribution of likelihood ratios

Most of the methods we will be concerned with, will use a likelihood ratio statistic

$$\delta = \ln \frac{L_a}{L_b} = \ln L_a / \ln L_b = l_a - l_b \tag{12.2}$$

to compare the difference of the log-likelihoods l_a and l_b for two trees T_a and T_b. These likelihoods are obtained by maximum likelihood optimization of model parameters, branch lengths, etc. on a given sequence data set D, as described in Section 6.3 (Chapter 6).

To judge whether the obtained likelihoods are significantly different, we need information on how the "real" distribution of likelihood differences under the null hypothesis looks like.

In the ideal case one would like to draw further samples from the process that generated our data. Unfortunately, we cannot re-run the process of evolution, as we would be able to by tossing a coin or roll dice a couple of additional times. Usually, we only have a limited data set, the alignment, where each column is usually regarded as an independent sample from the "true" process of evolution, to determine the desired distribution.

A common way to determine such distributions from limited data sets if further samples from the original process cannot be obtained are bootstrap re-sampling methods (Efron, 1979; Efron & Tibshirani, 1994; Goldman, 1993).

12.4.1 Non-parametric bootstrap

The **non-parametric bootstrap** has been mentioned in various chapters of this book. This bootstrap randomly re-samples columns from the alignment D with replacement to produce a number of pseudo-samples $D^{(i)}$ from the processes of evolution. In each of these pseudo-samples some columns might be included several times, while others have not been chosen at all (Felsenstein, 1985; Efron *et al.*, 1996). Each generated pseudo-sample $D^{(i)}$ is then used to compute the values of interest and to determine their distribution.

Here, based on each pseudo-sample alignment $D^{(i)}$ the maximum log-likelihood values $l_x^{(i)}$ of each tree T_x in the set of M trees $\mathcal{T}$ of interest are computed by complete optimization of branch lengths and model parameters.

For the use of bootstrap in a hypothesis testing scenario it is necessary for the values computed via the bootstrap to reflect the assumed null-distribution, although the pseudo-sample data might not. Several steps available to ensure this null-hypothesis conformity have been described (Hall & Wilson, 1991; Westfall & Young, 1993). The method of choice to adjust the log-likelihood values is the so-called *centering*, where each log-likelihood value $l_x^{(i)}$ of each tree T_x on pseudo-sample $D^{(i)}$ is shifted by the mean value $\bar{l}_x = \frac{1}{B} \sum_{i=1}^{B} l_x^{(i)}$ leading to a centered log-likelihood $\tilde{l}_x^{(i)} = l_x^{(i)} - \bar{l}_x$. From the centered log-likelihoods $\tilde{l}_x^{(i)}$ the log-likelihood differences $\delta^{(i)}$ are computed between pairs of trees according to (12.2) for each sample $D^{(i)}$. The obtained values $\delta^{(i)}$ are then used to infer the mean μ and the standard deviation σ of the respective normal (sample) distribution to test the observed ratio δ.

The re-optimization of the likelihood values for all the bootstrap samples is computationally very intense. Hence, Kishino *et al.* (1990) suggested **resampling** *estimated log-likelihoods* (*RELL*), a variant of the non-parametric bootstrap, that is computationally less demanding but not necessarily as accurate. We have seen in Chapter 6 that the likelihood values are computed by multiplying the *site-likelihoods* of each column D_j (6.12) or the log-likelihood as the sum of all site-log-likelihoods (6.18). Kishino *et al.* (1990) keep the site-log-likelihoods fixed and only "bootstrap" the pre-estimated site-log-likelihoods. This RELL method saves the time-consuming likelihood re-estimation, but it assumes some asymptotic conditions such as sufficiently large data and correctly specified models of evolution to produce valid results.

12.4.2 Parametric bootstrap

A different way to infer the distribution of δ is the ***parametric bootstrap*** (also called ***Monte Carlo simulation***). Here, the bootstrap samples are not drawn from the alignment but simulated along a tree with branch lengths and model parameters. That means, a tree with branch lengths and model parameters has to be inferred first from the original alignment D which then serve as input for Monte-Carlo simulation performed by sequence generation programs such as SEQ-GEN (Rambaut & Grassly, 1997). From the simulated bootstrap samples, one again estimates trees and their likelihoods which are, in turn, used to determine the distribution of δ.

Differently to non-parametric bootstrapping no adjustment step like centering (Section 12.4.1) is necessary, since the given tree, model, and parameters act as null-model according to which the bootstrap samples are generated by Monte-Carlo simulation.

For detailed descriptions of parametric bootstrap approaches refer to Goldman (1993) and Huelsenbeck and Crandall (1997).

12.5 Testing tree topologies

A large number of test variants exist by combining different approaches in the various steps of a test, like different bootstrap methods to generate the samples, the amount of optimization to compute the likelihoods, the choice of the trees of interest, or the assumptions made on the kind of normal distribution. To get an extensive overview on such variants, discussions about the possible ways, problems, and pitfalls of tree topology testing, we recommend Goldman *et al.* (2000) and Huelsenbeck and Crandall (1997) and references therein.

We only review a limited number of tests which are commonly used. To that end, we will mostly use the same notation as Goldman *et al.* (2000).

12.5.1 Tree tests – a general structure

First, the null hypothesis H_0 and the alternative hypothesis H_A have to be stated, since they determine the results of the test and also determine whether a test is applicable at all for the available data and the question a researcher is asking.

Second, testing trees with likelihoods follows a global structure:

(i) Compute the log-likelihood values l_x for all trees $T_x \in \mathcal{T}$ by fully optimizing all parameters. Also, all site-likelihoods are kept for bootstrapping.

(ii) Generate many ($B \geq 1000$) bootstrap samples $D^{(i)}$ ($i = 1 \ldots B$). Re-estimate the log-likelihood values $l_x^{(i)}$ (with optimization) for each tree T_x and each bootstrap sample $D^{(i)}$.

(iii) Adjust for each tree topology T_x all log-likelihoods $l_x^{(i)}$ to conform to the null hypothesis, if the bootstrap samples have been generated by non-parametric bootstrap or RELL. This is typically done by *centering* the log-likelihoods with the mean log-likelihood $\bar{l}_x^{(i)} = \frac{1}{B} \sum_{i=1}^{B} l_x^{(i)}$ across all bootstrap samples i:

$$\tilde{l}_x^{(i)} = l_x^{(i)} - \bar{l}_x^{(i)} \tag{12.3}$$

Refer to Hall and Wilson (1991) for more details on the necessity of centering.

(iv) Compute the log-likelihood differences $\delta^{(i)} = \tilde{l}_a^{(i)} - \tilde{l}_b^{(i)}$ between the relevant pair(s) of trees T_a and T_b. Use the $\delta^{(i)}$ values to determine their distribution.

Note, that the number and specification of the relevant pairs of trees depend on the respective null hypothesis H_0 (see following sections).

(v) Use the distribution of $\delta^{(i)}$ to test whether the null hypothesis is to be rejected. Obtain the p-value for the observed δ.

12.5.2 The original Kishino–Hasegawa (KH) test

Kishino and Hasegawa (1989) devised a test based on the RELL method to compare two *a priori* selected trees T_a and T_b, e.g. produced by a Markov chain.

The null and alternative hypotheses to be compared are (two-sided test):

H_0:   The two trees are equally supported, i.e. the expected value $E[\delta] = \mu = 0$
H_A: The two trees are not supported equally, i.e. the expected value $E[\delta] = \mu \neq 0$

The KH test itself follows the following procedure (see also Fig. 12.3):

(i) Infer the log-likelihood values l_a and l_b for trees T_a and T_b. Compute $\delta = l_a^{(i)} - l_b^{(i)}$.

(ii) Generate many ($B \geq 1000$) bootstrap samples i and the respective log-likelihood values $l_a^{(i)}$ and $l_b^{(i)}$ with the RELL method.

(iii) Center the obtained likelihood values of each tree with the mean log-likelihood $\bar{l}_x^{(i)}$ across all samples i, as $\tilde{l}_x^{(i)} = l_x^{(i)} - \bar{l}_x^{(i)}$.

(iv) Determine the distribution of differences $\delta^{(i)} = \tilde{l}_a^{(i)} - \tilde{l}_b^{(i)}$.

(v) Use the distribution inferred from $\delta^{(i)}$ to test whether your trees are equally supported in a two-sided test. Obtain the p-value for the observed δ.

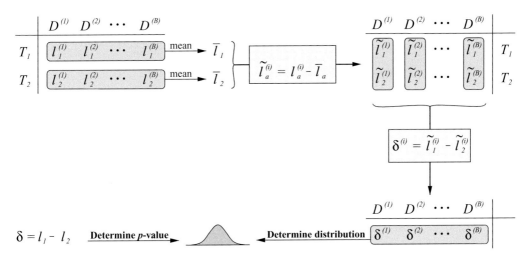

Fig. 12.3   Sketch of the Kishino–Hasegawa test. For both trees T_1 and T_2 the log-likelihoods l_a^i are computed for each bootstrap sample $D^{(i)}$. The log-likelihoods are subsequently "centered" by the trees mean log-likelihood $\bar{l}_a$ across the bootstrap samples. The log-likelihood difference $\delta_a^{(i)}$ between $\tilde{l}_1^{(i)}$ and $\tilde{l}_2^{(i)}$ is computed. Finally, the distribution of δ-values is determined and used to determine the p-value of the observed δ-value.

Since the two trees T_a and T_b are selected *a priori* both have an equal chance to gain the higher log-likelihood and, hence, δ might be positive or negative. Thus a two-sided test is applied as in Fig. 12.1a.

12.5.3 One-sided Kishino–Hasegawa test

Although the KH test was devised for *a priori* selected trees, in the majority of its applications it is (mis-)used to test whether sets of suboptimal trees are equally supported or significantly worse than the best tree, or to compare all trees against the one having the highest likelihood among all *a priori* selected trees.

The problem arises that, if T_a is the maximum likelihood tree or the tree with highest likelihood in the set and all trees in $\mathcal{T}$ are tested against this one tree T_{ML}, then $\delta = l_{ML} - l_b$ can hardly be negative. The above hypotheses are thus not tested properly by the original KH approach. However, the KH test has often been applied this way (see Goldman *et al.*, 2000 and Shimodaira & Hasegawa, 1999 for extensive discussion).

The only way to adjust the KH test to some extent to this scenario would be to use a one-sided test as indicated in Fig. 12.1b (cf. Goldman *et al.*, 2000). However, the null hypothesis $E[\delta] = 0$ might still be violated.

Many published conclusions based on wrongly applied KH tests might not be valid. Goldman *et al.* (2000) stated that the only possible adjustment to correct for this mistake might be to adjust the p-value to $p/2$. This has a similar effect as having performed a one-sided test instead.

12.5.4 Shimodaira–Hasegawa (SH) test

Shimodaira and Hasegawa (1999) devised a valid test to assess a set of *a posteriori* selected trees when the maximum likelihood tree is among the tested trees.

The null and alternative hypotheses tested by the Shimodaira–Hasegawa (SH) test look different in this case:

H_0:   All trees $T_x \in \mathcal{T}$ (including the ML tree T_{ML}) are equally good explanations of the data.

H_A: Some or all trees $T_x \in \mathcal{T}$ are not equally good explanations of the data.

The test itself follows the following procedure (see also Fig. 12.4):

(i)   Estimate the log-likelihood values l_{ML} and l_x for all trees $T_x \in \mathcal{T}$. Compute all $\delta_x = \ell_{ML} - l_x$.

(ii)   Generate many ($B \geq 1000$) bootstrap samples i and compute the respective log-likelihood values $l_{ML}^{(i)}$ and $l_x^{(i)}$ (using the RELL method).

(iii)   For each tree T_x, center the log-likelihood values with the mean log-likelihood $\bar{l}_x^{(i)}$ across all samples i, as $\tilde{l}_x^{(i)} = l_x^{(i)} - \bar{l}_x^{(i)}$.

(iv)   For each bootstrap sample i, find the maximal log-likelihood $\tilde{l}_{ML}^{(i)}$ over all trees $T_x \in \mathcal{T}$ and compute the differences $\delta_x^{(i)} = \tilde{l}_{ML}^{(i)} - \tilde{l}_x^{(i)}$.

(v)   For each tree T_x separately test whether the obtained δ_x value is in the rejection area beyond 95%. If so, reject the null hypothesis for T_x. If not, H_0 cannot be rejected. Obtain the p-value for the observed δ_x.

The one-sided test is appropriate here, since the log-likelihood $\tilde{l}_x^{(i)}$ of any tree T_x can only be smaller or equal to $\tilde{l}_{ML}^{(i)}$.

When applying the SH test, one has to keep in mind that the maximum likelihood tree is required to be among the tested trees, otherwise the estimated significance levels will be inaccurate (Goldman *et al.*, 2000 and Westfall & Young, 1993, p. 48).

Furthermore, it has been pointed out by Strimmer and Rambaut (2002) that the number of trees selected in the SH test is strongly correlated with the number of tested input trees, meaning, the more tree topologies are included in the test, the more trees are accepted. This conservative behavior makes the use of the SH test problematic for large sets of trees.

12.5.5 Weighted test variants

Shimodaira and Hasegawa (1999, comment 4) have suggested weighted variants of the SH and also the KH test, namely WSH and WKH, for cases where one wants

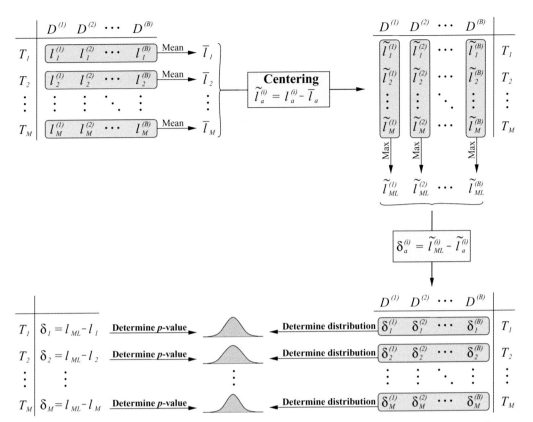

Fig. 12.4   Sketch of the Shimodaira–Hasegawa test. Log-likelihoods l_a^i are computed for each tree T_a and each bootstrap sample $D^{(i)}$ and subsequently "centered" by the trees mean log-likelihood $\bar{l}_a$ across the bootstrap samples. For each bootstrap sample $D^{(i)}$ the tree with maximal log-likelihood $\tilde{l}_{ML}^{(i)}$ is determined. Then log-likelihood difference $\delta_a^{(i)}$ between $\tilde{l}_{ML}^{(i)}$ and the corresponding $\tilde{l}_a^{(i)}$ is computed. Finally, the distribution of δ-values is determined and used to determine the p-value of the corresponding trees' observed δ-values.

to be less conservative. In these variants the likelihood ratio $l_a - l_b$ is weighted by the square root of its of variance $\sigma^2(l_a - l_b)$.

This is straightforward for the KH test in step (i) and (iv). In the SH test, however, all instances of $\delta_x = l_{ML} - l_x$ (step (i)) have to be substituted by

$$\delta_x = \max_{a \neq x} \left(\frac{\bar{l}_a - \bar{l}_x}{\sigma\left(\bar{l}_a - \bar{l}_x\right)} \right) \tag{12.4}$$

while $\delta_x^{(i)} = \tilde{l}_{ML}^{(i)} - \tilde{l}_x^{(i)}$ (step (iv)) is replaced by

$$\delta_x^{(i)} = \max_{a \neq x} \left(\frac{\tilde{l}_a^{(i)} - \tilde{l}_x^{(i)}}{\sigma\left(\tilde{l}_a^{(i)} - \tilde{l}_x^{(i)}\right)} \right) \tag{12.5}$$

Note, due to the weighting the maximal δ-value is not necessarily gained between the current tree and the ML tree, which is the case in the unweighted test. By weighting the likelihood ratio depending on its variance, the tests are less conservative. Although this compensates for some of the above-mentioned conservative behavior of the SH test, it does not completely correct for it (Shimodaira, 2002). Furthermore, both the WSH and WKH tests rely on the same assumptions like the presence or absence of the ML trees in the set of compared trees as their un-weighted counterparts.

12.5.6 The approximately unbiased (AU) test

Shimodaira (2002) explains the correlation of the number of input trees and the size of the confidence set returned by the SH test by the fact that the SH test is heavily biased. On the one hand, SH is very good in controlling its type I error, but it overestimates the selection bias and, thus, acts more conservative as the number of input trees grows.

Zharkikh and Li (1995) have suggested a method that is based on a complete, as well as a partial, bootstrap to enable the inference of the selection bias. Shimodaira (2002) later devised an **approximately unbiased (AU)** test based on a multiscale bootstrap to be able to better correct for the selection bias. The *multiscale bootstrap* works as follows.

From our input alignment D of length N, the multiscale bootstrap draws bootstrap replicates for a number of different lengths N_r. Some are smaller but also some are larger than the original sequence length N. For each length N_r, many bootstrap samples are drawn ($B \geq 10\,000$). The log-likelihood $l_x^{(i)*}$ obtained by the RELL method for the sequence length N_k are scaled with the factor N/N_k to the same virtual length N:

$$l_x^{(i,r)} = \frac{N}{N_k} l_x^{(i,r)*} \tag{12.6}$$

Using the results from the different sequence lengths N_r, the method is able to infer the unknown curvature of the selection bias needed for a proper correction. Thus, the AU test is approximately unbiased if an appropriate set of sequence lengths is used (see Shimodaira, 2002, for details).

According to Shimodaira (2002), the AU test tests for each tree $T_a \in \mathcal{T}$ the following null hypothesis.

$H_0(T_a)$:   the expected value $E[l_a]$ of T_a is larger or equal to the expected values $E[l_x]$, for all $T_x \in \mathcal{T}$.

Although the AU test is not susceptible to the increase of trees, one has to be careful if many of the best trees are almost equally well supported – one might miss the true tree, since there is over-confidence in the wrong trees (Shimodaira, 2002). Furthermore, the method might be computationally infeasible if the tree set $\mathcal{T}$ contains several thousand trees. Shimodaira (2002) suggests a prefiltering with the KH test using a very conservative significance value (e.g. $\alpha = 0.001$) to reduce the tree set before applying the AU test.

12.5.7 Swofford–Olsen–Waddell–Hillis (SOWH) test

Swofford *et al.* (1996) suggested an approach (SOWH test) which – different from the above tests – applies parametric bootstrapping to compare the trees. The SOWH test tests the following hypotheses (cf. Goldman *et al.*, 2000):

H_0:   The tree T_a is the true topology.
H_A: Some other topology is the true one.

To test each tree T_a from a set $\mathcal{T}$ the SOWH test proceeds as follows:

(i) Estimate the log-likelihood values l_{ML} and l_a and compute the test statistic $\delta = l_{ML} - l_a$.
(ii) Generate parametric bootstrap samples with Monte Carlo simulation along tree T_a with the ML parameters $\hat{\theta}_a$ estimated for tree T_a.
(iii) For each bootstrap sample, re-estimate the model parameters $\theta_a^{(i)}$ and the log-likelihood value $l_a^{(i)}$ for tree T_a (under the null hypothesis).
(iv) For each bootstrap sample, also re-estimate the model parameters $\theta_x^{(i)}$ and the log-likelihood value $l_x^{(i)}$ for all other trees $T_x \in \mathcal{T}$ to find the ML log-likelihood $l_{ML}^{(i)}$ for this bootstrap sample.
(v) Compute the difference values $\delta^{(i)} = l_{ML}^{(i)} - l_a^{(i)}$ which are interpreted as samples according to the distribution of δ under the null hypothesis H_0. Due to this assumption, no estimation of distribution parameters is performed.
(vi) Obtain the border of the rejection area directly from the generated distribution of $\delta^{(i)}$ values. To this end, empirically sum the $\delta^{(i)}$ values in ascending order until you have passed 95% of all $\delta^{(i)}$ values. Use this $\delta^{(i)}$ value as the critical value beyond which the null hypothesis is rejected.
(vii) Repeat this procedure for all different trees $T_a \in \mathcal{T}$.

The SOWH test utilizes the same test statistic δ as the KH and the SH test. Due to using the ML tree in the computation of δ, the assumption of $E[\delta] = 0$ would be inappropriate, however. Therefore, a one-sided test is used.

The repeated parametric bootstrap based on the respective T_a produces data conforming to the null hypothesis. Hence, no centering is necessary.

The main problem with tests based on parametric bootstrap is that they are computationally very demanding, often making extensive tests unfeasible. Furthermore, no straightforward implementation of the SOWH test seems available. Yet Goldman *et al.* (2000) give some advice at *http://www.ebi.ac.uk/*

goldman/tests/ how to implement SOWH tests using Paup* (see Chapter 8) and Seq-Gen (Rambaut & Grassly, 1997).

12.6 Confidence sets based on likelihood weights

Strimmer and Rambaut (2002) approach the problem of comparing trees from a different perspective. Instead of significance testing, they devised a method that generates a **confidence set** of trees based on **expected likelihood weights** (ELW). They define a confidence set as the smallest subset of models – here trees – which together obtain a pre-defined probability C to be selected based on some random data set D (with length N) drawn from the true distribution of the evolutionary process. Note that this concept is related to *credible sets* of trees in Chapter 7.

Given the likelihoods L_x of each tree $T_x \in \mathcal{T}$, the likelihood weight w_a of a single tree T_a is computed as the fraction of the total likelihood summed over all trees $T_x \in \mathcal{T}$:

$$w_a = \frac{L_a}{\sum_x L_x} \tag{12.7}$$

with all likelihood weights w_a adding up to 1.0. One way of constructing a confidence set would be to collect all trees T_a in descending order of their weights until the sum of collected weights meets the pre-defined threshold value C, typically 0.95. This view is related to significance testing (see above) where the $1 - \alpha$ confidence region corresponds to the coverage of our confidence set by the cumulative level of confidence C.

To compute the precise selection probability, the *expected* likelihood weight $E[w_a]$, the true model has to be known which is hardly ever the case in reality. Hence, estimating the expected weights is based on a non-parametric bootstrap as in the previous sections (Fig. 12.5):

(i) Generate B bootstrap samples $D^{(i)}$. Estimate the corresponding likelihood values $L_x^{(i)}$ for each tree $T_x \in \mathcal{T}$ (e.g. with the RELL method).
(ii) Compute the likelihood weights $w_a(i)$ for all $T_a \in \mathcal{T}$ according to (12.7), within each bootstrap sample separately.
(iii) For each tree T_x derive its expected likelihood weight $E[w_a]$ by averaging over all bootstrap samples, assuming $E[w_a] \approx \overline{w}_a$:

$$\overline{w}_a = \frac{1}{B} \sum_{i=1}^{B} w_a^{(i)} \tag{12.8}$$

(iv) Construct the confidence set by selecting trees T_x in descending orders of their (inferred) expected weights $\overline{w}_x$ until their accumulated sum meets the pre-set level of confidence $C = 0.95$.

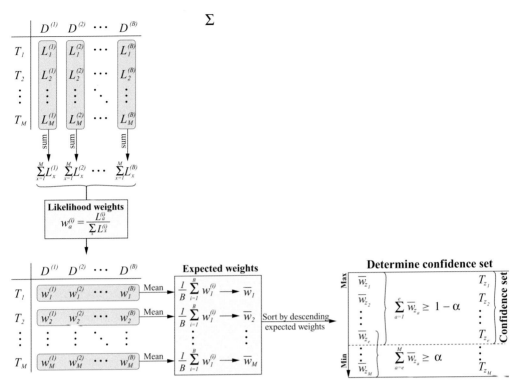

Fig. 12.5   Sketch of the confidence set generation from expected likelihood weights. From the bootstrap samples $D^{(1)} \ldots D^{(B)}$ likelihoods $L_1^{(1)} \ldots L_M^{(B)}$ are computed for each tree $T_1 \ldots T_M$. Based on each bootstrap sample, each likelihood $L_a^{(i)}$ is converted to a likelihood weight $w_a^{(i)}$. From these weights, the expected likelihood weight $\overline{w}_a$ for each tree T_a across the corresponding bootstrap samples is computed. The trees are sorted by their expected likelihood weights $\overline{w}_a$. The confidence set collects trees in descending order such that the cumulative expected weights $\sum_{a=1}^{e} \overline{w}_{z_a}$ just contains the fraction $1 - \alpha$.

This method for selecting a confidence set seems not to be affected by the problem of SH, i.e. extending the constructed confidence set as more and more trees are added as input (Strimmer & Rambaut, 2002). It is also independent of whether or not the true best tree is among the input trees. Nevertheless, the simplifications made need long sequence data sets D to correct for possible model mis-specification, and large enough numbers of bootstrap samples to get valid estimates from the parametric bootstrap (especially if the RELL method is used). The impact of model mis-specification with data sets, however, remains unclear.

12.7 Conclusions

All methods we have examined above provide us with a kind of confidence set of trees, a subset from our input set $\mathcal{T}$. The trees within this confidence set cannot be classified statistically as significantly better, worse, or different (depending on the hypotheses tested) by the means of their likelihood values. That means, when two trees are selected for the confidence set, we cannot discuss their differences as significant even though their likelihoods might differ and their topologies might substantially contradict each other.

The trees in the confidence set are usually assumed to be close to the true tree. This conclusion is difficult to confirm, however, since the true tree might not be among those tested. Furthermore, model mis-specifications and violations of basic assumptions might render the test results invalid.

We have seen that it is of utmost importance to take into account the hypotheses and assumptions a test is based on. Knowing these limitations allows us to draw valid conclusions from tests we apply and, vice versa, to determine what tests are appropriate to answer certain questions we want to ask about our data.

PRACTICE

Heiko A. Schmidt

12.8 Software packages

While LRTs (Chapters 10 and 11) for nested models are very abundant, only a few packages allow the execution of various tests like Tree-Puzzle, Paup*, or the Consel software package. More programs are available which only implement the KH and SH test. A detailed list of phylogenetic software which also might contain other test software or different tests can be found at Joe Felsenstein's website, *http://evolution.genetics.washington.edu/PHYLIP/ software.html*.

Because such program packages are emerging at a rapid pace, the reader is advised to visit this website and the software pages for updates.

12.9 Testing a set of trees with Tree-Puzzle and Consel

In this example we will compare a set of 15 trees contained in `hivALN.15trees` based on the alignment in the `hivALN.phy` file. Both files are available from *http://www.thephylogenetichandbook.org*. We will use Tree-Puzzle 5.3 (Schmidt *et al.*, 2002) to perform a number of tests (KH, SH, ELW) and to compute the site-log-likelihoods for the 15 trees. Subsequently, we will use the Consel (Shimodaira & Hasagawa, 2001) software to perform the AU test and a number of other tests. Both programs can be run under Linux/Unix, MacOS X, and Windows. Consel does need an input file with the site-log-likelihood values for the assessed trees, which we will also generate with Tree-Puzzle. The tests implemented in both programs are based on bootstrapping with the RELL method. Hence, no re-estimation of parameters or branch lengths will be performed. For the following we will assume a significance limit $\alpha = 0.05$.

The maximum likelihood analysis of the `hivALN.phy` dataset in Chapter 6 has shown that the relationships between five subtrees could not be resolved reliably (Fig. 6.8b). Those subtrees were the outgroup (comprising HIV-1 group O and SIV sequences), a subtree joining HIV-1 group M subtypes B and D, and the three separate group M subtypes A, C, and G. The five disjoint subtrees can be combined in a fully resolved tree in exactly 15 possible ways. These 15 possible trees are contained in `hivALN.15trees` and also sketched in Fig. 12.6. The branching orders within the subtrees were kept, since there was no contradiction between the two trees reconstructed with the two different methods (see Fig. 6.8).

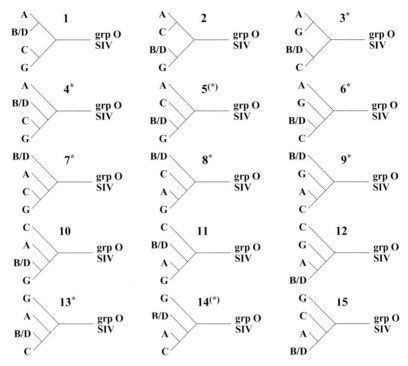

Fig. 12.6   The 15 possible tree topologies relating the four HIV-1 group M subtype-clades A, B/D, C, and G with the outgroup. The numbers give the order in the tree file. Tree 6 was the reconstructed ML tree from Fig. 6.8a. The stars * indicate the acceptance by the AU test. Those stars in brackets (*) had been excluded by the ELW method although accepted by the AU test.

12.9.1 Testing and obtaining site-likelihood with TREE-PUZZLE

The usage of TREE-PUZZLE has been explained in more detail in Chapter 6. Therefore, we will only highlight the relevant details and refer to the practical part of Chapter 6 for more information.

For analysis in the subsequent section we require TREE-PUZZLE to write the estimated site-log-likelihood values to a file for later use with CONSEL. To that end, we start TREE-PUZZLE with the command–line option -wsl (write site-likelihoods). It is recommended to execute TREE-PUZZLE with the following command from a terminal for this example:

```
puzzle -wsl hivALN.phy hivALN.15trees
```

where the first filename after the option is the alignment and the second the tree file. (An ".exe" extension is required for the windows executable.)

Then we will run TREE-PUZZLE with the following settings (changing options k, x, w, and c:

```
GENERAL OPTIONS
b                    Type of analysis? Tree reconstruction
k            Tree search procedure? Evaluate user defined trees
z   Compute clocklike branch lengths? No
e              Parameter estimates? Approximate (faster)
x       Parameter estimation uses? Neighbor-joining tree
SUBSTITUTION PROCESS
d       Type of sequence input data? Auto: Nucleotides
h          Codon positions selected? Use all positions
m          Model of substitution? HKY (Hasegawa et al., 1985)
t   Transition/transversion parameter? Estimate from data set
f          Nucleotide frequencies? Estimate from data set
RATE HETEROGENEITY
w        Model of rate heterogeneity? Gamma distributed rates
a   Gamma distribution parameter alpha? Estimate from data set
c        Number of Gamma rate categories? 4
Quit [q], confirm [y], or change [menu] settings: y
```

After typing "y", TREE-PUZZLE will compute the maximum-likelihood values for all trees in hivALN.15trees, output the site-likelihood values to a file called hivALN.15trees.sitelh, and then proceeds with the implemented KH and SH tests and the ELW method, indicated by

```
[...]
Computing maximum likelihood branch lengths (without clock) for tree # 1
Computing maximum likelihood branch lengths (without clock) for tree # 2
Computing maximum likelihood branch lengths (without clock) for tree # 3
Computing maximum likelihood branch lengths (without clock) for tree # 4
Computing maximum likelihood branch lengths (without clock) for tree # 5
Computing maximum likelihood branch lengths (without clock) for tree # 6
Computing maximum likelihood branch lengths (without clock) for tree # 7
Computing maximum likelihood branch lengths (without clock) for tree # 8
Computing maximum likelihood branch lengths (without clock) for tree # 9
Computing maximum likelihood branch lengths (without clock) for tree # 10
Computing maximum likelihood branch lengths (without clock) for tree # 11
Computing maximum likelihood branch lengths (without clock) for tree # 12
Computing maximum likelihood branch lengths (without clock) for tree # 13
Computing maximum likelihood branch lengths (without clock) for tree # 14
Computing maximum likelihood branch lengths (without clock) for tree # 15
Performing single sided KH test.
```

```
Performing ELW test.
Performing SH test.


All results written to disk:
   Puzzle report file: hivALN.15trees.puzzle
   Likelihood distances: hivALN.15trees.dist
   Phylip tree file: hivALN.15trees.tree
   Site likelihoods (PHYLIP): hivALN.15trees.sitelh


The parameter estimation took 9.00 seconds (= 0.15 minutes = 0.00 hours)
The computation took 28.00 seconds (= 0.47 minutes = 0.01 hours)
      including input 28.00 seconds (= 0.47 minutes = 0.01 hours)
```

Note, that when analyzing tree files, the tree file's name hivALN.15trees is used as prefix for the output files. Besides other information, the resulting output file hivALN.15trees.puzzle shows the results of the ML analysis and the test results for the different topologies at the end of the file:

```
COMPARISON OF USER TREES (NO CLOCK)
```

Tree	log L	difference	S.E.	p-1sKH	p-SH	c-ELW	2sKH
1	-17405.05	12.13	9.0392	0.0960 +	0.1870 +	0.0051 -	+
2	-17405.90	12.99	8.9989	0.0780 +	0.1760 +	0.0027 -	+
3	-17395.02	2.11	3.4895	0.2600 +	0.7860 +	0.1147 +	+
4	-17401.24	8.33	8.2551	0.1580 +	0.3830 +	0.0704 +	+
5	-17404.03	11.12	7.4308	0.0720 +	0.2290 +	0.0097 -	+
6	-17392.91	0.00	<---- best	1.0000 +	1.0000 +	0.4437 +	best
7	-17401.49	8.58	9.7675	0.1780 +	0.3760 +	0.0587 +	+
8	-17396.14	3.22	6.8145	0.3160 +	0.7170 +	0.1770 +	+
9	-17401.98	9.07	9.7895	0.1700 +	0.3400 +	0.0460 +	+
10	-17408.52	15.61	8.3014	0.0380 -	0.0780 +	0.0003 -	+
11	-17399.72	6.81	5.7552	0.1170 +	0.4840 +	0.0085 -	+
12	-17408.66	15.75	8.3151	0.0250 -	0.0740 +	0.0003 -	+
13	-17396.12	3.21	2.9334	0.1280 +	0.6930 +	0.0580 +	+
14	-17405.43	12.52	8.9404	0.0670 +	0.1910 +	0.0042 -	+
15	-17408.24	15.33	8.2263	0.0370 -	0.0860 +	0.0009 -	+

This table contains the trees in the same order as in the tree file. The columns show the index of the tree, the log-likelihoods, the log-likelihood differences δ_x to the best tree, their standard error, the p-values of the one-sided Kishino–Hasegawa test (p-1sKH), the Shimodaira–Hasegawa test (p-SH) and the expected likelihood weights (c-ELW). The results of the two-sided KH test (2sKH) are only included for historical reason and should not be used. Due to the random generation of the bootstrap samples the p-values might differ slightly between repeated analyses. The

"+" signs after the values indicate whether a method has chosen the corresponding tree for the confidence set or (equivalently) where the null hypothesis could not be rejected.

We already see that the three relevant columns show different sizes of their confidence sets (SH > KH > ELW), which reflects the different levels of conservativeness. The Shimodaira–Hasegawa test has even included all 15 trees; that means it could not reject a single null hypothesis.

12.9.2 Testing with CONSEL

The CONSEL program does not estimate the likelihoods for the trees itself. Hence we generated a site-likelihood file `hivALN.15trees.sitelh` with the `-wsl` option in TREE-PUZZLE in the previous section.

Note that CONSEL version 0.1i does have problems with filenames that contain more than one dot (" . "). In such cases CONSEL discards the last file extension, but fails to attach its own ones, which in this example would lead to the loss of the contents of `hivALN.15trees`, being overwritten by the CONSEL output.

Therefore, rename the site-likelihood file accordingly, say to `hivALN.sitelh`. CONSEL generates the multiscaled bootstrap samples from the TREE-PUZZLE `.sitelh` file in CONSEL's own native `.rmt` format by running:

```
makermt --puzzle hivALN.sitelh
```

While `makermt` performs the conversion and the bootstrapping, it indicates its progress:

```
# $Id: makermt.c,v 1.14 2005/09/20 07:58:03 shimo Exp $
# seed:0 (MT19937 generator)
# K:10
# R:0.5 0.6 0.7 0.8 0.9 1 1.1 1.2 1.3 1.4
# B:10000 10000 10000 10000 10000 10000 10000 10000 10000 10000
# reading hivALN.sitelh
# M:15 N:2352
# writing hivALN.vt
# writing hivALN.rmt
# start generating total 100000 replicates for 15 items.........
# time elapsed for bootstrap t=15.89 sec
# exit normally
```

K indicates the number of differently scaled bootstrap samples, R lists the different alignment length scale factors $\frac{N}{N_r}$, and B the numbers of bootstrap samples for each N_r (Section 12.5.6).

The command (omitting the `.rmt` extension)

```
consel hivALN
```

then performs the actual tests on the multiscaled bootstrap samples and writes the
p-values to file, indicating as it proceeds

```
# $Id: consel.c,v 1.19 2004/11/11 08:14:09 shimo Exp $
# reading hivALN.rmt.........
# K:10
# R:0.5 0.599915 0.69983 0.799745 0.89966 1 1.09991 1.19983 1.29974 1.39966
# B:10000 10000 10000 10000 10000 10000 10000 10000 10000 10000
# M:15
# generate the identity association
# CM:15
# MC-TEST STARTS
# centering the replicates
# calculating kh-pvalue...............
# calculating mc-pvalue...............
# calculating the variances...............
# calculating weighted kh-pvalue...............
# calculating weighted mc-pvalue...............
# MC-TEST DONE
# calculate replicates of the statistics..........
# BP-TEST STARTS - DONE
# AU-TEST STARTS
# sorting the replicates..........
# calculating approximately unbiased p-values by MLE (fast) fit-
ting...............
# time elapsed for AU test is t=0.05 sec
# ALPHA:0.05 0.1 0.5 0.9 0.95
# calculating confidence intervals...............
# AU-TEST DONE
# writing hivALN.pv
# writing hivALN.ci
# exit normally
```

To extract the resulting *p*-values from Consel's data files, we run the
command

```
catpv -s 1 hivALN
```

This generates a table of *p*-values in the following format:

```
# rank item obs     au      np |   bp     pp      kh     sh      wkh     wsh |
#   1   1   12.1   0.030   0.001 |  0.001   4e-06  0.094  0.193  0.094   0.419 |
#   2   2   13.0   0.044   0.001 |  0.001   2e-06  0.076  0.159  0.076   0.396 |
#   3   3    2.1   0.433   0.095 |  0.097   0.101  0.272  0.793  0.272   0.768 |
#   4   4    8.3   0.235   0.060 |  0.062   2e-04  0.164  0.384  0.164   0.520 |
```

```
#   5   5   11.1   0.072   0.009 |   0.009   1e-05   0.072   0.227   0.072   0.322 |
#   6   6   -2.1   0.826   0.484 |   0.475   0.831   0.728   0.960   0.681   0.961 |
#   7   7    8.6   0.281   0.066 |   0.063   2e-04   0.190   0.370   0.190   0.578 |
#   8   8    3.2   0.475   0.184 |   0.191   0.033   0.319   0.721   0.319   0.745 |
#   9   9    9.1   0.272   0.070 |   0.067   1e-04   0.175   0.339   0.175   0.551 |
#  10  10   15.6   0.005   5e-05 |   2e-04   1e-07   0.034   0.072   0.034   0.216 |
#  11  11    6.8   0.042   0.003 |   0.004   0.001   0.121   0.487   0.089   0.380 |
#  12  12   15.7   0.040   0.002 |   1e-04   1e-07   0.034   0.064   0.034   0.225 |
#  13  13    3.2   0.141   0.026 |   0.025   0.034   0.135   0.695   0.135   0.492 |
#  14  14   12.5   0.074   0.006 |   0.005   3e-06   0.082   0.177   0.082   0.393 |
#  15  15   15.3   0.033   0.001 |   3e-04   2e-07   0.037   0.075   0.037   0.227 |
```

The -s 1 option causes the trees to be output in the order of the tree file instead of their likelihood (the default). The output contains the index (rank, which would differ when ordered by likelihood), the tree number in the input (item), the log-likelihood difference δ_x to the best tree (obs), except for the best itself, which shows the negative distance of the second best. Among other statistics and information values it prints the p-values of the AU test (au), KH test (kh), the SH test (sh), and weighted variants of the two (wkh and wsh). Due to the random generation of the bootstrap samples the p-values might differ slightly between repeated analyses. The confidence sets based on a significance level $\alpha = 0.05$ are the largest for the SH test(s) and their size is decreasing via the KH test(s) down to the AU test. While the p-values sometimes differ substantially between the weighted and unweighted versions of the SH and KH tests, the confidence sets remain the same.

12.10 Conclusions

Comparing the results of TREE-PUZZLE and the CONSEL program which are collected in Table 12.1, one can easily see that, although the p-values of the tests computed by both programs differ slightly, the resulting confidence sets are the same. The differences of the p-values can be attributed to two reasons. First, the stochastic nature of the bootstrap samples causes fluctuations in the estimated values. Furthermore, rounding errors are introduced during the export of the site-log-likelihoods. Only TREE-PUZZLE itself has the advantage of the full precision, since the export to a text file for CONSEL does automatically decrease the precision of exported values.

Note that, although the 15 trees were constructed manually, their selection was still *a priori*, since we know that tree 6 was the reconstructed ML tree (see Chapter 6). That tree remained the one with highest likelihood even among the constructed trees in the analysis.

The KH test has been performed as a one-sided test, but still the assumptions implied allow one to use this test only with caution. The SH tests nicely

Table 12.1 *p*-values estimated with Tree-Puzzle and Consel

			Tree-Puzzle			Consel			
	l	*δ*	KH	SH	ELW	AU	KH	SH	WKH	WSH
1	−17405.05	12.13	0.096 +	0.187 +	0.0051	0.030	0.094 +	0.193 +	0.094 +	0.419 +
2	−17405.90	12.99	0.078 +	0.176 +	0.0027	0.044	0.076 +	0.159 +	0.076 +	0.396 +
3	−17395.02	2.11	0.260 +	0.786 +	0.1147 +	0.433 +	0.272 +	0.793 +	0.272 +	0.768 +
4	−17401.24	8.33	0.158 +	0.383 +	0.0704 +	0.235 +	0.164 +	0.384 +	0.164 +	0.520 +
5	−17404.03	11.12	0.072 +	0.229 +	0.0097	0.072 +	0.072 +	0.227 +	0.072 +	0.322 +
6	−17392.91	0.00	1.000 +	1.000 +	0.4437 +	0.826 +	0.728 +	0.960 +	0.681 +	0.961 +
7	−17401.49	8.58	0.178 +	0.376 +	0.0587 +	0.281 +	0.190 +	0.370 +	0.190 +	0.578 +
8	−17396.14	3.22	0.316 +	0.717 +	0.1770 +	0.475 +	0.319 +	0.721 +	0.319 +	0.745 +
9	−17401.98	9.07	0.170 +	0.340 +	0.0460 +	0.272 +	0.175 +	0.339 +	0.175 +	0.551 +
10	−17408.52	15.61	0.038	0.078 +	0.0003	0.005	0.034	0.072 +	0.034	0.216 +
11	−17399.72	6.81	0.117 +	0.484 +	0.0085	0.042	0.121 +	0.487 +	0.089 +	0.380 +
12	−17408.66	15.75	0.025	0.074 +	0.0003	0.040	0.034	0.064 +	0.034	0.225 +
13	−17396.12	3.21	0.128 +	0.693 +	0.0580 +	0.141 +	0.135 +	0.695 +	0.135 +	0.492 +
14	−17405.43	12.52	0.067 +	0.191 +	0.0042	0.074 +	0.082 +	0.177 +	0.082 +	0.393 +
15	−17408.24	15.33	0.037	0.086 +	0.0009	0.033	0.037	0.075 +	0.037	0.227 +

The "+" signs mean that a tree belongs to a confidence set or the null hypothesis could not be rejected.

demonstrated its biasedness being the only test (together with its weighted variant WSH) not rejecting any of the trees.

The ELW and the AU test show very similar confidence sets, but the AU test is a bit more conservative. Still one has to keep in mind that the impact of model mis-specification on ELW remains unclear.

The interpretation of these results are, that there is no tree to be clearly chosen over the others. But also the topological features of the accepted or rejected trees do not show an obvious trend. It seems clear, however, that the subtype C is not located close to the root, since all such trees (trees 10–12 in Fig. 12.6) have been rejected by ELW and the approximately unbiased AU test.

Furthermore, all those topologies are in the confidence sets of AU and ELW, where subtype A and G are located closely in the tree (Fig. 12.6, trees 3, 6, 8, 9, and 13) except if subtype C is close to the root (trees 11 and 12). This might reflect the possible recombinant history of subtype G with subtype A being one of its parents as reported by Abecasis *et al.* (2007).

The tests showed that a number of rather different tree topologies could not be rejected. Other subsequent analyses have to be performed to clarify the possible influence of inter-subtype recombination (e.g. Chapter 16) or of other phenomena like noise, lateral gene transfer, or directed selection which can also obscure the true phylogenetic history.

Section V

Molecular adaptation

## 13

Natural selection and adaptation of molecular sequences

Oliver G. Pybus and Beth Shapiro

Adaptation by natural selection is the most important process in biology – nothing else can explain the staggering complexity and diversity of organisms, cells, enzymes, and proteins. From the skeleton of a blue whale to the 100 million times smaller flagellum of a bacterium, all living structures result from the repeated fixation and elimination of genetic variants within populations. It should come as no surprise, then, that a key goal of modern evolutionary biology is the identification of genes or genome regions that have been targeted by natural selection, thereby pinpointing the genotypic variation that causes some individuals to live longer and reproduce faster than others. Great efforts have been made over recent decades to understand the "molecular footprint" left in nucleotide and protein sequences by the past action of natural selection, efforts that have resulted in a substantial framework of theory and an impressive array of statistical methods, of which we give only the barest introduction in this chapter. We begin by considering the dynamics through time of genetic variants within a population, then review the statistical approaches most commonly used to identify natural selection at the molecular level.

13.1 Basic concepts

To illustrate the essential concepts of molecular adaptation, it is convenient to consider the fate of a new genetic variant (*mutant*) present in a single individual belonging to a population whose other members are genetically homogeneous and carry a "*wild-type*" genotype. This approach is, of course, a deliberate oversimplification, but it does allow us to consider independently the different evolutionary forces that affect the frequency of the mutant in the population: *mutation*, *natural selection*, and **genetic drift**. Although mutation is the ultimate source of all genetic

The Phylogenetic Handbook: a Practical Approach to Phylogenetic Analysis and Hypothesis Testing,
Philippe Lemey, Marco Salemi, and Anne-Mieke Vandamme (eds.). Published by Cambridge
University Press. © Cambridge University Press 2009.

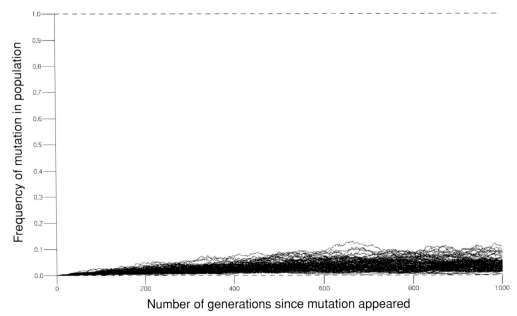

Fig. 13.1   Rapid mutation and no selection in a large population. Mutation, by itself, is a very slow
and inefficient process for changing mutation frequencies. The plot shows the frequency of
100 mutations through time. The x-axis represents the number of generations elapsed since
each mutation first appeared in a single individual. The y-axis represents the frequency of
the mutation in the population since then, with the dashed line denoting fixation. Because
selection is absent and genetic drift is very weak, the mutation frequency remains low, even
after 1000 generations (compare this with the timescale used in Fig. 13.2). The simulated
populations contained 10 000 individuals and the mutation rate was 0.00001 per individual
per generation. The simulations were performed using the program PopG, available from
http://evolution.gs.washington.edu/popgen/popg.html.

variation, it is by far the weakest of these evolutionary forces, and by itself cannot
rapidly change the frequency of the mutant in the population (see Fig. 13.1). This
is because the only way for the mutation to increase in frequency is for it to be
acquired independently over and over again, a very slow process even if mutation is
frequent and one-way (i.e. from wild type to mutant but not vice versa). Observed
differences among gene or protein sequences are therefore unlikely to result from
mutation alone, although it may be more important in organisms that have exceptionally high rates of mutation, such as RNA viruses. Putting mutation aside, we
are left with two processes capable of rapidly changing mutant frequency – natural
selection and genetic drift.

Mutation generates multiple genetic variants in a population, and the existence
of such variability is termed **polymorphism**. However, genetic polymorphisms do
not persist indefinitely and therefore a new mutation must eventually succumb to

one of two evolutionary fates: it either increases in frequency, eventually going to *fixation*, or it is *eliminated* from the population. The probability that a recently generated mutation is fixed or eliminated depends on two factors: (i) the degree to which the mutation increases or decreases an individual's ability to survive and reproduce in the current environment (i.e. *fitness*), and (ii) the size of the population concerned, typically denoted *N*. *Beneficial* or *advantageous* mutations are those that increase an individual's fitness relative to the wild-type, whilst *deleterious* mutations decrease relative fitness, and *neutral* mutations have no appreciable fitness effect. In population genetics this property of a mutation is quantified as *s*, the **selection coefficient**, which is positive for beneficial mutations, zero for neutral mutations, and negative for deleterious mutations. The vast majority of mutations are deleterious to some extent, and through careful experimentation it has been possible to estimate selection coefficients for some organisms (e.g. Crill *et al.*, 2000; Sanjuán *et al.*, 2004) . Although the recent generation of huge genomic data sets has led to improvements in the estimation of mutational fitness effects (e.g. Eyre-Walker *et al.*, 2006; Nielsen *et al.*, 2005), it is generally unknown whether the remaining non-deleterious mutations are mostly neutral or beneficial, and so far it has proved difficult to directly estimate selection coefficients from sequence data.

The effect of selection is to increase the frequency of a beneficial mutation until it becomes fixed in the population (**positive selection**), or to decrease the frequency of a deleterious mutation until it is eliminated (**negative selection**). Selection does not affect the frequency of neutral mutations. The greater the selection coefficient of a mutation – either positive or negative – the more rapidly the mutation frequency will change (see Fig. 13.2). In contrast, the effect of genetic drift is to cause the frequencies of all types of mutations to fluctuate randomly through time, with no net tendency towards increase or decrease, until the mutations become either fixed or eliminated. The size of the random fluctuations is greater if *N* is small and lesser if *N* is large (Fig. 13.2). Genetic drift arises because, in the absence of selection, each member of a generation is equally likely to be the parent of any member of the subsequent generation, and the random sampling effect this creates is greater if the number of potential parents involved is smaller. Although these fluctuations are very small in large populations, they are still of critical importance, because new mutations begin at very low frequencies (i.e. in single individuals) and are therefore highly susceptible to elimination by the tiniest of random effects. Hence most mutations, whether beneficial, neutral or deleterious, are swiftly lost in populations large or small (Fig. 13.2). Another important consequence of drift is that it takes much longer for a newly generated mutation to reach fixation by genetic drift than by selection; in the simplest of cases, it takes on average $4N$ generations for a neutral mutation to be fixed via drift in a **diploid** population (or $2N$ generations in a population of **haploid** individuals).

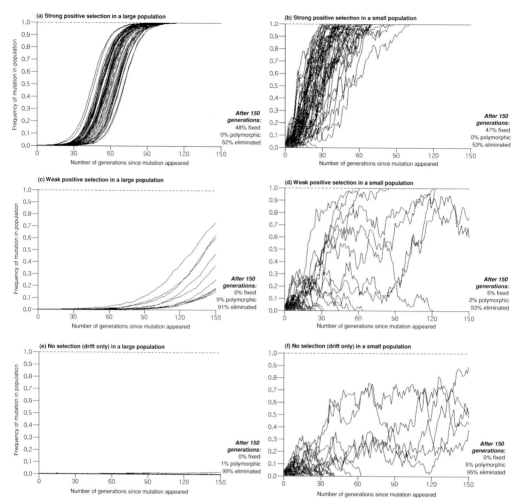

Fig. 13.2   The effects of positive selection and genetic drift on the frequency of mutations through time. As in Fig. 13.1, the x-axis represents the number of generations that have elapsed since each mutation initially arose in one individual in the population. The left-hand plots show the trajectories of 100 mutations in large populations (10 000 individuals) and the right-hand plots show the equivalent trajectories for small populations (50 individuals). Note the much greater random fluctuations that occur in the small populations. Plots (a) and (b) show the rapid fixation of strongly beneficial mutations in less than 150 generations. Plots (c) and (d) show the relatively slower process of fixation for mutations that are only slightly advantageous. For neutral mutations, fixation is very rare in large populations (e) and much slower than selection in small populations (f). In all cases, note that very many mutations are immediately eliminated very soon after they first appear. For more details, see the Fig. 13.1 legend.

In real populations, mutation frequencies are affected by natural selection and genetic drift at the same time, so that the evolutionary behavior of each mutation will depend on the product $2Ns$ (or Ns for haploids). If $2Ns \gg 1$ or $2Ns \ll -1$, then a mutation will act as if it were under strong selection and will be rapidly fixed or quickly eliminated, even if the relative fitness benefit or cost to the individual is small (i.e. s is close to zero). However, if $-1 < 2Ns < 1$, then the mutant will fluctuate randomly in frequency as if natural selection were absent, even if the fitness effect of the mutation is quite significant (i.e. s is not close to zero). Although these conclusions are based on population genetic models that make somewhat simplistic assumptions, they do provide a useful framework for understanding the complex dynamics of genes in real biological populations. For example, individual mutations do not evolve in isolation, but in the context of a chromosome or genome that likely contains many other sites under positive or negative selection. The presence of genetically-linked selected sites has the effect of generating evolutionary "interference," typically reducing the overall speed and efficiency of natural selection. However, a significant amount of recombination among loci reduces this interference and increases selection's potency (Hill & Robertson, 1966). Adding these interactions necessarily complicates the simple models introduced above (for further exploration of these effects see, for example, Birky & Walsh, 1988; Gerrish & Lenski, 1998; Miralles *et al.*, 1999; McVean & Charlesworth, 2000; Comeron & Guthrie, 2005; Desai *et al.*, 2007).

In addition to positive and negative selection, ***balancing selection*** also operates in real populations and acts to maintain multiple genetic variants in a population for very long periods of time (for review, see Hedrick, 2007). In diploids, balancing selection can be caused by ***overdominance***, which occurs when the heterozygote at a particular locus has greater fitness than both the homozygotes, thereby maintaining both alleles. Alternatively, both haploids and diploids may display ***frequency-dependent selection***, another form of balancing selection that occurs when a rare variant has greater fitness than a more common one. Frequency-dependent selection may be a common consequence of immune responses to infectious disease; for example, immune "escape" mutants are repeatedly selected during chronic viral infections (e.g. Frost *et al.*, 2005; Holmes *et al.*, 1992; Simmonds *et al.*, 1991; Wei *et al.*, 2003; Williamson, 2003; Yamaguchi & Gojobori, 1997). Lastly, if a population's environment fluctuates sufficiently frequently then multiple variants may be maintained because no single advantageous mutation has sufficient time to reach fixation before the environment within which it is beneficial changes, a process sometimes called ***fluctuating selection***. Spatial heterogeneity is one such cause of environmental variability, for example, when viruses infect multiple tissue types (e.g. Cuevas *et al.*, 2003).

13.2 The molecular footprint of selection

To detect natural selection and adaptation from molecular sequences we must first characterize the ways in which sequences differ. If each sequence in an alignment is representative of a different species or population, then the observed differences signify past fixation events, whereas differences among sequences sampled from individuals within a population represent current polymorphisms. Of course, both fixation and polymorphism are driven by the same causal processes – selection and genetic drift – but the distinction is important because the statistical analyses introduced in this book differ in their interpretation when applied to different types of data. Fixed changes will most likely be neutral or beneficial, whereas polymorphisms may be neutral, beneficial, or slightly deleterious. (Lethal or strongly deleterious changes are effectively never observed in natural populations, for obvious reasons.) The discovery in the twentieth century of copious variability among molecular sequences initiated a protracted debate, between those who considered the vast majority of polymorphisms to be under selection (e.g. Gillespie, 1991; Ohta, 1973, 1992, 2000, 2002) and those who ascribed their existence to genetic drift (e.g. Kimura, 1968, 1969, 1983; Kimura & Ohta, 1971; King & Jukes, 1969). The debate is now largely of historical interest (see Ohta & Gillespie, 1996 for detailed discussion), partly because the sequence analysis methods now available provide a way to evaluate the relative contribution of selection and drift for any data set of interest.

Molecular sequences contain three different types of information that can be used, separately or in combination, to infer the past action of selection.

(1) The frequency of observed polymorphisms

> The proportion of the population in which a mutation is found is its *frequency*, or **site-frequency**, and depends on the action of selection and drift at that site. For example, deleterious mutations are more likely to be found at low frequencies because they are typically removed by negative selection before they become common. Mutations that have become fixed are much more likely to represent neutral or beneficial changes.

(2) The relative rate of silent and replacement fixations

> *Replacement* (or **non-synonymous**) nucleotide changes are those that alter the encoded amino acid, whereas *silent* (or **synonymous**) mutations leave the amino acid unchanged due to the redundancy inherent in the genetic code. A greater rate of fixation for replacement changes, relative to the rate of silent change, can only be explained by the action of positive selection, given certain assumptions.

(3) Differences in genetic variation among genomic loci or among populations

Fixation of a mutation by positive selection not only leads to the loss of genetic variation at the selected locus, but also at genetically linked loci that may be evolving neutrally. Hence the pattern of genetic variability among genomic loci can be used to infer selection in a recombining population (see also Chapter 15). Similarly, differences in genetic variation at the same locus in different populations can also indicate the action of natural selection.

We can categorize methods for detecting selection into three groups, which approximately follow the three types of information above: **summary statistic methods**, which primarily use source (1), d_N/d_S **methods** that concentrate on source (2), and **comparative methods** that use information type (3), often in combination with (1) or (2). These labels are somewhat arbitrary and are used here to frame discussion rather than to provide a definitive classification – some methods utilize more than one source of information and these combined methods are likely to become increasingly common in the future. In the remainder of this chapter we introduce those approaches that can be applied to sequence alignments that represent single genomic regions within a single population, namely *summary statistic methods* and *d_N/d_S methods*. Comparative approaches, such as the *extended haplotype test* (Sabeti *et al.*, 2002), often extend these basic methods, either to multiple loci within a genome (e.g. Bakker *et al.*, 2006; Bustamante *et al.*, 2005; Sabeti *et al.*, 2006), or to multiple populations (e.g. Bersaglieri *et al.*, 2004; Evans *et al.*, 2006; Sabeti *et al.*, 2002; Tishkoff *et al.*, 2007; Toomajian & Kreitman, 2002; Voight *et al.*, 2006).

13.2.1 Summary statistic methods

The simplest way to investigate selection using sequences is to calculate a *summary statistic* from an alignment of sequences. Typically, the sequences represent different individuals from a single population, but sometimes represent multiple individuals from two closely related species. In order to detect the presence of selection, statistics that summarize the relative frequency of polymorphic sites are calculated from the observed alignment. These statistics are then compared against the values expected to occur under a "null model" of neutral evolution. If the observed statistics are significantly different from their expected values, then the neutral null model can be rejected. In addition to assuming that all mutations are either lethal (and therefore unobserved) or neutral, the null "standard neutral model" also assumes a constant *effective population size*, a constant mutation rate, no recombination, no migration, random mating, and lastly, that each mutation occurs at a different nucleotide site (this is the so-called "*infinite sites*" assumption).

The advantage of summary statistics lies in their simplicity; however, a rejection of the null "neutral model" does not necessarily denote the action of selection, because rejection can be caused by violation of any one of the model assumptions listed above.

One of the first and most commonly used summary statistics is **Tajima's D** (Tajima, 1989), which for n observed sequences is defined as:

$$D = \frac{\pi - \left(S/k\right)}{\sqrt{V}}, \quad \text{where} \quad k = \sum_{i=1}^{n-1} 1/i \tag{13.1}$$

π equals the average number of sites that differ between each pair of sequences and S equals the number of polymorphic or **segregating** sites in the alignment. V is a bit of mathematical jiggery-pokery representing variance, and is less relevant here than the equation's top line. Under the null "neutral model" both population diversity (S) and pairwise divergence (π) are generated by the same process – genetic drift – and as a result D is expected to equal zero if all the assumptions of the neutral model are met. Because S represents the number of polymorphic sites, but ignores their frequency in the population, it is sensitive to the existence of rare, low-frequency variants. Conversely, π is affected more by the existence of intermediate-frequency variants, because these generate differences between more pairs of sequences in the sample. Tajima's D therefore summarizes the distribution of site frequencies of polymorphic sites, with significantly negative D values resulting from an excess of rare variants, and significantly positive values reflecting an excess of intermediate frequency sites. As a rough rule-of-thumb, negative D values might signify a recent bout of positive selection or a rapidly growing population, whereas positive D values might signify the presence of population structure, balancing selection, or a population declining in size.

Other test statistics closely related to Tajima's D have been developed that concentrate on slightly different properties of polymorphic sites. For example, Fu and Li's D^* and F^* (1993) statistics compare π and S against the number of "*singletons*" (polymorphisms found only in one sequence) in an alignment. In contrast, the related H statistic introduced by Fay and Wu (2000) considers the abundance of very high-frequency variants relative to that of intermediate-frequency variants, and is thought to be more powerful in detecting the signature of genetic hitchhiking associated with a selective sweep. The H statistic uses an "**outgroup**" sequence from a closely related population or species to identify sites that have become fixed in the main study population.

When using Tajima's D or related statistics to test for the presence of selection, it is vital to remember that they all test whether the distribution of polymorphism is different from that expected under the null "neutral model". Deviation from this model could be due to natural selection, but this is by no means the only assumption of the neutral model. For example, population structure or changes in population size will also affect the distribution of polymorphisms in a population, and considerable further work must be done to rule out these alternative explanations before selection becomes the most likely or sole remaining explanation. Furthermore, summary statistic methods assume that the sampled sequences have been taken randomly from the population under investigation. The d_N/d_S methods described in the following section do not require this assumption.

13.2.2 d_N/d_S methods

Mutations in protein coding sequences can be classified as either synonymous (silent) or non-synonymous (replacement). Due to the inherent redundancy of the genetic code, silent changes do not alter the amino acid that they encode, whereas replacement changes do (Li et al., 1985; Miyata & Yasunaga, 1980; Nei & Gojobori, 1986). If we assume that selection acts less strongly on silent mutations, then observed differences between the patterns of silent and replacement changes should reflect the action of natural selection.

The application of d_N/d_S methods depends on the manner by which the sequence data was collected. If each sequence is from a different species or population, then we calculate the ratio of the rate of replacement fixations to the rate of silent fixations (denoted d_N/d_S or ω). If all replacement fixations are neutral then, by definition, the ratio of the two rates must equal one and hence a statistically significant deviation from $d_N/d_S = 1$ signifies selection. Note that d_N and d_S are calculated per replacement or silent site, respectively, thereby taking into account the fact that random mutations generate more replacement changes than silent changes due to the structure of the genetic code. If $d_N/d_S < 1$, then the rate of replacement fixation is slower, indicating that negative selection has acted more strongly on replacement changes. If $d_N/d_S \approx 0$, then the amino acid sequence encoded must be under exceptional selective constraint. Crucially, if d_N/d_S is significantly greater than one, then positive selection must have occurred, because no other obvious process can result in a faster fixation rate for replacement changes.

The modern interpretation of d_N/d_S is that it reflects the *difference* in the selective forces acting on silent versus replacement changes; that is, it is not strictly necessary to assume that all silent changes are selectively neutral. Silent changes in coding sequences may be selected because the sequence contains overlapping reading frames or contains functional secondary DNA or RNA structure, or because there is a fitness advantage in using some codons in preference to others. Although these factors should always be borne in mind, it is often reasonable to make the simplifying assumption that all silent mutations are effectively neutral. Users of d_N/d_S methods should avoid making the common mistake of equating the absolute value of d_N/d_S with the "strength" of selection or the average selection coefficient of replacement fixations. As an example, consider a codon within which an advantageous replacement mutation arose and rapidly got fixed some time in the past – a codon that could not be bettered by any subsequent mutation. After a sufficient period of evolution, several silent/neutral changes will have accumulated within the selected codon, leading to an observed d_N/d_S close to zero! Hence high d_N/d_S values most commonly occur when genes undergo recurrent rounds of selection, rather than single selective sweeps. As a result d_N/d_S methods are most successful in detecting adaptation when they are applied to genes that are under antagonistic co-evolution, such as that generated by sexual conflict, predator–prey interactions, or host–parasite interactions (see Nielsen & Yang, 2003; Yang & Bielawski, 2000). In summary, the general feature of d_N/d_S ratio methods is that they are statistically weak and will fail to detect many instances of selection (Endo *et al.*, 1996; Sharp, 1997; Wong *et al.*, 2004). However, unlike the summary statistics introduced above, d_N/d_S methods do not require us to make strong assumptions about the sampled population and are therefore considered to be more robust. This trade-off between "power" and "robustness" is nothing special and is common to many problems in statistics.

Later chapters will describe in detail how d_N/d_S ratios are computed and how significant deviations from $d_N/d_S = 1$ are statistically tested. The first d_N/d_S methods calculated d_N and d_S for each pair of sequences in a sample and averaged the resulting ratio across all codons in the alignment (e.g. Ina, 1995; Li, 1993; Li *et al.*, 1985; Miyata & Yasunaga, 1980; Nei & Gojobori, 1986; Pamilo & Bianchi, 1993). More recent approaches explicitly incorporate the phylogenetic history of the sampled sequences and also use statistical models of the process of codon substitution (see Chapter 14), analogous to the models of nucleotide substitution introduced earlier in Chapter 4. More recent methods are also capable of calculating d_N/d_S separately for each codon, or for any defined subset of codons within an alignment, enabling the precise location of molecular adaptation within a gene to be identified. Modern phylogenetic approaches are also capable of estimating d_N/d_S separately for different clades or sets of branches, allowing us to investigate within which common ancestors adaptation actually occurred (for a recent detailed review, see Yang, 2006).

If sequences have been sampled from different individuals within a population, then d_N/d_S reflects the ratio of replacement to silent polymorphism in the population. As before, if positive and negative selection are absent, then this ratio should equal one, and statistically significant deviations indicate the action of natural selection. The observation that, for neutrally evolving sequences, the ratio of silent to replacement polymorphism equals the ratio of silent to replacement fixation (Maynard-Smith, 1970), has been used as a basis of an entire class of selection-detection methods. The simplest such method, called the **McDonald–Kreitman test** (McDonald & Kreitman, 1991), uses an outgroup sequence to distinguish fixations from polymorphisms, then tests for a significant difference between the d_N/d_S ratios of polymorphic and fixed sites. Such methods combine information on polymorphism frequency with information on silent/replacement ratios, and have been refined and extended in several ways (e.g. Andolfatto, 2005; Bustamante et al., 2002; Rand & Kann, 1996; Sawyer & Hartl, 1992; Smith & Eyre-Walker, 2002; Williamson et al., 2004). For example, if sequences from a rapidly evolving pathogen population have been sampled at different timepoints, then earlier sequences can provide an "outgroup" to later sequences, allowing us to directly observe fixation events between timepoints (e.g. Williamson, 2003). The phylogenetic equivalent of the McDonald–Kreitman approach is to partition phylogenetic branches into different sets (e.g. intra-species branches versus inter-species branches, or internal branches versus external branches) and then calculate d_N/d_S separately for each set (e.g. Hasegawa et al., 1998; Pybus et al., 2007).

The rationale that lies at the heart of all d_N/d_S methods is that there exists a set of sites that evolves more neutrally than another set of sites, specifically the set of silent sites and the set of replacement sites. This argument can, of course, be extended to any two sets of sites that have the same properties, for example, sites in exons vs. sites in introns (e.g. Hughes & Yeager, 1997; Metz et al., 1998), or sites in coding regions vs. those in duplicate genes (e.g. Castillo-Davis et al., 2004), related pseudogenes (e.g. Bustamante et al., 2002; Li et al., 1981), or other non-coding regions (e.g. Wolfe et al., 1987; Wong & Nielsen, 2004). By comparing such sets of sites, it becomes possible to investigate the action of natural selection on non-coding regions, which are not amenable to analysis using standard d_N/d_S methods.

13.2.3 Codon volatility

Recently, Plotkin and co-workers introduced a method based on **codon volatility** to detect selection using only a single genome (Plotkin & Dushoff, 2003; Plotkin et al., 2004, 2006). This approach obviously presents a substantial departure from the sample-based methods described previously, in that it promises to detect molecular adaptation across entire genomes, whilst requiring far less data. *Codon volatility* represents the likelihood that a random point mutation in a codon will result in an amino acid change. The volatility of codon X is thus calculated as the number of other codons that are reachable from X by a single point mutation and that code for a different amino acid than X, divided by the total number of non-stop codons that are one nucleotide different to codon X. Selection is inferred when the average codon volatility for a given gene is high relative to that of other genes in the same genome. Plotkin et al. argue that, if the average volatility of codons in a sequence is high relative to that in other regions in the same genome, then there is an increased likelihood that the previous substitution was replacement.

Codon volatility makes some sense if we argue there are occasions when an individual's fitness may be increased by raising the phenotypic variability of off-spring (akin to the evolution of "mutator" bacterial strains), and that high codon volatility is one means to achieve such variability. However, it then follows that the empirical observation of high codon volatility can only be used to detect selection for "increased diversity," which is one phenotypic trait among the thousands that we may wish to investigate using sequence data. In addition, codon volatility has disappointingly been shown not to correlate with increased rates of replacement substitution, in both real (Friedman & Hughes, 2005; Pillai *et al.*, 2005) and sim-ulated data (Chen *et al.*, 2005; Dagan & Graur, 2005; Zhang, 2005), and instead correlates with nucleotide content and codon usage bias. As a result, sample-based analyses remain the most robust methods for detecting adaptive evolution in gene sequences.

13.3 Conclusion

In concluding this introduction, we stress the importance of fully understanding the assumptions and limitations of each method before interpreting analysis results. Summary statistic methods typically make strong assumptions about the nature of the sampled population and about the sample taken from it. Although d_N/d_S methods make fewer assumptions, they are typically statistically weaker and more indirect – the relationship between d_N/d_S values and the past action of positive and negative selection is more complex than many users realize, and absolute d_N/d_S values should not be equated directly with the "strength" of selection. Finally, if our aim is a comprehensive biological understanding of organismal evolution and adaptation, then sequence-based methods should form but one component of a larger research program. The detection of positive selection from gene sequences tells us nothing about *why* adaptation has taken place and *what* the nature of the fitness benefit to the organism was, nor *how* the new variant might interact epistatically with other genes or traits. These are questions that may require further functional experiments and investigations to be undertaken in order to be answered. We therefore urge you to consider the statement "neutrality was rejected ($p < 0.05$)" as a starting point, not a finishing line!

Estimating selection pressures on alignments of coding sequences

THEORY

Sergei L. Kosakovsky Pond, Art F. Y. Poon, and Simon D. W. Frost

14.1 Introduction

Understanding the selective pressures that have shaped genetic variation is a central goal in the study of evolutionary biology. As **non-synonymous** mutations can directly affect protein function, they are more likely to influence the **fitness** of an organism than mutations that leave the amino acid sequence unchanged (i.e. **synonymous** mutations). Under **negative** or **purifying selection**, less "fit" non-synonymous substitutions accumulate more slowly than synonymous substitutions, and under **diversifying** or **positive selection**, the converse is true. Therefore, an important concept in the analysis of coding sequences is that the comparison of relative rates of non-synonymous (β) and synonymous (α) substitutions can provide information on the type of selection that has acted on a given set of protein-coding sequences. The ratio $\omega = \beta/\alpha$ (also referred to as dN/dS or K_A/K_S) has become a standard measure of selective pressure (Nielsen & Yang, 1998); $\omega \approx 1$ signifies neutral evolution, $\omega < 1$ – negative selection and $\omega > 1$ – positive selection.

There are five fundamental questions which can be answered with existing methods and software tools that estimate such substitution rates.

- Is there evidence of selection operating on a gene? We can address this question by testing whether $\beta \neq \alpha$ for some regions in the sequence.

The Phylogenetic Handbook: a Practical Approach to Phylogenetic Analysis and Hypothesis Testing, Philippe Lemey, Marco Salemi, and Anne-Mieke Vandamme (eds.). Published by Cambridge University Press. © Cambridge University Press 2009.

- Where did selection happen? We can identify sequence regions or individual codons under selection, and determine the level of statistical significance.
- When did selection happen? We can estimate at what point in the evolutionary past (i.e. along which branch of the phylogenetic tree) non-neutral evolution occurred.
- What types of substitutions were selected for or against? We can classify amino acid substitutions (e.g. leucine↔valine) into those which were preferred and those which were suppressed.
- Are selective pressures different between genes/samples? Given two genes from the same set of taxa, or two samples of taxa covering the same gene, we can determine whether they evolved under similar or different selective pressures.

Our ability to address these questions depends on the accurate estimation of non-synonymous and synonymous rates, which was recently facilitated by the adoption of codon models of evolution within a phylogenetic *maximum likelihood* framework. We begin by providing a simple justification for why such complex models are necessary, despite the steep computational cost.

The unit of evolution

The structure of the genetic code forces realistic *evolutionary models* to consider triplets of nucleotides, i.e. codons, to be the basic unit of evolution. For example, the common assumption that the rates of evolution of the third codon position can serve as a proxy for synonymous substitution rates is a rather crude approximation. If x_i and y_i denote nucleotides in the ith position in a codon, then among all possible substitutions at the third codon position starting from codon $n_1 n_2 n_3$ and ending in codon $n_1 n_2 m_3$ ($n_3 \neq m_3$ and $n_1 n_2 m_3$ is not a stop codon), 50 are non-synonymous and 126 are synonymous, based on the universal genetic code[†]. More importantly, whether or not $x_3 \rightarrow y_3$ is a synonymous or a non-synonymous substitution, depends on the *context* of $n_1 n_2$. For example, $GG x_3 \rightarrow GG y_3$ is always synonymous, whereas $CA x_3 \rightarrow CA y_3$ is synonymous if $x_3 \rightarrow y_3$ is a *transition* ($A \leftrightarrow G$ or $C \leftrightarrow T$), and non-synonymous otherwise. Probabilistic codon substitution models (Muse & Gaut, 1994; Goldman & Yang, 1994) offer a natural and formal way to take into account such context dependency.

Estimating the neutral expectation

Before one can begin testing for selection, it is necessary to correctly assess what would happen under the neutral scenario. In the simplest case, one would like to know what proportion of random substitutions (measured per codon, to

---

[†] Run the HYPHY script CountSubstitutions.bf to tabulate various kinds of substitutions given a genetic code. You will need to download and install HYPHY (*http://www.hyphy.org/*), execute the Analysis>Standard Analyses menu option, choose Phylohandbook.bf from Miscella-neous rubrik and finally select the appropriate analysis from the list.

correct for the length of the alignment) would be synonymous and what proportion would be non-synonymous. Simply estimating the raw numbers of synonymous and non-synonymous substitutions and comparing them to detect selection will almost always fail because: (i) the structure of the genetic code is such that a larger proportion of random substitutions are non-synonymous rather than synonymous; (ii) some kinds of substitutions (e.g. transitions) are more frequent than others; and (iii) some codons are more frequent than others. The effect of each factor can be significant. For example, using the universal genetic code, assuming neutral evolution, equal codon frequencies (1/61 for each non-stop codon) and no biases in nucleotide substitution rates (i.e. the Jukes–Cantor model; Jukes & Cantor, 1969; see Chapter 4), 25.5% substitutions are expected to be synonymous and 74.5% non-synonymous (`NeutralExpectation.bf`). If transitions are assumed to happen at the rate five times that of *transversions* (such ratios are quite common in biological data), these numbers change to 30.9% and 69.1%, respectively. Furthermore, taking codon frequencies to be unequal (e.g. the distribution estimated from the human immunodeficiency virus type 1 [HIV-1] *pol* gene alters the counts to 27.1% and 72.9%. Hence, if one were to infer that a given sequence sample contained, on average, two non-synonymous and one synonymous substitutions per codon, one would have to account for all the factors influencing the neutral expectation before making a deduction about what selection mode – neutral, positive or negative – acted on the sample.

There are relatively sophisticated methods based on *codon distances*, which attempt to estimate the neutral expectation and *evolutionary rates* by comparing pairs of *homologous* sequences (Tzeng *et al.*, 2004), and in certain cases, these estimates can be quite accurate. However, there are several statistical issues inherent to all such methods (Muse, 1996), in particular they are difficult to generalize to comparing more than two sequences at a time. The use of codon substitution models can easily account for all the above confounding factors when estimating substitution rates, and can represent neutral evolution simply by setting the synonymous and non-synonymous rates to be the same ($\alpha = \beta$), or, equivalently, setting their ratio $\omega = \beta/\alpha$ to one.

Taking phylogenies into account

Genetic variation found in extant sequences is a combination of independent substitutions in different lineages and substitutions inherited by related sequences from a common ancestor. Correct estimation of substitution rates depends on our being able to separate these two effects. Consider a simple example shown in Fig. 14.1, where the same five extant codons at site 142 in Influenza A/H5N1 hemagglutinin are analyzed using two different phylogenies, one a maximum likelihood tree and the other a star topology, which is equivalent to the assumption

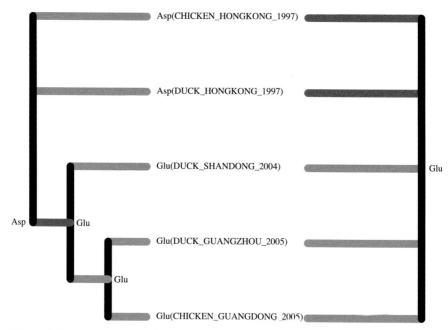

Fig. 14.1   Effect of phylogeny on estimating synonymous and non-synonymous substitution counts in a data set of Influenza A/H5N1 hemagglutinin sequences. Using the maximum likelihood tree on the left, the observed variation can be parsimoniously explained with one non-synonymous substitution, whereas the star tree on the right involves at least two.

made for naïve pairwise sequence comparison (e.g. no substitutions are shared by descent). Even for such a small sample, some of the conclusions (e.g. the minimal number of substitutions needed to explain the observed pattern) clearly may be affected by which phylogeny was used to relate the sequences. Without going into much detail, however, we note that, in practice, the inference of substitution rates using codon substitution models tends to be "robust" to some errors in the phylogenetic tree, i.e. so long as the tree is not "too wrong," rate estimates will not change much from tree to tree. As a simple illustration, EffectOfTopology.bf examines how the estimate of ω changes over all 15 possible trees relating the five influenza sequences (other small data sets can also be examined with this file). The *likelihood* scores differ by over 100 points between the best- and the worst-fitting trees, yet the ω estimate ranges only from 0.22 to 0.26. Applying grossly incorrect phylogenies (or assuming the lack of phylogenetic relatedness via pairwise comparisons) may have a much more profound effect on other types of inference, or larger data sets, as we will show later.

Different types of selection

The term "positive selection" encompasses several different evolutionary processes. It is critically important to distinguish between the two primary ones: ***directional*** and *diversifying selection*, because specific comparative methods must be used to identify each kind. Directional selection operating at a given position in a gene is manifested by concerted substitution towards a particular residue, which, given enough time, will result in a *selective sweep*, i.e. fixation of the new allele in the population. For example, when wildtype HIV-1 infects a number of different patients receiving the same antiretroviral drug, there will be strong selective pressure on the virus to independently acquire those mutations that confer drug resistance. For many early antiretroviral drugs, a single non-synonymous substitution was often sufficient for the acquisition of strong drug resistance, explaining why early HIV treatments were ineffectual (Frost *et al.*, 2000). If one were to sample these viruses after a complete selective sweep, when the virus population in each host has fixed the resistance mutation, there would be no remaining evidence of any selection having taken place. Diversifying selection, on the other hand, results from a selective regime whereby amino acid diversity at a given site is maintained in the population (Moore *et al.*, 2002). In HIV-1, this might occur at those codon positions that are the targets of host immune response. As immune systems in different hosts generally vary in their ability to recognize and target specific viral antigens, some viruses may be under selective pressure to evolve immune escape, while others may maintain wild-type residues.

In the rest of the chapter, we show how probabilistic models of codon substitution can be used to identify various types of selection pressures. While these methods are powerful and represent the current state-of-the-art in coding sequence analysis, it is important to realize that these models are able to recapitulate only some of the actual evolutionary processes that shape sequence evolution. There are many remaining assumptions and simplifications, which are often made to retain computational feasibility, and in certain cases the methods can be misled by recombination, small sample size, or biological processes not included in the model. We take great care to highlight such possible shortcomings, because it is our firm belief that the knowledge of a methods' limitations is as important as the knowledge of their power.

14.2 Prerequisites

In order to conduct an analysis of selection on a gene, one needs a multiple sequence alignment (see Chapter 3), and an underlying phylogenetic tree (Section III of this book), or in the case of recombination (Section VI), multiple phylogenetic trees (one for each non-recombinant segment).

When preparing alignments for codon analyses, one should ensure that the alignment process does not introduce frameshifts and preserves codons (i.e. by only inserting/deleting nucleotides in multiples of three). Hence it is advisable to carry out sequence alignment on *translated* protein sequences, and then map aligned residues to codons.

A number of algorithms have been proposed in order to estimate a phylogenetic tree, including distance-based methods such as ***neighbor-joining*** (Saitou & Nei, 1987), maximum likelihood-based methods (Felsenstein, 1981), and Bayesian approaches. Most of the time, rate estimates derived with a substitution model are robust to the details of the phylogenetic tree. An important exception to this occurs when recombinant sequences are present. Recombination has relatively little impact on estimates of global rates, but can have a profound effect on estimates of site-to-site and branch-to-branch variation in selection pressure. In order to accommodate recombination in a phylogenetic context, it is necessary to split the alignment into non-recombinant sequence fragments first. We have implemented a method called Genetic Algorithms for Recombination Detection (**GARD**; Kosakovsky Pond *et al.*, 2006; *http://www.datamonkey.org/GARD*), that uses a genetic algorithm to identify non-recombinant fragments; several other programs can be employed to do the same (see Chapters 15 and 16). Once these have been identified, selection analyses can be run separately on each fragment, or jointly by assuming that some parameters are shared between fragments.

14.3 Codon substitution models

The first tractable codon models were proposed independently by Goldman and Yang (1994) and Muse and Gaut (1994), and published in the same issue of *Molecular Biology and Evolution*. The process of substituting a non-stop codon $x = n_1 n_2 n_3$ with another non-stop codon $y = m_1 m_2 m_3$ over a time interval $t > 0$ is described by a continuous time, homogeneous, ***stationary*** and time-reversible ***Markov process***, described by the *transition matrix* $T(t)$, whose (i, j) entry contains the probability of replacing codon i with codon j over time interval $t \geq 0$. Stop codons are disallowed as evolutionary states since their random introduction in an active gene is overwhelmingly likely to destroy the function of the translated protein.

The (i, j) element or the rate matrix (Q^{MG94}) for the generalized Muse–Gaut (MG94) model defines the *instantaneous rate* of replacing codon i with codon j ($i \neq j$).

$$
q_{ij}^{MG94} = \begin{cases} \theta_{mn}\alpha_s^b \pi_n^p, & i \to j \text{ is a one–nucleotide synonymous substitution} \\ & \text{from nucleotide } m \text{ to nucleotide } n \text{ in codon position } p, \\ \theta_{mn}\beta_s^b \pi_n^p, & i \to j \text{ is a one–nucleotide non–synonymous substitution} \\ & \text{from nucleotide } m \text{ to nucleotide } n \text{ in codon position } p, \\ 0, & \text{otherwise.} \end{cases}
$$

π_n^p denotes the frequency of nucleotide $n \in \{A, C, G, T\}$ in codon position $p = 1, 2, 3$. Synonymous and non-synonymous substitution rates α_s^b and β_s^b may depend both on the alignment site (s) and the branch of the tree (b), as denoted by the sub/superscript. For example, the synonymous $ACG \to ACT$ substitution involves the change $G \to T$ in the third codon position, and its corresponding rate is $q_{ACG, ACT} = \theta_{GT}\alpha_s^b\pi_T^3$. For the remainder of the chapter, we assume that these frequencies (π's) are estimated by counts from the data. Although it is easy to estimate these frequencies by maximum likelihood, in most practical situations the observed frequencies are used, as this approximation (which is usually very good) saves computational time. θ_{mn} corrects for the nucleotide substitution bias, and because of time reversibility $\theta_{mn} = \theta_{nm}$. In the simplest case, all θ_{mn} are equal to 1, reducing the model to the original Muse–Gaut model, and in the most general, six rates can be specified; however, because the phylogenetic **likelihood function** depends only on *products* of rates and evolutionary times $q_{xy}t$, only five of those can be estimated, hence we arbitrarily set one of the rates (we choose θ_{AG}) to one, and all other nucleotide rates are estimated *relative* to the θ_{AG} rate. Diagonal entries of the rate matrix are defined by $q_{ii} = -\sum_{j \neq i} q_{ij}$, ensuring that each row of the transition matrix $T(t)$ forms a valid probability distribution.

The model assumes that point mutations alter one nucleotide at a time, hence most of the instantaneous rates (3134/3761 or 84.2% in the case of the universal genetic code) are 0. This restriction, however, does not mean that the model disallows any substitutions that involve multiple nucleotides (e.g. $ACT \to AGG$). Such substitutions must simply be realized via several single nucleotide steps. In fact, the (i, j) element of $T(t) = \exp(Qt)$ sums the probabilities of all such possible pathways of length t. For a model which does allow multiple nucleotides to be substituted at once, we refer the reader to a recent paper by Whelan and Goldman (2004).

Stationary codon frequencies for the MG94 model are given by $\pi (x = n_1 n_2 n_3) = \pi_{n_1}^1 \pi_{n_2}^2 \pi_{n_3}^3 / N$ and include nine parameters (also referred to as the $F3 \times 4$ estimator for those familiar with the PAML (Yang, 1997) package). N is the normalizing constant, which accounts for the absence of stop codons and is defined as $N = 1 - \sum_{(m_1 m_2 m_3 \text{ is a stop codon})} \pi_{m_1}^1 \pi_{m_2}^2 \pi_{m_3}^3$. In the original MG94 paper (Muse & Gaut, 1994), nucleotide frequencies were pooled from all three codon positions, i.e. $\pi_n^1 = \pi_n^2 = \pi_n^3$ for all four nucleotides ($F1 \times 4$ estimator), yielding three frequency parameters.

The GY94 model, first implemented in the PAML package (Yang, 1997) is similar to MG94, but it differs in two choices of model parameterization. First, the synonymous evolutionary rate is set to 1, making $\omega = \beta/\alpha = \beta$. As we will point out later, it is important to tease apart the effects of both α and β on the estimates of their ratio (or difference). From a statistical perspective, ratios are notoriously difficult to estimate, especially when the denominator is small. Second, in GY94, rates are proportional not to the frequency of target nucleotides (π_n^p), but rather to the frequency of target *codons*. In most practical cases, this distinction has a minor effect on the estimation of substitution rates α and β, and for the remainder of the chapter we will primarily focus on the MG94 model, albeit nearly everything we discuss can be run with the GY94 model instead.

The rest of the section concerns itself mostly with describing different methods for estimating α_s^b and β_s^b. There is no biological reason to assume that the selective pressures are the same for any two branches or any two sites, however, one's ability to estimate these quantities from finite data demands that we consider simplified models. Indeed, for N sequences on S sites, there would be $2S(2N-3)$ parameters to estimate, if each branch/site combination has its own rate – a number greater than the most liberally counted number of samples (each branch/site combination: $S(2N-3)$) available in the alignment. Clearly, a model with more parameters than available data points is going to grossly overfit the data, and no measure of statistical sophistication can do away with this fundamental issue.

14.4 Simulated data: how and why?

Biological data are very complex, and even the most sophisticated models that we can currently propose are at best scraping the surface of biological realism. In order to evaluate the statistical properties of parameter estimates and various testing procedures, it is therefore imperative to be able to generate sequence data that evolved according to a pre-specified Markov process with known rates. At the very basic level, if enough data with known rates are generated, the inference procedure (with the same model as the one used to generate the data) should be able to return correct parameter estimates on average (***consistency***) and render accurate estimates of rate parameters (***efficiency***). The powerful statistical technique of ***bootstrapping*** is dependent on simulated data generated either by resampling the original data (***non-parametric bootstrapping***; Felsenstein, 1985; Efron *et al.*, 1996) or by simulating a substitution model on a given tree (***parametric bootstrapping***; Goldman, 1993). Very large codon data sets can be simulated quickly, enabling the study and validation of various statistical properties of different estimation procedures.

14.5 Statistical estimation procedures

14.5.1 Distance-based approaches

A number of distance-based approaches have been proposed to estimate the relative rate of non-synonymous to synonymous substitution, some of which incorporate biased nucleotide frequencies and substitution rates. We use a heavily cited (over

1600 references) method proposed by Nei and Gojobori (1986) to illustrate the concepts. The cornerstone idea has already been mentioned in Introduction: one estimates the expected ratio of non-synonymous/synonymous substitutions under the neutral model and compares it to the one inferred from the data (we reframe several of the concepts using stochastic models in the following sections). Consider an alignment of two homologous sequences on C codons. For each codon $c = 1 \ldots C$ in every sequence, we consider how many of the nine single-nucleotide substitutions leading away from the codon are synonymous (f_c^s), and how many are non-synonymous (f_c^n). For example, $f_{GAA}^n = 8$, because 8/9 one nucleotide substitutions (AAA, CAA, GCA, GGA, GTA, GAC, GAG, GAT, TAA) are non-synonymous (compare with Section 14.4.3, for a note on substitutions to a stop codon) and $f_{GAA}^s = 1$ (GAG is the only synonymous change). For every position in the alignment, we average these quantities for the corresponding codon in each sequence, and sum over all positions to arrive at the estimates S (number of synonymous sites) and N (number of non-synonymous sites). N/S provides an estimate of the expected ratio of non-synonymous to synonymous substitutions under the neutral model for the given codon composition of the two sequences. The actual number of synonymous (D_s) and non-synonymous (D_n) substitutions between two sequences is estimated by counting the differences codon by codon, assuming the shortest evolutionary path between the two. If the codons differ by more than one nucleotide, all shortest paths, obtained by considering all possible orderings of the substitutions (two if two substitutions are needed, six – if three are needed), are identified, and the numbers of synonymous and non-synonymous substitutions are averaged over all pathways (see Section 14.7.3 for further insight). One can now estimate mean $dS = D_s/S$ and $dN = D_n/N$ for the entire sequence, and the corresponding ratio $dN/dS = (D_n/D_s)/(N/S)$. The effect of multiple substitutions at a site can be approximated by setting $dS_c = -3/4 \log(1 - 4/3 dS)$, and applying an analogous correction to dN. A careful reader will recognize this as the standard Jukes–Cantor (Jukes & Cantor, 1969) estimate of the genetic distance between two sequences, assuming a single substitution rate between all synonymous (and non-synonymous) codons, but it is merely an approximation. For example, it cannot, in principle, handle the case when all evolutionary paths between two observed codons x and y involves both synonymous and non-synonymous substitutions, since this would imply that different substitution rates apply in parts of the evolutionary past of the sequences. To decide if dN/dS is statistically different from 1, one can, for example, obtain a confidence interval around dN/dS (by bootstrap or using a variance estimator) and examine whether or not the confidence interval overlaps with 1.

While these approaches are useful exploratory tools, especially since they can be run very quickly, they are poorly suited to hypothesis testing, because statistical significance may be difficult to assess, and the effect of phylogenetic relatedness on estimated rates can be strong when rate estimates are based on pairwise sequence comparisons (see Section 14.7.3 regarding the adaptation of these ideas to account for phylogenetic relatedness). The *de facto* standard package for distance-based sequence analysis is MEGA (*http://www.megasoftware.net/*), a mature and feature-rich program whose main limitation is that the software can only be run under Microsoft Windows. The multi-platform HYPHY package discussed in the Practice section also provides a number of features for distance based estimation.

14.5.2 Maximum likelihood approaches

The very basic – global, or single rate – model posits that α and β do not vary from site to site or branch to branch. Clearly, this assumption cannot be expected to hold in most biological systems and the model must, therefore, be treated as a rough approximation. As with all phylogenetic maximum likelihood methods, a variant of Felsenstein's pruning algorithm (Felsenstein, 1981) is used to evaluate the probability of extant codon sequences given all model parameters, i.e. the likelihood function, and independent parameters are adjusted using a numerical optimization technique to obtain their **maximum likelihood estimates (MLEs)**. There are numerous statistical and philosophical reasons to use maximum likelihood methods for parameter estimation (e.g. Edwards, 1972). For example, assuming that the model which generated the data is the same as the one being fitted, and given enough data (alignment columns), MLEs will be consistent (i.e. converge to the true values) and efficient (have minimum variance among all unbiased estimators). For example, Rogers (1997) has demonstrated the consistency of maximum likelihood phylogenetic tree estimation for any reversible substitution model with a finite number of states.

Global α and β models are the simplest computationally and contain the fewest number of estimable parameters, hence they are suitable for coarse data characterization (exploratory analysis), analyses of small samples (a few sequences or very short alignments) or when substitution rates are a **nuisance parameter** (i.e. there are used as a means to estimate something else, e.g. phylogeny or ancestral sequences), although more complex models may provide a better result in the latter case. Lastly, the global model serves as a useful null hypothesis to form the basis of testing for spatial or temporal rate heterogeneity.

When α and β are the primary object of evolutionary analysis, the global model is nearly always a poor choice for inference. Selective regimes may differ from site to site in a gene due to varying functional and structural constraints, selective pressures, and other biological processes. Since the global model is only able to estimate the mean, it reveals remarkably little about the unknown distribution of rates. Two genes with mean $\omega = 0.5$ may, for example, have dramatically different distributions of ω across sites (Fig. 14.2), hence it might be erroneous to state,

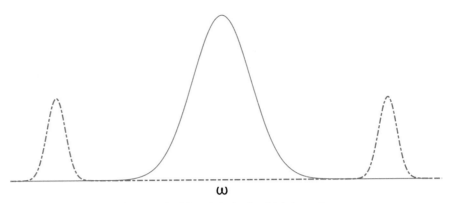

Fig. 14.2 Two different rate distributions (solid and dashed) which have the same mean ω.

based solely on the equality of the means, that two genes are evolving under similar selective pressures. WhatsInTheMean.bf shows how simulated alignments with vastly different distributions of ω yield very similar estimates of mean ω.

14.5.3 Estimating dS and dN

Quite often one may be interested in estimating *evolutionary distances* between coding sequences, usually measured as the expected substitutions per site per unit time, or % divergence. As codon-substitution Markov processes are time-homogeneous, one can use the property that the distribution of waiting times (i.e. the time for a process to change its state from codon i to some other codon) is an exponential distribution with rate parameter defined by the off-diagonal entries of the rate matrix Q, as $r_i = \sum_{j \neq i} q_{ij}$. Recalling that the diagonal elements of the rate matrix Q were defined as $q_{ii} = -r_i$, the expected time to change from i to some other state is $-1/q_{ii}$, i.e. an average of q_{ii} changes from i to some other state given over a unit length of time. The total expected number of changes per codon per unit time can be obtained by taking a weighted average over all possible codons

$$E[subs] = -\sum_i \pi_i \hat{q}_{ii},$$

where $\hat{q}$ denotes that the rate matrix Q is evaluated using maximum likelihood estimates for all model parameters. To make codon-based distances directly comparable with those obtained from nucleotide models, it is customary to divide the estimates by 3, reflecting the fact that there are three nucleotides in a codon.

The total expected number of substitutions can be decomposed into the sum of synonymous and non-synonymous changes per codon, by summing rate matrix entries which correspond to synonymous and non-synonymous substitutions only

---

Box 14.1 Calculating the number of non-synonymous and synonymous sites

---

The calculation of the number of non-synonymous and synonymous sites is performed as described previously (Muse, 1996).

- Given a genetic code, for each codon i compute three numbers: T_i – the total number of one-nucleotide substitutions that do not lead to a stop codon, S_i – those substitution which are synonymous and N_i^t – those which are non-synonymous. Clearly, $T_i = S_i + N_i$. For example, $T_{GAA} = 8$, because 8/9 one nucleotide substitutions (AAA, CAA, GCA, GGA, GTA, GAC, GAG, GAT) do not lead to a stop codon, but one (TAA) does, $S_{GAA} = 1$ (GAG is the only synonymous change) and $N_{GAA} = 7$. This step depends only on the genetic code, and not on the alignment being analyzed.

- Compute the expected values of the three quantities for a given alignment, by averaging over the stationary distribution of codons π:

$$T = \sum_i \pi_i T_i, \quad S = \sum_i \pi_i S_i, \quad N = \sum_i \pi_i N_i$$

Note, that $T = S + N$.

- Define $dS = E[syn]\, T/S$, $dN = E[nonsyn]\, T/N$, which can now be interpreted as the expected number of synonymous substitutions per synonymous site, and the non-synonymous analog, respectively.

as follows:

$$q_{ii} = q_{ii}^s + q_{ii}^{ns} = \sum_{j \neq i,\ j \text{ and } i \text{ are synonymous}} q_{ij} + \sum_{j \neq i,\ j \text{ and } i \text{ are non-synonymous}} q_{ij}$$

and

$$E[subs] = E[syn] + E[nonsyn] = -\sum_i \pi_i \hat{q}_{ii}^s - \sum_i \pi_i \hat{q}_{ii}^{ns}$$

In order to convert the expected numbers of substitutions *per codon* to a more customary dN and dS, one must normalize the raw counts by the proportions of synonymous and non-synonymous sites (see Box 14.1), allowing us to compensate for unequal codon compositions in different alignments.

It is important to realize that $\omega = \beta/\alpha$ is, in general, *not equal* to dN/dS as defined in the box, although the two quantities are proportional, with the constant dependent upon base frequencies and other model parameters, such as nucleotide substitution biases. When more than two sequences are involved in an analysis, the computation of genetic distances between any pair of sequences can be carried out by summing the desired quantities (e.g. dS and dN) over all the branches in the phylogenetic tree which lie on the shortest path connecting the two sequences in question. An alternative approach is to estimate the quantities directly from a two-sequence analysis, which implicitly assumes that the sequences are unrelated (e.g. conform to the star topology). Depending on the strength of phylogenetic signal and the assumptions of the model (e.g. variable selective pressure over the tree), the estimates obtained by the two methods can vary a great deal, and, generally, phylogeny-based estimates should be preferred. dSdN.bf provides a numerical example of generating dN and dS estimates from biological sequence data.

14.5.4 Correcting for nucleotide substitution biases

As we noted in the introduction, biased nucleotide substitutions, e.g. the preponderance of transitions over transversions, can have a significant effect on the proportions of synonymous and non-synonymous substitutions, and, by extension, they can affect the estimates of α and β. The MG94 model incorporates parameters (θ_{mn}) to correct for such biases. These parameters, in most cases, are not of primary interest to selection analyses, and, often they are indeed nuisance parameters.

There are several possible strategies for selecting one of the 203 possible nucleotide bias models; having chosen a "nuisance" nucleotide substitution model (or models), we can incorporate (or "cross") this model with a codon substitution model in order to estimate β/α. The NucleotideBiases.bf example evaluates the effect of nucleotide biases on the β/α estimate. For five H5N1 influenza sequences, β/α ranges from 0.148 to 0.233. The estimate for REV is 0.216, and that for the best fitting model (which happens to be 010023 as determined by the lowest *Akaike's information criterion* score, see Box 14.2) is 0.214. Lastly, a model averaged estimate for β/α is 0.221.

Hypothesis testing

Hypothesis testing concerns itself with selecting, from a number of *a priori* available models, or *hypotheses*, the one that explains the observed data best, or minimizes a loss function (e.g. squared distance). For example, one might test for evidence of non-neutral evolution across the gene on average by comparing an MG94 model which enforces $\beta = \alpha$ (neutral evolution) with one that estimates both β and α independently.

In the likelihood framework, all the information about how well the data D support any given model H is contained in the likelihood function $L(H|D) = Pr\{D|H\}$, i.e. the probability of generating the data given the model. When comparing two models H_0 (null) and H_A (alternative), the strength of support for model H_A relative to H_0 is often assessed using the **likelihood ratio statistic** (often abbreviated as LR), defined as $LR = 2(\log L(H_A|D) - \log L(H_0|D))$. A classical

---

Box 14.2 Choosing a nucleotide model

Consider the symmetric matrix form of nucleotide substitution biases:

$$B = \begin{pmatrix} - & \theta_{AC} & \theta_{AG}(=1) & \theta_{AT} \\ - & - & \theta_{CG} & \theta_{CT} \\ - & - & - & \theta_{GT} \\ - & - & - & - \end{pmatrix}$$

Reading this matrix left to right and top to bottom arranges the six rates as $\theta_{AC}, \theta_{AG}(=1), \theta_{AT}, \theta_{CG}, \theta_{CT}, \theta_{GT}$. Any of the models of nucleotide biases can be defined by specifying some constraints of the form $\theta_{ab} = \theta_{cd}$ (e.g. $\theta_{AG} = \theta_{CT} = 1, \theta_{AC} = \theta_{AT} = \theta_{CG} = \theta_{GT}$ for HKY85). A convenient shorthand (adapted from PAUP* and the first exhaustive model search publication (Muse, 1999)) for defining such constraints is a string of six characters where each character corresponds to a θ rate in the above ordering, and if two characters are equal, then the two corresponding rates are constrained to be equal as well. The shorthand for HKY85 is 010010, and the model specified by 010121 defines the constraints $\theta_{AC} = \theta_{AT}, \theta_{CG} = \theta_{GT} = \theta_{AG}(=1)$.

- A "named model", such as HKY85 (the "hard-wired" choice in GY94). Generally, this is a poor choice, because many organisms/genes seem to have non-standard substitution biases (Huelsenbeck *et al.*, 2004), unless the alignment is small or a model selection procedure suggests that a "named" model is appropriate.
- The general reversible model (REV), which estimates five biases from the data as a part of the model. While this is a good "default" model, some of the biases may be difficult to estimate from small data sets, and, as is the case with overly parameter rich models, overfitting is a danger. Overfitted models can actually increase the error in some parameter estimates, because instead of reflecting a real biological process, they may be fitting the noise.
- A nucleotide bias correction based on a model selection procedure (e.g. MODELTEST; Posada & Crandall, 1998; or as described in Practice). Generally, this is the preferred approach, because it allows the correction of substitution biases without overfitting. This approach has two drawbacks: additional computation expense (although it is usually small compared to the cost of fitting a codon model), and the fact that most model selection procedures are based on nucleotide inference, and may incorrectly deduce nucleotide biases because they fail to account for codon constraints and selective pressures.
- Model averaged estimates (see Chapter 10). The most robust, yet computationally expensive, process is to fit all 203 models, obtain α and β estimates from each model, and then compute a weighted sum, where the contribution from each model is determined by its Akaike Weight (defined in Section 14.5.3), which can be interpreted as the probability that the model is the "best-fitting" model given the data. Such exhaustive model fitting is an overkill except for small and/or short data sets, where several models may provide nearly equally good fits to the data. Fortunately, fitting all 203 models is practical precisely for smaller data sets, where model uncertainty is likely to occur.

likelihood ratio test decides between H_0 and H_A by comparing the LR to a fixed number c, selecting H_A if $LR > c$, and selecting H_0 otherwise. The Neyman–Pearson lemma gives theoretical grounds to prefer the likelihood ratio test to all other procedures for simple hypothesis testing, because for all tests of given *size* α (defined as the probability of selecting H_A when H_0 is true, i.e. making a false positive/Type I error), the likelihood ratio test is the most powerful test, i.e. it has the highest probability of selecting H_A when it is true, and hence the lowest Type II/false negative error rate.

An important particular case of the likelihood ratio test arises when H_0 and H_A are *nested*. In phylogenetics, H_0 and H_A can almost always be defined as a parametric model family (e.g. MG94), where some of the parameters are free to vary (e.g. branch lengths), and some may be constrained (e.g. $\beta = \alpha$). When H_0 can be obtained from H_A by adding B well-behaved constraints on model parameters, then the distribution of the likelihood ratio test statistic LR when the data are generated under H_0 follows the χ^2 distribution with B degrees of freedom, if the data sample is sufficiently large[†]. Given a significance level α, which describes the willingness to tolerate false positive results, one computes the critical level c which solves $Pr\{\chi_B^2 \geq c\} = \alpha$, and rejects H_0 if $LR \geq c$. Otherwise, one *fails to reject* H_0, which may be either because H_0 is true, or because the data sample is not sufficiently large to distinguish the two (lack of power), thus the hypothesis testing framework is geared towards rejecting the null hypothesis.

Some intuition for how hypothesis testing using this framework is justified can be gained by means of this simple example. Consider some sequence data that evolves according to the neutral MG94 model (i.e. $\alpha = \beta$). Here $H_0 : \beta = \alpha$ and H_A estimates β and α separately. H_0 is nested in H_A, with one fewer degree of freedom. The log-likelihood of H_0 can always be matched or improved by H_A because it contains H_0 and one finds $LR \geq 0$ in this case. If one were to consider a large number of independent data sets (of the same size) for which the correct model was H_0, compute the LR for each one and tabulate a histogram, then the histogram of LR would approximate the density function of χ^2, if the size of each sample was large enough. It is possible for the LR to exceed any arbitrarily large cutoff by chance, but with vanishing probabilities, hence one settles for the value large enough that only in α proportion of the cases does one falsely reject H_0 by estimating the location of the appropriate tail of the χ_1^2 distribution.

[†] If some parameters are bounded, e.g. by 0, then the distribution of the LR follows a mixture of χ^2 distributions and a point mass at 0, with the proportion of each generally being dependent on the data at hand. In the simple case of a single one-sided constraint, the appropriate mixture is 50% χ_1^2 and 50% point mass at 0.

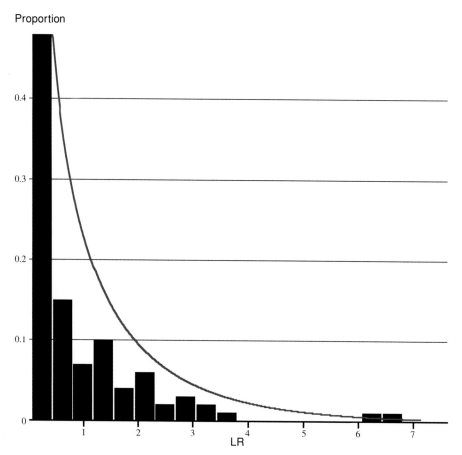

Fig. 14.3   Simulated distribution of the likelihood ratio test statistic based on 100 iterates, and the asymptotic χ_1^2 distribution density (solid line). Note that the simulated distribution is skewed towards smaller values compared to χ_1^2, suggesting that the data sample is too small to have achieved the asymptotic approximation, and tests based on χ_1^2 are likely to be slightly conservative.

The example LRT.bf tests the hypothesis $H_0 : \alpha = \beta$ (neutral evolution) vs. the alternative of non-neutral evolution on average across an entire gene. For instance, in the Influenza A/H5N1 HA data set, the hypothesis of neutrality is rejected ($LR = 65.44$, $p \ll 0.001$), with $\omega = 0.23$, suggesting overall purifying selection. Based on 100 random parametric replicates under H_0, we found the distribution of LR (Fig. 14.3) to range from 0.0004 to 6.81.

Hypothesis testing is a powerful tool if applied judiciously, but its findings can easily be over-interpreted. Indeed, it is crucially important to remember that, *when H_0 is rejected in favor of H_A, all we can be certain of is that the data being analyzed are unlikely to have been produced by H_0 not that H_A is the best explanation for the* *data.* For example, it is often tempting to reject an overly simplistic H_0 in favor of an overly complicated H_A, when in fact, a model intermediate in complexity may be the best choice. We will return to this topic in a later section when we describe a general procedure for model selection.

Confidence intervals on parameter estimates

As most sequence samples have relatively few independent observations (alignment columns) per model parameter, there tends to be a fair amount of stochastic "noise" in model parameter estimates. This noise is usually referred to as *sampling variance*, and it gives one an idea of how variable the estimate of a particular parameter would have been, had independent data samples of the same size been available. Sometimes, it may also be beneficial to estimate the entire *sampling distribution* of one or more model parameters, especially if maximum likelihood estimates are being used for *post-hoc* inference, and ignoring the errors in such estimates can lead to patently erroneous conclusions (Suzuki & Nei, 2004). We note that very few studies that employ codon substitution models (or indeed, any substitution model) report confidence intervals for the model parameters, hence we devote a section to how these intervals can be obtained, in order to encourage their use. Within a maximum likelihood framework, there are at least three different approaches to deducing how much error there is in a given parameter estimate.

Asymptotic normality

For sufficiently large samples, the joint sampling distribution of all model parameters approaches a multivariate normal distribution with the variance-covariance matrix given by the inverse of the Fisher information matrix, i.e. the log likelihood surface is quadratic in the vicinity of the maximum. Briefly, if $l(\theta_1, \ldots \theta_n) = \log L(H(\theta_1, \ldots \theta_n)|D)$ is the log likelihood function of all estimable model parameters, the information matrix I is defined as

$$I(\theta_1, \ldots \theta_n) = \begin{pmatrix} \dfrac{\partial^2 l}{\partial \theta_1^2} & \dfrac{\partial^2 l}{\partial \theta_1 \partial \theta_2} & \cdots & \dfrac{\partial^2 l}{\partial \theta_1 \partial \theta_n} \\ \vdots & & & \vdots \\ \dfrac{\partial^2 l}{\partial \theta_n^2} & \dfrac{\partial^1 l}{\partial \theta_n \partial \theta_2} & \cdots & \dfrac{\partial^2 l}{\partial \theta_n^2} \end{pmatrix}$$

evaluated at the maximum likelihood values of the parameters. The advantage of using this method is that it is able, given enough sample data, to correctly model possible codependencies between model parameter estimates in addition to yielding parameter errors, because the entire joint distribution of model parameters is approximated. However, it may be difficult to check whether the sample is large enough to allow the normal approximation to hold, especially if one or more of the parameter values are near the boundary of a parameter space. In addition, the information matrix has to be estimated numerically, which turns out to be computationally expensive and numerically challenging, because one has to estimate $\approx n^2/2$ partial derivatives with sufficient accuracy to obtain an accurate matrix inverse. As in phylogenetic models n grows linearly with the number of sequences, the only reliable way to achieve asymptotic normality is to analyze very long sequences, which may be impossible due to biological constraints (e.g. gene lengths).

Profile likelihood

If only confidence intervals around a parameter estimate, or a collection of parameters, is desired, especially for smaller samples and if asymptotic normality may be in doubt, then component-wise **profile likelihood** intervals may be used instead. A $1 - \alpha$ level confidence interval around a maximum likelihood estimate $\hat{\theta}_i$ is defined as all those values of t for which the hypothesis $\theta_i = t$ cannot be rejected in favor of the hypothesis $\theta_i = \hat{\theta}_i$ at significance level α, when all other model parameters are fixed at their maximum likelihood estimates. Profile confidence intervals are easy to understand graphically: if one plots the log-likelihood function as a function of parameter θ_i only, then (assuming the likelihood function is unimodal) a confidence interval is obtained simply by bracketing the $\hat{\theta}_i$ using the line c_α units below the maximum, where c_α is a critical value for the χ^2 distribution: $Pr\{\chi_1^2(x) \geq c_\alpha\} = \alpha$ (Fig. 14.4). Profile likelihood intervals can be found quickly, and can handle cases when the likelihood surface is not well approximated by a quadratic surface over a long range of parameter values. For instance, profile likelihood can produce asymmetric confidence intervals, and handle values near the boundary of the parameter space. However, because profile likelihood works with only one parameter at a time, the intervals it finds may be too small.

Sampling techniques

If MLE values of model parameters are used for subsequent inference (e.g. in the REL method for site-by-site selection analysis, Section 14.7.1), it may be desirable to incorporate parameter estimation errors and see how they might influence the conclusions of an analysis. One possible way to accomplish this is to average over values from an approximate joint distribution of parameter estimates. The **sampling importance resampling** (SIR) algorithm is a simple technique for sampling from probability distributions. Firstly, parameter values are sampled from the likelihood surface. In order to sample parameter values that have high likelihood, one

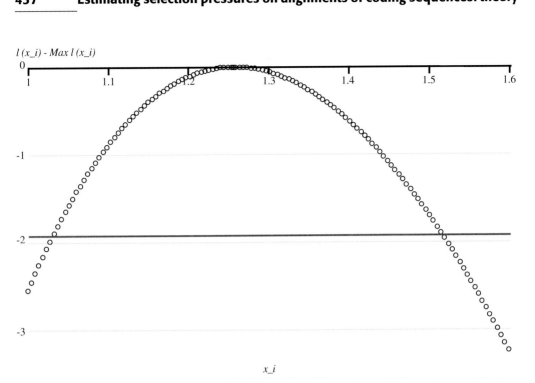

Fig. 14.4   Graphical interpretation for profile likelihood confidence intervals. The maximum of the log-likelihood function l is set at 0 and, as the value of the x_i parameter is taken further away from the MLE, we seek the points where the likelihood curve intersects the $c_{0.05} = -1.92$ line. For this parameter, the MLE is 1.26, and the 95% confidence interval turns out to be 1.03, 1.52). Note that the curve is slightly asymmetric, suggesting that asymptotic normality has not yet been achieved.

could estimate 95% profile confidence intervals for each model parameter: (θ_i^l, θ_i^u) (perhaps enlarging them by an inflation factor > 1), arriving at an n-dimensional rectangle from which one draws a large number $N \gg n$ (e.g. 1000) of samples. A technique called Latin Hypercube Sampling (LHS) can sample this space very efficiently. For each parameter i, LHS draws points by sequentially picking a random index r_i from 0 to N, with the constraint that all r_i are distinct and forming a sample point $\theta_i = \theta_i^l + (\theta_i^u - \theta_i^l)r_i/N$. Last, a resampling step is applied, where a much smaller number of points $M < N$ are drawn from the original sample, but now in proportion to their likelihood score. Statements about the sampling properties of all θ_i, and derivative quantities can now be made based on the resampled set of points. The closer the original distribution is to the true distribution, the larger the *effective sample size* will be.

ErrorEstimates.bf applies all three methods to estimating the sampling properties of ω ratios for the MG94 model with a separate ω for each branch.

Note that short branches tend to provide highly unreliable estimates of ω when compared to longer branches.

14.5.5 Bayesian approaches

Instead of a maximum likelihood approach, which works with $L(H|D) = Pr\{D|H\}$, i.e. the probability of generating the data given the model, a Bayesian approach works with $Pr\{H|D\}$, i.e. the probability of the model given the data. In this way, a Bayesian approach is much more similar to the way that many people interpret statistical results, and the output of a Bayesian model, called a **posterior distribution**, can be interpreted as true probabilities. However, a Bayesian approach requires the assumption of **prior distributions** for the parameters. Not only may the choice of priors affect the results, the use of certain prior distributions necessitates the use of computationally intensive **Markov chain Monte Carlo (MCMC)** techniques to sample from the posterior distribution. Although different from a philosophical point, Bayesian approaches and maximum likelihood approaches should arrive at the same results, given sufficient data. The two major phylogenetic software packages based on the Bayesian paradigm are MRBAYES (*http://mrbayes.csit.fsu.edu/*) and BEAST (*http://evolve.zoo.ox.ac.uk/beast/*), and we refer interested readers to Chapter 7 and Chapter 18, respectively.

14.6 Estimating branch-by-branch variation in rates

As selection pressures almost certainly fluctuate over time, it may be unreasonable to use models that assume a constant selective pressure for all branches in the phylogenetic tree. For instance, in a tree of influenza sequences from various hosts, one might expect to find elevated selection pressure on branches separating sequences from different hosts, because they are reflective of adaptation to different evolutionary environments.

We have already mentioned the model which allows a separate ω in every branch of the tree – the "local" model, or to follow the nomenclature of the original paper (Yang, 1998), the "free ratio" model. Other possibilities are the global (single-ratio) model, which posits the same ω for all branches and a large array of intermediate complexity models, where some branches are assigned to one of several classes, with all branches within a single class sharing one ω value. Formally, this model can be described as

$$\beta^b = \omega^{I(b)}\alpha^b$$

where $I(b)$ is the assignment of branch b to an ω class. For the global model $I(b) = 1$ and for the local model $I(b) = b$.

14.6.1 Local vs. global model

A naive approach to test for branch-to-branch rate variation is to fit the global model as H_0, the local model as H_A, and declare that there is evidence of branch-by-branch rate heterogeneity if H_0 can be rejected. Since the models are nested, one can use the likelihood ratio test with $B - 1$ (B is the total number of branches in the tree) degrees of freedom. `LocalvsGlobal.bf` (Standard Analyses> Miscellaneous> Phylohandbook.bf) performs this test using the MG94 model.

This procedure, however is lacking in two critical areas. Firstly, it may lack power if only a few branches in a large tree are under strong selective pressure, because the signal from a few branches may be drowned out by the "background." Secondly, it lacks specificity, in that the real question a biologist may want to ask is "Where in the tree did selection occur?" and not "Did selection occur somewhere in the tree?." A rough answer to this question may be gleaned from examining the confidence intervals on branch by branch ω and saying that two branches are under different selective pressures if their confidence intervals do not overlap. However, these confidence intervals are suitable only for data exploration, as they may not achieve appropriate coverage. For instance, there is an implicit multiple comparison problem and the intervals may be too narrow in some cases, and they may be difficult to interpret for large trees where many pairwise comparisons would have to be made.

14.6.2 Specifying branches *a priori*

The first likelihood-based procedure for identifying different selective regimes on tree branches (Yang, 1998) relied on an *a priori* specification of some branches of interest. For example, if a branch separates taxa from different evolutionary environments (e.g. virus in different hosts, geographically separate habitats, taxa with and without a specific phenotype), one may be interested in studying the strength of selection on that branch. The *a priori* branch model separates all B branches into a few ($F < B$) of interest (foreground), for which the ω parameter is estimated individually, and all other branches (background), which share a common ω_b – background selection regime. To test for significance, one conducts a LRT with F degrees of freedom. This analysis boosts the detection power because the number of model parameters is significantly reduced, and focuses on specific branches.

The main drawback of such a test is that it assumes that the rest of the branches have a uniform selective pressure. This assumption is less likely to hold as the number of taxa (and tree branches) is increased, and the model can be easily misled into claiming selection on a "foreground" branch if the background is strongly non-uniform. A simple example in Fig. 14.5 shows that the likelihood

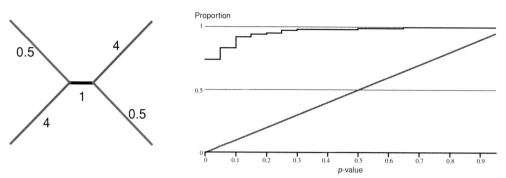

Fig. 14.5   The effect of model mis-specification on the likelihood ratio test. We simulated 100 long
data sets (with 5000 codons each) using the tree on the left, with branch ω shown for every
branch and then tested whether $\omega = 1$ along the short middle branch using the likelihood
ratio test, which makes an incorrect assumption that all other branches of the tree have
the same ω. The panel on the right tabulates the cumulative distribution of p-values for
rejecting the null hypothesis $\omega = 1$ along the internal branch. A correct test for this model
would reject neutrality at level p in approximately $p \times 100$ cases (the solid line), whereas
this test has uncontrollable rates of false positives. It rejects neutrality 74 times at $p = 0.05$,
for example. In addition, the point estimate of ω along the internal branch is strongly biased
by the incorrect model (mean of 1.94, instead of the correct value of 1).

ratio test can perform very poorly if the assumptions of the model are violated. The
assumption of neutrality along a given branch was rejected at $P = 0.05$, i.e. very
strongly, for 74/100 data sets simulated with the neutral "foreground" branch.

A test which is more robust to a non-uniform background may be the following:
to decide whether a given branch b_0 is under positive selection (i.e. has $\omega^{b_0} > 1$),
one fits $H_0 : \omega^{b_0} \leq 1$ and $H_A : \omega^{b_0}$ is unconstrained, allowing all other branches to
have their own ω and conducts an LRT. This is a one-sided test (e.g. the constraint
is an inequality, rather than an assignment), and the appropriate distribution of
the LR test statistic to check against is a mixture of χ_1^2 and a point mass at 0
(Self & Liang, 1987).

`BranchAPriori.bf` can be used to conduct both types of tests. To conduct
one of the tests of (Yang, 1998), select Node1 and Node5 as foreground branches
in the Primate Lysozyme data set.

14.6.3 Data-driven branch selection

However, in many instances there may exist no *a priori* evidence to suspect selection
at a specific branch in the tree. There are several naive approaches which we would
like to caution against. First, it may be tempting to begin by fitting a local model,
inspecting the values of ω estimated for each branch, and then selecting some
branches which appear to be under selection for further testing. Using the data

to formulate a hypothesis to test is referred to as "data-dredging" and this should generally be avoided. Hypotheses formulated using a data set and then tested with the same data set will nearly always be highly biased, i.e. appear significant when in fact they are not. Second, one might attempt to test every branch one at a time, and declare all those which appear under selection in individual tests to be significant. However, this method consists of multiple tests, and as such, requires a multiple test correction (e.g. Bonferroni or false discovery rate; Benjamini & Hochberg, 1995). Intuitively, if each test has the probability of making a Type I error 5% of the time, at least one of the 17 tests (each branch in a 10 sequence tree) would, in the worst case scenario make a Type I error in $1 - (1 - 0.05)^{17} = 58.2\%$ of the time, i.e. every other significant finding could be a false positive! There is another issue, in that each individual test assumes that every branch, except the foreground branch being tested, is under a uniform (background), presumably negative, selective pressure. Hence, if two of the tests return positive results (i.e. two foreground branches are positively selected), these results are incompatible.

Since the model of rate variation is a nuisance parameter in this case, we advocate the idea of *searching* the space of many possible models, selecting those which fit well, and averaging over models to achieve robust inference. For full details of the methods we refer the reader to Kosakovsky Pond & Frost (2005a), but the basic idea is intuitive. Let us consider models with up to C different ω assigned to branches. A model like this is completely specified by assigning each branch in a phylogenetic tree to one of C classes, with the total number of models on B branches given by the Stirling numbers of the second kind – the number of unique ways to assign B objects to C bins:

$$S(C; B) = \frac{1}{C!} \sum_{k=1}^{C} (-1)^{C-k} \frac{C!}{k!(C-k)!} k^B$$

The number of models grows combinatorially fast with B, even if C is small (e.g. 3). The models considered during the search will no longer always be nested, hence a new model comparison technique is called for. We chose a *small sample Akaike information criterion* score of each model, defined as

$$AIC_c(M) = -2 \log L + 2p \frac{s}{s - p - 1}$$

where L is the maximum log-likelihood score of the model, p is the number of model parameters and s is the number of independent samples available for inference (the number of sites in an alignment). AIC_c rewards a model for a good likelihood score and penalizes it for the number of parameters, progressively more so as the number of parameters approaches the number of independent samples. AIC_c minimizes the expected Kullback–Liebler divergence between model M and the true model that generated the data, and there exist fundamental results supporting its use. We use a ***genetic algorithm (GA)*** to search the space of possible models, measuring the fitness of each by its AIC_c score. GAs have proven very adept at rapidly finding good solution in complex, poorly understood optimization problems. As an added benefit, the availability of AIC_c scores allows one to compute Akaike weights for each model, defined as $w_M = \exp(\min_M AIC_c(M) - AIC_c(M))/2$, normalized to sum to one. w_M can be interpreted as the probability that model M is the best (in the Kullback–Liebler divergence sense) of all those considered, given the data. Now, instead of basing inference solely on the best fitting model, one can compute the model averaged probability of finding $\beta^b > \alpha^b$ for every branch in the tree. The GA search is a computationally expensive procedure, and we recommend that the reader first try our web interface (***http://www.hyphy.org/gabranch***).

14.7 Estimating site-by-site variation in rates

Often, we are most interested in positive or diversifying selection, which may be restricted to a small number of sites. Several methods to detect site-specific selection pressure have been proposed; for a review and detailed discussion of these methods, see Kosakovsky Pond and Frost (2005b). There are two fundamentally different approaches to estimating site-specific rates. The first approach, originally proposed by Nielsen and Yang (1998) assumes that the true distribution g of α_s and β_s can be well represented by some predefined, preferably simple distribution f, uses the data to infer the parameters of f, integrates the values of unobserved model parameters out of the likelihood function, and then assigns each site to a rate class from f to make inference about what occurs at a single site. This class of models – ***random effects likelihood (REL)*** models – was first implemented in the PAML (Yang, 1997) package and has since been widely adopted in the field of molecular evolution. Bayesian versions of these models have been developed by Huelsenbeck and colleagues (Huelsenbeck & Dyer, 2004; Huelsenbeck *et al.*, 2006). The second approach, first proposed by Suzuki and Gojobori (1999) estimates site-specific rates directly from each site independently, either using maximum likelihood methods (***fixed effects likelihood, FEL***), or counting heuristics. We now describe how these methods differ in their approach to estimating site-specific rates.

14.7.1 Random effects likelihood (REL)

When the objective of an analysis is to estimate, for each codon $c = 1 \ldots S$ in the alignment, a pair of *unobserved* rates α_s and β_s, the random effects approach posits that there exists a distribution of rates, which is almost always assumed to be discrete with D categories for computational feasibility, with values (α^d, β^d), and the probability of drawing each pair of values is (p^d), subject to $\sum_d p^d = 1$.

Examples of such distributions might be the general discrete distribution (all parameters are estimated), or the M8 model of PAML, where $D = 11, \alpha_d = 1$ and β is sampled from a discrete beta distribution (10 bins) or a point mass $\omega > 1$. The value D is fixed *a priori*, and all other distribution parameters are usually estimated by maximum likelihood. To compute the likelihood function at codon site c, one now has to compute an expectation over the distribution of rates

$$L(\text{site } c) = \sum_{d=1}^{D} L(\text{site } c | \alpha_c = \alpha_d, \beta_c = \beta_d) p^d,$$

where each of the conditional probabilities in the sum can be computed using the standard pruning algorithm (Felsenstein, 1981). Finally, the likelihood of the entire alignment, making the usual assumption that sites evolve independently, can be found as the product of site-by-site likelihoods.

The REL approach can be used to test whether positive selection operated on a proportion of sites in an alignment. To that end, two nested REL models are fitted to the data: one which allows $\beta > \alpha$, and one that does not. The most cited test involves the M8a (or M7) and M8b models of Yang and colleagues (Swanson *et al.*, 2003), which each assume a constant synonymous rate $\alpha \equiv 1$ and use a beta distribution discretized into ten equiprobable bins to model negative selection ($\beta_i \leq 1$ for $i \leq 10$), but M8a forces $\beta_{11} = 1$ while M8b allows $\beta_{11} \geq 1$. If M8a can be rejected in favor of M8b using the likelihood ratio test with a one-sided constraint, this provides evidence that a p_{11} proportion of sites are evolving under positive selection. Another test, allowing for variable synonymous rates has been proposed by Sorhannus and Kosakovsky Pond (2006) and involves fitting two D (e.g. $D = 2$ or $D = 3$) bin general discrete distributions with one of them constraining $\beta_d \leq \alpha_d$ for all rate classes.

To find individual codon sites under positive selective pressure, REL methods can use an **empirical Bayes** approach, whereby the posterior probability of a rate class at every site is computed. A simple application of *Bayes' rule*, treating the inferred distribution of rates as a prior shows that

$$Pr(\alpha_s = \alpha_d, \beta_s = \beta_d | \text{site } c) = \frac{L(\text{site } c | \alpha_s = \alpha_d, \beta_s = \beta_d) p^d}{L(\text{site } c)}$$

A site can be classified as selected if the posterior probabilities for all those rate classes with $\beta_d > \alpha_d$ exceed a fixed threshold (e.g. 0.95). While it may be tempting to interpret this value as an analog of a *p*-value, it is not one. The Bayesian analog of a *p*-value is the **Bayes factor** for some event E, defined as

$$BF(E) = \frac{\text{posterior odds } E}{\text{prior odds } E} = \frac{Pr_{posterior}(E)/(1 - Pr_{posterior}(E))}{Pr_{prior}(E)/(1 - Pr_{prior}(E))}$$

A Bayes factor measures how much more confident (in terms of odds) one becomes about proposition E having seen the data. If $BF(\beta > \alpha)$ at site c exceeds some pre-defined value (e.g. 20), a site can be called selected. Our simulations (Kosakovsky Pound & Frost, 2005b) indicate that in phylogenetic REL methods $1/BF$ is approximately numerically equivalent to a standard p-value.

REL methods are powerful tools if applied properly, and can detect selection when direct site-by-site estimation is likely to fail. For example, if there are 100 sites in an alignment with $\omega = 1.1$, unless the alignment consists of hundreds of sequences, methods which estimate rates from individual sites (discussed below) are unlikely to identify any *given* site with $\omega = 1.1$ to be positively selected, because selection is very weak. Hence they may miss evidence for positive selection altogether. On the other hand, REL methods can pool small likelihood contributions from all 100 sites to a single class with $\omega > 1$ and use *cumulative* evidence to conclude that there is selection somewhere in the sequence. However, REL methods may also fail to identify any individual site with any degree of confidence. The ability to pool information across multiple sites is a key advantage of REL.

However, one needs to be keenly aware of two major shortcomings of REL. First, there is a danger that the distribution of rates chosen *a priori* to model α and β is inadequate. For example, there is *no compelling biological reason* to support the mixture of a beta and a point mass distribution (PAML's M8). In extreme cases (see Section 14.7.5) this can lead to positively misleading inference, whereas in others "smoothing" – the underestimation of high rates and overestimation of low rates – can occur, and result in loss of power. Second, posterior empirical Bayes inference assumes that rate estimates are exact. For example, an $\omega = 1.05$ may be estimated and used to compute Bayes factors, but if the confidence interval on that estimate is $(0.5, 1.5)$, then one cannot be certain whether or not the contribution from this rate class should be counted for or against evidence of positive selection at a given site. REL methods have received somewhat undeserved criticisms (e.g. Suzuki & Nei, 2004), mostly arising from the application of these methods to very sparse (small and low divergence) data, where all inferred parameter values had large associated errors. Yang and colleagues (2004) partially addressed the issue of incorporating parameter errors into REL analyses using a *Bayes empirical Bayes (BEB)* technique, which uses a series of approximations to integrate over those parameters which influence the distribution of α_d, β_d. More generally, sampling methods (such as the SIR algorithm described in Section 14.5.4) can also be drawn upon to average over parameter uncertainty, by drawing a sample of parameter values, computing a Bayes factor for positive selection at every site, and then reporting a site as positively selected if a large (e.g. >95%) proportion of sampled Bayes factors exceeded the preset threshold.

Comparing distributions of rates in different genes

One of the advantages of using a random effects approach is that it is straightforward to test whether two sequence alignments (which may be totally unrelated) have the same distribution of substitution rates across sites. The easiest way to test for this is to select a distribution family (e.g. M8 or a general discrete distribution), $f(\nu)$, where ν denotes the full vector of distribution parameters. Several tests can be readily conducted.

Are the rate distributions different between two data sets?
To answer this question, we fit the null model H_0, which forces the ν parameter vector to be the same on two data sets (letting all other model parameters, e.g. base frequencies, branch lengths and nucleotide biases to vary freely in both data sets), and the alternative model H_A, where each data set is endowed with its own vector of distribution parameters (ν_1 and ν_2). A likelihood ratio test with the degrees of freedom equal to the number of estimable parameters in ν can then be used to decide whether the distributions are different. This test is better than simply comparing the means (see an earlier section), but it is effectively qualitative, in that no insight can be gleaned about *how* the distributions are different if the test is statistically significant.

Is the extent or strength of selection the same in two data sets?
If the choice of distribution f is fixed, it may be possible to pose a more focused series of questions regarding how the distributions might be different between two data sets. Let us consider, for instance, a 4 bin general discrete distribution (GDD_4) with rates (α_1, β_1), (α_2, β_2), (α_3, β_3), (α_4, β_4), probabilities p_1, p_2, p_3 and $p4 = 1 - p_1 - p_2 - p_3$, with the further restriction (explained in 14.7.5) that $\sum \alpha_i p_i = 1$. Furthermore, let us assume that the first two bins represent negative selection ($\alpha_1 > \beta_1$ and $\alpha_2 > \beta_2$), bin three reflects neutral evolution ($\alpha_3 = \beta_3$), and bin four – positive selection ($\beta_4 > \alpha_4$). Distributions with more bins, or a different allocation of selective pressures between bins can be readily substituted. Using the most general model, where two independent GDD_4 distributions are fitted to each data set as the alternative hypothesis, the following likelihood ratio tests can be carried out: (i) are the proportions of positively selected sites the same in both data sets ($H_0 : p_4^1 = p_4^2$); (ii) are the strengths of selection the same in both data sets ($H_0 : \beta_4^1/\alpha_4^1 = \beta_4^2/\alpha_4^2$); (iii) are the proportions and strengths of selection the same in both data sets (both constraints)?

14.7.2 Fixed effects likelihood (FEL)

With a FEL approach, instead of drawing (α, β) from a distribution of rates, one instead estimates them directly at each site. First, the entire data set is used to infer global alignment parameters, such as nucleotide substitution biases, codon frequencies and branch lengths; those values are fixed afterwards for all individual site fits. Second, FEL considers each codon site c as a number of independent realizations of the substitution process (now depending only on the site specific rates (α_c, β_c)) operating on the tree – roughly one realization per branch, yielding a sample size on the order of N – the number of sequences in the alignment. This observation underscores the need to have a substantial number of sequences (e.g. $N \geq 20$) before reliable estimates of (α_c, β_c) can be obtained. Two models are fitted to every codon site c: $H_0 : \alpha_c = \beta_c$ (the neutral model) and H_A, where (α_c, β_c) are estimated independently (the selection model), and, because the models are nested, the standard LRT can be applied to decide if site c evolves non-neutrally. When the LRT is significant, if $\alpha_c < \beta_c$, site c is declared to be under positive selection, otherwise site c is under negative selection.

FEL has several advantages over REL. First, no assumptions about the underlying distribution of rates are made, making the estimation of (α, β) potentially more accurate. Second, a traditional p-value is derived as a level of significance at every site, automatically taking into account errors in the estimates of α_c and β_c. Third, FEL can be trivially parallelized to run quickly on a cluster of computers, because each site is processed independently of others.

The drawbacks of FEL are that it may require a substantial number (e.g. 20 – 30) of sequences to gain power, that it cannot pool information across sites to look for evidence of alignment-wide selection, and that it is not well suited for rate comparisons between unrelated data sets. In addition, nuisance parameters such as nucleotide substitution biases, codon frequencies and branch lengths are treated as known rather than subject to error, however simulations suggest that this assumption does not lead to high type I (i.e. false positives) errors, and FEL appears to be extremely good at capturing "true" substitution rates, at least in simulated data (Kosakovsky Pond & Frost, 2005b).

14.7.3 Counting methods

Counting methods provide a "quick-and-dirty" alternative to FEL, and perform nearly as well for sufficiently large (e.g. 50 or more sequences) (Kosakovsky Pond & Frost, 2005b). A counting method consists of the following steps.

(i) Unobserved ancestral codons, i.e. those which reside at internal tree branches, are reconstructed. The original counting method of Suzuki and Gojobori (1999) used nucleotide-based parsimony reconstruction, which led to problems such as the possibility of inferring stop codons at internal branches or hundreds of equally good reconstructions. In our Single Likelihood Ancestor Counting (SLAC) method (Kosakovsky Pond & Frost, 2005b), a global MG94 model is fitted to the entire alignment, and used for maximum likelihood reconstruction of ancestral codons.

(ii) Based on a given ancestral reconstruction, the number of observed synonymous and non-synonymous substitutions (NS and NN) are counted, averaging over all possible shortest paths when multiple substitutions are required, e.g. to handle the $ACT \rightarrow AGA$ substitution one would average over $ACT \rightarrow ACA, ACA \rightarrow AGA$ and $ACT \rightarrow AGT, AGT \rightarrow AGA$ pathways.

(iii) Using the same ancestral state reconstruction, one computes the mean (over all branches) proportion of synonymous and non-synonymous sites (see Section 14.5.3) at a given site, ES and EN. $p_e = ES/(ES + EN)$ can then serve as the *expected proportion* of synonymous changes at that site under neutrality.

(iv) One then tests whether the "observed" (strictly speaking, inferred) proportion of synonymous substitutions $p_o = NS/(NS + NN)$ is significantly different from p_e, using an extended (to deal with non-integer counts) binomial distribution with p_e probability of success and $NS + NN$ outcomes to determine the level of significance. If $p_e < p_o$ and the result is statistically significant, then a site is called negatively selected, and if $p_e > p_o$ – positively selected.

The method is very fast and intuitively attractive (Fig. 14.6) and performs nearly identically with FEL and REL on large simulated data sets. However, there are many assumptions implicit in the method, which may mislead it in certain cases. First, the ancestral state reconstruction is treated as known, where in fact it is inferred (with possible errors) from the data. This shortcoming can be dealt with by averaging over all possible ancestral states or using a sampling procedure to study the effect of ancestral uncertainty on inference of selection (Kosakovsky Pond & Frost, 2005b). Secondly, parsimonious evolution is assumed, hence multiple possible substitutions along a branch are discounted. This is generally not a major issue, unless the phylogeny possesses many long branches. Third, the binomial distribution is only a rough approximation to neutral evolution. For example, even though averaged over all codons at a site, a random substitution will be synonymous with probability p_e, codon usage variation throughout the tree may bias this proportion quite strongly.

14.7.4 Which method to use?

The availability of multiple methods to estimate site-specific non-synonymous and synonymous substitution rates has led to discussion about which method is most appropriate. Fortunately, given enough sequence data, we have shown that all methods (properly applied) tend to perform similarly (Kosakovsky Pond & Frost, 2005b), hence the choice of a method is mostly a matter of preference or computational expediency. Difficult cases arise when inference must be made from limited sequence data, in which case we advocate the use of every available method for a consensus-based inference.

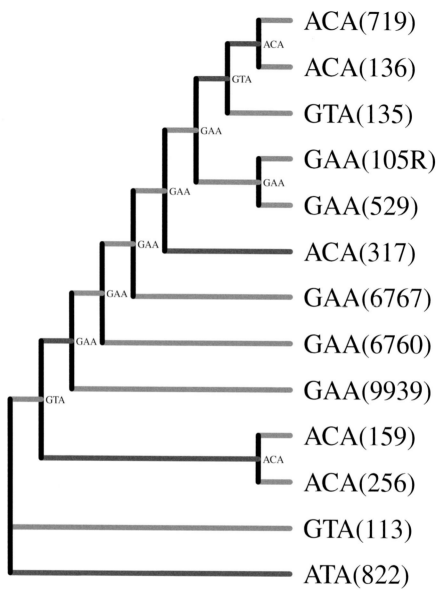

Fig. 14.6   An illustration of SLAC method, applied to a small HIV-1 envelope V3 loop alignment. Sequence names are shown in parentheses. Likelihood state ancestral reconstruction is shown at internal nodes. The parsimonious count yields 0 synonymous and 9 non-synonymous substitutions at that site. Based on the codon composition of the site and branch lengths (not shown), the expected proportion of synonymous substitutions is $p_e = 0.25$. Under an extended binomial distribution on 9 substitutions with the probability of success of 0.25, the probability of observing 0 synonymous substitutions is 0.07, hence the site is borderline significant for positive selection.

14.7.5 The importance of synonymous rate variation

When REL methods were first introduced in 1998 (Nielsen & Yang, 1998), the assumption that synonymous rates α were constant for the entire length of the gene (proportional to the underlying neutral mutation rate) was made. If this assumption is incorrect, however, then an elevated ω could be due to either lowered α, that could be a result of a functional constraint on synonymous substitutions, such as selection on exon splicing enhancement elements (Chamary *et al.*, 2006) or secondary RNA structure (Tuplin *et al.*, 2002). In addition, when a site is hypervariable, e.g. has both high α and β (but $\beta \leq \alpha$), models which assume a constant α are likely to misclassify such sites as strongly selected.

SLAC, FEL, and REL (with an appropriate rate distribution) are all able to model variation in both synonymous and non-synonymous rates, and we cannot emphasize enough how important it is to do so. We strongly encourage routine testing for synonymous rate variation, using the procedure described by Kosakovsky Pond and Muse (2005). In the same paper, it was demonstrated that a wide variety of sequence alignments, sampled from different genes and different types of organisms show a near universal strong support for variation in synonymous substitution rates. The test fits two models: H_A – a three bin general discrete distribution to β and a separate three bin general discrete distribution to α (constrained to have mean one, see Kosakovsky Pond and Muse (2005) for technical details), and H_0, which imposes the restriction $\alpha = 1$. The models are nested, with four constraints, hence the usual likelihood ratio test applies.

14.8 Comparing rates at a site in different branches

One of the most complex models considers both site-to-site variation and branch-to-branch variation in selection pressures. Since rates α_s^b, β_s^b now depend both on the branch and the site, the models must adopt some restrictions on how this dependency is modeled, otherwise too many parameters will be fitted to too few data points leading to unreliable inferences.

If one is interested in selection in a *group* of branches, for example, a specific subtree of the phylogeny, or internal branches of the tree, then FEL methods can be readily adapted. Branches in the tree are split into the group of interest (B_I) and the rest of the branches (B_B). The alternative model fits three rates to each codon c in the alignment: α_c, β_c^I, β_c^B, where β_c^I operates on branches of interest, and β_c^B – on the rest of tree branches. The null model forces neutral evolution on the branches of interest $\beta_c^I = \alpha_c$, and to test for significance, a one degree of freedom LRT is employed. This method has previously been applied to the study of population level selective forces on the HIV-1 virus, where each sequence was sampled from a different individual, which fell into one of two genetically distinct populations (Kosakovsky Pond *et al.*, 2006). The branches of interest (all internal branches), represent the evolution of successfully transmitted viral variants, and can be used as a proxy for population level selective forces. Based on simulation studies, this approach has low error rates, but needs many sequences to gain power, and would not be advisable with a group of interest branches that comprised only a few branches.

The REL take on the problem was advanced in 2002 by Yang and his colleagues (Yang & Nielsen, 2002), who proposed what has since become known as "branch-site" methods. The most recent version of the method (Zhang *et al.*, 2005) requires that a branch or branches of interest be specified *a priori*, assumes a constant synonymous rate for all sites, and samples β_s^b from a four bin distribution defined as follows:

(i) *Class 0.* Negative selection with $\beta = \omega_0 < 1$ on all tree branches. Weight p_0.
(ii) *Class 1.* Neutral evolution with $\beta = 1$ on all tree branches. Weight p_1.
(iii) *Class 2a.* Negatively selected background $\beta^B = \omega_0 < 1$, positively selected foreground $\beta^I = \omega_2 \geq 1$. Weight $(1 - p_0 - p_1)p_0/(p_0 + p_1)$.
(iv) *Class 2b.* Neutrally evolving background $\beta^B = 1$, positively selected foreground $\beta^I = \omega_2 \geq 1$. Weight $(1 - p_0 - p_1)p_1/(p_0 + p_1)$.

The alternative model fits all parameters independently, whereas the null model fixes $\omega_2 = 1$. A one-sided LRT can be used for evidence of selection on foreground branches somewhere in the sequence, and empirical Bayes inference is used to detect sites with strong Bayes factors (or posterior probabilities) of being in Classes 2a or 2b. As with other REL methods, the main advantage gained from adopting a heavily parameterized form of possible rate classes is the pooling of information from many sites. Hence, it may be possible to use a "branch-site" REL to look for evidence of episodic selection along a single branch. One has to be aware of issues similar to those highlighted in Section 14.6.2, when the assumption of uniform background selection may be violated, leading to incorrect inference. In addition, it is unclear what effect synonymous rate variability would have on this method.

14.9 Discussion and further directions

To conclude, currently available methodology allows the fitting of codon substitution models to characterize patterns of non-synonymous and synonymous substitution in multiple sequence alignments of coding sequences. Extensions of these models allow us to identify particular sites or branches in a phylogenetic tree that are evolving under positive or negative selection. These models can incorporate biological features such as biased nucleotide substitution patterns, and recombination through the assumption of different phylogenies for different parts of the sequence.

While the codon models discussed in this chapter represent the current state of the art, it is important to remember that they still make simplifying assumptions. Perhaps the most important assumption is that sites are assumed to evolve independently from one another. We cannot simply look for associations between codons in extant sequences. Simple "counting based" methods have been applied to look for associations while taking the shared evolutionary history into account (Dutheil *et al.*, 2005), but ideally, a process-based model is more desirable. While "covarion" models of codon substitution have been proposed (Guindon *et al.*, 2004), strictly speaking, these are heterotachy models, that allow the substitution rate to vary over time. Other models allow correlations between the evolutionary rates at nucleotide sites (Zang, 1995; Felsenstein & Churchill, 1996; Stern & Pupko, 2006), which can be extended to consider codon substitution models; however, these do not consider how the rates may change at one site depending on a particular state at another site. Recently, models have been proposed that explicitly consider interactions between sites by considering the entire sequence as a single state (Robinson *et al.*, 2003; Rodrigue *et al.*, 2005); these approaches are highly computationally intensive, due to the large number of possible states. Another approach would be to consider interactions between a small number of sites (2–3).

PRACTICE

Sergei L. Kosakovsky Pond, Art F. Y. Poon, and Simon D. W. Frost

In the practice section, we briefly review software packages for the analysis of selection pressures on coding sequences, before embarking on a detailed walk-through of some of the analyses available within our HyPhy software package.

14.10 Software for estimating selection

There is currently a wide variety of software packages available for inferring patterns of selection from protein-coding sequences, and the majority are freely-downloadable from the web. In most cases, there are compiled binaries available for "conventional" operating systems (i.e. Macintosh and Windows), and can otherwise be compiled from publicly released source code. As is often the case with public-domain software, however, the source code and documentation is often unavailable or poorly maintained. Here, we will review some of the software packages that have been used to estimate selection.

14.10.1 Paml

Paml (an abbreviation of "Phylogenetic Analysis by Maximum Likelihood") was developed by Ziheng Yang and originally released in 1997, providing the first publicly available implementation of codon model-based methods of selection inference. Paml can estimate selection on a site-by-site basis, modeling variation by random effects likelihood (REL). It has since become widely adopted as the gold standard for estimating selection from sequence alignments, having reached over 1000 citations in the literature. A large number of nested models are available within Paml for likelihood-based model selection. However, the programs that make up Paml are command–line executable binaries, meaning that no graphical user interface is provided. Program settings are modified by editing plain-text (i.e. ASCII) files in which various analytical options are listed. Furthermore, the programs in Paml cannot be easily customized to implement other related models. Source code and pre-compiled binaries for Macintosh and Windows can be obtained for free at the Paml website (*http://abacus.gene.ucl.ac.uk/software/paml.html*). A substantial manual written by Ziheng Yang is currently included in distributions of Paml as a PDF file.

14.10.2 ADAPTSITE

ADAPTSITE was developed by Yoshiyuki Suzuki, Takashi Gojobori, and Masatoshi Nei and was originally released in 2001 (Suzuki *et al.*, 2001). This program was the first implementation of a method for estimating selection by counting the number of inferred non-synonymous and synonymous substitutions throughout the tree (Suzuki & Gojobori, 1999). The source code for ADAPTSITE is distributed for free from the website (*http://www.cib.nig.ac.jp/dda/yossuzuk*), but there are no pre-compiled binaries available. With the exception of basic instructions for compiling and installing ADAPTSITE, there is no additional documentation provided.

14.10.3 MEGA

MEGA (an abbreviation of "Molecular Evolutionary Genetic Analysis") was developed by Sudhir Kumar, Koichiro Tamura, Masatoshi Nei, and Ingrid Jacobsen, and originally released in 1993 (Kumar *et al.*, 1994); it is currently at version 4. This software package provides a comprehensive array of methods for sequence alignment, reconstructing phylogenetic trees, and hypothesis testing. However, it is distributed as a Windows executable binary only (from the MEGA homepage, *http://www.megasoftware.net*). For estimating selection, MEGA implements distance-based methods for estimating the number of non-synonymous and synonymous substitutions (Nei & Gojobori, 1986), and then evaluates the hypothesis $\beta = \alpha$ using one of several available statistical tests.

14.10.4 HYPHY

HYPHY (an abbreviation of the phrase "HYpothesis testing using PHYlogenies") was developed by Sergei Kosakovsky Pond, Spencer Muse, and Simon Frost, and was first released publicly in 2000 (Kosakovsky Pond *et al.*, 2005). It is a free and actively-maintained software package that can be downloaded from the web-page (*http://www.hyphy.org*) as either a pre-compiled binary executable for Macintosh or Windows, or as source code. HYPHY provides a broad array of tools for the analysis of genetic sequences by maximum likelihood, and features an intuitive graphical user interface. It is uniquely flexible in that all of the models and statistical tests implemented in the software package are scripted in a custom batch language (instead of compiled source code), which can be freely modified to suit one's needs. Furthermore, many of the default methods can be executed in parallel, allowing HYPHY to handle exceptionally difficult problems by parallel computing, using either multiple threads (useful for multiple-core processors) or multiple processes (for clusters of computers). A user manual and batch language reference guide are included with distributions of HYPHY.

14.10.5 DATAMONKEY

DATAMONKEY (*http://www.datamonkey.org*) is actually a web-server application of specific programs in the HyPhy package, hosted on a high-performance computing cluster that currently consists of 80 processors. As the HyPhy programs for estimating selection on a site-by-site basis are designed to take advantage of parallel computing, DATAMONKEY provides a very fast method for analyzing large sequence data sets within an easy-to-use interface. In the first 3 years of its existence, DATAMONKEY has been used to analyze over 15 000 sequence alignments. DATAMONKEY also provides a parallelized implementation of genetic algorithms for rapid detection of recombination breakpoints in a sequence alignment (Kosakovsky Pond *et al.*, 2006), which, if unaccounted for, could lead to spurious results in a selection analysis.

14.11 Influenza A as a case study

We will demonstrate the various methods in HyPhy for measuring selection pressures in protein-coding sequences, by applying these methods to a series of data sets based on 357 sequences of the influenza A virus (serotype H3) haemagglutinin gene that was originally analyzed by Robin Bush and colleagues (Bush *et al.*, 1999). We will henceforth refer to this data set as "Influenza/A/H3(HA)." Influenza A can infect a broad range of animal hosts (e.g. waterfowl, swine, and horses) and currently accounts for about 30 000 human deaths a year in the United States alone (Webster *et al.*, 1992; Thompson *et al.*, 2003). The influenza A virus possesses a single-stranded RNA genome and lacks proof-reading upon replication, leading to an exceptionally rapid rate of evolution and abundant genetic variation (Webster *et al.*, 1982). The hemagglutinin gene (HA) encodes several antigenic sites that can elicit an immune response and is therefore likely to be responsible for much of the virus' adaptability. This data set also provides an example of branch-by-branch variation in substitution rates, which was attributed by Bush *et al.* (1999) to selection in a novel environment caused by the serial passaging of influenza A virus isolates in chicken eggs in the laboratory. Furthermore, it demonstrates the importance of modeling site-by-site variation in rates, due to the immunological roles of specific amino acids in the HA protein.

We have also prepared several example data sets containing a smaller number of influenza A virus (H3) HA gene sequences. Although sample sizes in this range are only marginally sufficient for measuring selection with confidence, such samples are more representative of what investigators tend to employ for this purpose. They are also more manageable and can be analyzed using the methods described below on a conventional desktop computer quickly. An archive with example alignments can be downloaded from *http://www.hyphy. org/phylohandbook/data.zip*, and some of the analytic results from *http://www.hyphy.org/phylohandbook/results.zip*. Alternatively, these files can also be downloaded from *http://www.thephylogenetichandbook.org*.

14.12 Prerequisites

14.12.1 Getting acquainted with HYPHY

HYPHY is a software package for molecular evolutionary analysis that consists of three major components:

(i) a formal scripting language (HYPHY Batch Language or HBL) designed to implement powerful statistical tools for phylogenetic inference by maximum likelihood;

(ii) a pre-packaged set of templates in HBL that implement a comprehensive array of standard phylogenetic analyses; and

(iii) a graphical interface for Mac OS, Windows, and the GTK toolkit for X11 which runs on nearly all Unix and Linux distributions, that provides quick and intuitive access to these templates alongside visual displays of sequences, trees, and analytical results.

As a result, there is frequently more than one way to do things in HYPHY. In most cases, we will use the template files (i.e. "standard analyses") to demonstrate how to carry out various analyses of selection. A standard analysis is activated by selecting the menu option "`Analysis > Standard Analyses....`"

Installing the HYPHY package will create a folder containing the executable file and a number of subfolders. Running the executable from your desktop will activate the graphical user interface (GUI), while running the executable from a command–line will activate a text-based analog. Upon activation of the HYPHY GUI, a console window will appear on your desktop, which may be accompanied by an optional greeting window (which can be dismissed by clicking "OK").

The console window consists of two parts: the log window, in which most results from standard analyses will be displayed; and the input window, in which you may be prompted by HYPHY to specify various options during an analysis. When a standard analysis is being executed, the name of the corresponding template batch file is displayed in the lower-left corner of the console window. There is also a status indicator at the base of the console window, which is most useful for tracking the progress of a long-term analysis (i.e. optimization of a likelihood function). At the bottom right corner, there are icons that activate various functions such as web-updating of the software package; however, most users will not need to use these functions in general. Depending on which platform you use, there may be menu options at the top of the console window.

Any object (i.e. sequence alignment, tree, likelihood function) in HYPHY can be viewed in a type-specific window. The most important of these is the data window, which displays sequences and provides a number of tools for setting up an analysis of selection. To open a window for any object stored in memory, access the object inspector window by selecting the menu option "Windows > Object Inspector". We will describe the basic aspects of each window type as we proceed in the following sections. In versions of HYPHY compiled for Microsoft Windows and the GTK toolkit for various flavors of Unix and Linux, different menu options are distributed between the various types of windows. For Macintosh versions of HYPHY, a single set of menu options that change in response to the active window type is displayed only at the top of the screen. For command–line versions of HYPHY, most menu options in the GUI are either irrelevant or have an analogous batch language command.

14.12.2 Importing alignments and trees

No matter what software package you choose, any implementation of a codon substitution model for measuring selection will require both an alignment of sequences and a corresponding tree. There are many programs available for generating either an alignment or inferring a tree. Although its main purpose is for analyzing alignments and trees provided by the user, HYPHY has limited capabilities built-in for preparing both an alignment and a tree from any given set of related sequences, which will be discussed in Sections 14.12.5 and 14.12.6, respectively. Here, we will demonstrate how to import an alignment and a tree from the respective files into HYPHY. We will also discuss how to use the graphical user interface (GUI) of HYPHY to visually inspect and manipulate alignments and trees. Although this is not usually a necessary step in a selection analysis, we consider it prudent to confirm that your data is being imported correctly, regardless of whichever software you use.

To import a file containing sequences in HYPHY, select the menu option "File > Open > Open Data File..." to bring up a window displaying the contents of the current working directory. By default, HYPHY does not necessarily assume that every file in the directory is readable, and might not allow your file to be selected as a result. For example, on Mac OS you may need to set the "Enable" tab located above the directory display to the option "All Documents". At this setting, HYPHY will attempt to open any file that you select, irrespective of type. In contrast to other packages, HYPHY is actually quite versatile in being able to handle FASTA, GDE, NEXUS, and PHYLIP-formatted sequences from files generated in Windows, Unix, or Macintosh-based systems equally well.

Nevertheless, like any other software package, HYPHY makes certain assumptions about how a file containing sequences or trees is formatted. For example, some non-standard characters that are occasionally inserted into sequences (e.g. the tilde character ~ which is used by BioEdit) will be ignored by HyPhy and can induce a frame-shift upon importing the sequences (but this does not affect standard IUPAC symbols for ambiguous nucleotides, e.g. "R" or "Y", or gaps, e.g. "–" or "."). In addition, HyPhy has some restrictions on how sequences can be named. Every sequence name must begin with a letter or a number, and cannot contain any punctuation marks or spaces (i.e. non-alphanumeric characters) with the exception of the underscore character, "_" (which is conventionally used to replace whitespace). Selecting the menu option "`Data > Name Display > Clean up sequence names`" will automatically modify the names of imported sequences to conform to these requirements, and also renames sequences with identical names. There is no restriction on the length of a sequence name. Having incompatible names in the sequence file will not prevent you from importing the file or displaying the sequences in the GUI, but may cause problems in subsequent analyses.

The procedure for opening a tree file in HyPhy is virtually identical to that for opening a sequence file, except that you select the menu option "`File > Open > Open Tree File...`". HyPhy will import trees generated in other software packages such as Paup* (Swofford, 2003) or Phylip (Felsenstein, 1989), but may fail to load a tree from a PHYLIP-formatted file (i.e. a Newick string) if the tree contains incompatible sequence names containing punctuation marks or spaces. If a tree containing incompatible sequence names is imported from a NEXUS-formatted file, however, then the tree will be accepted and displayed properly (but it is highly recommended to clean up the sequence names). Trees recovered from a NEXUS-formatted file will not be automatically displayed in HyPhy, but are available for subsequent analyses.

14.12.3 Previewing sequences in HyPhy

Importing sequences from a file will automatically spawn a data window in the HyPhy GUI (Fig. 14.7). A data window consists of several fields:

(i) **Name display.** Sequence names are listed in the leftmost field, and can be highlighted individually or in sequence. You can rename a sequence by double-clicking on the corresponding name in this field. A number of options affecting sequence names can be accessed in the menu "`Data > Name Display`" or by right-clicking on the field.

(ii) **Sequence display.** The nucleotide or amino acid sequence corresponding to each name is displayed in the field immediately to the right of the name display. Directly above this field, there is a modified horizontal scroll bar indicating what interval of the sequence alignment is currently being displayed in the window.

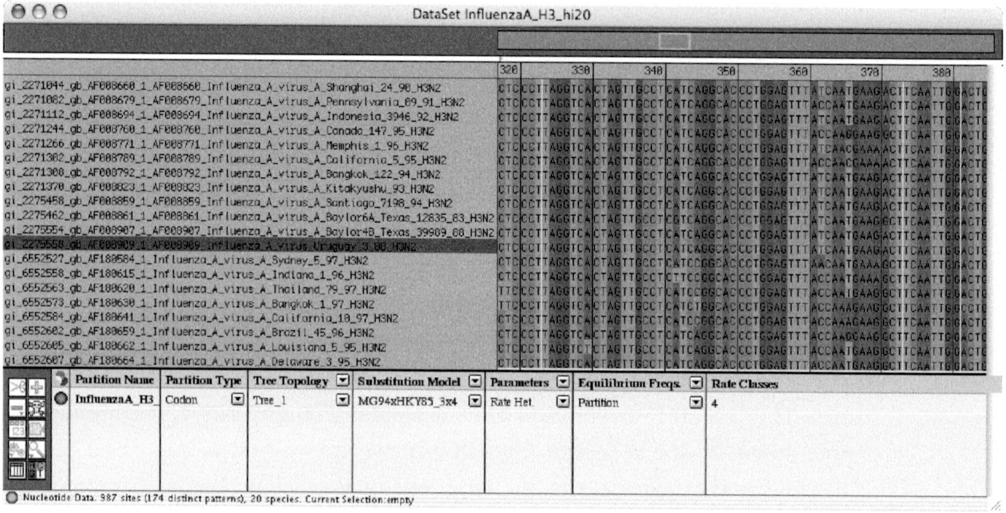

Fig. 14.7   Example of a data window in HyPhy.

(iii) **Partition display.** Located at the base of the data window, the partition display consists of a chart in which a new row is displayed for every data partition that is defined on the alignment. This display can be used for setting up a limited number of analyses.

A data partition is a fundamental object in HyPhy that defines which columns in the alignment will be passed onto an analysis. By specifying a data partition, for instance, we can perform a selection analysis on a specific region of a gene. You can create a partition from the entire alignment by selecting the menu option "Edit > Select All", followed by "Data > Selection->Partition". Every time a new partition is defined on a data set, a colored field will appear in the horizontal scroll bar at the top of the data window to indicate the range of the alignment in the partition, and the partition display will be updated.

Having an open data window provides a good opportunity to visually inspect your alignment. To make it easier to see discrepancies in the alignment, HyPhy can display nucleotide and amino acid alignments in a block-color mode that is activated by the "Toggle Coloring Mode" (colored AC/GT) icon located to the left of the partition display. If one of your sequences acts as a reference for aligning other sequences, you can click and drag it to the desired location in the name display window.

The Influenza/A/H3(HA) alignment extends beyond the region coding for the hemagglutinin gene to include non-coding nucleotides. In order to fit a codon substitution model to our data, we first must isolate the coding region in our alignment. You can select a range of columns in the sequence display by clicking at one location and dragging across the desired interval (this automatically selects every sequence in the alignment). Alternatively, you can click on one column of the alignment, and while holding the "shift"-key click on another column to select the interval contained in-between. Once the desired interval has been selected, create a partition as above so that a new data partition appears in the list. You can also specify a data partition by selecting "`Data > Input Partition`" from the menu and entering a partition specification string; e.g. "1–30" creates a partition of the first 30 nucleotide positions in the alignment.

One advantage of creating a data partition is that a number of additional tools specific to partitions can be activated using the icons to the left of the partition display in the data window. For example, you can search for identical sequences in the data partition by selecting the "`Data Operations`" (magnifying glass) icon. Removing identical sequences from the partition can provide a considerable savings in computational time for complex selection analyses. In our example, this operation identifies eight sequences that are identical to another sequence in the partition, which become highlighted in the name display field. To filter these sequences from the alignment, select "`Data > Invert Selection`" from the menu and right-click on the name display field to spawn a new data window. The contents of the new window can be saved to a NEXUS-formatted file (by selecting the menu option "`File > Save > Save As...`" and setting Format to "`Include Sequence Data, NEXUS`" in the window that appears). Any data partition can be written to a file by selecting the "Save Partition to Disk" (floppy disk) icon, which can then be re-imported for subsequent analyses in HYPHY. Box 14.3 provides an exercise on importing and manipulating a sequence data file in HYPHY.

14.12.4 Previewing trees in HYPHY

Importing a tree from a file will automatically open a tree display window (unless you have unflagged this option in the HYPHY preferences menu). Any tree that is stored in memory can also be viewed at any time by opening the object inspector window, selecting the "`Trees`" setting, and double-clicking the tree object in the list. An example of a tree viewing window is shown in Fig. 14.8. The main viewer window depicts a portion of the tree, which corresponds to the section indicated in the summary window in the upper left corner. By click-dragging the box in the summary window, you can quickly select what region of the tree to display in the main window. There are also a number of tools for modifying the main viewer perspective activated by various icons grouped under the label "`Zoom/Rotate`".

Box 14.3 Importing and manipulating sequence data files in HʏPʜʏ

Import sequences from `InfluenzaA_H3.fasta` into the data viewer using `File>Open>Open Data File` (on Mac OS X you need to enable `All Documents` in the file dialog to make the file "selectable"). Some 357 sequences with 987 nucleotides each will be displayed. As is often common with GenBank sequence tags, sequence names are very long and contain repetitive information. For example, the first sequence in the file is named `gi|2271036|gb|AF008656.1|AF008656 Influenza A virus (A/Perth/01/92(H3N2)) hemagglutinin (HA) gene, partial cds`. Invoke `Edit>Search and Replace` from the data viewer to access a search and replace dialog for sequence names. HyPhy incorporates support for *regular expressions* – a powerful mechanism to specify text patterns. Enter the search pattern `.+Influenza A virus \(` – a regular expression that will match an arbitrary number of characters followed by the substring `Influenza A virus` and an opening parenthesis " (" – and an empty pattern to replace with. Next, repeat this with the search pattern `\(H3N2\).+`, specifying (H3N2) followed by an arbitrary number of any characters. Now, the first sequence name is a lot more manageable `A/Perth/01/92`, with others following in kind. Scroll down to examine other sequence names, and notice that two of the names are still long. One of them is `gi|2271150|gb|AF008713.1|AF008713 Influenza A virus (A/Fukushima/114/96)`. This is because two sequences would have been marked as `A/Fukushima/114/96` had the second one been processed, thus HʏPʜʏ skipped that sequence during renaming. Double click on the long sequence name, and manually edit it to `A/Fukushima/114/96_2`. Finally, manually edit the last remaining long sequence name. Select `Data>Name Display>Clean up sequence names` to enforce valid HʏPʜʏ sequence names. Execute `Edit>Select All` followed with `Data>Selection→Partition` to create a partition with all the alignment columns. Click on the looking glass button, and choose `Identical Sequences Matching Ambiguities` to identify all sequences which are an identical copy (where ambiguities are treated as a match if they overlap in at least one resolution) of another sequence in the data set (8 sequences will be selected). Execute `Data>Invert Selection` followed by a right click (or Control-click on Macintosh computers with a single button mouse) in the sequence name part of the panel to display a context sensitive menu, from which you should select `Spawn a new datapanel with selected sequences`. This will create a new panel with 349 (unique) sequences. Create a new partition with all sequence data, and click on the disk button to save it to disk for later use (e.g. as a NEXUS file). The file generated by these manipulations can be found in `InfluenzaA_H3.nex`.

HʏPʜʏ can plot the tree in several conventional styles (e.g. slanted, radial) that are selected from the "`Tree Style`" tab, but defaults to plotting "rectangular" trees as shown in Fig. 14.8. The "`Branch Scaling`" tab allows you to select unscaled branches (i.e. ***cladogram***) or branches scaled according to the length values (***phylogram***) either provided in the original file, or estimated *de novo* by HʏPʜʏ. If the tree is associated with a substitution model, then the branch lengths can also be scaled to specific model parameters.

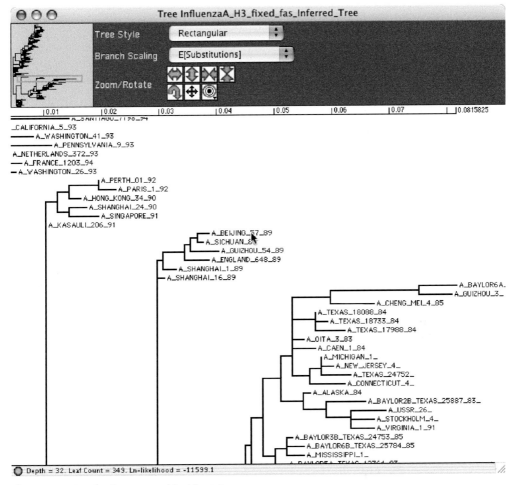

Fig. 14.8   Previewing a tree object in HyPhy.

14.12.5 Making an alignment

Generating a multiple sequence alignment is a computationally demanding problem that can require a long period of time to complete on conventional desktop computers. For alignment methods and tools we refer the reader to Chapter 3 in this book. HyPhy has a limited built-in capability for sequence alignment that is faster than the algorithms implemented in most alignment software, by making assumptions about genetic distance that are appropriate under certain circumstances. This alignment algorithm is implemented in a batch file called SeqAlignment.bf, which can be accessed in the Standard Analyses menu under the heading Data File Tools. The most common procedures for multiple sequence alignment are based on progressive algorithms, in which the addition of each sequence requires a pairwise comparison to every sequence in the

> **Box 14.4** Sequence alignment in HYPHY
>
> Open `InfluenzaA_H3.nex` in a data viewer and create a single partition with all
> sequence data. Change partition type from `Nucleotide` to `Codon`, selecting `Uni-versal` genetic code. Note that, instead of creating a solid partition (shown in the upper
> scroll bar), HYPHY skipped a number of positions, because they contained premature stop
> codons in some of the sequences. Select `Data>Partition→Selection` to highlight
> sites included in the alignment, and scroll around to see which sequences appear to have
> premature stop codons. A cursory examination will show that some of the sequences are
> simply missing a few starting nucleotides, and hence, are in another reading frame. Exe-cute `StandardAnalyses>Data File Tools>SeqAlignment.bf` to perform
> a simple clean-up alignment on `InfluenzaA_H3.nex`. Use BLOSUM62 scoring matrix,
> with `No penalty` for Prefix/Suffix Indels and `First in file` reference sequence,
> `No reference sequence` and `Universal` genetic code. The analysis finds 347 sequences
> in reading frame 1 and 2 sequences in frame 3. The output of `SeqAlignment.bf`
> consists of both a protein and a nucleotide alignment (`.nuc` extension) saved to a file
> of your choice. Import the two cleaned up alignments and check that the reading frame
> is now preserved throughout. For your reference, the output is available in `Influen-zaA_H3_cleaned.fas` and `InfluenzaA_H3_cleaned.fas.nuc`. Note that, if
> `SeqAlignment.bf` does not work well, another alignment program or manual edit-ing may be needed to prepare an alignment for codon model analyses.

alignment (i.e. with a time complexity $O(n^2)$ or greater for n sequences). In con-trast, `SeqAlignment.bf` performs pairwise alignments of each sequence to a
user-defined reference sequence, requiring linear time $O(n)$. This can provide a
sufficient alignment when the sequences can be assumed to be related by a star
phylogeny, as is often the case with HIV-1 sequences from within-patient isolates.
`SeqAlignment.bf` translates codons sequences to amino acids, aligns amino
acid sequences and then maps them back to nucleotides, enforcing the maintenance
of reading frame (see Box 14.4).

Regardless of which procedure is employed to generate an alignment, it is always
recommended to visually inspect your alignment before attempting a selection
analysis. For example, misaligned codons in a sequence may be interpreted as
spurious non-synonymous substitutions in the alignment that may cause your
analysis to overestimate β at that position.

14.12.6 Estimating a tree

Once an alignment has been prepared and inspected, it can be applied towards
the reconstruction of the evolutionary relationships among the sequences. Again,
there are many programs available for estimating the tree from an alignment.
HYPHY has a reasonably diverse suite of procedures built-in for phylogenetic
reconstruction, including clustering methods (e.g. *UPGMA*; Sokal & Michener

(1958), neighbor-joining; Saitou & Nei (1987), and maximum-likelihood-based tree search algorithms (e.g. *star decomposition*). Details on any of these procedures can be found in previous chapters on phylogenetic reconstruction. These procedures can be accessed from the `Standard Analyses` menu under the heading `Phylogeny Reconstruction`.

To generate a tree from an alignment by the neighbor-joining method in HYPHY, select the batch file `NeighborJoining.bf` from the menu. You will be prompted to specify the following:

(i) `Distance Computation` – Quantify the genetic distance between two related sequences by using predefined distance measure (e.g. Jukes–Cantor); estimating branch lengths by maximum likelihood given a standard substitution model; or using a user-defined matrix of pairwise distances[†].

(ii) `Data type` – Whether genetic distances are to be defined at the nucleotide, amino acid, or codon level.

(iii) `Data file` – Select a file containing the sequences from a window displaying the current working directory.

(iv) `Negative Branch Lengths` – Whether to allow branch lengths with negative values, or to reset these to zero.

(v) `Distance formula/standard model` – Select a distance formula or substitution model from a list of standard formulae/models. If using a model to compute distances, specify whether to estimate the model parameters locally or globally, and whether to model site-by-site rate variation (see sections below).

Once HYPHY has finished reconstructing a tree by the neighbor-joining method, the corresponding Newick tree string can be exported to a user-specified file for later use in a selection analysis. HYPHY will always ask whether a newly inferred tree should be saved to a file.

In practice, it turns out that the outcome of a selection analysis is usually fairly robust to the method used to reconstruct the tree; nevertheless, it is always a good idea to make a reasonable effort to provide an accurate reconstruction. Moreover, many tree search algorithms can require a long time on most desktop computers, especially when dealing with large alignments, although there exist very fast heuristic algorithms, which can handle even large alignments quickly, e.g. as implemented in PHYML (*http://atgc.lirmm.fr/phyml/*, see Chapter 6) and GARLI (*http://www.bio.utexas.edu/faculty/antisense/garli/Garli.html*). In most cases, the neighbor-joining method with an appropriate distance formula (e.g. Tamura–Nei (TN93) distance; Tamura & Nei (1993)) will provide a sufficiently accurate reconstruction for an analysis of selection. Estimating a neighbor-joining tree

---

† You will be prompted to select a file containing a HYPHY-formatted $n \times n$ matrix, where n is the number of sequences, e.g. "{{0, 0.1}, {0.1, 0}}" specifies a pairwise distance of 0.1 between two sequences.

Box 14.5 Inferring a neighbor-joining tree and estimating branch lengths

Open `InfluenzaA_H3_fixed.fas.nuc` in a data viewer and create a single partition with all sequence data. Set `Tree Topology` to "Infer Tree," `Substitution Model` to "REV," `Parameters` to "Global" and `Equilibrium Frequencies` to "Partition." Select `Likelihood>Inference>Infer Topology` choosing `Neighbor_Joining, Force Zero` and `TN93`. HYPHY will infer the tree in about 15 seconds and then proceed to fit the "REV" model to the alignment, which will take a few minutes. Inspect the fitted tree (`Window->Object Inspector`). A part of the resulting tree with branch lengths derived from the REV model is shown in Fig. 14.8. Use `MeanPairwiseDivergence` accessible from `User Actions` (the gears icon) button in the bottom right corner of the *console* window to compute mean pairwise nucleotide divergence of the Influenza tree (4.25% with the 95% confidence interval of 4.03%–4.48%). Save the results via `File>Save` choosing `Include sequence data`, NEXUS option from the pull down menu in the file dialog – this will create a self-contained document with the alignment, tree and the code needed to define the models and store parameter estimates. Finally, export the partition to a file (this will also include the inferred tree in the file).

using these settings for the Influenza/A/H3(HA) alignment (Box 14.5), for example, should only require about half a minute on a conventional desktop computer.

14.12.7 Estimating nucleotide biases

Different types of nucleotide substitutions rarely occur at exactly the same rate over time. For example, there is a well-known bias favoring transitions (e.g. G→A) over transversions (e.g. G→C). Failing to account for such biases can severely affect the accuracy of estimating non-synonymous and synonymous substitution rates, and the reliability of a selection analysis thereby. Biases in the nucleotide substitution rates can be estimated by fitting a general time-reversible model (GTR) of nucleotide substitution to the alignment given a tree (Lanave *et al.*, 1984). By definition, this is the most complex time-reversible model that requires the estimation of six parameters. However, simplified versions of the GTR model may often provide a sufficient approximation of the nucleotide substitution rates with fewer parameters, improving the efficiency of subsequent analyses and preventing over-fitting of the data. Although there are several standard or "named" models of nucleotide substitution (e.g. Hasegawa–Kishino–Yano or HKY85 (Hasegawa *et al.*, 1985)), the evolution of nucleotide sequences is often best explained by non-standard models (Muse, 1999; Huelsenbeck *et al.*, 2004). The GTR model contains six parameters, corresponding to the substitution rates between $A \leftrightarrow C/G/T$, $C \leftrightarrow G/T$, and $G \leftrightarrow T$. The parameters can be grouped in 203 different ways – with each group sharing a common rate, defining a range of models from F81 (Felsenstein, 1981) (a single group, all rates are equal), to HKY85 (Hasegawa *et al.*, 1985) (two groups, one for transitions and one for transversions) and ultimately GTR itself (six groups, one parameter in each).

An iterative procedure that evaluates every possible time-reversible model has been implemented in HYPHY in the `Standard Analyses` menu under the heading `Model Comparison`, called "NucModelCompare.bf." This procedure determines which model accomplishes the most accurate fit to the data with the fewest number of parameters according to Akaike's Information Criterion (AIC). Other than requiring you to select files containing sequences and the tree, the batch file will prompt for the following:

(i) `Model Options` – Whether to fit the model with local or global parameters and to model site-by-site variation in rates (see Section 14.13.2)
(ii) `Estimate Branch Lengths` – whether branch lengths are to be re-estimated for every model, or to re-use estimates from the GTR model;
(iii) whether to create a NEXUS-formatted file for every model fit (please note that this will create 203 files);
(iv) and the significance level at which models are rejected (in console window).

Fitting the models using global parameters and re-using branch length estimates can substantially reduce the amount of time required for the procedure to run, and should yield a sufficiently accurate assessment in almost all cases. Because a large number of models are being fit, this procedure can require a substantial amount of time. For example, the nucleotide model comparison for our Influenza/A/H3(HA) alignment required about 35 minutes to complete on a desktop computer (Box 14.6). (`NucModelCompare.bf` contains a parallel-computing implementation for running on a distributed cluster, and a version of it has also been implemented in *http://www.datamonkey.org*.)

It is usually a good idea to export the best fitting nucleotide model by writing the likelihood function object to a NEXUS-formatted file, which is carried out in HYPHY by selecting the menu option "`Analysis > Results > Save Results > Export Analysis`." In Section 14.15, we will describe how to import the fitted nucleotide model from this file into a selection analysis.

14.12.8 Detecting recombination

Many phylogenetic methods implicitly assume that all sites in a sequence share a common evolutionary history. However, recombination can violate this assumption by allowing sites to move freely between different genetic backgrounds, which may cause different sections of an alignment to lead to contradictory estimates of the tree (Posada & Crandall, 2002) and subsequently confuse model inferences

Box 14.6 Comparing models of nucleotide substitution

Use `Standard Analyses>Model Comparison>NucModelCompare.bf` to find the best nucleotide model for `InfluenzaA_H3_final.nex`. Use `Global` model options, estimate branch lengths `Once`, model rejection level of 0.0002 (a number this small ensures that even after the maximal possible number of tests, the overall error rate does not exceed 0.05, based on the conservative Bonferroni correction) and forego the saving of each of the 203 model fits. Based on AIC, the best fitting model should be `(012312)`. Repeat the analysis with a random subset of 35 sequences from the master alignment (`InfluenzaA_H3_Random35.nex`), to investigate how much of the signal is lost when only 10% of the sequences are included. Not surprisingly, the best fitting model is a slight simplification of the one we found from the large alignment: `(012212)`.

(Schierup & Hein, 2000). For example, failing to account for recombination can elevate the false positive error rate in positive selection inference (Anisimova *et al.*, 2003; Shriner *et al.*, 2003) (see Chapter 15). This problem in model inference can be accommodated by allowing different parts of an alignment to evolve independently, each according to their own unique tree. However, we are still required to specify the positions in the alignment at which recombination events have taken place (i.e. *breakpoints*). There is a large number of methods available for accomplishing this task (Posada & Crandall, 2002). A simple approach that performs at least as well as any other method available (Kosakovsky Pond *et al.*, 2006) specifies a given number of breakpoints ($B \geq 1$) at which recombination events have occurred between some number of sequences in the alignment. This method is implemented in HyPhy in the batch file `SingleBreakpointRecomb.bf` for the case $B = 1$, which can be accessed from the `Standard Analyses` menu under the heading `Recombination`.

The batch file will prompt you for the following options:

(i) `Data type` – Should HyPhy fit a nucleotide or codon model of substitution to the data? You will be prompted to select a file containing the sequence data.
(ii) `KH Testing` – Evaluate the fit of models using AIC (Akaike, 1974); or use Kishino–Hasegawa resampling of sequences (Hasegawa & Kishino, 1989) to estimate the confidence interval for the improvement of log-likelihood from the null model (either the model fitted without a recombination breakpoint, or by swapping topologies between sites on either side of the breakpoint).
(iii) `Standard model` – To fit to the data for evaluating incongruence of phylogenies on either side of the breakpoint. There may also be the usual additional options for model fitting (i.e. local vs. global, site-by-site variation);
(iv) and a file to write output from the analysis too.

The batch file will iterate through every potential recombination breakpoint in the alignment, and fit independent models to the sequences on either side of the

---

Box 14.7 Recombination detection in HYPHY

Run Standard Analyses>Recombination>SingleBreakpointRecomb.
bf on InfluenzaA_H3_Random35.nex using Nucleotide model, Skip KH
testing, use CUSTOM model with (012212) correction found in an earlier exercise,
Global model options, with Estimated rate parameters, using Observed equilibrium frequencies. The analysis will examine 189 potential breakpoint locations (all
variable sites in the alignment) using three information criteria (AIC_c is the default
one) and fail to find evidence of recombination in these sequences, because models with
two trees have worse scores (negative improvements) than the model without recombination. The analysis completes in about 5–10 minutes on a desktop. You can also
run this analysis via DATAMONKEY at *http://www.datamonkey.org/gard* (Kosakovsky Pond
et al., 2006).

---

breakpoint (Box 14.7). Consequently, the detection of recombination breakpoints
can require a long time to compute for a large alignment (i.e. over 50 sequences),
although other methods of recombination detection can be run more quickly (see
Chapter 15 and Goldman, 1993).

14.13 Estimating global rates

As mentioned in Section 14.3, comparing the global estimates of α and β averaged
over the entire alignment can provide a crude measure of the overall strength of
selection on the coding region. Global estimates of α and β can be obtained by
fitting a codon substitution model to a given alignment and corresponding tree
(Muse & Gaut, 1994). There are several procedures that are available in the HYPHY
package for fitting codon models. We will describe two methods of fitting a codon
model to the Influenza/A/H3(HA) alignment, first using a basic implementation
using the graphical interface, and then a more comprehensive implementation that
is available under the Standard Analyses menu.

14.13.1 Fitting a global model in the HYPHY GUI

Importing an alignment in the HYPHY graphical user interface (GUI) will automatically spawn a data window, containing a basic array of analytical tools. We
have already described how to create a data partition from an alignment in
Section 14.12.2, which will appear as a new row in the list of data partitions
located at the base of the data window. To set up an analysis on the data partition, you will need to specify a number of options under each column of the
partition list, which can be modified by clicking on the corresponding button.
(Note that each partition row has an associated set of buttons, distinct from the
buttons located beside each column heading that specify options for all partitions

in the list.) For example, under the column heading "`Partition Type`" there are three options available: nucleotide, dinucleotide, and codon. To obtain global estimates of non-synonymous and synonymous substitution rates, we specify that the partition contains codon data, which requires a user-specified reading frame and genetic code to be selected in the window that appears.

A partition must also be associated with a tree topology and substitution model, which are specified in the next two columns of the partition list. A tree can be imported from a file by selecting "`Read Tree From File...`" from the pop-up menu under the column heading "`Tree Topology`." (If a tree was included in the imported NEXUS file, then it will also appear as a selectable option.) Selecting the "`Infer Tree`" option allows you to access one of the many available methods in HyPhy to estimate a tree *de novo* (see Section 14.12.6). Finally, selecting the "`Create New Tree...`" option allows you to specify an arbitrary tree topology by either choosing one of several template topologies, or inputting a Newick tree string. In most cases, however, you will want to select one of the other two options.

A limited number of standard substitution models are available for each type of partition under the column heading "`Substitution Model`." For example, the substitution models that are available for a codon partition are either based on Goldman–Yang (GY94) (Golman & Yang, 1994) or Muse–Gaut (MG94) (Muse & Gaut, 1994) rate matrices. Many models based on MG94 matrices are further "crossed" by one of several standard nucleotide rate matrices, as indicated in HyPhy by the concatenation of "MG94" with a standard nucleotide model abbreviation (e.g. MG94xHKY85). Equilibrium codon frequencies are either estimated from the overall frequency of nucleotides, or from nucleotide frequencies specific to each codon position (indicated in the model specification string by the suffix "3×4"). The standard codon model generated by crossing MG94 with the REV nucleotide model (MG94xREV_3x4) provides a general model which can be constrained down to a simpler version as needed (see the Box 14.8).

For any model that depends on more than one rate parameter, HyPhy can either obtain local parameter estimates for each branch in the tree, or global parameter estimates that are shared by all branches in the tree. Moreover, HyPhy can allow global parameter estimates to vary across codon positions in the data partition (i.e. rate heterogeneity). If this option is selected, then a field will appear under the column heading `Rate Classes`, which can be edited by the user. Model parameter options can be specified for a data partition under the column heading `Parameters`. Whichever option is selected for analyzing a codon partition will determine the total number of parameters to be estimated, and thereby the length of time required to optimize the likelihood function. For example, estimating a local MG94 model effectively doubles the number of parameters because the rates α and β are being estimated for every branch in the tree.

Box 14.8 Estimating the global *dN/dS* ratio with confidence intervals

Estimate the global ω for `InfluenzaA_H3_Random35.nex` using the GUI. Import the alignment into a data panel viewer via `File>Open>Open Data File` and create a partition with the entire length of the sequence. Change data type of the partition to `Codon`. In the dialog which appears, ensure that `Universal` genetic code is selected and rename the partition to HA (for brevity). Select `InfluenzaA_H3_Random35_tree` for the tree topology (this tree was read automatically from the alignment file), `MG94xREV_3x4` substitution model with `Global` parameters and `Partition` based equilibrium frequencies. Note that the light icon in the bottom left corner of the data panel window has changed to yellow from red, indicating that we have provided enough information for HyPhy to build a likelihood function. Execute `Likelihood>Build Function` and examine console output for confirmation that the function has been built. Note that the HA partition name has been converted to boldface, to indicate that this partition is a part of an active likelihood function and the light icon has turned to green. Before we optimize the likelihood function, we need to constrain the REV model down to (012212) – a best fitting nucleotide model. Open the parameter viewer (the table button in the bottom left panel of the data viewer). In the parameter table, select the rows for global rate parameters HA_Shared_AT (see *http://www.hyphy.org/docs/HyphyDocs.pdf* for a HyPhy primer including variable naming conventions and many other useful tips, HA_Shared_CG and HA_Shared_GT (hold the Shift key down and click to select multiple rows), and click on the constrain button, forcing all parameters to share the value of HA_Shared_AT. Note how the display for two of the tree variables has changed to show the constraints. Also, double-click in the constraints window for HA_Shared_CT and enter "1" (to enforce $\theta_{AG} = \theta_{CT} = 1$). Finally, select `Likelihood>Optimize LF` to obtain MLE of parameter values. When the optimization has finished (1–3 minutes on a desktop), HyPhy will refresh the parameter table with derived parameter estimates (see Fig. 14.9). The global ω estimate, represented by HA_Shared_R has been estimated at 0.495. Select the row with this variable and execute `Likelihood>Covariance, Sampler and CI` to obtain profile likelihood error estimates on ω (`Likelihood Profile [chi2]` with significance of 0.95). The reported 95% confidence interval bounds are 0.419 and 0.579. Once a likelihood function has been fitted to the data, it is immediately possible to simulate data under the model(s) in the function via the `Data>Simulation` menu. Finally, save the likelihood function with all parameter estimates by switching back to the data panel, executing `File>Save>Save`, and choosing `Include sequence data, NEXUS` as a format option in the save dialog.

After an analysis has been set up for the codon partition, a likelihood function can be constructed by selecting the menu option "`Likelihood > Build Function`." When building a likelihood function, HyPhy will apply an algorithm in order to improve the efficiency of likelihood evaluation, which can substantially reduce the amount of time required to estimate the model parameters (Kosakovsky Pond & Muse, 2004).

Box 14.9 Estimating the global *dN/dS* ratio using a standard analysis batch file

Estimate the global ω for `InfluenzaA_H3_Random35.nex` using `Standard Analyses>Basic Analyses>AnalyzeCodonData.bf`. Select `MG94CUSTOM`, `Global`, `012212`, and use the tree provided in the file. When the analysis is finished, HYPHY will report global ω as $R = 0.495$. This value may be slightly different from the one found through the GUI, because of how initial guesses for the optimization procedure are obtained. Once an analysis has finished, look at the `Analysis>Results` menu – it contains a list of *post processors* which HYPHY can execute after most standard analyses. For instance, the `Syn and non-syn trees` option can be used to display the trees scaled on the expected numbers of synonymous and non-synonymous substitution per codon site, and on *dS* and *dN*. Note that because the model has a shared ω for all branches, all four trees have proportional branch lengths. Finally, many of the GUI tools can be applied to models fitted by standard analysis. To do this, open the likelihood function parameter table, using the `Object Inspector`.

14.13.2 Fitting a global model with a HYPHY `batch file`

There are also a number of template batch files listed in the `Standard Analyses` menu that will fit a codon substitution model to a data partition. The most appropriate procedure for estimating the global rates α and β, however, has been implemented in the batch file `AnalyzeCodonData.bf`, which can be selected from the `Basic Analyses` submenu (Box 14.9). A number of codon models are available in this batch file that have not been implemented in the HYPHY GUI.

As before, we have the option of fitting the model locally or globally, and can allow global parameter estimates to vary across codon positions in the data partition. In addition, this batch file can model variation across codon positions according to a hidden Markov model, which assumes that the evolutionary rates at adjacent positions are more similar, according to an autocorrelation parameter (λ) that is estimated by the model. There are also many more distributions available for modeling rate variation across codon positions in `AnalyzeCodonData.bf` (e.g. beta, log-normal, and mixture distributions); not all of these distributions are available via HYPHY GUI. If the data partition was imported from a NEXUS file containing a tree, HYPHY will ask if you wish to use this tree in fitting the model. If not, then you will be prompted to select a tree file.

14.14 Estimating branch-by-branch variation in rates

As noted in Section 14.6, it is unreasonable to assume that evolutionary rates remain constant over time. For example, a sudden change in the environment affecting one lineage may be manifested as episodic selection (Gillespie, 1984), which may become averaged out over the entire tree to an undetectable level. In

HYPHY, it is possible to assign a unique set of substitution rates to every branch in a tree, by locally fitting a codon substitution model. However, because the number of parameters in a local model is approximately proportional to the number of sequences, this may require an exceedingly long time to compute. It also does not provide a robust framework for hypothesis testing (e.g. whether a specific branch evolves at a significantly different rate than the rest of the tree). Hence, many procedures for selection inference that allow branch-by-branch variation in rates require the investigator to pre-specify which branches evolve under a particular model (Messier & Stewart, 1997; Yang, 1998), as discussed in Section 14.6. Several procedures of this type are implemented as batch files in HYPHY. In the batch file `SelectionLRT.bf`, a single branch is chosen to partition the other branches of the tree into two clades, for which the model parameters are estimated separately. Another batch file called `TestBranchDNDS.bf` allows you to test whether the strength of selection is significantly different for an arbitrary selection of branches in contrast to the rest of the tree.

To demonstrate the use of these methods in HYPHY, we will use a data set containing 20 sequences of influenza A serotype H3 viruses that have been isolated from waterfowl and mammals (mostly equine). Severe outbreaks in horse populations have been caused by equine influenza A virus, with mortality rates as high as 20%. The equine and waterfowl-isolated sequences form two distinct clades in a reconstructed phylogeny. As waterfowl are a main reservoir of influenza A virus infection, the transmission of virus populations to equine hosts may be accompanied by strong positive selection.

14.14.1 Fitting a local codon model in HYPHY

Fitting a local codon model is very similar to the procedure for fitting a global model, as demonstrated in Section 14.13. Whether using the partition list menus to setup an analysis in the HYPHY GUI, or while executing a batch file, you will almost always have the option to specify whether to fit a model globally or locally. Local models provide a useful exploratory tool, identifying branches in the tree in which strong selection has occurred and providing an alternative model for hypotheses testing. Unless a panel displaying the model parameter estimates was automatically spawned after optimization of the likelihood function, the most convenient means for viewing local parameter estimates is to access the `Object Inspector` panel (see Section 14.12.3) and select the corresponding likelihood function object. Figure 14.9 depicts a likelihood function panel in which all global and local parameters are displayed.

Each branch in the tree is associated with local parameter estimates of α and β, labeled as "`synRate`" and "`nonSynRate`," respectively. A quick procedure for calculating the ratio β/α for each branch in the tree requires you to open a tree

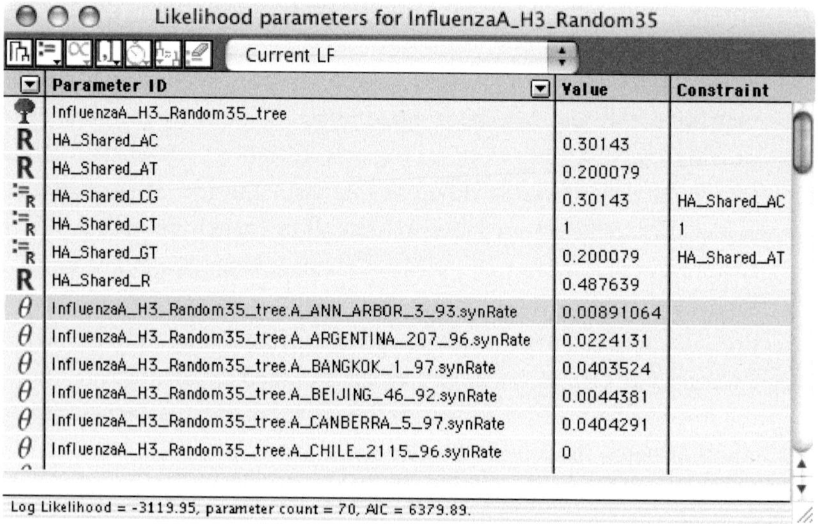

Parameter ID	Value	Constraint
InfluenzaA_H3_Random35_tree		
HA_Shared_AC	0.30143	
HA_Shared_AT	0.200079	
HA_Shared_CG	0.30143	HA_Shared_AC
HA_Shared_CT	1	1
HA_Shared_GT	0.200079	HA_Shared_AT
HA_Shared_R	0.487639	
InfluenzaA_H3_Random35_tree.A_ANN_ARBOR_3_93.synRate	0.00891064	
InfluenzaA_H3_Random35_tree.A_ARGENTINA_207_96.synRate	0.0224131	
InfluenzaA_H3_Random35_tree.A_BANGKOK_1_97.synRate	0.0403524	
InfluenzaA_H3_Random35_tree.A_BEIJING_46_92.synRate	0.0044381	
InfluenzaA_H3_Random35_tree.A_CANBERRA_5_97.synRate	0.0404291	
InfluenzaA_H3_Random35_tree.A_CHILE_2115_96.synRate	0	

Log Likelihood = -3119.95, parameter count = 70, AIC = 6379.89.

Fig. 14.9   Inspecting a likelihood function in HyPhy.

viewer window by "double-clicking" on the first row in the likelihood function panel, which is labeled by a tree icon. Select all branches by choosing the menu option "Edit > Select All," and then create a new parameter for each branch by selecting "Tree > Edit Properties," clicking on the "plus" icon in the panel that appears, and entering the string "nonSynRate/synRate" in the Formula window. Once this panel is closed, this new variable will be available for scaling branch lengths in the tree viewer window, in the Branch Scaling menu.

It is not unusual for estimates of α or β to converge to zero for one or more branches in a tree. Clearly, this outcome implies that there is insufficient sequence variation to support a branch of non-zero length in that part of the tree. If this occurs, then local estimates of the ratio β/α may assume the undefined value 0/0 or ∞. As a result, the tree will not be displayed properly in the HyPhy tree viewer window when it is scaled to β/α. Selection analyses performed on such trees, however, will not be affected by these poorly defined branch lengths.

Although locally fitting a codon substitution model can provide a detailed picture of branch-by-branch variation (Box 14.10), we strongly caution against re-applying the outcome from this method to other analyses. For example, it is tempting to use a local fit as an initial screen for branches with accelerated evolution, and then specify those branches in a hypothesis-testing procedure on the same data (e.g. to obtain a "p-value"). However, this unfairly biases the analysis towards obtaining a significant result.

Box 14.10 Comparing the fit of a global and a local *dN/dS* ratio model

---

Fit a local model to the sample of 12 bird and 8 mammalian Influenza/A/H3(HA) sequences (file `MammalsBirds.nex`). Import the alignment (`File>Open>Open Data File`) into a data viewer, create a partition with all sequence data, convert it to a codon partition, select the topology included with file, apply the `MG94xREV_3x4` substitution model with `Local` parameters, and frequencies estimates from `Partition`. Fit the model `Likelihood>Build Function` followed by `Likelihood>Optimize`. The procedure completes in a few minutes on a desktop, and should yield a likelihood score of -4837.85. Next, we will demonstrate how to test hypotheses with HYPHY GUI by establishing that this tree has variable dN/dS along its branches by comparing the local model fit with the global (single rate model fit). First, we save the likelihood function state describing the local model, by switching to the parameter table display, and choosing `Save LF state` from the pull-down menu at the top of that window. Enter `Local` when prompted to name the likelihood function state. Next, select this model as the alternative hypothesis, by choosing `Select as alternative` from the pulldown menu. Proceed to define the global model by constraining the β/α ratio for every branch to be the same for all branches in the tree. This is equivalent to setting *treeName.branchName.nonSynRate := globalRatio** *treeName.branchName.synRate* for every branch in the tree. The fastest way to apply this constraint is to invoke `Likelihood>Enter Command` to bring up an interface with the HYPHY batch language command parser, and type in `global globalRatio=1;ReplicateConstraint("this1.?.nonSynRate:= globalRatio*this2.?.synRate",MammalsBirds_tree,MammalsBirds _tree);`. This code instructs HYPHY to traverse the tree MammalsBirds_tree and apply the constraint to every branch ('?' matches all names). Note how the parameter display table changes to reflect newly imposed constraints. Optimize the likelihood function (`Likelihood>Optimize`) to obtain the global model fit. `Save LF state`, naming it `Global` and `Select as null`. Now that we have defined two hypotheses to compare, one can use the LRT (for nested models), and parametric or non-parametric bootstrap to evaluate support for the alternative hypothesis. Select LRT from the parameter table pull-down menu. The likelihood ratio test statistic in this case is 100.05, and there are 36 constraints, resulting in a very strong ($p \approx 10^{-7}$) support in favor of the local model. If you now save the likelihood function from the data panel window, all hypotheses defined in the parameter table will be preserved for later.

14.14.2 Interclade variation in substitution rates

We have *a priori* reasons to expect that the ratio β/α could vary significantly between the waterfowl and equine clades, because the virus is evolving in two distinct environments. We have previously investigated these phenomena in epidemiologically linked patient pairs of HIV patients (Frost *et al.*, 2005). To evaluate the support for this hypothesis, use the batch file `SelectionLRT.bf` specifying

Box 14.11 Compare selective pressures between different clades

Compare selective pressures on waterfowl and equine clade Influenza sequences from the file MammalsBirds.nex. Execute StandardAnalyses > Compartmental- ization > SelectionLRT.bf using the best fitting (012343) nucleotide model, and selecting Node1 – the root of the waterfowl clade to define one of the clades (Clade A) of interest. To verify that this is indeed the correct node, you can use the tree viewer. The equine–waterfowl data is best explained (based on AIC) by a fully unconstrained model (MG94x012343) in which β is estimated independently for each clade and the separating branch. For the equine clade, β/α was estimated to be 0.223 (95% CI: 0.158, 0.302). In contrast, β/α for the waterfowl clade was estimated to be 0.058 (95% CI: 0.038, 0.083), and the estimate for the internal branch separating the two clades was lower still ($\beta/\alpha = 0.033$; 95% CI: 0.026, 0.041).

the internal branch which separates the clades of interest (Box 14.11). This batch file is executed through the Standard Analyses menu, under the submenu Compartmentalization. Upon execution of this batch file, HyPhy prompts the user to specify:

  (i) a codon translation table (genetic code);
 (ii) the file containing the protein-coding sequences;
(iii) a PAUP* nucleotide model specification string (e.g. 010020 for TN93 (Tamura & Nei, 1993));
 (iv) the file containing the corresponding tree;
  (v) and the branch separating two clades in the tree.

Subsequently, HyPhy will globally fit the codon model (MG94 crossed by the specified nucleotide model) to the data, before iteratively evaluating models in which estimates of β for each clade and the internal branch are either independent or constrained to be equal (for a total of five phases of model fitting). To determine whether a nested model provides a significant improvement of fit to the data, HyPhy reports p-values from the likelihood ratio test (Yang, 1998) and the Akaike information criterion (AIC) values, which adjust likelihood ratios for the difference in the number of model parameters (Akaike, 1974). HyPhy also provides 95% confidence bounds, derived from the likelihood profile (see Section 14.5.4), for estimates of the ratio β/α for each clade and the internal branch.

14.14.3 Comparing internal and terminal branches

A useful hypothesis that can be tested in viral sequences is whether the relative rate of non-synonymous substitution varies between terminal and internal branches of the tree. Finding a significant difference between these classes would suggest that selection on a virus population within a host was distinct from selection for

---

Box 14.12 Testing for *dN/dS* variation among branches

Use `TestBranchDNDS.bf` from the `Positive Selection` rubrik of `Standard Analyses` to test whether terminal branches in `MammalsBirds.nex` evolve with *dN/dS* different from the rest of the tree (and each other). Use the (012343) nucleotide model, and begin by choosing `None` for site-to-site rate variation and `Default` for amino acid bias models. When prompted to select branches of interest, use shift-click (or control-click to highlight a range) to select all 20 terminal branches. Because there are 20 terminal branches, the alternative model has 20 additional parameters compared to the null model. Despite the addition of so many parameters, this model was significantly favored over the null model by a likelihood ratio test ($\chi^2_{20} = 68.87$, $p \approx 2 \times 10^{-7}$). The examination of ω values for each branch reported to the console indicate great variability from tip to tip, with several showing accelerated non-synonymous rates (but note wide 95% confidence intervals which highlight the limitations in our ability to infer tight ω estimates for a single branch). As an additional exercise, examine the effect of allowing site-to-site rate variation on the conclusions of the model.

---

transmission among hosts. Support for this hypothesis can be evaluated by the batch file `TestBranchDNDS.bf`, which is found in the `Standard Analyses` menu, under the heading `Positive Selection` (Box 14.12). The model selection procedure implemented in this file is similar to the previous file, attempting to fit a global model in which β is constrained to the same value (`SharedNS1`) on all branches, before relaxing this constraint for an arbitrary selection of branches.

However, there are several additional options that are available in `TestBranchDNDS.bf`. For example, you can also model site-by-site variation in rates (see Section 14.15), either in β only, or under a "complete" model in which α and β are both sampled from independent distributions[†]. As always, HYPHY will prompt you to specify the number of rate classes when modeling rate variation. In addition, you can weight the codon substitution rates according to amino acid classifications, such that substitutions between similar amino acids are assumed to be more common.

14.15 Estimating site-by-site variation in rates

So far, estimates of β/α have represented an average over all codon positions in the gene sequence. But there are many reasons to expect substantial site-by-site variation in these rates, as discussed in Section 14.7. Counting and fixed-effects

---

[†] In either case, rate variation for β and/or α is being modeled by separate gamma distributions, each partitioned into discrete classes according to a beta distribution whose parameters are also estimated (Kosakovsky Pond & Muse, 2005).

likelihood (FEL) methods for evaluating site-specific levels of selection have been implemented in HʏPʜʏ as a single batch file named `QuickSelectionDetec-tion.bf` (some of the options lead to methods that are not terribly quick!), which is found in the `Standard Analyses` menu under the heading `Positive Selection`. Random-effects likelihood (REL) methods have also been implemented in HʏPʜʏ and are described in the next section. In practice, however, all three classes of methods for site-by-site inference of selection converge to very similar results (Kosakovsky Pond & Frost, 2005b).

Counting methods (e.g. single likelihood ancestor counting, SLAC) are the most efficient and are well suited to analyzing large data sets (i.e. over 40 sequences), but can be more conservative than the other methods. On the other hand, FEL methods are far more time-consuming but more sensitive, and may be more successful at detecting selection in smaller data sets. REL methods are even slower and may suffer from a high false-positive rate for small data sets.

14.15.1 Preliminary analysis set-up

The batch file `QuickSelectionDetection.bf` was designed to provide a versatile and powerful array of methods for detecting site-by-site variation in selection. As a result, there are several options that need to be specified before an analysis can be carried out. This batch file proceeds in four phases: (i) fitting a nucleotide model; (ii) generating a codon model approximation; (iii) fitting the approximate codon model; and (iv) ancestral state reconstruction and substitution counting. A general outline for the preliminary set-up of an analysis follows:

(i) `Choose Genetic Code` – Select a genetic code for codon translation. A comprehensive list of codon translation tables covering a broad range of taxa has been built into HʏPʜʏ.
(ii) `New/Restore` – Restore nucleotide model parameter estimates from a previous analysis. Choosing "`Restore`" requires you to select a file that contains a previously exported likelihood function, which will also contain the sequence data and tree. This option allows you to iteratively run the batch file under different settings without having to re-optimize the same nucleotide model for every session. It also provides a convenient means of importing the best-fitting nucleotide model from the automated model selection procedure implemented in the batch file `Nuc-ModelCompare.bf` (see Section 14.12.7). Choosing "`New Analysis`" will prompt you to select a file containing the protein-coding sequences.
(iii) `Model Options` – (`New Analysis` only) Select a nucleotide model to fit to your data. Choose "`Default`" to fit an HKY85 nucleotide model. Otherwise, you will be prompted to enter a "custom" Pᴀᴜᴘ* model specification string. HʏPʜʏ subsequently prompts you to select a file containing a tree corresponding to your sequences, and second file for exporting the fitted nucleotide model.

14.15.2 Estimating β/α

Because a codon model contains a large number of parameters, it is impractical to optimize all of them for a large number of sequences. To accelerate estimation under codon models, HYPHY applies the branch lengths and nucleotide substitution rate parameter estimates from the nucleotide model to approximate the analogous parameters of the codon model (Kosakovsky Pond & Frost, 2005b). As a result, it is potentially important to select a nucleotide model that can provide a reasonably accurate fit to the data (see Section 14.12.7).

This approximation scheme introduces a global scaling parameter, called rConstr, that is shared by all branches in the tree. This scaling approximation is based on the observation that the joint distributions of branch lengths in nucleotide and codon models tend to be highly correlated (Yang, 2000), such that all branch lengths from the nucleotide model can be adjusted upwards by a given factor to approximate codon branch lengths. The scaling factor will be reported in the HYPHY console window during the analysis.

Thus, the next step in setting up our analysis is to specify how the global parameters β/α and rConstr are handled during optimization of the approximate codon model. These methods are listed under the heading "dN/dS bias parameter options" as:

- "Neutral," to constrain β/α to 1.
- "User," to prompt for a constant value (>0) for constraining β/α, if for instance it had been estimated in a previous analysis.
- "Estimate," to estimate β/α and rConstr from the data.
- "Estimate + CI," to estimate β/α and rConstr, and calculate confidence intervals for β/α.
- "Estimate dN/dS only," to estimate β/α and constrain rConstr to be calculated directly from β/α and nucleotide model parameters.

The first two options are provided for model selection, i.e. calculating the improvement of fit from incorporating β/α into the model as a global parameter.

At this point, it remains only to specify which counting or FEL method to use before proceeding with the selection analysis. The available methods are listed under the heading "Ancestor Counting Options." In the following sections, we will discuss the practical aspects of applying two of these methods; namely, single-likelihood ancestor counting (SLAC), and two-rate FEL.

14.15.3 Single-likelihood ancestor counting (SLAC)

As discussed in Section 14.7.3, the nucleotide and codon model parameter estimates are used to reconstruct the ancestral codon sequences at internal nodes of the tree. The single most likely ancestral sequences are then fixed as known variables, and applied to inferring the expected number of non-synonymous or synonymous substitutions that have occurred along each branch, for each codon position. This procedure requires you to specify the following:

(i) SLAC Options – Apply ancestral reconstruction and counting to the entire tree at once (Full tree), or as two separate analyses for terminal and internal branches of the tree, respectively (Tips vs. internals).

(ii) Treatment of Ambiguities – Ambiguous reconstructions of ancestral codons are averaged over all possible codon states (Averaged), or resolved into the most frequent codon (Resolved). The latter is more appropriate when ambiguous codons may have been due to sequencing errors, as opposed to being representative of sequence polymorphism (e.g. as in bulk viral sequences).

(iii) Test Statistic – Use the continuous extension of the binomial distribution (Approximate), or simulate a null distribution from the data (Simulated Null) – this is an experimental and very slow option. For assigning a p-value to whether β is significantly different from α at a given site, you will be prompted for a significance level in the HYPHY console window.

(iv) Output Options – Spool the output to the HYPHY console window (ASCII Table); write the output as a tab-separated table into a file (Export to File); or display the output as a graphical chart in a separate window (Chart). The last option is only available in GUI versions of HYPHY. You will be prompted for a file name if you select Export to File.

(v) Rate class estimator – Skip the estimation of the number of β and α rate classes (Skip), or approximate the number of classes from the data (Count).

The output of a SLAC analysis consists of 12 columns (see Box 14.13 for an explanation), and as many rows as there are codon positions in the alignment. All codon positions with significant positive or negative selection, according to the user-defined significance level, are automatically reported to the HYPHY console window.

14.15.4 Fixed effects likelihood (FEL)

The procedure "Two rate FEL" fits the rate parameters α and β to each codon position independently in order to accommodate site-by-site variation. User configurable options specific to FEL include the significance level (P-value) for the likelihood ratio test and which branches should be tested for selection.

(i) All. The entire tree is tested for evidence of non-neutral evolution.

(ii) Internal Only. Only interior branches are tested for evidence of non-neutral evolution, while terminal branches share an arbitrary β/α ratio.

(iii) A subtree only. Only branches in a given subtree (clade) are tested, while the rest of the branches share an arbitrary β/α ratio.

(iv) Custom subset. User selects a set of branches to be tested, while the rest of the branches share an arbitrary β/α ratio.

Box 14.13 A single-likelihood ancestor counting exercise

Conduct a SLAC analysis on the `InfluenzaA_H3_final.nex` file selecting the
`(012321)` nucleotide model, `Estimate` dN/dS `only` option for the estimation
of dN/dS, `Single likelihood ancestor` for the "Ancestor options" dialog,
selection analysis over the `Full tree`, `Averaged` for the treatment of ambiguities,
`Approximate` test statistic, p-value of 0.05, `Chart window` for result display and
`Skip` the *post-hoc* rate class counter. The entire analysis should take 5–10 minutes
on a desktop computer and identify 8 (128, 135, 138, 145, 194, 226, 275, and 276)
positively and 75 negatively selected codons. A chart displayed at the end of a SLAC
analysis shows detailed inference for each site, and a dS vs. dN plot codon by codon
(Fig. 14.10). HYPHY charts can be saved and reloaded (`File>Save>Save Chart`), and
their contents can be exported in a variety of formats (`File>Save>Save Table`). To
understand how to read the output for a given codon, look at the entries corresponding
to codon 226. Reading the columns left to right, we find out that SLAC inferred 2.5
synonymous and 44.5 non-synonymous substitutions at codon 226; the codon has an
expected 1.1 synonymous sites (out of 3) and 1.8 non-synonymous sites. The observed
proportion of synonymous substitutions was $2.5/47 = 0.05$, whereas the expectation was
much higher $1.10/(1.10 + 1.82) = 0.37$. $dS = 2.5/1.1 = 2.3$, $dN = 44.5/1.82 = 24.4$
and $dN - dS = 22.14$ are reported next (dN/dS would often be undefined or infinite
because of $dS = 0$ – hence the difference is reported). Based on the binomial distribution on 47 substitutions with the expected proportion of synonymous substitutions
under neutrality of 0.37, we find that the probability of observing 2.5 or fewer synonymous substitutions is very low (i.e. the p-value is 5×10^{-7}), suggesting that the site is
under strong positive selection. The probability of observing 2.5 or more synonymous
substitutions is ≈ 1 (arguing against negative selection). Lastly, $dN - dS$ scaled by the
total length of the tree is 15.7 (this scaling enables the comparison across different data
sets). Finally, experiment with the built-in data processing tools in the chart window. For
example, click on the "dN–dS" column heading to select the entire column, then execute `Chart>Data Processing>Descriptive Statistics`. This command
will report a number of standard descriptive statistics computed from the selected values
to the console, including the mean of -1.805.

While the FEL analysis iterates through every codon position in the alignment,
it will report estimates of the ratio β/α and the corresponding log-likelihood,
likelihood ratio test statistic, and p-value to the console window for each position (Box 14.14). Positions under significant positive or negative selection are
identified on the right margin by the symbol "*P" or "*N". At the end of the
analysis, GUI versions of HYPHY will show a chart window displaying analysis
results, and in all cases the user will be prompted to specify a file to save a
comma separated output that is suitable for subsequent analyses in a spread-
sheet program, HYPHY GUI – via the `Open>Open Table` menu, or a statistical
package.

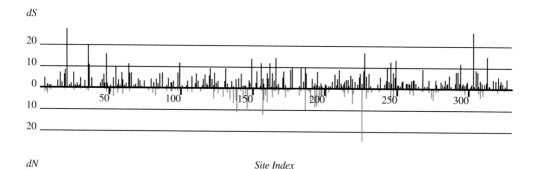

Fig. 14.10   *dS* and *dN* estimated from the 349 Influenza sequences using SLAC.

Box 14.14 A fixed effects likelihood exercise in HYPHY

Conduct a FEL analysis on a sample of 12 bird and 8 mammalian Influenza/A/H3(HA) sequences (file `MammalsBirds.nex`). Confirm that the best nucleotide model is (012343) and then use it to obtain the nucleotide model fit. First, run the `Two rate` FEL analysis on the entire tree using $p = 0.1$. Sample output for two of the codons is shown below.

```
| Codon: 175| dN/dS: 0.54| dN: 0.57| dS: 1.05| dS(=dN): 0.75| Log(L): -10.90|
LRT: 0.16| p: 0.69

| Codon: 176| dN/dS: inf | dN: 2.35| dS: 0.00| dS(=dN): 1.32| Log(L): -16.67|
LRT: 3.62| p: 0.06 *P
```

The analysis should finish in about 30 minutes on a modern desktop. Two codons (23 and 176) should be found under selection with $p < 0.1$. Repeat the FEL analysis on our cluster using *http://www.datamonkey.org*. Now the analysis completes in about a minute. Next, conduct a FEL analysis with the same settings except test for selection only within the mammalian clade, rooted at internal node `Node4` (see Fig. 14.11); this test will also include the long branch separating viral samples from birds and mammals. Codons 13, 209, 277 should be found under `selection`. Lastly, use FEL to do a "branch-site" type analysis on `Node4` (the separating branch). Codons 9, 11, 13, 15, 127, 176, 277, 458 are found to be under selection. Run the SLAC analysis using *http://www.datamonkey.org*, and use the `Inferred Substitutions` link from the results page to visualize the evolutionary history of each of the selected codons (see Fig. 14.11 for two stylized examples).

For every codon, the output from a FEL analysis consists of eight columns (or seven if the entire tree is used for testing): dN/dS – the ratio of β/α for the branches of interest (could be undefined or infinite if α is estimated to be 0); dN – the β estimate for the branches of interest; dN – the α estimate for the entire tree; $dS(=dN)$ – the $\alpha = \beta$ estimate for the branches of interest under the hypothesis of neutral evolution; $\texttt{Log(L)}$ – the likelihood score of a given site (only reported for

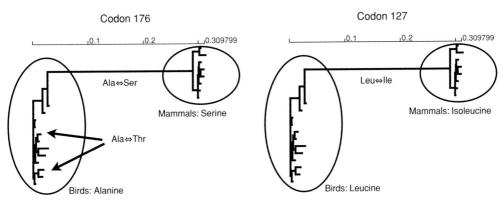

Fig. 14.11   The tree of sequences contained in `MammalsBirds.nex`, with 8 samples from mammals and 12 from birds. Codon 176 is an example of ongoing diversifying selection (in birds), while codon 127 shows what looks like a selective sweep in one of the populations, which can only be detected with a "branch-site" type method.

variable sites); LRT – the likelihood ratio test statistic for non-neutral evolution at a given site; P – the p-value for the likelihood ratio test; `dN_other` – the β estimate for the background branches (except if the `All` branches option is selected).

You can also use HyPhy to determine whether the site-by-site patterns of selection vary from the internal branches to the tips of the tree. A procedure for performing this analysis has been implemented in the batch file `Subtree-SelectionComparison.bf`, under the heading `Positive Selection` in the `Standard Analyses` menu. This analysis performs a site-by-site FEL estimation, but instead of testing whether $\alpha \neq \beta$ for some branches at a given site, it tests whether $\beta_1 \neq \beta_2$ at a given branch, where β_1 is applied to a set of selected branches, and β_2 – to the rest of the tree.

14.15.5 REL methods in HyPhy

Like the previous batch file, `dNdSRateAnalysis.bf` (located in the `Codon Selection Analyses` rubrik) requires the user to import files containing the sequences and a tree, and to specify a genetic code. (Recall that dN and dS are alternative notations for β and α, respectively.) In addition, it prompts the user to specify the following options (for complete details, see Kosakovsky Pond & Muse (2005)):

(i) `Branch Lengths` – Re-estimate branch lengths jointly with the codon rate parameters (`Codon Model`); or use the values proportional to those estimated in the nucleotide model before optimizing the codon model (`Nucleotide Model`), taking advantage of the strong correlation between nucleotide and codon branch lengths. For large data sets (e.g. >25 sequences), the `Nucleotide Model` option can cut computational time by several orders of magnitude.

(ii) `Rate Matrix Options` – Select one of several standard models of codon substitution rates to estimate from the data. `MG94Multi` permits multiple classes of β, adjusted according to one of several amino acid similarity matrices (selected at a later stage, under the heading `Amino Acid Class Model`).

(iii) `Rate Variation Options` – Evaluate all available models of rate variation (`Run All`), or an arbitrary selection from a subsequent list of models (`Run Custom`).

(iv) `Rate Variation Models` (Run Custom) – `Constant` assumes that neither α nor β vary across sites (simple global mean model); `Proportional` allows α and β to vary, but constrains β/α to be constant, assuming that the variability in substitution rates can be well explained by local mutation rates (this model is not expected to fit well); `Non-synonymous` constrains $\alpha = 1$ while allowing β to vary, making it very similar to the models implemented in **PAML**; `Dual` draws values of β and α from independent or correlated distributions. You can hold "shift" while clicking to select more than one option, so that multiple models will be evaluated in sequence. To run REL, select `Dual`. In addition, to test for evidence of synonymous rate variation, select `Non-synonymous` (together with `Dual`). A likelihood ratio test (Section 14.7.5) will then be used to derive a p-value to decide whether or not α is variable in a given data set.

(v) `Distribution Option` – Values of β and/or α are drawn from independent gamma distributions (`Syn:Gamma`, `Non-syn:Gamma`); β is instead drawn from a mixed gamma distribution with a class for invariant sites (`Syn:Gamma`, `Non-syn:Inv+Gamma`); β and α are drawn from independent general discrete distributions (GDDs[†], `Independent Discrete`), correlated GDDs (`Correlated Discrete`), or constrained GDDs such that $\beta \leq \alpha$ (`Non-positive Discrete`).

A good default setting is to use independent GDD distributions with three bins each (or two bins each for very small, or low divergence data sets). One can try several runs with various numbers of bins, starting with two each, and then increasing the number of bins by one (β followed by α) until the AIC score for the `Dual` model no longer decreases.

(vi) `Initial Value Options` – Use default or randomized initial values for parameters of the rate distributions. Repeated runs that make use of randomized initial values can be used to verify that the maximum likelihood found by **HYPHY** is indeed a global maximum.

The batch file will prompt you to specify a file (call it `ResultFile`) for saving the analysis results (Box 14.15). The analysis will generate two summary files: `ResultFile` containing a copy of console output and `ResultFile.distributions` listing maximum likelihood estimates for

---

† GDDs are useful non-parametric distributions that make a minimal number of assumptions about the true underlying distribution of rate variation across sites.

Box 14.15 A random effects likelihood analysis of variation in synonymous and non-synonymous rates

---

Conduct a REL analysis on a sample of 35 randomly chosen Influenza/A/H3(HA) sequences (file `InfluenzaA_H3_Random35.nex`) using nucleotide-based branch lengths, MG94×CUSTOM with (012212) correction model, Non-synonymous and Dual rate variation models, GDD distributions with three classes for α and three for β. Sample console output is shown below.

```
RUNNING MG94x012212 MODEL COMPARISONS on /home/sergei/Latest/HYPHY_Source/data/InfluenzaA_
H3_Random35.nex

########## 3x3 CLASSES ##########
```

	Log	Synonymous	NS Exp	N/S Exp				
Model	Likelihood	CV	and CV	and CV	P-Value	Prm	AIC	
Var. N.Syn.	-3053.54189	N/A	0.50446,	0.50446,	N/A	74	6255.08	
Rates			1.76495	1.76495				
Dual Variable	-3044.93374	1.04164118	0.49665,	1.02586,	0.00175455	78	6245.87	
Rates			1.81033	2.08251				

A large CV (coefficient of variation, defined as standard deviation/mean), for synonymous rates α and a low (0.002) p-value for the test that $CV(\alpha) = 0$ indicate that synonymous rates vary from codon to codon in this alignment. Use `dNdSResultProcessor.bf` analysis to find positively selected sites under both models, based on a Bayes Factor of 50, locating the appropriate `.marginals` file written by the REL analysis when prompted. The list from the Dual model should contain 13 codons (in particular, codons 135, 145, 194, 226, and 275, previously identified by SLAC on the complete alignment), and 20 codons under Non-synonymous model. Additional exercises might include: (i) investigating the effect of the number of rate classes on model fit and inference of sites under selection; (ii) identifying the list of sites which are found to be under selection by the Non-synonymous model, but not the Dual model and vice versa; (iii) generating various plots (e.g. α, β by codon position) using `dNdSResultProcessor.bf`.

α, β, β/α rate distributions inferred by each model. In addition, for each rate variation model a `ResultFile.model` (likelihood function fit with that model) and a `ResultFile.model.marginals` (a file containing rate distributions and conditional probabilities of each codon being in a given rate class) will be generated. The `.marginals` can be used as input to `dNdSResultProcessor.bf` (located in the `Codon Selection Analyses` rubrik) – a template file which can be used, in particular, to compute Bayes Factors (or posterior probabilities) of a site being under positive selection.

14.16 Estimating gene-by-gene variation in rates

Ultimately, the estimation of selection pressures on gene sequences is a comparative study, with the implicit objective of finding divergent patterns among genes of branch-by-branch or site-by-site variation in selection. For instance, copies of the same gene may experience strong positive selection at a codon position in one population, but negligible levels of selection in a second population. Drastic changes can easily be caused by environmental differences between populations. Different genes will also undergo different patterns of selection. However, it is more difficult to compare variation in selection at this level without having an overall sequence configuration in common between the two groups. We will discuss how this issue can be resolved through the application of data-driven model inference.

14.16.1 Comparing selection in different populations

Divergence in the site-by-site patterns of selection can be evaluated in HyPhy using batch files `CompareSelectivePressure.bf` and `CompareSelective-PressureIVL.bf`, under the heading `Positive Selection` in the `Standard Analyses` menu. These batch files are set up in much the same way as the preceding examples, except that it requires two sets of files containing sequences and trees from different populations. The former analysis tests the hypothesis that at a given site, β/α differs between the two samples along the entire tree, whereas the latter concerns itself only with *interior* tree branches. Please note, that both alignments must encode homologous sequences (i.e. two samples of the same gene from different populations or environments), and they must be aligned in the same codon coordinates, otherwise the comparisons are meaningless. Finally, large samples (at least 50 sequences in each sample) are required to gain power to discriminate differential evolution.

The output of these analyses will be displayed in a chart (GUI versions) and written to a comma separated file. `CompareSelectivePressure.bf` will report dS, dN, dN/dS estimates for both samples and those under the null model (joint, i.e. when dN/dS is shared between two samples), the LR statistic and the asymptotic p-value (Box 14.16). `CompareSelectivePressureIVL.bf` will print the estimates of dS, dN_{leaves}, $dN_{internal}$ for each sample, and the estimates under the shared $dN_{internal}/dS$ model, the LR statistic and the asymptotic p-value.

Neither analysis prompts for a specific significance level; sites of interest can be culled after the run is complete by selecting all those sites for which p is less than a desired value.

Box 14.16 Comparing site-by-site selective pressures in different populations

---

Compare selective pressures on Influenza neuraminidase gene from two different serotypes: H5N1 ("avian" flu) and H1N1 (the same serotype as the 1918
pandemic). Even though the antigenic properties of neuraminidase are similar, as it belongs to the N1 subtype in both samples, the evolutionary environments are different, because of the effect of other viral genes, and different host environments. Execute `Standard Analyses>Positive Selection>`
`CompareSelectivePressure.bf` on `H1N1_NA.nex` – an alignment of 186 H1N1
sequences and `H5N1_NA.nex` – an alignment of 96 H5N1 sequences. Select the
(`012345`) model for each alignment. The analysis can take a few hours on a desktop or about 15 minutes on a small computer cluster. Nine sites are evolving differentially
at $P \leq 0.01$: 3, 28, 44, 74, 135, 257, 259, 412, and 429 and in all nine cases, stronger
positive selection is indicated for the H5N1 sample. A complementary method that
can be used to compare `distributions` of rates, including proportions of sites
under selection (but not their location) and the strength of selection is outlined in
Section 14.16.2.

14.16.2 Comparing selection between different genes

As we noted in Section 14.5.2, comparing the means of β/α rate distributions
between two data sets to determine whether or not they are under similar selective
pressures can be misleading. `dNdSDistributionComparison.bf` (under
the `Codon Selection Analyses` rubrik in `Standard Analyses`) implements a more general procedure which fits an independent four bin distribu-
tion of (α_s^i, β_s^i) (with bin weights p_s^i) to data set i, with two negatively selected
($\alpha > \beta$) bins, one neutral ($\alpha = \beta$) bin and one positively selected ($\alpha < \beta$) bin
(the "Independent" model). To test for differences between the distributions,
four simpler models are also fitted: (a) same selection strength model with
$\beta_4^1/\alpha_4^1 = \beta_4^2/\alpha_4^2$ ("SharedStrength" model) ; (b) same selected proportion of sites:
$p_4^1 = p_4^2$ ("SharedProportion" model); (c) same selective regime: (a) and (b) combined ("SharedPositiveSelection" model); (d) same distributions – all distribution
parameters are shared between the data sets ("JointAll" model).

    The analysis prompts for two data sets and a nucleotide bias model for each
(Box 14.17). The user can select whether relative branch lengths should be approximated with a nucleotide model, or estimated directly. The former option greatly
accelerates convergence and, in most cases, has very little effect on the estimates of
rates. Lastly, starting parameter values can be set to default values, or chosen
randomly. Optimizing parameter-rich distributions can be numerically unstable, and to alleviate convergence problems, `dNdSDistributionCompari-`
`son.bf` incorporates a series of steps to ensure reliable convergence to a good
maximum.

Box 14.17 Comparing *dN/dS* distributions between two simulated data sets

Compare selective pressures on two 16 sequence simulated data sets Sim1.nex and Sim2.nex. The alignments each consist of 500 codons and were generated with different rate profiles.

	Sim1.nex				Sim2.nex		
dN/dS	dS	dN	Prob	dN/dS	dS	dN	Prob
0.5	1	0.5	0.4	0.5	1	0.5	0.3
0.1	0.5	0.05	0.2	0.2	0.5	0.1	0.3
1.0	1.5	1.5	0.2	1.0	1.5	1.5	0.3
4.0	1	4.0	0.2	10	1	10	0.1

Execute dNdSDistributionComparison.bf on files Sim1.nex and Sim2.nex, specifying Nucleotide branch lengths, (010010) models for both data sets, and Default starting values. The analysis takes about an hour to run on a desktop. The inferred distributions under the independent rates model are listed below:

```
Inferred rates for data set 1:
    dN/dS      dS        dN        Prob
    3.167      1.005     3.184     0.260
    1.000      0.038     0.038     0.003
    0.133      0.758     0.101     0.326
    0.684      1.195     0.818     0.411
Inferred rates for data set 2:
    dN/dS      dS        dN        Prob
   11.947      0.714     8.531     0.104
    1.000      0.000     0.000     0.019
    0.252      0.750     0.189     0.470
    0.847      1.410     1.193     0.407
```

Note that because of the small sample size, inferred distributions are not quite the same as those used for data generation. However, all four tests correctly report that the distributions are different, in general, and in the positive selection regime. We also note that because of the complexity of the models being fitted, it may be advisable to run the analysis several times (using both Default and Random starting values) to verify convergence.

```
Are the distributions different?
     LR = 89.843 DF = 10 p = 0.000
Are selective regimes (dN/dS and proportions) different?
     LR = 25.421 DF = 2 p = 0.000
Are selection strengths (dN/dS) different?
     LR = 15.554 DF = 1 p = 0.000
Are the proportions of codons under selection different?
     LR = 18.457 DF = 1 p = 0.000
```

At the end of the run, HYPHY will report the results of four tests discussed above, write five files with models fits, which can be examined and reused later, print a summary of model fits to the console and echo it to a file.

14.17 Automating choices for HYPHY analyses

Interactive, dialog driven analyses are useful for learning, exploring new options, and running analyses infrequently. However, if a large number of sequence files must be subjected to the same analysis flow, then a mechanism to automate making the same choices over and over again is desirable.

HYPHY provides a mechanism for scripting any standard analysis using *input redirection*. To instruct HYPHY to make any given set of selections automatically, one must first execute the standard analysis for which the selections must be made and make notes of the actual choices being made. For instance, to use the standard analysis `AnalyzeCodonData.bf` with a local $MG94 \times 012232$ substitution model, six choices must be made: genetic code to use (`Universal`), alignment file to import, substitution model (MG94CUSTOM), model options (`Local`), nucleotide bias (`012232`), and the tree to use. Having made these choices, one next creates a text file with a script in the HYPHY batch language which may look like this (assuming that the tree included in the file is to be used for step 6).

```
inputRedirect = {};
inputRedirect["01"]="Universal";
inputRedirect["02"]="/Users/sergei/Desktop/MyFiles/
                     somealignment.nex";
inputRedirect["03"]="MG94CUSTOM";
inputRedirect["04"]="Local";
inputRedirect["05"]="012232";
inputRedirect["06"]='y';
ExecuteAFile (HYPHY_BASE_DIRECTORY + "TemplateBatchFiles"+
DIRECTORY_SEPARATOR+"AnalyzeCodonData.bf", inputRedirect);
```

`inputRedirect` is a data structure (an associative array) which stores a succession of inputs, indexed by the order in which they will be used, and the `ExecuteAFile` command executes the needed standard analysis using some predefined variables to specify the path to that file, using `inputRedirect` to fetch user responses from. All standard analyses reside in the same directory, so this command can be easily adjusted for other analyses. The input for step "02" must, of course, refer to an existing file. Another option is to leave that option blank (""), and have HYPHY prompt just for the file, keeping other options as specified. To execute a file like this, invoke `File> Open> Open Batch File`.

14.18 Simulations

HYPHY has versatile data simulation capabilities. In particular, any combination of site and branch specific α_s^b, β_s^b can be employed to generate codon data.

There are two primary ways for specifying a model to simulate under. If a likelihood function has been defined and optimized then it can be simulated from with a single command. GUI based analyses should use `Data>Simulate` menu from the data panel. Likelihood functions defined otherwise, e.g. via a standard analysis, can be simulated from using the `SimulateFromLF` tool accessible via the `User Actions` button in the console window.

Often times, however, it may be easier to specify all the model parameters, such as base frequencies, branch lengths and selection profiles and simulate replicates from those parameters. We have written a customizable script for simulating both site-by-site rate variation and branch-and-site rate variation, available as HYPHY scripts from *http://www.hyphy.org/pubs/dNdS_Simulator.tgz*. Instructions on how to use and modify simulation scripts are included in configuration files.

14.19 Summary of standard analyses

Here we list all HYPHY standard analyses which can be used to conduct codon-based rate estimates and test selection hypotheses. Please note that the collection of HYPHY analyses is always growing, so new analyses may have been added since this chapter was written.

Analysis	Primary use	Reference
Miscellaneous Phylohandbook.bf	Run example/tutorial analyses referenced in the theory section.	
Basic Analyses AnalyzeCodonData.bf	Estimate mean ω for an alignment; fit a local model where each branch has a separate omega.	Goldman & Yang (1994), Muse & Gaut (1994), Yang (1998)
Codon Selection Analyses dNdSRateAnalysis.bf	REL site-by-site selection. Test for synonymous rate variation. Test for global evidence of positive selection ($\beta > \alpha$) when synonymous rates are variable.	Kosakovsky Pond & Frost (2005b), Kosakovsky Pond & Muse (2005), Sorhannus & Kosakovsky Pond (2006)

Codon Selection Analyses dNdSResultProcessor.bf	Process and visualize results generated with dNdSRateAnalysis.bf	Kosakovsky Pond & Muse (2005)
Codon Selection Analyses dNdSDistributionComparison.bf	Fit a series of REL models to two alignments and compare whether or not the distribution of rates differ between the alignments. Also compares the proportion of sites under and/or the strength of positive selection.	
Compartmentalization SelectionLRT.bf	Splits the tree into a clade of interest and the rest of the tree. Compare mean ω between the clade, the rest of the tree, and the branch separating the two and test for significant differences.	Frost et al. (2005)
Miscellaneous SlidingWindowAnalysis.bf	Fit a global (or local) model to a series of sliding windows. This is an obsolete distribution-free way to estimate spatial rate variation, which has been superseded by FEL to a large extent.	
Miscellaneous MGvsGY.bf	Investigate whether Muse–Gaut or Goldman–Yang codon model parameterization fits the data better.	Goldman & Yang (1994), Muse & Gaut (1994)
Miscellaneous NucModelCompare.bf	Select the best fitting nucleotide model which is an input option to most other analyses.	Kosakovsky Pond & Frost (2005)
Molecular Clock Multiple files	Carry out a molecular clock (or dated molecular clock) test using a codon model. The clock can be imposed on branch lengths (global models), or only on synonymous or non-synonymous distances (local models).	
Positive Selection TestBranchDNDS.bf	Test whether selected branches (foreground) have different ω when compared to the rest of the tree (uniform background). Site-to-site rate variation can also be handled.	Kosakovsky Pond & Frost (2005)
Positive Selection CompareSelectivePressure.bf CompareSelectivePressureIVL.bf	Use FEL to test for differential selection at a site between two samples of the same gene. The entire tree, or only internal branches can be tested.	Yang (1998)
Positive Selection SubtreeSelectionComparison.bf	Use FEL to check whether β/α differ significantly between a user-specified clade and the rest of the tree. Can also test internal versus terminal branches.	

Positive Selection NielsenYang.bf	Use a series of REL models (with constant α) to test for selection. This analysis (with the GY94 model) is identical to PAML analyses, e.g. M8a vs. M8b tests.	Kosakovsky Pond *et al.* (2006)
Positive Selection QuickSelectionDetection.bf	SLAC (and other counting based methods) and FEL site-by-site analyses. FEL also supports the analysis of a part of the tree, enabling branch-site type analyses.	Yang *et al.* (2000), Swanson *et al.* (2003)
Positive Selection YangNielsenBranchSite2005.bf	Use the improved branch-site REL method of Yang and Nielsen (2005) to look for episodic selection in sequences.	Kosakovsky Pond & Frost (2005b)
Recombination SingleBreakpointRecomb.bf	Screen an alignment for evidence of recombination using the single breakpoint method	Zhang *et al.* (2005)
http://www.datamonkey.org	A web interface to run model selection, SLAC, FEL, and REL analyses on our cluster.	Kosakovsky Pond & Frost (2005b)
http://www.datamonkey.org/Gard	A web interface to run recombination detection tools on our cluster.	Kosakovsky Pond *et al.* (2006)
http://www.hyphy.org/gabranch	Run a genetic algorithm to find good fitting models of temporal (branch-to-branch) variation in selection pressure	Kosakovsky Pond & Frost (2005a)

14.20 Discussion

While we have made every attempt to be thorough, this practice section only touches on what can be done in terms of selection analyses. HyPhy and DATAMONKEY have active user communities, and specific questions, can be posted on our bulletin board (*http://www.hyphy.org/cgi-bin/yabb/yabb.pl*). We make every effort to address these questions in a timely and informative manner. We also encourage users to develop their own batch files to implement specific analyses, and share them with the rest of the community. As further funding is secured, we hope to expand the capabilities of DATAMONKEY in order to offer a wider range of computationally intensive analyses to the research community.

Section VI

Recombination

Introduction to recombination detection

Philippe Lemey and David Posada

15.1 Introduction

Genetic exchange, henceforth recombination, is a widespread evolutionary force. Natural populations of many organisms experience significant amounts of recombination, sometimes with markedly different underlying molecular mechanisms. Although mutation is the ultimate source of genetic variation, recombination can easily introduce allelic variation into different chromosomal backgrounds, generating new genetic combinations. This genetic mixing can result in major evolutionary leaps, for example, allowing pathogens to acquire resistance against drugs, and ensures, as part of meiosis in eukaryotes, that offspring inherit different combinations of alleles from their parents. Indeed, when organisms exchange genetic information they are also swapping their evolutionary histories, and because of this, ignoring the presence of recombination can also mislead the evolutionary analysis. Because of its fundamental role in genomic evolution and its potential confounding effect in many evolutionary inferences, it is not surprising that numerous bioinformatic approaches to detect the molecular footprint of recombination have been developed. While the next chapter discusses methods to detect and characterize individual recombination events with a particular focus on genetically diverse viral populations, the aim of this chapter is to briefly introduce different concepts in recombination analysis and to position the approaches discussed in Chapter 16 in the large array of currently available recombination detection tools.

15.2 Mechanisms of recombination

Before we briefly review genetic recombination in different organisms, it should be pointed out that there are several scenarios for genetic exchange. On the one hand, the donor nucleotide sequence can neatly replace a homologous region in

The Phylogenetic Handbook: a Practical Approach to Phylogenetic Analysis and Hypothesis Testing,
Philippe Lemey, Marco Salemi, and Anne-Mieke Vandamme (eds.). Published by Cambridge
University Press. © Cambridge University Press 2009.

the acceptor molecule (***homologous recombination***). On the other hand, recombination can also occur as a result of crossovers at non-homologous sites or between unrelated nucleotide sequences (***non-homologous recombination***). From another perspective, the exchange can occur in both directions (*symmetrical recombination*) or there can be a donor organism and an acceptor (*non-symmetrical recombination*).

Different organisms have evolved and integrated various processes of genetic exchange in their life cycle. Biologists are generally most familiar with the process of genetic recombination that is part of the sexual reproduction cycle in eukaryotes. In males and females, a special type of cell division called ***meiosis*** produces *gametes* or sex cells that have halved the number of complete sets of chromosomes. During meiosis, crossing over between homologous chromosomes occurs as part of the segregation process. Whether genetic recombination occurs in a particular chromosome depends on how *chromatides* or chromosome arms are cut and ligated before the chromosome homologues are pulled apart. Half of the time, the original parental chromosome arms can be rejoined. This mechanism is not present in all eukaryotes; in a few systems, crossing does not occur during meiosis, and asexual reproduction is frequent. It is noteworthy that ***recombination rate*** variation in some human genome locations has been directly estimated from the analysis of sperm samples (Jeffreys *et al.*, 1998; Jeffreys *et al.*, 2000). Although the difficulty of carrying out such experiments prohibits a detailed and comprehensive picture of genomic recombination rates, this information can be extremely valuable in evaluating recombination rate profiles estimated from population genetic data (Stumpf & McVean, 2003), and in stimulating the development of more realistic statistical models of recombination (e.g. Wiuf & Posada, 2003).

In bacteria, unidirectional genetic exchange can be accomplished in at least three ways. A bacterium can pass DNA to another through a tube – the *sex pilus* – that temporarily joins to bacterial cells. This process of *lateral gene transfer* is called ***conjugation*** and only occurs between closely related bacteria. The second process, ***transduction***, occurs when *bacteriophages* transfer portions of bacterial DNA from one bacterial cell to another. In the third process, referred to as ***transformation***, a bacterium takes up free pieces of DNA from the surrounding environment. These mechanisms for lateral transfer and recombination are the bacterial equivalent of sexual reproduction in eukaryotes. Comparative analysis has revealed that many bacteria are genomic chimaeras, and that foreign genomic regions can be often associated with the acquisition of pathogenicity (Ochman *et al.*, 2000; Ochman & Moran, 2001). For bacteria, recombination rates can also be estimated in the laboratory and, interestingly, these are in general agreement with population genetic estimates (Awadalla, 2003; Ochman & Moran, 2001).

Viruses can exchange genetic material when at least two viral genomes co-infect the same host cell, and in the process of infection, they can even acquire genes from their hosts. Indeed, the physical process of recombination can occur between identical genomes, but the evolutionary impact of genetic recombination is only noticeable when these two genomes are genetically different. Physically exchanging genetic segments within nucleotide molecules can occur in both segmented and non-segmented genomes. Genetic mixing among viruses is greatly facilitated when their genomes are segmented (*multipartite* viruses). In these cases, these segments can simply be reshuffled during co-infection, a process called **reassortment**. Antigenic shift in influenza A is an important example of the evolutionary significance of reassortment. For DNA viruses, recombination is probably similar to that seen in other DNA genomes involving breakage and rejoining of DNA strands. For RNA genomes, it was long thought that recombination was absent until recombinant polioviruses were detected (Cooper *et al.*, 1974). The most supported model of RNA virus recombination nowadays is a copy-choice mechanism during replication, which involves mid replication switches of the RNA-dependent RNA polymerase between RNA molecules. A similar template-switching mechanism during reverse transcription has been invoked for retroviruses (discussed in more detail in the next chapter). However, alternatives to the copy-choice model of RNA virus recombination have been suggested (e.g. Negroni & Buc, 2001).

Viral recombination can have important biological implications. Recombination events have been demonstrated to be associated with viruses expanding their host range (Gibbs & Weiller, 1999; Vennema *et al.*, 1998) or increasing their virulence (Suarez *et al.*, 2004). In addition, for many virus examples, the theoretical advantages of recombination (discussed below) have been experimentally put to test (Chao *et al.*, 1992; Chao *et al.*, 1997), analyzed through simulation (e.g. Carvajal-Rodriguez *et al.*, 2007), or demonstrated by naturally occurring examples (Georgescu *et al.*, 1994). There appears to be a considerable variation in recombination rate estimates for different viruses (Chare *et al.*, 2003; Chare & Holmes, 2006). Different constraints in viral recombination will be, at least partially, responsible for this observation (for review see Worobey & Holmes, 1999). In addition, genome architecture and networks of interactions will shape the evolutionary consequences of recombination along the genome (Martin *et al.*, 2005).

15.3 Linkage disequilibrium, substitution patterns, and evolutionary inference

In population genetic data sets, the extent to which recombination breaks up linkage between loci is generally reflected in the pattern of **linkage disequilibrium**. Linkage disequilibrium is observed when the frequency of a particular multilocus haplotype is significantly different from that expected from the product of the observed allelic frequencies at each locus. Consider, for example, two loci, *A* and *B*, with alleles *A/a* and *B/b* respectively. Let P_A denote the frequency of allele *A*, and so forth. Similarly, let P_{AB} stand for the frequency of the *AB* haplotype. The classical linkage disequilibrium coefficient is then $D = P_{AB} - P_A P_B$. Recombination breaks up linkage disequilibrium. When a population is recombining "freely," segregating sites become independent, and the population is in *linkage equilibrium*. Measuring linkage disequilibrium as function of the distance between loci is the cornerstone of population genetic methods to estimate recombination rates, $4N_e r$, where N_e is the **effective population size** and *r* is the recombination rate per site per generation. It is also interesting to note that intermediate levels of linkage disequilibrium, and thus intermediate levels of recombination, assist us in identifying disease markers or drug resistance-genes in genome-wide association studies. Recombination ensures that only particular genome regions will be associated with a particular phenotype. Full independence among sites, however, will break up all the associations between markers and the phenotype (Anderson *et al.*, 2000).

When measuring linkage disequilibrium in sequence alignments, it is important to realize that haplotype structure can be shaped in a similar way by recombination and recurrent substitution (see Box 15.1). Because population genetic analyses have traditionally been performed on closely related genotypes, analytical methods were initially developed under an **infinite sites** model (Kimura, 1969), which constrains each mutation to occur at a different nucleotide site. Under this assumption, the possibility of recurrent substitution does not exist. For most populations, however, infinite site models are not appropriate and recurrent substitution needs to be accounted before attributing *incompatible sites* – sites for which the character evolution cannot be reconciled on a single tree, see Box 15.1 – to recombination. In particular, **homoplasies** as a result of convergent changes at the same site can be relatively frequent in sites under positive selection (especially under **directional selection**, but also under **diversifying selection**), even more than in neutral finite site models. In sequence analysis, the contribution of recombination and selection has been notoriously difficult to separate (Grassly & Holmes, 1997).

15.4 Evolutionary implications of recombination

Shuffling genes or parts of genes through recombination inevitably results in mosaic genomes that are composed of regions with different evolutionary histories, and in the extreme case, this will result in the complete independence of **segregating sites**. Conversely, sites in which allelic combinations are inherited together across generations, because recombination never separated them, are said to be linked, and share a single, common evolutionary history. In this way, recombination can be considered as a process that releases allelic variation at neutral loci from the action of selection at nearby, linked sites.

Box 15.1 Recombination and recurrent substitution

Consider four taxa with the following nucleotide sequences:

Taxon1 ATTG
Taxon2 AATG
Taxon3 AATC
Taxon4 ATTC

Both sites 2 and 4 are *parsimony informative sites*: they contain at least two types of nucleotides, and at least two of them occur with a frequency of two. However, these two site patterns are *incompatible* because there is no single tree that can represent the evolutionary history of these taxa. In order to reconcile the evolutionary history of two incompatible sites, either recombination or recurrent substitution needs to be invoked:

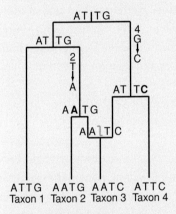

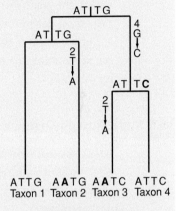

| Recombination | Recurrent mutation |

In this figure sequence characters are indicated at the internal and terminal nodes. Sites that have mutated along a branch are indicated in bold. In the left graph, the site patterns are explained by a *recombination event*, represented by the joining of two branches into the lineage of taxon 3, which carries a recombinant sequence. In the right graph, character evolution is explained by *recurrent substitution*: change T→A has evolved independently and in parallel at site 2 in taxa 2 and 3. Both recurrent substitution and recombination result in *homoplasies*.

To fully understand the implications of this, let us consider a population that recently experienced a severe population bottleneck or a founder effect. Such populations will only show limited genetic diversity because much ancient diversity has been lost due to the bottleneck event or through the founder effect. Similarly, a beneficial mutation can be rapidly fixed in the population through a *selective sweep*. In non-recombining populations, neutral mutations at all other linked sites in the variant harboring the beneficial mutation will reach a high frequency in the population (a process known as *genetic hitchhiking*), which will lead to a genome-wide low degree of genetic diversity. When recombination occurs, however, the fate of neutral variation in the genome can become independent from the action of selection at another site. And this is increasingly so for neutral variation that occurs at genomic sites more distantly located from the site where selective pressure operates because of the higher probability of a recombination event between them. This can be of major importance when, for example, a beneficial mutation occurs in an individual that has a less favorable genetic background. The presence of (slightly) deleterious mutations in this individual will seriously delay the time to fixation for the beneficial mutation in the population (if it will become fixed at all). However, recombination can remove deleterious mutations in the genetic background and thus speed up the fixation of the beneficial mutation. It has been predicted that in finite, asexual populations deleterious alleles can gradually accumulate because of the random loss of individuals with the fewest deleterious alleles (Felsenstein, 1974; Muller, 1932). Purging a population from lower fitness mutations has therefore been a longstanding theoretical explanation for the evolution of recombination and some aspects of sexual reproduction.

15.5 Impact on phylogenetic analyses

A first step in understanding the impact of recombination on phylogenetic inference is to consider the result of a single recombination event in the evolutionary history of a small sample of sequences. Figure 15.1 illustrates how the recombination event, described in Box 15.1 results in different phylogenetic histories for the left and right part of the alignment.

The impact of a single recombination event on the genetic make-up of the sampled sequences, and hence on further evolutionary inference, will strongly depend on how fast substitutions are accumulated in the evolutionary history and on which lineages recombine at what time point in the history. We illustrate this in Fig. 15.2 for three different evolutionary scenarios. Note that the top three diagrams represent the processes of recombination and coalescence of lineages as we go back in time; mutation will be added later. For simplicity, we will assume that all breakpoints occur in the middle of the sequences, represented by the blocks under the diagrams. In Fig. 15.2a, a recombination event occurred shortly after a lineage split into two lineages. None of the "pure" parental lineages persist and only the recombinant taxon B is sampled. In the second example, an additional splitting event has occurred before two lineages recombine (Fig. 15.2b). Taxon or sequence C represents a descendant of one of the parental lineages of recombinant D. In the last example in this row, two relatively distinct lineages recombined, after which the recombinant lineage split into two lineages leading to recombinant individuals B and C that share the same breakpoint (Fig. 15.2c). Note that we have described

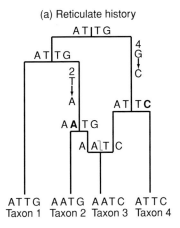

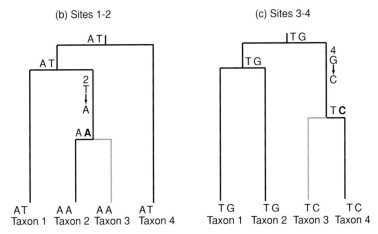

Fig. 15.1   Recombination generates different phylogenetic histories for different regions in the genome. (a) Reticulated history for the alignment discussed in Box 15.1. A recombination event between lineages 2 and 4 results in recombinant lineage 3. The recombination breakpoint was places between sites 2 and 3. (b) Phylogenetic tree for sites 1–2. (c) Phylogenetic tree for sites 3–4.

these histories of ancestral and descendent lineages without referring to mutation. In fact, these could represent extreme cases where no mutations have occurred. In this case, none of the recombination events resulted in any observable effect on the DNA sequences.

To illustrate an observable effect on the genetic sequences, we have added 8 (low substitution rate) and 20 (moderate substitution rate) non-recurrent substitution events to these histories (Fig. 15.2d–f, and Fig. 15.2g–i respectively). Horizontal lines in the evolutionary history represent the substitutions; a small circle on either side of the line indicates whether the substitution occurs on the left or right side of the genome, which is important to assess the presence and absence

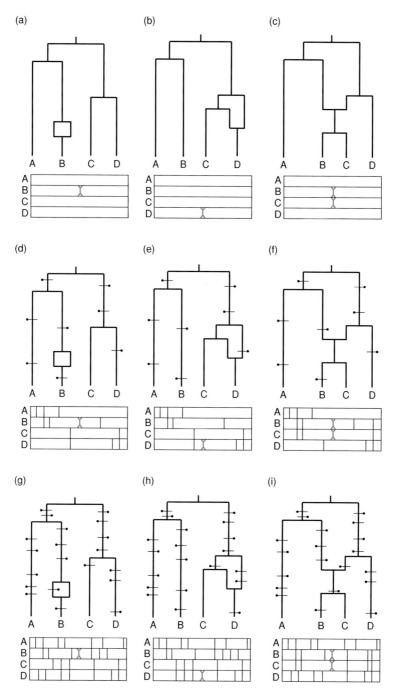

Fig. 15.2   Effect of different recombination events on the genetic make-up of molecular sequences. Three evolutionary histories with a different recombination event are represented in (a), (d), (g); (b), (e), (h), and (c), (f), (i), respectively. Different substitution processes are superimposed onto these histories resulting in different amount of substitutions in the sampled sequences: no mutation in (a), (b) and (c), a low mutation rate in (d), (e), and (f), a moderate mutation rate in (g), (h), and (i). A circle on either side of a horizontal line, representing a mutation in the evolutionary history, indicates whether a mutation occurs in the left or right part of the sequences, which are represented by boxes underneath the diagrams. Vertical lines in the sequence boxes represent inherited mutations.

of substitutions in recombinant lineages. For a recombinant, we assume that the left part of the genome was inherited from the leftmost pathway in the diagram and the right part from the rightmost pathway in the diagram (if the diagram could be disentangled into two evolutionary histories, similar to Fig. 15.1, then the left part of the genome evolved according to the left history, whereas the right part of the genome evolved according to the right history). For the first example (Fig. 15.2d and g), both low and moderate mutation rate leave no trace of incompatible sites or mosaic patterns and hence it will not impact our evolutionary analyses. Mutations that occur after the splitting event and before the recombination will be inherited by the recombinant depending on where in the genome they occur (Fig. 15.2g). In the second example, and under low substitution rate (Fig. 15.2e), the sequences have the same genetic make-up as in the first example (Fig. 15.2d). This is not the case for a higher substitution rate; sequence C and D have become more similar due to the recombination event (Fig. 15.2h). In this case, although it will impact the estimation of branch lengths, the recombination event does not generate incompatible sites, and estimates of the tree topology will not be affected. In the last example, incompatible sites are generated for both low and moderate substitution rates (Fig. 15.2h and i). Some mutations that were shared for A–B and C–D are now also shared for both recombinant lineages. The relative proportions of sites shared among A–B–C and B–C–D determine whether the recombinants will cluster with A or D. As a consequence, this recombination event will affect estimates of both tree topology and branch lengths.

The examples above illustrate the different impact of a single recombination event in the middle of the genome. The complexity of how recombination shapes genetic variation increases enormously as multiple recombination events occur during evolutionary history, each shuffling different parts of the genome. This is illustrated in Fig. 15.3, where different histories were simulated using a population genetic model with increasing recombination rates. The shaded boxes represented the sequence alignments, and alternate white and shaded areas indicate different phylogenetic histories. Recombination events similar to the three types we described in Fig. 15.3 can be observed, but now overlapping recombination events occur. If not only random substitutions, but also recurrent substitutions and selection processes generating convergent and parallel changes occur in these histories, it is not difficult to imagine how complex the impact on evolutionary inferences can be and how challenging the task of recombination detection will be.

Generalizing our arguments above, no single strictly bifurcating tree can accurately capture the true evolutionary relationships if different genome regions have evolved according to different phylogenetic histories. In this respect, phylogenetic network methods can be very useful to visualize complex evolutionary relationships (Posada & Crandall, 2001b)(Chapter 21). We illustrate some of the consequences

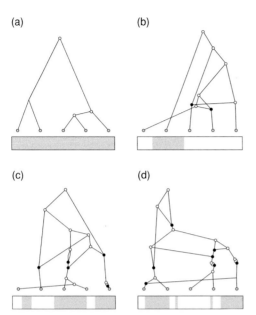

Fig. 15.3   Different population histories simulated for five sampled individuals using a population genetic model with increasing recombination rates. Each recombination event results in differently positioned breakpoint; *n* recombination events generate *n* + 1 different phylogenetic histories, represented by alternating white and shaded boxes. The histories were simulated using the HUDSON RECOMBINATION ANIMATOR (*http://www.coalescent.dk/*) with increasing population recombination rate: 0, 1, 2, 3 for a, b, c, and d, respectively.

of including a mosaic sequence in phylogenetic inference in Box 15.2. The first one is the effect on branch lengths and tree shape; phylogenetic simulation of recombinant data sets show a reduced ratio of internal branch lengths relative to external branch lengths. This agrees with population genetic simulations showing that relatively frequent recombination in the evolutionary history of larger data sets can create star-like trees (Schierup & Hein, 2000a). This effect on tree-shape is important to keep in mind when applying, for example, variable population size models that relate demography to tree shape (see the chapters on population genetic inference in this book). Phylogenetic simulations have also demonstrated the confounding effect of recombination on tree topology inference (Posada & Crandall, 2002), especially when the recombining lineages were divergent and the breakpoints divide the sequences in half (Fig. 15.2i). In some cases, a phylogeny can be inferred that is very different from any of the true histories underlying the data.

Figure 15.2 suggests that recombination homogenizes the alignments; this is also evident from the simulations showing a reduced variance in pairwise distances in Box 15.2. Not surprisingly, the variance of pairwise differences has been interpreted

Box 15.2 The impact of recombination on phylogenetic inference

To demonstrate some of the effects of recombination on phylogenetic analyses, 100 data sets of five sequences were simulated under three different scenarios A, B and C:

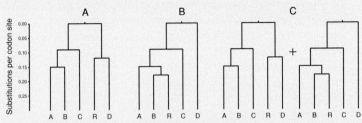

In A and B, sequences were simulated according to phylogenetic trees that only differ with respect to the position of taxon R. One hundred alignments of sequences encompassing 500 codons were simulated using a codon model of evolution with three discrete site classes: 40% with $d_N/d_S = 0.1$, 30% with $d_N/d_S = 0.5$ and 30% with $d_N/d_S = 1.0$, representing strong negative selection, moderate negative selection and neutral evolution (for more information on codon models and d_N/d_S, see Chapter 14). In the case of C, 100 data sets were constructed by concatenating the alignments obtained in A and B. As a consequence, these data sets contain a mosaic sequence R that has a different evolutionary history in the first and second 500 codons.

Tree shape and branch lengths

Maximum likelihood trees were inferred using the best-fit nucleotide substitution model. The inferred tree topologies were the same for all data sets within their specific simulation scenarios, but branch lengths varied. Topologies with branch lengths averaged over all data sets are shown below:

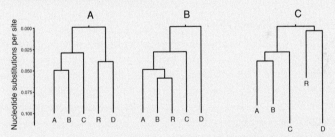

For the non-recombinant data sets in A and B, tree topologies were correctly reconstructed with mean branch lengths in nucleotide substitutions being almost exactly one third of the branch lengths in codon substitution units used for simulation. In the case of C, the best "compromise" tree is similar to the one used for simulation in A, however, with remarkably different branch lengths. Although R clusters with D, the branch lengths indicate a much larger evolutionary distance between R and D, but a lower evolutionary distance between R and A/B/C. This markedly reduces the ratio of internal branch lengths relative to external branch lengths; the mean ratio of external/internal branch lengths is 3.26 and 6.89 for A and C, respectively.

Box 15.2 (*cont.*)

Genetic distances

The effect on tree shape results from the homogenizing of sequences induced by recombination. This should also be apparent from pairwise genetic distances; histograms for the pairwise distances of all data sets are shown below:

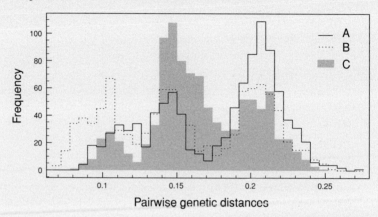

In general, the pairwise distance distributions appear to be trimodal, with data sets from A having a higher frequency of large pairwise genetic distances. For the data sets including the recombinant R sequence (C), the distances regress to the average value. This has little effect on the average pairwise genetic distance (0.174, 0.153, and 0.162 substitutions per site for A, B, and C, respectively), but it reduces the variance to some extent in the recombinant data sets (0.00175, 0.0022 and 0.00118 for A, B, and C, respectively).

Rate variation among sites

In addition to branch lengths, also parameters in the evolutionary model are affected when including mosaic gene sequences. In particular, the shape parameter of the gamma distribution to model rate heterogeneity among sites is lower in the recombinant data sets (median value of 1.96, 2.10, and 1.17 for A, B, and C, respectively), indicating higher rate variation in this case (see Chapter 4). This is because the tree with the mosaic sequence will require extra changes at some sites to account for the homoplasies introduced by recombination. So, although R clusters with D, assuming extra changes at some sites in the region where R is most closely related to sequence B will be required.

Rate variation among lineages

Sequences were simulated according to an ***ultrametric*** tree meaning that each tip is exactly at the same distance from the root. The inferred trees with mean branch lengths for A and B are perfect reproductions of such trees. Because the tree with mean branch lengths in C is far from ultrametric, the presence of a recombinant might be erroneously interpreted as rate variation among branches (Schierup & Hein, 2000). To demonstrate this more formally, we have performed a ***likelihood ratio test*** (*LRT*) for the ***molecular clock*** in each simulated data set and compared the distribution of the LRT statistic with the expected χ^2 distribution with three degrees of freedom:

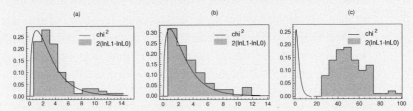

(a)   (b)   (c)

Both simulated and expected distributions should be very similar under conditions of ***asymptotic normality***. This appears to be the case for the non-recombinant data sets. In fact, the 95% cut-off values are very similar for simulated (9.1 and 7.6 for A and B, respectively) and expected distributions (7.8). The picture for the recombinant data sets is remarkably different. In this case, the LRT would strongly reject the molecular clock for each simulated data set based on the χ^2 distribution cut-off.

Positive selection

The codon sequences were simulated under a selection regime that did not allow for positively selected sites. Although a sample size of five sequences is not very powerful to estimate site-specific selection parameters, we can still compare the fit of a codon model representing neutral evolution (M7; Yang *et al.*, 2000) and a model that allows an extra class of positively selected sites (M8) (for codon models, see also Chapter 14). These models can be compared using a LRT, similar to the molecular clock test, and significance can be assessed using a χ^2 distribution with two degrees of freedom. For a 95% cut-off value of 5.99, a proportion of 0.06 and 0.02 simulated data sets in A and B, respectively rejects the neutral model in favor of a model that allows for positively selected sites, which are very close to the expected 5%. For the recombinant simulation, however, there is a proportion of 0.10 data sets that reject the neutral model. This tendency confirms a more elaborate simulation study showing that the LRT often mistakes recombination as evidence for positive selection (Anisimova *et al.*, 2003).

as a measure of linkage disequilibrium and this and other summary statistics have been employed to estimate recombination rates (Hudson, 1985). The simulations in Box 15.2 also indicate inflated apparent rate heterogeneity, an important parameter in phylogenetic reconstruction (see Chapters 4 and 5). This observation has motivated the development of a specific test for recombination (Worobey, 2001).

Finally, any evolutionary analysis that models shared ancestry using a single underlying bifurcating tree will be affected by recombination. The simulations in Box 15.2 clearly indicate that recombination results in apparent rate heterogeneity among lineages. This corroborates the claim that even small levels of recombination invalidate the ***likelihood ratio test*** of the ***molecular clock*** (Schierup & Hein, 2000b) (but see Posada, 2001). Likewise, likelihood ratio tests for detecting selection using codon models are also invalid in the presence of recombination (see also Anisimova *et al.*, 2003). Importantly, also ***non-synonymous/synonymous rate ratios*** (d_N/d_S) can be overestimated and high number of sites can be falsely identified as positively selected (Shriner *et al.*, 2003) (see Chapter 14 for inferences of this type). Because of this, several approaches have been recently developed to estimate this ratio and recombination rate simultaneously (e.g. OMEGAMAP: Wilson & McVean, 2006), or to accommodate the presence of obvious mosaic sequences when estimating d_N/d_S (Scheffler *et al.*, 2006). In summary, it is important to be aware that recombination affects many of the evolutionary analyses discussed in this book.

15.6 Recombination analysis as a multifaceted discipline

The myriad of available recombination analysis tools probably reflects the difficulty of evaluating recombination in molecular sequence data and the different questions we can pose about it (Posada *et al.*, 2002). Different bioinformatics strategies can be explored to tackle the problem, and very importantly, with different objectives. Identifying specific questions prior to data analyses will therefore be an important guidance in choosing among several approaches. In this chapter, we will distinguish four major goals: (i) detecting evidence of recombination in a data set, (ii) identifying the mosaic sequences, (iii) delineating their breakpoints, and (iv) quantifying recombination.

15.6.1 Detecting recombination

The first goal is determining whether recombination has occurred during the evolution of the sampled sequences. Although a judgment can be made based on different approaches, including graphical exploration tools, statistical tests are required to appropriately evaluate any hypothesis. Typically, *substitution distribution* and *compatibility* methods have been developed for this purpose and were among the earliest available recombination detection tools. These methods examine the uniformity of relatedness across gene sequences, measure the similarity or compatibility between closely linked sites or measure their composition in terms of homoplasies or two-state ***parsimony informative sites***. The null distribution of the test statistic, which determines the level of significance, can be obtained using specific statistical models; it can be generated by randomly reshuffling sites, or by simulating data sets according to a strictly bifurcating tree. The latter approach to determine *p*-values is often referred to as ***Monte Carlo simulation*** (or ***parametric bootstrapping***). Both permutation and simulation procedures have made different approaches amenable to statistical evaluation (including, for example, population genetic methods). Substitution distribution and compatibility methods are generally relatively powerful compared to methods that measure phylogenetic discordance (Posada & Crandall, 2001a).

15.6.2 Recombinant identification and breakpoint detection

If significant evidence for recombination can be detected in a data set, the question naturally arises which sequences are responsible for this signal. The methods that try to answer this question are usually, but not necessarily, methods that also attempt to localize recombination breakpoints. Obviously, if a mosaic pattern can be demonstrated for a particular sequence, then this also provides evidence for the recombinant nature of the sequence. Scanning methods employing distance or phylogenetic methods are generally well suited for this purpose, but also substitution distribution methods that examine the uniformity of relatedness across gene sequences can be used in this context. The next chapter expands on detecting and characterizing individual recombination events using these types of methods. It is, however, important to be aware of some caveats in using sliding window approaches for detecting recombinants and delineating breakpoints. First, scanning methods usually require *a priori* specification of a query sequence and putative parental sequences (or rather, the progeny thereof). So, if this approach is used in an attempt to identify recombinant sequences in a data set, the analysis needs to iterate through every sequence as possible recombinant, which obviously generates multiple testing problems. Second, Suchard *et al.* (2002) pointed out that scanning approaches fall into a "sequential testing trap," by first using the data to determine optimal breakpoints and parental sequences for a particular query sequence and then using the same data to assess significance conditional on the optimal solution. To overcome this, a Bayesian approach was developed to simultaneously infer recombination, breakpoints, and parental representatives, avoiding the sequential testing trap (Suchard *et al.*, 2002). Interestingly, such probabilistic models have also been extended to map recombination hotspots in multiple recombinants (Minin *et al.*, 2007).

15.6.3 Recombination rate

A different goal would be to quantify the extent to which recombination has shaped the set of sequences under study. Simply estimating the proportion of mosaic sequences in the data set does not provide adequate information because different sequences might share the same recombinant history and/or different recombinants might have different complexity resulting from a different number of recombinant events in their past. Some substitution distribution methods that measure homoplasies or two-state parsimony informative sites also determine the expected value for this compositional measure under complete linkage equilibrium. The ratio of the observed value over the expected value for complete independence of sites therefore provides some assessment of how pervasive recombination is (e.g. the *homoplasy ratio* and the *informative sites index*, Maynard Smith & Smith, 1998; Worobey, 2001). However, this is not always suitable for comparative analysis because the number of recombination events also depends on the time to The

Most Recent Common Ancestor (TMRCA) for a particular sample of sequences (and related to this, the effective population size of the population from which the sequences were sampled). The same criticism is true for methods that simply count the number of recombination events that have occurred in the history of a sample. This can be achieved by counting the occurrences where all allelic combinations are observed for two bi-allelic loci (AB, Ab, aB, ab). If an infinite sites model is assumed (see above), such an occurrence can only be explained by a recombination event. Because this *four-gamete test* only scores a recombination event if all four possible two-locus haplotypes can be observed in the sample, it is only a conservative estimate of the minimum number of recombination events that have occurred in the absence of recurrent mutation.

   A quantitative estimate of recombination that takes into account the evolutionary time scale of the sampled sequences would be an estimate of the **population recombination rate**. Very importantly, this implies a completely different scenario. Now we are trying to estimate a population parameter from a sample (*parametric* methods), while before we were trying to characterize the history of a given alignment (non-*parametric* methods). In this setting it is very important that the random sample provides a good estimate of the frequency of the different alleles present in the population. Note that now we will work with all the sequences, when before, we just used the different haplotypes. In Chapter 17, a population genetics model is introduced that describes the genealogical process for a sample of individuals from a population (**the coalescent**). Recombination can easily be incorporated into this model, in which case the coalescent process is represented using an **ancestral recombination graph** (**ARG**) instead of a single **genealogy**. (In Figs. 15.2 and 15.3, most graphs represent ARGs, whereas the graph in Fig. 15.3a represents a genealogy.) Parametric methods based on linkage disequilibrium attempt to estimate the population recombination rate, ρ, which is the product of the per-generation recombination rate, r, and the effective population size (N_e): $\rho = 4N_e r$ (for diploid populations). The ARGs in Fig. 15.3 were simulated using increasing population recombination rates. If the **population mutation rate** Θ, a fundamental population genetics quantity equal to $4N_e m$ in diploid populations, can also be estimated, then the recombination rate estimate can be expressed as $\rho/\Theta = r/m$, which represents importance of per generation recombination relative to per generation mutation. As discussed above (Box 15.1), applying such estimators on diverse sequences, like many bacterial and viral data sets, requires relaxing the infinite site assumption. Other assumptions that we need to make under the parametric approaches are also important to keep in mind. The standard, neutral coalescent process operates under a constant population size, no selection, random mating, and no population structure. Although this ideal might be far from biological reality, the model can still be useful for comparing recombination rates among genes and for the purpose of prediction (Stumpf & McVean, 2003). Nevertheless, it is clear that factors like population structure and demographic history may affect the ability of coalescent methods to correctly infer the rate of recombination (Carvajal-Rodriguez *et al.*, 2006). Despite the disadvantage of having to make simplifying assumptions, population genetic methods offer a powerful approach to estimating recombination rates across the genome, which can lead to a better understanding of the molecular basis of recombination, its evolutionary significance, and the distribution of linkage disequilibrium in natural populations (Stumpf & McVean, 2003).

15.7 Overview of recombination detection tools

Although we have attempted to associate different objectives to different recombination detection approaches, this does not result in an unequivocal classification of the available bioinformatics tools. Many tools share particular computational approaches or algorithmic details, but none of these are ideal classifiers. Both in terms of algorithms and objectives, the line has become blurred in many software packages. To provide a relatively detailed overview of different methods and software packages available, we have compiled a table that includes relevant features from a user perspective and particular algorithmic details (Table 15.1). To provide an approximate chronological overview, the methods are roughly ordered according to the year of publication. An updated version of this table will be maintained at www.thephylogenetichandbook.org and URLs for the software packages will be provided at *www.bioinf.manchester.ac.uk/recombination/*.

The criteria we list as important from a user perspective are statistical support, the ability to identify recombinant and parental sequences, the ability to locate breakpoints, and the speed of the application. A method is considered to provide statistical support if a *p*-value can support the rejection of the null hypothesis of clonal evolution. Methods that are designed to detect recombination signal in sequence data based on some test statistic are generally well suited to provide statistical support (e.g. through a Z-test, a chi^2-test, a binomial *p*-value or parametric, and non-parametric bootstrapping; see Section 15.5 and the next chapter). Statistical support can also be obtained in the form of a ***posterior probability*** (Suchard *et al.*, 2002) or using an incongruence test (e.g. the Shimodaira–Hasegawa test in GARD; Kosakovsky Pond *et al.*, 2006). Methods that are more graphical in nature do not always provide such support. For example, we do not consider bootscanning to be statistically supported, even though bootstrap values might give a good indication of the robustness of different clustering patterns in different gene regions. Bootscanning might not only fall into the sequential testing trap, but it also does

Table 15.1 Overview of different recombination detection tools

Method	Program	User perspective criteria				Algorithmic criteria			Method/Program reference
		Statistical support	Identifies recombinants/parentals	Locates breakpoints	Speed	Sliding window	Phylogenetic Incongruence	Run mode	
Sawyer's statistical test for gene conversion	GENECONV	yes	yes	yes	fast	no	no	pairs (against average)	(Sawyer, 1989)
	RDP3	.	.	.	.	.	.	triplet/query vs. reference[a]	(Martin et al., 2005)
	START	.	.	.	.	.	.	.	(Jolley et al., 2001)
Maximum chi-squared	MAXIMUM CHI-SQUARED	yes	no	yes	fast	yes	no	triplets	(Maynard Smith, 1992)
	RDP3 (maxchi)	.	yes	.	.	.	.	triplets/query vs. references[a]	(Martin et al., 2005)
	RDP3 (Chimaera)	.	yes	.	.	.	.	triplet	(Posada & Crandall, 2001)
	PHIPACK	.	.	no	.	.	.	.	(Bruen et al., 2006)
	START	.	.	.	.	.	.	.	(Jolley et al., 2001)
	4ISIS	.	.	.	.	no	.	quartet	(Robertson et al., 1995)
Heuristic parsimony analysis of recombination	RECPARS	no	no	yes	fast	no	yes	all sequences	(Hein, 1993)
Similarity/distance plot	SIMPLOT	no	no	yes	fast	yes	no	query vs. references	(Lole et al., 1999)
	RDP3	.	.	.	.	.	.	.	(Martin et al., 2005)
	RIP	.	.	.	.	.	.	.	(Siepel et al., 1995)
	FRAGDIST	.	.	.	.	.	.	.	(Gao et al., 1998; van Cuyck et al., 2005)
	BLAST genotyping	.	.	.	.	.	.	.	n.a.

Method	Program								Reference
								query vs. references[b]	
Bootscanning	SIMPLOT	no	no	yes	fast	yes	yes	query vs. references[b]	(Lole et al., 1999)
	RDP3								(Martin et al., 2005)
	REGA								(de Oliveira et al., 2005)
Compatibility matrix and neighbor similarity score	RETICULATE	yes	no	no	fast	no	no	all sequences	(Jakobsen & Easteal, 1996)
Partition matrices	PARTIMATRIX	no	no	yes	fast	no	no	all sequences	(Jakobsen et al., 1997)
Difference in Sums of Squares method	TOPALi/rDSS	yes	yes[c]	yes	slow[d]	yes	yes	all sequences	(McGuire et al., 1997)
Likelihood detection of Spatial Phylogenetic Variation	RDP3							quartets	(Martin et al., 2005)
	PLATO	yes	no	yes	fast	yes	yes	all sequences	(Grassly & Holmes, 1997)
Homoplasy test	(qbasic programs)	yes	no	no	fast	no	no	all sequences	(Maynard Smith & Smith, 1998)
	START								(Jolley et al., 2001)
	PHIPACK								(Bruen et al., 2006)
Modified Sherman Test	SNEATHST	yes	yes	no	fast	no	no	pairs	(Sneath, 1998)
Graphical recombination detection using Phylogenetic Profiles	PHYLPRO	no	yes	yes	fast	yes	no	query vs. references[e]	(Weiller, 1998)
	RDP3								(Martin et al., 2005)
Likelihood Analysis of Recombination in DNA	LARD	yes	no	yes	slow	yes	yes	triplet	(Holmes et al., 1999)
	RDP3		yes				yes			(Martin et al., 2005)

(cont.)

Table 15.1 (*cont.*)

Method	Program	User perspective criteria				Algorithmic criteria			Method/Program reference
		Statistical support	Identifies recombinants/ parentals	Locates breakpoints	Speed	Sliding window	Phylogenetic Incongruence	Run mode	
Sister scanning method	SISCAN	no	no	yes	fast	yes	no	quartet/triplet	(Gibbs et al., 2000)
	RDP3	yes	yes	.	.	.	.	quartet/triplet/ query vs. references[a]	(Martin et al., 2005)
the RDP method	RDP3	yes	yes	yes	fast	yes	no	triplet	(Martin & Rybicki, 2000)
Informative sites test	PIST	yes	no	no	fast	no	no	all sequences	(Worobey, 2001)
Phylogenetic hidden Markov model with Bayesian inference	SERAD (Matlab)[f]	yes	no	yes	slow	no	yes	quartet	(Husmeier & Wright, 2001a)
	BARCE[f]	.	.	.	.	.	.	.	(Husmeier & McGuire, 2003)
	TOPALi/rHMM[f]	.	no[c]	.	slow[d]	.	.	.	(Milne et al., 2004)
Probabilistic divergence method using MCMC	JAMBE[f]	yes	no	yes	slow	yes	yes	all sequences	(Husmeier & Wright, 2001b)
	TOPALi/rPDM	.	no[c]	.	slow[d]	.	.	.	(Milne et al., 2004)
Bayesian multiple change-point modelling	DUALBROTHERS	yes	no	yes	slow	no	yes	query vs. reference	(Suchard et al., 2002)
	cBROTHER	.	.	.	.	.	.	.	(Fang et al., 2007)
Distance-matrix calculation across breakpoints	BELLEROPHON	no	yes	yes	fast	yes	no	query vs. references[e]	(Huber et al., 2004)

Method	Program				Speed			Scope	Reference
Visual recombination detection using quartet scanning	VisRD	no	no	yes	fast	yes	yes	quartets	(Strimmer et al., 2003)
Distance-based recombination analysis tool	RAT	no	yes	yes	fast	yes	no	query vs. references[e]	(Etherington et al., 2005)
Automated bootscanning (Recscan)	RDP3	yes	yes	yes	fast	yes	yes	triplets	(Martin et al., 2005)
Recombination detection using multiple approaches	RDP3	yes	yes	yes	method dependent	yes	yes	triplets & quartets	(Martin et al., 2005)
Stepwise recombination detection	STEPWISE (R-package)			method dependent[g]					(Graham et al., 2005)
Pairwise homoplasy index	PHIPACK	yes	no	no	fast	no	no	all sequences	(Bruen et al., 2006)
	SPLITSTREE								(Huson, 1998)
Phylogenetic compatibility method	SIMMONICS	no	yes	yes	slow	yes	yes	all sequences	(Simmonds & Welch, 2006)
Recombination analysis using cost optimization	RECCO	yes	yes	yes	slow	no	no	all sequences	(Maydt & Lengauer, 2006)
Genetic algorithm for recombination detection	GARD	yes	no	yes	slow	no	yes	all sequences	(Kosakovsky Pond et al., 2006)
Jumping profile hidden markov models	JPHMM	no	no	yes	slow	no	yes	query vs. reference	(Schultz et al., 2006)
Recombination detection using hyper-geometric random walks	3SEQ	yes	yes	yes	fast	no	no	triplets	(Boni et al., 2007)
	RDP3	.	.	.	.	.	.	triplets/query vs. references[a]	(Martin et al., 2005)

(cont.)

Table 15.1 (*cont.*)

		User perspective criteria				Algorithmic criteria			
Method	Program	Statistical support	Identifies recombinants/ parentals	Locates breakpoints	Speed	Sliding window	Phylogenetic Incongruence	Run mode	Method/Program reference
Building evolutionary networks of serial samples	SLIDING MINPD	no	yes	yes	fast	yes	method dependent[h]	all sequences	(Buendia & Narasimhan, 2007)
Comparing trees using likelihood ratio testing	TOPALi/rLRT	yes	no[c]	yes	slow [d]	yes	yes	all sequences	(Milne et al., 2004)

+ A dot in the cells refers to the same content as the first software package implementing the same approach.

a For these approaches, RDP3 allows the user to perform a "manual" analysis by assigning a query and parental sequences or to perform an "automated" analysis that evaluates every possible quartet or triplet. The latter setting provides a useful approach to identify putative recombinant sequences in the data set.

b The bootscan approach analyzes all sequences simultaneously (phylogenetic tree inference), but uses the "query vs. reference" scheme *a posteriori* to trace the clustering of a particular query sequence.

c In TOPALi, a modified difference in sums of squares (DSS) method can be used to find *which* sequence(s) appear to be responsible for the recombination breakpoints. This "leave-one-out" method uses as windows the homogeneous regions between the breakpoints, identified using any method. The DSS scores for each breakpoint are calculated, leaving out one sequence at a time. To assess significance, 100 alignments are simulated. A sequence is a putative recombinant if removing it results in a non-significant recombination breakpoint. This algorithm can be applied after a recombinant pattern is identified using any method implemented in TOPALi.

d The methods in TOPALi are generally slow when run on a single processor, but when spread on multiple CPUs analyses will run significantly faster.

e Although these software packages compare a query sequence against all the other sequences, they can perform this comparison for every sequence in the data set being assigned as a query. In PHYLPRO, phylogenetic correlations, which are based on pairwise genetic distances, are computed for each individual sequence in the alignment at every position using sliding-window techniques. BELLEROPHON evaluates for each sequence the contribution to the absolute deviation between two distance matrices for two adjacent windows. RAT has an "auto search" option to evaluate the similarity of every sequence to all other sequences. Therefore these approaches can be useful in identifying putative recombinants.

f SERAD is the MATLAB precursor of BARCE (C++ program). Both BARCE and JAMBE have been integrated into TOPALI, which provides a user-friendly GUI and several on-line monitoring diagnostic tools. Note that the BARCE method may predict erroneous recombination events when a DNA sequence alignment shows strong rate heterogeneity. An improved method that addresses this problem via a factorial hidden Markov model has been described in Husmeier (2005). The method has been implemented in a MATLAB program (*http://www.bioss.ac.uk/~dirk/Supplements/phyloFHMM/*), but unfortunately, this implementation is slow and computationally inefficient for the time being.

g The stepwise approach can be applied to any recombination detection method that uses a permutation test and provides estimates of breakpoints. The criteria depend on the method that is used in the stepwise approach.

h SLIDING MINPD implements three different methods to identify recombinants: a percentage identity method (as implemented in the Recombination Detection Program, RIP), a standard bootscanning method and a distance bootscanning method. Only the standard bootscanning method infers trees for each alignment window.

not assess how much topological variability can be expected as a result of chance alone (under the null hypothesis).

The ability to detect recombinant sequences and breakpoint locations has been discussed above (see Section 15.5). It should be noted that methods examining every possible triple or quartet combination in a data set (e.g. RDP3 and 3SEQ), are generally designed to identify (a) combination(s) of taxa for which the null hypothesis can be rejected, with appropriate multiple testing correction (see Section 15.5 and next chapter), but cannot always discriminate between the recombinant and parental sequences within that triplet/quartet. In terms of speed, we have only used a slow/fast classification based on user-experience or input from the original authors. If the analysis of a moderate size data set was thought to take no more than a coffee break, we classified it as "fast." Of course, coffee breaks are stretchable, software speed can heavily depend on the settings involved, and a more objective evaluation on benchmark data sets is required to provide a more accurate and quantitative classification.

The algorithmic criteria in Table 15.1 include the use of a sliding window approach, "phylogenetic incongruence" and "run mode." Methods are classified as using a phylogenetic incongruence criterion if phylogenetic trees form the cornerstone of the inference. Whereas sliding window refers to a particular form of alignment partitioning, the "run mode" refers to a taxa partition used by the method. Several methods do not partition the taxa in their analysis strategy and detect recombination in the complete data set ("all sequences"). Other methods focus on subsets like pairs, triplets or quartets. In some cases, these methods analyze all possible combinations of this partitioning scheme in an attempt to pinpoint putative recombinants and their parental sequences (e.g. RDP3 and 3SEQ) or in an attempt to provide a graphical visualization of phylogenetic incongruence (e.g. VISRD). In other software programs, the application of the algorithm is simply restricted to such subsets (e.g. a quartets in BARCE and TOPALI/rHMM and a triplet in the original MAXIMUM CHI-SQUARED program). Methods that use the "query vs. reference" setup focus on the relationship – frequently inferred from pairwise genetic distances (e.g. PHYLPRO, SISCAN, SIMPLOT, BELLEROPHON and RAT) – between one particular sequence and all the other sequences in the alignment. Some of these programs iterate through all sequences when evaluating these relationships (see footnote *e* in Table 15.1), while others restrict themselves to the *a priori* assigned query sequence.

Because population genetic inferences are considered as a different class of methods with the primary objective of quantifying recombination (see Section 15.5), they are not included in Table 15.1. A list of population genetic methods is provided in Stumpf and McVean (2003), which includes software packages like SEQUENCELD and FINS (Fearnhead & Donnelly, 2001), LDHAT (McVean *et al.*, 2002) and its four-allele extension that implements more complex evolutionary models (Carvajal-Rodriguez *et al.*, 2006) (*http://darwin.uvigo.es*), RECMIN (Myers & Griffiths, 2003) and LAMARC (Kuhner *et al.*, 2000) (see Chapter 19).

15.8 Performance of recombination detection tools

Probably the most important user criterion to make an objective choice is lacking in Table 15.1; how good are these methods at achieving their goal? Two aspects are important in their evaluation: *power* (or false negative rate) and *false positive rate*. The power and false positive rate of detecting recombination signal in molecular data has been evaluated for different methods (Brown *et al.*, 2001; Posada & Crandall, 2001a; Smith, 1999; Wiuf *et al.*, 2001). For this purpose, coalescent-based simulations appear to be very useful. To evaluate power, two variables are important in the simulation procedures: recombination rate and mutation rate. Increasing recombination rates result in an increasing number of recombination events in the history of the sampled sequences (Fig. 15.3). Although not all events leave a molecular footprint of recombination (Fig. 15.2), the frequency of simulated data sets in which recombination can be detected should increase towards 100% as higher recombination rates are used in the simulation. However, also the mutation/substitution process that is superimposed onto the ancestral recombination graphs to simulate the sequences will impact the power of recombination detection (Fig. 15.2). The higher the mutation rate per site per generation, the larger the genetic difference between two randomly drawn sequences and the higher the frequency of incompatible sites and conflicting phylogenetic information (compare Fig. 15.2c, f, and i). For every tool, it is hoped that the sensitivity in detecting recombination is not at the expense of the rate of false positive detection (Type I error). To evaluate this, genealogies are simulated without recombination events. In the mutational process, not only increasing genetic divergence is now important, but also increasing rate heterogeneity among sites (Posada & Crandall, 2001a) (see Chapter 4 on how this is modeled in the nucleotide substitution process). The latter increases the probability of recurrent substitution, which also increases the frequency of incompatible sites in the generated sequences (see Box 15.1).

The most comprehensive study comparing 14 different methods using such simulations revealed that recombination detection tools are generally not very powerful, but they do not seem to infer many false positives either (Posada & Crandall, 2001a). Methods that examine substitution patterns or incompatibility among sites appeared more powerful than phylogenetic methods, a conclusion also shared by smaller-scale studies (Brown *et al.*, 2001; Wiuf *et al.*, 2001). This might not be surprising in case of low sequence diversity. If the boxes in Fig. 15.2 would be translated to real sequences, no well-supported trees could probably be inferred

from either side of the breakpoint, even for those representing "moderate" mutation rates. However, the substitution patterns and the amount of incompatible
sites can still exhibit significant deviations from clonality. In real molecular data,
such deviations might also result from other evolutionary process. So comparisons
on empirical data sets (Posada, 2002), and those for which recombination is well
characterized in particular (Drouin *et al.*, 1999), can provide important insights.
Particular processes that can lead to false positive results are now also being implemented in simulation procedures (e.g. substitution rate correlation, Bruen *et al.*,
2006). Finally, it is worth noting that methods have only been systematically evaluated for their performance in revealing the presence of recombination. Similar
studies to evaluate the accuracy of identifying recombinant sequences within a
data set and locating breakpoints will greatly assist the selection among available
methods (e.g. Chan *et al.*, 2006).

Acknowledgment

We thank David Robertson and Darren Martin for comments and suggestions on
how to classify recombination detection methods.

Detecting and characterizing individual recombination events

THEORY

Mika Salminen and Darren Martin

16.1 Introduction

In addition to point mutation, the most important mechanisms whereby organisms generate genomic diversity are undoubtedly nucleic acid recombination and chromosomal **reassortment**. Although this chapter will mostly deal with the characterization of true recombination events, many of the methods described here are also applicable to the detection and description of reassortment events. Moreover, we will focus on **homologous recombination**, which is defined as the exchange of nucleotide sequences from the same genome coordinates of different organisms. Although **heterologous recombination** (i.e. recombination resulting in either the physical joining of segments in unrelated genes or gross insertion/deletion events) will not be considered, many of the methods described in this chapter may also be applicable to the detection and analysis of this form of genetic exchange.

Whereas the genomes of all cellular organisms and some viruses are encoded in DNA, many viruses use RNA as their genetic material. The importance of recombination in the evolution of life on earth is underlined by the fact that various mechanisms for both DNA and RNA genomic recombination have evolved. Well-studied recombination mechanisms mediating double-stranded DNA break repair and/or chromosomal recombination during meiotic cell division are extremely common amongst cellular organisms and it is believed that DNA viruses also access these mechanisms during nuclear replication within infected host cells.

The Phylogenetic Handbook: a Practical Approach to Phylogenetic Analysis and Hypothesis Testing, Philippe Lemey, Marco Salemi, and Anne-Mieke Vandamme (eds.). Published by Cambridge University Press. © Cambridge University Press 2009.

Conversely, the recombination mechanisms used by RNA viruses and retroviruses (whose genomes pass intermittently through DNA and RNA phases), are generally encoded by the viruses themselves. For example, two features of retrovirus life cycles that greatly facilitate recombination are packaging of two RNA genomes within each virus particle (i.e. they are effectively *diploid*), and the predisposition of retroviral reverse transcriptases to periodically drop on and off these RNA molecules during DNA synthesis. If the reverse transcriptase drops off one of the RNA molecules and reinitiates DNA strand elongation on the other, and the two RNA molecules are genetically different, then the newly synthesized DNA molecule will have a mixed ancestry and will therefore be effectively recombinant.

Given the biological and evolutionary significance of recombination and the real probability that recombination features in the recent histories of most DNA sequences on Earth, it is perhaps surprising that so many (if not most) evolutionary analysis methods in common use today assume that nucleotide sequences replicate without recombining. The inescapable fact that a recombinant nucleic acid sequence has more than one evolutionary history implies that recombination will have at least some effect on any sequence analysis method that assumes correctly inferred evolutionary relationships (see Box 15.2 in the previous chapter). It is probably for this reason that an enormous number of recombination detection and analysis methods have been devised (Table 16.1 and see *http://www.bioinf.manchester.ac.uk/recombination/programs.shtml* for a reasonably up-to-date list of the computer software that implements most of these methods). This chapter will briefly discuss how some of these methods can be practically applied to identify and characterize evidence of individual recombination events from multiple alignments of recombining nucleotide sequences.

16.2 Requirements for detecting recombination

The vast majority of recombination events leave no trace on the recombinant molecule that is generated. For an individual recombination event to be detectable, the recombining or parental sequences must differ at two or more nucleotide positions. In practice, however, proof that a recombination event has occurred usually involves a statistical or phylogenetic test to determine whether or not a potential recombinant has a non-clonal ancestry. To detect and characterize individual recombination events, such tests demand considerable amounts of sequence data that must meet certain criteria. Generally they require: (1) sampling of a recombinant sequence and at least one sequence resembling one of the recombinant's parental sequences; (2) that the parental sequences are different enough that at least one of the two sequence tracts inherited by the recombinant contains sufficient

Table 16.1 Available software tools for characterizing individual recombination events

Program	Method(s) implemented	References
3Seq	3Seq	Boni *et al.*, 2007
Barce	Barce	Husmeier & McGuire, 2003
DualBrothers	DualBrothers	Minin *et al.*, 2005
Gard	Gard	Kosakovsky Pond *et al.*, 2006
Geneconv	Geneconv	Sawyer, 1989
Jambe	Jambe	Husmeier & Wright, 2001
JpHMM	Jphmm	Shultz *et al.*, 2007
Lard	Lard	Holmes *et al.*, 1999
Maxchi	Maxchi	Posada & Crandall, 2001; Maynard Smith, 1992
Phylpro	Phylpro	Weiller, 1998
Pist	Pist	Grassly & Holmes, 1997
Plato	Plato	Woroby, 2001
Rat	Rat	Etherington *et al.*, 2005
Recpars	Recpars	Hein, 1993
Rega	Rega	de Oliveira *et al.*, 2005
Rdp3	RDP, Geneconv, 3Seq, Bootscan, Maxchi, Chimaera, Dss, Siscan, Phylpro, Lard	Martin *et al.*, 2004
Rip	Rip	Siepel *et al.*, 1995
Simplot	Simplot, Bootscan	Lole *et al.*, 1999; Salminen *et al.*, 1995
Siscan	Siscan	Gibbs *et al.*, 2000
Topal	Dss	McGuire & Wright, 2000
TOPALi	Dss, Barce, Jambe	Milne *et al.*, 2004

polymorphisms to unambiguously trace its origin to a parental lineage; (3) that the distribution of polymorphisms inherited by the recombinant from its parental sequences cannot be credibly accounted for by convergent point mutation (see Box 15.1 in the previous chapter); (4) that the recombination event has not occurred so long ago that the distinguishing pattern of polymorphisms created by the event has not been erased by subsequent mutations.

A good illustration of the importance of each of these factors can be seen when attempting to detect recombination events in HIV sequences. In the case of detecting recombination between HIV-1M subtypes (the viruses responsible for the vast majority of HIV infections worldwide), all of these requirements are met. (1) The over 600 publicly available full HIV-1M genome sequences provide ample data for recombination analysis; (2) following the introduction of HIV-1M into humans, founder effects in the epidemiological history of HIV-1 group M allowed enough distinguishing genetic variation to occur between the subtype lineages

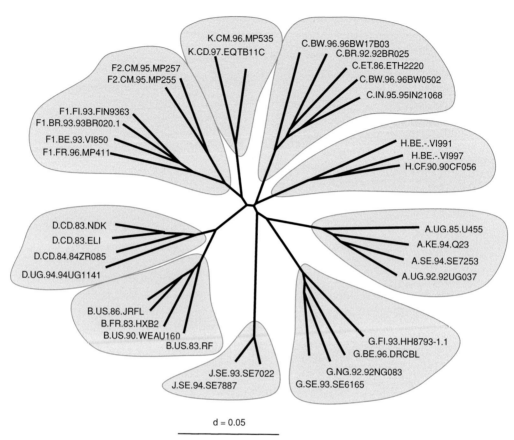

Fig. 16.1   HIV subtypes. K2P model Neighbor-Joining phylogenetic tree using the 1999/2000 Los Alamos HIV database complete genome reference sequence alignment (*http://hiv-web.lanl.gov*). Strain-name coding: X.CC.00.YYYY with X = Subtype, CC = two-letter country code, 00 = year of sampling, YYYY = original isolate identifier. Note the nine discrete groups of sequences.

(Fig. 16.1) that the origins of many sequence tracts found in today's inter-subtype recombinants can be quite easily traced; (3) the recombination mechanism yielding HIV recombinants often results in "exchanges" of sequence tracts large enough to contain sufficient polymorphisms that clusters of polymorphisms characteristic of different subtypes within a single sequence cannot be reasonably explained by convergent evolution; (4) mosaic polymorphism patterns that characterize the many inter-subtype recombinants that have emerged in the wake of relatively recent widespread epidemiological mixing of HIV-1 subtypes are still largely unobscured by subsequent substitutions.

The ease with which inter-subtype HIV-1M recombinants can be detected is starkly contrasted with the difficulty of characterizing individual recombination events that have occurred between sequences belonging to the same HIV-1M sub-types – so-called intra-subtype recombination. While there are many publicly available sequences for most of the subtypes and the sequences within each sub-type are sufficiently divergent for intra-subtype recombinants to be detectable, it is still very difficult to accurately characterize intra-subtype recombination events because: (1) within subtypes, phylogenetic trees generally show star-like struc-tures lacking clusters with enough distinguishing genetic variation. Such trees can be expected for exponentially growing viral epidemics. However, also recombina-tion makes structured trees appear more star-like, which aggravates the problem. (2) Many of the mosaic polymorphism patterns of older intra-subtype recombi-nants may have been obscured by subsequent mutations.

16.3 Theoretical basis for recombination detection methods

Again, using HIV recombination as an example, inter-subtype recombination can be graphically illustrated using phylogenetic analyses. If separate phylogenetic trees are constructed using sequences corresponding to the tracts of a recombinant sequence inherited from its different parents, the recombinant sequence will appar-ently "jump" between clades when the two trees are compared (Fig. 16.2). Most methods developed to detect specific recombination events and/or map the posi-tions of recombination breakpoints apply distance- or phylogenetic-based meth-ods to identify such shifting relationships along the lengths of nucleotide sequence alignments. We will hereafter refer to these shifts or jumps in sequence relatedness as "recombination signals."

There are a variety of ways in which recombination signals are detected. The most common are variants of an average-over-window-based scanning approach. Some measure of relatedness is calculated for a set of overlapping windows (or alignment partitions) along the length of an alignment (Fig. 16.3) and evidence of recombination is obtained using some statistic that compares the relatedness measures calculated for different windows. Every scanning window approach is, however, a simplification of a more computationally intense analysis-of-every-possible-alignment-partition approach. Scanning window approaches are gener-ally quite fast because relatedness measures are only calculated for a relatively small proportion of all possible alignment partitions and sometimes only adjacent win-dows are compared. All recombination signal detection approaches that do not employ scanning windows have some other (occasionally quite complex) means of first selecting the partitions (from all those possible) for which relatedness measures are calculated, and then selecting which of the examined partitions are statistically compared with one another. Generalizing, therefore, recombination signal detec-tion involves partitioning of alignments, calculation of relatedness measures for

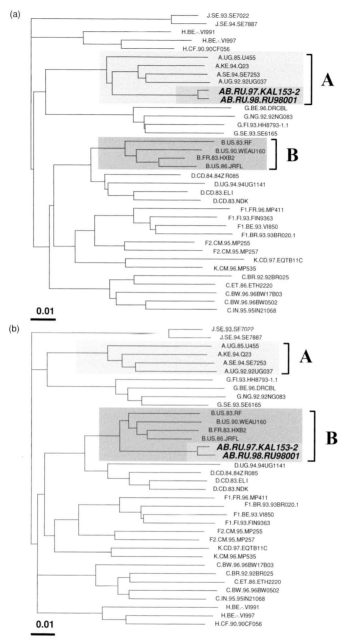

Fig. 16.2   "Jumping" of recombinant isolates. Two representatives (AB.RU.97.KAL153-2 and AB.RU.98.RU98001, in boldface) of a recombinant between HIV-1 M-group subtypes A and B were phylogenetically analyzed in two different genome regions. The trees represent a region clustering the recombinant isolates in (a) with subtype A and in (b) with subtype B. Reference sequences are from the 1999/2000 Los Alamos HIV-1 database.

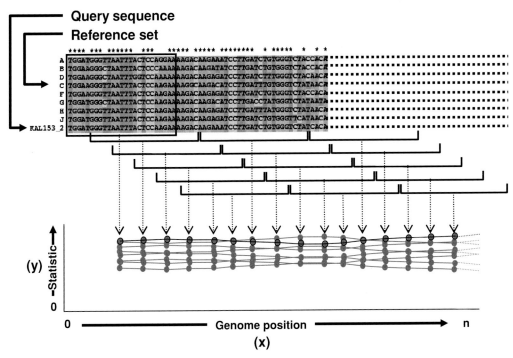

Fig. 16.3   Basic principle of average-over-window scanning methods. An alignment of reference sequences (A–J) and a potential recombinant (the query sequence) are sequentially divided into overlapping sub-regions (i.e. window, box, and brackets) for which a statistic/measure is computed. This statistic/measure is then plotted (broken arrow lines) in an x/y scatter plot using the alignment coordinates on the x-axis and the statistic/measure range on the y-axis. Each value is recorded at the midpoint of the window and connected by a line.

each partition, and use of some statistic determined from these measures to identify pairs or sets of partitions where recombination signals are evident.

There are many possible relatedness measures and statistics that could be used to compare partitions. The simplest and computationally quickest to compute measures are pairwise **_genetic distances_** between the sequences in an alignment. These are based on the reasonable (but not always true) assumption that any given sequence will be most similar to whichever sequence it shares a most recent common ancestor with. Therefore, if the genetic distances of every sequence pair are computed along an alignment using, for example, a sliding-window approach, the relative distances between non-recombinant sequences should remain consistent across all alignment partitions (Fig. 16.4). If recombination occurs, however, the relative distances between the recombinant sequence and sequences closely related to one or both of its parental sequences might shift from one partition to the next, with the point at which the shift occurs indicating the recombination breakpoint.

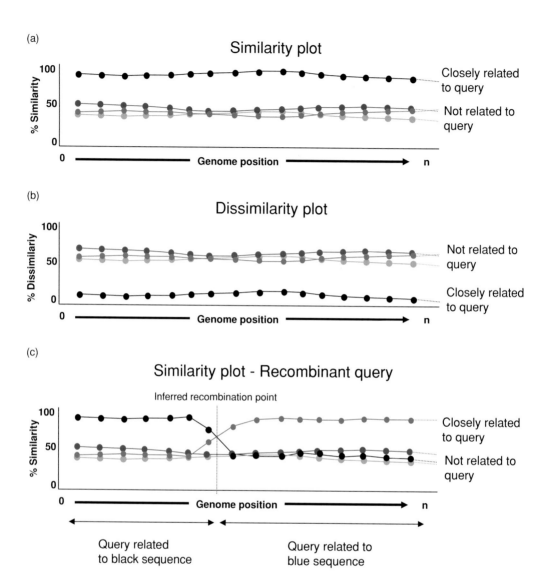

Fig. 16.4   Similarity and dissimilarity methods. (a) Similarity plot. In this type of analysis, the measure recorded is the similarity value (sim) between the query sequence and each of the reference sequences. (b) The same analysis, except that the inverse value to similarity (1−sim) or the dissimilarity is plotted. In various methods, similarity/dissimilarity values corrected by an evolutionary model, i.e. any of the JC69, TN93, K80, F81, F84, HKY85, or GTR models (see Chapter 4 and Chapter 10) may be used. (c) Schematic view of a plot of a recombinant sequence.

In this simple case, the calculated statistic would be something like the differences in genetic distances between the potential recombinant and its two potentially parental sequences in adjacent windows. Using raw pairwise genetic distances as a relatedness measure for identifying potential recombination signals is one component of many recombination detection methods including, for example, those implemented in programs such as SIMPLOT (Lole *et al.*, 1999), RECSCAN (Martin *et al.*, 2005), RAT (Etherington *et al.*, 2005), and RIP (Siepel *et al.*, 1995).

Although the relative pairwise genetic distances between sequences usually correlates well with their evolutionary relatedness, this is not always the case. This is important because more accurate estimates of evolutionary relatedness should enable more accurate identification of recombination signals. Therefore, using phylogenetic methods rather than pairwise distances to infer the relative relatedness of sequences in different alignment partitions has been extensively explored in the context of recombination signal detection. Recombination detection methods that employ phylogeny-based comparisons of alignment partitions include those implemented in programs such as TOPAL (McGuire & Wright, 1998; McGuire *et al.*, 1997), DUALBROTHERS (Minin *et al.*, 2005; Suchard *et al.*, 2002), PLATO (Grassly & Holmes, 1997), RECPARS (Hein, 1993), GARD (Kosakovsky Pond *et al.*, 2006), JAMBE (Husmeier & Wright, 2001; Husmeier *et al.*, 2005), and BARCE (Husmeier & McGuire, 2003). The most popular of these phylogenetic methods is the *bootscan* method (Fig. 16.5). Implementations of bootscan can be found in the programs RDP3 and SIMPLOT, which will be used later to demonstrate how recombination can be detected with the method. The bootscan method involves the construction of bootstrapped *neighbor joining* trees in sliding window partitions along an alignment. The relative relatedness of the sequences in each tree is then expressed in terms of *bootstrap* support for the phylogenetic clusters in which they occur. Recombination is detected when a sequence, defined as a query, "jumps" between different clusters in trees constructed from adjacent alignment partitions. This jump is detectable as a sudden change in bootstrap support grouping the potential recombinant with different sequences resembling its potential parental sequences in different genome regions.

A third, very diverse, set of recombination signal detection approaches employ a range of different relatedness measures and statistics and often draw on some phylogenetic information to compare relationship measures determined for different alignment partitions. Usually, these are based on identifying shifts in the patterns of sites shared by subsets of sequences within an alignment. These so-called *substitution distribution methods* use a statistic (e.g. a Z-score, a chi square value, Pearson's regression coefficient, or some other improvised statistic) that can be used to express differences in the relative relationship between sequences in different, usually adjacent, alignment partitions. The relatedness measures used

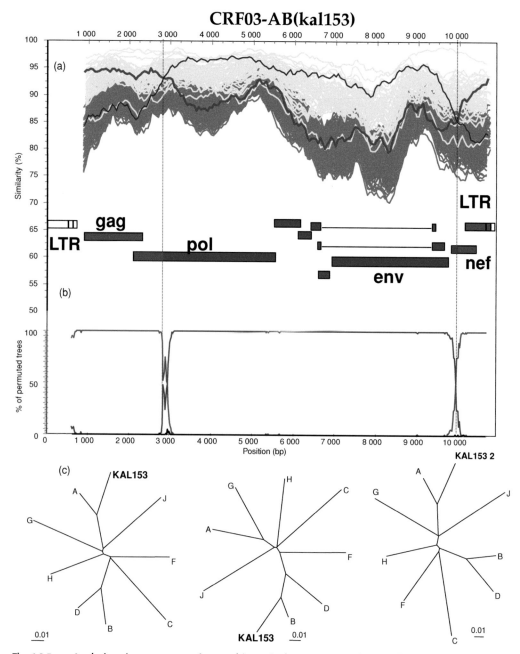

Fig. 16.5 Analysis using SIMPLOT of recombinant isolate KAL153 using (a) the distance-based similarity method and (b) the phylogenetic based bootscanning method. (c) Phylogenetic confirmation (K2P + NJ) of the identified recombination signal. The similarity analysis with subtypes A, B (parental subtypes), and C (outgroup) is superimposed on the ranges of intra- and inter-subtype variation.

are generally counts of nucleotide sites in common and/or different between pairs, triplets or quartets of sequences in the alignment being analyzed. The exact sites that are considered differ from method to method with some counting all sites of the alignment (e.g. the PHYLPRO method; Weiller, 1998) and others examining only those sites that differ in some specified way amongst the sequences being compared (e.g. the SISCAN method; Gibbs *et al.*, 2000). The major advantage of the substitution distribution methods over pure phylogenetic and distance based approaches is that they often allow detection of recombination events that cannot, for example, be visualized as sequences "jumping" between clades of phylogenetic trees constructed using different alignment partitions. By accessing information on overall patterns of nucleotide substitution within an alignment, many substitution distribution methods (such as those implemented in programs like 3SEQ, Boni *et al.*, 2007; GENECONV, Sawyer, 1989; MAXCHI, Maynard Smith, 1992; and CHIMAERA, Posada & Crandall, 2001) can detect when sequences are either more closely related or more distantly related in certain alignment partitions than would be expected in the absence of recombination. These methods are able to detect such recombination signals regardless of whether they have sufficient phylogenetic support.

One of the most powerful substitution distribution methods is the MAXCHI method and it will be demonstrated in the practical exercises later. Like the bootscan method mentioned above, the MAXCHI method involves moving a window across the alignment. However, before scanning the alignment, the MAXCHI method first discards all the alignment columns where all sequences are identical and then moves a window with a partition in its centre along this condensed alignment. For every pair of sequences in the alignment, nucleotide matches and mismatches are scored separately for the two halves of each window and then compared using a 2×2 chi square test. Whenever a window partition passes over a recombination breakpoint, the method records peaks in the chi square values calculated for sequence pairs containing the recombinant and one of its "parents" (they are actually usually just sequences resembling the recombinant's parents).

While substitution distribution methods such as MAXCHI are both extremely fast and amongst the most powerful ever devised, many of these methods generally also assume that sequence similarity is perfectly correlated with evolutionary relatedness – an assumption that is often violated and could potentially compromise their accurate inference of recombination.

Given that there are so many different methods with which recombination signals could be detected, it is important to realize that none of these methods has yet emerged as being best under all possible analysis conditions. Some extremely sophisticated and quite powerful methods, such as those implemented in GARD, DUALBROTHERS and TOPALI are also extremely slow and can currently only be applied to relatively small or simplified analysis problems. Some simpler methods, while less powerful, are capable of rapidly scanning enormous, extremely complex data sets. Also, certain methods are incapable of detecting certain types of recombination events, whereas others are prone to false positives if certain of their assumptions are violated. For this reason certain recombination analysis tools such as RDP (Martin & Rybicki, 2000; Martin *et al.*, 2005) and TOPALi (McGuire & Wright, 2000) provide access to multiple recombination signal detection methods that can be used in conjunction with one another. Not only can these methods be used to crosscheck potential recombination signals but they can, in the case of RDP3, also be collectively combined to analyze sequences for evidence of recombination. While this may seem like a good idea, it is still unclear whether detection of any recombination signal by more than one method should afford additional credibility.

16.4 Identifying and characterizing actual recombination events

Many of the methods aimed at identifying recombination signals are also suited to characterize certain aspects of the recombination events – such as localizing recombination breakpoint positions, identifying recombinant sequences and identifying sequences resembling parental sequences. How this information is accessed differs from one method to the next and also relies largely on the amount of prior knowledge one has on the potential recombinant status of the sequences being analyzed.

Currently available analysis tools use two main approaches to identify and characterize recombination events. Choosing an approach and the tools that implement it depends firstly on the types of recombination events one wants to analyze and, secondly, on knowledge of which of the sequences available for analysis are probably not recombinant. We will again use analysis of HIV-1 recombination as an example. Analysis of inter-subtype HIV-1M recombination is vastly simplified by the existence of a large publicly available collection of so-called "pure-subtype" full genome sequences. While not definitely non-recombinant (many of these sequences may be either intra-subtype recombinants or ancient inter-subtype recombinants in which the recombination signals have been degraded by subsequent mutations), this collection of sequences can be used to reliably scan potentially recombinant sequences for evidence of recent inter-subtype recombination events. Many recombination analysis tools such as SimPlot, Rega, Rip, DualBrothers, jpHMM, and RDP3 will allow one to analyze a potential recombinant, or *query sequence*, against a set of known non-recombinants, or *reference sequences*, and identify: (1) whether the query sequence is recombinant; (2) the locations of potential breakpoint positions; (3) the probable origins of different tracts of sequence within a recombinant (Fig. 16.5).

In cases where no reliable set of non-recombinant reference sequences is available, such as when attempting to analyze intra-subtype HIV-1 recombination, one is faced with two choices. Either a set of reference sequences must be constructed from scratch before applying one of the query vs. reference scanning methods, or use must be made of one of the many exploratory recombination signal detection methods that do not rely on a reference sequence set. Construction of a reference sequence data set will not be discussed here, but see Rosseau *et al.* (2007) for an example involving analysis of intra-subtype HIV recombination.

The exploratory recombination signal detection methods, such as those implemented in RDP3, GENECONV, 3SEQ, SISCAN, and PHYLPRO, all accept sequence alignments as input and, without any prior information on which sequences might be recombinant, will attempt to identify signals of recombination. Although these methods are completely objective and their use might seem more appealing than that of methods relying on a largely subjective query vs. reference scanning approach, one should be aware that there are two serious drawbacks to the exploratory analysis of recombination signals. First, when enumerating all the recombination signals evident in an alignment, the exploratory methods will often compare thousands or even millions of combinations of sequences. This can create massive multiple testing problems that must be taken into account when assessing the statistical support of every recombination signal detected. In extreme cases, such as when alignments containing hundreds of sequences are being analyzed, statistical power can become so eroded by multiple testing corrections, that even relatively obvious recombination signals are discounted.

The second, and probably most important, problem with exploratory recombination detection is that, even if one is provided with a very clear recombination signal and a set of sequences used to detect the signal (something that many of the exploratory methods will provide), it is often extremely difficult to determine which sequence is the recombinant. As a result, "automated" exploratory scanning of recombination often still requires a great deal of manual puzzling over which of the identified sequences is "jumping" most between clades of phylogenetic trees constructed using different alignment partitions. This can be a particularly serious problem because the exploratory methods do not, for better or worse, exclude the possibility of detecting recombination events between parental sequences that are themselves recombinant. The apparent objectivity of exploratory recombination detection is therefore ultimately compromised by a largely subjective process of recombinant identification.

PRACTICE

Mika Salminen and Darren Martin

16.5 Existing tools for recombination analysis

Software for analyzing recombination is available for all major operating systems (see *http://www.bioinf.manchester.ac.uk/recombination/programs.shtml* for a reasonably comprehensive list). The two programs that will be used here for demonstrating the detection and characterization of recombination events are the Windows programs SIMPLOT (downloadable from *http://sray.med.som.jhmi.edu/ RaySoft/SimPlot/*) and RDP3 (downloadable from *http://darwin.uvigo.es/rdp/ rdp.html*). Both programs will run on Apple Macs under Virtual PC emulation software. Both programs are capable of reading alignments in a wide variety of formats and performing similarity/dissimilarity plots of any of the sequences against all others in the alignment. They allow multiple analysis parameters to be varied, including window sizes, window overlaps and evolutionary-distance corrections. Both produce graphical outputs that can be exported in both bitmap (.bmp) and windows metafile formats (.wmf or .emf). The programs enable reasonably precise mapping of inferred recombination breakpoints and allow trees to be constructed for different alignment partitions; RDP3 implements various tree reconstruction methods to this purpose. SIMPLOT also allows sequence partitions to be exported for phylogenetic analysis by other programs. Both RDP3 and SIMPLOT also implement the popular bootscanning method (Salminen *et al.*, 1995), the use of which will be described in some detail here using SIMPLOT.

To install SIMPLOT, go to *http://sray.med.som.jhmi.edu/SCRoftware/simplot/* and download the zip-compressed installation file; for this exercise, we used SIM-PLOT version 3.5.1. Place the file in a temporary file folder (e.g. the C:\temp folder found on most systems) and uncompress the file. Install SIMPLOT using the installer file SETUP.EXE. By default, SIMPLOT is installed in Program Files/RaySoft/ folder and a link is added to the Start menu. More detailed instructions can be found on the Simplot website. Installation of RDP3 follows a similar process. Download the RDP3 installation files from *http://darwin.uvigo.es/rdp/rdp.html* to the temporary folder and uncompress it. Run the SETUP.EXE that is unpacked and RDP3 will add itself to the Start menu.

To successfully perform the SIMPLOT exercises, another program, TREEVIEW (see Section 5.5 in Chapter 5), is also needed.

16.6 Analyzing example sequences to detect and characterize individual recombination events

Several sets of aligned sequences will be used for the exercises. These are available at the website *http://www.thephylogenetichandbook.org* and have the following features:

(1) File A-J-cons-kal153.fsa: A single recombinant HIV-1 genome (KAL153) aligned with 50% consensus sequences derived for each HIV-1 subtype (proviral LTR-genome-LTR form).
(2) File A-J-cons-recombinants.fsa: two recombinant HIV-1 sequences in FASTA format aligned with 50% consensus sequences derived for each HIV-1 subtype (proviral LTR-genome-LTR form).
(3) File 2000-HIV-subtype.fsa: A reference set of virtually complete HIV genome sequences aligned in FASTA format (virion RNA R-U5-genome-U3-R form).

In the first three exercises SIMPLOT will be used to demonstrate the query vs. reference approach to detecting and characterizing recombination events. RDP3 will then be used in the last three exercises to demonstrate some exploratory approaches to recombination detection and analysis that can be used if one has no prior knowledge of which sequences in a data set are non-recombinant.

16.6.1 Exercise 1: Working with SIMPLOT

This exercise shows the use of SIMPLOT and its basic properties. To start SIMPLOT, select the program from its group on the `Start` Menu in Windows or double-click on the SIMPLOT icon in the SIMPLOT folder. The program will start with a window from which it is possible to select the `file` menu to open an alignment file. Open the file A-J-cons-kal153.fsa (the .fsa extension may or may not be visible, depending on the settings of the local computer). This file contains an alignment of the CRF03-AB strain Kal153 and 50% consensus reference sequences for subtypes A through J. The tree-like graph shown in Fig. 16.6 appears, which has all the sequences contained in a group with the same name as the sequence (by default). Groups can be automatically defined using any number of characters in the sequence names or using characters prior to a separator in the sequence names by clicking on "Use first character to identify groups". To reveal the sequences in each group, select "Expand Groups". Groups have a color code and are primarily used to quickly generate consensus sequences; they are discussed in another exercise.

Groups or individual sequences can be excluded from the analysis by deselecting them. Groups can be moved or individual sequences can be moved between groups using the options in the right panel. Select the KAL153 group (not the sequence) and try to move it to the top of the tree by pressing the "Move up" button repeatedly. Under the `File` and `Help` menus are four tabs; click on `SimPlot` to go to the page

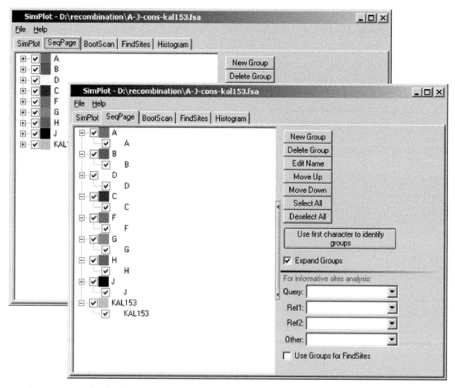

Fig. 16.6 Starting SIMPLOT. The back window represents the view after opening a sequence-alignment file. The front window represents the view after expanding the groups. Each sequence belongs to a group, which by default is named the same as the sequence itself.

where the similarity plots are performed. The other tabs, SeqPage, Bootscan, FindSites and Histogram, are the input page, the bootscanning-analysis page, the phylogenetically informative site-analysis page and the histogram page; the bootscanning analysis is discussed later, and the SIMPLOT documentation describes the FindSites and Histogram analysis.

To do the first analysis, go to the Commands menu and select Query and KAL153 in the pop-up list, which reflects the sequences in the order determined in the SeqPage window. This first operation determines which of the aligned sequences is compared to all the others (Fig. 16.7). To set the window size for the similarity scanning to 400 bp, click on "Window Size" in the lower left corner and choose 400 in the pop-up list. Keep the default settings for "Step Size". To perform the analysis, go to the Commands menu again and click on DoSimplot; the result should be similar to Fig. 16.7. The analysis indicates that the KAL153 strain is a recombinant of subtypes A and B, as reflected by the legend key. The parameters for this analysis can be changed in the Options (Preferences)

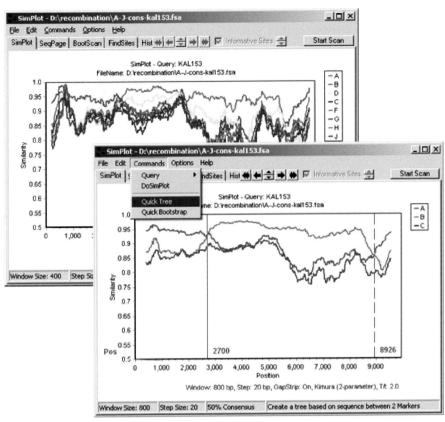

Fig. 16.7   Comparing KAL153 to the reference sequences with SIMPLOT. The back window represents the similarity plot for KAL153 as query compared to all reference sequences. The front window represents the similarity plot for KAL153 compared to subtypes A, B, and C. In this plot, the approximate positions of the breakpoints are indicated using red lines that appear by double clicking on the appropriate positions in the plot. When two breakpoint positions are indicated, a quick tree can be inferred for the alignment slice in between the two red lines.

menu. These include the options for the distance calculation (see Chapter 4 for models of evolution), the bootstrap options and miscellaneous options. Change the window size to 800 bases and compare the plot to the composite similarity plot shown in Fig. 16.5. Another useful option is whether to exclude gap regions from the analysis (usually recommended). Consult the SIMPLOT documentation, accessible from the Help (Contents) menu, for more information about the different options.

Return to the SeqPage tab and deselect groups so that only A, B, C, and KAL153 will be included. Go back to the SimPlot tab. When doing the analysis, it should be easy to pinpoint the recombination breakpoints. In fact, the user can click on

Mika Salminen and Darren Martin

the breakpoint positions in the plot (but not on the distance lines themselves) and a red line will appear with the position in the alignment where the breakpoint is approximately located. A second position, e.g. representing the second breakpoint position can also be selected and a quick tree can be reconstructed for the alignment segment in between the red lines using the "Quick Tree" command (Fig. 16.7). The red lines can be removed by double clicking on them. Clicking on the similarity lines will result in a pop-up window with the name of the relevant reference sequence, the position, the similarity score and the possibility to change the color of the line.

The next analysis we will perform is bootscanning. Make sure that in the Seq-Page tab only the four sequences are still included for this analysis. Click on the BootScan tab, select again KAL153 as query sequence and start the bootscan procedure by selecting DoBootscan from the Commands menu. The default options for this analysis are the Kimura-two-parameter model (see Chapter 4), a *transition/transversion ratio* of 2 and stripping positions that have over 50% of gaps in a column. These settings can be changed in the Options (Preferences) menu. As the bootscanning begins, a running dialog appears in the bottom right corner that shows for which window a tree and bootstrap analysis is being performed; simultaneously, the plot grows on the screen. The analysis is fairly slow but can be stopped using the Stop Bootscan button in the upper right corner (it is not possible to continue the analysis after prematurely stopping it).

16.6.2 Exercise 2: Mapping recombination with SIMPLOT

In this exercise, we will map the structure of two other HIV-1 recombinant strains. The sequences under study are in the file A-J-cons-recombinants. fsa. Open the file A-J-cons-recombinants.fsa and use the skills learned in the previous exercise to map the recombination breakpoints in the sequences. First, analyze the recombinants individually by deselecting the other group including a recombinant. Identify the subtypes of the parental sequences using all reference sequences and a relatively large window size (i.e. 600–800). Finally, map the recombination breakpoints with only the parental sequences and a smaller window size (i.e. 300–400). To verify the results, quick trees can be reconstructed for the putative recombinant regions identified by SIMPLOT or those regions can be exported and analyzed running separate phylogenetic analyses with PHYLIP or PAUP*. To export a region of an alignment from SIMPLOT, first select the region of interest (i.e. map the breakpoints) using two red lines. Go to the File menu and select "Save Marked Slice of Alignment as . . .". The format of the new alignment file can be chosen; for example, the PHYLIP interleaved format. Table 16.2 shows the correct result of the analysis.

Table 16.2 Key to recombinant strain structures of Exercise 2

Strain	Parental subtypes	Breakpoints[a]
UG266	AD	5430, 6130, 6750, 9630
VI1310	CRF06-DF	2920, 3640, 4300, 5260, 6100

[a] There may be some variation in the exact coordinates depending on the window settings used.

16.6.3 Exercise 3: Using the "groups" feature of SIMPLOT

To focus on recombinant events between major clades, e.g. HIV-1 group M subtypes, the "group" feature provides a way to specify groups of reference sequences and perform quick analyses on their consensus sequences. This is done by adding an extra NEXUS block at the end of a NEXUS-formatted alignment file (see Box 8.4 in Chapter 8 for an explanation on the NEXUS format). This block specifies groups identified by users or written by the SIMPLOT software. The following is an example of a group-specifying block at the end of an alignment:

begin RaySoft; [!This block includes information about groups identified by users of software written by Stuart Ray, M.D.] groups group1 = '"@A":clRed(@0, @1, @2, @3), "@B":clGreen(@4, @5, @6, @7), "@C":clYellow(@8, @9, @10, @11, @12), "@D":clBlue(@13, @14, @15, @16), "@F":clGray(@17, @18, @19, @20, @21, @22), "@G":clFuchsia(@23, @24, @25, @26), "@H":clTeal(@27, @28, @29), "@J":clNavy(@30, @31), "@K":clSilver(@32, @33), "@CRF01":clWhite(@37, @36, @35, @34), "@CRF02":clWhite(@41, @40, @39, @38), "@CRF03":clWhite(@43, @42), "@CRF04":clWhite(@46, @45, @44), "@CRF05":clWhite(@48, @47), "@CRF06":clMaroon(@49, @50)'; end;

The NEXUS-formatted alignment contains 51 sequences. The sequences in the group definitions are represented by "@" and a number according to their occurrence in the alignment (starting with 0). Sequences are grouped by bracket notation and the group is preceded by a "@name" and a color code.

To start the exercise, open the file 2000-HIV-subtype.nex in SIMPLOT and compare the SeqPage view to the example NEXUS block; this file contains the same groups as defined in the block. Expand the groups to see how the program groups all the sequences. When the user scrolls to the bottom of the window, six groups are visible: CRF01, CRF02, CRF03, CRF04, CRF05, and CRF06. They contain reference sequences for six currently recognized circulating recombinant forms of HIV-1 (CRF01_AE, CRF02_AG, CRF03_AB, CRF04_cpx, CRF05_DF, and CRF06_cpx). The F subtype actually consists of two groups, the F1 and F2 subsubtypes; split them into two groups. First, rename the F group to F1. Next, create a new group using the buttons on the right, name it F2, and move it right under the F1 group. Expand the F1 group and move the F2 sequences one by one to

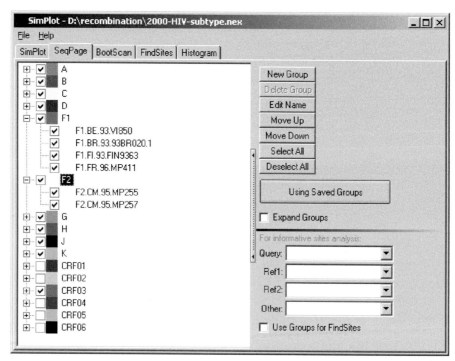

Fig. 16.8   Using and rearranging groups specified in the alignment file. In the 2000-HIV-subtype.nex
file groups have been specified by the inclusion of a NEXUS block. The F group has been
renamed to F1 and a new group, F2, has been created. Both groups are expanded. F2
sequences were moved to the F2 group.

the new F2 group using the buttons to the right or by dragging and dropping
(Fig. 16.8). Deselect all other CRF groups except the CRF03 group. Switch to the
SimPlot tab and perform a default parameter analysis with window settings
of 400 nucleotides/20 steps and CRF03 as the query. The program calculates by
default a 50% consensus for the groups and plots the similarity values to that
consensus. By clicking on the consensus panel in the lower part of the window, it
is possible to change the type of consensus used in the analysis. Explore the effect
of different types of consensus models (see the SIMPLOT documentation for more
details).

16.6.4 Exercise 4: Setting up RDP3 to do an exploratory analysis

This exercise demonstrates how to modify RDP3 analysis settings and do a basic
exploratory search for recombination signals in a simple data set. Start RDP3 and
open the file A-J-cons-kal153.fsa. Certain default RDP3 settings are not ideal for
the analysis of HIV sequences and should be changed. Press the "Options"
button at the top of the screen. All options for all methods implemented in RDP3

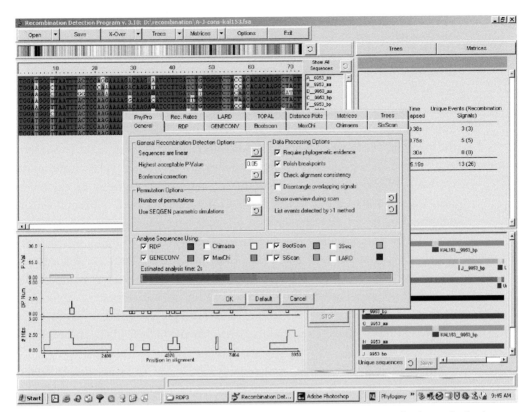

Fig. 16.9   Working with RDP3. The top left panel represents the alignment. The boxes in the bottom left panel indicate the positions of the breakpoints and the sizes of recombinant tracts that have been detected. The top right panel provides information on the duration of the analysis and the number of recombination events identified. The bottom right panel shows the recombinant tracts identified in the different sequences. The middle window is the options window in which the sequences have been set to linear.

(including tree drawing, matrix drawing and recombination detection methods) can be changed using the form that is displayed (Fig. 16.9). The main page of the options form contains details of the general exploratory recombination detection settings. The only thing that needs to be changed here is that sequences should be handled as though they are linear. Although it would not invalidate the analysis if sequences were handled as though they were circular, this setting will make analysis results a little harder to interpret. Press the button with the circular arrow besides the "Sequences are circular" caption.

At the bottom of the general settings form are a series of colored rectangles with names next to them and a colored strip beneath them (Fig. 16.9). Note that some of the names next to the rectangles ("RDP", "GENECONV", "MaxChi" etc.) have ticks next to them. Each colored rectangle represents a different recombination signal detection method. These are all of the methods implemented in RDP3 that can be used to automatically explore and enumerate recombination signals in an alignment. If you wish to explore an alignment for recombination with any of the methods you should click on the "check box" next to the method's name. A tick in this box means it will be used, along with all the other methods with ticks next to them, to explore for recombination signals in the alignment. Note that the BOOTSCAN and SISCAN methods each have two boxes, only one of which is ticked at the moment. These methods are far slower than all the other methods (except LARD) and the option is given to use these as either primary or secondary exploration methods. Primary exploration methods will thoroughly examine an alignment searching for and counting all detectable recombination signals. Whenever signals are found by the primary exploration methods, secondary exploration methods will be used to thoroughly re-examine sequences similar to those in which the initial recombination signal was found. All of the listed methods except BOOTSCAN, SISCAN and LARD will automatically be used as secondary exploration methods. The LARD method is so slow that RDP3 only permits its use as a secondary exploration method. The colored strip at the bottom of the form gives an estimate of the relative execution times of the different methods. An estimate of total analysis time is given above the bar.

For the moment all analysis options on this form will be left to their default values. However, some analysis settings need to be changed for some of the particular exploratory recombination detection methods. Click on the "RDP" tab and change the window size setting from 30 to 60. Click on the "MAXCHI" tab and change this window size setting to 120. Click on the "CHIMAERA" tab and also change this window size setting to 120. Click on the "BOOTSCAN" tab and change the window size setting to 500. Click on the "SISCAN" tab and also change this window size setting to 500. Window sizes were increased from their default settings because HIV sequences are relatively unusual in that they experience mutation rates that are so exceptionally high that recombination signals are rapidly obscured. Increasing window sizes increases the ratio of signal relative to mutational "noise." It is important to note that increasing window sizes also makes detection of smaller recombination events more difficult.

Now that the analysis settings have been made, press the "OK" button at the bottom of the form. If you shut the program down now these settings will be saved. The saved settings will be used whenever you start the program and you will not have to reset them at the start of every analysis.

16.6.5 Exercise 5: Doing a simple exploratory analysis with RDP3

To start an exploratory analysis with the RDP, GENECONV, and MAXCHI methods in primary exploratory mode and the CHIMAERA, SISCAN, BOOTSCAN, and

3SEQ methods in secondary exploratory mode press the "X-Over" button at the top of the screen. A series of colored lines will flash past on the bottom of the screen and will be almost immediately replaced by a set of black and pinkish boxes (Fig. 16.9). These are indicating the positions of breakpoints and sizes of recombinant tracts that have been detected and provide some indication of the *p*-values associated with detected events. These graphics are intended to give you something to look at when the analysis is taking a long time, so don't worry about exactly what they mean right now. In the top right-hand panel you will notice that information is given on how long each analysis method took to explore the data (Fig. 16.9). You will also notice that under the "Unique events (recombination signals)" heading there are some numbers. The first number indicates the number of unique recombination events detected by each method and the second number (in parentheses) indicates the number of unique recombination signals detected. The second number will always be equal to or larger than the first number. If the second number is larger it will indicate that two or more sequences in the alignment carry traces of the same ancestral recombination event(s).

In the bottom right panel you will notice a series of colored rectangles (Fig. 16.9). Move the mouse pointer around this panel a bit. You will notice that when it moves over some of the rectangles, information flashes onto the top right-hand panel. The information displayed here relates to the rectangle that the mouse pointer is positioned over. The rectangles that are "sensitive" to the mouse pointer are graphical representations of individual recombination signals detected in the alignment. Use the scroll-bar on the right-hand side of the bottom right panel to move down so that the rectangles representing the sequence, KAL153, are visible (move the scroll bar right down to the bottom). This is the recombinant sequence analyzed using the query vs. reference method in exercise 1 above.

Click on the button with the circular arrow at the bottom of the bottom right panel displaying the graphical representation of recombination signals. The caption beside the button should now read "Methods" (Fig. 16.10). You will notice that the color of all the rectangles has changed. They are now either grey, red, orange or blue. The long rectangle at the bottom, that is intermittently dark and light gray, represents sequence KAL153. The dark parts represent the "background" sequence and the light bits represent tracts of sequence that possibly have a recombinant origin (Fig. 16.10). As mentioned before, the colored rectangles directly beneath the light gray bits represent the recombination signal. These rectangles also represent recombination hypotheses delimiting the bounds of tracts of horizontally inherited sequence. The labels to the right of the colored rectangles indicate sequences in the alignment resembling the donor parents. As the "Methods" coloring scheme is selected, the red, blue and orange rectangles denote recombination events detected by the RDP, GENECONV, and MAXCHI methods, respectively. Pressing the left mouse

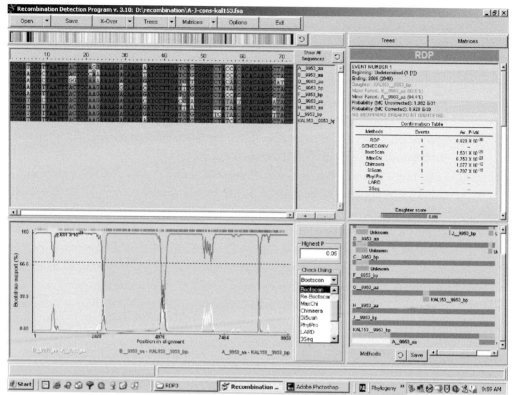

Fig. 16.10   Detecting recombination events using RDP3. The top right panel provides information that relates to the recombination signal, the rectangle in a sequence in the lower right panel, over which the mouse has been positioned. In the bottom right panel, the graphical information representing recombination signals has been changed from "Unique sequences" to "Methods." The bottom left panel shows the bootscan plot for the recombination event identified for KAL153. (After clicking on the first recombinant tract identified by RDP in KAL153 at the bottom of the lower right panel, the "Check Using" option was changed to "Bootscan".)

button in the window when the mouse pointer is not over a colored rectangle gives you a legend explaining the color coding.

Move the mouse pointer over the red rectangle to the left of the panel and press the left mouse button. What should now be displayed in the bottom left panel is a graphical representation of the actual recombination signal used to detect the recombination event depicted by the red rectangle. This plot can be interpreted in much the same way as the BOOTSCAN plots described in earlier exercises. To see a BOOTSCAN version of this plot look for the label "Check using" to the right of the plot and press the little arrow besides the label, "RDP". On the menu that appears select either "Bootscan" or "Re-Bootscan" (Fig. 16.10). Now

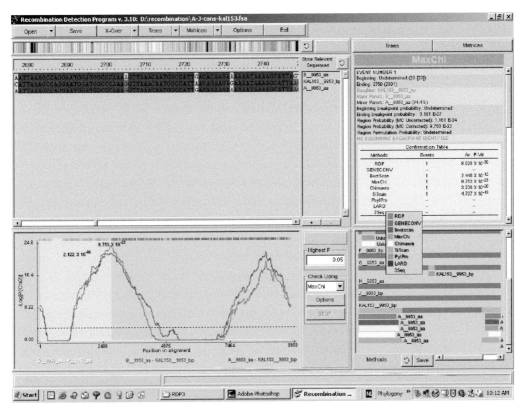

Fig. 16.11   Investigating breakpoint positions using RDP3. The top right panel provides information that relates to the recombination signal identified by MAXCHI for KAL153. The bottom right panel shows "All Events for All Sequences", the option selected after right clicking in the panel. The color key for the different methods is shown (by pressing the left mouse button on an open part of the panel). The bottom left panel shows the MAXCHI plot for KAL153. By clicking on the right border of the first pink region, the alignment in the top left panel jumps to around position 2800. By clicking "Show all sequences" next to the alignment panel, the option changes to "Show relevant sequences", which results in the differently colored three sequences in the alignment panel.

try the GENECONV, MAXCHI, CHIMAERA, SISCAN, PHYLPRO, 3SEQ, TOPAL, and DISTANCE options. (Be warned not to select the LARD option unless you are prepared to wait a few minutes.) The distance plot is similar to the "simplots" described in previous exercises.

Notice that most of the plots involve three lines (Fig. 16.11). This is because most of the exploratory methods scan through the alignment successively examining all possible combinations of three sequences. These three-sequence sub-alignments are examined in two basic ways, which are reflected in the two different color schemes used in the plots. The green–purple–yellow schemes indicate pairwise comparisons of sequences in the three sequence sub-alignments whereas the green–blue–red schemes indicate triplet analyses of the three sequences in the sub-alignment where the pattern of sites in each sequence is compared to the other two. For LARD and TOPAL a single line is plotted because only a single statistic is used to compare partitions of the three-sequence sub-alignment.

Point the mouse cursor at a region of the bottom right panel that has no colored rectangles in it and press the right mouse button. One of the options on the menu that appears is "Show All Events for All Sequences." Select this option and use the scroll bar to get back to the representation of sequence KAL153. You should see that some more colored rectangles have appeared (Fig. 16.11). These include light green, dark green, yellow, and purple ones. Use the color key to see what methods these represent (Fig. 16.11, press the left mouse button on an open part of the bottom right panel).

Going back to the left-hand recombinant tract of sequence KAL153, you may notice that, of the five methods detecting this recombination signal, only the BOOTSCAN and MAXCHI methods (dark green and orange rectangles, respectively) agree on the position of the breakpoint to the right (or "ending breakpoint"). This indicates that there is some degree of uncertainty associated with the placement of this breakpoint. Click on the orange rectangle. The recombination signal detected by the MAXCHI method is displayed (Fig. 16.11). Notice that the purple and green peaks in the plot coincide with the right breakpoint position. It is worthwhile pointing out here that the peaks in MAXCHI, CHIMAERA, TOPAL, and PHYLPRO plots all indicate estimated breakpoint positions (note, however that for PHYLPRO plots the peaks face downward). These methods are geared to breakpoint detection.

Look at the upper left panel where the alignment is displayed and look for the caption "Show all sequences." Beside the caption there is a button with a circular arrow on it. Press this until the caption says "Show relevant sequences" (Fig. 16.11). You will notice that the color-coding of the alignment has changed. On the MAXCHI plot double click on the right border of the pink region. You will notice that this causes the alignment to jump to around position 2800. Use the horizontal scroll bar beneath the alignment to scan backwards. Whereas most nucleotides are grayed out, some are colored green, purple, and yellow (Fig. 16.11). The grayed positions indicate nucleotides ignored by the MAXCHI analysis. Nucleotide pairs labeled yellow, green, and purple represent nucleotide positions contributing to the recombination signal plots in the same colors below. As the peaks in the plot below are green and purple, one would expect to see more purple colored nucleotide pairs on one side of the breakpoint (in this case it is the left) and more green nucleotide pairs on the other side of the breakpoint (in this case it is the right). Scan the sequence left of the breakpoint and you will notice that although purple sites are

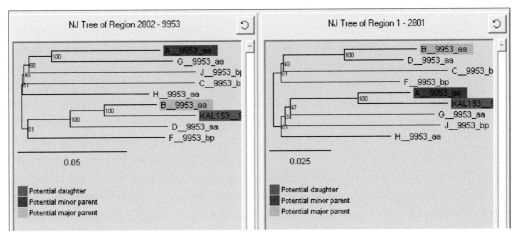

Fig. 16.12   Tree reconstruction using RDP3. On the right, the neighbor-joining (NJ) tree is shown for the first 2801 nucleotides in the alignment. In the NJ tree on the left, the KAL153 is clustering with subtype B.

most common there are also a lot of yellow sites. There is a particularly high ratio of yellow : purple sites between alignment positions 2570 and 2780. These yellow sites may represent either post-recombination mutations or sites at which the sequence in the alignment identified as resembling a parent (A in this case) differs from the KAL153's actual parent. RDP3's uncertainty over the breakpoint position is due to sequence A being a poor match for KAL153 in the immediate vicinity of the breakpoint, as indicated by the different breakpoint positions estimated by the different methods.

At the top of the screen press the "Trees" button. A window should appear with two phylograms or trees in it (Fig. 16.12). Now whenever the left mouse button is pressed on one of the colored rectangles in the bottom right panel of the main program window, two trees will be automatically drawn. The left tree is drawn using an alignment partition corresponding with the "background" sequence of the recombinant sequence being analyzed (in this case it is KAL153). The right tree is drawn from the alignment partition corresponding with the currently selected "recombinant region" – i.e. the alignment partition bounded by the ends of the tract of sequence represented by the colored rectangle currently selected. The default tree that is displayed is an *UPGMA* without any bootstrap replicates. While not necessarily the best type of tree for describing evolutionary relationships, it is very fast to construct and allows a quick preliminary assessment of whether there is any obvious phylogenetic evidence of recombination, i.e. you should look and see whether the proposed recombinant does indeed "jump" between clades of the two trees.

To more rigorously explore the phylogenetic evidence in favor of a recombination hypothesis you should at the very least draw a bootstrapped neighbor-joining tree. To do this, right click on each of the trees and select the "Change tree type > Neighbor-joining" option that is displayed on the menu that appears. Notice that whereas there is 100% bootstrap support for a KAL153-B clade in the left tree, there is 100% support for a KAL153-A clade in the right tree (Fig. 16.12). This is good phylogenetic support that KAL153 has arisen following a recombination event between A- and B-like viruses. These trees are constructed using the neighbor joining method found in the NEIGHBOR component of PHYLIP (Felsenstein, 1996). You could also have constructed least squares trees (using the FITCH component of PHYLIP; see Chapter 5), maximum likelihood trees (using PHYML; Guindon & Gascuel, 2003; see Chapter 6) or Bayesian trees (using MRBAYES; Ronquist & Huelsenbeck, 2003; see Chapter 7), but be warned that large trees can take a very long time to construct with these methods.

16.6.6 Exercise 6: Using RDP3 to refine a recombination hypothesis

In this exercise we will use what you have learned to complete our analysis of the A-J-cons-KAL153.fsa data set. It should become clear that the exploratory analyses carried out by RDP3 yield a set of recombination hypotheses that are not necessarily consistent among different methods or not necessarily the absolute truth about the recombination events that have left detectable traces in a set of aligned sequences.

RDP3 formulates its recombination hypotheses in a stepwise fashion. It firstly scans the alignment for all the recombination signals detectable with the primary exploratory methods selected, and then, starting with the clearest signal (i.e. the one with the best associated p-value), it attempts to identify which of the sequences used to detect the recombination signal is the recombinant. This is not a trivial process and the program uses a variety of phylogenetic and distance-based tests to try to figure this out. While the results of these tests are a completely objective assessment of which sequence is recombinant, the program will incorrectly identify the recombinant from time to time. When it has identified a putative recombinant sequence, RDP3 looks for other sequences in the alignment that might share evidence of this same recombination event (i.e. the recombination event may have occurred in some common ancestor of two or more sequences in the alignment and RDP3 tries to see if there is any evidence for this). From time to time it will miss a sequence, but RDP3 will usually be a bit overzealous grouping sequences it thinks are descended from a common recombinant ancestor. It then takes the recombinant sequence (or family of recombinant sequences) and splits it into two sections – one piece corresponding to each of the bits inherited from the recombinant's two parents. The smaller bit (from the "minor" parent) is then added to the alignment as an extra sequence. The alignment of the split sequences with the rest of the alignment is maintained by marking deleted bits of sequence with a "missing data" character (as opposed to the conventional "A,C,G,T,-" characters). This process effectively "erases" the recombination signal from the alignment and ensures it is not counted again when RDP3 next rescans the alignment from scratch for the remaining signals. Once the remaining signals are detected, it again selects the best and restarts the process.

The most important point of this explanation of how RDP3 explores and counts unique recombination signals is that the program does it in a stepwise fashion. You can see the order in which signals were analyzed by moving the mouse cursor over one of the colored rectangles at the bottom right of the screen. On the first line of the information appearing in the top right panel you will see the caption "EVENT NUMBER." This number tells you the order in which this event was analyzed. The number is very important because if the program has made any judgment errors when analyzing earlier events, it might affect the validity of the conclusions it has reached regarding the currently examined event.

When refining RDP3's recombination hypotheses after the automated phase of its exploratory analysis, it is strongly advisable to trace the exact path the program took to derive its hypotheses. You can do this by first clicking the left mouse button when the mouse pointer is over an empty part of the bottom right panel and then using the "pg up" (page up) and "pg dn" (page down) buttons on your computer keyboard to go through events in the same order that the program went through them. Do this and navigate to event 1 using the "pg up" and "pg dn" buttons. Don't be surprised if some events are skipped – these represent recombination signals that were detected by fewer methods than the cut-off specified in the "general" section of the options form you looked at earlier.

Event one is the recombination signal we analyzed earlier and persuaded ourselves was, indeed, genuine evidence of a recombination event. Move the mouse cursor over one of the colored rectangles representing the event and press the right mouse button. On the menu that appears select the "Accept all similar" option. This allows you to specify to RDP3 that you are happy to accept that this is genuine evidence of recombination and that it does not need to concern itself with this event during any subsequent reanalysis cycles. Press the "pg dn" button and you will be taken to event 2. Assess the evidence for this event the same way that you did event 1. Notice the blue area of the recombination signal plot. This specifies that one of the three sequences being examined in the plot contains missing data characters – these missing data is in the "blued-out" region of the plot. The missing data characters are those introduced by RDP3 into the background KAL153 sequence to keep it in alignment with the rest of the sequences after the recombination signal yielding the event 1 was analyzed.

After looking at event 2, move on to subsequent events. One of the following events has red capitalized text in the top right panel, "POSSIBLE MISSALIGN-MENT ARTIFACT", which is particularly worrying. In the plot window for this event double click on the pinkish region specifying the recombinant tract. If you are not on the "Show relevant sequences" view of the alignment display, change the alignment display to this setting now (see the previous exercise). It is quite plain that there has been unreliable alignment of the three displayed sequences in the area identified as being a potential recombinant tract. Recombination signal detection is particularly error prone in poorly aligned sequences and it would be advisable to mark this evidence of recombination as being of dubious quality. To do this, move the mouse pointer over the flashing rectangle, press the right mouse button and select the "Reject all similar" option on the menu that appears. The rectangle and its associated caption should become gray.

If you navigate through the remainder of the signals you will notice both that their associated *p*-values are just barely significant and that they are only detectable by the CHIMAERA and MAXCHI methods (see the RDP3 manual for specific differences between both methods). These two methods are the most similar of the methods implemented in RDP3 and supporting evidence for one by the other should perhaps not carry as much weight as if it was provided by another of the methods. Nevertheless these remaining signals may indicate some genuine evidence of recombination. It is not entirely improbable that trace signals of ancient recombination events might be detectable in the sequences identified.

If you are feeling particularly brave, you may want to attempt an exploratory analysis of the more complicated recombinants in A-J-cons-recombinants.fsa. Besides some obvious inter-subtype HIV recombination events, an exploratory analysis will reveal that these contain some pretty good evidence of other, currently uncharacterized, recombination events.

Section VII

Population genetics

The coalescent: population genetic inference using genealogies

Allen Rodrigo

17.1 Introduction

Most readers will know that *genealogies* are family trees which depict the ancestors and descendents of individuals in a population. In a *diploid* population, each individual has two ancestors in the preceding generation, four in the generation before that, eight in the generation before that, and so on. With *haploid* populations, each individual's lineage can be traced back through a single line of ancestors, one in each generation. In the same way that we can construct genealogies of individuals, we can also construct genealogies of genes within individuals. In diploid individuals, each copy of a *homologous* gene has a different pattern of inheritance and, consequently, a different genealogy. We can think of the genealogies of individual genes as intra-specific gene phylogenies.

Interestingly, genealogies contain information about historical demography and the processes that have acted to shape the diversity of populations. Imagine selecting two people at random from a large city, and two people from a small town. Intuitively, we would guess that the two individuals from the small town would share a common ancestor only a few generations in the past, perhaps a great-grandparent or a great-great-grandparent, whereas the two individuals from the city may have to dig back several generations before finding a common ancestor. We would, of course, realize that the number of generations that separate the two individuals from their common ancestor would depend on the numbers of people that immigrated or emigrated to/from the city or the small town – if we are told that there are large numbers of people coming in or leaving the town, for instance, we would revise our estimate on the time to common ancestry. Similarly, if we are told that what is now a small town had been a thriving metropolis some 40 or

The Phylogenetic Handbook: a Practical Approach to Phylogenetic Analysis and Hypothesis Testing, Philippe Lemey, Marco Salemi, and Anne-Mieke Vandamme (eds.). Published by Cambridge University Press. © Cambridge University Press 2009.

50 years ago, we would not be so confident that these two individuals have a recent common ancestor.

In these examples, three factors determine the time to common ancestry: the *size of the population, rates of migration,* and *change in population size.* Although the examples are simple thought experiments, they capture the essence of how genealogies are used to make inferences about historical population processes and demographies. In the absence of any selective bias that confers differential reproductive success amongst individuals, the numbers of generations that separate individuals from their common ancestors – the times that lineages on a genealogy take to *coalesce* – are functions of historical population size and, if it happens, migration between different groups of individuals. In 1982, John Kingman described this process formally in mathematical terms. He called it the **coalescent** (Kingman, 1982a, b).

17.2 The Kingman coalescent

The simplest formulation of the coalescent begins with a **panmictic** haploid population where the number of individuals, N, has remained constant over time, generations are discrete so that at each generation, only the offspring of the preceding generation survive, there are no selective forces acting on the population and all individuals have an equal chance of producing offspring. This is the very well-known **Wright–Fisher population** model (Fisher, 1930; Wright, 1931). If we sample two individuals from such a population at the present time, the probability that both will share a common ancestor in the preceding generation is $1/N$. To make it a little easier to keep track of time, we number the present generation t_0, the preceding generation t_1, the generation before that t_2, etc., so that t_k indicates a time k generations before the present. The probability that two individuals will share a common ancestor at t_2 is the probability that they will not share an ancestor at t_1 (this probability is $1-1/N$), multiplied by the probability that their respective parents will share an ancestor at t_2 (this probability is $1/N$, since the population size has remained unchanged). This is equal to $1/N(1-1/N)$. We can generalize this, and calculate the probability that any pair of randomly sampled individuals will have their *most recent common ancestor* (MRCA) at time t_k; this probability is given by:

$$P(t_k) = \left(\frac{1}{N}\right)\left(1 - \frac{1}{N}\right)^{k-1} \tag{17.1}$$

Suppose, instead of sampling only two individuals, we sample n individuals (where n is larger than or equal to 2, but much smaller than N). Now there are $n(n-1)/2$ possible pairs of individuals in the present generation that may share a common ancestor in the preceding generation, but each of these $n(n-1)/2$ pairs only has a $1/N$ chance that both individuals will have the same parent, i.e. the probability that there will be one common ancestor in the preceding generation is

$$P(t_1) = \frac{n(n-1)}{2N} \tag{17.2}$$

and the probability that the first MRCA of any of these $n(n-1)/2$ pairs will be at time t_k is

$$P(t_k) = \left(\frac{n(n-1)}{2N}\right)\left(1 - \frac{n(n-1)}{2N}\right)^{k-1} \tag{17.3}$$

The assumption that N is large and n is much smaller than N allows us to use some quite nice mathematical approximations. The net result is that we can move from talking about time in discrete generations, to continuous time, and we can express (17.3) more neatly as:

$$P(t_k) = \left(\frac{n(n-1)}{2N}\right)\exp\left(\frac{n(n-1)}{2N}k\right)dt \tag{17.4}$$

(technical note: since time is now continuous, the function $P(t_k)$ is strictly a **probability density function** – the "dt" at the end of the equation indicates that we are calculating the probability of a coalescent event in the infinitesimally small interval t_k to $t_k + dt$). Equation (17.4) is the density function of the exponential distribution, and has a mean of $\frac{2N}{n(n-1)}$ and variance of $\frac{4N^2}{[n(n-1)]^2}$. In a genealogy of n individuals, as time moves backwards from present to past, each successive coalescent event is a merger of two lineages randomly chosen from those that remain. From tips to root, there are $n-1$ coalescent events. With the approximations used to obtain (17.4), it is assumed that no two coalescent events occur at exactly the same point in time (Fig. 17.1). The expected time to the MRCA of all n individuals tends to $2N$ generations, as n gets large, and this is the expectation obtained under the Wright–Fisher model. The coalescent, then, is a continuous-time approximation to the Wright–Fisher model when population size, N, is very large.

Instead of speaking of genealogies of individuals, it is useful from this point forward to focus on the genealogies of gene sequences, because these are likely to be the primary source of data for most biologists. For the moment, we will only consider homologous and non-recombining genes at a single locus. With haploid organisms, there are only as many homologous genes as there are individuals in the population (each individual possesses only a single copy of the gene in question). In contrast, with diploid organisms, in a population of N individuals, the size of the *population of genes* is $2N$. All this does is change the denominators in (17.4),

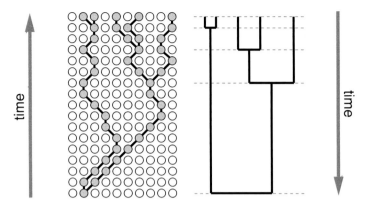

Fig. 17.1   On the left, a genealogy of five sequences, sampled at present from a constant-sized Wright–Fisher population of 10 individuals. At some point in the past, all lineages coalesce to a common ancestor. On the right, the lineages on the genealogy are shown as continuous lines, because the Kingman coalescent is a continuous-time description of the times-to-ancestry of a genealogy of a small sample of sequences obtained from a large population. With the coalescent, time is measured from present to past.

so that instead of $2N$, we now write $4N$; the mean and variance of the distribution also change to $\frac{4N}{n(n-1)}$ and $\frac{16N^2}{[n(n-1)]^2}$, respectively.

17.3 Effective population size

Of course, most biological populations fail to satisfy one or more of the assumptions of the Wright–Fisher model. For instance, it is impossible to find a natural population that remains at a constant size from generation to generation. How then can the coalescent be applied to real-world situations? The elegant solution is to use, in place of N, an abstract parameter called the **effective population size**, N_e (the quantity N is often called the *census population size*). The effective size, N_e, of a real biological population is proportional to the rate at which genetic diversity is lost or gained – if one obtains a measure of this rate in a real biological population, the effective size of that population is equal to the size of an ideal Fisher–Wright population that loses or gains genetic diversity at exactly that rate. There are several ways of calculating effective population size, but for the coalescent, arguably the most appropriate is the "**coalescent effective size**" proposed by Sjödin *et al.* (2005).

Essentially, the coalescent effective size of a real population is obtained by finding the value of N in (17.4) that delivers the same distribution of times-to-ancestry that one would obtain with a sample from the real population in question. This has a certain intuitive appeal – if two independently evolving lineages coalesce into a single ancestral lineage, then we have lost one lineage-worth of genetic diversity.

With the coalescent, this happens at an average rate of $1/N_e$, which is exactly the expectation of (17.4) when $n = 2$. This means that, when N_e is small, lineages are lost very quickly as one moves from present to past along a genealogy. Conversely, if one moves forward along a genealogy, the rate at which the number of lineages with shared recent ancestry increases or decreases due to stochastic effects alone (= *genetic drift*) is faster in populations with smaller effective sizes.

Effective population size is a useful parameter because it allows biologists to compare populations using a common measure. Lineages in a population with a larger effective size drift more slowly than one with a smaller effective size. Two populations with the same effective size have the same rate of genetic drift. Genetic drift can be quantified by measuring genetic diversity or **heterozygosity** (under the assumption of neutrality – see below). For the rest of the chapter, we will use the term "population size" and N to mean effective population size.

As I have noted above, and as we shall see in a later section, the coalescent tells us about historical population processes. Consequently, our ability to relate the rate of genetic drift to the coalescent, via the coalescent effective size, means that we have the opportunity to determine whether the historical dynamics of two or more populations are similar, as these relate to the accumulation or loss of genetic diversity. This is tremendously useful, as one can imagine, and these techniques have been applied in fields as diverse as conservation genetics, the reconstruction of colonization histories of humans and animals, and the epidemiology of infectious diseases.

17.4 The mutation clock

As noted in Chapter 11 of this book, for most phylogenies of gene sequences, we are not able to separate **mutation rate** from time (except, as we shall see, when we have calibration points or sequences collected at different times). Branch length and the time-to-ancestry of sequences is measured as a composite variable, $b = \mu t$, where μ is the mutation rate measured as the expected number of mutations (across the entire gene or per sequence position) per generation, and t is time in generation units. The branch length (or time-to-ancestry) measures the number of mutations that is expected to have accumulated between any pair of ancestral and descendant nodes. If we can rely on the constancy of μ across the genealogy of a sample of sequences, we can rescale time, so that instead of asking, "How many generations have elapsed since these two lineages separated from a common ancestor?" we ask, "How many substitutions have accumulated since these two lineages separated?." Under a **molecular clock**, each generation is equivalent to μ substitutions, t' generations is equivalent to $b = \mu t_k$ substitutions, and N generations is equivalent to $N\mu$ substitutions. With a little calculus, we can re-derive (17.4) so that, instead of measuring time in generations (t), it is now measured in substitutions (t'):

$$P(t'_b) = \left(\frac{n(n-1)}{2N\mu}\right) \exp\left(\frac{n(n-1)}{2N\mu}b\right) dt' \qquad (17.5)$$

By convention, instead of writing $2N\mu$ (or $4N\mu$ for diploid populations), population geneticists use θ in its place. The parameter θ is quite interesting: for one thing, it is equal to the average number of mutational differences between any two randomly sampled pair of sequences from a population with constant effective size (note that this is not quite the same as saying that θ is equal to the average number of *observable* mutational differences, because there may be hidden mutations, or rather, substitutions; see Chapter 4). Amongst its many other uses in population genetics, θ also features as a fundamental quantity in Ewens' sampling formula (Ewens, 1972), which tells us about the distribution of the number of allelic haplotypes in a population. Not surprisingly, as a rescaled measure of effective size, θ is also a measure of genetic diversity.

17.5 Demographic history and the coalescent

Since the effective size of a population correlates with the expected intervals between coalescent events, changes in population size (provided they are not too rapid; see Sjödin *et al.*, 2005) will result in changes to the distributions of these times. Consider, for example, a population that has grown in size. If we sample a set of genes from the population now, when it has a large effective size (say, N_0), we expect to find that the time to the first coalescent event between a pair of sequences will be large. However, after that first coalescent event, some generations in the past, we encounter a population with an effective size, N_t, that is smaller than N_0. Two randomly sampled lineages will coalesce at a faster rate, proportional to N_t. The effect of this process on the genealogy is to produce a tree with long terminal branches and shorter internal branches compared to a genealogy from a constant-sized population (Fig. 17.2).

We can use a similar thought experiment to work out the properties of a gene genealogy from a declining population. In this case, effective size at present, N_0, is small relative to population sizes in the past. Coalescent events tend to occur with greater rapidity, but as one moves past a series of coalescences backwards in time, population sizes grow ever larger and, concomitantly, coalescent intervals get longer. The *rate of growth or decline*, g, of a population where size changes exponentially, is the expected number of offspring per individual per generation. Growth rate is estimated as the composite parameter N_0g or g/μ, since it is not possible to obtain a measure of g that is not scaled in units of substitutions.

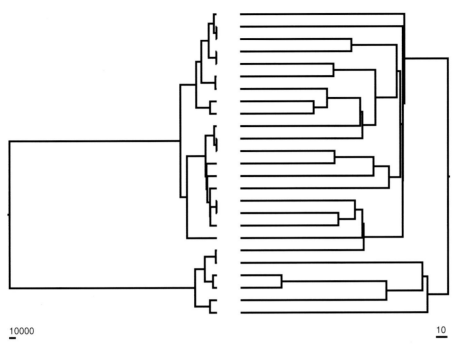

10000   10

Fig. 17.2   Two examples of genealogies. On the left, a genealogy of 25 sequences from a population
of constant size 10^5 and, on the right, a genealogy of the same number of sequences from
a population with present size 10^5 that has been growing at an exponential growth rate of
0.05. Note the very different timescales, measured in the number of generations: whereas
the genealogy on the right has a root only a few hundred generations in the past, the
genealogy on the left goes back hundreds of thousands of generations.

Up to this point, we have concentrated on describing the coalescent process for
panmictic populations. However, most biological populations have some degree
of geographical structure, and are not panmictic. The coalescent can be modified
to accommodate different types of geographical structure; for instance, the **island
model** of migration breaks the population into subpopulations or *demes*, with
migration of individuals between demes. A special type of island model is the
stepping-stone model of migration, where demes are arranged linearly or in a two-
dimensional grid, and migration only takes place between neighboring demes. The
stepping-stone model is often used to approximate a continuous-space model of
migration, where **migration rates** are a function of distance. For both the island
model and the stepping stone model of migration, the distributions of times-to-
ancestry depend on the rates of migration between demes and the effective sizes
of the population within demes. This is because two lineages can only coalesce if
they are in the same deme. If a population is divided into several demes, each with
a small effective size, and there are low rates of migration between demes, lineages

within demes will coalesce relatively quickly, leaving a single ancestral lineage in each deme. These lineages take a long time to coalesce, because to do so requires a rare migration event to another deme.

We can write down expressions for the distribution of coalescent times when populations are subdivided. These expressions extend (17.5) to allow for migration events on a genealogy, and involve another parameter, $M = Nm$, which is simply the proportion of migrating individuals per generation, m ($0 < m < 1$), scaled by the total number of individuals, N (note that some authors set $M = m/\mu$, instead; this represents the proportion of migrating individuals per substitution).

17.6 Coalescent-based inference

If we have at our disposal a collection of gene sequences sampled from a non-recombining population, we are able to construct a phylogeny of these sequences. If we know both the topology and branch-lengths of this (clock-constrained) phylogeny perfectly, then we are able to estimate the rescaled effective population size θ, the rates of growth or decline in substitution units, or the rate(s) of migration between subpopulations (again scaled by μ) and the rescaled effective sizes of the demes. This is because, as noted in the previous section, the coalescent intervals (and, therefore, the branch lengths on a genealogy) are functions of these rates. For a constant-sized population, for instance, we may ask what value of θ gives us the greatest relative probability of observing the series of coalescent intervals obtained with our sample genealogy – this type of estimator is called a ***maximum likelihood estimator*** (***MLE***) because it chooses the parameter estimate that has the highest probability of delivering the observed data. Because the genealogy of a sample is just one of many genealogies that may be obtained from the population, we can also obtain standard errors of our parameter estimates.

However, the topology and coalescent intervals of a genealogy are rarely, if ever, known with absolute certainty. In fact, this phylogeny that one obtains is an estimate of the genealogy of the sampled genes. We call it an estimate for several reasons: (1) we rarely know that the reconstructed genealogy corresponds to the true genealogy, (2) unless the sequences are very long, the data may provide similar support for a number of different genealogies, (3) if an explicit ***model of evolution*** is used to reconstruct the genealogy, it is unlikely that the model will fully describe the complexity of the molecular evolution of the sequences, and (4) the method used to reconstruct the phylogeny may not do a good job. Of these, the nastier side effects of (3) and (4) on phylogenetic reconstruction may be avoided by good scholarship (reading the chapters on phylogenetic analysis and models of evolution in this book is probably a good start!). The length of the sequences (2) is determined by experimental design and logistics; additionally, if the genomes recombine and/or are short (as in the case of many viral genomes), there is a limit to just how much sequence information one can collect. Finally, (1) is simply a fact of life.

Therefore, whereas we may derive parameter estimates and their standard errors using a single reconstructed genealogy, to do so is to ignore the added variation that these estimates truly have, added variation that is a result of the uncertainty in our reconstruction of the genealogy. Since we are typically more interested in the evolutionary parameters associated with the dynamics of the population than in the genealogy itself, the genealogy is a *"nuisance" parameter*, and we can free our parameter estimates from their conditional dependence on a specified genealogy by integrating or averaging over very many possible genealogies. The methods described in the next two chapters use this approach to obtain MLEs or Bayesian *posterior distributions* of parameters that do not depend on genealogy.

17.7 The serial coalescent

The coalescent has been a particularly useful tool, and has led to the development of a host of powerful methods in evolutionary and population genetics. One recent extension of the coalescent involves its use when samples of gene sequences are obtained at different times from *Measurably Evolving Populations* (*MEPs*; Drummond *et al.*, 2003). A MEP is any population that permits the detection of a statistically significant accumulation of substitutions when samples are obtained at different times. Rapidly evolving pathogens, for example RNA viruses including HIV, Human Influenza virus, Dengue virus, the SARS coronavirus, Hepatitis B and C viruses, are MEPs, because their relatively high rates of mutation mean that one is able to detect mutational and genetic change from samples collected a few months or years apart. Populations which leave subfossil traces of their ancient past from which DNA may subsequently be extracted may also be classed as MEPs because, although rates of mutation in these (generally, eukaryotic) populations are low, the times between sequence samples are large. The definition of a MEP is an operational one, and serves to differentiate those populations where we need to take sampling time into account from those for which differences in sampling times will not influence our evolutionary inferences (Drummond *et al.*, 2003).

Rodrigo and Felsenstein (1998) developed the *serial coalescent* (*or s-coalescent*) to describe the distributions of coalescent intervals on a genealogy of samples obtained serially in time (Fig. 17.3). The serial extension to the Kingman coalescent is not a difficult one, but there are interesting differences that arise as a consequence of sampling sequences over time. First, it is possible to obtain a direct estimate of mutation rate simply by estimating the expected number of substitutions that accumulate over each sampling interval, and dividing by the amount of time between samples. However, a mutation rate derived in this manner is expressed in

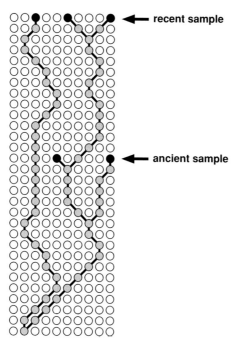

recent sample

ancient sample

Fig. 17.3   A genealogy of sequences sampled at two different times. Note that some lineages from the recent sample do not coalesce before they reach the ancient sample.

units of chronological time, for example, days, months or years. This, of course, is different from the way mutation rate is generally expressed, that is, in units of generations. There is a correspondence between the timescales of mutation rates in generations, and the coalescent process (where we measure the times-to-ancestry in generations). To accommodate our new measure of mutation rate in chronological time units within the inferential framework of the coalescent, we need to know how many chronological units correspond to a generation (= the **generation time**). Generation time permits us to scale our genealogies in chronological time by describing the distributions of chronological times-to-ancestry (t'') as follows:

$$P(t'') = \left(\frac{n(n-1)}{2N\tau} \right) \exp \left(\frac{n(n-1)}{2N\tau} \omega \right) dt'' \tag{17.6}$$

where τ is generation time, and $\omega = \tau \times t$, that is, the product of generation time and time measured in generations. Notice that effective size is now scaled by generation time. Therefore, although we are able to estimate mutation rate directly, we still have not freed ourselves completely from the burden of estimating effective size as a composite parameter: whereas with non-serial samples, we estimate $\theta = 2N\mu$, with serial samples we estimate $\theta' = 2N\tau$. Nonetheless, it is easy to obtain independent estimates of generation time for most populations so it is reasonable to say that, for all intents and purposes, with serial samples we are able to estimate both mutation rate and effective size separately.

A second important difference between the Kingman coalescent and the *s*-coalescent is that with serial samples, the number of lineages can increase as one moves back in time. Whereas this does not have a profound effect on the mathematics of the coalescent (nor, indeed, on the inferential framework *per se*) it can influence the extent to which we are able to make statements about historical population dynamics. In particular, with the standard coalescent, the inevitable decrease in the number of lineages backwards in time, means that we have less and less certainty about the lengths of our coalescent intervals and consequently, greater variance in our estimates of evolutionary parameters. This is a particular problem when there have been changes in the dynamics of the population over time. For instance, if glaciation and subsequent recolonization have changed the migration dynamics of a population, we may have difficulty recovering these changes in a genealogy of contemporaneous sequences. In contrast, with serial samples, our ability to add sequences as we move back along a genealogy means that we can increase the efficiency of our estimates of times-to-ancestry. This in turn means that we have more power to detect changes in the dynamics of the population. Consequently, a researcher can – with judicious choice of sampling times – increase the power of his/her analysis.

17.8 Advanced topics

In my description of the coalescent above, I have assumed that (a) the sequences that are available are not subject to recombination or selection, and (b) substitutions accrue at a clock-like rate. In this section, I discuss these assumptions as they apply to coalescent-based inference.

Recombination amongst homologous genes or sequences result in changes in the topology of the genealogy and/or differences in the lengths of coalescent intervals as one moves along an alignment (see also Chapter 15). The **recombination rate**, c, is the expected number of crossover events between any pair of adjacent sites per generation. If you have read this far, you will not be surprised to know that the recombination rate is estimated as the composite parameter c/μ using the coalescent. In essence, current coalescent-likelihood estimates of recombination rate are based on the extent to which genealogical conflicts across different regions of the locus explain different partitions (or site patterns) of the sequence data.

Natural selection presents a challenge to the coalescent because its action on a locus means that different individuals may have statistically different propensities to leave offspring in the next generation. Consequently, the random "choice" of parents in each preceding generation, predicated by the neutral process on which the coalescent is based, no longer holds. One method that uses a coalescent-based approach to model genealogies in the presence of selection was developed by Kaplan *et al.* (1985). With this method, sequences are classified into groups, each with a different *fitness* (or propensity to reproduce). The historical dynamics of the groups are known, within each group there is no difference in fitness, and mutations lead to "migrations" between classes. In this case, the method effectively reduces to the coalescent with migration, except that in the place of demes we have selective categories.

A second approach developed by Neuhauser and Krone (Krone & Neuhauser, 1997; Neuhauser & Krone, 1997) involves a new representation of the genealogical process, one that incorporates selective events as branches that occur at a rate proportional to the difference in fitness between groups. Here, as time proceeds backwards, we encounter both coalescences and branching events, representing potential lineage trajectories. With the superimposition of mutations on this coalescing and branching diagram, we are able to remove paths that correspond to less fit genotypes. What remains is a genealogy where favored mutations and lineages have a tendency to remain, and less fit lineages have a tendency to die out.

None of the methods described in the following two chapters incorporate selection within their inferential scheme. It is also difficult to describe the effects of selection on the coalescent concisely, because there are so many ways that selection can manifest itself. *Selective sweeps* (e.g. Jensen *et al.*, 2005), *background* and *hitchhiking selection* (e.g. Barton & Etheridge, 2004; Pfaffelhuber *et al.*, 2006), *directional* (e.g. Przeworski *et al.*, 2005), or *purifying* (e.g. Williamson & Orive, 2002) selection, all have some effect on the genealogies of gene sequences. Perhaps the best that a researcher can do, if he/she is intent on using coalescent-based methods, is (a) test for the presence of selection (there are a variety of tests available; see Chapters 14 and 15); (b) choose a locus that is unlikely to be subject to significant selection, and/or (c) determine the extent to which the particular type of selection that is likely to apply to one's locus of choice may confound the results of one's analysis.

Another troublesome requirement of coalescent-based inference is the need to quantify times-to-ancestry. Traditionally, researchers have used a molecular clock to measure these times, and have either assumed that these clocks are appropriate for their sequences, or have applied tests of rate constancy, before using coalescent-based methods. Over the last few years, there have been several methods that allow the rates of substitution to vary in some constrained way. These "*relaxed-clock*" approaches (Pybus, 2006) typically assign to each lineage a rate that is either correlated or uncorrelated to an ancestral rate, and is drawn from some plausible distribution of rates. The value of these approaches is that they permit times to be assigned to coalescent events, so that one is still able to use the coalescent to make population genetic inferences.

As noted above, different unlinked loci will have different genealogical histories. However, population parameters including effective size, growth, and migration rates are expected to be identical across loci. If we are able to obtain multiple independent loci, then, we have the opportunity to obtain multiple independent estimates of our parameters of interest, and in so doing, reduce the variance of these estimates. However, there is obviously a cost incurred with more sequencing – should we allocate costs to obtain more sequences per locus, more loci per individuals or longer sequences? With some reasonable assumptions about sequencing costs, it turns out that to estimate effective size, θ, the theoretical optimum sample size per locus is around eight (Pluzhnikov & Donnelly, 1996; Felsenstein, 2006), allowing us to sequence between 10 and 12 loci, each 600 bases long (Felsenstein, 2006). Felsenstein (2006) offers some suggestions on what type of sampling strategy is required if recombination, growth or migration rates are to be estimated, but cautions that these strategies still need to be worked out rigorously.

This chapter is meant to provide an introduction to the next two chapters. It is not meant to be an exhaustive review of coalescent theory. For an excellent and very readable account of the coalescent and its application to current evolutionary genetics, readers should consult Rosenberg and Nordborg (2002). Kingman (2000) has written a personal perspective on the development of the coalescent that is worth reading. For an in-depth treatment of *genealogy-based population genetics*, the text by Hein *et al.* (2005) is recommended.

Bayesian evolutionary analysis by sampling trees

THEORY

Alexei J. Drummond and Andrew Rambaut

18.1 Background

The BEAST software package is an ambitious attempt to provide a general framework for parameter estimation and hypothesis testing of evolutionary models from molecular sequence data. BEAST is a Bayesian statistical framework and thus provides a role for *prior* knowledge in combination with the information provided by the data. Bayesian *Markov chain Monte Carlo* (*MCMC*) has already been enthusiastically embraced as the state-of-the-art method for phylogenetic reconstruction, largely driven by the rapid and widespread adoption of MRBAYES (Huelsenbeck & Ronquist, 2001) (see Chapter 7). This enthusiasm can be attributed to a number of factors. First, Bayesian methods allow the relatively straightforward implementation of extremely complex *evolutionary models*. Second, there is an often erroneous perception that Bayesian estimation is "faster" than heuristic optimization based on the *maximum likelihood* criterion.

BEAST can be compared to a number of other software packages with similar goals, such as MRBAYES (Huelsenbeck & Ronquist, 2001), which currently focuses on phylogenetic inference and LAMARC (Kuhner, 2006) (discussed in the next chapter) and BATWING (Wilson *et al.*, 2003), which focus predominantly on *coalescent*-based population genetics. Like these software packages, the core algorithm implemented in BEAST is *Metropolis–Hastings MCMC* (Metropolis *et al.*, 1953; Hastings, 1970). MCMC is a stochastic algorithm that produces sample-based estimates of a target

The Phylogenetic Handbook: a Practical Approach to Phylogenetic Analysis and Hypothesis Testing, Philippe Lemey, Marco Salemi, and Anne-Mieke Vandamme (eds.). Published by Cambridge University Press. © Cambridge University Press 2009.

distribution of choice. For our purposes the target distribution is the ***posterior distribution*** of a set of evolutionary parameters given an alignment of molecular sequences.

Possibly the most distinguishing feature of BEAST is its firm focus on calibrated phylogenies and genealogies, that is, rooted trees incorporating a timescale. This is achieved by explicitly modeling the ***rate of molecular evolution*** on each branch in the tree. On the simplest level this can be a uniform rate over the entire tree (i.e. the strict ***molecular clock model*** (Zuckerkandl & Pauling, 1965)) with this rate known in advance or estimated from calibration information. However, one of the most promising recent advances in molecular phylogenetics has been the introduction of ***relaxed molecular clock*** models that do not assume a constant rate across lineages (Thorne *et al.*, 1998; Yoder & Yang, 2000; Kishino *et al.*, 2001; Sanderson, 2002; Thorne & Kishino, 2002; Aris-Brosou & Yang, 2003). BEAST was the first software package that allows phylogenetic inference under such models (Drummond *et al.*, 2006).

In the context of ***genealogy***-based population genetics (see previous chapter), the target distribution of interest is the posterior probability of the population genetic parameters (ϕ) given a multiple sequence alignment (D):

$$p(\phi|D) = \frac{1}{Z} \int_{g,\omega} Pr\{D|g, \omega\} p(g|\phi) p(\phi) p(\omega) dg d\omega \qquad (18.1)$$

In order to estimate the posterior probability distribution of ϕ it is necessary to average over all possible genealogies (g) and substitution model parameters (ω) proportional to their probabilities. This integration is achieved by MCMC. In the above equation $Pr\{D|g, \omega\}$ is the ***likelihood*** of genealogy g given the sequence data and the substitution model (Felsenstein, 1981) and $p(g|\phi)$ is the coalescent prior of the genealogy given the population parameters ϕ. In the original formulation of the Kingman coalescent (Kingman, 1982) (see also previous chapter), there is a single population size, $\phi = \{\theta\}$ and the coalescent prior takes the form:

$$p(g|\phi) = \frac{1}{\theta^{n-1}} \prod_{i=2}^{n} \exp \frac{-i(i-1)u_i}{2\theta} \qquad (18.2)$$

where u_i is the length of time over which the genealogy g has exactly i lineages. This formulation assumes that the units of time are substitutions per site and that all sequences are sampled from the same time. Both of these assumptions can be relaxed (Drummond *et al.*, 2002). It is also possible to devise more complex coalescent models so that the population size is a function of time. BEAST supports a number of demographic models including constant size, exponential growth, logistic growth, expansion, and the highly parameteric *Bayesian skyline plot* (Drummond *et al.*, 2005). Currently, BEAST does not include coalescent models of migration or recombination but these processes will be included in a future version. For the case of contemporaneously sampled sequences, these processes can be investigated using LAMARC (see next chapter).

The purpose behind the development of BEAST is to bring a large number of complementary evolutionary models (e.g. substitution models, demographic tree priors, relaxed clock models, node calibration models) into a single coherent framework for evolutionary inference. This building-block principle of constructing a complex evolutionary model out of a number of simpler model components provides powerful new possibilities for molecular sequence analysis. The motivation for doing this is: (1) to avoid the unnecessary simplifying assumptions that currently exist in many evolutionary analysis packages; and (2) to provide new model combinations and a flexible system for model specification so that researchers can tailor their evolutionary analyses to their specific set of questions.

18.2 Bayesian MCMC for genealogy-based population genetics

The integration in (18.1) is achieved by constructing a chain of parameter/genealogy combinations in such a way that they form a (correlated) sample of states from the full posterior distribution:

$$p(g, \omega, \phi | D) = \frac{1}{Z} Pr\{D | g, \omega\} \, p(g | \phi) \, p(\phi) \, p(\omega) \tag{18.3}$$

We summarize the **marginal density** $p(\phi | D)$ by using samples $(g, \omega, \phi) \sim p(g, \omega, \phi | D)$. The sampled genealogies and substitution model parameters can be thought of as uninteresting **nuisance parameters**.

To construct the **Markov chain** we begin with an initial state $x_0 = (g^{(0)}, \omega^{(0)}, \phi^{(0)})$. At each step i in the chain we begin by proposing a new state y. An operator (m) proposes this state by copying the previous state x_{i-1} and making a small alteration (to the genealogy, or the parameter values, or both). The probability of the previous state and the newly proposed state are then compared in an accept/reject step. The proposed state is accepted as the new state in the Markov chain with probability:

$$\alpha = \min\left(1, \frac{p(y | D)}{p(x_{i-1} | D)}\right) \tag{18.4}$$

If the proposed state y is accepted, then state $x_i = y$ otherwise the previous state is kept ($x_i = x_{i-1}$). Notice that, if the posterior probability of y is greater than x_{i-1}, then y will definitely be accepted, whereas when y has lower probability than x_{i-1} it will only be accepted with a probability proportional to the ratio of their posterior probabilities. The above acceptance probability assumes that the operator is symmetric, so that the probability of proposing state y from state x,

$q(y|x)$, is the same as proposing state x from state y, $q(x|y)$. BEAST uses a mixture of symmetric and asymmetric operators. At each step in the chain an operator (m) is chosen at random (with weights). When operator m is not symmetric, then $q_m(y|x) \neq q_m(x|y)$ and the acceptance probability becomes

$$\alpha = \min \left(1, \frac{p(y|D)}{p(x_{i-1}|D)} \frac{q(x_{i-1}|y)}{q(y|x_{i-1})} \right) \tag{18.5}$$

The additional ratio of proposal probabilities is called the *Hastings ratio* (Hastings, 1970).

18.2.1 Implementation

The overall architecture of the BEAST software package is a file-mediated pipeline. The core program takes, as input, an XML file describing the data to be analyzed, the models to be used and technical details of the MCMC algorithm such as the proposal distribution (defined by the operators), the length of the Markov chain (chain length) and the output options. The output of a BEAST analysis is a set of tab-delimited plain text files that summarize the estimated posterior distribution of parameter values and trees.

A number of additional software programs assist in generating the input and analyzing the output:

- BEAUTI is a software package written in Java and distributed with BEAST that provides a graphical user interface for generating BEAST XML input files for a number of simple model combinations.
- TRACER is a software package written in Java and distributed separately from BEAST that provides a graphical tool for MCMC output analysis. It can be used for the analysis of the output of BEAST as well as the output of other common MCMC packages such as MRBAYES (Huelsenbeck & Ronquist, 2001) and BALI-PHY (Suchard & Redelings, 2006).

Because of the combinatorial nature of the BEAST XML input format, not all models can be specified through the graphical interface of BEAUTI. Indeed, the sheer number of possible combinations of models mean that, inevitably, some combinations will be untested. It is also possible to create models that are inappropriate or meaningless for the data being analyzed. BEAUTI is therefore intended as a way of generating commonly used and well-understood analyses. For the more adventurous researcher, and with the above warnings in mind, the XML file can be directly edited. A number of online tutorials are available to guide users on how to do this.

18.2.2 Input format

One of the primary motivations for providing a highly structured XML input format is to facilitate reproducibility of complex evolutionary analyses. While an interactive graphical user interface provides a pleasant user experience, it can be time-consuming and error-prone for a user to record and reproduce the full sequence of choices that are made, especially with the large array of options typically available for MCMC analysis. By separating the graphical user interface (BEAUTi) from the analysis (BEAST), we accommodate an XML layer that captures the exact details of the MCMC analysis being performed. We strongly encourage the routine publication of XML input files as supplementary information with publication of the results of a BEAST analysis. Because of the non-trivial nature of MCMC analyses and the need to promote reproducibility, it is our view that the publication of the exact details of any Bayesian MCMC analysis should be made a prerequisite for publication of all MCMC analysis results.

18.2.3 Output and results

The output from BEAST is a simple tab-delimited plain text file format with one a row for each sample. When accumulated into frequency distributions, this file provides an estimate of the marginal posterior probability distribution of each parameter. This can be done using any standard statistics package or using the specially written package, TRACER (Rambaut & Drummond, 2003). TRACER provides a number of graphical and statistical ways of analyzing the output of BEAST to check performance and accuracy. It also provides specialized functions for summarizing the posterior distribution of population size through time when a coalescent model is used.

The phylogenetic tree of each sample state is written to a separate file as either NEWICK or NEXUS format (see Section 5.5 in Chapter 5). This can be used to investigate the posterior probability of various phylogenetic questions such as the *monophyly* of a particular group of organisms or to obtain a consensus phylogeny (e.g. a *majority-rule consensus tree*).

18.2.4 Computational performance

Although there is always a trade-off between a program's flexibility and its computational performance, BEAST performs well on large analyses (e.g. Shapiro *et al.*, 2004). A Bayesian MCMC algorithm needs to evaluate the likelihood of each state in the chain and thus performance is dictated by the speed at which these likelihood evaluations can be made. BEAST attempts to minimize the time taken to evaluate a state by only recalculating the likelihood for parts of the model that have changed from the previous state. Furthermore, the core computational functions have been implemented in the C programming language. This can be compiled into a highly optimized library for a given platform providing an improvement in speed. If this library is not found, BEAST will use its Java version of these functions, thereby retaining its platform independence.

18.3 Results and discussion

BEAST provides considerable flexibility in the specification of an evolutionary model. For example, consider the analysis of a multiple sequence alignment of protein-coding DNA. In a BEAST analysis, it is possible to allow each codon position to have a different rate, a different amount of rate heterogeneity among sites, and a different amount of rate heterogeneity among branches, while, at the same time, sharing the same intrinsic ratio of **transitions** to **transversions** with the other codon positions. In fact, all parameters can be shared or made independent among partitions of the sequence data.

An unavoidable feature of Bayesian statistical analysis is the specification of a prior distribution over parameter values. This requirement is both an advantage and a burden. It is an advantage because relevant knowledge such as palaeontological calibration of phylogenies is readily incorporated into an analysis. However, when no obvious prior distribution for a parameter exists, a burden is placed on the researcher to ensure that the prior selected is not inadvertently influencing the posterior distribution of parameters of interest.

In BEAST, all parameters (whether they be substitutional, demographic or genealogical) can be given *informative priors* (e.g. exponential, normal, lognormal or uniform with bounds, or combinations of these). For example, the age of the root of the tree can be given an exponential prior with a pre-specified mean.

The five components of an evolutionary model for a set of aligned nucleotides in BEAST are:

- *Substitution model* – The substitution model is a homogeneous **Markov process** that defines the relative rates at which different substitutions occur along a branch in the tree.
- *Rate model among sites* – The rate model among sites defines the distribution of relative rates of evolutionary change among sites.
- *Rate model among branches* – The rate model among branches defines the distribution of rates among branches and is used to convert the tree, which is in units of time, to units of substitutions. These models are important for divergence time estimation procedures and producing timescales on demographic reconstructions.
- *Tree* – A model of the phylogenetic or genealogical relationships of the sequences.
- *Tree prior* – The tree prior provides a parameterized prior distribution for the node heights (in units of time) and tree topology.

18.3.1 Substitution models and rate models among sites

For nucleotide data, all of the models that are nested in the general time-reversible (GTR) model (Rodriquez *et al.*, 1990), including the well-known HKY85 model (Hasegawa *et al.*, 1985) – can be specified (see Chapter 4). For the analysis of amino acid sequence alignments, all of the following replacement models can be used: Blosum62, CPREV, Dayhoff, JTT, MTREV, and WAG. When nucleotide data represents a coding sequence (i.e. an in-frame protein-coding sequence) the Goldman and Yang model (Goldman & Yang, 1994) can be used to model codon evolution (see Chapter 14).

In addition, both Γ-distributed rates among sites (Yang, 1994) and a proportion of invariant sites can be used to describe rate heterogeneity among sites.

18.3.2 Rate models among branches, divergence time estimation, and time-stamped data

Without calibration information, *substitution rate* (μ) and time (t) are confounded and thus branches must be estimated in units of substitutions per site, μt. However, when a strong prior is available for: (1) the time of one or more nodes, or (2) the overall substitution rate, then the genealogy can be estimated in units of time.

The basic model for rates among branches supported by BEAST is the strict molecular clock model (Zuckerkandl & Pauling, 1965), calibrated by specifying either a substitution rate or the date of a node or set of nodes. In this context, dates of divergence for particular clades can be estimated. The clades can be defined either by a monophyletic grouping of taxa or as the most recent common ancestor of a set of taxa of interest. The second alternative does not require monophyly of the selected taxa with respect to the rest of the tree. Furthermore, when the differences in the dates associated with the tips of the tree comprise a significant proportion of the age of the entire tree, these dates can be incorporated into the model providing a source of information about the overall rate of evolutionary change (Drummond *et al.*, 2002, 2003).

In BEAST, divergence time estimation has also been extended to include *relaxed phylogenetics* models, in which the rate of evolution is allowed to vary among the branches of the tree. In particular, we support a class of **uncorrelated relaxed clock** branch rate models, in which the rate at each branch is drawn from an underlying distribution such as exponential or lognormal (Drummond *et al.*, 2006).

If the sequence data are all from one time point, then the overall evolutionary rate must be specified with a strong prior. The units implied by the prior on the evolutionary rate will determine the units of the node heights in the tree (including the age of the most recent common ancestor) as well as the units of the demographic parameters such as the population size parameter and the growth rate. For example, if the evolutionary rate is set to 1.0, then the node heights (and root height) will be in units of substitutions per site (i.e. the units of branch lengths produced by common software packages such as MRBAYES 3.0). Similarly, for a *haploid* population, the coalescent parameter will be an estimate of $N_e\mu$. However, if, for example, the evolutionary rate is expressed in substitutions per site per year, then the branches in the tree will be in units of years. Furthermore, the population size parameter of the demographic model will then be equal to $N_e\tau$, where N_e is the *effective population size* and τ is the *generation time* in years. Finally, if the evolutionary rate is expressed in units of substitutions per site per generation then the resulting tree will be in units of generations and the population parameter of the demographic model will be in natural units (i.e. will be equal to the effective number of reproducing individuals, N_e).

18.3.3 Tree priors

When sequence data has been collected from a homogeneous population, various coalescent (Kingman, 1982; Griffiths & Tavare, 1994) models of demographic history can be used in BEAST to model population size changes through time. At present, the simple parametric models available include constant size $N(t) = N_e$ (1 parameter), exponential growth $N(t) = N_e e^{-gt}$ (two parameters), expansion or logistic growth (three parameters).

In addition, the highly parametric Bayesian skyline plot (Drummond *et al.*, 2005) is also available, but this model can only be used when the data are strongly informative about population history. All of these demographic models are parametric priors on the ages of nodes in the tree, in which the *hyperparameters* (e.g. population size, N_e, and growth rate, g) can be sampled and estimated. As well as performing single locus coalescent-based inference, two or more unlinked gene trees can be simultaneously analyzed under the same demographic model. Sophisticated multi-locus coalescent inference can be achieved by allocating a separate overall rate and substitution process for each locus, thereby accommodating loci with heterogeneous evolutionary processes.

At present, there are only a limited number of options for non-coalescent priors on tree shape and branching rate. Currently, a simple *Yule prior* on birth rate of new lineages (one parameter) can be employed. However, generalized birth–death tree priors are currently under development.

In addition to general models of branching times such as the coalescent and Yule priors, the tree prior may also include specific distributions and/or constraints on certain node heights and topological features. These additional priors may represent other sources of knowledge such as expert interpretation of the fossil record. For example, as briefly noted above, each node in the tree can have a prior distribution representing knowledge of its date. A recent paper on "relaxed phylogenetics" contains more information on calibration priors (Drummond *et al.*, 2006).

18.3.4 Multiple data partitions and linking and unlinking parameters

BEAST provides the ability to analyze multiple data partitions simultaneously. This is useful when combining multiple genes in a single multi-locus coalescent analysis (Lemey *et al.*, 2004), or to allocate different evolutionary processes to different regions of a sequence alignment, such as the codon positions (Pybus *et al.*, 2003). The parameters of the substitution model, the rate model among sites, the rate model among branches, the tree, and the tree prior can all be "linked" or "unlinked" in an analysis involving multiple partitions. For example, in an analysis of HIV-1 group O by Lemey *et al.* (2004), three loci (*gag, int, env*) were assumed to share the same substitution model parameters (GTR), as well as sharing the same demographic history of exponential growth. However, they were assumed to have different shape parameters for Γ-distributed rate heterogeneity among sites, different rate parameters for the strict molecular clock and the three tree topologies and sets of divergence times were also assumed to be independent and unlinked.

18.3.5 Definitions and units of the standard parameters and variables

Crucial to the interpretation of all BEAST parameters is an understanding of the units that the tree is measured in. The simplest situation occurs when no calibration information is available, either from knowledge of the rate of evolution of the gene region, or from knowledge of the age of any of the nodes in the tree. If this is the case, the rate of evolution is set to 1.0 (via the clock.rate or ucld.mean parameters) and the branch lengths in the tree are then in substitutions per site. However, if the rate of evolution is known in substitutions per site per unit time, then the genealogy will be expressed in the relevant time units. Likewise, if the age of one or more nodes (internal or external) are known, then this will also provide the units for the rest of the branch lengths and the rate of evolution. With this in mind, Box 18.1 lists the parameters that are used in the models that can be generated by BEAUTI, with their interpretation and units.

18.3.6 Model comparison

Considering the large number of models available in a Bayesian inference package like BEAST, a common question is "Which model should I use?." This is especially the case for the parts of the evolutionary model that are not of direct interest to the researcher (and responsible for the so-called nuisance parameters). If the research question is a question of demographic inference, then the researcher may not be interested in the substitution model parameters, but nevertheless some substitution model must be chosen. It is in these situations that it often makes sense to choose the substitution model which best fits the data.

Box 18.1 Parameters in models generated by BEAUTI

■ `clock.rate` – The rate of the strict molecular clock. This parameter only appears when you have selected the strict molecular clock in the model panel. The units of this parameter are in substitutions per site per unit time. If this parameter is fixed to 1.0, then the branch lengths in the tree will be in units of substitutions per site. However, if, for example, the tree is being calibrated by using fossil calibrations on internal nodes and those fossil dates are expressed in millions of years ago (Mya), then the `clock.rate` parameter will be an estimate of the evolutionary rate in units of substitutions per site per million years (Myr).

■ `constant.popSize` – This is the coalescent parameter under the assumption of a constant population size. This parameter only appears if you select a constant size coalescent tree prior. This parameter represents the product of effective population size (N_e) and the generation length in units of time (τ). If time is measured in generations, this parameter is a direct estimate of N_e. Otherwise it is a composite parameter and an estimate of N_e can be computed from this parameter by dividing it by the generation length in the units of time that your calibrations (or `clock.rate`) are defined in. Finally, if `clock.rate` is set to 1.0 then `constant.popSize` is an estimate of $N_e \mu$ for haploid data such as mitochondrial sequences and $2 N_e \mu$ for ***diploid*** data, where μ is the substitution rate per site per generation.

■ `covariance` – If this value is significantly positive, it means that within your phylogeny, branches with fast rates are followed by branches with fast rates. This statistic measures the covariance between parent and child branch rates in your tree in a relaxed molecular clock analysis. If this value spans zero, then branches with fast rates and slow rates are next to each other. It also means that there is no strong evidence of autocorrelation of rates in the phylogeny.

■ `exponential.growthRate` – This is the coalescent parameter representing the rate of growth of the population assuming exponential growth. The population size at time t is determined by $N(t) = N_e \exp(-gt)$ where t is in the same units as the branch lengths and g is the `exponential.growthRate` parameter. This parameter only appears if you have selected an exponential growth coalescent tree prior.

■ `exponential.popSize` – This is the parameter representing the modern day population size assuming exponential growth. Like `constant.popSize`, it is a composite parameter unless the timescale of the genealogy is in generations. This parameter only appears if you have selected an exponential growth coalescent tree prior.

■ `gtr.{ac,ag,at,cg,gt}` – These five parameters are the relative rates of substitutions for $A \leftrightarrow C$, $A \leftrightarrow G$, $A \leftrightarrow T$, $C \leftrightarrow G$ and $G \leftrightarrow T$ in the general time-reversible model of nucleotide substitution (Rodrignez *et al.*, 1990). In the default set up these parameters are relative to $r_{C \leftrightarrow T} = 1.0$. These parameters only appear if you have selected the GTR substitution model.

■ `hky.kappa` – This parameter is the ***transition/transversion ratio*** (κ) parameter of the HKY85 model of nucleotide substitution (Hasegawa *et al.*, 1985). This parameter only appears if you have selected the HKY substitution model.

Box 18.1 *(cont.)*

- `siteModel.alpha` – This parameter is the shape (α) parameter of the Γ-distribution of rate heterogeneity among sites (Yang, 1994). This parameter only appears when you have selected Gamma or Gamma+Invariant Sites in the site heterogeneity model.
- `siteModel.pInv` – This parameter is the proportion of invariant sites (p_{inv}) and has a range between 0 and 1. This parameter only appears when you have selected "Invariant sites" or "Gamma+Invariant Sites" in the site heterogeneity model. The starting value must be less than 1.0.
- `treeModel.rootHeight` – This parameter represents the total height of the tree (often known as the t_{MRCA}). The units of this parameter are the same as the units for the branch lengths in the tree.
- `ucld.mean` – This is the mean molecular clock rate under the uncorrelated lognormal relaxed molecular clock. This parameter can be in "real" space or in log space depending on the BEAST XML. However, under default BEAUTI options for the uncorrelated lognormal relaxed clock this parameter has the same units as `clock.rate`.
- `ucld.stdev` – This is the standard deviation (σ) of the uncorrelated lognormal relaxed clock (in log-space). If this parameter is 0 there is no variation in rates among branches. If this parameter is greater than 1 then the standard deviation in branch rates is greater than the mean rate. This is also the case for the *coefficient of variation*. When viewed in TRACER, if the coefficient of variation frequency histogram is abutting against zero, then your data can't reject a strict molecular clock. If the frequency histogram is not abutting against zero, then there is among branch rate heterogeneity within your data, and we recommend the use of a relaxed molecular clock.
- `yule.birthRate` – This parameter is the rate of lineage birth in the Yule model of speciation. If `clock.rate` is 1.0 then this parameter estimates the number of lineages born from a parent lineage per substitution per site. If the tree is instead measured in, for example, years, then this parameter would be the number of new lineages born from a single parent lineage per year.
- `tmrca(taxon group)` – This is the parameter for the t_{MRCA} of the specified taxa subset. The units of this variable are the same as the units for the branch lengths in the tree and will depend on the calibration information for the rate and/or dates of calibrated nodes. Setting priors on these parameters and/or `treeModel.rootHeight` parameter will act as calibration information.

In a Bayesian setting, the most theoretically sound method of determining which of two models is better is to calculate the ***Bayes Factor (BF)***, which is the ratio of their marginal likelihoods. Generally speaking calculating the BF involves a Bayesian MCMC that averages over both models (using a technique called ***reversible jump MCMC***), and this is not something that can currently be done in BEAST. However there are a couple of ways of approximately calculating the marginal likelihood of each model (and therefore the Bayes factor between them) that can be achieved by processing the output of two BEAST analyses. For example, a simple method first described by Newton and Raftery (1994) computes the Bayes factor via *importance sampling* (with the posterior as the importance distribution). With this importance distribution it turns out that the *harmonic mean* of the sampled likelihoods is an estimator of the marginal likelihood. So, by calculating the harmonic mean of the likelihood from the posterior output of each of the models and then taking the difference (in log space) you get the log BF and you can look up this number in a table to decide if the BF is large enough to strongly favor the better model (see Table 7.2 in Chapter 7). This method of calculating the BF is only approximate and in certain situations it is not very stable, so model comparison is an area of Bayesian evolutionary analysis that could certainly be improved.

18.3.7 Conclusions

BEAST is a flexible analysis package for evolutionary parameter estimation and hypothesis testing. The component-based nature of model specification in BEAST means that the number of different evolutionary models possible is very large and therefore difficult to summarize. However, a number of published uses of the BEAST software already serve to highlight the breadth of application the software enjoys (Pybus *et al.*, 2003; Lemey *et al.*, 2004; Shapiro *et al.*, 2004; Drummond *et al.*, 2005; Lunter *et al.*, 2005).

BEAST is an actively developed package and enhancements for the next version include: (1) birth–death priors for tree shape; (2) faster and more flexible codon-based substitution models; (3) the structured coalescent to model subdivided populations with migration; (4) models of continuous character evolution and; (5) new relaxed clock models based on random local molecular clocks.

PRACTICE

Alexei J. Drummond and Andrew Rambaut

18.4 The BEAST software package

This chapter provides a step-by-step tutorial for analyzing a set of virus sequences which have been isolated at different points in time (heterochronous data). The data are 35 sequences from the *G* (attachment protein) gene of human respiratory syncytial virus subgroup A (RSVA) from various parts of the world with isolation dates ranging from 1956–2002 (Zlateva *et al.*, 2004). The input file required for this exercise is available for download at *http://www.thephylogenetichandbook.org*. The aim is to obtain an estimate of the rate of molecular evolution, an estimate of the date of the most recent common ancestor and to infer the phylogenetic relationships with appropriate measures of statistical support.

The first step will be to convert a NEXUS file with a DATA or CHARACTERS block into a BEAST XML input file. This is done using the program BEAUTi (this stands for Bayesian Evolutionary Analysis Utility). This is a user-friendly program for setting the evolutionary model and options for the MCMC analysis. The second step is to actually run BEAST using the input file that contains the data, model and settings. The final step is to explore the output of BEAST in order to diagnose problems and to summarize the results.

To undertake this tutorial, you will need to download three software packages in a format that is compatible with your computer system (all three are available for Mac OS X, Windows, and Linux/UNIX operating systems):

- BEAST – this package contains the BEAST program, BEAUTi, and a couple of utility programs. At the time of writing, the current version is v1.4.6. It is available for download from *http://beast.bio.ed.ac.uk/*.
- TRACER – this program is used to explore the output of BEAST (and other Bayesian MCMC programs). It graphically and quantitively summarizes the distributions of continuous parameters and provides diagnostic information. At the time of writing, the current version is v1.4. It is available for download from *http://beast.bio.ed.ac.uk/*.
- FIGTREE – this is an application for displaying and printing molecular phylogenies, in particular those obtained using BEAST. At the time of writing, the current version is v1.1. It is available for download from *http://tree.bio.ed.ac.uk/*.

18.5 Running BEAUTi

The exact instructions for running BEAUTi differs depending on which computer you are using. Please see the README text file that was distributed with the version you downloaded. Once running, BEAUTi will look similar irrespective of

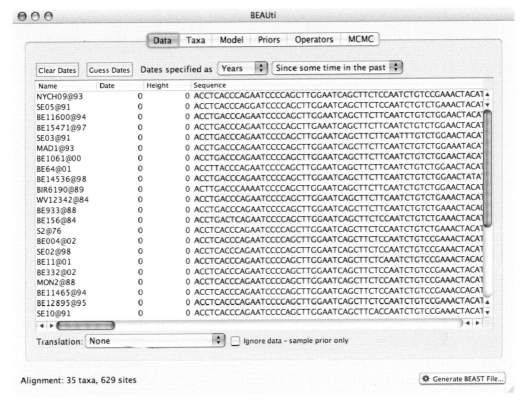

Fig. 18.1   The data panel in BEAUTI.

which computer system it is running on. For this tutorial, the Mac OS X version will be used in the Figures but the Linux and Windows versions will have exactly the same layout and functionality.

18.6 Loading the NEXUS file

To load a NEXUS format alignment, simply select the Import NEXUS...option from the File menu. The example file, called RSVA.nex, is available from *http://www.thephylogenetichandbook.org/*. This file contains an alignment of 35 sequences from the *G* gene of RSVA virus, 629 nucleotides in length. Once loaded, the list of taxa and the actual alignment will be displayed in the main window (Fig. 18.1).

18.7 Setting the dates of the taxa

By default, all the taxa are assumed to have a date of zero (i.e. the sequences are assumed to be sampled at the same time). In this case, the RSVA sequences have

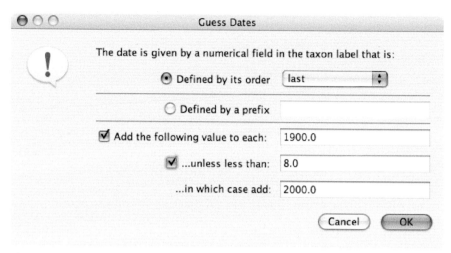

Fig. 18.2   The Guess Dates dialog.

been sampled at various dates going back to the 1950s. The actual year of sampling is given in the name of each taxon and we could simply edit the value in the Date column of the table to reflect these. However, if the taxa names contain the calibration information, then a convenient way to specify the dates of the sequences in BEAUTi is to use the "Guess Dates" button at the top of the Data panel. Clicking this will make a dialog box appear (Fig. 18.2).

This operation attempts to guess what the dates are from information contained within the taxon names. It works by trying to find a numerical field within each name. If the taxon names contain more than one numerical field (such as the RSVA sequences, above) then you can specify how to find the one that corresponds to the date of sampling. You can either specify the order that the date field comes (e.g. first, last or various positions in-between) or specify a prefix (some characters that come immediately before the date field in each name). For the RSVA sequences you can select "last" from the drop-down menu for the order or use the prefix option and specify "@" as the prefix ("@" is the prefix used for dates by PAML, see Chapter 11).

In this dialog box, you can also get BEAUTi to add a fixed value to each guessed date. In this case the value "1900" has been added to turn the dates from 2 digit years to 4 digit. Any dates in the taxon names given as "00" would thus become "1900." Some of the sequences in the example file actually have dates after the year 2000 so selecting the "unless less than:" option will convert them correctly, adding 2000 to any date less than 08. When you press OK the dates will appear in the appropriate column of the main window. You can then check these and edit them manually as required. At the top of the window you can set the units that the dates

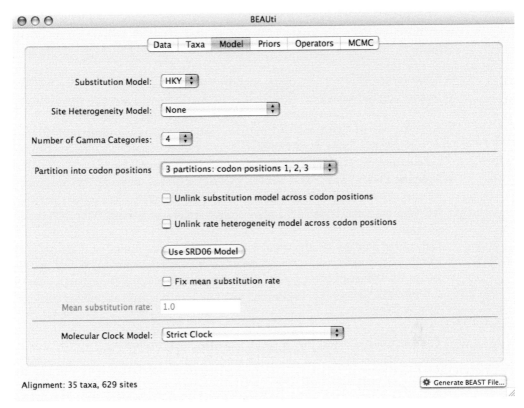

Fig. 18.3   The evolutionary model settings in BEAUTI.

are given in (years, months, days) and whether they are specified relative to a point in the past (as would be the case for years such as 1984) or backwards in time from the present (as in the case of radiocarbon ages).

18.7.1 Translating the data in amino acid sequences

At the bottom of the main window is the option to translate the data into amino acid sequences. This will be done using the genetic code specified in the associated drop down menu. If the loaded sequence are not nucleotides, then this option will be disabled.

18.8 Setting the evolutionary model

The next thing to do is to click on the Model tab at the top of the main window. This will reveal the evolutionary model settings for BEAST. Exactly which options appear depend on whether the data are nucleotides or amino acids (or nucleotides translated into amino acids). Figure 18.3 shows the settings that will appear after loading the RSVA data and selecting a codon partitioning.

This chapter assumes that you are familiar with the evolutionary models available, however there are a couple of points to note about selecting a model in BEAUTi:

- Selecting the `Partition into codon positions` option assumes that the data are aligned as codons. This option will then estimate a separate rate of substitution for each codon position, or for 1+2 vs. 3, depending on the setting.
- Selecting the `Unlink substitution model across codon positions` will specify that BEAST should estimate a separate transition–stransversion ratio or general time reversible rate matrix for each codon position.
- Selecting the `Unlink rate heterogeneity model across codon positions` will specify that BEAST should estimate separate rate heterogeneity parameters (gamma shape parameter and/or proportion of invariant sites) for each codon position.
- If there are no dates for the sequences (they are contemporaneous), then you can specify a fixed `mean substitution rate` obtained from another source. Setting this to 1.0 will result in the ages of the nodes of the tree being estimated in units of substitutions per site (i.e. the normal units of branch lengths in popular packages such as MrBayes).

For this tutorial, select the "`3 partitions: codon positions 1, 2 & 3`" option so that each codon position has its own rate of evolution.

18.9 Setting up the operators

Each parameter in the model has one or more "operators" (these are variously called *moves* and *proposals* by other MCMC software packages such as MrBayes and Lamarc). The operators specify how the parameters change as the MCMC runs. The `operators` tab in BEAUTi has a table that lists the parameters, their operators and the tuning settings for these operators. In the first column are the parameter names. These will be called things like `hky.kappa` which means the HKY model's kappa parameter (the transition-transversion bias). The next column has the type of operators that are acting on each parameter. For example, the scale operator scales the parameter up or down by a proportion, the random walk operator adds or subtracts an amount to the parameter and the uniform operator simply picks a new value uniformly within a range of the current value. Some parameters relate to the tree or to the divergence times of the nodes of the tree and these have special operators.

The next column, labelled `Tuning`, gives a tuning setting to the operator. Some operators don't have any tuning settings so have `n/a` under this column. The tuning parameter will determine how large a move each operator will make which will affect how often that change is accepted by the MCMC which will affect the efficency of the analysis. For most operators (like random walk and subtree slide operators) a larger tuning parameter means larger moves. However for the scale operator a tuning parameter value closer to 0.0 means bigger moves. At the top of the window is an option called Auto Optimize which, when selected, will automatically adjust the tuning setting as the MCMC runs to try to achieve maximum efficiency. At the end of the run a table of the operators, their performance and the final values of these tuning settings will be written to standard ouput. These can then be used to set the starting tuning settings in order to minimize the amount of time taken to reach optimum performance in subsequent runs.

The next column, labelled Weight, specifies how often each operator is applied relative to the others. Some parameters tend to be sampled very efficiently – an example is the kappa parameter – these parameters can have their operators downweighted so that they are not changed as often (this may mean upweighting other operators since the weights must be integers).

18.10 Setting the MCMC options

The MCMC tab in BEAUTI provides settings to control the MCMC chain (Fig. 18.4). Firstly we have the Length of chain. This is the number of steps the MCMC will make in the chain before finishing. How long this should be depends on the size of the data set, the complexity of the model, and the precision of the answer required. The default value of 10 000 000 is entirely arbitrary and should be adjusted according to the size of your data set. We will see later how the resulting log file can be analyzed using TRACER in order to examine whether a particular chain length is adequate.

The next couple of options specify how often the current parameter values should be displayed on the screen and recorded in the log file. The screen output is simply for monitoring the program's progress so can be set to any value (although if set too small, the sheer quantity of information being displayed on the screen will slow the program down). For the log file, the value should be set relative to the total length of the chain. Sampling too often will result in very large files with little extra benefit in terms of the precision of the estimates. Sample too infrequently and the log file will not contain much information about the distributions of the parameters. You probably want to aim to store no more than 10 000 samples so this should be set to the chain length / 10 000.

For this data set let's initially set the chain length to 100 000 as this will run reasonably quickly on most modern computers. Although the suggestion above would indicate a lower sampling frequency, in this case set both the sampling frequencies to 100.

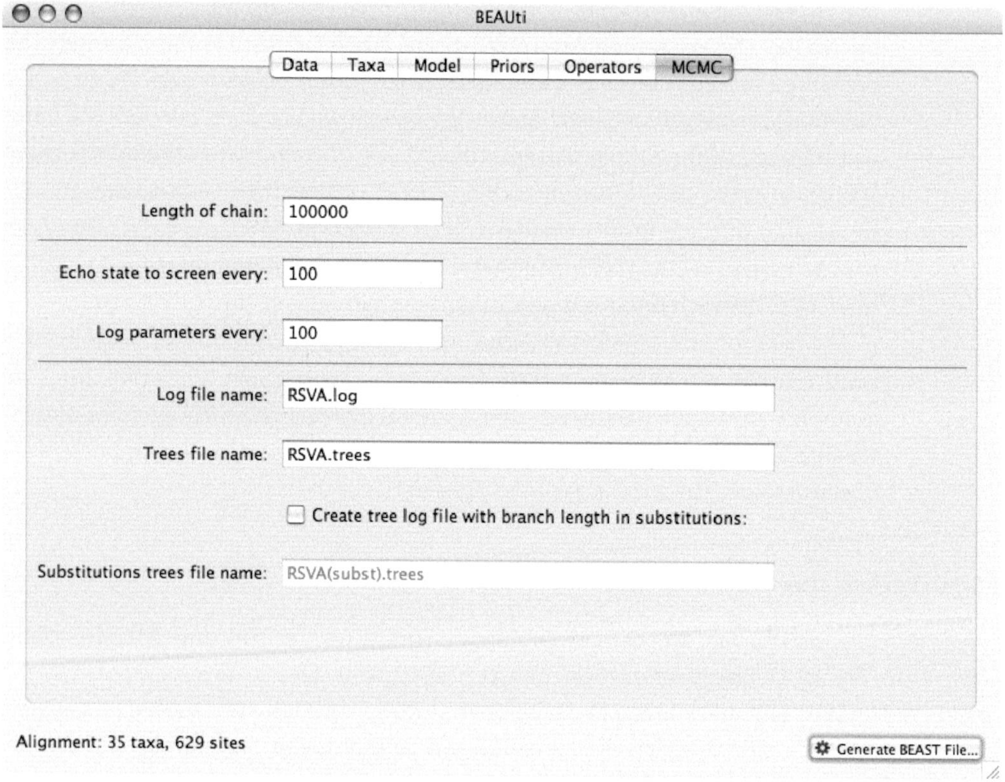

Fig. 18.4   The MCMC settings in BEAUTI

The final two options give the file names of the log files for the parameters and the trees. These will be set to a default based on the name of the imported NEXUS file but feel free to change these.

18.11 Running BEAST

At this point we are ready to generate a **BEAST** XML file and to use this to run the Bayesian evolutionary analysis. To do this, either select the Generate **BEAST** File... option from the File menu or click the similarly labeled button at the bottom of the window. Choose a name for the file (for example, RSVA.xml) and save the file. For convenience, leave the **BEAUTI** window open so that you can change the values and re-generate the **BEAST** file as required later in this tutorial.

Once the **BEAST** XML file has been created the analysis itself can be performed using **BEAST**. The exact instructions for running **BEAST** depends on the computer you are using, but in most cases a standard file dialog box will appear in which you select the XML file. If the command line version is being used, then the name of the XML file is given after the name of the BEAST executable. The analysis will then be performed with detailed information about the progress of the run being written to the screen. When it has finished, the log file and the trees file will have been created in the same location as your XML file.

18.12 Analyzing the BEAST output

To analyze the results of running BEAST we are going to use the program TRACER. The exact instructions for running TRACER differs depending on which computer you are using. Please see the README text file that was distributed with the version you downloaded. Once running, TRACER will look similar irrespective of which computer system it is running on.

Select the Open option from the File menu. If you have it available, select the log file that you created in the previous section. The file will load and you will be presented with a window similar to the one below (Fig. 18.5). Remember that MCMC is a stochastic algorithm so the actual numbers will not be exactly the same.

On the left-hand side is the name of the log file loaded and the traces that it contains. There are traces for the posterior (this is the log of the product of the tree likelihood and the prior probabilities), and the continuous parameters. Selecting a trace on the left brings up analyses for this trace on the right-hand side depending on tab that is selected. When first opened (Fig. 18.5), the "posterior" trace is selected and various statistics of this trace are shown under the Estimates tab.

In the top right of the window is a table of calculated statistics for the selected trace. The statistics and their meaning are described in the table below.

- *Mean* – The mean value of the samples (excluding the **burn-in**).
- *Stdev* – The standard error of the mean. This takes into account the effective sample size so a small ESS will give a large standard error.
- *Median* – The median value of the samples (excluding the burn-in).
- *95% HPD Lower* – The lower bound of the **highest posterior density (HPD) interval**. The HPD is the shortest interval that contains 95% of the sampled values.
- *95% HPD Upper* – The upper bound of the highest posterior density (HPD) interval.
- *Auto-Correlation Time (ACT)* – The average number of states in the MCMC chain that two samples have to be separated by for them to be uncorrelated (i.e. independent samples from the posterior). The ACT is estimated from the samples in the trace (excluding the burn-in).
- *Effective Sample Size (ESS)* – The **effective sample size (ESS)** is the number of independent samples that the trace is equivalent to. This is calculated as the chain length (excluding the burn-in) divided by the ACT.

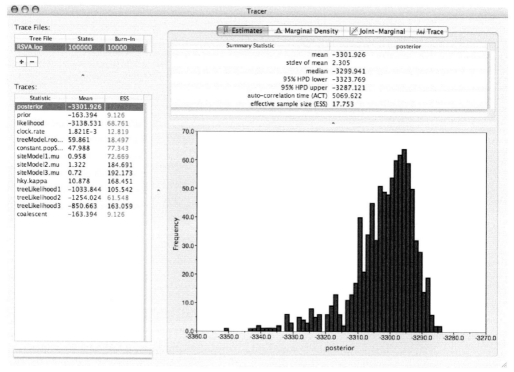

Fig. 18.5   The main TRACER window with a BEAST log file loaded.

Note that the effective sample sizes (ESSs) for all the traces are small (ESSs less than 100 are highlighted in red by TRACER). This is not good. A low ESS means that the trace contained a lot of correlated samples and thus may not represent the posterior distribution well. In the bottom right of the window is a frequency plot of the samples which is expected given the low ESSs is extremely rough (Fig. 18.5).

If we select the tab on the right-hand side labelled "Trace" we can view the raw *trace plot*, that is, the sampled values against the step in the MCMC chain (Fig. 18.6).

Here you can see how the samples are correlated. There are 1000 samples in the trace (we ran the MCMC for 100 000 steps sampling every 100) but it is clear that adjacent samples often tend to have similar values. The ESS for the age of the root (treeModel.rootHeight) is about 18 so we are only getting 1 independent sample to every 56 actual samples. It also seems that the default burn-in of 10% of the chain length is inadequate (the posterior values are still increasing over most of the chain). Not excluding enough of the start of the chain as burn-in will bias the results and render estimates of ESS unreliable.

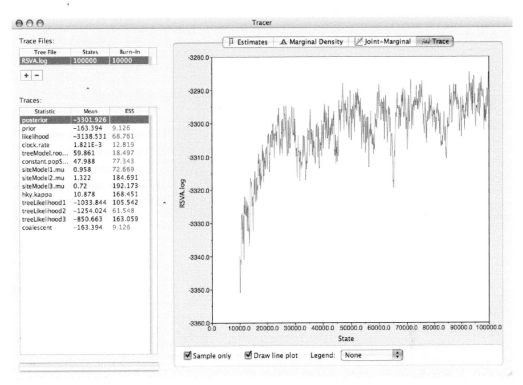

Fig. 18.6   The trace of posterior against chain length in TRACER for a run of 100 000 steps.

The simple response to this situation is that we need to run the chain for longer. Given the lowest ESS (for the `prior`) is 9, it would suggest that we have to run it at least 12 times longer to get ESSs that are >100. However it would be better to aim higher so let's go for a chain length of 5 000 000. Go back to Section 18.10 and create a new BEAST XML file with a longer chain length. Now run BEAST and load the new log file into TRACER (you can leave the old one loaded for comparison). Click on the "`Trace`" tab and look at the raw trace plot (Fig. 18.7).

Again we have chosen options that produce 1000 samples and with an ESS of about 500 there is still auto-correlation between the samples but 500 effectively independent samples will now provide a good estimate of the posterior distribution. There are no obvious trends in the plot, which would suggest that the MCMC has not yet converged, and there are no large-scale fluctuations in the trace which would suggest poor *mixing*. As we are happy with the behavior of log-likelihood we can now move on to one of the parameters of interest: substitution rate. Select `clock.rate` in the left-hand table. This is the average substitution rate across all sites in the alignment. Now choose the density plot by selecting the tab labeled

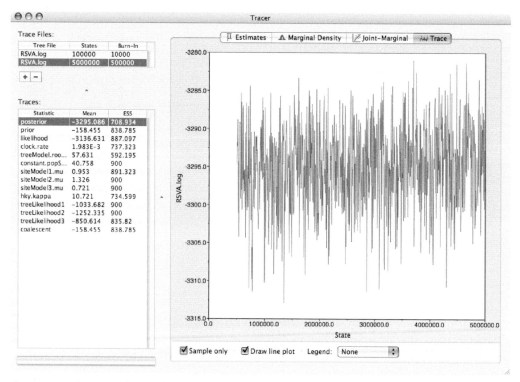

Fig. 18.7   The trace of posterior against chain length in TRACER for a run of 5 000 000 steps.

"Density". This shows a plot of the posterior probability density of this parameter. You should see a plot similar to Fig. 18.8.

As you can see the posterior probability density is roughly bell-shaped. There is some sampling noise which would be reduced if we ran the chain for longer, but we already have a good estimate of the mean and HPD interval. You can overlay the density plots of multiple traces in order to compare them (it is up to the user to determine whether they are comparable on the same axis or not). Select the relative substitution rates for all three codon positions in the table to the left (labeled siteModel1.mu, siteModel2.mu and siteModel3.mu). You will now see the posterior probability densities for the relative substitution rate at all three codon positions overlaid (Fig. 18.9).

18.13 Summarizing the trees

We have seen how we can diagnose our MCMC run using TRACER and produce estimates of the marginal posterior distributions of parameters of our model. However, BEAST also samples trees (either phylogenies or genealogies) at the

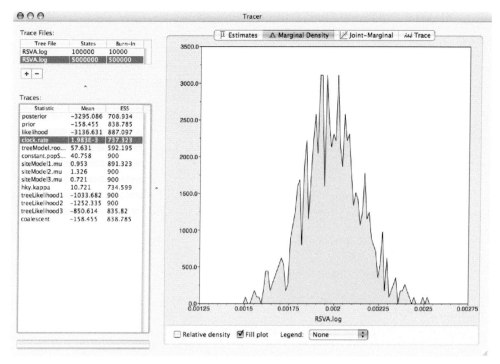

Fig. 18.8   The posterior density plot for the substitution rate.

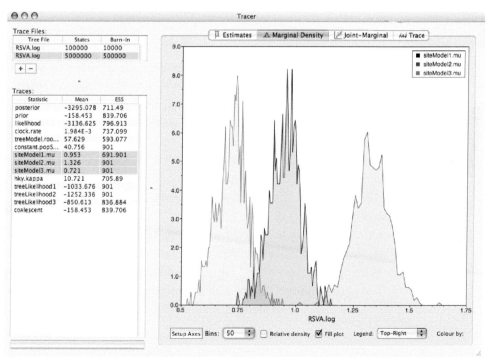

Fig. 18.9   The posterior density plots for the relative rate of evolution at each codon position.

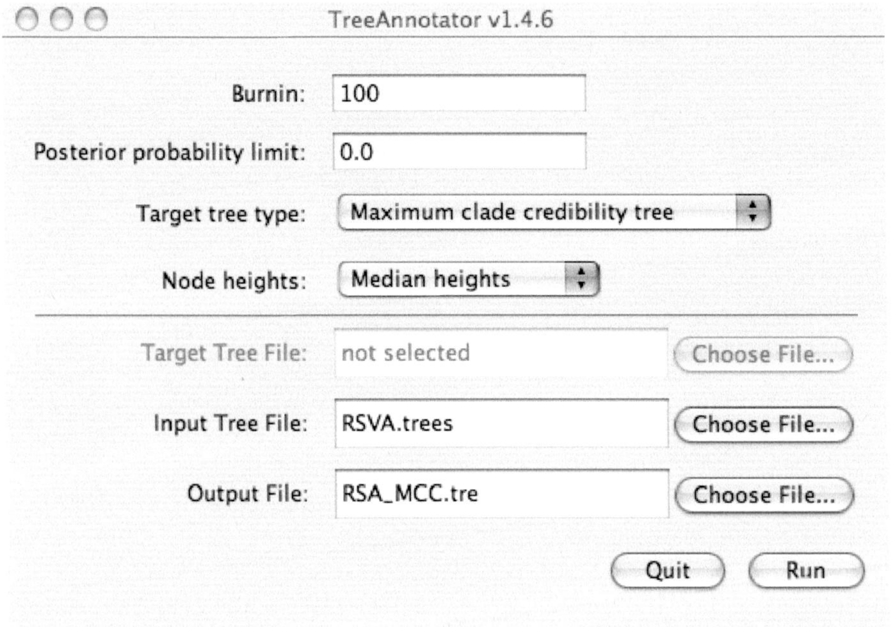

Fig. 18.10   The user-interface for the TREEANNOTATOR tool.

same time as the other parameters of the model. These are written to a separate file called the "trees" file. This file is a standard NEXUS format file. As such, it can easily be loaded into other software in order to examine the trees it contains. One possibility is to load the trees into a program such as PAUP* and construct a consensus tree in a similar manner to summarizing a set of bootstrap trees. In this case, the support values reported for the resolved nodes in the consensus tree will be the posterior probability of those clades.

In this tutorial, however, we are going to use a tool that is provided as part of the BEAST package to summarize the information contained within our sampled trees. The tool is called "TREEANNOTATOR" and once running, you will be presented with a window like the one in Fig. 18.10.

TREEANNOTATOR takes a single "target" tree and annotates it with the summarized information from the entire sample of trees. The summarized information includes the average node ages (along with the HPD intervals), the posterior support and the average rate of evolution on each branch (for models where this can vary). The program calculates these values for each node or clade observed in the specified "target" tree.

- *Burn-in* – This is the number of trees in the input file that should be excluded from the summarization. This value is given as the number of trees rather than the number of steps in the MCMC chain. Thus for the example above, with a chain of 1 000 000 steps, sampling every 1000 steps, there are 1000 trees in the file. To obtain a 10% burn-in, set this value to 100.

■ *Posterior probability limit* – This is the minimum posterior probability for a node in order for TREEANNOTATOR to store the annoted information. The default is 0.5 so only nodes with this posterior probability or greater will have information summarized (the equivalent to the nodes in a majority-rule consensus tree). Set this value to 0.0 to summarize all nodes in the target tree.

■ *Target tree type* – This has three options "Maximum clade credibility," "Maximum sum of clade credibilities" or "User target tree." For the latter option, a NEXUS tree file can be specified as the Target Tree File, below. For the first two options, TREEANNOTATOR will examine every tree in the Input Tree File and select the tree that has the highest product ("Maximum clade credibility") or highest sum ("Maximum sum of clade credibilities") of the posterior probabilities of all its nodes.

■ *Node heights* – This option specifies what node heights (times) should be used for the output tree. If the "Keep target heights" is selected, then the node heights will be the same as the target tree. The other two options give node heights as an average (mean or median) over the sample of trees.

■ *Target Tree File* – If the "User target tree" option is selected then you can use "Choose File . . ." to select a NEXUS file containing the target tree.

■ *Input Tree File* – Use the "Choose File . . ." button to select an input trees file. This will be the trees file produced by BEAST.

■ *Output File* – Select a name for the output tree file.

Once you have selected all the options, above, press the "Run" button. TREEANNOTATOR will analyse the input tree file and write the summary tree to the file you specified. This tree is in standard NEXUS tree file format so may be loaded into any tree drawing package that supports this. However, it also contains additional information that can only be displayed using the FIGTREE program.

18.14 Viewing the annotated tree

Run FIGTREE now and select the Open. . . command from the File menu. Select the tree file you created using TREEANNOTATOR in the previous section. The tree will be displayed in the FIGTREE window. On the left-hand side of the window are the options and settings which control how the tree is displayed. In this case we want to display the posterior probabilities of each of the clades present in the tree and estimates of the age of each node (see Fig. 18.11). In order to do this you need to change some of the settings.

First open the Branch Labels section of the control panel on the left. Now select posterior from the Display pop-up menu. The posterior probabilities won't actually be displayed until you tick the check-box next to the Branch Labels title.

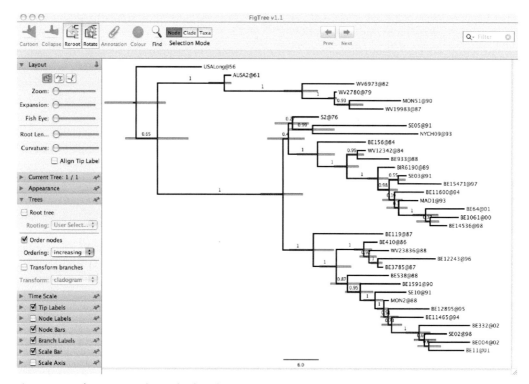

Fig. 18.11   The annotated tree displayed in FIGTREE.

We now want to display bars on the tree to represent the estimated uncertainty in the date for each node. TREEANNOTATOR will have placed this information in the tree file in the shape of the 95% highest posterior density (HPD) intervals (see the description of HPDs, above). Open the Node Bars section of the control panel and you will notice that it is already set to display the 95% HPDs of the node heights so all you need to do is to select the check-box in order to turn the node bars on.

Finally, open the Appearance panel and alter the Line Weight to draw the tree with thicker lines. None of the options actually alters the tree's topology or branch lengths in any way, so feel free to explore the options and settings. You can also save the tree and this will save all your settings so that, when you load it into FIGTREE again, it will be displayed exactly as you selected.

18.15 Conclusion and resources

This chapter only scratches the surface of the analyses that are possible to undertake using BEAST. It has hopefully provided a relatively gentle introduction to the fundamental steps that will be common to all BEAST analyses and provide a basis for more challenging investigations. BEAST is an ongoing development project with new models and techniques being added on a regular basis. The BEAST website provides details of the mailing list that is used to announce new features and to discuss the use of the package. The website also contains a list of tutorials and recipes to answer particular evolutionary questions using BEAST as well as a description of the XML input format, common questions, and error messages.

- The BEAST website: *http://beast.bio.ed.ac.uk/*
- Tutorials: *http://beast.bio.ed.ac.uk/Tutorials/*
- Frequently asked questions: *http://beast.bio.ed.ac.uk/FAQ/*

LAMARC: **Estimating population genetic parameters from molecular data**

THEORY

Mary K. Kuhner

19.1 Introduction

The LAMARC *programs* estimate parameters such as *effective population size*, *growth rate*, **migration rates**, and **recombination rate** using molecular data from a random sample of individuals from one or several populations (Felsenstein *et al.*, 1999). The relationship structure among sampled individuals, their *genealogy*, contains a great deal of information about the past history of the population from which those individuals were drawn. For example, in a population that has been large for a long time, most of the individuals in the sample will be distantly related; in a population that has been smaller, most of the individuals will be closely related.

The mathematical theory relating a genealogy to the structure of its underlying population, called **coalescent theory,** was first developed by Kingman (1982) and expanded by Hudson and Kaplan (1988) (see Chapter 17). However, use of coalescent theory to estimate parameters is hampered by the fact that the sample genealogy is almost never known with certainty and is difficult to infer accurately. Population samples are less likely to yield their correct genealogy than samples from multiple species since fewer informative mutations will be available. Additionally the possibility of recombination can make accurate genealogy reconstruction of a population almost impossibly daunting. Analysis of pairs of individuals can reveal some coalescent-based information – the genealogy connecting two individuals is relatively easy to infer. However, considerable information is lost in reducing a

The Phylogenetic Handbook: a Practical Approach to Phylogenetic Analysis and Hypothesis Testing, Philippe Lemey, Marco Salemi, and Anne-Mieke Vandamme (eds.). Published by Cambridge University Press. © Cambridge University Press 2009.

genealogy to a series of pairs, and the non-independence of the pairs also poses some problems in analysis.

The Lᴀᴍᴀʀᴄ programs work around our ignorance of the correct genealogy by considering many possible genealogies. This is done using a statistical method called ***Metropolis–Hastings Markov chain Monte Carlo sampling***. Working backwards through this imposing term, *Monte Carlo* means that there is a random element to the sampling. ***Markov Chain*** means that the genealogies are produced using a scheme of defining transition probabilities from one state to another. Metropolis and colleagues (1953) developed (for atom bomb research!) the specific type of sampler used today, one which starts with a plausible configuration, makes a small change, and accepts or rejects the new configuration. Hastings (1970) subsequently added statistical details to this procedure.

Markov Chain Monte Carlo (MCMC) methods are necessarily approximate, and they often demand a great deal of computer time and effort. However, methods considering the whole genealogy can potentially extract more information from the same amount of data than methods based on pairs of individuals; they provide not only estimates of the parameters, but also approximate error bars around those estimates.

19.2 Basis of the Metropolis–Hastings MCMC sampler

Since the correct genealogy is not known, especially in more complex cases such as those with recombination, the estimate should be based on a good sample of possible genealogies. To make the sampling as efficient as possible, only genealogies that are reasonably concordant with the data are chosen. Undirected random sampling ($\approx$ *Monte Carlo sampling*) is not efficient since the number of possible genealogies skyrockets as the number of sampled individuals increases. MCMC sampling, by imposing preferences on the random walk, allows movement through the space of possible genealogies in a purposeful fashion.

Two things are necessary to build up a Metropolis–Hastings MCMC coalescent sampler. First, a mathematical statement of how the parameters are expected to affect the shape of the genealogy is needed. For example, in the simple case of a constant size, ***haploid*** population without recombination or migration, Kingman's original work provides the necessary expectations (see Chapter 17). They take the general form of the probability of the genealogy G for a given value of the parameter (or parameters). In the example, the single parameter is Θ ($2N_e\mu$, where N_e is the ***effective population size*** and μ is the ***mutation rate***) and the expectation is written as the probability of observing genealogy G given Θ or $P(G|\Theta)$. Second, the relative fit of the data to the various genealogies must be assessed so that the sampler can concentrate on genealogies that explain the data well. This is the goal of phylogeny estimation as well; therefore similar methods may be used. *Likelihood* methods (see Chapter 6; Kishino & Hasegawa, 1989) are the most appropriate in this situation because they are accurate and flexible and because they can tell not only which genealogy is better, but also by how much. The fit of data to a genealogy can be expressed as the probability of the data, assuming an appropriate *model of molecular evolution*, with respect to any given genealogy (written $P(D|G)$). Combining the two parts it can be shown that $L(\Theta) = \Sigma_G P(D|G)P(G|\Theta)$, where the likelihood of any given value of Θ is the sum, over all possible genealogies, of the probability of the data on that genealogy multiplied by the probability of that genealogy for the given value of Θ. Unfortunately, the whole summation is not possible in any but trivial cases. To overcome this problem, the *Metropolis–Hastings sampler* generates a biased set of genealogies driven by an assumed value of Θ, called Θ_0, and then it corrects for that bias in evaluating the likelihood. The result is a relative likelihood:

$$L(\Theta)/L(\Theta_0) = \Sigma_{G^*}(P(D|G)P(G|\Theta))/(P(D|G)P(G|\Theta_0)) \qquad (18.1)$$

Here Σ_{G^*} is a sum over genealogies selected in proportion to $P(D|G)P(G|\Theta_0)$. If an infinitely large sample could be generated, then this approximation would give the same results as the straightforward likelihood. In practice, a sufficiently large sample must be considered so that the region of plausible genealogies is well explored. The algorithm will only efficiently explore the right region if Θ_0, which acts as a guide, is close to the true, unknown Θ. LAMARC's strategy is to make short runs of the program in order to obtain a preliminary estimate of Θ, and then feed that estimate back in as Θ_0. The final run will then have Θ_0 close to Θ, and will be more efficient (and less biased) than the earlier ones. The program generates its sample of genealogies by starting with some arbitrary or user-supplied genealogy and proposing small rearrangements to it. The choice of rearrangements is guided by $P(G|\Theta_0)$. Once a rearranged genealogy has been produced, its plausibility is assessed ($P(D|G)$) and compared to the plausibility of the previous genealogy. If the new genealogy is superior, it is accepted. If it is inferior, it still has a chance to be accepted: for example, genealogies that are ten times worse are accepted one time in ten that they occur. This behavior helps keeping the sampler from walking up the nearest "hill" in the space of genealogies and sticking there, even if there are better regions elsewhere. Given sufficient time, all areas of the space will be searched, though proportionally more time will be spent in regions where $P(D|G)P(G|\Theta_0)$ is higher.

Once a large sample of genealogies has been produced, it is then used to construct a likelihood curve showing $L(\Theta)/L(\Theta_0)$ for various values of Θ, which is normally displayed as a log-likelihood curve. The maximum of this curve is the *maximum likelihood estimate* of Θ; the region within two log-likelihood units of the maximum forms an approximate 95% confidence interval. Typically, the strategy is to run 5–10 "short chains" of a few thousand genealogies each, to get a good starting value of Θ, and then 1–2 "long chains" to generate the final estimate.

The most difficult part of creating such a Metropolis–Hastings sampler is working out a way to make rearrangements guided by $P(G|\Theta_0)$: this is particularly challenging in cases with recombination, where the genealogy becomes a tangled graph.

19.2.1 Bayesian vs. likelihood sampling

MCMC is a general tool that can be used in either a likelihood or a Bayesian context. The fundamental difference is that a Bayesian sampler searches not only among genealogy resolutions, but also among values of its model parameters. For example, a likelihood estimate of Θ involves sampling many possible genealogies guided by a driving value of Θ, and using those genealogies to estimate Θ, factoring out the influence of the driving value. A Bayesian estimate of Θ involves searching simultaneously among genealogies and values of Θ. The values of Θ must be drawn from a ***prior***, and this reliance on a prior is the critical feature of Bayesian estimation.

This chapter focuses on likelihood estimation using Lamarc, but the program can also perform Bayesian estimation. The previous chapter, describing Beast, is a good overview of how such an analysis is conducted. Since this is a recently added capability in Lamarc, little is known about comparative performance of likelihood and Bayesian methods. Kuhner and Smith (2007) found little difference between the two approaches on highly informative data sets. Beerli (2006) showed that because of the constraining prior, Bayesian analysis produces more reasonable results when the data are rather uninformative. However, like all Bayesian methods, Bayesian Lamarc will return its prior when data are lacking, and this should not be mistaken for an informative result. When using Bayesian Lamarc, as with any Bayesian analysis, it is critical to choose appropriate priors. A prior that disallows the true answer will obviously produce nonsense results; more subtly, a prior that puts little weight on reasonable values will lead to an inefficient search and potentially misleading results.

19.2.2 Random sample

Lamarc assumes that the sampled individuals were chosen at random from the population(s) under study. For Migrate and Lamarc, one may freely choose how intensively to sample each subpopulation, but within each subpopulation individuals must be chosen at random. Violating this assumption will lead to bias: for example, if the most diverse individuals are chosen then Θ will be overestimated. Researchers are sometimes tempted to delete "boring" identical individuals from their data sets in favor of non-identical ones, but doing so produces misleading results. If the data set is too large to analyze, individuals must be dropped at random.

19.2.3 Stability

LAMARC assumes that the population situation has been stable for "a long time" (longer than the history of the coalescent genealogy). MIGRATE and RECOMBINE assume that it has been of constant size; FLUCTUATE (Kuhner et al., 1998) and LAMARC (Kuhner, 2006) can instead assume that it has been growing or shrinking at a constant exponential rate. MIGRATE (Beerli & Felsenstein, 1999) and LAMARC assume that the subpopulation structure and migration rates have been stable; RECOMBINE and LAMARC assume that recombination rates have remained stable. If these assumptions are not met, the results will be misleading. For example, two populations which currently exchange no migrants, but which are recently derived from a single ancestral population, will give spurious evidence for migration. These assumptions can be relaxed by constructing more complex samplers, if a model is available for how things have changed: for example, LAMARC can relax the constant-size assumption, replacing it with a model of exponential growth.

19.2.4 No other forces

LAMARC assumes that no forces other than the ones modeled are acting on the locus. In particular, it assumes that the variation observed is neutral and is not being dragged by selection at a linked marker. Natural selection could bias the results in a variety of directions, depending on the type of selection. *Balancing selection* will lead to overestimates of Θ, and *directional selection*, in general, will lead to underestimates. It is difficult to predict the effects of selection on estimates of migration or recombination rates. Investigating selective pressure in molecular sequences is discussed in Chapters 13 and 14.

19.2.5 Evolutionary model

LAMARC's results can only be as good as its model of molecular evolution. For example, analyzing human mitochondrial DNA with a model of DNA sequence evolution that assumes *transitions* and *transversions* are equally frequent will lead to misleading results, since in fact mtDNA appears to greatly favor transitions. Fortunately, it is fairly easy to add more complex evolutionary models as the phylogenetic community develops them (see Chapter 4). LAMARC inherits any assumptions made by its evolutionary models: for example, a common assumption in DNA models is that all sites are independent and identically distributed. Elsewhere in this book (see Chapter 6) information about the assumptions of maximum likelihood phylogeny models can be found; those caveats also apply here.

19.2.6 Large population relative to sample

Coalescent theory is an approximation of the real genealogical process. To make the mathematics tractable, it is assumed that only one event happens at a time: for example, two individuals may trace back to a common ancestor at a particular point in time, but not three individuals (see Chapter 17). This assumption is safe when the population is much larger than the sample (it may occasionally be violated, but these rare violations will not affect the result). However, it can break down if the method is applied to a tiny population, most of which is being sampled. An extreme example would be a lineage of highly inbred mice, where the size of the whole population is just a few individuals each generation. Any sample of such a population will represent a large proportion of the whole, and the coalescent approximations will not be very good. In practice, there are few stable biological populations small enough to encounter breakdown of the coalescent approximation. Tiny populations are more likely to represent a violation of the stability assumption: they have probably not been tiny throughout their history.

19.2.7 Adequate run time

The sample of genealogies produced by a LAMARC program is meant to stand in for the entire space of possible genealogies, a tall order. Luckily, most genealogies are such poor explanations of the data that they contribute very little to the final estimate, so failure to sample such genealogies does not matter much. However, the program does need to run long enough to adequately sample the high-probability regions of genealogy space. This is particularly difficult if genealogy space contains multiple isolated peaks, rather than a single peak. For example, consider a data set which can support either high migration from A to B and low from B to A, or vice versa, but is less supportive of intermediate values. This corresponds to two peaks in the space of possible genealogies for these data. The program must run long enough to visit both peaks since an estimate based only on the first peak found may be misleading, for example, leading to a false conclusion that high migration from B to A can be rejected. Metropolis–Hastings samplers in other applications sometimes encounter cases where the "valley" between two peaks is infinitely deep and cannot be crossed, stalling the sampler. It is believed that such cases do not exist for the genealogy applications, but very deep valleys are possible. There are some ways to detect stalled-out samplers, suggested later in this chapter, but they are not guaranteed to work. Getting good results from the LAMARC programs will require patience and a degree of skepticism.

PRACTICE

Mary K. Kuhner

19.3 The LAMARC software package

Copies of the source code and executables for various systems can be obtained from *http://evolution.gs.washington.edu/lamarc/* or by anonymous ftp to *evolution.gs.washington.edu* in directory `/pub/lamarc`. See Table 19.1 for details of the available programs.

19.3.1 FLUCTUATE (COALESCE)

COALESCE, the first LAMARC program released, estimated Θ $(2N_e\mu)$ in a single constant-size, haploid population using DNA or RNA sequence data and a Kimura 2-parameter model of sequence evolution (Kuhner *et al.*, 1995). Its successor, FLUCTUATE, added estimation of an exponential growth (or decline) rate, *g*. COALESCE and FLUCTUATE are no longer supported, since they can easily be mimicked using LAMARC.

Estimation of Θ is quite robust and accurate even with fairly small data sets. Estimation of *g* is much more difficult. For many data sets, the likelihood surface resembles a long, flat ridge, so that many different values of *g* are almost equally likely. Each value of *g* has its own preferred value of Θ, but it is difficult to choose between various (Θ, *g*) pairs. There are two ways to overcome this problem. Using multiple loci can greatly improve the estimate of *g*. Most information about *g* comes from coalescences deep in the genealogy (towards the root) but any given genealogy will have only a few such coalescences. Adding a second, unlinked locus will nearly double the available information. Another option is to provide an estimate of Θ from external evidence. This will allow a much more stable estimate of *g*. If neither multiple loci nor an independent estimate of Θ are available, the estimate of *g* should be considered skeptically as it is likely to be biased upwards; its error bars, however, are reasonably trustworthy.

19.3.2 MIGRATE-N

This program estimates, for a subdivided population, Θ for each subpopulation and the migration rate into each subpopulation from each of the others. It can also analyze migration models with particular restrictions: for example, it can assume that migration rates are symmetrical, or that only certain subpopulations are capable of exchanging migrants (as in a ***stepping-stone model***). The more parameters estimated, the less power is available for each one: thus, if a stepping-stone model is a good representation of the data, then it should be used rather than

598

Table 19.1 LAMARC programs as of 2006

Program	Estimates	Data types supported
COALESCE[a]	Θ	DNA, RNA
FLUCTUATE[a]	Θ, growth rate	DNA, RNA
MIGRATE	Θ, migration rates	DNA, RNA, microsats, electrophoretic alleles
RECOMBINE[a]	Θ, recombination rate	DNA, RNA, SNPs
LAMARC	Θ, growth rates, migration rates, recombination rate	DNA, RNA, SNPs, microsats, electrophoretic alleles

SNPs = *single nucleotide polymorphisms*; microsats refer to *microsatellite DNA*.
[a] These programs are provided for reference since they are superseded by LAMARC and no longer supported.

the most general model. The converse, of course, is that, if a stepping-stone model is imposed and the biological situation is not consistent with it, estimates will be actively misleading.

MIGRATE needs multiple loci and reasonably large numbers of individuals per subpopulation, at least 20 or more, to perform well. It is often the case that the researcher is aware of many subpopulations, but has samples only from a few. For example, one might be studying two island populations of mink and have no samples from the adjacent mainland. In such cases, it may be useful to include a third population, containing no observed individuals, into the MIGRATE analysis. While the parameters of this unsampled population will not be estimated very accurately, its presence will allow more accurate estimation of parameters for the sampled populations.

When MIGRATE performs badly, failure of the steady-state assumption is often the cause. For example, if two populations recently diverged from a common ancestor but currently have little migration, MIGRATE will infer high migration in an attempt to explain the shared similarities due to common ancestry. Another potential difficulty is that if two subpopulations are mixing at an extremely high rate, so that they are in effect a single population, MIGRATE is theoretically capable of analyzing the situation and coming to correct conclusions, but in practice will slow down severely. The current version of MIGRATE can be run in parallel on a computer cluster, which significantly improves the speed of the analysis.

19.3.3 RECOMBINE

This program, which assumes a single population, estimates Θ and the recombination rate r, expressed as recombinations per site over mutations per site: C/μ. It also assumes that the recombination rate is constant across the region. RECOMBINE can successfully estimate recombination rate with short sequences only if Θ is high (in other words, data have to be very polymorphic) and the recombination rate is also high. If either parameter is relatively low, long sequences (thousands of base pairs) will be needed to make an accurate estimate. The space of possible genealogies searched by RECOMBINE is huge, and long runs are necessary to produce an accurate estimate. One practical consideration when running RECOMBINE is whether the "short chains" used to find initial values of Θ and r are long enough. The user should examine the values of r produced from the short chains. If they change greatly from one short chain to the next, the short chains are probably too short: it may be better to run half as many short chains as before, each one twice as long. RECOMBINE is no longer supported because it is now superseded by LAMARC.

19.3.4 LAMARC

The LAMARC program, not to be confused with the package to which it belongs, combines the basic capabilities of FLUCTUATE, RECOMBINE, and MIGRATE-N. It can simultaneously estimate migration rates, growth rates, and recombination rates. This is useful when all parameters are of interest, but it is also important when only one set is of interest, for example for a researcher trying to estimate migration rates using nuclear DNA, because acknowledging the presence of other evolutionary forces will avoid bias in the results. Attempts to estimate both recombination and migration in complex cases will demand extremely long runs; a good rule of thumb is to start by adding together the run length needed for each program individually. In the future the LAMARC program will be enhanced with additional evolutionary forces such as selection and population divergence, so that complex biological situations can be accurately analyzed.

19.4 Starting values

The closer the initial values are to the truth, the more efficiently the program will search. If a "quick and dirty" estimator is available (e.g. the **Watterson estimate** for Θ, see Section 19.8 below), it should be used to generate initial values. This estimator is provided by most of the LAMARC programs. If it is necessary to guess an initial value, the guess should err upwards rather than downwards. If that parameter is known to be zero, it should be fixed at that value; otherwise, never use zero as an initial guess. Trying more than one set of initial values can help spot problems in the analysis. The results of multiple runs with different starting values should be quite similar. If they are not, the reason could be twofold: (a) the likelihood surface is so flat that almost any value is equally plausible, or (b) the program is not running long enough for it to reach its final solution. Examining the shape of the likelihood surface will help distinguish between these two alternatives. If it is very flat, situation (a) is true and the data simply do not allow a sharp estimate. If the likelihood surface is sharply peaked, but varies hugely with starting values, the program is not running long enough.

19.5 Space and time

On many machines extra memory will have to be assigned to the LAMARC programs in order to run them on realistic data sets. The method for doing this is operating-system specific: ask the local system operator for help. This is worth trying even if the program does function with less memory, because additional memory may allow it to run more quickly. If the program still will not run, or runs intolerably slowly, consider reducing the size of the data set. Randomly delete sequence regions, or randomly delete individuals. In cases with multiple populations, deletion of individuals can be non-random between populations as long as it is random within populations. For example, the user can randomly drop three individuals from each population. Do not preferentially delete "uninteresting" sites or individuals as this will bias the results!

Increasing the interval between sampled genealogies will save space and a little time. The LAMARC programs do not normally use all of the genealogies they generate in their final estimate, but sample every 10th or 20th genealogy. Denser sampling does not help the estimate much, since successive genealogies are highly correlated. There is significantly more information in a sample of every 100th genealogy from 100 000 than every 10th genealogy from 10 000, even though the number of genealogies sampled and the amount of space used are the same. Of course, generating 100 000 genealogies is still ten times slower than generating 10 000.

19.6 Sample size considerations

If one must choose between studying more individuals, more sites, or more loci, almost always choose more loci for FLUCTUATE and MIGRATE, and almost always choose more sites for RECOMBINE. Users of LAMARC will have to decide whether they are more interested in recombination or migration. However, a minimum number of individuals is needed to get a good estimate. In general, single-population analysis should have at least 20 individuals, and multiple-population analysis should have at least 20 per subpopulation. Doubling of these numbers is reasonable, but beyond that point there are diminishing returns. Running LAMARC with 400 individuals will gain almost nothing over running it with 40; in fact, the larger search space may lead to a worse analysis because the sampler does not find all of the peaks. DNA or RNA sequences should be at least a few hundred base pairs long for analysis without recombination, but will need to be much longer, thousands to tens of thousands, for analysis with recombination unless both Θ and r are very high.

Microsatellite and electrophoretic data require multiple loci even more strongly than do sequences, since the amount of information per locus is smaller. The reason that adding more individuals does not help much is that the first few individuals already establish the basic structure of the genealogy. After that point, additional individuals tend to join the genealogy as twigs, and each twig carries only a tiny amount of new information. Adding a second locus, in contrast, gives a new trunk and main branches. The reason that longer sequences do not improve the estimate much is that uncertainty in the LAMARC estimates comes from three sources. First, the program introduces some uncertainty by its approximate nature, which can be overcome by longer runs. Second, the data is only an approximate representation of its underlying genealogy, which can be overcome by longer sequences. Finally, the genealogy is only an approximate representation of its underlying parameters, which can be slightly improved by a larger genealogy and significantly improved by adding a second unlinked locus, thereby examining a fully independent new genealogy. After a certain point, no amount of running the program longer or adding longer sequences will improve the estimate, because the remaining error comes from the relationship between genealogy and parameters. This is less true in the presence of recombination because a long recombining sequence, like multiple loci, provides additional information. RECOMBINE can accommodate multiple loci but is actually most efficient with a single, long, contiguous stretch of markers.

19.7 Virus-specific issues

Estimating population genetic parameters for viral data sets, especially for fast evolving viruses like HIV or HCV, needs a few additional considerations.

19.7.1 Multiple loci

Many viruses are too small to provide multiple independent loci for analysis. RECOMBINE or LAMARC can extract additional information out of partially linked loci. Another alternative is to examine several different isolates of the virus, using the same locus from each isolate but entering them into the program as separate loci. If each isolate traces back to a common ancestor before any shared ancestry with any other isolate, then they are independent samples of the evolutionary process. For example, if within-patient recombination rate or growth rate are the parameters of interest, different patients can be entered as separate loci with all the variants within a patient representing a single locus. However, this method assumes that the evolutionary and population genetic processes are the same within each patient, which is not necessarily true.

19.7.2 Rapid growth rates

Experience from using Lᴀᴍᴀʀᴄ to estimate HIV growth rates shows that no sensible estimates are produced when Θ and g are co-estimated. HIV genealogies often resemble stars, and a star genealogy is equally consistent with any high value of Θ and corresponding high value of g. Therefore, the likelihood surface is flat along the Θg diagonal, and the program will wander about on this surface until the estimate becomes so huge that the program overflows. If there is a star-like genealogy, it is only possible to estimate g if an external estimate of Θ can be provided. Conversely, if g is known, then Θ can be estimated.

19.7.3 Sequential samples

One external source of information about these parameters is sequential sampling from a ***measurably evolving population***, i.e. two or more samples taken at a reasonable time interval allowing a direct estimate of the population parameters (Rodrigo & Felsenstein, 1999; Rodrigo *et al.*, 1999). It would be desirable to adapt the Lᴀᴍᴀʀᴄ programs to handle sequential samples since they should be very powerful for estimating growth in particular. Currently, however, mixing samples from different times in a Lᴀᴍᴀʀᴄ analysis will result in confused estimates and analyzing them as separate data sets results in a problem of non-independence. The coalescent has been extended to serially sampled populations and is implemented in other software (Bᴇᴀsᴛ, see previous chapter).

19.8 An exercise with Lᴀᴍᴀʀᴄ

A data set will only be suitable for Lᴀᴍᴀʀᴄ analysis if it represents a random sample from one or several populations of the same species. It is essential that the sample be random. Selecting a few viruses from each serological subtype, for example, will produce a severe bias. As an example we will use a data set from Ward *et al.* (1991), containing 360 bp from the human mitochondrial genome D-loop region, sequenced in 63 members of the Nuu-Chah-Nulth people. The data originally came in the form of the 28 unique sequences only. The data were expanded into a full-sized data set with all 63 individuals by duplicating the sequences representing the same haplotype. (Note that this can only be done because Ward actually sequenced all 63 individuals. If he had sequenced only one representative of each haplotype group, we would miss private polymorphisms, leading to a bias.) The data can be found at *http://www.thephylogenetichandbook.org*.

The following exercise assumes you have a working version of LAMARC installed on your computer. LAMARC version 2.1.2b was used for the exercise below. Pre-compiled executables (binary distribution) can be found at *http://evolution. genetics.washington.edu/lamarc/lamarc.html*, and instructions to build the software executables on your computer can be found in doc/html/compiling. html in the distribution folder. Each software module of LAMARC works in the same basic way: it needs a file containing the input data to be placed in the same directory where the program resides (otherwise, the full path to the file will have to be specified); it produces an output file, by default called "outfile.txt", in text format, containing the results of its analysis. The user-interface looks like one of the applications of the PHYLIP package (see Chapter 4). The older programs in the package use an input format similar to that of PHYLIP, whereas the LAMARC program uses its own XML-based input format; a file conversion utility is provided.

19.8.1 Converting data using the LAMARC file converter

The Ward data are in the form of a "sequential" PHYLIP input file, a format available in many bioinformatics programs and databases (see Box 2.4 in Chapter 2). The older LAMARC package programs (FLUCTUATE, RECOMBINE, MIGRATE) could read this file with minimal change, but for the LAMARC program it needs to be converted to XML. This can be done by the graphical or text-based converter lam_conv, provided with the LAMARC program.

We will use the graphical version of lam_conv to create the xml file. Execute lam_conv and load the ward.phy data file using "Read Data File" in the File menu. Under "Data Partitions", four panels will appear after parsing the file: population, region, contiguous segment and parsed data (Fig. 19.1). "Regions" refer to unlinked areas, each representing an independent realization of the evolutionary process. For example, a nuclear DNA sample from the same population could be added as an additional region. Some regions need to be treated as containing multiple "contiguous segments" of data, which are close enough to be fairly linked but need to be modeled differently. Verify that 63 sequences (samples) and 360 base pairs (sites) have been found correctly. Settings and names can be changed by double clicking on the panels. Name the population "Nuu-Chah-Nulth", the region "mtdna", and the segment "D-loop" for mnemonic value. The names are particularly important if you have multiple regions or populations, so that you will be able to interpret your output reports. While changing the contiguous segment name, also select the check box for DNA Data Type and click Apply. Now you can write the file in xml format to the LAMARC folder using "Write Lamarc File" in the File menu. Use the default "infile.xml" file name. Inspecting this file with

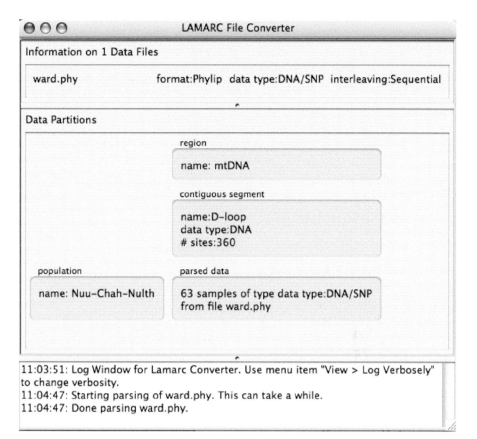

Fig. 19.1   The L AMARC file converter utility. The Ward data set is loaded and names of the population, region, and contiguous segment are changed to "Nuu-Chah-Nulth", "mtdna", and "D-loop", respectively.

a text editor shows that it is indeed a L AMARC XML file containing the sequences. More information on converting data files using lam_conv can be found in doc/html/converter.html.

19.8.2 Estimating the population parameters

Now we can execute L AMARC itself, which prompts for the data file; type return and the program will read the default "infile.xml" file in the same directory. A text menu appears:

```
Main Menu

D  Data options (tell us about your data)
A  Analysis (tell us what you want to know)
S  Search Strategy menu (tell us how to do it)
```

```
I   Input and Output (tell us where to keep it)
O   Overview of current settings (what you told us)
    ----------
    . = Run | Q = Quit
    >  Generate Lamarc input file for these settings
```

For the first exercise, we will co-estimate Θ and growth rate. The "D" (Data options) menu entry is used to set the substitution model. Moving downward through the D menus (typing "D") and the Edit data model(s) menu (typing "1"), we find that the data will be analyzed by default under an F84 substitution model with one rate category, a ***transition/transversion ratio*** of 2.0, and base frequencies calculated from the data:

```
Edit data model for Region 1: mtDNA

    Datatype for this region                          DNA
M   Data Model                                        F84
D   Use default data model for this region/segment    Yes
C   Number of Categories                              1
A   Auto-Correlation                                  1
T   TT Ratio                                          2
B   Base Frequencies (A C G T)     0.3297 0.3371 0.1120
                                          0.2213 <calc>
R   Relative mutation rate                            1
```

Change the transition/transversion ratio to a more reasonable mtDNA value of 30.0 (typing "t" and entering the value). Other programs, such as PAUP* or TREE-PUZZLE, can be used to estimate the best value here (see Chapters 8 and 6, respectively). These programs could also be used to explore multiple rate categories for these data. Hitting return repeatedly will bring you back up to the main menu.

The "A" (Analysis) section allows you to specify which evolutionary forces need to be measured. By default, on this single-population data set LAMARC will try to estimate only Θ. Enter the Growth Rate sub-menu to turn on Growth estimation (by typing "G" and then "X"). Note that you are not allowed to turn on Migration, since there is only one population. Accept the default starting values for the population parameters, since the starting values are of little importance if the run is sufficiently long. Return again to the main menu (by typing return repeatedly).

The "S" (Search Strategy) menu determines how much effort LAMARC will put into searching among genealogies, and how that effort will be distributed. For our purposes, the most important sub-menu here is "S" (Sampling `strategy`). Entering this submenu, we find that the default is to run 1 replicate, with 10 initial chains and 2 final chains:

```
Sampling strategy (chains and replicates)

R   Number of replicates                               1

    Initial Chains
1   Number of chains (initial)                        10
2   Number of recorded genealogies (initial)         500
3   Interval between recorded items (initial)         20
4   Number of samples to discard (initial burn-in)  1000
    --------------
    Final Chains
5   Number of chains (final)                           2
6   Number of recorded genealogies (final)         10000
7   Interval between recorded items (final)           20
8   Number of samples to discard (final burn-in)    1000
```

Initial chains are used to quickly refine the initial values of the parameters, and final chains are used to make a robust final estimate. Normally, final chains should be at least 10 times longer than initial chains. Here, we are taking 500 samples (sampling every 20th genealogy) from the initial chains and 10 000 samples from the final chains. The program is set to discard 1000 genealogies at the beginning of each chain, since the first sampled genealogies may be atypical. These defaults have deliberately been set low so that the program will run quickly. For a publication-quality run, the number of samples in both initial and final chains should be increased at least fivefold, and the output should be examined to make sure that the run was long enough. For now, we will accept these defaults and return to the Main Menu.

The "I" (Input and Output) menu allows the user to control the wordiness of the output and the file to which it will be written. The defaults are adequate for now. The "O" (Overview) menu can be used to quickly check that the settings are what you expect them to be. Type " . " to start the program. A progress bar appears showing how the run is progressing. (If it is not visible within a minute, the program is not working and it's time to troubleshoot.) As it turns out, this run takes about 4 hours on a modern desktop computer. Realistic LAMARC runs take hours, if not days. You may want to experiment with a small subset of your data first.

19.8.3 Analyzing the output

The output is by default written to a file named "outfile.txt". Table 19.2 shows a selected section of the output. Note that your results might be slightly different because of the stochastic nature of the algorithm. The two most important parts in

Table 19.2 LAMARC maximum likelihood estimates and confidence intervals

Population Best Val (MLE) Percentile	Theta Theta1 0.060816	GrowthRate Growth1 246.0408
99%	0.005	0.033104	21.63394
95%	0.025	0.037098	71.87632
90%	0.050	0.039484	97.41261
75%	0.125	0.043920	137.8771
50%	0.250	0.049423	178.5456
	MLE	0.060816	246.0408
50%	0.750	0.075975	334.9648
75%	0.875	0.088914	454.1918
90%	0.950	0.105897	591.3140
95%	0.975	0.117563	640.8869
99%	0.995	0.140588	720.0954

Theta1: Theta for Nuu-Chah-Nulth
Growth1: Growth for Nuu-Chah-Nulth

the output are the very first table, showing the estimates of Θ and g with their error bars, and the progress reports at the very end. These progress reports are your best indication of whether the run was adequate to produce reliable answers.

Table 19.2 shows that the maximum likelihood estimate (MLE) of Θ was 0.060816 and of g was 246.0408. The table also indicates approximate confidence intervals around these estimates. Note that, despite the large positive estimate of g, the confidence interval is relatively large. In order to interpret these results, it would be very helpful to have an external estimate of the mitochondrial D loop mutation rate. Siguroardottir *et al.* (2000) provide a rate of 0.32/site/million years. This needs to be converted to substitutions/site/generations; if we assume 20 years/generation in humans, this gives a μ of 6.4×10^{-6} substitutions/site/generation. In mtDNA, $\Theta = N_e\mu$, so our estimate of N_e is around 10 000. Knowledge of the relationship between N_e and N would be needed to convert this to census size. Note that this estimate is affected by any uncertainty in the mutation rate and ***generation time*** estimates.

This number is overly large for the Nuu-Chah-Nulth tribe, probably because LAMARC is sampling back into the tribe's shared ancestry with other American and Eurasian peoples. These data violate the steady-state assumptions of LAMARC in two ways: the assumption that there has been no migration (unless it is explicitly modeled) and that the population growth regime has been stable. The answer should therefore be interpreted cautiously.

The growth rate estimate is positive, indicating that this population has been growing. We can ask questions such as "What would the population size have been 700 generations ago, at the end of the last Ice Age?" $\Theta_t = \Theta_0 \exp(-gt)$, but the time in this equation is scaled in terms of expected substitutions per site. We must rescale (using the mutation rate given above) into generations before solving for N_t. The answer is about 3156, so our best estimate is that the population size has tripled since the Ice Age. However, using the upper or lower confidence interval boundaries in this calculation shows us that we have very poor certainty. A single locus generally gives a poor estimate of the growth rate. To refine this one, we would need nuclear DNA sequences – not additional mtDNA, as the entire mitochondrial genome is completely linked and gives little additional information. Even in cases where we don't have the external information to estimate N_e, the value of Θ can be compared to other species or populations with a similar μ; we can observe, for example, that chimpanzee populations typically have a Θ much larger than this, suggesting bottlenecks in the human lineage.

The next section of the output file shows *profile likelihoods*, indicating how each parameter's best values vary as the other parameter is systematically varied. They show us a strong correlation between Θ and g. High values of Θ (modern-day population size) require higher values of g to adequately explain the observed data.

At the end of the output file, a copy of the progress reports is included that also appeared during the run. They can contain some warnings, but probably only in the early chains for this data. These generally come from use of a starting value far from the truth (in this case, the starting value of $g = 1$ was far from the eventual $g = 246$) and are usually not cause for concern. In assessing the progress reports, there are five things to be noted:

"Accepted" – How often did the genealogy sampler find a genealogy it could accept? If this number drops below about 5%, the program will have to run for a very long time in order to do an adequate search. If it climbs very high, 90% or more, there may be an error in the data (such as failure to align it), causing all trees to be equally bad. If the *acceptance rate* is very low, "heating" should be considered in the next run to improve the program's ability to search. In this case, the acceptance rate is around 10%–13%, which is quite satisfactory.

"Posterior lnL" – How well did the genealogies fit their final estimate of the parameters, compared to how well they fit their initial estimate for that chain? This number should always be positive, but not very large – a rule of thumb is that it should not be greater than twice the number of parameters, at least in the final chains. We have two parameters (Θ and g), so we should be unhappy to see a posterior lnL greater than 4. Already after the first initial chain, the posterior lnL decreases to values much lower than this, which is good. Note that this number is the ratio of two likelihoods, and cannot be used in a *likelihood ratio test*, nor

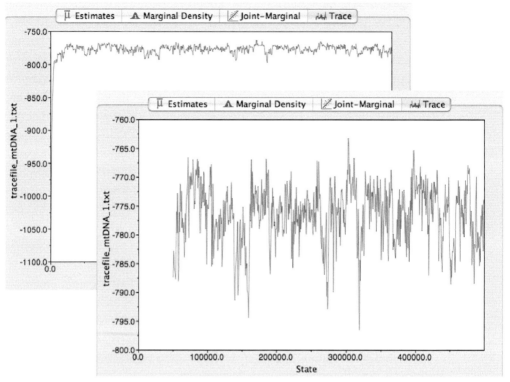

Fig. 19.2
Trace plots in TRACER. Whereas the plot in the back has no samples discarded, a burn-in of 10% has been specified for the plot in the front.

compared meaningfully among runs. It is supplied only because very high values are a strong indication that the run is too short.

"Data lnL" – This is the log-likelihood of the data on the last tree of each chain. It should go up at first, and then fluctuate around a value. If it is still going up at the end of the run, better and better trees are still being found and the run was too short. Note that the magnitude of this number depends mainly on the data set; a data set that generates a hugely negative data log-likelihood is actually a good, information-rich data set. In this run, the best data log-likelihood was found in the second of the two final chains, which is later than one would like it to be. This indicates that a longer run is advisable. Again, this can slightly vary among runs because of the stochastic nature of MCMC. The behavior of the log-likelihood can be investigated in more detail using a *trace plot*. LAMARC will write the log-likelihood for every sample in the initial and final chains to a file for which the name also refers to the region; in our case this is "tracefile_mtDNA_1.txt". This file can be read using the program TRACER (see previous chapter on how to use this application). Figure 19.2 shows the trace plots that can be visualized using TRACER;

the plot in the back is without ***burn-in***, the plot in the front has the first 10% of the samples discarded. The full sample clearly shows that the log-likelihood steeply increases to values around -775 and then indicates **stationary** behavior. With a burn-in of 10%, the trace plot indicates that there is a relatively large degree of autocorrelation. In fact, the ***effective sample size*** (ESS) in this example is only 91.6 and flagged in red by TRACER. Runs with ESS <200 have not obtained a reasonable sample, which again indicates that a longer run for this data set is required.

"`Trees discarded`" – LAMARC sets internal limits on how many migrations or recombinations a tree can contain, and how extreme branch lengths in a tree can become. Trees are discarded if they reach the limits. A few discarded trees are no problem, but high numbers here suggest that LAMARC is bashing against its limits and its estimates will be questionable. This run has none.

"`Estimated parameter values`" – After each chain LAMARC estimates the values of the parameters. These may change directionally at first, but if they are still systematically increasing or decreasing by the end of the run, the run is too short. If they are bouncing hugely, the individual chains are too short. This run is satisfactory.

In conclusion, this run is a bit short, and for publication a fivefold increase – leaving the program to run overnight – would be a good solution. Estimating further parameters – recombination, or migration in a multiple-population case – would need to run much longer. It's best to explore the program with a small data set initially, and then use the large one for final production runs.

19.9 Conclusions

With LAMARC programs, as with phylogeny estimation programs, it is important to know when not to use them. Results from inadequate or inappropriate data will have no meaning, and although they will look impressive, they are potentially misleading. In order to assess the appropriateness of LAMARC analysis, users should consider the following questions:

- Is the population really likely to be in a stable state?
- Are forces other than the ones analyzed likely to affect the population?
- Is the data set large enough?
- Were the individuals chosen at random?
- Are the results stable from chain to chain? From run to run?
- What is the uncertainty of auxiliary values used (such as external estimates of the mutation rate)?
- How wide are the confidence intervals of the estimates?

People often balk at the slowness of modern phylogenetic and population-genetic estimation programs, but considering that it could take several years to collect samples, it is worthwhile to spend a few weeks making a thoughtful and careful analysis. Slapping together an analysis can greatly reduce the value of the results. There are currently no methods that can be applied mechanically which will guarantee good results. A degree of good judgment is always essential!

Section VIII

Additional topics

Assessing substitution saturation
with DAMBE

THEORY

Xuhua Xia

20.1 The problem of substitution saturation

The accuracy of phylogenetic reconstruction depends mainly on (1) the sequence quality, (2) the correct identification of **homologous** sites by sequence alignment, (3) regularity of the substitution processes, e.g. **stationarity** along different lineages, absence of **heterotachy** and little variation in the substitution rate over sites, (4) **consistency, efficiency** and little **bias** in the estimation method, e.g. not plagued by the **long-branch attraction** problem, and (5) sequence divergence, i.e. neither too conserved as to contain few substitutions nor too diverged as to experience substantial **substitution saturation**. This chapter deals with assessing substitution saturation with software DAMBE, which is a Windows program featuring a variety of analytical methods for molecular sequence analysis (Xia, 2001; Xia & Xie, 2001).

Substitution saturation decreases phylogenetic information contained in sequences, and has plagued the phylogenetic analysis involving deep branches, such as major arthropod groups (Lopez et al., 1999; Philippe & Forterre, 1999; Xia et al., 2003). In the extreme case when sequences have experienced full substitution saturation, the similarity between the sequences will depend entirely on the similarity in nucleotide frequencies (Lockhart et al., 1992; Steel et al., 1993; Xia, 2001, pp. 49–58; Xia et al., 2003), which often does not reflect phylogenetic relationships.

The Phylogenetic Handbook: a Practical Approach to Phylogenetic Analysis and Hypothesis Testing, Philippe Lemey, Marco Salemi, and Anne-Mieke Vandamme (eds.). Published by Cambridge University Press. © Cambridge University Press 2009.

Other than simple suggestions such as avoiding sequences with many pairwise JC69 (Jukes & Cantor, 1969) distances larger than 1 (Nei & Kumar, 2000, p. 112) and plotting *transitions* and *transversions* against a corrected *genetic distance*, as implemented in DAMBE (Xia, 2001; Xia & Xie, 2001)(see also practice in Chapter 4), there are currently five main approaches for testing whether molecular sequences contain phylogenetic information. The first approach involves the randomization or permutation tests (Archie, 1989; Faith, 1991). The second employs the standard g_1 statistic for measuring the skewness of tree lengths of alternative trees (Swofford, 1993). These approaches, in addition to having difficulties with sequences with divergent nucleotide frequencies (Steel *et al.*, 1993), suffer from the problem that, as long as we have two closely related species, the tests will lead us to conclude that significant phylogenetic information is present in the data set even if all the other sequences have experienced full substitution saturation. This problem is also shared by the third approach, a tree-independent measure based on relative apparent *synapomorphy*, implemented in the RASA program (Lyons-Weiler *et al.*, 1996). The fourth approach (Steel *et al.*, 1993, 1995) is based on the *parsimony* method, proposed specifically to alleviate the problem of sequence convergence due to similarity in nucleotide frequencies. The convergence would become increasingly serious with increasing substitution saturation. Indeed, sequence similarity will depend entirely on similarity in nucleotide frequencies with full substitution saturation. The fifth is an *information entropy*-based index of substitution saturation (Xia *et al.*, 2003). DAMBE (Xia, 2001; Xia & Xie, 2001) implements the last two approaches.

In what follows, I will (1) outline the method by Steel *et al.* (1993) to highlight its potential problems and its implementation in DAMBE with extensions, and (2) introduce the entropy-based index (Xia *et al.*, 2003) and its implementation in DAMBE with extensions not covered in the original paper. This is followed by a "Practice" section on how to use DAMBE to carry out the assessment of substitution saturation with molecular sequences by using one or both of the implemented methods. For simplicity, I will refer to the two implemented methods as *Steel's method* and *Xia's method*.

20.2 Steel's method: potential problem, limitation, and implementation in DAMBE

Steel *et al.* (1993) presented two statistical tests for testing phylogenetic hypotheses in a maximum parsimony (MP) framework for four species. The first evaluates the relative statistical support of the three possible unrooted topologies. The second

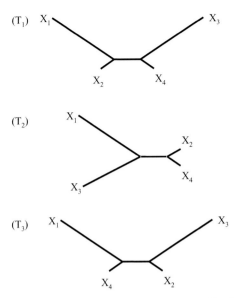

Fig. 20.1   The true topology (T$_1$) and the two possible alternative topologies: T$_2$ and T$_3$.

tests whether the distribution of the phylogenetically informative sites differs from the null model in which sequence variation is generated randomly, i.e. equivalent to sequences having experienced full substitution saturation.

Steel's method has a serious problem involving long-branch attraction mediated by substitution saturation. Long-branch attraction refers to the estimation bias (or tendency) of grouping highly divergent taxa as **sister taxa** even when they are not. While it is historically associated with the MP method (Felsenstein, 1978), other methods, such as distance-based methods with the **minimum evolution** criterion and with a distance index that severely underestimates the true genetic distance between divergent taxa, also suffer from the problem (Felsenstein, 2004).

To illustrate the problem in the MP context, let us focus on the topology in Fig. 20.1, with four species (four nucleotide sequences) designated X$_i$ (i = 1, 2, 3, 4) and three possible unrooted topologies designated T$_i$ (i = 1, 2, 3, Fig. 20.1, with T$_1$ being the true topology). Let X$_{ij}$ be the nucleotide at site j for species X$_i$, and L be the sequence length. For simplicity, assume that nucleotide frequencies are all equal to 0.25. Suppose that the lineages leading to X$_1$ and X$_3$ have experienced full substitution saturation, so that

$$\Pr(X_{1j} = X_{ij,i\neq1}) = \Pr(X_{3j} = X_{ij,i\neq3}) = 0.25 \tag{20.1}$$

The lineages leading to X_2 and X_4 have not experienced substitution saturation and have

$$Pr(X_{2j} = X_{4j}) = P \tag{20.2}$$

where $P > 0.25$ (Note that P approaches 0.25 with increasing substitution saturation). For simplicity, let us set $P = 0.8$, and $L = 1000$.

We now consider the expected number of informative sites, designated n_i ($i = 1, 2, 3$) as in Steel et al. (1993), favoring T_i. Obviously, site j is informative and favors T_1 if it meets the following three conditions: $X_{1j} = X_{2j}$, $X_{3j} = X_{4j}$, $X_{1j} \neq X_{3j}$. Similarly, site j favors T_2 if $X_{1j} = X_{3j}$, $X_{2j} = X_{4j}$, $X_{1j} \neq X_{2j}$. Thus, the expected numbers of informative sites favoring T_1, T_2, and T_3, respectively, are

$$\begin{aligned}
E(n_1) &= Pr(X_{1j} = X_{2j}, X_{3j} = X_{4j}, X_{1j} \neq X_{3j}) L \\
&= 0.25 \times 0.25 \times 0.75 \times 1000 \approx 47 \\
E(n_2) &= Pr(X_{1j} = X_{3j}, X_{2j} = X_{4j}, X_{1j} \neq X_{2j}) L \\
&= 0.25 \times 0.8 \times 0.75 \times 1000 = 150 \\
E(n_3) &= E(n_1) \approx 47
\end{aligned} \tag{20.3}$$

Designating c as the total number of informative sites, we have $c = \Sigma n_i = 244$. Equation (20.3) means that, given the true topology T_1, $P = 0.8$, $L = 1000$, and the condition specified in equations (20.1)–(20.2), we should have, on average, about 47 informative sites favoring T_1 and T_3, but 150 sites supporting T_2. Thus, the wrong tree receives much stronger support (150 informative sites) than the true topology (T_1) and the other alternative topology (T_3). This is one of the many causes for the familiar problem of long-branch-induced attraction (Felsenstein, 1978), popularly known as long-branch attraction, although short-branch attraction would seem more appropriate.

Suppose we actually have such sequences and observed $n_1 = n_3 = 47$, and $n_2 = 150$. What would Steel's method tell us? Steel et al. (1993) did not take into account the problem of long-branch attraction. They reasoned that, if the sequences are randomly generated, then some sites will be informative by chance and it is consequently important to assess whether the number of informative sites supporting a particular topology exceeds the number expected by chance. They designate the number of informative sites favoring T_i by chance alone as N_i ($i = 1, 2, 3$). The probability that N_i is at least as large as n_i, according to Steel et al. (1993), is

$$Pr(N_i \geq n_i) = \sum_{k \geq n_i}^{c} \binom{c}{k} s_i^k (1 - s_i)^{c-k} \tag{20.4}$$

where s_i is defined in Steel *et al.* (1993) as the expected proportion of informative sites supporting T_i by chance. Because the nucleotide frequencies are all equal to 0.25 in our fictitious case, $s_i = 1/3$, so that

$$\Pr(N_2 \geq n_2) = \sum_{k \geq 150}^{244} \binom{244}{k} \left(\frac{1}{3}\right)^k \left(\frac{2}{3}\right)^{244-k} \approx 0$$

$$\Pr(N_1 \geq n_1) = \Pr(N_3 \geq n_3) = \sum_{k \geq 47}^{244} \binom{244}{k} \left(\frac{1}{3}\right)^k \left(\frac{2}{3}\right)^{244-k} \approx 1 \qquad (20.5)$$

These equations mean that, by chance alone, it is quite likely to have $N_1 \geq 47$ and $N_3 \geq 47$. That is, there is little statistical support for T_1 and T_3. However, there is significant support for T_2 since it is quite unlikely to have $N_2 \geq 150$ by chance alone. The same conclusion is reached by using the normalized values of n_i as recommended in Steel *et al.* (1993). So T_2, which is a wrong topology, is strongly supported by Steel's method.

In addition to the test above for evaluating the relative support of alternative topologies, Steel *et al.* (1993) also presented a statistic for testing whether the distribution of the informative sites differs from what is expected from random sequences (the null model):

$$X^2 = \sum_{i=1}^{3} \frac{(n_i - \mu_i)^2}{\mu_i} \qquad (20.6)$$

where $\mu_i = cs_i = c/3 \approx 81.3$ and X^2 follows approximately the χ^2 distribution with two degrees of freedom. In our case,

$$X^2 = \sum_{i=1}^{3} \frac{(n_i - \mu_i)^2}{\mu_i} \approx \frac{2(47 - 81.3)^2}{81.3} + \frac{(150 - 81.3)^2}{81.3} \approx 87.17 \qquad (20.7)$$

With two degrees of freedom, the null model is conclusively rejected ($p = 0.0000$). In applying the tests, one typically would test the null model first by the χ^2-test above to see if there is any phylogenetic information left in the sequences. If the null model is rejected, then one proceeds to evaluate the relative statistical support for the three alternative topologies. In our case, we would conclude that there is a very strong phylogenetic signal in the sequences and, after the next step of evaluating the statistical support of the three alternative topologies, reject T_1 and T_3, and adopt T_2. This seemingly flawless protocol would lead us to confidently reject the true tree (T_1) and adopt the wrong tree (T_2).

Steel's method is limited to four OTUs, and its extension to more than four species (Steel *et al.*, 1995) is not well described for efficient computer implementation. One way to circumvent the problem is to take a heuristic approach by sampling all possible combinations of four OTUs (quartets), performing Steel's test by calculating X^2, and checking which species are most frequently involved in tests that fail to reject the null hypothesis of no phylogenetic information. For example, with five OTUs, there are five possible combinations of four OTUs, i.e. $\{1, 2, 3, 4\}$, $\{1, 2, 3, 5\}$, $\{1, 2, 4, 5\}$, $\{1, 3, 4, 5\}$, and $\{2, 3, 4, 5\}$. In general, given N OTUs, the number of possible quartets are

$$N_{quartet} = \frac{N(N-1)(N-2)(N-3)}{4 \times 3 \times 2 \times 1} \tag{20.8}$$

For each quartet, we apply Steel's method to perform the χ^2-test. If the null hypothesis is rejected in all five tests, then we may conclude that the sequences have experienced little substitution saturation. On the other hand, if some tests fail to reject the null hypothesis, then one can check the OTU combination in the quartet and which OTU is involved most often in such cases. An OTU that is involved in a large number of tests that fail to reject the null hypothesis (designated as $N_{insignificant}$) may be intuitively interpreted as one with little phylogenetic information useful for resolving the phylogenetic relationships among the ingroup OTUs. $N_{insignificant}$ does not include tests with $c \leq 15$ because, with a small c, the failure to reject the null hypothesis is not due to substitution saturation but is instead due to lack of sequence variation. However, it is important to keep in mind that such intuitive interpretation may be misleading given the long-branch attraction problem outlined above.

DAMBE generates two indices to rank the OTUs. The first is $N_{insignificant}$. The second is

$$\phi = \sqrt{\frac{\chi^2}{c}} \tag{20.9}$$

which is often used in contingency table analysis as a measure of the strength of association. The value of ϕ ranges from 0 to 1 in contingency table analysis, but can be larger than 1 in a goodness-of-fit test when χ^2 is calculated according to (20.6). However, the scaling with c renders ϕ relatively independent of the number of informative sites and consequently more appropriate for inter-quartet comparisons. With five OTUs, each OTU is involved in four quartets and associated with four ϕ values. The mean of the four φ values for an OTU should be correlated with phylogenetic information of the OTU.

The interpretation of both $N_{insignificant}$ and ϕ are problematic with the long-branch attraction problem mentioned above. For illustration, suppose we have four sequences that have experienced substitution saturation (designated as group 1 sequences) and four sequences that are conserved with few substitutions among them (designated as group 2 sequences). Define a bad quartet as the combination of two group 1 sequences and two group 2 sequences, i.e. where long-branch attraction will happen. Such a bad quartet will always generate a large χ^2 and ϕ.

The total number of bad quartets is 36 out of a total of 70 possible quartets in this fictitious example. This means that group 1 sequences will often be involved in tests rejecting the null hypothesis with a large χ^2 and ϕ and be interpreted as containing significant phylogenetic information according to the two indices. On the other hand, if there are eight group 1 sequences and eight group 2 sequences, then a large number of quartets will be made of group 1 sequences only to allow substitution saturation among group 1 sequences to be revealed. Based on my own unpublished computer simulation, the indices are useful when there are more group 1 sequences than group 2 sequences or when there are at least four group 1 sequences.

20.3 Xia's method: its problem, limitation, and implementation in Dambe

Xia's method (Xia *et al.*, 2003) is based on the concept of entropy in information theory. For a set of N aligned sequences of length L, the entropy at site i is

$$H_i = -\left(\sum_{j=1}^{4} p_j \log_2 p_j \right) \tag{20.10}$$

where $j = 1, 2, 3, 4$ corresponding to nucleotide A, C, G, and T, and p_j is the proportion of nucleotide j at site i. The maximum value of H_i is 2 when nucleotide frequencies at each site are represented equally. The mean and variance of H for all L sites are simply

$$\overline{H} = \frac{\sum_{i=1}^{L} H_i}{L}, \quad Var(H) = \frac{\sum_{i=1}^{L} (H_i - \overline{H})^2}{L-1} \tag{20.11}$$

When sequences have experienced full substitution saturation, then the expected nucleotide frequencies at each nucleotide site are equal to the global frequencies P_A, P_C, P_G, and P_T. The distribution of the nucleotide frequencies at each site then follows the multinomial distribution of $(P_A + P_C + P_G + P_T)^N$, with the expected entropy and its variance expressed as follows:

$$H_{FSS} = -\left(\sum_{N_A=0}^{N} \sum_{N_C=0}^{N} \sum_{N_G=0}^{N} \sum_{N_T=0}^{N} \frac{N!}{N_A! N_C! N_G! N_T!} P_A^{N_A} P_C^{N_C} P_G^{N_G} P_T^{N_T} \sum_{j=1}^{4} p_j \log_2 p_j \right) \tag{20.12}$$

$$Var(H_{FSS}) =$$
$$\sum_{N_A=0}^{N} \sum_{N_C=0}^{N} \sum_{N_G=0}^{N} \sum_{N_T=0}^{N} \frac{N!}{N_A! N_C! N_G! N_T!} P_A^{N_A} P_C^{N_C} P_G^{N_G} P_T^{N_T} \left(\sum_{j=1}^{4} p_j \log_2 p_j - H_{FSS} \right)^2 \tag{20.13}$$

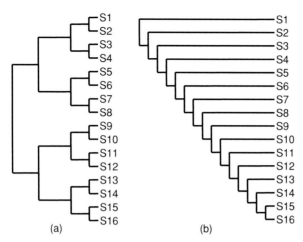

Fig. 20.2   Two extreme topologies used in simulation. (a) – symmetrical, (b) – asymmetrical.

where N_A, N_C, N_G, and N_T are smaller or equal to N and subject to the constraint of $N = N_A + N_C + N_G + N_T$, $j = 1, 2, 3$, and 4 corresponding to A, C, G, and T, and $p_j = N_j/N$. The subscript FSS in H_{FSS} stands for full substitution saturation.

Theoretically, the test of substitution saturation can be done by simply testing whether the observed $\overline{H}$ value in (20.11) is significantly smaller than H_{FSS}. If $\overline{H}$ is not significantly smaller than H_{FSS}, then the sequences have experienced severe substitution saturation. This leads to a simple index of substitution saturation and is defined as

$$I_{ss} = \overline{H}/H_{FSS} \tag{20.14}$$

We can see intuitively that the sequences must have experienced severe substitution saturation when I_{ss} approaches 1, i.e. when $\overline{H}$ equals H_{FSS}. However, the test of $\overline{H} = H_{FSS}$ is only theoretically useful because the sequences often will fail to recover the true phylogeny long before the full substitution saturation is reached, i.e. long before I_{ss} reaches 1. For this reason, we need to find the critical I_{ss} value (referred to hereafter as $I_{ss.c}$) at which the sequences will begin to fail to recover the true tree. Once $I_{ss.c}$ is known for a set of sequences, then we can simply calculate the I_{ss} value from the sequences and compare it against the $I_{ss.c}$. If I_{ss} is not smaller than $I_{ss.c}$, then we can conclude that the sequences have experienced severe substitution saturation and should not be used for phylogenetic reconstruction.

Computer simulation (Xia *et al.*, 2003) suggests that $I_{ss.c}$ depends on N, L, and the topology, with larger $I_{ss.c}$ associated with more symmetrical topologies (Fig. 20.2). The ability of phylogenetic methods in recovering the true tree decreases with the degree of substitution saturation, but the effect of substitution saturation is alleviated by increasing L. The relation between P_{true} (the probability of the true tree being recovered) and the tree length (TL) appears to be sufficiently described by the following equation:

$$P_{true} = 1 - e^{-e^{(a-TL)/b}} \qquad (20.15)$$

which was graphically plotted for various combinations of N and L and for symmetric and asymmetric topologies (Xia *et al.*, 2003). The last term in (20.15), with an exponential of an exponential, is a special form of the ***extreme value distribution (EVD)*** or Gumbel distribution (Gumbel, 1958). For the symmetrical topology, the fit of the equation to the data is almost perfect in all cases, with r^2 values greater than 0.965.

Defining the critical tree length (TL_c) as the tree length when $P_{true} = 0.95$, $I_{ss.c}$ is naturally the I_{ss} value corresponding to TL_c. When an observed I_{ss} value is significantly smaller than $I_{ss.c}$, we are confident that substitution saturation is not serious. This will be illustrated later with the elongation factor-1α sequences.

The computer simulation in Xia *et al.* (2003) is limited to $N \leq 32$. Because the $I_{ss.c}$ is based on simulation result, there is a problem with more than 32 species. To circumvent this problem, DAMBE will randomly sample subsets of 4, 8, 16, and 32 OTUs multiple times and perform the test for each subset to see if substitution saturation exists for these subsets of sequences.

PRACTICE

Xuhua Xia and Philippe Lemey

The results in this section are generated with DAMBE version 4.5.56, which better reflects the content in this chapter than previous versions. The new version of DAMBE can be found at *http://dambe.bio.uottawa.ca/dambe.asp.* The installation of DAMBE involves only a few mouse clicks.

Three sets of sequences will be used for practice: (1) 8 aligned cytochrome oxidase subunit I (COI) sequences from vertebrate mitochondrial genomes in the VertebrateMtCOI.FAS file, (2) 16 aligned EF-1α sequences (Xia *et al.*, 2003) from major arthropod groups and putative outgroups in the InvertebrateEF1a.FAS file, and (3) 41 aligned simian immunodeficiency virus (SIV) genomes, restricted to a single, non-overlapping reading frame for the coding genes that could be unambiguously aligned, in the SIV.fas file (Paraskevis *et al.*, 2003). These files come with DAMBE installation and can be found at the DAMBE installation directory (C:\Program Files\DAMBE by default) or they can be downloaded from *www.thephylogenetichandbook.org.*

Start DAMBE and Click "Tools|Options" to set the default input and output directories to the directory where you have downloaded and saved these files. Set the default input file format to the FASTA format (DAMBE can read and convert sequence files in almost all commonly used sequence formats).

20.4 Working with the VertebrateMtCOI.FAS file

The VertebrateMtCOI.FAS file contains the mitochondrial COI sequences from *Masturus lanceolatus* (sunfish), *Homo sapiens* (human), *Bos taurus* (cow), *Balaenoptera musculus* (blue whale), *Pongo pygmaeus* (Bornean orangutan), *Pan troglodytes* (chimpanzee), *Gallus gallus* (chicken), and *Alligator mississippiensis* (American alligator). The third codon position will be analyzed separately from the other two codon positions.

Protein-coding genes consist of codons, in which the third codon position is the most variable, and the second the most conserved (Xia *et al.*, 1996; Xia, 1998). The third codon position is often not excluded from the analysis, mainly for two reasons. First, excluding the third codon position would often leave us with few substitutions to work on. Second, substitutions at the third codon position should conform better to the neutral theory of molecular evolution than those at the other two codon positions. Consequently, the former may lead to better phylogenetic estimation than the latter, especially in estimating divergence time (Yang, 1996).

However, these two potential benefits of using substitutions at the third codon position may be entirely offset if the sites have experienced substitution saturation and consequently contain no phylogenetic information.

(1) Click "File|Open standard sequence file" to open VertebrateMtCOI. FAS. When prompted for sequence type, choose "Protein-coding Nuc. Seq", select "VertMtDNA (Trans_Table = 2)" in the dropdown box (DAMBE has implemented all known genetic codes), and click the "Go!" button. The sequences will be displayed, with identical sites indicated by a "*".

(2) Click "Sequence|Work on codon position 1 and 2". Codon positions 1 and 2 are highly conserved, with many "*'s" below the sequences indicating many identical sites.

(3) Click "Seq.Analysis|Measure substitution saturation|Test by Xia *et al*." A dialog appears where you can specify the proportion of invariant sites (P_{inv}) with the default being 0. P_{inv} is important for sequences with very different substitution rates over sites. For example, the first and second codon positions of functionally important protein-coding genes are often nearly invariant relative to the third codon position. The effect of substitution saturation at highly variable third codon positions may therefore go unrecognized without specifying P_{inv} because, at nearly two thirds of the sites, hardly any substitutions may be observed. In DAMBE, one can estimate P_{inv} by clicking "Seq.Analysis|Substitution rates over sites|Estimate proportion of invariant sites". So, "cancel" the test for the moment, and estimate P_{inv} using this analysis option. By specifying to "Use a new tree", a window appears providing a choice of tree-building algorithm and options. Choose the Neighbor-Joining algorithm, keep the default settings, click "Run" and then "Go!". At the end of the text output, the estimated P_{inv} is shown (P(invariate) = 0.73769). So, go back to the Test by Xia *et al*. and enter "0.74" as proportion of invariant sites. Clicking "Go!" results in the following text output:

```
Part I. For a symmetrical tree.
===========================================
Prop. invar. sites               0.7400
Mean H                           0.5550
Standard Error                   0.0298
Hmax                             1.6642
Iss                              0.3335
Iss.c                            0.7873
T                               15.2523
DF                          261
Prob (Two-tailed)                0.0000
95% Lower Limit                  0.2749
95% Upper Limit                  0.3920
```

```
Part II. For an extreme asymmetrical (and generally very
unlikely) tree.
===============================================
Iss.c                              0.6817
T                                  11.7056
DF                                 261
Prob (Two-tailed)                  0.0000

95% Lower Limit                    0.2749
95% Upper Limit                    0.3920
```

In this example, we obtain $I_{ss} = 0.3335$, much smaller than $I_{ss.c}$ ($= 0.7873$ assuming a symmetrical topology and 0.6817 assuming an asymmetrical topology). The sequences obviously have experienced little substitution saturation.

(4) We will build a tree to serve as a reference against the tree built with the third codon position. Click "Phylogenetics|Maximum likelihood|Nucleotide sequence|DNAML" and have a look at the options that you can specify. Click the "Run" button and you will see the tree topology shown in Fig. 20.3. You may use distance-based methods such as the **neighbor-joining** (Saitou & Nei, 1987), *Fitch–Margoliash* (Fitch & Margoliash, 1967) or *FastME* method (Desper & Gascuel, 2002; Desper & Gascuel, 2004) implemented in DAMBE and generate exactly the same topology with virtually any genetic distances. To obtain a distance-based tree from aligned nucleotide sequences with DAMBE, click "Phylogenetics|distance method|nucleotide sequence", optionaly set of the options, and click the "Run" button.

(5) Click "Sequence|Restore sequences" to restore the sequences to its original form with all three codon positions (or just re-open the file). Click "Sequence|Work on codon position 3". The 3rd codon positions in vertebrate mitochondrial genes evolve extremely fast (Xia *et al.*, 1996).

(6) Click "Seq.Analysis|Measure substitution saturation|Test by Xia et al.". For the 3rd codon position, P_{inv} can be left at its default value since

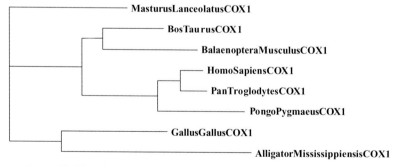

Fig. 20.3   Maximum likelihood tree from DNAML based on the first and second codon positions of the vertebrate mitochondrial COI sequences.

there are very few invariable sites. The resulting $I_{ss} = 0.7145$ is only marginally smaller than $I_{ss.c}$ $(= 0.7518)$ assuming a symmetrical topology and substantially larger than $I_{ss.c}$ $(= 0.6401)$ assuming an asymmetrical topology. This means that the sequences may still be useful if the true topology is not very asymmetrical. To verify this, Click "Phylogenetics|Maximum likelihood|Nucleotide sequence|DNAML" and click the "Run" button. The resulting tree has exactly the same topology as in Fig. 20.3. Note that, for this set of sequences consisting of the 3rd codon positions only, genetic distances based on substitution models more complicated than K80 (Kimura, 1980) cannot be calculated for all pairs of OTUs. For example, TN93 distance (Tamura & Nei, 1993) cannot be computed if $Q/(2\pi_Y) + P_1\pi_Y/(2\pi_T\pi_C) \geq 1$, or $Q/(2\pi_R\pi_Y) \geq 1$, or $Q/(2\pi_R) + P_2\pi_R/(2\pi_A\pi_G) \geq 1$, where Q, P_1 and P_2 are proportions of transversions, T↔C transitions and A↔G transitions, respectively and π_Y, π_R, π_T, π_C, π_A, and π_G are the frequencies of pyrimidines, purines, T, C, A, and G, respectively. This highlights one of the limitations for distance-based methods involving highly diverged sequences. For the ***p-distance***, the Poisson-corrected p-distance and K80-distance that can still be calculated with this set of sequences, only the **UPGMA** method, but not the neighbor-joining (Saitou & Nei, 1987), Fitch–Margoliash (Fitch & Margoliash, 1967) or FastME method (Desper & Gascuel 2002; Desper & Gascuel 2004), will recover the topology in Fig. 20.3. Interestingly, the maximum parsimony tree from this set of sequences is the same as that in Fig. 20.3. It is therefore important to keep in mind that establishing the existence of phylogenetic information in the sequences does not mean that the true tree can be recovered by any tree-building algorithm, and that we still know poorly as to which method is better than others.

(7) To perform a test using Steel's method, click "Seq.Analysis|Measure substitution saturation|Test by Steel *et al.*" and click the "OK" button. The output is in three parts. The first is the nucleotide frequencies. The second is the output for individual tests of each quartet. The third part shows which OTU might be problematic:

```
Sequences ranked from the best to the worst.
==============================================
Seq_Name                    Mean_Phi   Num_Insignif
----------------------------------------------------
PanTroglodytesCOX1           0,1584       10
HomoSapiensCOX1              0,1495       12
PongoPygmaeusCOX1            0,1262       12
GallusGallusCOX1             0,1131       20
AlligatorMississippiensis    0,1124       20
BosTaurusCOX1                0,1096       20
BalaenopteraMusculusCOX1     0,1085       20
MasturusLanceolatusCOX1      0,0897       22
==============================================
Num_Insignif conditional on c > 15.
```

The output for the third codon positions only shows that *Masturus lanceolatus* (sunfish) has the smallest mean ϕ value and is involved in the largest number of tests that fail to reject the null hypothesis of no phylogenetic information, indicating that it might be too divergent from the rest of the OTUs. One may be interested to learn that the JC69 distances between *M. lanceolatus* and other OTUs are all greater than 1. This reminds us of the suggestion (Nei & Kumar, 2000, p. 112) to avoid such sequences.

20.5 Working with the InvertebrateEF1a.FAS file

The elongation factor-1α (EF-1α) is one of the most abundant proteins in eukaryotes (Lenstra *et al.*, 1986) and catalyzes the GTP-dependent bindings of charged tRNAs to the ribosomal acceptor site (Graessmann *et al.*, 1992). Because of its fundamental importance for cell metabolism in eukaryotic cells, the gene coding for the protein is evolutionarily conserved (Walldorf & Hovemann, 1990), and consequently has been used frequently in resolving deep-branching phylogenies (Cho *et al.*, 1995; Baldauf *et al.*, 1996; Regier & Shultz, 1997; Friedlander *et al.*, 1998; Lopez *et al.*, 1999).

The InvertebrateEF1a.FAS file contains the EF-1α from four chelicerate species (CheU90045, CheU90052, CheU90047, CheU90048), four myriapod species (MyrU90055, MyrU90053, MyrU90057, MyrU90049), two branchiopod species (BraASEF1A, BraU90058), two hexapod species (HexU90054, HexU90059), one molluscan species (MolU90062), one annelid species (AnnU90063) and two malacostracan species (MalU90046, MalU90050). The phylogenetic relationship among major arthropod taxa remains controversial (Regier & Shultz 1997).

(1) Click "File|Open standard sequence file" to open InvertebrateEF1a.FAS as before, choose the default "standard" genetic code and click the "Go!" button.
(2) Click "Sequence|Work on codon position 3".
(3) Click "Seq.Analysis|Measure substitution saturation|Test by Xia et al." P_{inv} can be left at its default value. The resulting $I_{ss} = 0.6636$, not significantly ($p = 0.1300$) smaller than $I_{ss.c}$ ($= 0.7026$) assuming a symmetrical topology and substantially larger than $I_{ss.c}$ ($= 0.4890$) assuming an asymmetrical topology. This means that the sequences consisting of 3rd codon positions only have experienced so much substitution saturation that they are no longer useful in phylogenetic reconstruction. To verify this, click "Phylogenetics|Maximum likelihood|Nucleotide sequence|DNAML" and click the "Run" button. The resulting tree, with no consistent clustering of EF1-α from the same species, is absurd and totally different from the tree one would obtain by using the 1st and 2nd codon positions.

(4) To apply Steel's method to the analysis of the 3rd codon positions, click "Seq.Analysis|Measure substitution saturation|Test by Steel et al" and click the "OK" button. The last part of the output shows the mean ϕ values ranging from 0.028 to 0.0376, in dramatic contrast to the mean ϕ values for the 3rd codon position of the mitochondrial COI gene (between 0.0897 and 0.1584). An OTU with a mean ϕ value smaller than 0.04 may be taken as lacking phylogenetic information based on computer simulation. The mean ϕ values range from 0.0774 to 0.1061 when Steel's method is applied to the 1st and 2nd codon positions of the EF-1α sequences, but range from 0.2296 to 0.3133 when applied to the 1st and 2nd codon positions of the vertebrate mitochondrial COI gene. In short, all indications suggest that the set of invertebrate EF-1α sequences have experienced much greater substitution saturation than the set of vertebrate mitochondrial COI sequences.

20.6 Working with the SIV.FAS file

The test with Xia's method involving more than 32 OTUs is different from those with 32 or fewer OTUs, and is illustrated with this set of 41 SIV sequences obtained from various African primates.

(1) Click "File|Open standard sequence file" to open SIV.FAS as before. Since this file has unresolved bases, a window pops up asking you how to deal with them. DAMBE presents three options for dealing with ambiguous codes. The first is to explicitly mark them as unresolved. The second will treat them in a probabilistic manner depending on what computation is involved. Take R (coding for either A or G) for example: if 80% of the purines in the input sequences are As, then an R is counted as 0.8 A and 0.2 G in computing frequencies. In computing nucleotide substitutions, such an R facing a homologous A on another sequence will be treated as identical with a probability of 0.8 and a transition with a probability of 0.2. The final option keeps the ambiguities in the sequences. Choose option 2 by entering "2" and clicking the "Go!" button.

(2) Click "Seq.Analysis|Measure substitution saturation|Test by Xia et al." and set P_{inv} to "0.17" before performing the analysis. The output table shows that the average I_{ss} for subsets of 4, 8, 16, and 32 are significantly smaller than the corresponding $I_{ss.c}$ if the true topology is symmetrical:

NumOTU	Iss	Iss.cSym	T	DF	P Iss.	cAsym	T	DF	P
4	0.573	0.850	35.922	5001	0.0000	0.845	35.283	5001	0.0000
8	0.558	0.847	36.663	5001	0.0000	0.767	26.534	5001	0.0000
16	0.575	0.832	33.524	5001	0.0000	0.680	13.686	5001	0.0000
32	0.576	0.814	31.387	5001	0.0000	0.568	1.058	5001	0.2902

Note: two-tailed tests are used.

While substitution saturation becomes a problem when the true topology is extremely asymmetrical and when the number of OTUs (N) is greater than 16 (e.g. $P = 0.2902$ for $N = 32$), such asymmetrical trees are probably not realistic for these SIV sequences. We can conclude that there is still sufficient phylogenetic information in the complete SIV data set. However, analyzing only the 3rd codon position (keeping the $P_{inv} = 0$) will reveal that the average I_{ss} values are already considerably higher.

Acknowledgment

We thank Stephane Aris-Brosou, Pinchao Ma, and Huiling Xiong for comments, and NSERC Discovery, RTI, and Strategic grants for financial support.

Split networks. A tool for exploring complex evolutionary relationships in molecular data

THEORY

Vincent Moulton and Katharina T. Huber

21.1 Understanding evolutionary relationships through networks

The standard way to represent evolutionary relationships between a given set of taxa is to use a *bifurcating leaf-labeled tree*, in which internal nodes represent hypothetical ancestors and leaves are labeled by present-day species (see Chapter 1). Using such a tree presumes that the underlying evolutionary processes are *bifurcating*. However, in instances where this is not the case, it is questionable whether a bifurcating tree is the best structure to represent phylogenetic relationships. For example, the phenomena of explosive evolutionary radiation, e.g. when an AIDS virus infects a healthy person, might be best modeled not by a bifurcating tree, but by a *multifurcating tree* (see Chapter 1). In addition, it may be necessary to label internal nodes by taxa if ancestors and present-day species co-exist, as has also been observed with fast evolving viruses.

In certain cases, one might want to allow even more general structures than multifurcating trees to represent evolutionary histories. For example, certain viruses/plants/bacteria are known to exhibit **recombination/hybridization/gene transfer**, and this process might not always be best represented by a tree. In particular, a tree implicitly assumes that once two lineages are created they subsequently never interact with one another later on. However, if it is assumed that such

The Phylogenetic Handbook: a Practical Approach to Phylogenetic Analysis and Hypothesis Testing, Philippe Lemey, Marco Salemi, and Anne-Mieke Vandamme (eds.). Published by Cambridge University Press. © Cambridge University Press 2009.

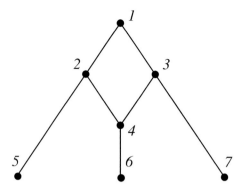

Fig. 21.1   A simple network relating three taxa.

interactions might have occurred, then a simplistic representation of this might look something like the **network** (or *labeled-graph*) presented in Fig. 21.1. In this network, the nodes (or *vertices*) are the dots labeled by *1, 2, . . . , 7*, and the branches (or *edges*) are the lines connecting the nodes. In this example, *1, 2, 3, 4* can be thought of as being potential hypothetical ancestors, and *5, 6, 7* as extant taxa. As with a rooted tree, this structure indicates that all taxa have evolved from *1*, but that at a certain point in time taxa *2* and *3* interacted (or their ancestors), resulting in taxon *4*. The key difference between networks and trees illustrated by this example is that *cycles* are allowed, paths that begin and start at the same node. For example, the nodes labeled *1, 2, 3, 4* form a cycle (of length four).

Recombination/hybridization/gene transfer is sometimes regarded as being quite special and hence it might be reasonable to suppose that networks are only useful in studying certain types of evolution. However, networks can also serve other purposes. Since they do not implicitly assume a tree-like evolutionary process they will not "force" the data onto a tree. Most tree building programs will output a tree for any input even when it is evidently not tree-like. Even when the taxa in question did arise from a tree-like evolutionary process, it may be that **parallel evolution,** model heterogeneity, or sampling errors in the phylogenetic analysis have led to a data set which is no longer best described by a tree.

In this situation, the use of networks can allow the possibility of spotting any deviation from tree-likeness or at least confirm that a tree is still probably the best structure to use. This is analogous to some tree-building programs, which allow one to explore the possibility of using various trees for representing the same data. Consequently, networks can provide a useful complementary technique to building trees, allowing one to explore and better understand the data, as well as derive the sought-after evolutionary histories. Moreover, they provide the possibility to visualize the structure of the data and hence to identify any underlying patterns which a tree representation might miss.

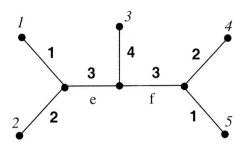

Fig. 21.2   A weighted tree relating five taxa.

At this point, it is natural to ask, "What kinds of networks are there, and how can they be built?" Various networks have been considered, usually with certain special types of data in mind. For example, **reticulation networks** (Legendre & Makarenkov, 2002) and **statistical parsimony networks** (Templeton *et al.*, 1992) have been developed for use within the field of phylogeography. Several reviews have recently appeared concerning networks (Posada & Crandall, 2001; Huber & Moulton, 2005; Morrison, 2005). However, the present chapter will focus on only one kind of network used in evolutionary analysis called *split-networks*, which were originally introduced in Bandelt and Dress (1992b), and can be computed using the program SPLITSTREE (Huson, 1998; Huson & Bryant, 2006).

Split-networks were until quite recently primarily constructed from distances using the mathematics of *split decomposition* theory (Bandelt & Dress, 1992a, b, 1993). In the first part of this chapter we will present an intuitive introduction to split decomposition theory and how it is used to construct networks. Other descriptions of the split decomposition technique for phylogenetic analysis may be found in (Dress *et al.*, 1996; Page & Holmes, 1998; Swofford *et al.*, 1996). We will also briefly describe the recently introduced **NeighborNet** method for computing split-networks (Bryant & Moulton, 2004), and the other types of split-networks called **median networks** (Bandelt, 1995; Bandelt *et al.*, 1995) and **consensus/super networks** (Holland *et al.*, 2004; Huson *et al.*, 2004).

21.2 An introduction to split decomposition theory

The cornerstone of split decomposition theory is the notion of **splits**. Consider the weighted tree in Fig. 21.2, which is labeled by the set of taxa $\{1, 2, 3, 4, 5\}$ with branch lengths denoted in bold. Note that the removal of any branch from this tree results in a partition of the data into two non-empty, disjoint sets, each of which labels one of the two resulting trees. For example, removing the branch *e* results in the partition $A = \{1, 2\}$, $B = \{3, 4, 5\}$. Such a partition is called a split, and it is usually denoted by $S = \{A, B\}$.

Now, notice that any pair of splits given by this tree satisfy a special condition: For example, consider the two splits $\{A = \{1, 2\}, B = \{3, 4, 5\}\}$ and $\{C = \{1, 2, 3\}, D = \{4, 5\}\}$ corresponding to the branches e and f, respectively. Then A and D do not intersect, i.e. $A \cap D = \emptyset$, whereas the intersections $A \cap C$, $B \cap C$, $B \cap D$ are all non-empty. With this example in mind, two splits $S = \{A, B\}$ and $T = \{C, D\}$ of a set X are said to be *compatible* if precisely one of the intersections $A \cap C$, $A \cap D$, $B \cap C$, $B \cap D$ is empty. Otherwise, if all of these intersections are non-empty, then S and T are *incompatible* (convince yourself that these are really the only two possibilities!).

In 1971, Buneman presented a ground-breaking paper in which he proved that a collection of splits corresponds to the branches of a phylogenetic tree, which is not necessarily a strictly bifurcating tree, if and only if every pair of splits in this collection is compatible. Thus, searching for trees to represent the evolutionary relationships between a given set of taxa is, in fact, equivalent to looking for compatible collections of splits of the taxa. However, even for relatively small data sets there are many possible splits. For example, there are 15 possible splits of five taxa and, for a set of n taxa there are $2^{(n-1)} - 1$ splits. Searching for a bifurcating tree is equivalent to trying to find a collection of $2n - 3$ compatible splits amongst all of these splits. For example, when $n = 15$ we are looking for a collection of 27 compatible splits within 16 383 possible; a very computational-intensive task! One possible key to solving this problem is to use clever ways to search for collections of compatible splits that are well supported by the data. Although many solutions have been proposed, we only look in detail at the one provided by Buneman (1971) since it also provides a good warm up for understanding split decomposition.

21.2.1 The Buneman tree

As shown in Chapter 4, there are various ways to estimate the **genetic distance** d on a set of taxa X. Once this has been done, d can be used to select significant splits using the Buneman technique. Given a split $S = \{A, B\}$ of X, some x, y in A and some u, v in B put

$$\beta(xy|uv) = min\,\{d(x, u) + d(y, v), d(x, v) + d(y, u)\}$$
$$- (d(x, y) + d(u, v)) \tag{21.1}$$

and define the **Buneman index** β_S of S as $1/2\ min\ \beta(xy|uv)$ taken over all x, y in A and u, v in B.

For example, in Fig. 21.2 the **phenetic distance** d_T between any pair of taxa $\{1, 2, 3, 4, 5\}$ on the weighted tree T is defined as the sum of the weights of the branches on the path between them (e.g. $d_T(2, 5) = 2 + 3 + 3 + 1 = 9$). Consider the split

$S = \{\{1, 2\}, \{3, 4, 5\}\}$. Then calculating β for all possible pairs of this split gives $\beta(12|34) = 6$, $\beta(12|35) = 6$, and $\beta(12|45) = 12$. Therefore, $1/2\,\beta_S = 6/2 = 3$, i.e. exactly the weight of the branch corresponding to S.

The remarkable fact that Buneman noticed was that, if any distance d is taken on a set of taxa X, then the collection of splits S, for which $\beta_S > 0$ holds, is compatible and hence corresponds to a tree. Thus, considering β_S as being a measure of the support of the split S, Buneman's method tells us which splits to keep (those with positive Buneman index) and which to discard (those with non-positive Buneman index), so that the resulting set of splits corresponds to a tree. This tree, with weight β_S assigned to that branch of the tree corresponding to split S, is called the **Buneman tree** corresponding to the input genetic distance d.

At this point, it is worthwhile briefly discussing some properties that any distance tree-building method should probably satisfy.

(1) The method applied to a phenetic distance d arising from a labeled, weighted tree T should give back the tree T.
(2) The method applied to genetic distances d should depend "continuously" on d, i.e. small changes in d should not result in large changes in the topology of the output tree.
(3) It should be possible to perform the method efficiently (so that the computer can handle it!).
(4) The tree returned by the method should not depend on the order in which the taxa are input.

Although these seem like reasonable properties to demand, some of the well-known distance-based phylogeny methods do not satisfy them. For example, **UPGMA** does not satisfy 1 and **Neighbor-Joining (NJ)** does not always satisfy 2 or 3 (see Chapter 5; Moulton & Steel, 1999 for more details).

The Buneman tree does satisfy all of these demands but, as a result, it is quite conservative and usually elects to discard too many splits. The reason for this is simple: to compute the Buneman index of a split $\{A, B\}$, the *minimum* is taken of the values of $\beta(xy|uv)$ over all x, y in A and u, v in B. Therefore, just one of the $\beta(xy|uv)$ values has to be non-positive in order for $S = \{A, B\}$ to be rejected (as in this case the Buneman index will be non-positive), and for splits $\{A, B\}$ of a large set this is quite likely to happen. One possibility for addressing this problem is to take some kind of average of the $\beta(xy|uv)$'s paying attention, of course, to the fact that a tree should still be obtained. Although this idea can be made to work, leading to refined Buneman trees (Moulton & Steel, 1999), the next section will focus on another possible solution, which is at the heart of the split decomposition method.

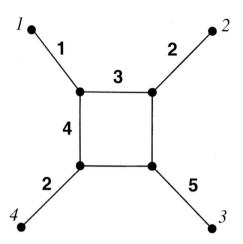

Fig. 21.3   A weighted network relating four taxa.

21.2.2 Split decomposition

The difference between split decomposition and Buneman's approach is a very simple but fundamental alteration to the definition of the index β_S of a split S. Given a set X of taxa, a distance matrix on X, and some split $S = \{A, B\}$ of X, for each x, y in A and u, v in B define the quantity

$$\alpha(xy|uv) = max\,\{d(x, u) + d(y, v), d(x, v) + d(y, u)\}$$
$$- (d(x, y) + d(u, v)). \qquad (21.2)$$

Note that this is almost exactly the same formula as the one given for the quantity $\beta(xy|uv)$ above, except that a *maximum* rather than a *minimum* is taken. Proceeding as with the definition of β_S, the **isolation index** α_S of the split S is defined to be one half the minimum value of $\alpha(xy|uv)$ taken over *all x, y* in A and u, v in B (Bandelt & Dress, 1992a). For example, consider the weighted network presented in Fig. 21.3, labeled by the taxa $\{1, 2, 3, 4\}$ with weights indicated in bold. In this network N, as with trees, the phenetic distance d_N between any pair of the taxa is the sum of the weights of the branches on a shortest path between the taxa. Note that there may be several such shortest paths as opposed to a tree in which there is always precisely one. Thus, for example, $d_N(1, 3) = 1 + 3 + 4 + 5 = 13$. Here there are two possible ways to go between *1* and *3*, either going around the top of the central square or around the bottom. Then $\alpha(14|23) = 6$, so that for the split $S = \{\{1, 4\}, \{2, 3\}\}$, we obtain $\alpha_S = 3$. Similarly, for the split $T = \{\{1, 2\}, \{3, 4\}\}$, we obtain $\alpha_T = 4$. It is no accident that the isolation indices just computed correspond to the weights on the two sets of parallel branches of the square in this network.

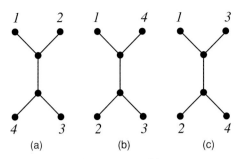

Fig. 21.4   Three possible binary trees of four taxa.

This example illustrates several other important points. First, the removal of any set of parallel branches results in a division of the network into two networks each labeled by one part of the split, and the isolation index determines the weight of these branches. This will be true in general for split networks and mimics the behavior of trees discussed in the previous section. Second, the splits $S = \{\{1, 4\}, \{2, 3\}\}$ and $T = \{\{1, 2\}, \{3, 4\}\}$ are incompatible (see Section 21.2) and hence cannot correspond to two branches of *any* tree. Thus, as opposed to the Buneman index, the collection of splits which have positive isolation index, i.e. those splits S for which $\alpha_S > 0$, is no longer necessarily compatible. It is also worth mentioning here the technique of **spectral analysis** which, rather than choosing any particular collection of splits, instead assigns a significance to every split and proceeds from there (Hendy & Penny, 1993; Lockhart *et al.*, 1995).

Now, consider the example in Fig. 21.3 once more. The isolation index of the split $U = \{\{1, 3\}, \{2, 4\}\}$ is $\alpha_U = 0$. Thus, this split is not one of those in the set $\{1, 2, 3, 4\}$ with a positive isolation index, and consequently this split is discarded. In other words, for the three possible binary tree topologies of the four taxa pictured in Fig. 21.4, the isolation index would keep alternatives (A) and (B), but discard (C). On the other hand, the Buneman index would keep alternative (A) and throw (B) and (C) away. Thus, the network in Fig. 21.3 in some sense represents a combination of the two trees (A) and (B), neither being less well supported than the other.

Buneman showed that the collection of all splits S with β_S positive is compatible; however, even though the set of splits S with positive isolation index α_S is not necessarily compatible, it does satisfy some relaxation of the notion of compatibility. In particular, it is *weakly compatible* (Bandelt & Dress, 1992a, b), i.e. for every *three* splits $S = \{A, B\}$, $T = \{C, D\}$, $U = \{E, F\}$ in this collection at least one of the intersections $A \cap C \cap E$, $A \cap D \cap F$, $B \cap C \cap F$, $B \cap D \cap E$ is empty.

Even though collections of weakly compatible splits are fairly complicated mathematical objects, the most important consequences of weak compatibility can be summarized as follows (Bandelt & Dress, 1992a): if X has n elements, then the

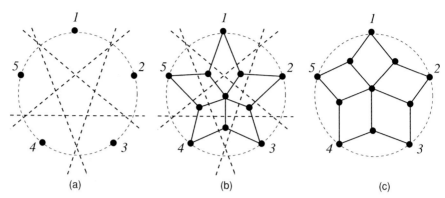

Fig. 21.5   Deriving outer planar split-networks from collections of circular splits.

number of splits with positive isolation index is at most $n(n - 1)/2$, and, as with the Buneman index, these can be computed efficiently. Moreover, the desired properties (1)–(4) given above for tree building methods also hold for the split decomposition method.

21.3 From weakly compatible splits to networks

Having computed a collection of weakly compatible splits, each one with its associated isolation index, it is now necessary to represent it by some weighted network. In general, this can always be achieved using a **median network** (Dress *et al.*, 1997). However, such networks have the problem that they may not, in general, be planar, i.e. it may not be possible to draw them in the plane without crossing branches. This is in contrast to trees, which are always planar. However, if the collection of splits under consideration is *circular*, then it can always be represented by a so-called *outer planar split-network* (Huson & Dress, 2004).

To illustrate these concepts, consider the set $\{1, 2, 3, 4, 5\}$ of taxa arranged on a circle as in Fig. 21.5. The splits of the taxa are represented by lines in the plane, e.g. the split $\{\{1, 2\}, \{3, 4, 5\}\}$ is represented by one of the dotted lines in Fig. 21.5b. In general, a collection of splits is circular if the taxa can be arranged on a circle so that every split in the collection is represented by a line. Note that weakly compatible collections of splits are not necessarily circular (although in practice, this is often the case).

Now, consider the collection of circular splits represented by the lines in Fig. 21.5a. If a node is placed in each of the bounded regions created by the lines, and if two nodes are connected by a branch if they are in regions separated by exactly one line, then the network pictured in Fig. 21.5b is obtained. The split network representing the original collection of splits, which is shown in

Fig. 21.5c, can now be obtained by adjusting the network in Fig. 21.5b so that splits are represented by equal length parallel branches. Note that the removal of any parallel set of branches in the split network results in two networks each labeled by the two parts of the split corresponding to the line passing through these branches.

The technique for constructing the split network used in this example is based on the mathematical principle of *DeBruijn dualization*, and, in general, the program SPLITSTREE computes networks using a related construction (Dress *et al.*, 1996). For non-circular split systems, non-planar networks are usually generated using techniques that we won't go into here.

Finally, note that the split network can be weighted by giving the parallel branches corresponding to a split weights equal to the isolation index of that split. The phenetic distance between taxa in the resulting *weighted network N*, computed using shortest paths, gives a representation d_N of the original distance. In general, the phenetic distance d_N will be an approximation of the input genetic distance d. Hence, in essence the isolation indices are telling us how to factor out as many splits as possible from the distance d to give the representation d_N of d, whilst possibly leaving some *split-prime residue* $(d - d_N)$ which is zero precisely when the distance d is represented exactly by the network (since in this case $d = d_N$) and that is disregarded when this is not the case.

In order to measure how well the distance d_N approximates d in case the network N is computed using the split decomposition method, it is common to compute a simple quantity called the *fit index*, which is defined as $\Sigma\ [d_N\ (x,\ y)\ /\ \Sigma\ d(x,\ y)]^*$ 100%, where both sums are taken over all pairs of elements in X. Then, the fit is 100% precisely when all $d = d_N$. This will be described in more detail below.

21.4 Alternative ways to compute split networks

The split networks that we constructed using split decomposition are in essence the end result of two main steps. First, we use the structure of the data to derive a collection of splits (with weights corresponding to the isolation indices of splits), and second we use these splits to construct the split network, in which parallel branches correspond to splits (and have lengths that are proportional to the split weights). We now describe some additional methods for constructing split networks that employ some variants of these two steps.

21.4.1 NeighborNet

As we have pointed out above, the Buneman method for tree reconstruction is quite conservative since it usually elects to throw out too many splits. The split decomposition method suffers from a similar problem, since the isolation index α_S of the split S is defined to be one-half the *minimum* value of $\alpha(xy|uv)$ taken over *all x, y* in *A* and *u, v* in *B*. As a result, split networks computed for large data sets tend to be highly unresolved.

To try to circumvent this systematic bias in the split decomposition method, in (Bryant & Moulton, 2004), a new approach for computing split networks from genetic distances is described called *NeighborNet*, which can be loosely thought of as a "hybrid" between the NJ and split decomposition methods. Essentially, it works by using the given distances to iteratively construct an ordering for the taxa on the circle. Once the ordering is constructed, *least squares estimates* are then computed for the associated collection of circular splits. This is analogous to the computation of branch lengths for a tree using least squares as described in Chapter 5. Finally, the collection of weighted circular splits is represented by an outer-planar split network as described above.

In general, it has been found that NeighborNet produces more resolved split networks for large data sets than the split decomposition method (for example, see the Salmonella data set described below). However, NeighborNets do tend to contain many small rectangles, which may need filtering out. Moreover, it should be noted that, by their very construction, NeighborNets are planar, and in some cases this does not necessarily allow one to capture the full complexity of the data in question, which may be high dimensional in nature (Bryant & Moulton, 2004). In such circumstances, the split decomposition method can still provide a useful tool for exploring data; as mentioned above, split networks corresponding to collections of weakly compatible splits are not necessarily planar.

21.4.2 Median networks

Median networks are a special class of split networks that are commonly used in the study of intraspecific and population data. They were introduced as a tool for analyzing mitochondrial data (Bandelt *et al.*, 1995), and this is still the main type of molecular data that median networks are used for. Median networks can be constructed directly from *any* collection of splits using a *convex expansion procedure* (Bandelt *et al.*, 1995). In the resulting network, a collection of *n* pairwise *incompatible* splits is represented by an *n*-dimensional hypercube. As a result, median networks can become quite complex, and various methods have been described to reduce this complexity (Bandelt *et al.*, 1995; Huber *et al.*, 2001). For a more detailed review of median networks see Huber and Moulton (2005).

21.4.3 Consensus networks and supernetworks

Quite often phylogenetic methods produce collections of trees as well as an estimate of the best tree (according to some predefined optimality criterion). For example, *maximum parsimony* (Chapter 8), *maximum likelihood* (Chapter 6), **Bayesian** (Chapter 7), and *bootstrapping* (Chapter 5) approaches can all produce collections of trees. For such methods it can be helpful to summarize the information contained in the trees, since collections of trees can be difficult to interpret and draw conclusions from. To do this *consensus trees* can be constructed, and many methods have been devised for their construction. However, all such methods have a common limitation: by summarizing the trees in a single tree, information contained in the trees concerning conflicting information is lost.

In an attempt to circumvent this problem, Holland and Moulton (2003) introduced the concept of a *consensus network* (extending a similar idea proposed in Bandelt (1995)). Essentially, this method generalizes the **majority rule consensus tree** method. For a collection of trees, it chooses those splits that are contained in some fraction of the trees (the majority rule consensus method takes one-half), and then represents them by a split network (in particular, a median network). More details may be found in Holland *et al.* (2005).

More recently, the use of consensus networks has been proposed for simultaneously representing a collection of gene trees (Holland *et al.*, 2004). This can help elucidate species phylogenies in case gene trees conflict (which might occur in case processes such as gene transfer and hybridization have occurred). In this situation, *super networks* (Huson *et al.*, 2004) can also be helpful. These are constructed from collections of partial gene trees, that is, trees that do not necessarily have identical leaf sets.

PRACTICE

Vincent Moulton and Katharina T. Huber

21.5 The SPLITSTREE program

21.5.1 Introduction

SPLITSTREE4 is an interactive and exploratory tool for performing phylogenetic analyses. It incorporates a range of phylogenetic tree and network methods, inference tools, data management utilities, and validation methods. It evolved from SPLITSTREE3, which was basically an implementation of the split decomposition method (Bandelt & Dress, 1992a, b). It is easy to use, portable and flexible. A user can either click his/her way through a split network analysis or control the entire program from a command–line.

21.5.2 Downloading SPLITSTREE

Several versions of the program SPLITSTREE are available. The most current one, SPLITSTREE4, was written in Java and extends all earlier versions (SPLITSTREE1 – 3), which were written in C++. It is formatted for Windows, Mac OS, Linux, and Unix and comes together with a set of examples and a user manual. You can also run a recent version of SPLITSTREE directly from within your web-browser by clicking on the WebStart link. Before installing SPLITSTREE4, it is necessary to install JAVA (Java runtime version 1.4.2 or newer), which is freely available from *http://www.java.org*. To install SPLITSTREE4, go to http://www.splitstree.org, click on the link for the current version and download, you will need to obtain a license key online. For the examples presented in this chapter, we used version 4.8 of SPLITSTREE4 for Mac OS.

21.6 Using SPLITSTREE on the mtDNA data set

We now go through a typical preliminary SPLITSTREE analysis using the mtDNA example (available at *http://www.thephylogenetichandbook.org*). Note that we will only go through some of the available menu items – full details for the remaining items may be found in the user manual.

Note that to use SPLITSTREE, your data needs to be in a variant of the NEXUS format (see Box 8.4 in Chapter 8). The user manual contains all of the details required for converting the input file into NEXUS format.

21.6.1 Getting started

To start, double click on the SPLITSTREE icon. A menu bar labeled SPLITSTREE4 will appear followed by two windows, the SPLITSTREE4 window and a Message window. The SPLITSTREE4 window is the main window; the Message window contains program status information and is also used to print error messages and/or general information about the current data. Go to the File menu and select Open. A new window entitled Open files appears. Select the mtDNA data set file by double clicking on it. In general, if SPLITSTREE cannot read an input NEXUS-file an appropriate error message or warning will appear. The Message window informs you what computations have been performed and completed and a split network appears in the main window.

The default split network produced is a NeighborNet. To produce a split network using split decomposition, click on the menu option Networks and select the item SplitDecomposition. A window appears called the *processing pipeline window*. This allows the user to organize their calculations in a systematic way (see user manual for more details). Click on Apply. This results in the computation of a split decomposition network as pictured in Fig. 21.6. At the bottom of the window is information concerning the input data; for example, how the network is computed (Uncorrected_Pdistances and SplitDecomposition), how it is displayed (equal angle), and the fit index. The three tabs at the top of the main window called Network, Source, and Data allow you to flip between the computed split-network/tree, the NEXUS blocks computed during an analysis, and the source NEXUS file, respectively.

21.6.2 The fit index

We now pause to explain the fit index. For split networks computed using the SplitDecomposition item (and also for trees computed using the Buneman trees item), this index is defined as the sum of all pairwise phenetic distances (i.e. distances in the network) divided by the sum of all pairwise genetic distances (i.e. distances in the input distance matrix), times 100 (see Section 21.3). This is the default setting for SPLITSTREE.

For all other split networks computed by SPLITSTREE from distances, an alternative index called the least squares fit should be computed. Denoting by L the sum of the squares of the difference between the pairwise genetic and phenetic distances, this quantity is $1 - L$ divided by the sum of the squared pairwise genetic distances (Winkworth *et al.*, 2005). To compute the least squares fit index, one has to select Edit in the menu bar and then click on the item Preferences. Within the popped up dialogue box, the tab Status line has to be selected, the item Show LSfit has to be ticked, and the item Show fit has to be unticked. Note that

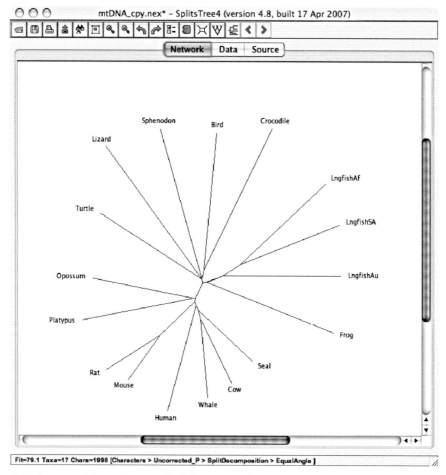

Fig. 21.6   Split decomposition split network of the mtDNA data set.

the change only applies to the data set whose name is given in the header of the dialogue box, but these setting can also be saved as defaults for all other analyses.

Returning to our example, the fit index for the split network based on split decomposition has value 79.1%, which is quite a reasonable fit, i.e. the split network represents about 80% of the original distance, whereas 20% – hopefully including any noise – has been thrown away. Although it is not possible to give precise values for a good or bad fit, based on past experience, anything with fit index above 80% is quite reasonable, and above 90% split networks can be considered as quite robust. Anything much below 60%–70% should be treated carefully. Note that the split network in Fig. 21.6 is quite well resolved and tree-like. Thus, in view of the high fit index, one would expect other distance-based methods, such as NJ, to produce similar looking trees (which is in fact the case, as can be easily checked using SPLITSTREE – see below).

21.6.3 Laying out split networks

The SPLITSTREE program has many useful options for laying out a split-network, which can be very important for publication purposes. We describe some of these options in the SPLITSTREE's menu bar. Note that several of these options are also included as icons in SPLITSTREE's main window. Note that some of the layout options involve the setting of parameters within the processing pipeline window, which will automatically appear when such an option is chosen.

Begin by enlarging the split network that you have produced by going to the `View` menu and repeatedly selecting the item `Zoom in`. This enlarges the split network, but can result in part of the network no longer fitting into the main window. However, by using the scroll bars that appear to the right and at the bottom of the window during the enlargement process, the network can be moved around to view any part of interest. For example, considerably enlarging the split network depicted in Fig. 21.6 reveals that the frog taxon and the three lungfish taxa are attached to the split network via a thin rectangle. Note that the `View` menu contains various additional easy-to-use items for laying out the network.

To interactively change the layout of a split network, move the pointer with the mouse to a branch in the network. Clicking on it will result in that branch (plus all the branches that are parallel with it) becoming red, and also in some nodes and taxa to become circumscribed with red lines (corresponding to one part of the split represented by the red branches). By clicking on any circumscribed node, it is now possible to move all red branches simultaneously and independently from the rest of the network, so that they can be positioned at any angle. Similarly, by clicking on a taxon name it is possible to move the name to any position. Note that more advanced options for laying out split networks can be found in the `Draw` menu and are described in the user manual.

Once you are happy with the layout of the split network, both taxon and branch labels can be edited by double clicking on them (the latter can be added by double clicking on a branch). The `Edit` menu provides further items to edit a split network. The most important are the `Find/replace`, `Select All`, and `Copy` items. The former allows you to globally edit taxa labels in a split network without having to modify the underlying nexus file. By first clicking on the `Select All` item, and then on the `Copy` item it is possible to directly include a split network in a word document (see below for alternative ways to export a split network).

21.6.4 Recomputing split networks

Using items in the `Data` menu it is possible to revise the computed split network. For example, the user is allowed to exclude some taxa so that a new split network will be computed with these taxa left out of the calculation. This option can be very helpful when exploring the effect of a particular subset of taxa on the topology of the network. By selecting the `Restore all taxa` item, the original taxa set can easily be restored. Also characters such as gap sites, constant sites and parsimonious uninformative sites can be excluded by toggling the respective items on or off.

A number of standard distance transformations are available in the `Distances` menu. The *Hamming-distance* computes the proportion of positions at which two sequences differ and can be calculated by using the `UncorrectedP` item. In addition, the following distance transformations are implemented: *Hasegawa, Kishino Yano-85, Jukes–Cantor, Kimura 2P, Kimurar-3ST, Felsenstein-81*, and *Felsesnstein-84* (see Chapters 4, 10, or Swofford *et al.* (1996)). These can be called by clicking on the correspondingly labeled item.

21.6.5 Computing trees

The `Trees` menu can be used for computing and processing trees. Tree building techniques that are implemented include, for example, UPGMA, NJ (see Chapter 5), BioNJ (a variant of NJ introduced by Gascuel, 1997), and Buneman trees. All of them can be called by clicking on the correspondingly labeled item. In particular, Buneman trees can be computed with the `Buneman tree` item using the formula discussed in Section 21.2.1. Figure 21.7 shows a Buneman tree for the mtDNA data. Note that this tree is less resolved than the split network obtained using split decomposition. This is precisely for the reasons discussed in Section 21.2.1; the Buneman tree is more conservative. Note that the items `Phylip-Parsimony` and PHYML can only be used in conjunction with the PHYLIP and the PHYML programs.

21.6.6 Computing different networks

The `Networks` menu is used to generate split networks using different algorithms. By default SPLITSTREE computes a NeighborNet, as described in Section 21.4.1. We will give an example of a NeighborNet below. The other distance-based methods implemented are split decomposition (as described above) and parsimony splits (Bandelt & Dress, 1992a). Character-based methods implemented include median networks (see Section 21.4.2) and spectral split (Hendy & Penny, 1993). The consensus network technique can also be accessed by selecting the `Consensus` item and, in case the input trees only overlap on subsets, the `Supernetwork` and `FilteredSupernetwork` option can be used (as described in Section 21.4.3).

21.6.7 Bootstrapping

The `Analysis` menu is used for analyzing computed trees and split networks. Bootstrap values can be computed for both of them by selecting the `Bootstrap` item and then specifying the number of bootstrap replicates. Once SPLITSTREE has computed these bootstraps, each branch in the network will be labeled by

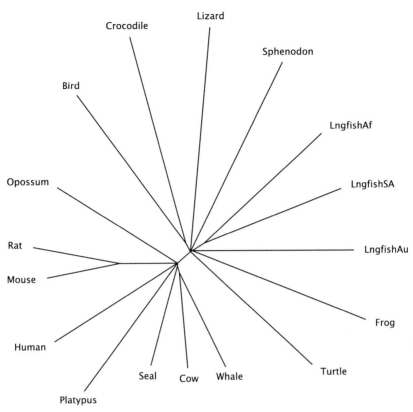

Fig. 21.7   Buneman tree of the mtDNA data set.

a value, which gives the percentage of computed networks in which the split corresponding to that branch occurred. For the split network depicted in Fig. 21.6, all pendant branches have bootstrap support 100% and, except for the five long internal branches all of which have support above 90%, all other branches in the split network have support values between about 35.0% and 78.1%. Note that the short branch in the thin rectangle (including the lungfish taxa) has bootstrap support of about 30%. In addition to usual bootstrap values, a bootstrap network can be computed that represents all splits appearing in any of the bootstrap replicates using the Show Bootstrap network item. (See Huson & Bryant, 2006 for more details.)

21.6.8 Printing

Once you have completed your analysis, the computed split networks can be saved, printed, or exported in various formats (including eps, gif, jpg formats) by choosing the appropriate items from the File menu.

Note that information on how to cite SPLITSTREE is contained in the How to cite item which can be found under the Window menu.

21.7 Using SPLITSTREE on other data sets

Running SPLITSTREE on the HIV sample data set and selecting SplitDecomposition (under Networks menu) produces a network that is basically a tree, except for a very thin box with two pendant branches leading to nodes labeled by the taxa U27399 and U43386 (Fig. 21.8a). The fit index is 88.53%, which is quite high, indicating that the data set is quite tree-like. The resolution at the central node from which the thin box emanates is very poor. The fact that there are *polytomies* indicate – in contrast to what is expected in a strictly bifurcating tree (see Chapter 1) – an explosive, star-like evolution of HIV-1 subtypes in this data set (Dress & Wetzel, 1993). Note that the NeighborNet for this example is slightly more resolved (least squares fit is very high at 99.99%) but still relatively tree-like (Fig. 21.8b). The NJ tree for this data set (Fig. 21.8c) looks very similar to both of the split networks, which is probably not too surprising in view of their tree-like behavior.

Figure 21.9 illustrates a split decomposition arising from a non-tree-like data set. The network was constructed using molecular data from a study on Salmonella strains characterized in Kotetishvili *et al.* (2002). The example shows that phylogenetic relationships in this data set are probably best represented by a network rather than a tree. Note that the split network generated for this data set using split decomposition is highly unresolved (Fig. 21.9). This is almost certainly due to the fact that this is a relatively large data set (see Section 21.4.1). For this reason, the original study by Kotetishvili *et al.* (2002) had to perform split decomposition on subsets of the taxa.

The NeighborNet in Fig. 21.10a can be used as a preliminary snapshot of the data, allowing us to explore it and also guide further analyses. To illustrate this, suppose we want to perform a more computationally intensive window-based analysis to investigate the possibility of recombination, which would be infeasible for the whole data set. Then, we can use NeighborNet to help select taxa for additional analysis. For example, the taxa She49, Sse94, UND8, UND79, Smb17, Sty85, UND64 are "well distributed" across the network and may constitute a useful subset to investigate evidence for recombination. Indeed, after running the LikeWin recombination detection algorithm (Archibald & Roger, 2002) on these taxa, the possibility of recombination was detected between sites 110 and 250 (see Bryant & Moulton, 2004 for more details). Running NeighborNet once more on sites 1–109 plus 251–660, all sites, and sites 110–250 results in the NeighborNets depicted in Fig. 21.10b–d. We see that the NeighborNet on sites 1–109 plus 251–660

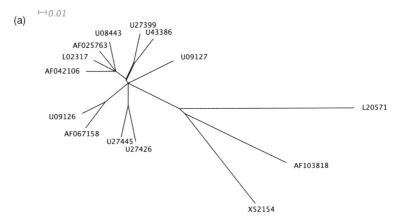

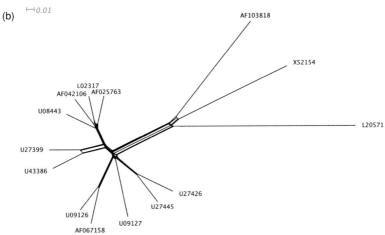

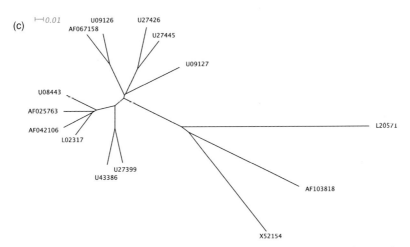

Fig. 21.8   (a) Split decomposition network, (b) NeighborNet, and (c) NJ tree of the HIV data set.

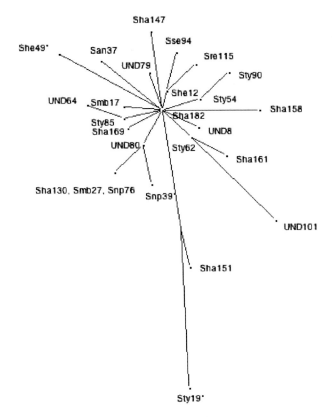

Fig. 21.9   Split decomposition network of the Salmonella data set.

(Fig. 21.10c) is quite tree-like, whereas for all sites (Fig. 21.10b) and sites 110–250 (Fig. 21.10d), we get non-tree-like NeighborNets, confirming that more complex evolutionary processes are probably at work for sites 110–250.

Note that, in case recombination is suspected to have occurred, some care has to be taken in interpreting split networks: a non-tree-like split network does not necessarily indicate recombination. For example, in Worobey *et al.* (2002), for the 1918 influenza virus it is shown that model heterogeneity can also lead to non-tree-like split networks (see also Huson & Bryant, 2006). Thus, it is important to make further tests for recombination in case non-tree-like split networks arise (see Section VI in this book).

In conclusion, we have presented some systematic methods for the construction of a special class of networks called ***split networks***, all of which are implemented

(a)

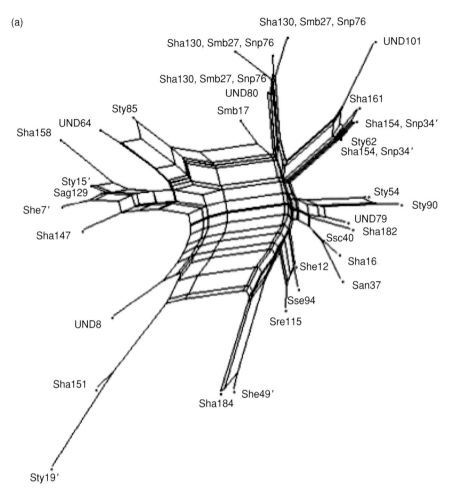

Fig. 21.10   NeighborNet for (a) the complete Salmonella alignment, (b) for all sites, (c) sites 1–109 plus 251–660, and (d) sites 110–250 (see Bryant & Moulton, 2004).

in the user-friendly package SPLITSTREE. This program is under constant develop-ment and new functions should be treated with some caution in case there are bugs (the authors Bryant and Huson are very helpful with fixing these). Also, as indicated above, some care needs to be taken in interpreting split networks, espe-cially when trying to unravel complicated underlying evolutionary processes such as recombination.

In general, the study of phylogenetic networks is a burgeoning area of phyloge-netics, and we can expect some exciting developments within this area in the next few years. These will include improved methods for constructing consensus and

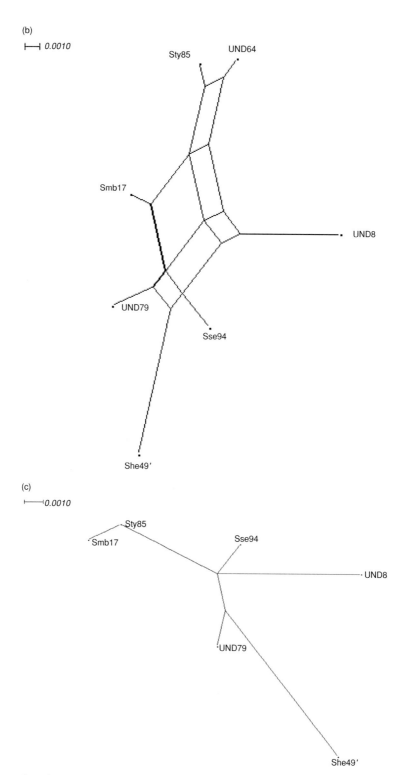

(b)

Fig. 21.10 (*cont.*)

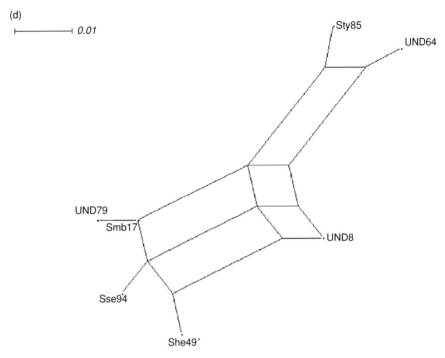

Fig. 21.10   (*cont.*)

supernetworks, new methods for constructing networks that represent explicit evolutionary scenarios, and improved tools for testing the significance of networks. Hopefully this chapter has served as a useful introduction to some of these topics, and also to the alternative possibilities to trees that networks can provide for analyzing complex data sets.

Glossary

Additive distances/additive distance trees Additive distances are distances that satisfy the *four-point metric condition*. Additive distance trees are trees inferred from additive distances. Additive distances will precisely fit a unique, additive phylogenetic tree.

Affine gap penalties Affine gap penalties are length dependent gap (*indel*) penalties used for alignment scoring. The penalty combines a gap-opening and gap-extension penalty.

AIC See *Akaike Information Criterion*.

Akaike Information Criterion (AIC) The Akaike Information Criterion is an estimate of the amount of information lost when we use a model to represent a stochastic process. The AIC for a model is $-2*\ln$ *likelihood* $+ 2*k$, where k is the number of estimated parameters. Models with smaller AIC provide a better fit to the data.

Allopatric speciation Allopatric speciation occurs due to geographic isolation. In this mode of speciation, something extrinsic to a population of organisms, for example island formation, mountain formation, prevents two or more groups from regularly mating with each other, eventually causing that lineage to speciate.

Ancestral recombination graph An ancestral recombination graph describes the genealogical relationships of *homologous* sequences undergoing recombination.

Asymptotic normality The property of a test statistic (function of the data) to approach the normal distribution with increasing sample sizes. For many estimation problems, the distribution of model parameter estimates are asymptotically normal.

Autapomorphy A derived character state (with reference to another, ancestral state) that is observed in only one of the taxa under consideration.

Autocorrelated relaxed clock In an autocorrelated relaxed clock model, the rate on each branch in a phylogeny is drawn from a distribution that is centered on the rate of the ancestral branch. Depending on the model specifications, the strength of rate correlation between ancestral and descendent branches can be or cannot be contingent on the time between the two rates. Additional assumptions must be made for the rate at the root node.

Background selection Background selection is the reduction in genetic variation linked to a gene frequently experiencing deleterious mutations.

Balancing selection The evolutionary force that maintains multiple genetic variants in a population indefinitely.

Bayes factor (BF) Given an event E (e.g. site X is under ***positive selection***), Bayes Factor (BF) of E is the ratio of ***posterior*** odds for E = Prob (E)/(1− Prob(E)) to the ***prior*** odds for E. BF estimates the factor by which one's prior belief in E was changed by observing the data. Large BF is evidence in favor of E.

Bayesian Information Criterion (BIC) BIC is an approximation to the log of the ***Bayes factor***. The *BIC for a model is* −2*ln ***likelihood*** + *k**ln(n), where *k* is the number of estimated parameters and n is the sample size of the alignment. Models with smaller BIC are preferred. The BIC is also known as the Schwarz criterion. Despite its name, the BIC is rarely used by Bayesian statisticians, who typically prefer to estimate Bayes factors using more precise methods, such as ***reversible-jump MCMC***, the harmonic mean estimator, or thermodynamic integration methods.

BF See ***Bayes factor***.

Bias Bias is the average deviation of an estimate from the true value. Biased estimators will consistently over- or under-estimate parameter values.

BIC See ***Bayesian Information Criterion***.

BLAST Basic Local Alignment Search Tool: software suite with programs for searching a sequence against a database in order to find similar sequences, freeware developed at the NCBI (Bethesda, Maryland, USA).

BLOSUM BLOCKS Substitution Matrices: tables with pairwise scores for amino acids, based on the frequencies of occurrence of amino acids in corresponding positions in ***homologous*** proteins, derived by the Henikoffs from the local alignments in the Blocks database, used for scoring protein alignments and for estimating evolutionary distance.

Bootstrap analysis (non-parametric) or bootstrapping Bootstrap analysis is a widely used sampling technique for estimating the statistical error in situations where the underlying sampling distribution is either unknown or difficult to derive analytically (Efron & Gong, 1983). The bootstrap method (Efron, 1979) offers a useful way to approximate the underlying distribution by resampling from the original data set. In phylogenetic analysis, non-parametric bootstrapping involves creating replicate data sets of the same size as the original alignment by randomly resampling alignment columns with replacement from the original alignment and reconstruction phylogenetic trees for each. The proportion of each clade among all the bootstrap replicates can be considered as measure of robustness of the ***monophyly*** of the taxa subset.

Branch-and-bound method Branch-and-bound methods are class algorithms for finding optimal solutions of various optimization problems. A branch-and-bound algorithm keeps the best solution found so far and abandons partial solutions that cannot improve the score or for which no globally optimal solution can exist. Therefore they are non-heuristic, in the sense that they maintain a provable upper and lower bound on the (globally) optimal value for the objective criterion.

Bremer support See ***Decay Index***.

Buneman index An index that can be computed for a ***split*** of a set of taxa on which a distance has been defined. It is computed using quartets, and measures how much the distance supports the split. It is related to the ***isolation index***, and is used to define branch lengths in ***Buneman trees***.

Buneman tree A tree corresponding to a set of ***splits*** for which the ***Buneman index*** computed using ***genetic distances*** resulted in a positive value. In a Buneman tree, the weight of each branch representing a particular split corresponds to the Buneman index for that split.

Burn-in Burn-in is a term that describes the initial phase of an ***MCMC*** run, when the sampled values are still influenced by the starting point. The samples collected during this phase are typically discarded.

Cladogram A branching or tree diagram representing the estimated evolutionary relationships among a set of taxa. In contrast to a ***phylogram***, branch lengths in a cladogram are not proportional to the amount of inferred evolutionary change.

The coalescent or coalescent theory A mathematical framework describing the times at which extant lineages had their most recent common ancestors as a function of population size. First developed by J.F.C. Kingman and later expanded by other researchers to include migration, recombination, growth, selection, population divergence, and other forces.

Coalescent effective size The coalescent effective size of a real population is obtained by finding the value of the ***effective population size*** that delivers the same distribution of times-to-ancestry that one would obtain with a sample from the real population in question.

Codon usage bias Codon usage bias is the phenomenon of different organisms having particular preferences for one of the several codons that encode the same amino acid.

Codon volatility The likelihood that a random point mutation in a codon will result in an amino acid change. Codon volatility tests have been used to detect natural selection within a single genome, which is inferred when the average codon volatility for a given gene is high relative to that of other genes in the same genome.

Cold chain The ***Markov chain*** sampling from the distribution of interest when using ***Metropolis coupling***. A cold chain is coupled with other Markov chains sampling from *heated* (flattened out) versions of the same distribution.

Conjugation In biology, conjugation is a bacterial mating process that involves transfer of genetic information from one bacterial cell to another, which requires physical contact between the two bacteria involved.

Consistency An estimator $T(N)$ of quantity T is consistent if $T(N)$ converges to T in probability as sample size N is increased. In other words, for any fixed estimation error C one can select a sufficiently large sample size so that the probability that $|T(N) - T| > C$ is arbitrarily small. In particular, this implies that the mean of $T(N)$ converges to T, and the variance of $T(N)$ converges to 0.

Consistency-based scoring A technique for multiple sequence alignment that makes a consistency measure among a set of pairwise sequence alignments before the ***progressive alignment*** steps.

Convergent evolution Convergent evolution is the independent evolution of similar traits (at the genotypic or phenotypic level) in separate lineages, starting from a different ancestral condition. ***Parallel evolution*** on the other hand, starts from a similar ancestral state.

Credibility interval An interval containing a specified probability. For instance, the probability is 0.95 that a value is within a 95% credibility interval. A credibility interval is typically estimated from a sample by removing the most extreme samples from each tail of the distribution, the same number of samples from each tail. Alternatively, one can find the shortest continuous interval containing the specified percentage of the samples (See *Highest posterior density interval*).

Credible sets A credible set is like a *credibility interval* but for discrete variables, such as the topology parameter in phylogenetic problems, where the values cannot be easily ordered. A credible set contains a specified probability and is often constructed by starting with the most probable value and then adding values in order of decreasing probability until the desired total probability is reached.

Decay Index The Decay Index or Bremer support is a measure to evaluate the support of clades in *maximum parsimony* analysis. The number is the difference between the shortest tree and the shortest tree that is incompatible with a pre-specified clade and thus reflects the number of steps that need to be taken in order to break the internal branch leading to the clade.

Delete-half jackknifing See *jackknifing*.

Diploid A diploid organism has cells possessing chromosomes in *homologous* pairs and thus two copies of each autosomal genetic locus, one from each parental set.

Directional selection An evolutionary process which increases the frequency of a specific beneficial allele in a population with several allelic variants, often leading to the fixation of the beneficial variant (a *selective sweep*).

Disruptive selection Disruptive selection occurs when extreme phenotypes have a fitness advantage over more intermediate phenotypes. A simple form of disruptive selection is a single locus with two alleles where the heterozygous individuals are at a disadvantage relative to the two homozygous classes. This will result in two peaks in the distribution of a particular trait. Disruptive selection on mutations in the reproductive machinery can lead for example to sympatric species pairs. Disruptive selection is also thought to have played a role in sexual dimorphisms.

Diversifying selection The evolutionary process which maintains several allelic variants in a population or among divergent taxa. This type of selection may arise for example, when different individuals (or taxa) are exposed to different environments, such as viruses in different hosts.

d_N/d_S The ratio of the rate of replacement fixations to the rate of silent fixations or *non-synonymous/synonymous rate ratio*, frequently used to evaluate the nature of selective pressure in molecular sequences. Note that d_N and d_S are usually calculated per replacement or silent site, respectively, thereby taking into account the fact that random mutations generate more replacement changes than silent changes due to the structure of the genetic code. $d_N / d_S \approx 1$ signifies neutral evolution, $d_N / d_S < 1$ – negative selection and $d_N / d_S > 1$ – positive selection.

Dynamic programming Dynamic programming is an algorithmic technique to solve an optimization problem efficiently by sorting/assembling sub-problem solutions in a way that guarantees optimality for the full problem.

Effective population size The size of an idealized *Wright–Fisher population* which would have the same carrying capacity for genetic variation as the given (census) population and thus loses or gains genetic diversity at the same rate as the census population size.

Effective sample size In *MCMC* analysis, the effective sample size is a measure used to evaluate *mixing behavior*. It is defined as the number of independent samples that the trace is equivalent to. This is calculated as the chain length (excluding the ***burn-in***) divided by the *auto-correlation time* (ACT, the average number of states in the MCMC chain that two samples have to be separated by for them to be uncorrelated). Low ESSs reflect a high degree of correlation among samples. More generally, when estimating a parameter, the effective sample size is the size of (an idealized) i.i.d. sample from the distribution of the estimator which would result in a comparable variance of the estimator. In practice, the sample does not often contain independent observations resulting in an effective sample size that is smaller than the number of points.

Efficiency An efficient estimator has the lowest variance (or squared mean error) among all estimators with a given set of properties. For unbiased estimators the minimum achievable variance is inversely proportional to the amount of information in the data.

Electrophoretic alleles Allelic variants that have amino acid substitutions causing a change in a protein's electrical charge, which can be detected via gel electrophoresis.

Empirical Bayes Empirical Bayes generally refers to Bayesian approaches that use the data to provide an empirical estimate of the prior distribution.

Evolutionary model See ***Model of evolution.***

Evolutionary rate See ***Rate of (molecular) evolution.***

Expressed Sequence Tags (***ESTs***) ESTs are obtained by extraction of total RNA, followed by reverse transcription and single-pass automated DNA sequencing. They are useful to localize genes (including intron/exon boundaries) on the genome and to show under which conditions genes are effectively transcribed.

Extreme value distribution The extreme value distribution or Gumbel distribution, ze^{-z}/b with $z = e^{-(x-m)/b}$, is used in statistics to find the minimum or the maximum value of a variable x in a sample taken from a population where x is normally or exponentially distributed or depends on many independent factors. In bioinformatics, database searching programs often rely on the assumption that the scores of the false positives follow an extreme value distribution.

FastA Software suite with programs for searching a sequence against a database in order to find similar sequences, freeware developed by Professor Pearson at the University of Virginia (Charlottesville, USA). FASTA is also a simple data format for sequences.

Fit index A measure for how well a ***split network*** that has been computed from a distance, using the ***split decomposition*** algorithm, represents the distance. A 100% fit implies that the split network represents the distance exactly; lower fit indices indicate that some part of the distance, the split prime component, is not represented in the network.

Fitness A measure of an individual's ability to survive and reproduce in the current environment.

Fixation rate The rate at which alleles or mutations become *fixed* in the population. This is not necessarily equivalent to ***evolutionary rate.***

Fixed effect likelihood A class of statistical models that model an unobserved quantity (e.g substitution rate at a site), which is fitted directly at each site as a model parameter. This approach is used for identifying positively or negatively selected sites in sequence alignments as an alternative approach to ***random effect likelihood.***

Fluctuating selection A process that is characterized by a sufficiently frequently fluctuating environment resulting in multiple variants being maintained because no single advantageous mutation has sufficient time to reach fixation before the environment within which it is beneficial changes.

Four-point metric condition   *Additive trees* satisfy the four-point metric condition, which states that for any four taxa A, B, C, and D, $d_{AB} + d_{CD} \leq \max(d_{AC} + d_{BD}, d_{AD} + d_{BC})$.

Frequency-dependent selection   A form of *balancing selection* that occurs when a rare variant has greater fitness than a more common one.

Genealogy   The pattern of common ancestry within a population. For non-recombining sequences this will resemble a phylogenetic tree, but recombining sequences will produce a tangled graph, also called the *ancestral recombination graph.*

Generation time   The generation time of an organism refers to the length of a generation expressed in time units, which is sometimes defined as the average age of mothers of newborn offspring. For bacteria and viruses, generation time is sometimes defined as the reciprocal of the growth rate and the reciprocal of the replication rate, respectively.

Genetic algorithm   Genetic Algorithms (GAs) are adaptive heuristic search algorithms that use the principles of evolution and selection to produce several solutions to a given problem. Classified as global search heuristics, the range of problems to which they can be applied to is quite broad.

Genetic bottleneck   A large reduction in genetic variation due to large reduction in population size, thus associated with a large effect of *genetic drift*, caused, for example, by a catastrophe or a founder effect.

Genetic distance   In evolutionary biology, genetic distance is a measure of the evolutionary divergence or dissimilarity between the genetic material of different species or individual of the same species. Genetic distances estimated from nucleotide sequences are generally based on *models of evolution*. In genetics, genetic distance sometimes refers to distance between the loci of two allele pairs known to reside on the same chromosome.

Genetic drift   The stochastic fluctuation in allele frequency due to random sampling of gametes from generation to generation in a population.

Genetic hitchhiking   See *Hitchhiking selection.*

Genome survey sequences (*GSS*)   GSSs are short genomic sequences obtained by single-pass automated DNA sequencing (unassembled fragments from genome sequencing projects, ends of cosmids/BACs/YACs, PCR products ...).

GSS   See *Genome survey sequences.*

Haploid   Haploid organisms have a single copy of each genetic locus per nucleus, cell or particle.

Hastings ratio   The ratio of proposal probabilities in *Metropolis–Hastings MCMC.*

Heterologous recombination   See *Non-homologous recombination.*

Heterotachy   In phylogenetics heterotachy, refers to the situation where the substitution rate of sites in a gene changes through time.

Heterozygosity   Heterozygosity at a locus is the proportion of individuals for which two distinct alleles can be detected. This definition obviously implies that heterozygosity can only be measured in *diploid* individuals.

Hidden Markov model or *HMM*   A hidden Markov model is used to statistically model a system that is assumed to be a *Markov process* with unknown parameters. The model is composed of a set of *states* associated with a probability distribution and transitions among the states are governed by a set of probabilities called *transition probabilities.* In a *hidden* Markov model, the state is not directly visible, but variables influenced by the state, the observable parameters, are visible. The challenge is to determine the hidden parameters from the observable parameters.

Highest posterior density (HPD) interval An interval representing the uncertainty of a parameter estimate in Bayesian inference. The HPD is the shortest interval in parameter space that contains x% of the sampled values, with x usually being 95.

Hitchhiking selection The process by which allelic neutral variants become fixed in the population because their locus is linked to a locus with beneficial alleles that are positively selected.

Homogeneity Homogeneity in *time-homogeneous, time-continuous stationary* **Markov models** refers to assumption that substitution rates – the parameters in the **substitution model** – do not change over time.

Homology/Homologous In biology, homology refers to similarity due to shared ancestry.

Homologous recombination Homologous recombination occurs when the donor nucleotide sequence replaces a homologous region in the acceptor molecule.

Homoplasy Sharing of identical character states that cannot be explained by inheritance from the common ancestor of a group of taxa.

HPD See **Highest posterior density interval.**

Hyperprior A hyperprior is a *prior distribution* on a parameter (also called a hyperparameter) of another distribution used in a hierarchical Bayesian model. For instance, rate variation across sites is often modeled using a gamma distribution with a shape parameter. If allowed to vary, the shape parameter is a hyperparameter and its prior is a hyperprior.

Infinite sites model A model assuming that each mutation occurs at a different nucleotide site; "multiple hits" are not accounted for.

Information entropy (or Shannon's entropy) Information entropy quantifies the amount of information in a discrete signal defined as $H(x) = -\sum_{i=1}^{n} p_i \log_2 p_i$. In biology, it is frequently used to quantify the information in a locus or a particular nucleotide or amino acid site in a set of aligned sequences. Consider a locus with two alleles, A and a, with A dominant over a, and their associated genotypic frequencies being $P_{AA} = 0.25$, $P_{Aa} = 0.5$ and $P_{aa} = 0.25$. The information entropy of the three genotypes is then 1.5 (bits). In contrast, there are only two recognizable phenotypes, with $p_1 = P_{AA} + P_{Aa}$ and $p_2 = P_{aa}$. The corresponding information entropy is only 0.8112781245 (bits), i.e. much information is lost at the phenotypic level.

Informative site See **Parsimony informative site.**

Island model An island model is a structured population model where migration between sub-populations or demes is random with respect to distance.

Isolation index An index that can be computed for a *split* of a set of taxa on which a distance has been defined. It is computed using quartets, and measures how much the distance supports the split. It is related to the **Buneman index**, and can be used to define branch lengths in *split networks*.

Jackknifing (or delete-half jackknifing) Jackknifing is a resampling technique often used to evaluate the reliability of specific clades in phylogenetic trees (similar to **bootstrap analysis**). The jackknife generates replicate data sets by randomly purging half of the sites from the original alignment. Trees are reconstructed for each replicate and the proportion of each clade among all the bootstrap replicates can be considered a measure of robustness of the ***monophyly*** of the taxa subset.

Joint probability The probability of a particular combination ***of two or more random variables.***

LAMARC programs Program package, available at http://evolution.gs.washington.edu/lamarc/, which uses ***Metropolis–Hastings Markov chain Monte Carlo*** sampling to make ***likelihood*** or Bayesian estimates of population parameters including Θ, *growth rate*, ***migration rate***, and ***recombination rate***. Suitable for DNA, RNA, SNP, microsatellite or electrophoretic data.

Likelihood Likelihood is proportional to the probability of observing given data under a known probabilistic model. For continuous probability models, likelihood is proportional to the density function evaluated at the observed values. The likelihood value can be considered as the goodness-of-fit of our model and parameters to the underlying data.

Likelihood function Likelihood function is ***likelihood*** considered as a function of model parameters. For N independent samples x_i and from a proposed density function $f(x; p)$, where p is a vector of parameters, the likelihood function is $f(x_1; p)\, f(x_2; p) \ldots f(x_N; p)$. In phylogenetics, x_i are most typically alignment columns, and p encompasses branch lengths and substitution rate parameters.

Likelihood mapping A method to visualize either the support at a single inner branch of a tree or to evaluate the "phylogenetic content" in a data set. The methods visualizes the amount of unresolved, partly resolved, and fully resolved trees, by plotting the vector of posterior weights from the ***likelihoods*** of each of the three possible unrooted topologies for each relevant set of four sequences (quartets).

Likelihood ratio statistic The likelihood ratio statistic is defined as $LR = 2 \log (L(H_A)/L(H_0))$, where $L(H_0)$ and $L(H_A)$ represent the ***likelihood functions*** for the null and the alternative model. In a ***Likelihood ratio test (LRT)***, the LR is used to decide between the two competing models or hypotheses.

Likelihood ratio test The likelihood ratio test can be used to decide between two competing models or hypotheses: the null hypothesis (H_0) and the alternative model (H_A). The likelihood functions for the null $(L(H_0))$ and the alternative $(L(H_A))$ models are maximized separately and the ***likelihood ratio test statistic*** $LR = 2 \log (L(H_A) / L(H_0))$ is used to choose a model. When H_0 is *nested* in H_A, i.e. it can be formulated by imposing D constraints on the parameters of H_A (e.g. of the form $a = 1$ or $b = c$), statistical significance is assessed by comparing the value of LR to a chi^2 distribution with D degrees of freedom. In other words, if H_0 is true, then LR is distributed as chi^2_D, and the probability (*p*-value) of falsely rejecting H_0 (false positive) if $LR = C$ can be approximated by $\text{Prob} (\text{chi}^2_D > = C)$. Additional considerations apply if some of the constraints are not regular, e.g. fall on the boundary of the parameter space.

Linkage disequilibrium Linkage disequilibrium refers to a situation where certain combinations of linked alleles occur in greater proportion than expected from the allele frequencies at their respective loci. This indicates that the loci are physically close on the DNA strand.

Local clock A local clock model allows different rates in different parts of the tree. In this model, different substitution rate parameters are assigned to different branches or collections of branches. Local clocks represent intermediates between the ***unrooted model*** and the global ***molecular clock*** model.

Log odds matrices Amino acid tables with pairwise scores, calculated as the logarithm of the proportion between the observed frequency and the expected frequency (e.g. the *PAM* and *BLOSUM* matrices for comparing amino acids).

Long-branch attraction The estimation bias (or tendency) of grouping highly divergent taxa as *sister taxa* even when they are not. While it is historically associated with the maximum *parsimony* method, other methods, such as distance-based *minimum evolution* methods with a distance index that severely underestimates the true *genetic distance* between divergent taxa, also suffer from the problem.

Majority-rule consensus tree A majority-rule consensus tree is a consensus tree of multiple trees. For instance, the consensus tree can summarize the information contained in all the trees from a posterior sample, from a bootstrap analysis or from (different) tree inference procedures. The majority-rule consensus tree is defined by its threshold x, with $x > 50\%$; only splits or clades that are present in $x\%$ of the trees are included in the consensus.

Marginal probability (density/distribution) The probability (density/distribution) of one random variable, obtained by summing or integrating probabilities over the entire range of the other random variables.

Markov chain Monte Carlo (MCMC) sampling A statistical technique for integrating a function by drawing samples at random ("Monte Carlo") from the function, basing each sample on the previous one ("*Markov chain*"). This stochastic technique is useful when the function cannot be integrated directly, but can fail if the sample drawn is not big enough or does not explore all important regions of the function.

Markov model A model describing a *Markov process*.

Markov process or *Markov chain* A random process that has the property that the states depend only on their direct predecessors, but not on the values of the past (hence, it is memory-less).

Maximum likelihood A principle of statistical inference developed by R. A. Fisher in the 1920s. Essentially, it is a generalization of least-squares to non-normal data, and can be shown to lead to optimal estimators, at least for large sample size. Moreover, it is fully automatic once a model is specified, and allows computing confidence bands by means of the so-called Fisher information.

Maximum likelihood estimate (MLE) The maximum likelihood estimate (MLE) of a vector of parameters p is the value of p that maximizes the *likelihood function* over the space of allowable values. This value is not necessarily unique and is frequently obtained using numerical optimization techniques.

Maximum parsimony A criterion for estimating a parameter from observed data based on the principle of minimizing the number of events needed to explain the data. In phylogenetic analysis, the optimal tree under the maximum parsimony criterion is the tree that requires the fewest number of character-state changes.

MCMC See *Markov chain Monte Carlo sampling*.

MCMCMC or *MC³* See *Metropolis Coupling*.

McDonald–Kreitman test A method for detecting selection in molecular sequences sampled within populations. The method uses an *outgroup* sequence or population to distinguish fixations (substitutions that occur on the branches separating the ingroup and the outgroup) from polymorphisms (mutations that occur within the ingroup), and then tests for a significant difference between the d_N/d_S ratios of polymorphic and fixed sites.

Measurably Evolving Population (MEP) A MEP is any population that permits the detection of a statistically significant accumulation of substitutions when samples are obtained at different times.

Meiosis In biology, meiosis is the process by which one ***diploid*** eukaryotic cell divides to generate four ***haploid*** cells, often called gametes.

MEP See ***Measurably Evolving Population.***

Metropolis Coupling or ***MCMCMC or MC³*** Metropolis coupling refers to the technique of coupling a ***Markov chain***, sampling from a distribution of interest, with other Markov chains sampling from *heated* (flattened out) versions of the same distribution. The heated chains explore the distribution more widely and can find multiple peaks more readily than the *cold* chain. The cold and heated chains are run in parallel and, at regular intervals, one attempts to swap the states of two chains using a Metropolis step. Inference is based solely on the samples from the cold chain. When the procedure works well, it can accelerate the mixing of the cold chain dramatically.

Metropolis–Hastings sampling/MCMC A form of ***Markov chain Monte Carlo*** sampling in which each newly proposed sample (such as a ***genealogy***) is made by proposing a small change to the previous sample and accepting or rejecting it based on how well it fits the data.

Microsatellites Short tandem DNA repeats; the repeat unit is generally 2–4 base pairs. The number of repeats in a microsatellite can change due to unequal crossing over or replication slippage, making them useful markers in within-population studies.

Midpoint rooting The midpoint rooting method places the root of a tree at the midpoint of the longest distance between two taxa in a tree.

Migration rate In population genetics, the migration rate is always the rate of immigration into a population. It can be measured either as $4N_e m$ ($2N_e m$ in ***haploids***) or as m/μ, where N_e is the ***effective population size***, μ is the ***mutation rate*** per site per generation, and m is the chance of immigration per lineage per generation. When $4N_e m$ is much greater than 1, it will generally homogenize populations; when it is much less than 1, it will allow differentiation. $4N_e m$ has an intuitive interpretation as the number of migrants per generation.

Minimum evolution (ME) The name applied by Rzhetsky and Nei (1992) to a phylogenetic optimality criterion that was originally described by Kidd and Sgaramella-Zonta (1971). The best tree under this criterion is the tree with the smallest sum of branch lengths.

Mixing behavior The speed with which the ***MCMC*** covers the interesting regions of the ***posterior*** or in other words, the efficiency with which the MCMC algorithm samples the posterior probability distribution of a parameter or set of parameters. If an MCMC chain is mixing well, it implies that autocorrelation in the chain is low, the ***effective sampling size*** (***ESS***) is high, and the estimates obtained are accurate.

Model averaging A technique that provides a way to estimate a parameter using several models at once, and thus includes model selection uncertainty in the estimate. The contribution of each model's estimate is weighted according to its ***AIC, BIC*** or ***posterior probability***. Sometimes it is called multi-model inference.

Model of (sequence/molecular) evolution A statistical description of the stochastic process of substitution in nucleotide or amino acid sequences.

Model of substitution See *Model of evolution*.

Molecular clock Constancy of *evolutionary rate* among lineages in a *genealogy* or phylogeny. In a genealogy with a molecular clock, any two lineages sampled at the same time should show about the same amount of genetic divergence from their common ancestor.

Monophyly/monophyletic In phylogenetics, a group of taxa is monophyletic or represents a monophyly if the group includes all descendants from its inferred common ancestor.

Monte Carlo simulation A stochastic technique to study the properties of random variables by simulating many instances of a variable and analyzing the results. This simulation procedure can be used to test phylogenetic hypotheses and *evolutionary models* using the *parametric bootstrap*.

Mutation rate The rate at which one nucleotide is replaced by another, usually expressed as the number of nucleotide (or amino acid) changes per site per replication cycle. Although it would be preferable to make a distinction between the biochemical mutation rate and number of mutations per generation per nucleotide site in the entire population (*population mutation rate*), mutation rate is frequently used for both processes. In case of population mutation rate, be careful to distinguish between studies and programs that estimate mutation rate per gene and those that measure mutation rate per base, and also between per-year and per-generation estimates.

Nearest Neighbor Interchange (NNI) A heuristic algorithm for searching through treespace by rearranging tree topologies. It starts from a particular topology and proceeds by juxtaposing the positions of neighbors on a phylogenetic tree. If the resulting tree is better according to an optimality criterion, then it is retained.

Negative selection/selective pressure The evolutionary force that decreases the frequency of a deleterious mutation until it is eliminated from the population.

Neighbor-joining (NJ) A heuristic method for estimating the *minimum evolution* tree originally developed by Saitou and Nei (1987) and modified by Studier and Keppler (1988). NJ is conceptually related to clustering, but does not require the data to be *ultrametric*. The principle of NJ is to find pairs of operational taxonomic units (OTUs) that minimize the total branch length at each stage of clustering of OTUs starting with a star-like tree. The neighbor-joining method is therefore a special case of the *star decomposition* method.

NNI See *Nearest Neighbor Interchange*.

Non-homologous recombination The process of non-homologous recombination involves crossovers at non-homologous sites or between unrelated nucleotide sequences.

Non-parametric bootstrap(ping). See *bootstrapping*.

Non-synonymous substitutions Nucleotide substitutions in coding sequences that alter the encoded amino acid.

Non-synonymous/synonymous rate ratio See d_N/d_S.

Nuisance parameter Most generally, an estimated model parameter that is needed to define the *likelihood function*, but is not of interest in and of itself. For example, when reconstructing a phylogenetic tree, the *substitution model* is a nuisance parameter. Conversely, when estimating selective pressures on sequences, the phylogenetic tree is a nuisance parameter.

Observed distance See *p-distance.*

Ontology In informatics an ontology means a set of names for objects, their attributes and their mutual relations, people have agreed about in order to reliably share information.

Orthologous Orthologous genes are homologues in different species that coalesce to a common ancestral gene without gene duplication or horizontal transmission.

Outgroup A taxon that is used to root a phylogenetic tree and thus providing directionality to the evolutionary history. An outgroup taxon is not considered to be part of the group in question (*the ingroup*), but preferably, it is closely related to that group. In a cladistic analysis, an outgroup is used to help resolve the polarity of characters, which refers to their state being original or derived.

Overdominance An alternative term for heterozygote advantage: the heterozygote at a particular locus has greater *fitness* than both the homozygotes.

PAM Point Accepted Mutation: tables with pairwise scores for amino acids, based on the frequencies of occurrence of amino acids in corresponding positions in homologous proteins, derived by M. Dayhoff from reliable global protein alignments. They are used for scoring protein alignments and for estimating evolutionary distance.

Parametric bootstrap In phylogenetics, parametric bootstrapping is an alternative for the *non-parametric bootstrap* procedure. Instead of generating pseudo-samples by sampling alignment columns with replacement from the initial alignment, parametric bootstrap generates pseudo-sample alignments by simulation along a given tree and using a particular *evolutionary model*, the parameters of which were estimated from the real data.

p-distance *p-distance* or *observed distance* is the proportion of different homologous sites between two sequences.

Panmictic Referring to unstructured, random-mating populations.

Parallel evolution Parallel evolution is the independent evolution of similar traits (at the genotypic or phenotypic level) in separate lineages, starting from a similar ancestral condition.

Paralogous Paralogous genes are homologues that diverged after a duplication event.

Parapatric speciation In contrast to *allopatric speciation*, there is no specific extrinsic barrier to gene flow. The population is continuous, but nonetheless, the population does not mate randomly. Individuals are more likely to mate with their geographic neighbors than with individuals in a different part of the population's range. In this mode, divergence may happen because of reduced gene flow within the population and varying selection pressures across the population's range.

Paraphyly/paraphyletic In phylogenetics, a group of taxa is paraphyletic or represents a paraphyly if the group does not include all descendants from its inferred common ancestor.

Parsimony See *Maximum parsimony.*

Parsimony informative site A site is parsimony informative if, and only if, it has at least two different characters (nucleotides or amino acids) that are represented by at least two different taxa or sequences each.

Patristic distance A patristic distance is the sum of the lengths of the branches that link two nodes in a tree, where those nodes are typically terminal nodes that represent extant gene sequences or species.

Phenetic distance   The distance given by the shortest paths between taxa labeling a phylogenetic tree (or network) with branch lengths.

Phenogram   A diagram or tree depicting the taxonomic relationships among organisms based on overall phenotypic similarity.

Phylogram   A branching or tree diagram representing the estimated evolutionary relationships among a set of taxa. In contrast to a *cladogram*, branch lengths are proportional to the inferred *genetic or evolutionary distance.*

Polytomy   Polytomies are multifurcating (as opposed to bifurcating) relationships in phylogenetic trees. On the one hand, they can represent the literal hypothesis that a common ancestral population split into multiple lineages (*hard polytomies*). On the other hand, they can reflect the uncertainty as to which resolved pattern is the best hypothesis (*soft polytomies*).

Population mutation rate   See *theta.*

Population recombination rate   In population genetics the population recombination rate is $4N_e r$, where N_e is the *effective population size* and r is the recombination rate per site per generation. In *LAMARC*, the recombination rate r is the probability of recombination per base per generation divided by the probability of mutation per base per generation (m), so that a recombination rate of 1 indicates that recombination and mutation events are equally frequent.

Positive selection/selective pressure   The evolutionary force that increases the frequency of a beneficial mutation until it becomes fixed in the population.

Posterior (probability/distribution)   The posterior is the probability distribution over the parameter space, given the data and the prior under the chosen model.

Potential Scale Reduction Factor   An index used to assess the convergence of *Markov chains* sampling from a continuous variable. It is based on the comparison of the variance within and among samples from independent Markov chains. Among other things, it assumes that the starting points are over-dispersed relative to the target distribution.

Prior (probability/distribution)   The prior is the probability distribution over the parameter space, prior to seeing the data. It represents the prior belief or prior assumptions about the probabilities of different parameter values before analyzing the data.

Probability density function   A function describing the probability of a continuous variable.

Probability mass function   A function describing the probability of a discrete random variable.

Profile likelihood   Profile likelihoods, also called *maximum relative likelihoods*, are found by replacing the *nuisance parameters* by their *maximum likelihood estimates* at each value of the parameters of interest. Component-wise profile likelihood intervals are frequently used to obtain confidence intervals around a parameter estimate. Because profile likelihood works with only one parameter at the time, the intervals it finds may be too small.

Progressive alignment   A heuristic method for multiple sequence alignment that progressively aligns sequences according to their relatedness as inferred using a clustering technique like *neighbor-joining.*

Purifying selection   See *Negative selection.*

Q matrix   Substitution rate matrix providing an infinitesimal description of the substitution process. The non-diagonal elements Q_{ij} of this matrix can be interpreted as the flow from nucleotide *i* to *j*. For a reversible substitution process, the rate matrix can be decomposed into rate parameters and nucleotide frequencies.

Quartet puzzling Quartet puzzling is a fast tree search algorithm that is employed as a heuristic strategy in ***maximum likelihood*** analysis. The method is based on the analysis of four-taxon trees. The quartet tree yielding the maximum likelihood is constructed from every possible four-taxon combination in a data set or a random subset thereof. A "puzzling" step is subsequently used to assemble the quartets in a tree for the complete data set.

Quasispecies Quasispecies theory describes the evolution of an infinite population of asexual replicators at high mutation rate. The theory predicts that the quasispecies is the target of selection since ensembles of mutants rather than individual genomes rise to dominance. From this perspective, individual genomes may have only a fleeting existence. Because of the high mutation rates and population turnover in RNA viruses, virologists have adopted this concept but it remains questionable to what extent they obey the mathematical laws governing quasispecies dynamics, in particular the "group" selection scenario.

Random effects likelihood A class of statistical models which model an unobserved quantity (e.g. substitution rate at a site) as a random variable independent of a particular sample. For example, standard gamma models of site-to-site rate variation are random effects models.

Rate of (molecular) evolution The rate of molecular evolution or *substitution rate* refers to the rate at which organisms accrue genetic differences over time, frequently expressed in units of substitutions per site per time unit. When sequences from different species or populations are compared, the genetic differences or substitutions represent the number of new mutations per unit time that become fixed in a species or population. Hence, substitution rate is equivalent to ***fixation rate***. However, when applied to sequences representing different individuals within a population, not all observed mutational differences among individuals will eventually become fixed in the population. In this case, evolutionary rate is not equivalent to fixation rate.

Reassortment Reassortment is the mixing of segmented genetic material of two viruses that are infecting the same cell.

Recombination rate See ***Population recombination rate***.

Relative rate test The relative rate test evaluates the ***molecular clock*** hypothesis for two taxa, A and B, by testing whether the genetic distance between taxa A and an outgroup O (d_{AO}) is not significantly different from the genetic distance between taxa B and an outgroup O (d_{BO}). The "triplet-based" relative test has been extended to multiple sequences by comparing the average or weighted-average distances of members of two clades to an outgroup.

Relaxed clock A relaxed-clock model relaxes the assumption of a strict ***molecular clock*** and typically assigns to each lineage a rate that is either correlated or uncorrelated to an ancestral rate, and is drawn from some plausible distribution of rates.

Resampling estimated log-likelihoods (RELL) This technique is used to avoid re-optimizing ***likelihood*** values for ***bootstrap*** samples in tree topology test procedures. Usually all parameters like branch lengths have to be re-optimized to obtain the likelihood value for a certain bootstrap sample alignment which is very time consuming. To reduce the running time, the RELL method resamples from the pre-estimated log-likelihoods for each site in the alignment instead of the alignment columns, hence avoiding the re-estimation of the parameters and site log-likelihood.

Reversible jump MCMC A *Markov chain Monte Carlo* technique used to sample across models that have different dimensionality (number of parameters).

Sampling distribution The distribution of a statistic – a function (e.g. sample mean, model parameter estimate) of the data for a given sample size.

Sampling importance resampling (SIR) A simple technique for approximating the (unknown) shape of the *sampling distribution* (e.g. of model parameters) by randomly drawing a large number of values using one of several possible schemes, and then using this sample to draw a smaller sample (resampling) in proportion to the weight of each point, e.g. the *likelihood* of a given set of model parameters (importance).

Sampling variance Variance of a sampling distribution.

Saturation See *Substitution saturation*.

Segregating sites Another name for polymorphic sites in a sequence alignment.

Selection coefficient Generally denoted "*s*", the selection coefficient is a measure of the change in an organism's *fitness* associated with a particular mutation, compared with the best genotype in the population.

Selective sweep An evolutionary process that leads to the fixation of a beneficial variant allele in a population with several allelic variants.

Serial coalescent (*or s-coalescent*) The mathematical framework describing the distributions of coalescent intervals on a genealogy of samples obtained serially in time.

Shannon's entropy See *Information entropy*.

Simulated annealing A technique to find a good solution to an *optimization problem* by trying random variations of the current solution. A worse variation is accepted as the new solution with a probability that decreases as the computation proceeds. As the process continues, the acceptance probability for a worse solution is decreased (the so-called *cooling* schedule).

Single Nucleotide Polymorphisms (SNPs) Sequence positions at which multiple bases (usually two) are segregating in the population. When data are collected by identifying SNPs in a trial sample and then sequencing only those positions in a larger sample, care must be taken to avoid bias in the resulting estimates, since this procedure systematically misses low-frequency alleles.

Sister taxa Sister taxa are any taxa derived from a common ancestral node, and therefore, they are the most closely related groups of organisms in a phylogeny.

Site-frequency The proportion of the population in which a mutation is found.

Split A bipartition of a set of taxa, that is, a partition of the set X into two non-empty sets A and B, say, so that the union of A and B equals X and the intersection of A and B is empty.

Split decomposition A canonical decomposition of a distance on a set of taxa that is given by computing *isolation indices* for all of the *splits* of the taxa. The split decomposition can be used to give a representation of the distance in terms of a *split network*.

Split network A network that represents a given collection of weighted splits. It is analogous to a phylogenetic tree in that the collection of splits can be obtained by removing certain "parallel" branches in the network.

SPR See *Subtree Pruning Regrafting*.

Star decomposition Star decomposition is an algorithmic approach to tree-building that starts with an unresolved (star) tree, with all taxa emanating from the central node. The star is gradually decomposed by inserting inner branches at the most probable bipartitions, according some optimality criterion. The most common star-decomposition method is the **neighbor-joining** algorithm of Saitou and Nei (1987).

Stationarity The property of a stochastic process to preserve the expected distribution of values that the process can take. For character substitution processes, this usually means that the expected nucleotide or amino acid composition of sequences does not vary across the phylogenetic tree. Stationarity is also a concept to describe the behavior of a **Markov chain** in **MCMC** analyses. Regardless of starting point, a Markov chain tends towards an equilibrium state determined by its transition probabilities. In MCMC sampling of a distribution, one makes sure that the transition probabilities are such that the equilibrium state is the distribution of interest. When the chain has reached its equilibrium state, it is said to be at stationarity.

Stepping-stone model In this model of geographical structure subpopulations are arranged linearly or in a two-dimensional grid, and migration is restricted to adjacent populations.

Stepwise addition Stepwise addition is a greedy tree-building algorithm that adds taxa sequentially, starting from a three-taxa tree (a strategy opposite to **star decomposition**). The algorithm uses an optimality criterion to evaluate each possible topology that can be generated by adding an additional taxon and commits to build upon the tree that looks most promising at the moment.

Substitution model See **Model of sequence/molecular evolution.**

Substitution rate See **Rate of (molecular) evolution.**

Substitution saturation A state of aligned, presumably **orthologous**, sequences that have experienced so many repeated substitutions that the number of substitutions between two sequences has become a poor predictor of their time of divergence.

Subtree Pruning Regrafting (SPR) A heuristic search algorithm for searching through treespace by rearranging tree topologies (see also **NNI**). It starts from a particular topology and proceeds by breaking off part of the tree and attaching it to another part of the tree. If it finds a better tree, then the new tree is used as a starting tree for another round of SPR. If the maximum branches to be crossed by SPR is set to one, then SPR reduces to **NNI**.

Sympatric speciation Sympatric speciation does not require large-scale geographic distance to reduce gene flow between parts of a population. Exploiting a new ecological niche may reduce gene flow between individuals of both populations.

Synapomorphy A synapomorphy is an ancestral character shared among two or more descendants and serving as the only valid information for phylogenetic reconstruction in cladistics.

Synonymous substitutions Nucleotide substitutions in coding sequences that leave the amino acid unchanged due to the redundancy inherent in the genetic code.

Tajima's D One of the first and most commonly used summary statistics to detect natural selection in gene sequences. Tajima's D compares the number of low- and intermediate-frequency mutations in an alignment, and can be used to predict recent selective sweeps or balancing selection in a constant-sized population.

TBR See **Tree bisection and reconnection.**

Theta (Θ) A fundamental population genetics quantity also referred to as the *population mutation rate*, which determines how much neutral genetic variation a population can carry. Equal to $4N_e\mu$ in **diploids**, $2N_e\mu$ in **haploids**, $2N_f\mu$ in mitochondrial DNA, where N_e is **effective**

population size and μ represents the mutation rate. Be careful to distinguish between analyses and programs that use *mutation rate* per gene to define μ and those that use mutation rate per base.

Trace plot A plot of the (posterior) likelihood values or parameter values against the generation of the chain, which is frequently used to evaluate *stationarity* and *mixing behavior* in *MCMC* analysis.

Transduction The process by which bacteriophages (viruses that infect bacteria) tranfer DNA between bacteria.

Transformation The process by which bacteria take up and incorporate DNA from their environment.

Transition A nucleotide substitution that exchanges a purine by a purine or a pyrimidine by a pyrimidine (A↔G, C↔T).

Transition/transversion ratio In nucleotide *substitution models*, the ratio between *transition* changes (purine to purine or pyrimidine to pyrimidine) and *transversion* changes (purine to pyrimidine or pyrimidine to purine). Since there are twice as many possible transversions and transitions, a ratio of 0.5 indicates that all base changes are equally likely.

Transversion A nucleotide substitution that exchanges a purine by a pyrimidine or vice versa (A↔C, A↔T, C↔G, G↔T).

Tree bisection and reconnection (TBR) This heuristic search algorithm is similar to *SPR*, but with a more extensive branch-swapping scheme. The tree is divided into two parts, and these are reconnected through every possible pair of branches in order to find a shorter tree. This is done after each taxon is added, and for all possible divisions of the tree.

Ultrametric/Ultrametricity Ultrametricity refers to the condition that is satisfied when, for any three taxa, A, B, and C, the distance between A and C (d_{AC}) is smaller or, at most, as large as the largest distance for the other pairs of taxa (d_{AB}, d_{BC}). Ultrametric trees are *rooted* trees in which all the tips are equidistant from the root of the tree, which is only possible by assuming a *molecular clock*.

Uncorrelated relaxed clock In an uncorrelated relaxed clock model, the rate on each branch in a phylogeny is drawn independently and identically from an underlying rate distribution, like an exponential or lognormal distribution.

Unrooted model / Unrooted tree The unrooted model of phylogeny assumes that every branch has an independent *rate of molecular evolution*. Under this assumption, branch lengths can be estimated as genetic distances but not the separate contributions of rate and time. The unrooted model is also referred to as the *different rates* model. The term unrooted refers to the fact that the *likelihood* of a tree is independent of the position of the root under this model given that the *substitution model* is time-reversible, which is typically the case.

Unweighted-pair group method with arithmetic means (UPGMA) UPGMA is a sequential clustering algorithm that builds a distance-based tree in a stepwise manner, by grouping sequences or groups of sequences that are most similar to each other, that is, for which the genetic distance is the smallest. When two operational taxonomic units (OTUs) are grouped, they are treated as a new single OTU. The new OTU replaces the two clustered OTUs in the distance matrix and the distance between the new OTU and a remaining OTU is calculated as the average of distances of the clustered OTUs to that remaining OTU. In contrast to weighted-pair group method with arithmetic means (WPGMA), UPGMA takes into account the number of OTUs in the different clusters when averaging the distances of these clusters. From the new group of OTUs, the pair for which the similarity is highest is again identified, clustered and so on, until only two OTUs are left. The method produces ultrametric trees and thence assumes that the rate of nucleotide or amino acid substitution is the same for all evolutionary lineages.

UPGMA See ***Unweighted-pair group method with Arithmetic means.***

Watterson estimate Watterson's estimate is a population genetics estimate of ***theta***, $\theta = 4N_e\mu$ (where N_e is the ***effective population size*** and μ is the ***mutation rate***). The estimate depends only on the number of segregating sites and the sample size: $\theta = K/a_n$, where K is the number of segregating sites and $a_n = \sum_{i=1}^{n-1} \frac{1}{i}$, and n is the number of sequences.

Wright–Fisher population A idealized population that is of constant size with non-overlapping generations, and in which each individual contributes an infinite number of gametes to the large gene pool and is replaced in every generation.

References

Abascal, F., Posada, D., & Zardoya, R. (2007). MtArt: a new model of amino acid replacement for Arthropoda. *Molecular Biology and Evolution*, **24**(1), 1–5.

Abascal, F., Zardoya, R., & Posada, D. (2005). ProtTest: selection of best-fit models of protein evolution. *Bioinformatics*, **21**(9), 2104–2105.

Abdo, Z., Minin, V. N., Joyce, P., & Sullivan, J. (2005). Accounting for uncertainty in the tree topology has little effect on the decision-theoretic approach to model selection in phylogeny estimation. *Molecular Biology and Evolution*, **22**(3), 691–703.

Abecasis, A. B., Lemey, P., Vidal, N. *et al.* (2007). Recombination is confounding the early evolutionary history of HIV-1: subtype G is a circulating recombinant form. *Journal of Virology*, **81**, 8543–8551.

Adachi, J. & Hasegawa, M. (1996). Model of amino acid substitution in proteins encoded by mitochondrial DNA. *Journal of Molecular Evolution*, **42**, 459–468.

Adachi, J. & Hasegawa, M. (1996a). MOLPHY version 2.3: programs for molecular phylogenetics based on maximum likelihood. *Computer Science Monographs of Institute of Statistical Mathematics*, **28**, 1–150.

Adachi, J. & Hasegawa, M. (1996b). Model of amino acid substitution in proteins encoded by mitochondrial DNA. *Journal of Molecular Evolution*, **42**(4), 459–468.

Adachi, J., Waddell, P. J., Martin, W., & Hasegawa, M. (2000). Plastid genome phylogeny and a model of amino acid substitution for proteins encoded by chloroplast DNA. *Journal of Molecular Evolution*, **50**(4), 348–358.

Akaike, H. (1974). A new look at the statistical model identification. *IEEE Transactions on Automatic Control*, **19**(6), 716–723.

Aldous, D. J. (2001). Stochastic models and descriptive statistics for phylogenetic trees, from Yule to today. *Statistical Science*, **16**, 23–34.

Allison, A. C. (1956). Sickle cells and evolution. *Scientific American*, 87–94.

Altschul, S. F., Gish, W., Miller, W., Myers, E. W., & Lipman, D. J. (1990). Basic local alignment search tool. *Journal of Molecular Biology*, **215**, 403–410.

Altschul, S. F., Madden, T. L., Schäffer, A. A. *et al.* (1997). Gapped BLAST and PSI-BLAST: a new generation of protein database search programs. *Nucleic Acids Research*, **25**, 3389–3402.

Anderson, T. J., Haubold, B., Williams, J. T. *et al.* (2000). Microsatellite markers reveal a spectrum of population structures in the malaria parasite *Plasmodium falciparum*. *Molecular Biology and Evolution*, **17**, 1467–1482.

Andolfatto, P. (2005). Adaptive evolution of non-coding DNA in *Drosophila*. *Nature*, **437**(7062), 1149–1152.

Anisimova, M., Nielsen, R., & Yang, Z. (2003). Effect of recombination on the accuracy of the likelihood method for detecting positive selection at amino acid sites. *Genetics*, **164**, 1229–1236.

Archibald, J. M. & Rogers, A. J. (2002). Gene conversion and the evolution of euryarchaeal chaperonins: a maximum likelihood-based method for detecting conflicting phylogenetic signals. *Journal of Molecular Evolution*, **55**, 232–245.

Archie, J. W. (1989). A randomization test for phylogenetic information in systematic data. *Systematic Zoology*, **38**, 219–252.

Aris-Brosou, S. & Yang, Z. (2003). Bayesian models of episodic evolution support a late Precambrian explosive diversification of the Metazoa. *Molecular Biology and Evolution*, **20**(12), 1947–1954.

Armougom, F., Moretti, S., Poirot, O. *et al.* (2006). Expresso: automatic incorporation of structural information in multiple sequence alignments using 3D-Coffee. *Nucleic Acids Research*, **34**(Web Server issue), W604–W608.

Awadalla, P. (2003). The evolutionary genomics of pathogen recombination. *National Review in Genetics*, **4**, 50–60.

Baake, E. (1998). What can and what cannot be inferred from pairwise sequence comparison? *Mathematical Biosciences*, **154**, 1–21.

Bäck, T. & Schwefel, H.-P. (1993). An overview of evolutionary algorithms for parameter optimization. *Evolutionary Computation*, **1**, 1–23.

Bailey, N. T. J. (1964). *The Elements of Stochastic Processes with Application to the Natural Sciences*. New York: Wiley.

Bakker, E. G., Stahl, E. A., Toomajian, C., Nordborg, M., Kreitman, M., & Bergelson, J. (2006). Distribution of genetic variation within and among local populations of *Arabidopsis thaliana* over its species range. *Molecular Ecology*, **15**(5), 1405–1418.

Baldauf, S. L., Palmer, J. D., & Doolittle, W. F. (1996). The root of the universal tree and the origin of eukaryotes based on elongation factor phylogeny. *Proceedings of the National Academy of Sciences, USA*, **93**, 7749–7754.

Bandelt, H.-J. (1995). Combination of data in phylogenetic analysis. *Plant systematics and Evolution*, **9**, S355–S361.

Bandelt, H.-J. & Dress, A. (1992a). A canonical decomposition theory for metrics on a finite set. *Advances in Mathematics*, **92**, 47–105.

Bandelt, H.-J. & Dress, A. (1992b). Split decomposition: a new and useful approach to phylogenetic analysis of distance data. *Molecular Phylogenetics and Evolution*, **1**, 242–252.

Bandelt, H.-J. & Dress, A. (1993). A relational approach to split decomposition. In *Information and Classification*, ed. O. Opitz, B. Lausen, & R. Klar, pp. 123–131. Springer.

Bandelt, H.-J., Forster, P., Sykes, B., & Richards, M. (1995). Mitochondrial portraits of human population using median networks. *Genetics*, **141**, 743–753.

Barnes, I., Matheus, P., Shapiro, B., Jensen, D., & Cooper, A. (2002). Dynamics of Pleistocene population extinctions in Beringian brown bears. *Science*, **295**(5563), 2267–2270.

Barton, N. & Etheridge, A. M. (2004). The effect of selection on genealogies. *Genetics*, **166**, 1115–1131.

Bateman, A., Birney, E., Durbin, R., Eddy, S. R., Howe, K. L., & Sonnhammer, E. L. L. (2000). The Pfam protein families database. *Nucleic Acids Research*, **28**, 263–266.

Beaumont, M. A. (1999). Detecting population expansion and decline using microsatellites. *Genetics*, **153**(4), 2013–2029.

Beerli, P. (2006). Comparison of Bayesian and maximum likelihood inference of population genetic parameters. *Bioinformatics*, **22**, 341–345.

Beerli, P. & Felsenstein, J. (1999). Maximum likelihood estimation of migration rates and effective population numbers in two populations using a coalescent approach. *Genetics*, **152**, 763–773.

Benjamini, Y. & Hochberg, Y. (1995). Controlling the false discovery rate: a practical and powerful approach to multiple testing. *Journal Royal Statistics Society B*, **57**(1), 289–300.

Bersaglieri, T., Sabeti, P. C., Patterson, N. *et al.* (2004). Genetic signatures of strong recent positive selection at the lactase gene. *American Journal of Human Genetics*, **74**(6), 1111–1120.

Bininda-Emonds, O. R. (2005). transAlign: using amino acids to facilitate the multiple alignment of protein-coding DNA sequences. *BMC Bioinformatics*, **6**, 156.

Birky, C. W. & Walsh, J. B. (1988). Effects of linkage on rates of molecular evolution. *Proceedings of the National Academy of Sciences, USA*, **85**(17), 6414–6418.

Birney, E., Thompson, J. D., & Gibson, T. J. (1996). PairWise and SearchWise: finding the optimal alignment in a simultaneous comparison of a protein profile against all DNA translation frames. *Nucleic Acids Research*, **24**, 2730–2739.

Bollback, J. P. (2002). Bayesian model adequacy and choice in phylogenetics. *Molecular Biology and Evolution*, **19**(7), 1171–1180.

Boni, M. F., Posada, D., & Feldman, M. W. (2007). An exact nonparametric method for inferring mosaic structure in sequence triplets. *Genetics*, **176**, 1035–1047.

Bray, N. & Pachter, L. (2004). MAVID: constrained ancestral alignment of multiple sequences. *Genome Research*, **14**(4), 693–699.

Bremer, K. (1988). The limits of amino-acid sequence data in angiosperm phylogenetic reconstruction. *Evolution*, **42**, 795–803.

Brenner, S. E., Chothia, C., & Hubbard, T. P. J. (1998). Assessing sequence comparison methods with reliable structurally identified distant evolutionary relationships. *Proceedings of the National Academy of Sciences, USA*, **95**, 6073–6078.

Britten, R. J. (1986). Rates of DNA sequence evolution differ between taxonomic groups. *Science*, **231**, 1393–1398.

Brocchieri, L. & Karlin, S. (1998). A symmetric-iterated multiple alignment of protein sequences. *Journal of Molecular Biology*, **276**(1), 249–264.

Brodie, R., Smith, A. J., Roper, R. L., Tcherepanov, V., & Upton, C. (2004). Base-By-Base: single nucleotide-level analysis of whole viral genome alignments. *BMC Bioinformatics*, **5**, 96.

Bromham, L. & Penny, D. (2003). The modern molecular clock. *National Reviews in Genetics*, **4**(3), 216–224.

Bromham, L., Penny, D., Rambaut, A., & Hendy, M. D. (2000). The power of relative rates tests depends on the data. *Journal of Molecular Evolution*, **50**(3), 296–301.

Brown, C. J., Garner, E. C., Keith Dunker, A., & Joyce, P. (2001). The power to detect recombination using the coalescent. *Molecular Biology of Evolution*, **18**, 1421–1424.

Brudno, M. & Morgenstern, B. (2002). Fast and sensitive alignment of large genomic sequences. *Proceedings of the IEEE Computer Society Bioinformatics Conference*, **1**, 138–147.

Brudno, M., Do, C. B., Cooper, G. M. *et al.* (2003). LAGAN and Multi-LAGAN: efficient tools for large-scale multiple alignment of genomic DNA. *Genome Research*, **13**(4), 721–731.

Bruen, T. C., Philippe, H., & Bryant, D. (2006). A simple and robust statistical test for detecting the presence of recombination. *Genetics*, **172**, 2665–2681.

Bruno, W. J., Socci, N. D., & Halpern, A. L. (2000). Weighted neighbor-joining: a likelihood-based approach to distance-based phylogeny reconstruction. *Molecular Biology and Evolution*, **17**, 189–197.

Bryant, D. & Moulton, V. (2004). NeighborNet: an agglomerative method for the construction of phylogenetic networks. *Molecular Biology and Evolution*, **21**, 255–265.

Buendia, P. & Narasimhan, G. (2007). Sliding MinPD: Building evolutionary networks of serial samples via an automated recombination detection approach. *Bioinformatics*, **23**, 2993–3000.

Buneman, P. (1971). The recovery of trees from measures of dissimilarity. In *Mathematics in the Archeological and Historical Sciences*, ed. F. R. Hodson, D. G. Kendall, & P. Tautu, pp. 387–395. Edinburgh, UK: Edinburgh University Press.

Burnham, K. P. & Anderson, D. R. (1998). *Model Selection and Inference: A Practical Information-Theoretic Approach*. New York, NY: Springer-Verlag.

Burnham, K. P. & Anderson, D. R. (2003). *Model Selection and Multimodel Inference: A Practical Information-theoretic Approach*. New York, NY: Springer-Verlag.

Bush, R. M., Bender, C. A., Subbarao, K., Cox, N. J., & Fitch, W. M. (1999). Predicting the evolution of human Influenza A. *Science*, **286**(5446), 1921–1925.

Bustamante, C. D., Nielsen, R., & Hartl, D. L. (2002). A maximum likelihood method for analyzing pseudogene evolution: implications for silent site evolution in humans and rodents. *Molecular Biology and Evolution*, **19**(1), 110–117.

Bustamante, C. D., Nielsen, R., Sawyer, S. A., Olsen, K. M., Purugganan, M. D., & Hartl, D. L. (2002). The cost of inbreeding in Arabidopsis. *Nature*, **416**(6880), 531–534.

Bustamante, C. D., Fledel-Alon, A., Williamson, S. *et al.* (2005). Natural selection on protein-coding genes in the human genome. *Nature*, **437**(7062), 1153–1157.

Cao, Y., Adachi, J., & Hasegawa, M. (1998). Comments on the quartet puizlling method for finding maximum-likelihood tree topologies. *Molecular Biology and Evolution*, **15**(1), 87–89.

Cao, Y., Adachi, J., Janke, A., Paabo, S., & Hasegawa, M. (1994). Phylogenetic relationships among eutherian orders estimated from inferred sequences of mitochondrial proteins: instability of a tree based on a single gene. *Journal of Molecular Evolution*, **39**, 519–527.

Carroll, S. B. (1995). Homeotic genes and the evolution of arthropods and chordates. *Nature*, **376**, 479–485.

Carvajal-Rodriguez, A., Crandall, K. A., & Posada, D. (2006). Recombination estimation under complex evolutionary models with the coalescent composite-likelihood method. *Molecular Biology and Evolution*, **23**, 817–827.

Carvajal-Rodriguez, A., Crandall, K. A., & Posada, D. (2007). Recombination favors the evolution of drug resistance in HIV-1 during antiretroviral therapy. *Infections and Genetic Evolution*, **7**, 476–483.

Castillo-Davis, C. I., Bedford, T. B., & Hartl, D. L. (2004). Accelerated rates of intron gain/loss and protein evolution in duplicate genes in human and mouse malaria parasites. *Molecular Biology and Evolution*, **21**(7), 1422–1427.

Cavalli-Sforza, L. L. & Edwards, A. W. F. (1967). Phylogenetic analysis: Models and estimation procedures. *Evolution*, **32**, 550–570.

Chakrabarti, S., Lanczycki, C. J., Panchenko, A. R., Przytycka, T. M., Thiessen, P. A., & Bryant, S. H. (2006). Refining multiple sequence alignments with conserved core regions. *Nucleic Acids Research*, **34**(9), 2598–2606.

Chamary, J. V., Parmley, J. L., & Hurst, L. D. (2006). Hearing silence: non-neutral evolution at synonymous sites in mammals. *National Reviews in Genetics*, **7**, 98–108.

Chan, C. X., Beiko, R. G., & Ragan, M. A. (2006). Detecting recombination in evolving nucleotide sequences. *BMC Bioinformatics*, **7**, 412.

Chao, L., Tran, T. R., & Matthews, C. (1992). Muller's ratchet and the advantage of sex in the RNA virus $\varphi6$. *Evolution*, **46**, 289–299.

Chao, L. & Tran, T. T. (1997). The advantage of sex in the RNA virus phi6. *Genetics*, **147**, 953–959.

Chare, E. R. & Holmes, E. C. (2006). A phylogenetic survey of recombination frequency in plant RNA viruses. *Archives of Virology*, **151**, 933–946.

Chare, E. R., Gould, E. A., & Holmes, E. C. (2003). Phylogenetic analysis reveals a low rate of homologous recombination in negative-sense RNA viruses. *Journal of General Virology*, **84**, 2691–2703.

Charleston, M. A., Hendy, M. D., & Penny, D. (1994). The effect of sequence length, tree topology, and number of taxa on the performance of phylogenetic methods. *Journal of Computational Biology*, **1**, 133–151.

Chen, Y., Emerson, J. J., & Martin, T. M. (2005). Evolutionary genomics: codon volatility does not detect selection. *Nature*, **433**(7023), E6–E7; discussion E7–E8.

Cho, S., Mitchell, A., Regier, J. C. *et al.* (1995). A highly conserved nuclear gene for low-level phylogenetics: elongation factor-1a recovers morphology-based tree for heliothine moths. *Molecular Biology and Evolution*, **12**, 650–656.

Clamp, M., Cuff, J., Searle, S. M., & Barton, G. J. (2004). The Jalview Java alignment editor. *Bioinformatics*, **20**(3), 426–427.

Comeron, J. M. & Guthrie, T. B. (2005). Intragenic Hill-Robertson interference influences selection intensity on synonymous mutations in *Drosophila*. *Molecular Biology and Evolution*, **22**(12), 2519–2530.

Corpet, F. (1988). Multiple sequence alignment with hierarchical clustering. *Nucleic Acids Research*, **16**(22), 10881–10890.

Cox, D. R. (1961). Tests of separate families of hypotheses. In *Proceedings of the 4th Berkeley Symposium on Mathematical Statistics and Probability*, pp. 105–123. Berkeley, CA, USA: UCB Press.

Cox, D. R. (1962). Further results on tests of separate families of hypotheses. *Journal of the Royal Society of Statistics B*, **24**, 406–424.

Crill, W. D., Wichman, H. A., & Bull, J. J. (2000). Evolutionary reversals during viral adaptation to alternating hosts. *Genetics*, **154**(1), 27–37.

Cuevas, J. M., Moya, A., & Elena, S. F. (2003). Evolution of RNA virus in spatially structured heterogeneous environments. *Journal of Evolutionary Biology*, **16**(3), 456–466.

Dagan, T. & Graur, D. (2005). The comparative method rules! Codon volatility cannot detect positive Darwinian selection using a single genome sequence. *Molecular Biology and Evolution*, **22**(3), 496–500.

Darwin, C. (1859). *On the Origin of Species by Means of Natural Selection.* London: Murray.

Dayhoff, M. O. (ed.) (1978). *Atlas of Protein Sequence and Structure*, vol. 5, Silver Spring, MD: National Biomedical Research Foundation.

Dayhoff, M. O., Schwartz, R. M., & Orcutt, B. C. (1978). A model of evolutionary change in proteins. In *Atlas of Protein Sequence and Structure*, vol. 5, suppl. 3, ed. M. O. Dayhoff, pp. 345–352. Washington DC, USA: National Biomedical Research Foundation.

de Oliveira, T., Miller, R., Tarin, M., & Cassol, S. (2003). An integrated genetic data environment (GDE)-based LINUX interface for analysis of HIV-1 and other microbial sequences. *Bioinformatics*, **19**(1), 153–154.

de Oliveira, T., Deforche, K., Cassol, S. *et al.* (2005). An automated genotyping system for analysis of HIV-1 and other microbial sequences. *Bioinformatics*, **21**, 3797–3800.

de Oliveira, T., Pybus, O. G., Rambaut, A. *et al.* (2006). Molecular epidemiology: HIV-1 and HCV sequences from Libyan outbreak. *Nature*, **444**(7121), 836–837.

de Queiroz, K. & Poe, S. (2001). Philosophy and phylogenetic inference: A comparison of likelihood and parsimony methods in the context of Karl Popper's writings on corroboration. *Systematic Biology*, **50**, 305–321.

DeBry, R. W. (2001). Improving interpretation of the decay index for DNA sequence data. *Systematic Biology*, **50**, 742–752.

Debyser, Z., Van Wijngaerden, E., Van Laethem, K. *et al.* (1998). Failure to quantify viral load with two of the three commercial methods in a pregnant woman harbouring an HIV-1 subtype G strain. *AIDS Research and Human Retroviruses*, **14**, 453–459.

Depiereux, E. & Feytmans, E. (1992). MATCH-BOX: a fundamentally new algorithm for the simultaneous alignment of several protein sequences. *Computer Applications in the Biosciences*, **8**(5), 501–509.

Desai, M. M., Fisher, D. S., & Murray, A. W. (2007). The speed of evolution and maintenance of variation in asexual populations. *Current Biology*, **17**(5), 385–394.

Desper, R. & Gascuel, O. (2002). Fast and accurate phylogeny reconstruction algorithms based on the minimum-evolution principle. *Journal of Computational Biology*, **9**, 687–705.

Desper, R. & Gascuel, O. (2004). Theoretical foundation of the balanced minimum evolution method of phylogenetic inference and its relationship to weighted least-squares tree fitting. *Molecular Biology and Evolution*, **21**, 587–598.

Dimmic, M. W., Rest, J. S., Mindell, D. P., & Goldstein, R. A. (2002). rtREV: an amino acid substitution matrix for inference of retrovirus and reverse transcriptase phylogeny. *Journal of Molecular Evolution*, **55**(1), 65–73.

Do, C. B., Mahabhashyam, M. S., Brudno, M., & Batzoglou, S. (2005). ProbCons: probabilistic consistency-based multiple sequence alignment. *Genome Research*, **15**(2), 330–340.

Do, C. B., Woods, D. A., & Batzoglou, S. (2006). *The Tenth Annual International Conference on Computational Molecular Biology*, vol. 3909, CONTRAlign: discriminative training for protein sequence alignment. In *Computational Molecular Biology*, pp. 160–174. Berlin/Heidelberg: Springer.

Domingo, E. (2006). Quasispecies: concept and implications for virology. *Current Topics in Microbiology and Immunology*, **299**, 51–92.

Donoghue, M. J., Olmstead, R. G., Smith, J. F., & Palmer, J. D. (1992). Phylogenetic relationships of Dipsacales based on rbcL sequences. *Annals of the Missouri Botanical Garden*, **79**, 333–345.

Doolittle, R. F. (1987). *Of URFs and ORFs*. University Science Books.

Dopazo, J., Dress, A., & von Haeseler, A. (1993). Split decomposition: A technique to analyze viral evolution. *Proceedings of the National Academy of Sciences, USA*, **90**, 10320–10324.

Dorit, R. L., Akashi, H., & Gilbert, W. (1995). Absence of polymorphism at the ZFY locus on the human Y chromosome. *Science*, **268**, 1183–1185.

Dress, A. & Krüger, M. (1987). Parsimonious phylogenetic trees in metric spaces and simulated annealing. *Advances in Applied Mathematics*, **8**, 8–37.

Dress, A. & Wetzel, R. (1993). The human organism – a place to thrive for the immuno-deficiency virus. In *Proceedings of IFCS*, Pairs.

Dress, A., Huson, D., & Moulton, V. (1996). Analyzing and visualizing sequence and distance data using SplitsTree. *Discrete and Applied Mathematics*, **71**, 95–109.

Dress, A., Hendy, M., Huber, K., & Moulton, V. (1997). On the number of vertices and branches in the Buneman graph. *Annals in Combinatorics*, **1**, 329–337.

Drouin, G., Prat, F., Ell, M., & Clarke, G. D. (1999). Detecting and characterizing gene conversions between multigene family members. *Molecular Biology and Evolution*, **16**, 1369–1390.

Drummond, A. & Strimmer, K. (2001). PAL: an object-oriented programming library for molecular evolution and phylogenetics. *Bioinformatics*, **17**(7), 662–663.

Drummond, A. J., Ho, S. Y. W., Phillips, M. J., & Rambaut, A. (2006). Relaxed phylogenetics and dating with confidence. *PLoS Biology*, **4**(5).

Drummond, A. J., Nicholls, G. K., Rodrigo, A. G., & Solomon, W. (2002). Estimating mutation parameters, population history and genealogy simultaneously from temporally spaced sequence data. *Genetics*, **161**(3), 1307–1320.

Drummond, A. J., Pybus, O. G., Rambaut, A., Forsberg, R., & Rodrigo, A. G. (2003). Measurably evolving populations. *Trends in Ecology and Evolution*, **18**(9), 481–488.

Drummond, A. J., Rambaut, A., Shapiro, B., & Pybus, O. G. (2005). Bayesian coalescent inference of past population dynamics from molecular sequences. *Molecular Biology and Evolution*, **22**, 1185–1192.

Durbin, R., Eddy, S., Krogh, A., & Mitchison, G. (1998). *Biological Sequence Analysis*. Cambridge, UK: Cambridge University Press.

Dutheil, J., Pupko, T., Alain, J.-M., & Galtier, N. (2005). A model-based approach for detecting coevolving positions in a molecule. *Molecular Biology and Evolution*, **22**(9), 1919–1928.

Edgar, R. C. & Batzoglou, S. (2006). Multiple sequence alignment. *Current Opinions on Structural Biology*, **16**(3), 368–373.

Edgar, R. C. & Sjolander, K. (2003). SATCHMO: sequence alignment and tree construction using hidden Markov models. *Bioinformatics*, **19**(11), 1404–1411.

Edgar, R. C. (2004a). MUSCLE: a multiple sequence alignment method with reduced time and space complexity. *BMC Bioinformatics*, **5**, 113.

Edgar, R. C. (2004b). MUSCLE: multiple sequence alignment with high accuracy and high throughput. *Nucleic Acids Research*, **32**(5), 1792–1797.

Edwards, A. W. F. (1972). *Likelihood*. Cambridge, UK: Cambridge University Press.

Efron, B. (1979). Bootstrap methods: another look at the jackknife. *Annals of Statistics*, **7**, 1–26.

Efron, B. & Gong, G. (1983). A leisurely look at the bootstrap, the jackknife, and cross-validation. *American Statistician*, **37**, 36–48.

Efron, B. & Tibshirani, R. (1994). *An Introduction to the Bootstrap*. New York: Chapman and Hall.

Efron, B., Halloran, E., & Holmes, S. (1996). Bootstrap confidence levels for phylogenetic trees. *Proceedings of the National Academy of Sciences, USA*, **93**(23), 13429–13434.

Eigen, M. & Biebricher, C. (1988). Sequence space and quasispecies distribution. In *RNA Genetics*, vol. 3, ed. E. Domingo, J. J. Holland, & P. Ahlquist, pp. 211–245. Boca Raton, Fl: CRC Press.

Endo, T., Ikeo, K., & Gojobori, T. (1996). Large-scale search for genes on which positive selection may operate. *Molecular Biology and Evolution*, **13**(5), 685–690.

Etherington, G. J., Dicks, J., & Roberts, I. N. (2005). Recombination Analysis Tool (RAT): a program for the high-throughput detection of recombination. *Bioinformatics*, **21**, 278–281.

Evans, J., Sheneman, L., & Foster, J. (2006). Relaxed neighbor joining: a fast distance-based phylogenetic tree construction method. *Journal of Molecular Evolution*, **62**(6), 785–792.

Evans, P. D., Mekel-Bobrov, N., Vallender, E. J., Hudson, R. R., & Lahn, B. T. (2006). Evidence that the adaptive allele of the brain size gene microcephalin introgressed into *Homo sapiens* from an archaic Homo lineage. *Proceedings of the National Academy of Sciences, USA*, **103**(48), 18178–18183.

Ewens, W. (1972). The sampling theory of selectively neutral alleles. *Journal of Theoretical Biology*, **3**, 87–112.

Eyre-Walker, A., Woolfit, M., & Phelps, T. (2006). The distribution of fitness effects of new deleterious amino acid mutations in humans. *Genetics*, **173**(2), 891–900.

Faith, D. P. (1991). Cladistic permutation tests for monophyly and nonmonophyly. *Systematic Zoology*, **40**, 366–375.

Fang, F., Ding, J., Minin, V. N., Suchard, M. A., & Dorman, K. S. (2007). cBrother: relaxing parental tree assumptions for Bayesian recombination detection. *Bioinformatics*, **23**, 507–508.

Farris, J. S. (1970). Estimating phylogenetic trees from distance matrixes. *American Nature*, **106**, 645–668.

Farris, J. S. (1977). On the phenetic approach to vertebrate classification. In *Major Patterns in Vertebrate Evolution*, ed. M. K. Hecht, P. C. Goody, & B. M. Hecht, pp. 823–850. New York: Plenum Press.

Fay, J. C. & Wu, C. I. (2000). Hitchhiking under positive Darwinian selection. *Genetics*, **155**(3), 1405–1413.

Fearnhead, P. & Donnelly, P. (2001). Estimating recombination rates from population genetic data. *Genetics*, **159**, 1299–1318.

Felsenstein, J. (1974). The evolutionary advantage of recombination. *Genetics*, **78**, 737–756.

Felsenstein, J. (1978a). Cases in which parsimony and compatibility methods will be positively misleading. *Systematic Zoology*, **27**, 401–410.

Felsenstein, J. (1978b). The number of evolutionary trees. *Systematic Zoology*, **27**, 27–33.

Felsenstein, J. (1981). Evolutionary trees from DNA sequences: a maximum likelihood approach. *Journal of Molecular Evolution*, **17**, 368–376.

Felsenstein, J. (1982). Numerical methods for inferring evolutionary trees. *Quarterly Review of Biology*, **57**, 379–404.

Felsenstein, J. (1984). Distance methods for inferring phylogenies: a justification. *Evolution*, **38**, 16–24.

Felsenstein, J. (1985). Confidence limits on phylogenies: an approach using the bootstrap. *Evolution*, **39**, 783–791.

Felsenstein, J. (1988). Phylogenies from molecular sequences: inference and reliability. *Annual Review of Genetics*, **22**, 521–565.

Felsenstein, J. (1989). Phylip – phylogeny inference package (version 3.2). *Cladistics*, **5**, 164–166.

Felsenstein, J. (1993). PHYLIP (Phylogeny Inference Package) version 3.5c. Distributed by the author. Department of Genetics, University of Washington, Seattle.

Felsenstein, J. (1996). PHYLIP: Phylogeny Inference Package, Version 3.572c. Seattle, WA: University of Washington.

Felsenstein, J. (2004). *Inferring Phylogenies*. Sunderland, MA: Sinauer.

Felsenstein, J. (2006). Accuracy of coalescent likelihood estimates: do we need more sites, more sequences, or more loci? *Molecular Biology and Evolution*, **23**, 691–700.

Felsenstein, J. & Churchill, G. A. (1996). A Hidden Markov model approach to variation among sites in rate of evolution. *Molecular Biology and Evolution*, **13**(1), 93–104.

Felsenstein, J. & Kishino, H. (1993). Is there something wrong with the bootstrap on phylogenies? A reply to Hillis and Bull. *Systematic Biology*, **42**, 193–200.

Felsenstein, J., Kuhner, M. K., Yamato, J., & Beerli, P. (1999). Likelihoods on coalescents: a Monte Carlo sampling approach to inferring parameters from population samples of molecular data. In *Statistics in Genetics and Molecular Biology*, ed. F. Seillier-Moiseiwitsch, IMS Lecture Notes-Monograph Series, pp. 163–185.

Feng, D. F. & Doolittle, R. F. (1987). Progressive sequence alignment as a prerequisite to correct phylogenetic trees. *Journal of Molecular Evolution*, **25**(4), 351–360.

Fernandes, A. P., Nelson, K., & Beverley, S. M. (1993). Evolution of nuclear ribosomal RNAs in kinetoplastid protozoa: perspectives on the age and origins of parasitism. *Proceedings of the National Academy of Sciences, USA*, **90**, 11608–11612.

Fisher, R. A. (1930). *The Genetical Theory of Natural Selection*. UK: Clarendon Press.

Fisher, R. A. (1971). *Design of Experiments*. 9th edn., Macmillan.

Fitch, W. (1981). A non-sequential method for constructing trees and hierarchical classifications. *Journal of Molecular Evolution*, **18**, 30–37.

Fitch, W. M. (1971). Toward defining the course of evolution: minimum change for a specific tree topology. *Systematic Zoology*, **20**, 406–416.

Fitch, W. M. & Margoliash, E. (1967). Construction of phylogenetic trees: A method based on mutation distances as estimated from cytochrome c sequences is of general applicability. *Science*, **155**, 279–284.

Fitch, W. M. & Margoliash, E. (1967). Construction of phylogenetic trees. *Science*, **155**, 279–284.

Fleissner, R., Metzler, D., & von Haeseler, A. (2005). Simultaneous statistical multiple alignment and phylogeny reconstruction. *Systematic Biology*, **54**, 548–561.

Friedlander, T. P., Horst, K. R., Regier, J. C., Mitter, C., Peigler, R. S., & Fez, Q. Q. (1998). Two nuclear genes yield concordant relationships within Attacini (Lepidoptera: Saturniidae) *Molecular Phylogenetics and Evolution*, **9**, 131–140.

Friedman, R. & Hughes, A. L. (2005). Codon volatility as an indicator of positive selection: data from eukaryotic genome comparisons. *Molecular Biology and Evolution*, **22**(3), 542–546.

Frost, S. D. W., Nijhuis, M., Schuurman, R., Boucher, C. A. B., & Leigh Brown, A. J. (2000). Evolution of lamivudine resistance in human immunodeficiency virus type 1-infected individuals: the relative roles of drift and selection. *Journal of Virology*, **74**(14), 6262–6268.

Frost, S. D. W., Liu, Y., Kosakovsky Pond, S. L. *et al.* (2005a). Characterization of human immunodeficiency virus type 1 (HIV-1) envelope variation and neutralizing antibody responses during transmission of HIV-1 subtype B. *Journal of Virology*, **79**(10), 6523–6527.

Frost, S. D. W., Wrin, T., Smith, D. M. *et al.* (2005b). Neutralizing antibody responses drive the evolution of human immunodeficiency virus type 1 envelope during recent HIV infection. *Proceedings of the National Academy of Sciences, USA*, **102**(51), 18514–18519.

Fu, Y. X. & Li, W. H. (1993). Statistical tests of neutrality of mutations. *Genetics*, **133**(3), 693–709.

Galtier, N., Gouy, M., & Gautier, C. (1996). SeaView and Phylo_win, two graphic tools for sequence alignment and molecular phylogeny. *Computer Applications in the Biosciences*, **12**, 543–548.

Gao, F., Robertson, D. L., Carruthers, C. D. *et al.* (1998). A comprehensive panel of near-full-length clones and reference sequences for non-subtype B isolates of human immunodeficiency virus type 1. *Journal of Virology*, **72**, 5680–5698.

Gascuel, O. (1997). BIONJ: an improved version of the NJ algorithm based on a simple model of sequence data. *Molecular Biology and Evolution*, **14**(7), 685–695.

Gelman, A. & Rubin, D. B. (1992). Inference from iterative simulation using multiple sequences. *Statistical Science*, **7**(4), 457–511.

Georgescu, M. M., Delpeyroux, F., Tardy-Panit, M. *et al.* (1994). High diversity of poliovirus strains isolated from the central nervous system from patients with vaccine-associated paralytic poliomyelitis. *Journal of Virology*, **68**, 8089–8101.

Gerrish, P. J. & Lenski, R. E. (1998). The fate of competing beneficial mutations in an asexual population. *Genetica*, **103**, 127–144.

Geyer, C. J. (1991). Markov chain Monte Carlo maximum likelihood. In *Computing Science and Statistics: Proceedings of the 23rd Symposium on the Interface,* ed. E. M. Keramidas & S. M. Kaufman, pp. 156–163.

Gibbs, M. J., Armstrong, J. S., & Gibbs, A. J. (2000). Sister-scanning: a Monte Carlo procedure for assessing signals in recombinant sequences. *Bioinformatics*, **16**, 573–582.

Gibbs, M. J. & Weiller, G. F. (1999). Evidence that a plant virus switched hosts to infect a vertebrate and then recombined with a vertebrate-infecting virus. *Proceedings of the National Academy of Sciences, USA*, **96**, 8022–8027.

Gillespie, J. H. (1984). The molecular clock may be an episodic clock. *Proceedings of the National Academy of Sciences, USA*, **81**(24), 8009–8013.

Gillespie, J. H. (1991). *The Causes of Molecular Evolution*. Oxford, UK: Oxford University Press.

Gogarten, J. P., Kibak, H., Dittrich, P. *et al.* (1989). Evolution of the vacuolar H+ -ATPase: implications for the origin of eukaryotes. *Proceedings of the National Academy of Sciences, USA*, **86**, 6661–6665.

Goldman, N. (1993). Statistical tests of models of DNA substitution. *Journal of Molecular Evolution*, **36**(2), 182–198.

Goldman, N. & Yang, Z. (1994). A codon-based model of nucleotide substitution for protein-coding DNA sequences. *Molecular Biology and Evolution*, **11**(5), 725–736.

Goldman, N., Anderson, J. P., & Rodrigo, A. G. (2000). Likelihood-based tests of topologies in phylogenetics. *Systematic Biology*, **49**, 652–670.

Goloboff, P. (1999). Analyzing large data sets in reasonable times: Solutions for composite optima. *Cladistics*, **15**, 415–428.

Gonnet, G. H., Cohen, M. A., & Benner, S. A. (1992). Exhaustive matching of the entire protein sequence database. *Science*, **256**, 1443–1445.

Goodman, S. N. (1999). Toward evidence-based medical statistics. 1: The P value fallacy. *Annals of Internal Medicine*, **130**, 1019–1021.

Gotoh, O. (1982). An improved algorithm for matching biological sequences. *Journal of Molecular Biology*, **162**, 705–708.

Gotoh, O. (1995). A weighting system and algorithm for aligning many phylogenetically related sequences. *Computer Applications in the Biosciences*, **11**(5), 543–551.

Gotoh, O. (1996). Significant improvement in accuracy of multiple protein sequence alignments by iterative refinements as assessed by reference to structural alignments. *Journal of Molecular Biology*, **264**, 823–838.

Gotoh, O. (1999). Multiple sequence alignment: algorithms and applications. *Advances in Biophysics*, **36**, 159–206.

Graessmann, M., Graessmann, A., Cadavid, E. O. *et al.* (1992). Characterization of the elongation factor 1-α gene of *Rhynchosciara americana*. *Nucleic Acids Research*, **20**, 3780.

Graham, J., McNeney, B., & Seillier-Moiseiwitsch, F. (2005). Stepwise detection of recombination breakpoints in sequence alignments. *Bioinformatics*, **21**, 589–595.

Grassly, N. C. & Holmes, E. C. (1997). A likelihood method for the detection of selection and recombination using nucleotide sequences.*Molecular Biology and Evolution*, **14**, 239–247.

Gribskov, M., McLachlan, A. D., & Eisenberg, D. (1987). Profile analysis: detection of distantly related proteins. *Proceedings of the National Academy of Sciences, USA*, **84**, 4355–4358.

Griffiths, R. C. & Tavare, S. (1994). Sampling theory for neutral alleles in a varying environment. *Philosophical Transactions of the Royal Society London B Biological Sciences*, **344**, 403–410.

Guindon, S. & Gascuel, O. (2003). A simple, fast, and accurate algorithm to estimate large phylogenies by maximum likelihood. *Systematic Biology*, **52**(5), 696–704.

Guindon, S., Rodrigo, A. G., Dyer, K. A., & Huelsenbeck, J. P. (2004). Modeling the site-specific variation of selection patterns along lineages. *Proceedings of the National Academy of Sciences, USA*, **101**(35), 12957–12962.

Gumbel, E. J. (1958). *Statistics of Extremes*. New York, NY: Columbia University Press.

Hall, P. & Wilson, S. R. (1991). Two guidelines for bootstrap hypothesis testing. *Biometrics*, **47**, 757–762.

Hall, T. A. (1999). BioEdit: a user-friendly biological sequence alignment editor and analysis program for Windows 95/98/NT. *Nucleic Acids Symposium Series*, **41**, 95–98.

Hancock, J. M. & Armstrong, J. S. (1994). SIMPLE34: an improved and enhanced implementation for VAX and Sun computers of the SIMPLE algorithm for analysis of clustered repetitive motifs in nucleotide sequences. *Computer Applications in the Biosciences*, **10**, 67–70.

Hannaert, V., Blaauw, M., Kohl, L., Allert, S., Opperdoes, F. R., & Michels, P. A. (1992). Molecular analysis of the cytosolic and glycosomal glyceraldehyde-3-phosphate dehydrogenase in *Leishmania mexicana*. *Molecular Biochemical Parasitology*, **55**, 115–126.

Hannaert, V., Saavedra, E., Duffieux, F. *et al.* (2003). Plant-like traits associated with metabolism of *Trypanosoma* parasites. *Proceedings of the National Academy of Sciences, USA*, **100**, 1067–1071.

Hartigan, J. A. (1973). Minimum mutation fits to a given tree. *Biometrics*, **29**, 53–65.

Hartl, D. L. & Clark, A. G. (1997). *Principles of Population Genetics*. Sunderland, MA: Sinauer Associates, Inc.

Hasegawa, M. & Kishino, H. (1989). Confidence limits on the maximum-likelihood estimate of the hominid tree from mitochondrial-DNA sequences. *Evolution*, **43**(3), 672–677.

Hasegawa, M., Cao, Y., & Yang, Z. (1998). Preponderance of slightly deleterious polymorphism in mitochondrial DNA: nonsynonymous/synonymous rate ratio is much higher within species than between species. *Molecular Biology and Evolution*, **15**(11), 1499–1505.

Hasegawa, M., Kishino, H., & Yano, T. A. (1985). Dating of the human ape splitting by a molecular clock of mitochondrial-DNA. *Journal of Molecular Evolution*, **22**, 160–174.

Hastings, W. K. (1970). Monte Carlo sampling methods using Markov chains and their applications. *Biometrika*, **57**, 97–109.

Hastings, W. K. (1970). Monte Carlo sampling methods using Markov chains and their applications. *Biometrika*, **57**, 97–109.

Hedges, S. B. (1992). The number of replications needed for accurate estimation of the bootstrap P value in phylogenetic studies. *Molecular Biology and Evolution*, **9**, 366–369.

Hedges, S. B. (1994). Molecular evidence for the origin of birds. *Proceedings of the National Academy of Sciences, USA*, **91**, 2621–2624.

Hedrick, P. W. (2007). Balancing selection. *Current Biology*, **17**(7), R230–R231.

Hein, J. (1993). A heuristic method to reconstruct the history of sequences subject to recombination. *Journal of Molecular Evolution*, **36**, 396–406.

Hein, J., Schierup, M. H., & Wiuf, C. (2005). *Gene Genealogies, Variation and Evolution*. Oxford, UK: Oxford University Press.

Hendy, M. D. & Penny, D. (1982). Branch-and-bound algorithms to determine minimal evolutionary trees. *Mathematical Biosciences*, **59**, 277–290.

Hendy, M. D. & Penny, D. (1993). Spectral analysis of phylogenetic data. *Journal of Classification*, **10**, 5–24.

Henikoff, S. & Henikoff, J. G. (1992). Amino acid substitution matrices from protein blocks. *Proceedings of the National Academy of Sciences, USA*, **89**(22), 10915–10919.

Hill, W. G. & Robertson, A. (1966). The effect of linkage on limits to artificial selection. *Genetical Research*, **8**, 269–294.

Hillis, D. M. (1996). Inferring complex phylogenies. *Nature*, **383**, 130.

Hillis, D. M. & Bull, J. J. (1993). An empirical test of bootstrapping as a method for assessing confidence in phylogenetic analysis. *Systematic Biology*, **42**, 182–192.

Hillis, D. M., Huelsenbeck, J. P., & Swofford, D. L. (1994). Hobgoblin of phylogenetics. *Nature*, **369**, 363–364.

Hillis, D. M., Moritz, C., & Mable, B. K. (1996). *Molecular Systematics*. Sunderland, MA: Sinauer Associates.

Hoeting, J. A., Madigan, D., & Raftery, A. E. (1999). Bayesian model averaging: a tutorial. *Statistical Science*, **14**(4), 382–417.

Hogeweg, P. & Hesper, B. (1984). The alignment of sets of sequences and the construction of phylogenetic trees. An integrated method. *Journal of Molecular Evolution*, **20**, 175–186.

Holland, B. & Moulton, V. (2003). Consensus networks: a method for visualising incompatibilities in collections of trees. In *Proceedings of the 3rd Workshop on Algorithms in Bioinformatics (WABI'03), Volume 2812 of Lecture Notes in Bioinformatics*, ed. G. Benson & R. Page, pp. 165–176. Berlin/Heidelberg, Germany: Springer-Verlag.

Holland, B., Huber, K., Moulton, V., & Lockhart, P. (2004). Using consensus networks to visualize contradictory evidence for species phylogeny. *Molecular Biology and Evolution*, **21**, 1459–1461.

Holland, B., Delsuc, F., & Moulton, V. (2005). Visualizing conflicting evolutionary hypotheses in large collections of trees using consensus networks. *Systematic Biology*, **54**, 56–65.

Holmes, E. C., Worobey, M., & Rambaut, A. (1999). Phylogenetic evidence for recombination in dengue virus. *Molecular Biology and Evolution*, **16**, 405–409.

Holmes, E. C., Zhang, L. Q., Simmonds, P., Ludlam, C. A., & Brown, A. J. L. (1992). Convergent and divergent sequence evolution in the surface envelope glycoprotein of human-immunodeficiency-virus type-1 within a single infected patient. *Proceedings of the National Academy of Sciences, USA*, **89**(11), 4835–4839.

Hordijk, W. & Gascuel, O. (2006). Improving the efficiency of SPR moves in phylogenetic tree search methods based on maximum likelihood. *Bioinformatics*, **21**, 4338–4347.

Huang, X. (1994). On global sequence alignment. *Computer Applications in the Biosciences*, **10**(3), 227–235.

Huber, K. & Moulton, V. (2005). Phylogenetic networks. In *Mathematics of Evolution and Phylogeny*, ed. O. Gascuel, pp. 178–204. Oxford, UK: Oxford University Press.

Huber, K., Moulton, V., Lockhart, P., & Dress, A. (2001). Pruned median networks: a technique for reducing the complexity of median networks. *Molecular Phylogenetics and Evolution*, **19**, 302–310.

Huber, T., Faulkner, G., & Hugenholtz, P. (2004). Bellerophon: a program to detect chimeric sequences in multiple sequence alignments. *Bioinformatics*, **20**, 2317–2319.

Hudson, R. R. (1985). The sampling distribution of linkage disequilibrium under an infinite allele model without selection. *Genetics*, **109**, 611–631.

Hudson, R. R. & Kaplan, N. L. (1988). The coalescent process in models with selection and recombination. *Genetics*, **120**, 831–840.

Huelsenbeck, J. P. & Crandall, K. A. (1997). Phylogeny estimation and hypothesis testing using maximum likelihood. *Annual Reviews of Ecology Systems*, **28**, 437–466.

Huelsenbeck, J. P. & Dyer, K. A. (2004). Bayesian estimation of positively selected sites. *Journal of Molecular Evolution*, **58**(6), 661–672.

Huelsenbeck, J. P. & Hillis, D. M. (1993). Success of phylogenetic methods in the four-taxon case. *Systematic Biology*, **42**, 247–264.

Huelsenbeck, J. P. & Rannala, B. (2004). Frequentist properties of Bayesian posterior probabilities of phylogenetic trees under simple and complex substitution models. *Systematic Biology*, **53**(6), 904–913.

Huelsenbeck, J. P. & Ronquist, F. (2001). MrBayes: Bayesian inference of phylogenetic trees. *Bioinformatics*, **17**, 754–755.

Huelsenbeck, J. P. & Ronquist, F. (2005). Bayesian analysis of molecular evolution using MrBayes. In *Statistical Methods in Molecular Evolution*, ed. R. Nielsen, pp. 183–232. New York: Springer.

Huelsenbeck, J. P., Jain, S., Frost, S. W. D., & Kosakovsky Pond, S. L. (2006). A Dirichlet process model for detecting positive selection in protein-coding DNA sequences. *Proceedings of the National Academy of Sciences, USA*, **103**(16), 6263–6268.

Huelsenbeck, J. P., Larget, B., & Alfaro, M. E. (2004). Bayesian phylogenetic model selection using reversible jump Markov chain Monte Carlo. *Molecular Biology and Evolution*, **21**(6), 1123–1133.

Hughes, A. L. & Yeager, M. (1997). Comparative evolutionary rates of introns and exons in murine rodents. *Journal of Molecular Evolution*, **45**(2), 125–130.

Hughey, R. & Krogh, A. (1996). Hidden Markov models for sequence analysis: extension and analysis of the basic method. *Computer Applications in the Biosciences*, **12**(2), 95–107.

Husmeier, D. (2005). Discriminating between rate heterogeneity and interspecific recombination in DNA sequence alignments with phylogenetic factorial hidden Markov models. *Bioinformatics*, **21**, 166–172.

Husmeier, D. & McGuire, G. (2003). Detecting recombination in 4-taxa DNA sequence alignments with Bayesian hidden Markov models and Markov chain Monte Carlo. *Molecular Biology and Evolution*, **20**, 315–337.

Husmeier, D. & Wright, F. (2001a). Detection of recombination in DNA multiple alignments with hidden Markov models. *Journal of Computing Biology*, **8**, 401–427.

Husmeier, D. & Wright, F. (2001b). Probabilistic divergence measures for detecting interspecies recombination. *Bioinformatics*, **17**, S123–S131.

Huson, D. (1998). SplitsTree: a program for analyzing and visualizing evolutionary data. *Bioinformatics*, **14**, 68–73.

Huson, D. & Dress, A. (2004). Constructing split graphs, *IEEE Transactions on Computational Biology and Bioinformatics*, **1**, 109–115.

Huson, D. & Bryant, D. (2006). Application of phylogenetic networks in evolutionary studies. *Molecular Biology and Evolution*, **23**, 254–267.

Huson, D., Dezulian, T., Kloepper, T., & Steel, M. (2004). Phylogenetic super-networks from partial trees. *IEEE Transactions on Computational Biology and Bioinformatics*, **1**, 151–158.

Ina, Y. (1995). New methods for estimating the numbers of synonymous and nonsynonymous substitutions. *Journal of Molecular Evolution*, **40**(2), 190–226.

Ingman, M., Kaessmann, H., Paabo, S., & Gyllensten, U. (2000). Mitochondrial genome variation and the origin of modern humans. *Nature*, **408**(6813), 708–713.

Jakobsen, I. B. & Easteal, S. (1996). A program for calculating and displaying compatibility matrices as an aid in determining reticulate evolution in molecular sequences. *Computing in Applied Biosciences*, **12**, 291–295.

Jakobsen, I. B., Wilson, S. R., & Easteal, S. (1997). The partition matrix: exploring variable phylogenetic signals along nucleotide sequence alignments. *Molecular Biology and Evolution*, **14**, 474–484.

Jeffreys, A. J., Murray, J., & Neumann, R. (1998). High-resolution mapping of crossovers in human sperm defines a minisatellite-associated recombination hotspot. *Molecular Cell*, **2**, 267–273.

Jeffreys, A. J., Ritchie, A., & Neumann, R. (2000). High resolution analysis of haplotype diversity and meiotic crossover in the human TAP2 recombination hotspot. *Human Molecular Genetics*, **9**, 725–733.

Jenkins, G. M., Rambaut, A., Pybus, O. G., & Holmes, E. C. (2002). Rates of molecular evolution in RNA viruses: a quantitative phylogenetic analysis. *Journal of Molecular Evolution*, **54**(2), 156–165.

Jensen, J. D., Kim, Y., DuMont, V. B., Aquadro, C. F., & Bustamante, C. D. (2005). Distinguishing between selective sweeps and demography using DNA polymorphism data. *Genetics*, **170**, 1401–1410.

Johnson, J. B. & Omland, K. S. (2004). Model selection in ecology and evolution. *TREE*, **19**, 101–108.

Jolley, K. A., Feil, E. J., Chan, M. S., & Maiden, M. C. (2001). Sequence type analysis and recombinational tests (START). *Bioinformatics*, **17**, 1230–1231.

Jones, D. T., Taylor, W. R., & Thornton, J. M. (1992). The rapid generation of mutation data matrices from protein sequences. *Computer Applications in the Biosciences*, **8**, 275–282.

Jukes, T. H. & Cantor, C. R. (1969). Evolution of protein molecules. In *Mammalian Protein Metabolism*, ed. H. H. Munro, Vol. III, pp. 21–132. New York: Academic Press.

Kaplan, N. L. & Hudson, R. R. (1985). The use of sample genealogies for studying a selectively neutral m-loci model with recombination. *Theoretical Population Biology*, **28**, 382–396.

Karlin, S. & Altschul, S. F. (1990). Methods for assessing the statistical significance of molecular sequence features by using general scoring schemes. *Proceedings of the National Academy of Sciences, USA*, **87**, 2264–2268.

Karlin, S. & Altschul, S. F. (1993). Applications and statistics for multiple high-scoring segments in molecular sequences. *Proceedings of the National Academy of Sciences, USA*, **90**, 5873–5877.

Kass, R. E. & Raftery, A. E. (1995). Bayes factors. *Journal of the American Statistical Association*, **90**(430), 773–795.

Katoh, K., Misawa, K., Kuma, K., & Miyata, T. (2002). MAFFT: a novel method for rapid multiple sequence alignment based on fast Fourier transform. *Nucleic Acids Research*, **30**(14), 3059–3066.

Katoh, K., Kuma, K., Toh, H., & Miyata, T. (2005). MAFFT version 5: improvement in accuracy of multiple sequence alignment. *Nucleic Acids Research*, **33**(2), 511–518.

Kececioglu, J. (1993). The maximum weight trace problem in multiple sequence alignment. *Proceedings of the 4th Symposium on Combinatorial Pattern Matching, Springer-Verlag Lecture Notes in Computer Science*, **684**, 106–119.

Kent, W. J. (2002). BLAT – the BLAST-like alignment tool. *Genome Research*, **12**, 656–664.

Kidd, K. K. & Sgaramella-Zonta, L. A. (1971). Phylogenetic analysis: Concepts and methods. *American Journal of Human Genetics*, **23**, 235–252.

Kim, H., Feil, I. K., Verlinde, C. L. M. J., Petra, P. H., & Hol, W. G. J. (1995). Crystal structure of glycosomal glyceraldehyde-3-phosphate dehydrogenase from *Leishmania mexicana*: implications for structure-based drug design and a new position for the inorganic phosphate binding site. *Biochemistry*, **34**, 14975–14986.

Kimura, M. (1968). Evolutionary rate at the molecular level. *Nature*, **217**, 624–626.

Kimura, M. (1969a). The number of heterozygous nucleotide sites maintained in a finite population due to steady flux of mutations. *Genetics*, **61**, 893–903.

Kimura, M. (1969b). The rate of molecular evolution considered from the standpoint of population genetics. *Proceedings of the National Academy of Sciences, USA*, **63**, 1181–1188.

Kimura, M. (1980). A simple method for estimating evolutionary rate of base substitutions through comparative studies of nucleotide sequences. *Journal of Molecular Evolution*, **16**, 111–120.

Kimura, M. (1981). Estimation of evolutionary distances between homologous nucleotide sequences. *Proceedings of the National Academy of Sciences, USA*, **78**(1), 454–458.

Kimura, M. (1983). *The Neutral Theory of Molecular Evolution*. Cambridge, UK: Cambridge University Press.

Kimura, M. & Ohta, T. (1971). Protein polymorphism as a phase of molecular evolution. *Nature*, **229**, 467–479.

Kimura, M. & Ohta, T. (1972). On the stochastic model for estimation of mutational distance between homologous proteins. *Journal of Molecular Evolution*, **2**, 87–90.

King, J. L. & Jukes, T. H. (1969). Non-Darwinian evolution. *Science*, **164**, 788–798.

Kingman, J. F. C. (1982a). The coalescent. *Stochastic Processes and Their Applications*, **13**, 235–248.

Kingman, J. F. C. (1982b). On the genealogy of large populations. *Journal of Applied Probability*, **19A**, 27–43.

Kingman, J. F. C. (2000). Origins of the coalescent, 1974–1982. *Genetics*, **156**, 1461–1463.

Kirkpatrick, S., Gelatt, Jr., C. D., & Vecchi, M. P. (1983). Optimisation using simulated annealing. *Science*, **220**, 671–680.

Kiryu, H., Tabei, Y., Kin, T., & Asai, K. (2007). Murlet: a practical multiple alignment tool for structural RNA sequences. *Bioinformatics*, **23**, 1588–1589.

Kishino, H. & Hasegawa, M. (1989). Evaluation of the maximum likelihood estimate of the evolutionary tree topologies from DNA sequence data, and the branching order in Hominoidea. *Journal of Molecular Evolution*, **29**, 170–179.

Kishino, H., Miyata, T., & Hasegawa, M. (1990). Maximum likelihood inference of protein phylogeny and the origin of chloroplasts. *Journal of Molecular Evolution*, **31**, 151–160.

Kishino, H., Thorne, J. L., & Bruno, W. J. (2001). Performance of a divergence time estimation method under a probabilistic model of rate evolution. *Molecular Biology of Evolution*, **18**(3), 352–361.

Klotz, L. C., Blanken, R. L., Komar, N., & Mitchell, R. M. (1979). Calculation of evolutionary trees from sequence data. *Proceedings of the National Academy of Sciences, USA*, **76**, 4516–4520.

Kluge, A. G. & Farris, J. S. (1969). Quantitative phyletics and the evolution of anurans. *Systematic Zoology*, **18**, 1–32.

Korber, B., Muldoon, M., Theiler, J. *et al.* (2000). Timing the ancestor of the HIV-1 pandemic strains. *Science*, **288**(5472), 1789–1796.

Korf, I., Yandell, M., & Bedell, J. (2003). *BLAST.* Sebastopol CA, USA: O'Reilly.

Korostensky, C. & Gonnet, G. H. (1999). *Proceedings of the String Processing and Information Retrieval Symposium and International Workshop on Groupwave, Cancun, Mexico*, p. 105.

Kosakovsky Pond, S. L. & Frost, S. D. W. (2005a). A simple hierarchical approach to modeling distributions of substitution rate. *Molecular Biology and Evolution*, **22**, 223–234.

Kosakovsky Pond, S. L. & Frost, S. D. W. (2005b). A genetic algorithm approach to detecting lineage-specific variation in selection pressure. *Molecular Biology and Evolution*, **22**(3), 478–485.

Kosakovsky Pond, S. L. & Frost, S. D. W. (2005c). Not so different after all: a comparison of methods for detecting amino-acid sites under selection. *Molecular Biology and Evolution*, **22**, 1208–1222.

Kosakovsky Pond, S. L. & Muse, S. V. (2004). Column sorting: rapid calculation of the phylogenetic likelihood function. *Systems in Biology*, **53**(5), 685–692.

Kosakovsky Pond, S. L. & Muse, S. V. (2005). Site-to-site variation of synonymous substitution rates. *Molecular Biology and Evolution*, **22**(12), 2375–2385.

Kosakovsky Pond, S. L., Frost, S. D. W., & Muse, S. V. (2005). HyPhy: Hypothesis testing using phylogenies. *Bioinformatics*, **21**(5), 676–679.

Kosakovsky Pond, S. L., Frost, S. D. W., Grossman, Z., Gravenor, M. B., Richman, D. D., & Leigh Brown, A. J. (2006a). Adaptation to different human populations by HIV-1 revealed by codon-based analyses. *PLoS Comparative Biology*, **23**, 2993–3000.

Kosakovsky Pond, S. L., Frost, S. D., & Muse, S. V. (2005). HyPhy: hypothesis testing using phylogenies. *Bioinformatics*, **21**(5), 676–679.

Kosakovsky Pond, S. L., Posada, D., Gravenor, M. B., Woelk, C. H., & Frost, S. D. W. (2006b). GARD: a genetic algorithm for recombination detection. *Bioinformatics*, **22**, 3096–3098.

Kosakovsky Pond, S. L., Posada, D., Gravenor, M. B., Woelk, C. H., & Frost, S. D. W. (2006c). Automated phylogenetic detection of recombination using a genetic algorithm. *Molecular Biology and Evolution*, **23**(10), 1891–1901.

Kotetishvili, M., Syine, O. C., Kreger, A., Morris, Jr, J. G., & Sulakvelidze, A. (2002). Multilocus sequence typing for characterization of clinical and environmental *Salmonella* strains. *Journal of Clinical Microbiology*, **40**(5), 1626–1635.

Krogh, A., Brown, M., Mian, I. S., Sjolander, K., & Haussler, D. (1994). Hidden Markov models in computational biology: applications to protein modeling. *Journal of Molecular Biology*, **235**, 1501–1531.

Krone, S. M. & Neuhauser, C. (1997). Ancestral processes with selection. *Theoretical Population Biology*, **51**, 210–237.

Kuhner, M. K. (2006). LAMARC 2.0: Maximum likelihood and Bayesian estimation of population parameters. *Bioinformatics*, **22**(6), 768–770.

Kuhner, M. K. & Felsenstein, J. (1994). A simulation comparison of phylogeny algorithms under equal and unequal evolutionary rates. *Molecular Biology and Evolution*, **11**, 459–468.

Kuhner, M. K. & Smith, L. P. (2007). Comparing likelihood and Bayesian coalescent estimators of population parameters. *Genetics*, **175**, 155–165.

Kuhner, M. K., Yamato, J., & Felsenstein, J. (1995). Estimating effective population size and mutation rate from sequence data using Metropolis–Hastings sampling. *Genetics*, **140**, 1421–1430.

Kuhner, M. K., Yamato, J., & Felsenstein, J. (1998). Maximum likelihood estimation of population growth rates based on the coalescent. *Genetics*, **149**, 429–434.

Kuhner, M. K., Yamato, J., & Felsenstein, J. (2000). Maximum likelihood estimation of recombination rates from population data. *Genetics*, **156**, 1393–1401.

Kullback, S. & Leibler, R. A. (1951). On information and sufficiency. *Annals of Mathematical Statistics*, **22**(1), 79–86.

Kumar, S. (1996). A stepwise algorithm for finding minimum-evolution trees. *Molecular Biology and Evolution*, **13**, 584–593.

Kumar, S., Tamura, K., & Nei, M. (1994). Mega: molecular evolutionary genetics analysis software for microcomputers. *Bioinformatics*, **10**(2), 189–191.

Kyte, J. & Doolittle, R. F. (1982). A simple method for displaying the hydropathic character of a protein. *Journal of Molecular Biology*, **157**, 105–132.

Lanave, C., Preparata, G., Saccone, C., & Serio, G. (1984). A new method for calculating evolutionary substitution rates. *Journal of Molecular Evolution*, **20**, 86–93.

Lanciotti, R. S., Gubler, D. J., & Trent, D. W. (1997). Molecular evolution and phylogeny of dengue-4 viruses. *Journal of General Virology*, **78**, 2279–2284.

Lartillot, N. & Philippe, H. (2006). Determining Bayes factors using thermodynamic integration. *Systematic Biology*, **55**, 195–207.

Lassmann, T. & Sonnhammer, E. L. (2005). Kalign – an accurate and fast multiple sequence alignment algorithm. *BMC Bioinformatics*, **6**, 298.

Lee, C., Grasso, C., & Sharlow, M. F. (2002). Multiple sequence alignment using partial order graphs. *Bioinformatics*, **18**(3), 452–464.

Legendre, P. & Makarenkov, V. (2002). Reconstruction of biogeographic and evolutionary networks using reticulograms. *Systematic Biology*, **51**, 199–216.

Lemey, P., Pybus, O. G., Rambaut, A. *et al.* (2004). The molecular population genetics of HIV-1 group O. *Genetics*, **167**, 1059–1068.

Lemey, P., Rambaut, A., & Pybus, O. G. (2006). HIV evolutionary dynamics within and among hosts. *AIDS Reviews*, **8**, 155–170.

Lemmon, A. R. & Milinkovitch, M. C. (2002). The metapopulation genetic algorithm: an efficient solution for the problem of large phylogeny estimation. *Proceedings of the National Academy Sciences, USA*, **99**, 10516–10521.

Lemey, P., Kosakovsky Pond, S. L., Drummond, A. J., Pybus, O. *et al.* (2007). Synonymous substitution rates predict HIV disease progression as a result of underlying replication dynamics. *PLoS Comparative Biology*, **3**(2).

Lenstra, J. A., Van Vliet, A., Carnberg, A. C., Van Hemert, F. J., & Mler, W. (1986). Genes coding for the elongation factor EF-1α in *Artemia*. *European Journal of Biochemistry*, **155**, 475–483.

Lewis, P. O. (1998). A genetic algorithm for maximum likelihood phylogeny inference using nucleotide-sequence data. *Molecular Biology and Evolution*, **15**, 277–283.

Li, W. H. (1981). Simple method for constructing phylogenetic trees from distance matrixes. *Proceedings of the National Academy of Sciences of the USA*, **78**, 1085–1089.

Li, W. H. (1993). Unbiased estimation of the rates of synonymous and nonsynonymous substitution. *Journal of Molecular Evolution*, **36**(1), 96–99.

Li, W. H. (1997). *Molecular Evolution*. Sunderland, MA: Sinauer Associates.

Li, W. H., Gojobori, T., & Nei, M. (1981). Pseudogenes as a paradigm of neutral evolution. *Nature*, **292**(5820), 237–239.

Li, W. H., Wu, C. I., & Luo, C. C. (1985). A new method for estimating synonymous and non-synonymous rates of nucleotide substitution considering the relative likelihood of nucleotide and codon changes. *Molecular Biology and Evolution*, **2**(2), 150–174.

Lindgren, B. W. (1976). *Statistical Theory*. 3rd edn., New York: Macmillan.

Linnaeus, C. (1758). *Systema Naturae*. 10th edn. Stockholm.

Liò, P. & Goldman, N. (1998). Models of molecular evolution and phylogeny. *Genome Research*, **8**(12), 1233–1244.

Lipman, D. J., Altschul, S. F., & Kececioglu, J. D. (1989). A tool for multiple sequence alignment. *Proceedings of the National Academy of Sciences, USA*, **86**, 4412–4415.

Lockart, P. J., Steel, M. D., Hendy, M. D., & Penny D. (1994). Recovering evolutionary trees under a more realistic model of evolution. *Molecular Biology and Evolution*, **11**, 605–612.

Lockhart, P., Penny, D., & Meyer, A. (1995). Testing the phylogeny of swordtail fishes using split decomposition and spectral analysis. *Journal of Molecular Evolution*, **41**, 666–674.

Lockhart, P. J., Penny, D., Hendy, M. D., Howe, C. J., Beanland, T. J., & Larkum, A. W. (1992). Controversy on chloroplast origins. *FEBS Letters*, **301**, 127–131.

Lockhart, P. J., Steel, M. A., Hendy, M. D., & Penny, D. (1994). Recovering evolutionary trees under a more realistic model of sequence evolution. *Molecular Biology and Evolution*, **11**, 605–612.

Lole, K. S., Bollinger, R. C., Paranjape, R. S. *et al.* (1999). Full-length human immunodeficiency virus type 1 genomes from subtype C-infected seroconverters in India, with evidence of intersubtype recombination. *Journal of Virology*, **73**, 152–160.

Lopez, P., Forterre, P., & Philippe, H. (1999). The root of the tree of life in the light of the covarion model. *Journal of Molecular Evolution*, **49**, 496–508.

Lundy, M. (1985). Applications of the annealing algorithm to combinatorial problems in statistics. *Biometrika*, **72**, 191–198.

Lunter, G., Miklos, I., Drummond, A., Jensen, J. L., & Hein, J. (2005). Bayesian coestimation of phylogeny and sequence alignment. *BMC Bioinformatics*, **6**(1), 83.

Lyons-Weiler, J., Hoelzer, G. A., & Tausch, R. J. (1996). Relative Apparent Synapomorphy Analysis (RASA) I: the statistical measurement of phylogenetic signal. *Molecular Biology and Evolution*, **13**, 749–757.

Maddison, D. R. (1991). The discovery and importance of multiple islands of most parsimonious trees. *Systematic Zoology*, **40**, 315–328.

Maddison, W. P. & Maddison, D. R. (1989). Interactive analysis of phylogeny and character evolution using the computer program MacClade. *Folia Primatologia (Basel)*, **53**(1–4), 190–202.

Maddison, D. R., Swofford, D. L., & Maddison, W. P. (1997). NEXUS: an extensible file format for systematic information. *Systematic Biology*, **46**(4), 590–621.

Madigan, D. M. & Raftery, A. E. (1994). Model selection and accounting for model uncertainty in graphical models using Occam's Window. *Journal of the American Statistical Association*, **89**, 1335–1346.

Mailund, T. & Pedersen, C. N. (2004). QuickJoin – fast neighbour-joining tree reconstruction. *Bioinformatics*, **20**(17), 3261–3262.

Martin, D. & Rybicki, E. (2000). RDP: detection of recombination amongst aligned sequences. *Bioinformatics*, **16**, 562–563.

Martin, D. P., Posada, D., Crandall, K. A., & Williamson, C. (2005a). A modified bootscan algorithm for automated identification of recombinant sequences and recombination breakpoints. *AIDS Research Human Retroviruses*, **21**, 98–102.

Martin, D. P., Van Der Walt, E., Posada, D., & Rybicki, E. P. (2005b). The evolutionary value of recombination is constrained by genome modularity. *PLoS Genetics*, **1**(4), e51.

Martin, D. P., Williamson, C., & Posada, D. (2005). RDP2: recombination detection and analysis from sequence alignments. *Bioinformatics*, **21**, 260–262.

Matsuda, H. (1995). Construction of phylogenetic trees from amino acid sequences using a genetic algorithm. In *Proceedings of the Genome Informatics Workshop VI*, ed. M. Hagiya, A. Suyama, T. Takagi, K. Nakai, S. Miyano, & T. Yokomori, pp. 19–28. Tokyo: Universal Academy Press.

Maydt, J. & Lengauer, T. (2006). Recco: recombination analysis using cost optimization. *Bioinformatics*, **22**, 1064–1071.

Maynard Smith, J. & Smith, N. H. (1998). Detecting recombination from gene trees. *Molecular Biology and Evolution*, **15**, 590–599.

Maynard Smith, J. (1992). Analyzing the mosaic structure of genes. *Journal of Molecular Evolution*, **34**, 126–129.

Maynard-Smith, J. (1970). Population size, polymorphism, and the rate of non-Darwinian evolution. *American Naturalist*, **104**, 231–236.

McDonald, J. H. & Kreitman, M. (1991). Adaptive protein evolution at the Adh locus in *Drosophila*. *Nature*, **351**(6328), 652–654.

McGinnis, S. & Madden, T. L. (2004). BLAST: at the core of a powerful and diverse set of sequence analysis tools. *Nucleic Acids Research*, **32**, W20–W25.

McGuire, G. & Wright, F. (2000). TOPAL 2.0: Improved detection of mosaic sequences within multiple alignments. *Bioinformatics*, **16**, 130–134.

McGuire, G., Wright, F., & Prentice, M. J. (1997). A graphical method for detecting recombination in phylogenetic data sets. *Molecular Biology and Evolution*, **14**, 1125–1131.

McLaughlin, P. J. & Dayhoff, M. O. (1973). Eukaryote evolution: a view based on cytochrome c sequence data. *Journal of Molecular Evolution*, **2**, 99–116.

McVean, G. A. T. & Charlesworth, B. (2000). The effects of Hill–Robertson interference between weakly selected mutations on patterns of molecular evolution and variation. *Genetics*, **155**(2), 929–944.

McVean, G., Awadalla, P., & Fearnhead, P. (2002). A coalescent-based method for detecting and estimating recombination from gene sequences. *Genetics*, **160**, 1231–1241.

Messier, W. & Stewart, C.-B. (1997). Episodic adaptive evolution of primate lysozymes. *Nature*, **385**, 151–154.

Metropolis, N., Rosenbluth, A. W., Rosenbluth, M. N., Teller, A. H., & Teller, E. (1953). Equation of state calculations by fast computing machines. *Journal of Chemical Physics*, **21**, 1087–1092.

Metz, E. C., Robles-Sikisaka, R., & Vacquier, V. D. (1998). Nonsynonymous substitution in abalone sperm fertilization genes exceeds substitution in introns and mitochondrial DNA. *Proceedings of the National Academy of Sciences, USA*, **95**(18), 10676–10681.

Milne, I., Wright, F., Rowe, G., Marshall, D. F., Husmeier, D., & McGuire, G. (2004). TOPALi: software for automatic identification of recombinant sequences within DNA multiple alignments. *Bioinformatics*, **20**, 1806–1807.

Minin, V., Abdo, Z., Joyce, P., & Sullivan, J. (2003). Performance-based selection of likelihood models for phylogeny estimation. *Systematic Biology*, **52**(5), 674–683.

Minin, V. N., Dorman, K. S., Fang, F. *et al.* (2005). Dual multiple change-point model leads to more accurate recombination detection. *Bioinformatics*, **21**, 3034–3042.

Minin, V. N., Dorman, K. S., Fang, F., & Suchard, M. A. (2007). Phylogenetic mapping of recombination hotspots in human immunodeficiency virus via spatially smoothed change-point processes. *Genetics*, **175**, 1773–1785.

Miralles, R., Gerrish, P. J., Moya, A., & Elena, S. F. (1999). Clonal interference and the evolution of RNA viruses. *Science*, **285**(5434), 1745–1747.

Miyata, T. & Yasunaga, T. (1980). Molecular evolution of mRNA: a method for estimating evolutionary rates of synonymous and amino acid substitutions from homologous nucleotide sequences and its application. *Journal of Molecular Evolution*, **16**(1), 23–36.

Moilanen, A. (2001). Simulated evolutionary optimization and local search: Introduction and application to tree search. *Cladistics*, **17**, S12–S25.

Moore, C. B., John, M., James, I. R., Christiansen, F. T., Witt, C. S., & Mallal, S. A. (2002). Evidence of HIV-1 adaptation to HLA-restricted immune responses at a population level. *Science*, **296**(5572), 1439–1443.

Morgenstern, B. (1999). DIALIGN2: improvement of the segment-to-segment approach to multiple sequence alignment. *Bioinformatics*, **15**, 211–218.

Morgenstern, B. (2004). DIALIGN: multiple DNA and protein sequence alignment at BiBiServ. *Nucleic Acids Research*, **32**(Web Server issue), W33–W36.

Morgenstern, B., Dress, A., & Werner, T. (1996). Multiple DNA and protein sequence alignment based on segment-to-segment comparison. *Proceedings of the National Academy of Sciences, USA*, **93**(22), 12098–12103.

Morgenstern, B., Goel, S., Sczyrba, A., & Dress, A. (2003). AltAVisT: comparing alternative multiple sequence alignments. *Bioinformatics*, **19**(3), 425–426.

Morgenstern, B., Prohaska, S. J., Pohler, D., & Stadler, P. F. (2006). Multiple sequence alignment with user-defined anchor points. *Algorithms Molecular Biology*, **1**(1), 6.

Morrison, D. (2005). Networks in phylogenetic analysis: new tools for population biology, *International Journal of Parasitology*, **35**, 567–582.

Moulton, V. & Steel, M. (1999). Retractions of finite distance functions onto tree metrics. *Discrete Applied Mathematics*, **91**, 215–233.

Muller, H. J. (1932). Some genetic aspects of sex. *American Naturalist*, **66**, 118–138.

Muller, T. & Vingron, M. (2000). Modeling amino acid replacement. *Journal of Computational Biology*, **7**(6), 761–776.

Muse, S. V. (1996). Estimating synonymous and nonsynonymous substitution rates. *Molecular Biology and Evolution*, **13**(1), 105–114.

Muse, S. V. (1999). Modeling the molecular evolution of HIV sequences. In *The Evolution of HIV*, ed. K. A. Crandall, Chapter 4, pp. 122–152. Baltimore, MD: The Johns Hopkins University Press.

Muse, S. V. & Gaut, B. S. (1994). A likelihood approach for comparing synonymous and nonsynonymous nucleotide substitution rates, with application to the chloroplast genome. *Molecular Biology and Evolution*, **11**, 715–724.

Myers, S. R. & Griffiths, R. C. (2003). Bounds on the minimum number of recombination events in a sample history. *Genetics*, **163**, 375–394.

Needleman, S. B. & Wunsch, C. D. (1970). A general method applicable to the search for similarities in the amino acid sequence of two proteins. *Journal of Molecular Biology*, **48**, 443–453.

Negroni, M. & Buc, H. (2001). Mechanisms of retroviral recombination. *Annual Reviews in Genetics*, **35**, 275–302.

Nei, M. (1985). *Molecular Evolutionary Genetics*. New York: Columbia University Press.

Nei, M. & Gojobori, T. (1986). Simple methods for estimating the numbers of synonymous and nonsynonymous nucleotide substitutions. *Molecular Biology and Evolution*, **3**(5), 418–426.

Nei, M. & Kumar, S. (2000). *Molecular Evolution and Phylogenetics*. New York: Oxford University Press.

Nei, M., Kumar, S., & Takahashi, K. (1998). The optimization principle in phylogenetic analysis tends to give incorrect topologies when the number of nucleotides or amino acids used is small. *Proceedings of the National Academy of Sciences, USA*, **95**, 12390–12397.

Neuhauser, C. & Krone, S. M. (1997). The genealogy of samples in models with selection. *Genetics*, **145**, 519–534.

Newton, M. A. & Raftery, A. E. (1994). Approximate Bayesian inference with the weighted likelihood bootstrap. *Journal of the Royal Statistics Society Series B*, **56**, 3–48.

Nickle, D. C., Heath, L., Jensen, M. A. *et al.* (2007). HIV-specific probabilistic models of protein evolution. *PLoS ONE*, **2**(6): e503. doi:10.1371/journal.pone.0000503.

Nielsen, R. & Yang, Z. H. (1998). Likelihood models for detecting positively selected amino acid sites and applications to the HIV-1 envelope gene. *Genetics*, **148**, 929–936.

Nielsen, R. & Yang, Z. (2003). Estimating the distribution of selection coefficients from phylogenetic data with applications to mitochondrial and viral DNA. *Molecular Biology and Evolution*, **20**(8), 1231–1239.

Nielsen, R., Williamson, S., Kim, Y., Hubisz, M. J., Clark, A. G., & Bustamante, C. (2005). Genomic scans for selective sweeps using SNP data. *Genome Research*, **15**(11), 1566–1575.

Nieselt-Struwe, K. & von Haeseler, A. (2001). Quartet-mapping, a generalization of the likelihood-mapping procedure. *Molecular Biology and Evolution*, **18**, 1204–1219.

Ning, Z. A., Cox, J., & Mullikin, J. C. (2001). SSAHA: a fast search method for large DNA databases. *Genome Research*, **11**, 1725–1729.

Nixon, K. C. (1999). The parsimony ratchet, a new method for rapid parsimony analysis. *Cladistics*, **15**, 407–414.

Notredame, C. & Higgins, D. G. (1996). SAGA: Sequence alignment by genetic algorithm. *Nucleic Acids Research*, **24**, 1515–1524.

Notredame, C., Holm, L., & Higgins, D. G. (1998). COFFEE: an objective function for multiple sequence alignments. *Bioinformatics*, **14**, 407–422.

Notredame, C., Higgins, D. G., & Heringa, J. (2000). T-Coffee: a novel method for fast and accurate multiple sequence alignment. *Journal of Molecular Biology*, **302**, 205–217.

Nuin, P. A., Wang, Z., & Tillier, E. R. (2006). The accuracy of several multiple sequence alignment programs for proteins. *Bioinformatics*, **7**, 471.

Nylander, J. A. A., Ronquist, F., Huelsenbeck, J. P., & Nieves-Aldrey, J. L. (2004). Bayesian phylogenetic analysis of combined data. *Systematic Biology*, **53**, 47–67.

O'Sullivan, O., Suhre, K., Abergel, C., Higgins, D. G., & Notredame, C. (2004). 3DCoffee: combining protein sequences and structures within multiple sequence alignments. *Journal of Molecular Biology*, **340**(2), 385–395.

Ochman, H. & Moran, N. A. (2001). Genes lost and genes found: evolution of bacterial pathogenesis and symbiosis. *Science*, **292**, 1096–1099.

Ochman, H., Lawrence, J. G., & Groisman, E. A. (2000). Lateral gene transfer and the nature of bacterial innovation. *Nature*, **405**, 299–304.

Ohta, T. (1973). Slightly deleterious mutant substitutions in evolution. *Nature*, **246**, 96–98.

Ohta, T. (1992). The nearly neutral theory of molecular evolution. *Annual Review of Ecology and Systematics*, **23**, 263–286.

Ohta, T. (2000). Mechanisms of molecular evolution. *Philosophical Transactions Royal Society of London B Biological Sciences*, **355**(1403), 1623–1626.

Ohta, T. (2002). Near-neutrality in evolution of genes and gene regulation. *Proceedings of the National Academy of Sciences, USA*, **99**, 16134–16137.

Ohta, T. & Gillespie, J. H. (1996). Development of neutral and nearly neutral theories. *Theory of Population Biology*, **49**(2), 128–142.

Olsen, G. J. (1987). Earliest phylogenetic branchings: Comparing rRNA-based evolutionary trees inferred with various techniques. *Cold Spring Harbor Symposia on Quantitative Biology*, **LII**, 825–837.

Olsen, G. J., Matsuda, H., Hagstrom, R., & Overbeek, R. (1994). fastDNAml: a tool for construction of phylogenetic trees of DNA sequences using maximum likelihood. *Computer Applications in the Biosciences*, **10**, 41–48.

Ota, S. & Li, W. H. (2000). NJML: a hybrid algorithm for the neighbor-joining and maximum-likelihood methods. *Molecular Biology and Evolution*, **17**(9), 1401–1409.

Page, R. D. M. & Holmes, E. C. (1998). *Molecular Evolution: A Phylogenetic Approach*. Oxford, UK: Blackwell Science.

Pagel, M. & Meade, A. (2004). A phylogenetic mixture model for detecting pattern-heterogeneity in gene sequence or character-state data. *Systematic Biology*, **53**(4), 571–581.

Pamilo, P. & Bianchi, N. O. (1993). Evolution of the Zfx and Zfy genes: rates and interdependence between the genes. *Molecular Biology and Evolution*, **10**(2), 271–281.

Papadopoulos, J. S. & Agarwala, R. (2007). COBALT: COnstraint Based ALignment Tool for Multiple Protein Sequences. *Bioinformatics*, **23**, 1073–1079.

Paraskevis, D., Lemey, P., Salemi, M., Suchard, M., Van De Peer, Y., & Vandamme, A. M. (2003). Analysis of the evolutionary relationships of HIV-1 and SIVcpz sequences using Bayesian inference: implications for the origin of HIV-1. *Molecular Biology and Evolution*, **20**, 1986–1996.

Parida, L., Floratos, A., & Rigoutsos, I. I. (1998). MUSCA: an Algorithm for Constrained Alignment of Multiple Data Sequences. *Genome Informatics Series Workshop Genome Information*, **9**, 112–119.

Pascarella, S. & Argos, P. (1992). Analysis of insertions deletions in protein structures. *Journal of Molecular Biology*, **224**, 461–471.

Patterson, C. (ed.) (1987). *Molecules and Morphology in Evolution: Conflict or Compromise?* Cambridge, UK: Cambridge University Press.

PAUP* 4.0 – Phylogenetic Analysis Using Parsimony (*and Other Methods): Sunderland, MA: Sinauer Associates.

Pearson, W. R. (1998). Empirical statistical estimates for sequence similarity searches. *Journal of Molecular Biology*, **276**, 71–84.

Pearson, W. R. & Lipman, D. J. (1988). Improved tools for biological sequence comparison. *Proceedings of the National Academy of Sciences, USA*, **85**, 2444–2448.

Pearson, W. R., Wood, T., Zhang, Z., & Miller, W. (1997). Comparison of DNA sequences with protein sequences. *Genomics*, **46**, 24–36.

Pearson, W. R., Robins, G., & Zhang, T. (1999). Generalized neighbor-joining: More reliable phylogenetic tree reconstruction. *Molecular Biology and Evolution*, **16**, 806–816.

Pei, J. & Grishin, N. V. (2006). MUMMALS: multiple sequence alignment improved by using hidden Markov models with local structural information. *Nucleic Acids Research*, **34**(16), 4364–4374.

Pei, J. & Grishin, N. V. (2007). PROMALS: towards accurate multiple sequence alignments of distantly related proteins. *Bioinformatics*, **23**(7), 802–808.

Pei, J., Sadreyev, R., & Grishin, N. V. (2003). PCMA: fast and accurate multiple sequence alignment based on profile consistency. *Bioinformatics*, **19**(3), 427–428.

Pfaffelhuber, P., Haubold, B., & Wakolbinger, A. (2006). Approximate genealogies under genetic hitchhiking. *Genetics*, **174**, 1995–2008.

Philippe, H. & Forterre, P. (1999). The rooting of the universal tree of life is not reliable. *Journal of Molecular Evolution*, **49**, 509–523.

Phuong, T. M., Do, C. B., Edgar, R. C., & Batzoglou, S. (2006). Multiple alignment of protein sequences with repeats and rearrangements. *Nucleic Acids Research*, **34**(20), 5932–5942.

Pillai, S. K., Kosakovsky Pond, S. L., Woelk, C. H., Richman, D. D., & Smith, D. M. (2005). Codon volatility does not reflect selective pressure on the HIV-1 genome. *Virology*, **336**(2), 137–143.

Plotkin, J. B. & Dushoff, J. (2003). Codon bias and frequency-dependent selection on the hemagglutinin epitopes of influenza A virus. *Proceedings of the National Academy of Sciences, USA*, **100**(12), 7152–7157.

Plotkin, J. B., Dushoff, J., & Fraser, H. B. (2004). Detecting selection using a single genome sequence of *M. tuberculosis* and *P. falciparum. Nature,* **428**(6986), 942–945.

Plotkin, J. B., Dushoff, J., Desai, M. M., & Fraser, H. B. (2006). Codon usage and selection on proteins. *Journal of Molecular Evolution,* **63**(5), 635–653.

Pluzhnikov, A. & Donnelly, P. (1996). Optimal sequencing strategies for surveying molecular genetic diversity. *Genetics,* **144**, 1247–1262.

Pol, D. (2004). Empirical problems of the hierarchical likelihood ratio test for model selection. *Systematic Biology,* **53**(6), 949–962.

Posada, D. & Crandall, K. (2001). Intraspecific gene genealogies: trees grafting into networks. *Trends in Ecology and Evolution,* **16**, 37–45.

Posada, D. (2001). Unveiling the molecular clock in the presence of recombination. *Molecular Biology of Evolution,* **18**(10), 1976–1978.

Posada, D. (2002). Evaluation of methods for detecting recombination from DNA sequences: empirical data. *Molecular Biology and Evolution,* **19**, 708–717.

Posada, D. (2003). Using Modeltest and PAUP* to select a model of nucleotide substitution. In *Current Protocols in Bioinformatics,* ed. A. D. Baxevanis, D. B. Davison, R. D. M. Page *et al.,* pp. 6.5.1–6.5.14. Chichester, UK: John Wiley & Sons, Inc.

Posada, D. (2006). ModelTest Server: a web-based tool for the statistical selection of models of nucleotide substitution online. *Nucleic Acids Research,* **34**, W700–W703.

Posada, D. & Buckley, T. R. (2004). Model selection and model averaging in phylogenetics: advantages of Akaike Information Criterion and Bayesian approaches over likelihood ratio tests. *Systematic Biology,* **53**(5), 793–808.

Posada, D. & Crandall, K. A. (1998). Modeltest: testing the model of DNA substitution. *Bioinformatics,* **14**(9), 817–818.

Posada, D. & Crandall, K. A. (2001a). Evaluation of methods for detecting recombination from DNA sequences: computer simulations. *Proceedings of the National Academy of Sciences, USA,* **98**, 13757–13762.

Posada, D. & Crandall, K. A. (2001b). Intraspecific gene genealogies: trees grafting into networks. *Trends in Ecology and Evolution,* **16**, 37–45.

Posada, D. & Crandall, K. A. (2001c). Selecting the best-fit model of nucleotide substitution. *Systematic Biology,* **50**(4), 580–601.

Posada, D. & Crandall, K. A. (2002). The effect of recombination on the accuracy of phylogeny estimation. *Journal of Molecular Evolution,* **54**, 396–402.

Posada, D., Crandall, K. A., & Holmes, E. C. (2002). Recombination in evolutionary genomics. *Annual Reviews in Genetics,* **36**, 75–97.

Press, W. H., Flannery, B. P., Teukolsky, S. A., & Vetterling, W. T. (1992). *Numerical Recipes in C: The Art of Scientific Computing,* 2nd edn. Cambridge, UK: Cambridge University Press.

Przeworski, M., Coop, G., & Wall, J. D. (2005). The signature of positive selection on standing genetic variation. *Evolution,* **59**, 2312–2323.

Pybus, O. G. (2006). Model selection and the molecular clock. *PLoS Biology,* **4**(5), e151.

Pybus, O. G., Drummond, A. J., Nakano, T., Robertson, B. H., & Rambaut, A. (2003). The epidemiology and iatrogenic transmission of hepatitis C virus in Egypt: a Bayesian coalescent approach. *Molecular Biology and Evolution,* **20**(3), 381–387.

Pybus, O. G., Rambaut, A., Belshaw, R., Freckleton, R. P., Drummond, A. J., & Holmes, E. C. (2007). Phylogenetic evidence for deleterious mutation load in RNA viruses and its contribution to viral evolution. *Molecular Biology and Evolution*, **24**(3), 845–852.

Quenouille, M. H. (1956). Notes on bias in estimation. *Biometrika*, **43**, 353–336.

Raftery, A. E. (1996). Hypothesis testing and model selection. In *Markov Chain Monte Carlo in Practice*, ed. W. R. Gilks, S. Richardson, & D. J. Spiegelhalter, pp. 163–187. London; New York: Chapman & Hall.

Rambaut, A. (2000). Estimating the rate of molecular evolution: incorporating non-contemporaneous sequences into maximum likelihood phylogenies. *Bioinformatics*, **16**(4), 395–399.

Rambaut, A. & Bromham, L. (1998). Estimating divergence dates from molecular sequences. *Molecular Biology of Evolution*, **15**(4), 442–448.

Rambaut, A. & Drummond, A. J. (2003). `Tracer`: Available from *http://evolve.zoo.ox. ac.uk/software/*.

Rambaut, A. & Grassly, N. C. (1997). Seq-Gen: an application for the Monte Carlo simulation of DNA sequence evolution along phylogenetic trees. *Computers Applications Biosciences*, **13**, 235–238.

Rand, D. M. & Kann, L. M. (1996). Excess amino acid polymorphism in mitochondrial DNA: contrasts among genes from *Drosophila*, mice, and humans. *Molecular Biology and Evolution*, **13**(6), 735–748.

Rannala, B. & Yang, Z. H. (2003). Bayes estimation of species divergence times and ancestral population sizes using DNA sequences from multiple loci. *Genetics*, **164**(4), 1645 1656.

Raphael, B., Zhi, D., Tang, H., & Pevzner, P. (2004). A novel method for multiple alignment of sequences with repeated and shuffled elements. *Genome Research*, **14**(11), 2336–2346.

Redelings, B. D. & Suchard, M. A. (2005). Joint Bayesian estimation of alignment and phylogeny. *Systematic Biology*, **54**(3), 401–418.

Regier, J. C. & Shultz, J. W. (1997). Molecular phylogeny of the major arthropod groups indicates polyphyly of crustaceans and a new hypothesis for the origin of hexapods. *Molecular Biology and Evolution*, **14**, 902–913.

Roberts, G. O. & Rosenthal, J. S. (1998). Optimal scaling of discrete approximations to Langevin diffusions. *Journal of the Royal Statistical Society: Series B*, **60**, 255–268.

Roberts, G. O. & Rosenthal, J. S. (2001). Optimal scaling for various Metropolis–Hastings algorithms. *Statistical Science*, **16**, 351–367.

Roberts, G. O. & Rosenthal, J. S. (2006). Examples of adaptive MCMC. Preprint available from http://www.probability.ca/jeff/ftpdir/adaptex.pdf.

Roberts, G. O., Gelman, A., & Gilks, W. R. (1997). Weak convergence and optimal scaling of random walk Metropolis algorithms. *Annals of Applied Probability*, **7**, 110–120.

Robertson, D. L., Sharp, P. M., McCutchan, F. E., & Hahn, B. H. (1995). Recombination in HIV-1. *Nature*, **374**, 124–126.

Robinson, D. M., Jones, D. T., Kishino, H., Goldman, N., & Thorne, J. L. (2003). Protein evolution with dependence among codons due to tertiary structure. *Molecular Biology and Evolution*, **20**(10), 1692–1704.

Robinson, M., Gouy, M., Gautier, C., & Mouchiroud, D. (1998). Sensitivity of the relative-rate test to taxonomic sampling. *Molecular Biology of Evolution*, **15**(9), 1091–1098.

Rodríguez, F., Oliver, J. F., Marín, A., & Medina, J. R. (1990). The general stochastic model of nucleotide substitution. *Journal of Theoretical Biology*, **142**, 485–501.

Rodrigo, A. G. & Felsenstein, J. (1999). Coalescent approaches to HIV population genetics. In *The Evolution of HIV*, ed. K. A. Crandall. Baltimore, MD: Johns Hopkins University Press.

Rodrigo, A. G., Shaper, E. G., Delwart, E. L. *et al.* (1999). Coalescent estimates of HIV-1 generation time in vivo. *Proceedings of the National Academy of Sciences, USA*, **96**, 2187–2191.

Rodrigue, N., Lartillot, N., Bryant, D., & Philippe, H. E. (2005). Site interdependence attributed to tertiary structure in amino acid sequence evolution. *Gene*, **347**(2), 207–217.

Rodriguez, F., Oliver, J. L., Marin, A., & Medina, J. R. (1990). The general stochastic model of nucleotide substitution. *Journal of Theoretical Biology*, **142**, 485–501.

Rogers, J. S. (1997). On the consistency of maximum likelihood estimation of phylogenetic trees from nucleotide sequences. *Systems in Biology*, **46**(2), 354–357.

Ronquist, F. & Huelsenbeck, J. P. (2003). MRBAYES 3: Bayesian phylogenetic inference under mixed models. *Bioinformatics*, **19**, 1572–1574.

Rosenberg, N. A. & Nordborg, M. (2002). Genealogical trees, coalescent theory, and the analysis of genetic polymorphisms. *Nature Reviews Genetics*, **3**, 380–390.

Roshan, U. & Livesay, D. R. (2006). Probalign: multiple sequence alignment using partition function posterior probabilities. *Bioinformatics*, **22**(22), 2715–2721.

Rousseau, C. M., Learn, G. H., Bhattacharya, T. *et al.* (2007). Extensive intrasubtype recombination in South African human immunodeficiency virus type 1 subtype C infections. *Journal of Virology*, **81**, 4492–4500.

Russo, C. A. M., Takezaki, N., & Nei, M. (1996). Efficiencies of different genes and different tree-building methods in recovering a known vertebrate phylogeny. *Molecular Biology and Evolution*, **13**, 525–536.

Rzhetsky, A. & Nei, M. (1992). A simple method for estimating and testing minimum-evolution trees. *Molecular Biology and Evolution*, **9**, 945–967.

Rzhetsky, A. & Nei, M. (1993). Theoretical foundation of the minimum-evolution method of phylogenetic inference. *Molecular Biology and Evolution*, **10**, 1073–1095.

Sabeti, P. C., Reich, D. E., Higgins, J. M. *et al.* (2002). Detecting recent positive selection in the human genome from haplotype structure. *Nature*, **419**(6909), 832–837.

Sabeti, P. C., Schaffner, S. F., Fry, B. *et al.* (2006). Positive natural selection in the human lineage. *Science*, **312**(5780), 1614–1620.

Saitou, N. & Nei, M. (1987). The neighbor-joining method: a new method for reconstructing phylogenetic trees. *Molecular Biology and Evolution*, **4**(4), 406–425.

Saitou, N. & Imanishi, T. (1989). Relative efficiencies of the Fitch–Margoliash, maximum-parsimony, maximum-likelihood, minimum-evolution, and neighbor-joining methods of phylogenetic tree construction in obtaining the correct tree. *Molecular Biology and Evolution*, **6**, 514–525.

Salemi, M., Desmyter, J., & Vandamme, A. M. (2000). Tempo and mode of human and simian T-lymphotropic virus (HTLV/STLV) evolution revealed by analyses of full-genome sequences. *Molecular Biology and Evolution*, **17**, 374–386.

Salemi, M., Strimmer, K., Hall, W. W. *et al.* (2001). Dating the common ancestor of SIVcpz and HIV-1 group M and the origin of HIV-1 subtypes using a new method to uncover clock-like molecular evolution. *FASEB Journal,* **15**, 267–268.

Salter, L. A. & Pearl, D. K. (2001) Stochastic search strategy for estimation of maximum likelihood phylogenetic trees. *Systematic Biology,* **50**, 7–17.

Sanderson, M. J. (2002). Estimating absolute rates of molecular evolution and divergence times: a penalized likelihood approach. *Molecular Biology and Evolution,* **19**, 101–109.

Sanjuán, R., Moya, A., & Elena, S. F. (2004). The distribution of fitness effects caused by single-nucleotide substitutions in an RNA virus. *Proceedings of the National Academy of Sciences, USA,* **101**, 8396–8401.

Sanjuan, R., Cuevas, J. M., Moya, A., & Elena, S. F. (2005). Epistasis and the adaptability of an RNA virus. *Genetics,* **170**(3), 1001–1008.

Sankoff, D. & Rousseau, P. (1975). Locating the vertixes of a Steiner tree in an arbitrary metric space. *Mathematic Progress,* **9**, 240–276.

Sankoff, D. (1985). Simultaneous solution of the RNA folding, alignment and protosequence problems. *SIAM Journal on Applied Mathematics,* **45**, 810–825.

Sawyer, S. (1989). Statistical tests for detecting gene conversion. *Molecular Biology and Evolution,* **6**, 526–538.

Sawyer, S. A. & Hartl, D. L. (1992). Population genetics of polymorphism and divergence. *Genetics,* **132**(4), 1161–1176.

Sawyer, S. L., Wu, L. I., Emerman, M., & Malik, H. S. (2005). Positive selection of primate TRIM5alpha identifies a critical species-specific retroviral restriction domain. *Proceedings of the National Academy of Sciences, USA,* **102**(8), 2832–7.

Schaffer, A. A., Aravind, L., Madden, T. L. *et al.* (2001). Improving the accuracy of PSI-BLAST protein database searches with composition-based statistics and other refinements. *Nucleic Acids Research,* **29**, 2994–3005.

Scheffler, K., Martin, D. P., & Seoighe, C. (2006). Robust inference of positive selection from recombining coding sequences. *Bioinformatics,* **22**, 2493–2499.

Schierup, M. H. & Hein, J. (2000a). Consequences of recombination on traditional phylogenetic analysis. *Genetics,* **156**, 879–891.

Schierup, M. H. & Hein, J. (2000b). Recombination and the molecular clock. *Molecular Biology and Evolution,* **17**, 1578–1579.

Schmidt, H. A., Strimmer, K., Vingron, M., & von Haeseler, A. (2002). TREE-PUZZLE: maximum likelihood phylogenetic analysis using quartets and parallel computing. *Bioinformatics,* **18**(3), 502–504.

Schuler, G. D., Altschul, S. F., & Lipman, D. J. (1991). A workbench for multiple alignment construction and analysis. *Proteins,* **9**(3), 180–190.

Schultz, A. K., Zhang, M., Leitner, T. *et al.* (2006). A jumping profile Hidden Markov Model and applications to recombination sites in HIV and HCV genomes. *BMC Bioinformatics,* **7**, 265.

Schwartz, A. S. & Pachter, L. (2007). Multiple alignment by sequence annealing. *Bioinformatics,* **23**(2), e24–e29.

Schwarz, G. (1978). Estimating the dimension of a model. *The Annals of Statistics,* **6**(2), 461–464.

Self, S. G. & Liang, K.-Y. (1987). Asymptotic properties of maximum likelihood estimators and likelihood ratio tests under nonstandard conditions. *Journal of the American Statistical Association*, **82**(398), 605–610.

Seo, T. K., Kishino, H., & Thorne, J. L. (2004). Estimating absolute rates of synonymous and nonsynonymous nucleotide substitution in order to characterize natural selection and date species divergences. *Molecular Biology of Evolution*, **21**(7), 1201–1213.

Shapiro, B., Drummond, A. J., Rambaut, A. *et al.* (2004). Rise and fall of the Beringian steppe bison. *Science*, **306**(5701), 1561–1565.

Sharp, P. M. (1997). In search of molecular Darwinism. *Nature*, **385**(6612), 111–112.

Shimodaira, H. (2002). An approximately unbiased test of phylogenetic tree selection. *Systematic Biology*, **51**, 492–508.

Shimodaira, H. & Hasegawa, M. (1999). Multiple comparisons of log-likelihoods with applications to phylogenetic inference. *Molecular Biology and Evolution*, **16**, 1114–1116.

Shimodaira, H. & Hasegawa, M. (2001). CONSEL: for assessing the confidence of phylogenetic tree selection. *Bioinformatics*, **17**, 1246–1247.

Shriner, D., Nickle, D. C., Jensen, M. A., & Mullins, J. I. (2003). Potential impact of recombination on sitewise approaches for detecting positive natural selection. *Genetics Research*, **81**, 115–121.

Siegel, S. & Castellan, Jr, N. J. (1988). *Nonparametric Statistics for the Behavioral Sciences*. 2nd edn. New York: McGraw-Hill.

Siepel, A. C., Halpern, A. L., Macken, C., & Korber, B. T. (1995). A computer program designed to screen rapidly for HIV type 1 intersubtype recombinant sequences. *AIDS Research Human Retroviruses*, **11**, 1413–1416.

Siguroardottir, S., Helgason, A., Gulchar, J. R., Stefansson, K., & Donnelly, P. (2000). The mutation rate in the human mtDNA control region. *American Journal of Human Genetics*, **66**, 1599–1609.

Sikes, D. S. & Lewis, P. O. (2001). *Beta Software, Version 1. PAUPRat: PAUP* Implementation of the Parsimony Ratchet*. Distributed by the authors. Storrs: University of Connecticut, Department of Ecology and Evolutionary Biology. (*http://viceroy.eeb.uconn.edu/paupratweb/pauprat.htm*).

Silva, A. E., Villanueva, W. J., Knidel, H., Bonato, V. C., Reis, S. F., & Von Zuben, F. J. (2005). A multi-neighbor-joining approach for phylogenetic tree reconstruction and visualization. *Genetics and Molecular Research*, **4**(3), 525–534.

Simmonds, P. & Welch, J. (2006). Frequency and dynamics of recombination within different species of human enteroviruses. *Journal of Virology*, **80**, 483–493.

Simmonds, P., Zhang, L. Q., McOmish, F., Balfe, P., Ludlam, C. A., & Brown, A. J. L. (1991). Discontinuous sequence change of human-immunodeficiency-virus (HIV) type-1 env sequences in plasma viral and lymphocyte-associated proviral populations in vivo – implications for models of HIV pathogenesis. *Journal of Virology*, **65**(11), 6266–6276.

Simossis, V. A. & Heringa, J. (2005). PRALINE: a multiple sequence alignment toolbox that integrates homology-extended and secondary structure information. *Nucleic Acids Research*, **33**(Web Server issue), W289–W294.

Sjödin, P., Kaj, I., Krone, S., Lascoux, M., & Nordborg, M. (2005). On the meaning and existence of an effective population size. *Genetics*, **169**, 1061–1070.

Smith, J. M. (1999). The detection and measurement of recombination from sequence data. *Genetics*, **153**, 1021–1027.

Smith, N. G. & Eyre-Walker, A. (2002). Adaptive protein evolution in *Drosophila*. *Nature*, **415**(6875), 1022–1024.

Smith, R. F. & Smith, T. F. (1992). Pattern-induced multi-sequence alignment (PIMA) algorithm employing secondary structure-dependent gap penalties for use in comparative protein modelling. *Protein Engineering*, **5**(1), 35–41.

Smith, T. F & Waterman, M. S. (1981). Identification of common molecular subsequences. *Journal of Molecular Biology*, **147**, 195–197.

Sneath, P. H. (1998). The effect of evenly spaced constant sites on the distribution of the random division of a molecular sequence. *Bioinformatics*, **14**, 608–616.

Sneath, P. H. A. & Sokal, R. R. (1973). *Numerical Taxonomy*. San Francisco: W. H. Freeman.

Sokal, R. R. & Michener, C. D. (1958). A statistical method for evaluating systematic relationships. *University of Kansas Science Bulletin*, **38**, 1409–1438.

Sorhannus, U. & Kosakovsky Pond, S. L. (2006). Evidence for positive selection on a sexual reproduction gene in the diatom genus Thalassiosira (Bacillariophyta). *Journal of Molecular Evolution*, **63**, 231–239.

Sourdis, J. & Krimbas, C. (1987). Accuracy of phylogenetic trees estimated from DNA sequence data. *Molecular Biology and Evolution*, **4**, 159–166.

Stamatakis, A. (2005). An efficient program for phylogenetic inference using simulated annealing. In *Online Proceedings of the 4th IEEE International Workshop on High Performance Computational Biology (HICOMB 2005)*, p. 8, Denver.

Stamatakis, A. (2006). RAxML-VI-HPC: maximum likelihood-based phylogenetic analyses with thousands of taxa and mixed models. *Bioinformatics*, **22**, 2688–2690.

Stamatakis, A. P., Ludwig, T., & Meier, H. (2005). RAxML-III: a fast program for maximum likelihood-based inference of large phylogenetic trees. *Bioinformatics*, **21**, 456–463.

Steel, M. (1994). Recovering a tree from the Markov leaf colourations it generates under a Markov model. *Applied Mathematics Letters*, **7**, 19–23.

Steel, M. & Penny, D. (2000). Parsimony, likelihood, and the role of models in molecular phylogenetics. *Molecular Biology and Evolution*, **17**, 839–850.

Steel, M. A., Lockhart, P. J., & Penny, D. (1993). Confidence in evolutionary trees from biological sequence data. *Nature*, **364**, 440–442.

Steel, M., Lockhart, P. J., & Penny, D. (1995). A frequency-dependent significance test for parsimony. *Molecular Phylogenetics and Evolution*, **4**, 64–71.

Stern, A. & Pupko, T. (2006). An evolutionary space-time model with varying among-site dependencies. *Molecular Biology and Evolution*, **23**(2), 392–400.

Stewart, C. A., Hart, D., Berry, D. K., Olsen, G. J., Wernert, E. A., & Fischer, W. (2001). Parallel implementation and performance of fastDNAml – a program for maximum likelihood phylogenetic inference. In *Proceedings of the International Conference on High Performance Computing and Communications – SC2001*, pp. 191–201.

Stoye, J. (1998). Multiple sequence alignment with the Divide-and-Conquer method. *Gene*, **211**(2), GC45–GC56.

Stoye, J., Moulton, V., & Dress, A. W. (1997). DCA: an efficient implementation of the divide-and-conquer approach to simultaneous multiple sequence alignment. *Computer Applications in the Biosciences*, **13**, 625–626.

Stremlau, M., Owens, C. M., Perron, M. J., Kiessling, M., Autissier, P., & Sodroski, J. (2004). The cytoplasmic body component TRIM5alpha restricts HIV-1 infection in Old World monkeys. *Nature*, **427**(6977), 848–853.

Strimmer, K. & Moulton, V. (2000). Likelihood analysis of phylogenetic networks using directed graphical models. *Molecular Biology and Evolution*, **17**, 875–881.

Strimmer, K. & Rambaut, A. (2002). Inferring confidence sets of possibly misspecified gene trees. *Proceedings of the Royal Society of London B*, **269**, 137–142.

Strimmer, K. & von Haeseler, A. (1996). Quartet-puzzling: a quartet maximum-likelihood method for reconstructing tree topologies. *Molecular Biology and Evolution*, **13**, 964–969.

Strimmer, K. & von Haeseler, A. (1997). Likelihood-mapping: a simple method to visualize phylogenetic content of a sequence alignment. *Proceedings of the National Academy of Sciences, USA*, **94**, 6815–6819.

Strimmer, K. & von Haeseler, A. (2003). Phylogeny inference based on maximum likelihood methods with tree-puzzle. In *The Phylogenetic Handbook*, ed. M. Salemi & A.-M. Vandamme, pp. 137–159, Cambridge, UK: Cambridge University Press.

Strimmer, K., Forslund, K., Holland, B., & Moulton, V. (2003). A novel exploratory method for visual recombination detection. *Genome Biology*, **4**, R33.

Strimmer, K., Goldman, N., & von Haeseler, A. (1997) Bayesian probabilities and quartet puzzling. *Molecular Biology and Evolution*, **14**, 210–213.

Studier, J. A. & Keppler, K. J. (1988). A note on the neighbor-joining algorithm of Saitou and Nei. *Molecular Biology and Evolution*, **5**, 729–731.

Stumpf, M. P. & McVean, G. A. (2003). Estimating recombination rates from population-genetic data. *Nature Review Genetics*, **4**, 959–968.

Suarez, D. L., Senne, D. A., Banks, J. *et al.* (2004). Recombination resulting in virulence shift in avian influenza outbreak, Chile. *Emergency Infections Diseases*, **10**, 693–699.

Subramanian, A. R., Weyer-Menkhoff, J., Kaufmann, M., & Morgenstern, B. (2005). DIALIGN-T: an improved algorithm for segment-based multiple sequence alignment. *BMC Bioinformatics*, **6**, 66.

Suchard, M. A. & Redelings, B. D. (2006). `Bali-Phy`: simultaneous Bayesian inference of alignment and phylogeny. *Bioinformatics*, **22**, 2047–2048.

Suchard, M. A., Weiss, R. E., Dorman, K. S., & Sinsheimer, J. S. (2002). Oh brother, where art thou? A Bayes factor test for recombination with uncertain heritage. *Systems in Biology*, **51**, 715–728.

Sullivan, J. & Joyce, P. (2005). Model selection in phylogenetics. *Annual Review of Ecology, Evolution and Systematics*, **36**, 445–466.

Sullivan, J. & Swofford, D. L. (2001). Should we use model-based methods for phylogenetic inference when we know that assumptions about among-site rate variation and nucleotide substitution process are violated? *Systematic Biology*, **50**, 723–729.

Sullivan, J., Holsinger, K. E., & Simon, C. (1996). The effect of topology on estimates of among-site rate variation. *Journal of Molecular Evolution*, **42**, 308–312.

Sullivan, J., Abdo, Z., Joyce, P., & Swofford, D. L. (2005). Evaluating the performance of a successive-approximations approach to parameter optimization in maximum-likelihood phylogeny estimation. *Molecular Biology and Evolution*, **22**, 1386–1392.

Suzuki, Y. & Gojobori, T. (1999). A method for detecting positive selection at single amino acid sites. *Molecular Biology and Evolution*, **16**, 1315–1328.

Suzuki, Y. & Nei, M. (2004). False-positive selection identified by ML-based methods: examples from the *sig*1 gene of the diatom *Thalassiosira weissflogii* and the *tax* gene of a human T-cell lymphotropic virus. *Molecular Biology and Evolution*, **21**, 914–921.

Suzuki, Y., Gojobori, T., & Nei, M. (2001). ADAPTSITE: detecting natural selection at single amino acid sites. *Bioinformatics*, **17**, 660–661.

Swanson, W. J., Nielsen, R., & Yang, Q. (2003). Pervasive adaptive evolution in mammalian fertilization proteins. *Molecular Biology and Evolution*, **20**(1), 18–20.

Swofford, D. (1993) PAUP (Phylogenetic Analysis Using Parsimony). Smithsonian Institution, Washington, Version 4.0beta, MA: Sinauer Associates of Sunderland.

Swofford, D. L. (2002). PAUP*. *Phylogenetic Analysis Using Parsimony (* and other methods)*. *Version 4.0b10*. Sunderland, MA (USA): Sinauer Associates, Inc.

Swofford, D. L. & Maddison, W. P. (1987). Reconstructing ancestral character states under Wagner parsimony. *Mathematical Biosciences*, **87**, 199–229.

Swofford, D. L., Olsen, G. J., Waddell, P. J., & Hillis, D. M. (1996). Phylogenetic inference. In *Molecular Systematics*, 2nd edn., ed. D. M. Hillis, C. Moritz, & B. K. Mable, pp. 407–514. Sunderland, Massachusetts, USA: Sinauer Associates, Inc.

Swofford, D. L. (1998). *PAUP*. Phylogenetic Analysis Using Parsimony (* and other methods)*. Version 4. Sunderland, MA: Sinauer Associates.

Sze, S. H., Lu, Y., & Yang, Q. (2006). A polynomial time solvable formulation of multiple sequence alignment. *Journal of Computer Biology*, **13**(2), 309–319.

Tajima, F. (1989). Statistical method for testing the neutral mutation hypothesis by DNA polymorphism. *Genetics*, **123**(3), 585–595.

Takezaki, N., Rzhetsky, A., & Nei, M. (1995). Phylogenetic test of the molecular clock and linearized trees. *Molecular Biology of Evolution*, **12**(5), 823–833.

Tamura, K. & Nei, M. (1993). Estimation of the number of nucleotide substitutions in the control region of mitochondrial DNA in humans and chimpanzees. *Molecular Biology and Evolution*, **10**, 512–526.

Tavaré, S. (1986). Some probabilistic and statistical problems in the analysis of DNA sequences. *Lectures in Mathematics and Life Sciences*, **17**, 57–86.

Templeton, A., Crandall, K., & Sing, C. (1992). A cladistic analysis of phenotypic associations with haplotypes inferred from restriction endonuclease mapping and DNA sequence data. III. Cladogram estimation, *Genetics*, **132**, 619–633.

Thompson, J. D., Higgins, D. G., & Gibson, T. J. (1994). CLUSTAL W: improving the sensitivity of progressive multiple sequence alignment through sequence weighting, position-specific gap penalties and weight matrix choice. *Nucleic Acids Research*, **22**, 4673–4680.

Thompson, J. D., Gibson, T. J., Plewniak, F., Jeanmougin, F., & Higgins, D. G. (1997). The CLUSTAL_X windows interface: flexible strategies for multiple sequence alignment aided by quality analysis tools. *Nucleic Acids Research*, **25**, 4876–4882.

Thompson, J. D., Plewniak, F., & Poch, O. (1999a). A comprehensive comparison of multiple sequence alignment programs. *Nucleic Acids Research*, **27**, 2682–2690.

Thompson, J. D., Plewniak, F., & Poch, O. (1999b). BaliBase: a benchmark alignment database for the evaluation of multiple alignment programs. *Bioinformatics*, **15**, 87–88.

Thompson, W. W., Shay, D. K., Weintraub, E. *et al.* (2003). Mortality associated with influenza and respiratory syncytial virus in the united states. *Journal of the American Medical Association*, **289**, 179–186.

Thomson, R., Pritchard, J. K., Shen, P., Oefner, P. J., & Feldman, M. W. (2000). Recent common ancestry of human Y chromosomes: Evidence from DNA sequence data. *Proceedings of the National Academy of Science, USA*, **97**(13), 7360–7365.

Thorne, J. L. & Kishino, H. (2002). Divergence time and evolutionary rate estimation with multilocus data. *Systematic Biology*, **51**(5), 689–702.

Thorne, J. L., Kishino, H., & Felsenstein, J. (1992). Inching toward reality: an improved likelihood model of sequence evolution. *Journal of Molecular Evolution*, **34**, 3–16.

Thorne, J. L., Kishino, H., & Painter, I. S. (1998). Estimating the rate of evolution of the rate of molecular evolution. *Molecular Biology of Evolution*, **15**(12), 1647–1657.

Thorne, J. L., Kishino, H., & Felsenstein, J. (1991). An evolutionary model for maximum likelihood alignment of DNA sequences. *Journal of Molecular Evolution*, **33**, 114–124.

Tishkoff, S. A., Reed, F. A., Ranciaro, A. *et al.* (2007). Convergent adaptation of human lactase persistence in Africa and Europe. *Nature Genetics*, **39**(1), 31–40.

Toomajian, C. & Kreitman, M. (2002). Sequence variation and haplotype structure at the human HFE locus. *Genetics*, **161**(4), 1609–1623.

Tuffley, C. & Steel, M. (1997). Links between maximum likelihood and maximum parsimony under a simple model of site substitution. *Bulletin of Mathematical Biology*, **59**, 581–607.

Tuplin, A., Wood, J., Evans, D. J., Patel, A. H., & Simmonds, P. (2002). Thermodynamic and phylogenetic prediction of RNA secondary structures in the coding region of hepatitis C virus. *RNA*, **8**, 824–841.

Tzeng, Y.-H., Pan, R., & Li, W.-H. (2004). Comparison of three methods for estimating rates of synonymous and nonsynonymous nucleotide substitutions. *Molecular Biology and Evolution*, **21**(12), 2290–2298.

Uzzel, T. & Corbin, K. W. (1971). Fitting discrete probability distributions to evolutionary events. *Sciences*, **172**, 1089–1096.

van Cuyck, H., Fan, J., Robertson, D. L., & Roques, P. (2005). Evidence of recombination between divergent hepatitis E viruses. *Journal of Virology*, **79**, 9306–9314.

Van de Peer, Y. & De Wachter, R. (1994). TREECON for Windows: A software package for the construction and drawing of evolutionary trees for the Microsoft Windows environment. *Computer Applications in the Biosciences*, **10**, 569–570.

Van de Peer, Y., De Rijk, P., Wuyts, J., Winkelmans, T., & De Wachter, R. (2000a). The European small subunit ribosomal RNA database. *Nucleic Acids Research*, **28**, 175–176.

Van de Peer, Y., Rensing, S., Maier, U.-G., & De Wachter, R. (1996). Substitution rate calibration of small ribosomal subunit RNA identifies chlorarachniophyte endosymbionts as remnants of green algae. *Proceedings of the National Academy of Sciences, USA*, **93**, 7732–7736.

Van de Peer, Y., Baldauf, S., Doolittle, W. F., & Meyer, A. (2000b). An updated and comprehensive rRNA phylogeny of crown eukaryotes based on rate-calibrated evolutionary distances. *Journal of Molecular Evolution*, **51**, 565–576.

Van Walle, I., Lasters, I., & Wyns, L. (2004). Align-m – a new algorithm for multiple alignment of highly divergent sequences. *Bioinformatics*, **20**(9), 1428–1435.

Van Walle, I., Lasters, I., & Wyns, L. (2005). SABmark – a benchmark for sequence alignment that covers the entire known fold space. *Bioinformatics*, **21**(7), 1267–1268.

Vennema, H., Poland, A., Foley, J., & Pedersen, N. C. (1998). Feline infectious peritonitis viruses arise by mutation from endemic feline enteric coronaviruses. *Virology*, **243**, 150–157.

Vigilant, L., Stoneking, M., Harpending, H., Hawkes, K., & Wilson, A. C. (1991). African populations and the evolution of human mitochondrial DNA. *Science*, **253**(5027), 1503–1507.

Vinh, L. S. & von Haeseler, A. (2004). IQPNNI: Moving fast through tree space and stopping in time. *Molecular Biology and Evolution*, **21**, 1565–1571.

Voight, B. F., Kudaravalli, S., Wen, X., & Pritchard, J. K. (2006). A map of recent positive selection in the human genome. *PLoS Biology*, **4**(3), e72.

Wakeley, J. (1993). Substitution rate variation among sites in hypervariable region 1 of human mitochondrial DNA. *Journal of Molecular Evolution*, **37**, 613–623.

Walldorf, U. & Hovemann, B. T. (1990). *Apis mellifera* cytoplasmic elongation factor 1α (EF-1α) is closely related to *Drosophila melanogaster* EF-1α. *FEBS*, **267**, 245–249.

Ward, R. H., Frazer, B. L., Dew-Jager, K., & Pääbo, S. (1991). Extensive mitochondrial diversity within a single Amerindian tribe. *Proceedings of the National Academy of Sciences, USA*, **88**, 8720–8724.

Wasserman, L. (2000). Bayesian model selection and model averaging. *Journal Mathematical Psychology*, **44**(1), 92–107.

Watterson, G. A. (1975). On the number of segregating sites in genetical models without recombination. *Theoretical Population Biology*, **7**, 256–276.

Webster, R. G., Laver, W. G., Air, W. G., & Schild, G. C. (1982). Molecular mechanisms of variation in influenza viruses. *Nature*, **296**(5853), 115–121.

Webster, R. G., Bean, W. J., Gorman, O. T., Chambers, T. M., & Kawaoka, Y. (1992). Evolution and ecology of influenza A viruses. *Microbiology Reviews*, **56**, 152–179.

Wei, X. P., Decker, J. M., Wang, S. Y. *et al.* (2003). Antibody neutralization and escape by HIV-1. *Nature*, **422**(6929), 307–312.

Weiller, G. F. (1998). Phylogenetic profiles: a graphical method for detecting genetic recombinations in homologous sequences. *Molecular Biology and Evolution*, **15**, 326–335.

Wernersson, R. & Pedersen, A. G. (2003). RevTrans: multiple alignment of coding DNA from aligned amino acid sequences. *Nucleic Acids Research*, **31**(13), 3537–3539.

Westfall, P. H. & Young, S. S. (1993). *Resampling-based Multiple Testing: Examples and Methods for P-value Adjustment*. New York, USA: John Wiley and Sons.

Whelan, S. & Goldman, N. (1999). Distributions of statistics used for the comparison of models of sequence evolution in phylogenetics. *Molecular Biology and Evolution*, **16**(9), 1292–1299.

Whelan, S. & Goldman, N. (2001). A general empirical model of protein evolution derived from multiple protein families using a maximum-likelihood approach. *Molecular Biology and Evolution*, **18**(5), 691–699.

Whelan, S. & Goldman, N. (2004). Estimating the frequency of events that cause multiple-nucleotide changes. *Genetics*, **167**, 2027–2043.

Williamson, S. (2003). Adaptation in the env gene of HIV-1 and evolutionary theories of disease progression. *Molecular Biology and Evolution*, **20**(8), 1318–1325.

Williamson, S. & Orive, M. E. (2002). The genealogy of a sequence subjected to purifying selection at multiple sites. *Molecular Biology and Evolution*, **19**, 1376–1384.

Williamson, S., Fledel-Alon, A., & Bustamante, C. D. (2004). Population genetics of polymorphism and divergence for diploid selection models with arbitrary dominance. *Genetics*, **168**(1), 463–475.

Wilson, D. J. & McVean, G. (2006). Estimating diversifying selection and functional constraint in the presence of recombination. *Genetics*, **172**, 1411–1425.

Wilson, I. J., Weale, M. E., & Balding, D. J. (2003). Inferences from DNA data: population histories, evolutionary processes and forensic match probabilities. *Journal of the Royal Statistics Society A–Statistics in Society*, **166**, 155–188.

Wiuf, C. & Posada, D. (2003). A coalescent model of recombination hotspots. *Genetics*, **164**, 407–417.

Wiuf, C., Christensen, T., & Hein, J. (2001). A simulation study of the reliability of recombination detection methods. *Molecular Biology and Evolution*, **18**, 1929–1939.

Wolfe, K. H., Li, W. H., & Sharp, P. M. (1987). Rates of nucleotide substitution vary greatly among plant mitochondrial, chloroplast, and nuclear DNAs. *Proceedings of the National Academy of Sciences, USA*, **84**(24), 9054–9058.

Wong, W. S. & Nielsen, R. (2004). Detecting selection in noncoding regions of nucleotide sequences. *Genetics*, **167**(2), 949–958.

Wong, W. S., Yang, Z., Goldman, N., & Nielsen, R. (2004). Accuracy and power of statistical methods for detecting adaptive evolution in protein coding sequences and for identifying positively selected sites. *Genetics*, **168**(2), 1041–1051.

Wootton, J. C. & Federhen, S. (1993). Statistics of local complexity in amino acid sequences and sequence databases. *Computers and Chemistry*, **17**, 149–163.

Worobey, M. (2001). A novel approach to detecting and measuring recombination: new insights into evolution in viruses, bacteria, and mitochondria. *Molecular Biology and Evolution*, **18**, 1425–1434.

Worobey, M. & Holmes, E. C. (1999). Evolutionary aspects of recombination in RNA viruses. *Journal of General Virology*, **80**, 2535–2543.

Worobey, M., Rambaut, A., Pybus, O. G., & Robertson, D. L. (2002). Questioning the evidence for genetic recombination in the 1918 "Spanish flu" virus. *Science*, **296**, 211a.

Wright, S. (1931). Evolution in Mendelian populations. *Genetics*, **16**, 97–159.

Xia, X. & Xie, Z. (2001). DAMBE: data analysis in molecular biology and evolution. *Journal of Heredity*, **92**, 371–373.

Xia, X. (1998). The rate heterogeneity of nonsynonymous substitutions in mammalian mitochondrial genes. *Molecular Biology and Evolution*, **15**, 336–344.

Xia, X., Hafner, M. S., & Sudman, P. D. (1996). On transition bias in mitochondrial genes of pocket gophers. *Journal of Molecular Evolution*, **43**, 32–40.

Xia, X. H., Xie, Z., & Kjer, K. M. (2003). 18S ribosomal RNA and tetrapod phylogeny. *Systematic Biology*, **52**, 283–295.

Xia, X. H., Xie, Z., Salemi, M., Chen, L., & Wang, Y. (2003). An index of substitution saturation and its application. *Molecular Phylogenetics and Evolution*, **26**, 1–7.

Xia, X. & Xie, Z. (2001). DAMBE: Software package for data analysis in molecular biology and evolution. *Journal of Heredity*, **92**, 371–373.

Yamada, S., Gotoh, O., & Yamana, H. (2006). Improvement in accuracy of multiple sequence alignment using novel group-to-group sequence alignment algorithm with piecewise linear gap cost. *BMC Bioinformatics*, **7**, 524.

Yamaguchi, Y. & Gojobori, T. (1997). Evolutionary mechanisms and population dynamics of the third variable envelope region of HIV within single hosts. *Proceedings of the National Academy of Sciences, USA*, **94**(4), 1264–1269.

Yang, Z. (1994a). Estimating the pattern of nucleotide substitution. *Journal of Molecular Evolution*, **39**, 105–111.

Yang, Z. (1994b). Maximum-likelihood phylogenetic estimation from DNA sequences with variable rates over sites: approximate methods. *Journal of Molecular Evolution*, **39**, 306–314.

Yang, Z. (1995). A space-time process model for the evolution of DNA sequences. *Genetics*, **139**(2), 993–1005.

Yang, Z. (1996). Maximum-likelihood models for combined analyses of multiple sequence data. *Journal of Molecular Evolution*, **42**, 587–596.

Yang, Z. (1997). PAML: a program package for phylogenetic analysis by maximum likelihood. *Computing in Applied Biosciences*, **13**(5), 555–556.

Yang, Z. (1998). Likelihood ratio tests for detecting positive selection and application to primate lysozyme evolution. *Molecular Biology and Evolution*, **15**, 568–573.

Yang, Z. (2000). *Phylogenetic Analysis by Maximum Likelihood (PAML), version 3.0.* University College, London, UK.

Yang, Z. & Bielawski, J. P. (2000). Statistical methods for detecting molecular adaptation. *Trends in Ecology and Evolution*, **15**(12), 496–503.

Yang, Z. & Nielsen, R. (1998). Synonymous and nonsynonymous rate variation in nuclear genes of mammals. *Journal of Molecular Evolution*, **46**, 409–418.

Yang, Z. & Nielsen, R. (2002). Codon-substitution models for detecting molecular adaptation at individual sites along specific lineages. *Molecular Biology and Evolution*, **19**(6), 908–917.

Yang, Z., Goldman, N., & Friday, A. E. (1995). Maximum likelihood trees from DNA sequences: a peculiar statistical estimation problem. *Systematic Biology*, **44**, 384–399.

Yang, Z., Wong, W. S. W., & Nielsen, R. (2005). Bayes empirical Bayes inference of amino acid sites under positive selection. *Molecular Biology and Evolution*, **22**(4), 1107–1118.

Yang, Z. H. (2000). Maximum likelihood estimation on large phylogenies and analysis of adaptive evolution in human influenza virus A. *Journal of Molecular Evolution*, **51**, 423–432.

Yang, Z. H. (2006). *Computational Molecular Evolution*. Oxford, UK: Oxford University Press.

Yang, Z. H., Nielsen, R., Goldman, N., & Pedersen, A. M. K. (2000). Codon-substitution models for heterogeneous selection pressure at amino acid sites. *Genetics*, **155**, 431–449.

Yoder, A. D. & Yang, Z. (2000). Estimation of primate speciation dates using local molecular clocks. *Molecular Biology of Evolution*, **17**(7), 1081–1090.

Zhang, J. (2005). On the evolution of codon volatility. *Genetics*, **169**(1), 495–501.

Zhang, J., Nielsen, R., & Yang, Z. (2005). Evaluation of an improved branch-site likelihood method for detecting positive selection at the molecular level. *Molecular Biology and Evolution*, **22**(12), 2472–2479.

Zharkikh, A. (1994). Estimation of evolutionary distances between nucleotide sequences. *Journal of Molecular Evolution*, **39**(3), 315–329.

Zharkikh, A. & Li, W.-H. (1992a). Statistical properties of bootstrap estimation of phylogenetic variability from nucleotide sequences. I. Four taxa with a molecular clock. *Molecular Biology and Evolution*, **9**, 1119–1147.

Zharkikh, A. & Li, W.-H. (1992b). Statistical properties of bootstrap estimation of phylogenetic variability from nucleotide sequences. II. Four taxa without a molecular clock. *Journal of Molecular Evolution*, **35**, 356–366.

Zharkikh, A. & Li, W.-H. (1995). Estimation of confidence in phylogeny: The complete-and-partial bootstrap technique. *Molecular Phylogenetic Evolution*, **4**, 44–63.

Zhou, H. & Zhou, Y. (2005). SPEM: improving multiple sequence alignment with sequence profiles and predicted secondary structures. *Bioinformatics*, **21**(18), 3615–3621.

Zlateva, K. T., Lemey, P., Vandamme, A. M., & Van Ranst, M. (2004). Molecular evolution and circulation patterns of human respiratory syncytial virus subgroup A: positively selected sites in the attachment g glycoprotein. *Journal of Virology*, **78**, 4675–4683.

Zuckerkandl, E. & Pauling, L. (1962). Molecular disease, evolution, and genetic heterogeneity. In *Horizons in Biochemistry*, ed. M. Kasha & B. Pullman, pp. 189–225. New York: Academic Press.

Zuckerkandl, E. & Pauling, L. (1965). Evolutionary divergence and convergence in proteins. In *Evolving Genes and Proteins*, ed. V. Bryson & H. J. Vogel, pp. 97–166. New York: Academic Press: Academic Press.

Zwickl, D. J. (2006). Genetic algorithm approaches for the phylogenetic analysis of large biological sequence datasets under the maximum likelihood criterion. Ph.D. thesis, University of Texas, Austin, USA.

Index

Locators for headings with subheadings refer to general or introductory aspects of that topic.
Locators in *italics* refer to pages/page ranges containing relevant figures and diagrams.

Printed in the United States
By Bookmasters